# Fundamentals of Logic Design

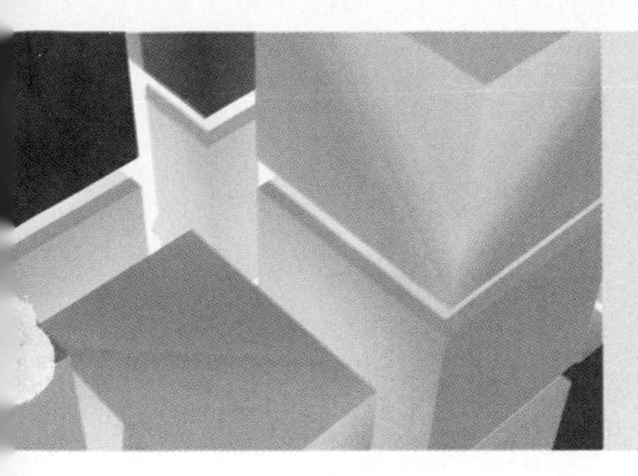

# Fundamentals of Logic Design

**Charles H. Roth, Jr.**
*University of Texas at Austin*

**Larry L. Kinney**
*University of Minnesota, Twin Cities*

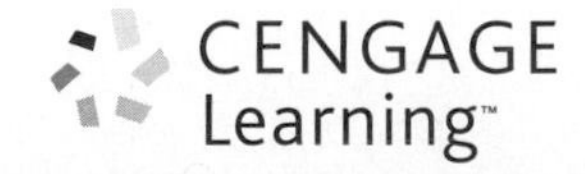


Australia • Brazil • Japan • Korea • Mexico • Singapore • Spain • United Kingdom • United States

**Fundamentals of Logic Design, Sixth Edition**

**Charles H. Roth, Jr. and Larry L. Kinney**

Director, Global Engineering Program:
Chris Carson

Senior Developmental Editor:
Hilda Gowans

Editorial Assistant:
Jennifer Dismore

Marketing Services Coordinator:
Lauren Bestos

Director, Content and Media Production:
Barbara Fuller-Jacobsen

Content Project Manager:
Cliff Kallemeyn

Production Service:
RPK Editorial Services, Inc.

Copyeditor:
Fred Dahl

Proofreader:
Harlan James

Indexer:
Ron Prottsman

Compositor:
Integra

Senior Art Director:
Michelle Kunkler

Internal Designer:
Carmela Periera

Cover Designer:
Andrew Adams

Cover Image:
© Shutterstock/guattièro boffi

Senior First Print Buyer:
Doug Wilke

Printed in the United States of America
1 2 3 4 5 6 7 12 11 10 09



Library of Congress Control Number: 2009920814

Student Edition with CD:
ISBN-13: 978-0-495-47169-1
ISBN-10: 0-495-47169-0

Student Edition:
ISBN-13: 978-0-495-66804-6
ISBN-10: 0-495-66804-4

**Cengage Learning**
200 First Stamford Place, Suite 400
Stamford, CT 06902
USA

Cengage Learning is a leading provider of customized learning solutions with office locations around the globe, including Singapore, the United Kingdom, Australia, Mexico, Brazil, and Japan. Locate your local office at: **international.cengage.com/region**.

Cengage Learning products are represented in Canada by Nelson Education Ltd.

For your course and learning solutions, visit **www.cengage.com/engineering**.
Purchase any of our products at your local college store or at our preferred online store **www.ichapters.com**.

# Brief Contents

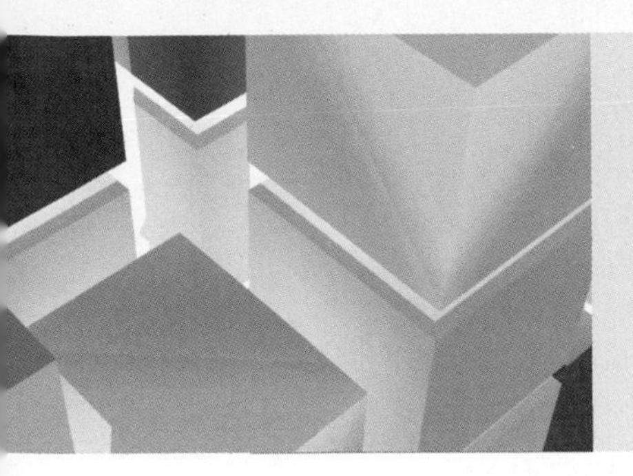

# Contents

## Unit 11 Latches and Flip-Flops



## Unit 12 Registers and Counters



## Unit 13 Analysis of Clocked Sequential Circuits

## Unit 20 VHDL for Digital System Design



## A Appendices

# Preface

After studying this text, you should be able to apply switching theory to the solution of logic design problems. This means that you will learn both the basic theory of switching circuits and how to apply it. After a brief introduction, you will study Boolean algebra, which is the basic mathematical tool needed to analyze and synthesize an important class of switching circuits. Starting from a problem statement, you will learn to design circuits of logic gates that have a specified relationship between signals at the input and output terminals. Then you will study the logical properties of flip-flops, which serve as memory devices in sequential switching circuits. By combining flip-flops with circuits of logic gates, you will learn to design counters, adders, sequence detectors, and similar circuits. You will also study the VHDL hardware description language and its application to the design of combinational logic, sequential logic, and simple digital systems.

The fifth edition offers a number of improvements over the fourth edition. Material in the text has been reorganized to provide a better teaching sequence, and obsolete material has been removed. The chapter on latches and flip-flops has been rewritten. Greater emphasis is placed on the use of programmable logic devices (PLDs), including programmable gate arrays and complex PLDs. New exercises and problems have been added to every unit, and several sections have been rewritten to clarify the presentation. Three chapters on the VHDL hardware description language have been added, and more emphasis is placed on the role of simulation and computer-aided design of logic circuits.

This text is designed so that it can be used in either a standard lecture course or in a self-paced course. In addition to the standard reading material and problems, study guides and other aids for self-study are included in the text. The content of the text is divided into 20 study units. These units form a logical sequence so that mastery of the material in one unit is generally a prerequisite to the study of succeeding units. Each unit consists of four parts. First, a list of objectives states precisely what you are expected to learn by studying the unit. Next, the study guide contains reading assignments and study questions. As you work through the unit, you should write out the answers to these study questions. The text material and problem set that follow are similar to a conventional textbook. When you complete a unit, you should review the objectives and make sure that you have met them.

The study units are divided into three main groups. The first 9 units treat Boolean algebra and the design of combinational logic circuits. Units 11 through 16, 18 and 19 are mainly concerned with the analysis and design of clocked sequential logic circuits, including circuits for arithmetic operations. Units 10, 17, and 20 introduce the VHDL hardware description language and its application to logic design.

Since the computer plays an important role in the logic design process, integration of computer usage into the first logic design course is very important. A computer-aided logic design program, called *LogicAid,* is included on the CD provided with this textbook. *LogicAid* allows the student easily to derive simplified logic equations from minterms, truth tables, and state tables. This relieves the student of some of the more tedious computations and permits the solution of more complex design problems in a shorter time. *LogicAid* also provides tutorial help for Karnaugh maps and derivation of state graphs.

Several of the units include simulation or laboratory exercises. These exercises provide an opportunity to design a logic circuit and then test its operation. The *SimUaid* logic simulator, provided on the CD, may be used to verify the logic designs. The lab equipment required for testing either can be a breadboard with integrated circuit flip-flops and logic gates or a circuit board with a programmable logic device. If such equipment is not available, the lab exercises can be simulated with *SimUaid* or just assigned as design problems. This is especially important for Units 8, 16, and 20 because the comprehensive design problems in these units help to review and tie together the material in several of the preceding units.

As integrated circuit technology continues to improve to allow more components on a chip, digital systems continue to grow in complexity. Design of such complex systems is facilitated by the use of a hardware description language such as VHDL. This text introduces the use of VHDL in logic design and emphasizes the relationship between VHDL statements and the corresponding digital hardware. VHDL allows digital hardware to be described and simulated at a higher level before it is implemented with logic components. Computer programs for synthesis can convert a VHDL description of a digital system to a corresponding set of logic components and their interconnections. Even though use of such computer-aided design tools helps to automate the logic design process, we believe that it is important to understand the underlying logic components and their timing before writing VHDL code. By first implementing the digital logic manually, students more fully can appreciate the power and limitations of VHDL.

This text is written for a first course in the logic design of digital systems. It is written on the premise that the student should understand and learn thoroughly certain fundamental concepts in a first course. Examples of such fundamental concepts are the use of Boolean algebra to describe the signals and interconnections in a logic circuit, use of systematic techniques for simplification of a logic circuit, interconnection of simple components to perform a more complex logic function, analysis of a sequential logic circuit in terms of timing charts or state graphs, and use of a control circuit to control the sequence of events in a digital system.

The text attempts to achieve a balance between theory and application. For this reason, the text does not overemphasize the mathematics of switching theory; however, it does present the theory that is necessary for understanding the fundamental concepts of logic design. After completing this text, the student should be prepared for a more advanced digital systems design course that stresses more intuitive concepts like the development of algorithms for digital processes, partitioning of digital systems into subsystems, and implementation of digital systems using currently available hardware. Alternatively, the student should be prepared to go on to a more advanced course in switching theory that further develops the theoretical concepts that have been introduced here.

Although the technology used to implement digital systems has changed significantly since the first edition of this text was published, the fundamental principles of logic design have not. Truth tables and state tables still are used to specify the behavior of logic circuits, and Boolean algebra is still a basic mathematical tool for logic design. Even when programmable logic devices are used instead of individual gates and flip-flops, reduction of logic equations is still desirable in order to fit the equations into smaller PLDs. Making a good state assignment is still desirable, because without a good assignment, the logic equations may require larger PLDs.

The text is suitable for both computer science and engineering students. Material relating to circuit aspects of logic gates is contained in Appendix A so that this material can conveniently be omitted by computer science students or other students with no background in electronic circuits. The text is organized so that Unit 6 on the Quine-McCluskey procedure may be omitted without loss of continuity. The three units on VHDL can be studied in the normal sequence, studied together after the other units, or omitted entirely.

Although many texts are available in the areas of switching theory and logic design, this text was originally developed to meet the needs of a self-paced course in which students are expected to study the material on their own. Each of the units has undergone extensive class testing in a self-paced environment and has been revised based on student feedback.

Study guides and text material have been expanded as required so that students can learn from the text without the aid of lectures and so that almost all of the students can achieve mastery of all of the objectives. Supplementary materials were developed as the text was being written. An instructor's manual is available that includes suggestions for using the text in a standard or self-paced course, quizzes on each of the units, and suggestions for laboratory equipment and procedures. The instructor's manual also contains solutions to problems, to unit quizzes, and to lab exercises.

To be effective, a book designed for self-study cannot simply be written. It must be tested and revised many times to achieve its goals. I wish to express my appreciation to the many professors, proctors, and students who participated in this process. Special thanks go to Dr. David Brown, who worked with me in teaching the self-paced course, and who made many helpful suggestions for improving the text. I am especially grateful to graduate teaching assistant, Mark Story, who developed many new problems and solutions for the fifth edition and who offered many suggestions for improving the consistency and clarity of the presentation.

Charles H. Roth, Jr.

# Preface to the Sixth Edition

The major change in the sixth edition of the text is the addition of over 150 new problems and the modification of several of the fifth edition problems. Substantial new discussion was added to the units on VHDL. Other topics receiving expanded discussion are hazards, latches and one-hot state assignments. In addition, the logic design and simulation software that accompanies the text has been updated and improved.

Larry L. Kinney Charles H. Roth, Jr.

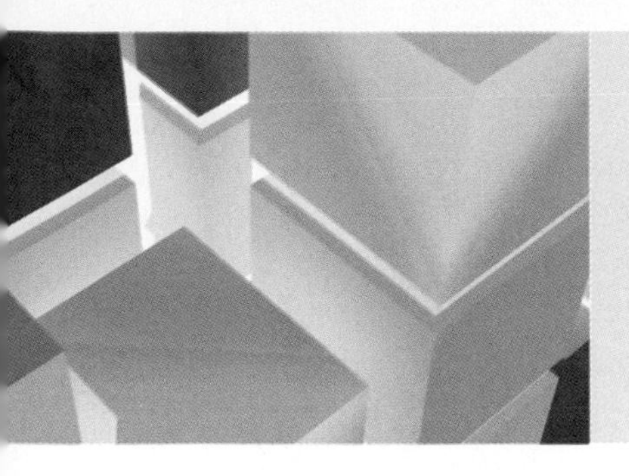

# How to Use This Book for Self-Study

If you wish to learn all of the material in this text to mastery level, the following study procedures are recommended for each unit:

1. Read the *Objectives* of the unit. These objectives provide a concise summary of what you should be able to do when you complete study of the unit.
2. Work through the *Study Guide.* After reading each section of the text, write out the answers to the corresponding study guide questions. In many cases, blank spaces are left in the study guide so that you can write your answers directly in this book. By doing this, you will have the answers conveniently available for later review. The study guide questions generally will help emphasize some of the important points in each section or will guide you to a better understanding of some of the more difficult points. If you cannot answer some of the study guide questions, this indicates that you need to study the corresponding section in the text more before proceeding. The answers to selected study guide questions are given in the back of this book; answers to the remaining questions generally can be found within the text.
3. Several of the units (Units 3, 5, 6, 11, 13, 14, and 18) contain one or more programmed exercises. Each programmed exercise will guide you step-by-step through the solution of one of the more difficult types of problems encountered in this text. When working through a programmed exercise, be sure to write down your answer for each part in the space provided before looking at the answer and continuing with the next part of the exercise.
4. Work the assigned *Problems* at the end of the unit. Check your answers against those at the end of the book and rework any problems that you missed.
5. Reread the *Objectives* of the unit to make sure that you can meet all of them. If in doubt, review the appropriate sections of the text.
6. If you are using this text in a self-paced course, you will need to pass a readiness test on each unit before proceeding with the next unit. The purpose of the readiness test is to make sure that you have mastered the objectives of one unit before moving on to the next unit. The questions on the test will relate directly to the objectives of the unit, so that if you have worked through the study guide and written out answers to all of the study guide questions and to the problems assigned in the study guide, you should have no difficulty passing the test.

# Fundamentals of Logic Design

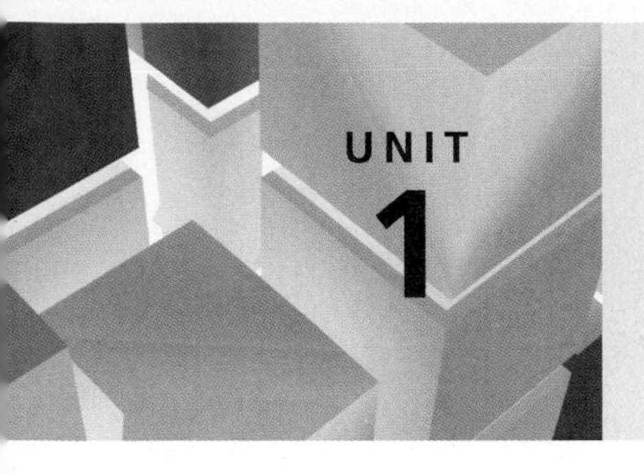


# Introduction
# Number Systems and Conversion

## Objectives

1. Introduction
   The first part of this unit introduces the material to be studied later. In addition to getting an overview of the material in the first part of the course, you should be able to explain
   a. The difference between analog and digital systems and why digital systems are capable of greater accuracy
   b. The difference between combinational and sequential circuits
   c. Why two-valued signals and binary numbers are commonly used in digital systems

2. Number systems and conversion
   When you complete this unit, you should be able to solve the following types of problems:
   a. Given a positive integer, fraction, or mixed number in any base (2 through 16); convert to any other base. Justify the procedure used by using a power series expansion for the number.
   b. Add, subtract, multiply, and divide positive binary numbers. Explain the addition and subtraction process in terms of carries and borrows.
   c. Write negative binary numbers in sign and magnitude, 1's complement, and 2's complement forms. Add signed binary numbers using 1's complement and 2's complement arithmetic. Justify the methods used. State when an overflow occurs.
   d. Represent a decimal number in binary-coded-decimal (BCD), 6-3-1-1 code, excess-3 code, etc. Given a set of weights, construct a weighted code.

# Study Guide

1. Study Section 1.1, *Digital Systems and Switching Circuits,* and answer the following study questions:

   (a) What is the basic difference between analog and digital systems?

   (b) Why are digital systems capable of greater accuracy than analog systems?

   (c) Explain the difference between combinational and sequential switching circuits.

   (d) What common characteristic do most switching devices used in digital systems have?

   (e) Why are binary numbers used in digital systems?

2. Study Section 1.2, *Number Systems and Conversion.* Answer the following study questions as you go along:

   (a) Is the first remainder obtained in the division method for base conversion the most or least significant digit?

   (b) Work through all of the examples in the text as you encounter them and make sure that you understand all of the steps.

   (c) An easy method for conversion between binary and hexadecimal is illustrated in Equation (1-1). Why should you start forming the groups of four bits at the binary point instead of the left end of the number?

   (d) Why is it impossible to convert a decimal number to binary on a digit-by-digit basis as can be done for hexadecimal?

(e) Complete the following conversion table.

| Binary (base 2) | Octal (base 8) | Decimal (base 10) | Hexadecimal (base 16) |
|---|---|---|---|
| 0 | 0 | 0 | 0 |
| 1 | | | |
| 10 | | | |
| 11 | | | |
| 100 | | | |
| 101 | | | |
| 110 | | | |
| 111 | | | |
| 1000 | | | |
| 1001 | | | |
| 1010 | | | |
| 1011 | | | |
| 1100 | | | |
| 1101 | | | |
| 1110 | | | |
| 1111 | | | |
| 10000 | 20 | 16 | 10 |

(f) Work Problems 1.1, 1.2, 1.3, and 1.4.

**3.** Study Section 1.3, *Binary Arithmetic.*

(a) Make sure that you can follow all of the examples, especially the propagation of borrows in the subtraction process.

(b) To make sure that you understand the borrowing process, work out a detailed analysis in terms of powers of 2 for the following example:

$$\begin{array}{r} 1100 \\ -\ 101 \\ \hline 111 \end{array}$$

**4.** Work Problems 1.5, 1.6, and 1.17(a).

**5.** Study Section 1.4, *Representation of Negative Numbers.*

(a) In digital systems, why are 1's complement and 2's complement commonly used to represent negative numbers instead of sign and magnitude?

(b) State two different ways of forming the 1's complement of an $n$-bit binary number.

(c) State three different ways of forming the 2's complement of an $n$-bit binary number.

(d) If the word length is $n = 4$ bits (including sign), what decimal number does $1000_2$ represent in sign and magnitude?
In 2's complement?
In 1's complement?

(e) Given a negative number represented in 2's complement, how do you find its magnitude?

Given a negative number represented in 1's complement, how do you find its magnitude?

(f) If the word length is 6 bits (including sign), what decimal number does $100000_2$ represent in sign and magnitude?

In 2's complement?

In 1's complement?

(g) What is meant by an overflow? How can you tell that an overflow has occurred when performing 1's or 2's complement addition?

Does a carry out of the last bit position indicate that an overflow has occurred?

(h) Work out some examples of 1's and 2's complement addition for various combinations of positive and negative numbers.

(i) What is the justification for using the end-around carry in 1's complement addition?

(j) The one thing that causes the most trouble with 2's complement numbers is the special case of the negative number which consists of a 1 followed by all 0's (1000 . . . 000). If this number is $n$ bits long, what number does it represent and why? (It is not negative zero.)

(k) Work Problems 1.7 and 1.8.

**6.** Study Section 1.5, *Binary Codes.*

(a) Represent 187 in BCD code, excess-3 code, 6-3-1-1 code, and 2-out-of-5 code.

(b) Verify that the 6-3-1-1 code is a weighted code. Note that for some decimal digits, two different code combinations could have been used. For example, either 0101 or 0110 could represent 4. In each case the combination with the smaller binary value has been used.

(c) How is the excess-3 code obtained?

(d) How are the ASCII codes for the decimal digits obtained? What is the relation between the ASCII codes for the capital letters and lowercase letters?

(e) Work Problem 1.9.

**7.** If you are taking this course on a self-paced basis, you will need to pass a readiness test on this unit before going on to the next unit. The purpose of the readiness test is to determine if you have mastered the material in this unit and are ready to go on to the next unit. Before you take the readiness test:

(a) Check your answers to the problems against those provided at the end of this book. If you missed any of the problems, make sure that you understand why your answer is wrong and correct your solution.

(b) Make sure that you can meet all of the objectives listed at the beginning of this unit.

# Introduction Number Systems and Conversion

## 1.1 Digital Systems and Switching Circuits

Digital systems are used extensively in computation and data processing, control systems, communications, and measurement. Because digital systems are capable of greater accuracy and reliability than analog systems, many tasks formerly done by analog systems are now being performed digitally.

In a digital system, the physical quantities or signals can assume only discrete values, while in analog systems the physical quantities or signals may vary continuously over a specified range. For example, the output voltage of a digital system might be constrained to take on only two values such as 0 volts and 5 volts, while the output voltage from an analog system might be allowed to assume any value in the range −10 volts to +10 volts.

Because digital systems work with discrete quantities, in many cases they can be designed so that for a given input, the output is exactly correct. For example, if we multiply two 5-digit numbers using a digital multiplier, the 10-digit product will be correct in all 10 digits. On the other hand, the output of an analog multiplier might have an error ranging from a fraction of one percent to a few percent depending on the accuracy of the components used in construction of the multiplier. Furthermore, if we need a product which is correct to 20 digits rather than 10, we can redesign the digital multiplier to process more digits and add more digits to its input. A similar improvement in the accuracy of an analog multiplier would not be possible because of limitations on the accuracy of the components.

The design of digital systems may be divided roughly into three parts—system design, logic design, and circuit design. System design involves breaking the overall system into subsystems and specifying the characteristics of each subsystem. For example, the system design of a digital computer could involve specifying the number and type of memory units, arithmetic units, and input-output devices as well as the interconnection and control of these subsystems. Logic design involves determining how to interconnect basic logic building blocks to perform a specific function. An example of logic design is determining the interconnection of logic gates and flip-flops required to perform binary addition. Circuit design involves specifying the interconnection of specific components such as resistors, diodes, and transistors to form a gate, flip-flop, or other logic building block. Most contemporary circuit design is done in integrated circuit form using appropriate computer-aided design tools to lay out and interconnect the components on a chip of silicon. This book is largely devoted to a study of logic design and the theory necessary for understanding the logic design process. Some aspects of system design are treated in Units 18 and 20. Circuit design of logic gates is discussed briefly in Appendix A.

Many of a digital system's subsystems take the form of a switching circuit (Figure 1-1). A switching circuit has one or more inputs and one or more outputs which take on discrete values. In this text, we will study two types of switching circuits—combinational and sequential. In a combinational circuit, the output values depend only on the present value of the inputs and not on past values. In a sequential circuit, the outputs depend on both the present and past input values. In other words, in order to determine the output of a sequential circuit, a sequence of input values must be specified. The sequential circuit is said to have memory because it must "remember" something about the past sequence of inputs, while a combinational circuit has no memory. In general, a sequential circuit is composed of a combinational circuit with added memory elements. Combinational circuits are easier to design than sequential circuits and will be studied first.

**FIGURE 1-1**
Switching Circuit

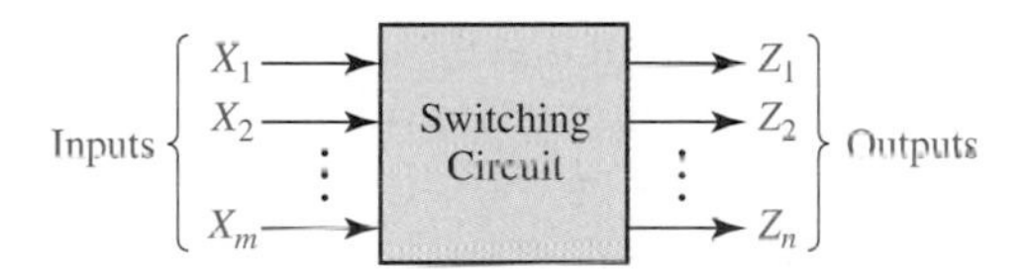


The basic building blocks used to construct combinational circuits are logic gates. The logic designer must determine how to interconnect these gates in order to convert the circuit input signals into the desired output signals. The relationship between these input and output signals can be described mathematically using Boolean algebra. Units 2 and 3 of this text introduce the basic laws and theorems of Boolean algebra and show how they can be used to describe the behavior of circuits of logic gates.

Starting from a given problem statement, the first step in designing a combinational logic circuit is to derive a table or the algebraic logic equations which describe the circuit outputs as a function of the circuit inputs (Unit 4). In order to design an economical circuit to realize these output functions, the logic equations which describe the circuit outputs generally must be simplified. Algebraic methods for this simplification are described in Unit 3, and other simplification methods (Karnaugh map and Quine-McCluskey procedure) are introduced in Units 5 and 6. Implementation of the simplified logic equations using several types of gates is described in Unit 7, and alternative design procedures using programmable logic devices are developed in Unit 9.

The basic memory elements used in the design of sequential circuits are called flip-flops (Unit 11). These flip-flops can be interconnected with gates to form counters and registers (Unit 12). Analysis of more general sequential circuits using timing diagrams, state tables, and graphs is presented in Unit 13. The first step in designing a sequential switching circuit is to construct a state table or graph which describes the relationship between the input and output sequences (Unit 14). Methods for going from a state table or graph to a circuit of gates and flip-flops are developed in Unit 15. Methods of implementing sequential circuits using programmable logic are discussed in Unit 16. In Unit 18, combinational and sequential design techniques are applied to the realization of systems for performing binary addition, multiplication, and division. The sequential circuits designed in this text are called synchronous sequential circuits because they use a common timing signal, called a clock, to synchronize the operation of the memory elements.

Use of a hardware description language, VHDL, in the design of combinational logic, sequential logic, and digital systems is introduced in Units 10, 17, and 20. VHDL is used to describe, simulate, and synthesize digital hardware. After writing VHDL code, the designer can use computer-aided design software to compile the hardware description and complete the design of the digital logic. This allows the completion of complex designs without having to manually work out detailed circuit descriptions in terms of gates and flip-flops.

The switching devices used in digital systems are generally two-state devices, that is, the output can assume only two different discrete values. Examples of switching devices are relays, diodes, and transistors. A relay can assume two states—closed or open—depending on whether power is applied to the coil or not. A diode can be in a conducting state or a nonconducting state. A transistor can be in a cut-off or saturated state with a corresponding high or low output voltage. Of course, transistors can also be operated as linear amplifiers with a continuous range of output voltages, but in digital applications greater reliability is obtained by operating them as two-state devices. Because the outputs of most switching devices assume only two different values, it is natural to use binary numbers internally in digital systems. For this reason binary numbers and number systems will be discussed first before proceeding to the design of switching circuits.

## 1.2 Number Systems and Conversion

When we write decimal (base 10) numbers, we use a positional notation; each digit is multiplied by an appropriate power of 10 depending on its position in the number. For example,

$$953.78_{10} = 9 \times 10^2 + 5 \times 10^1 + 3 \times 10^0 + 7 \times 10^{-1} + 8 \times 10^{-2}$$

Similarly, for binary (base 2) numbers, each binary digit is multiplied by the appropriate power of 2:

$$\begin{aligned} 1011.11_2 &= 1 \times 2^3 + 0 \times 2^2 + 1 \times 2^1 + 1 \times 2^0 + 1 \times 2^{-1} + 1 \times 2^{-2} \\ &= 8 + 0 + 2 + 1 + \tfrac{1}{2} + \tfrac{1}{4} = 11\tfrac{3}{4} = 11.75_{10} \end{aligned}$$

Note that the binary point separates the positive and negative powers of 2 just as the decimal point separates the positive and negative powers of 10 for decimal numbers.

Any positive integer $R$ ($R > 1$) can be chosen as the *radix* or *base* of a number system. If the base is $R$, then $R$ digits $(0, 1, \ldots, R-1)$ are used. For example, if $R = 8$, then the required digits are 0, 1, 2, 3, 4, 5, 6, and 7. A number written in positional notation can be expanded in a power series in $R$. For example,

$$\begin{aligned} N &= (a_4a_3a_2a_1a_0.a_{-1}a_{-2}a_{-3})_R \\ &= a_4 \times R^4 + a_3 \times R^3 + a_2 \times R^2 + a_1 \times R^1 + a_0 \times R^0 \\ &\qquad + a_{-1} \times R^{-1} + a_{-2} \times R^{-2} + a_{-3} \times R^{-3} \end{aligned}$$

where $a_i$ is the coefficient of $R^i$ and $0 \le a_i \le R-1$. If the arithmetic indicated in the power series expansion is done in base 10, then the result is the decimal equivalent of $N$. For examplc,

$$\begin{aligned} 147.3_8 &= 1 \times 8^2 + 4 \times 8^1 + 7 \times 8^0 + 3 \times 8^{-1} = 64 + 32 + 7 + \\ &= 103.375_{10} \end{aligned}$$

The power series expansion can be used to convert to any base. For example, converting $147_{10}$ to base 3 would be written as

$$147_{10} = 1 \times (101)^2 + (11) \times (101)^1 + (21) \times (101)^0$$

where all the numbers on the right-hand side are base 3 numbers. (*Note:* In base 3, 10 is 101, 7 is 21, etc.) To complete the conversion, base 3 arithmetic would be used. Of course, this is not very convenient if the arithmetic is being done by hand. Similarly, if $147_{10}$ is being converted to binary, the calculation would be

$$147_{10} = 1 \times (1010)^2 + (100) \times (1010)^1 + (111) \times (1010)^0$$

Again this is not convenient for hand calculation but it could be done easily in a computer where the arithmetic is done in binary. For hand calculation, use the power series expansion when converting from some base *into base 10*.

For bases greater than 10, more than 10 symbols are needed to represent the digits. In this case, letters are usually used to represent digits greater than 9. For example, in hexadecimal (base 16), $A$ represents $10_{10}$, $B$ represents $11_{10}$, $C$ represents $12_{10}$, $D$ represents $13_{10}$, $E$ represents $14_{10}$, and $F$ represents $15_{10}$. Thus,

$$A2F_{16} = 10 \times 16^2 + 2 \times 16^1 + 15 \times 16^0 = 2560 + 32 + 15 = 2607_{10}$$

Next, we will discuss conversion of a decimal *integer* to base $R$ using the division method. The base $R$ equivalent of a decimal integer $N$ can be represented as

$$N = (a_na_{n-1} \cdots a_2a_1a_0)_R = a_nR^n + a_{n-1}R^{n-1} + \cdots + a_2R^2 + a_1R^1 + a_0$$

If we divide $N$ by $R$, the remainder is $a_0$:

$$\frac{N}{R} = a_nR^{n-1} + a_{n-1}R^{n-2} + \cdots + a_2R^1 + a_1 = Q_1, \text{ remainder } a_0$$

Then we divide the quotient $Q_1$ by $R$:

$$\frac{Q_1}{R} = a_nR^{n-2} + a_{n-1}R^{n-3} + \cdots + a_3R^1 + a_2 = Q_2, \text{ remainder } a_1$$

Next we divide $Q_2$ by $R$:

$$\frac{Q_2}{R} = a_nR^{n-3} + a_{n-1}R^{n-4} + \cdots + a_3 = Q_3, \text{ remainder } a_2$$

This process is continued until we finally obtain $a_n$. Note that the remainder obtained at each division step is one of the desired digits and the least significant digit is obtained first.

**Example**

Convert $53_{10}$ to binary.

$$\begin{array}{rll}
2\,/\underline{53} & & \\
2\,/\underline{26} & \text{rem.} = 1 = a_0 & \\
2\,/\underline{13} & \text{rem.} = 0 = a_1 & \\
2\,/\underline{6} & \text{rem.} = 1 = a_2 & 53_{10} = 110101_2 \\
2\,/\underline{3} & \text{rem.} = 0 = a_3 & \\
2\,/\underline{1} & \text{rem.} = 1 = a_4 & \\
0 & \text{rem.} = 1 = a_5 &
\end{array}$$

---

Conversion of a decimal *fraction* to base $R$ can be done using successive *multiplications* by $R$. A decimal fraction $F$ can be represented as

$$F = (.a_{-1}\,a_{-2}\,a_{-3}\cdots a_{-m})_R = a_{-1}\,R^{-1} + a_{-2}R^{-2} + a_{-3}R^{-3} + \cdots + a_{-m}R^{-m}$$

Multiplying by $R$ yields

$$FR = a_{-1} + a_{-2}R^{-1} + a_{-3}R^{-2} + \cdots + a_{-m}R^{-m+1} = a_{-1} + F_1$$

where $F_1$ represents the fractional part of the result and $a_{-1}$ is the integer part. Multiplying $F_1$ by $R$ yields

$$F_1R = a_{-2} + a_{-3}R^{-1} + \cdots + a_{-m}R^{-m+2} = a_{-2} + F_2$$

Next, we multiply $F_2$ by $R$:

$$F_2R = a_{-3} + \cdots + a_{-m}R^{-m+3} = a_{-3} + F_3$$

This process is continued until we have obtained a sufficient number of digits. Note that the integer part obtained at each step is one of the desired digits and the most significant digit is obtained first.

***Example***

Convert $0.625_{10}$ to binary.

$$\begin{array}{rrr} F = .625 & F_1 = .250 & F_2 = .500 \\ \times\ 2 & \times\ 2 & \times\ 2 \\ \hline 1.250 & 0.500 & 1.000 \\ (a_{-1} = 1) & (a_{-2} = 0) & (a_{-3} = 1) \end{array} \qquad .625_{10} = .101_2$$

This process does not always terminate, but if it does not terminate, the result is a repeating fraction.

***Example***

Convert $0.7_{10}$ to binary.

$$\begin{array}{r} .7 \\ \underline{2} \\ (1).4 \\ \underline{2} \\ (0).8 \\ \underline{2} \\ (1).6 \\ \underline{2} \\ (1).2 \\ \underline{2} \\ (0).4 \\ \underline{2} \\ (0).8 \end{array}$$

← process starts repeating here because 0.4 was previously obtained (at the (0).4 step)

$0.7_{10} = 0.1\ \underline{0110}\ \underline{0110}\ \underline{0110} \ldots_2$

Conversion between two bases other than decimal can be done directly by using the procedures given; however, the arithmetic operations would have to be carried out using a base other than 10. It is generally easier to convert to decimal first and then convert the decimal number to the new base.

**Example**

Convert $231.3_4$ to base 7.

$$231.3_4 = 2 \times 16 + 3 \times 4 + 1 + \tfrac{3}{4} = 45.75_{10}$$

$$\begin{array}{rl} 7\,\underline{/45} & \\ 7\,\underline{/\ 6} & \text{rem. } 3 \\ 0 & \text{rem. } 6 \end{array} \qquad \begin{array}{r} .75 \\ \underline{7} \\ (5)\,.25 \\ \underline{7} \\ (1)\,.75 \\ \underline{7} \\ (5)\,.25 \\ \underline{7} \\ (1)\,.75 \end{array} \qquad 45.75_{10} = 63.5151\ldots_7$$

Conversion from binary to hexadecimal (and conversely) can be done by inspection because each hexadecimal digit corresponds to exactly four binary digits (bits). Starting at the binary point, the bits are divided into groups of four, and each group is replaced by a hexadecimal digit:

$$1001101.010111_2 = \underbrace{0100}_{4}\ \underbrace{1101}_{D}\ .\ \underbrace{0101}_{5}\ \underbrace{1100}_{C} = 4D.5C_{16} \tag{1-1}$$

As shown in Equation (1-1), extra 0's are added at each end of the bit string as needed to fill out the groups of four bits.

## 1.3 Binary Arithmetic

Arithmetic operations in digital systems are usually done in binary because design of logic circuits to perform binary arithmetic is much easier than for decimal. Binary arithmetic is carried out in much the same manner as decimal, except the addition and multiplication tables are much simpler.

The addition table for binary numbers is

$$\begin{aligned} 0 + 0 &= 0 \\ 0 + 1 &= 1 \\ 1 + 0 &= 1 \\ 1 + 1 &= 0 \quad \text{and carry 1 to the next column} \end{aligned}$$

Carrying 1 to a column is equivalent to adding 1 to that column.

*Example*

Add $13_{10}$ and $11_{10}$ in binary.

$$\begin{array}{rl} & 1111 \longleftarrow \text{carries} \\ 13_{10} = & 1101 \\ 11_{10} = & \underline{1011} \\ & 11000 = 24_{10} \end{array}$$

The subtraction table for binary numbers is

$$\begin{array}{ll} 0 - 0 = 0 & \\ 0 - 1 = 1 & \text{and borrow 1 from the next column} \\ 1 - 0 = 1 & \\ 1 - 1 = 0 & \end{array}$$

Borrowing 1 from a column is equivalent to subtracting 1 from that column.

*Examples of Binary Subtraction*

(a)
$$\begin{array}{rl} 1\longleftarrow & \text{(indicates} \\ 11101 & \text{a borrow} \\ -\underline{10011} & \text{from the} \\ 1010 & \text{3rd column)} \end{array}$$

(b)
$$\begin{array}{rl} 1111\longleftarrow & \text{borrows} \\ 10000 & \\ -\underline{\quad 11} & \\ 1101 & \end{array}$$

(c)
$$\begin{array}{rl} 111\longleftarrow & \text{borrows} \\ 111001 & \\ -\underline{\quad 1011} & \\ 101110 & \end{array}$$

Note how the borrow propagates from column to column in the second example. In order to borrow 1 from the second column, we must in turn borrow 1 from the third column, etc. An alternative to binary subtraction is the use of 2's complement arithmetic, as discussed in Section 1.4.

Binary subtraction sometimes causes confusion, perhaps because we are so used to doing decimal subtraction that we forget the significance of the borrowing process. Before doing a detailed analysis of binary subtraction, we will review the borrowing process for decimal subtraction.

If we number the columns (digits) of a decimal integer from right to left (starting with 0), and then we borrow 1 from column $n$, what we mean is that we subtract 1 from column $n$ and add 10 to column $n - 1$. Because $1 \times 10^n = 10 \times 10^{n-1}$, the value of the decimal number is unchanged, but we can proceed with the subtraction. Consider, for example, the following decimal subtraction problem:

column 2 ↘ column 1 ↙

$$\begin{array}{r} 205 \\ -\underline{\ 18} \\ 187 \end{array}$$

A detailed analysis of the borrowing process for this example, indicating first a borrow of 1 from column 1 and then a borrow of 1 from column 2, is as follows:

$$
\begin{aligned}
205 - 18 &= [2 \times 10^2 + 0 \times 10^1 + 5 \times 10^0] \\
&\quad - [\qquad\qquad 1 \times 10^1 + 8 \times 10^0] \\
&\qquad \text{note borrow from column 1} \\
&= [2 \times 10^2 + (0 - 1) \times 10^1 + (10 + 5) \times 10^0] \\
&\quad - [\qquad\qquad 1 \times 10^1 + \qquad 8 \times 10^0] \\
&\qquad \text{note borrow from column 2} \\
&= [(2 - 1) \times 10^2 + (10 + 0 - 1) \times 10^1 + 15 \times 10^0] \\
&\quad - [\qquad\qquad\qquad 1 \times 10^1 + 8 \times 10^0] \\
&= [1 \times 10^2 \quad + \quad 8 \times 10^1 \quad + \quad 7 \times 10^0] = 187
\end{aligned}
$$

The analysis of borrowing for binary subtraction is exactly the same, except that we work with powers of 2 instead of powers of 10. Thus for a binary number, borrowing 1 from column $n$ is equivalent to subtracting 1 from column $n$ and adding 2 ($10_2$) to column $n - 1$. The value of the binary number is unchanged because $1 \times 2^n = 2 \times 2^{n-1}$.

A detailed analysis of binary subtraction example (c) follows. Starting with the rightmost column, $1 - 1 = 0$. To subtract in the second column, we must borrow from the third column. Rather than borrow immediately, we place a 1 over the third column to indicate that a borrow is necessary, and we will actually do the borrowing when we get to the third column. (This is similar to the way borrow signals might propagate in a computer.) Now because we have borrowed 1, the second column becomes 10, and $10 - 1 = 1$. In order to borrow 1 from the third column, we must borrow 1 from the fourth column (indicated by placing a 1 over column 4). Column 3 then becomes 10, subtracting off the borrow yields 1, and $1 - 0 = 1$. Now in column 4, we subtract off the borrow leaving 0. In order to complete the subtraction, we must borrow from column 5, which gives 10 in column 4, and $10 - 1 = 1$.

The multiplication table for binary numbers is

$$
\begin{aligned}
0 \times 0 &= 0 \\
0 \times 1 &= 0 \\
1 \times 0 &= 0 \\
1 \times 1 &= 1
\end{aligned}
$$

The following example illustrates multiplication of $13_{10}$ by $11_{10}$ in binary:

$$
\begin{array}{r}
1101 \\
1011 \\
\hline
1101 \\
1101\phantom{0} \\
0000\phantom{00} \\
1101\phantom{000} \\
\hline
10001111
\end{array}
= 143_{10}
$$

Note that each partial product is either the multiplicand (1101) shifted over the appropriate number of places or is zero.

When adding up long columns of binary numbers, the sum of the bits in a single column can exceed $11_2$, and therefore the carry to the next column can be greater than 1. For example, if a single column of bits contains five 1's, then adding up the 1's gives $101_2$, which means that the sum bit for that column is 1, and the carry to the next column is $10_2$. When doing binary multiplication, a common way to avoid carries greater than 1 is to add in the partial products one at a time as illustrated by the following example:

```
    1111     multiplicand
    1101     multiplier
    ----
    1111     first partial product
   0000      second partial product
   -----
  (01111)    sum of first two partial products
  1111       third partial product
 --------
(1001011)    sum after adding third partial product
 1111        fourth partial product
 --------
11000011     final product (sum after adding fourth partial product)
```

The following example illustrates division of $145_{10}$ by $11_{10}$ in binary:

```
         1101
     ________
1011 |10010001
      1011
      ----
       1110
       1011
       ----
        1101     The quotient is 1101 with a remainder
        1011     of 10.
        ----
          10
```

Binary division is similar to decimal division, except it is much easier because the only two possible quotient digits are 0 and 1. In the above example, if we start by comparing the divisor (1011) with the upper four bits of the dividend (1001), we find that we cannot subtract without a negative result, so we move the divisor one place to the right and try again. This time we can subtract 1011 from 10010 to give 111 as a result, so we put the first quotient bit of 1 above 10010. We then bring down the next dividend bit (0) to get 1110 and shift the divisor right. We then subtract 1011 from 1110 to get 11, so the second quotient bit is 1. When we bring down the next dividend bit, the result is 110, and we cannot subtract the shifted divisor, so the third quotient bit is 0. We then bring down the last dividend bit and subtract 1011 from 1101 to get a final remainder of 10, and the last quotient bit is 1.

## 1.4 Representation of Negative Numbers

Up to this point we have been working with unsigned positive numbers. In most computers, in order to represent both positive and negative numbers the first bit in a word is used as a sign bit, with 0 used for plus and 1 used for minus. Several representations of negative binary numbers are possible. The *sign and magnitude* system is similar to that which people commonly use. For an $n$-bit word, the first bit is the sign and the remaining $n - 1$ bits represent the magnitude of the number. Thus an $n$-bit word can represent any one of $2^{n-1}$ positive integers or $2^{n-1}$ negative integers. Table 1-1 illustrates this for $n = 4$. For example, 0011 represents $+3$ and 1011 represents $-3$. Note that 1000 represents minus zero in the sign and magnitude system and $-8$ in the 2's complement system.

The design of logic circuits to do arithmetic with sign and magnitude binary numbers is awkward; therefore, other representations are often used. The 2's complement and 1's complement are commonly used because arithmetic units are easy to design using these systems. For the 2's complement number system, a positive number, $N$, is represented by a 0 followed by the magnitude as in the sign and magnitude system; however, a negative number, $-N$, is represented by its 2's complement, $N^*$. If the word length is $n$ bits, the 2's complement of a positive integer $N$ is defined as for a word length of $n$ bits.

$$N^* = 2^n - N \tag{1-2}$$

For $n = 4$, $-N$ is represented by $16 - N$ as shown in Table 1-1. For example, $-3$ is represented by $16 - 3 = 13 = 1101_2$. As is the case for sign and magnitude numbers, all negative 2's complement numbers have a 1 in the position furthest to the left (sign bit).

For the 1's complement system a negative number, $-N$, is represented by its 1's complement, $\overline{N}$. The 1's complement of a positive integer $N$ is defined as

$$\overline{N} = (2^n - 1) - N \tag{1-3}$$

**TABLE 1-1** Signed Binary Integers (word length: $n = 4$)

| $+N$ | Positive Integers (all systems) | $-N$ | Negative Integers | | |
|---|---|---|---|---|---|
| | | | Sign and Magnitude | 2's Complement $N^*$ | 1's Complement $\overline{N}$ |
| +0 | 0000 | −0 | 1000 | — | 1111 |
| +1 | 0001 | −1 | 1001 | 1111 | 1110 |
| +2 | 0010 | −2 | 1010 | 1110 | 1101 |
| +3 | 0011 | −3 | 1011 | 1101 | 1100 |
| +4 | 0100 | −4 | 1100 | 1100 | 1011 |
| +5 | 0101 | −5 | 1101 | 1011 | 1010 |
| +6 | 0110 | −6 | 1110 | 1010 | 1001 |
| +7 | 0111 | −7 | 1111 | 1001 | 1000 |
| | | −8 | — | 1000 | — |

Note that 1111 represents minus zero, and $-8$ has no representation in a 4-bit system. An alternate way to form the 1's complement is to simply complement $N$ bit-by-bit by replacing 0's with 1's and 1's with 0's. This is equivalent to the definition, Equation (1-3), because $2^n - 1$ consists of all 1's, and subtracting a bit from 1 is the same as complementing the bit. No borrows occur in this subtraction. For example, if $n = 6$ and $N = 010101$,

$$\begin{aligned} 2^n - 1 &= 111111 \\ N &= \underline{010101} \\ \overline{N} &= 101010 \end{aligned}$$

From Equations (1-2) and (1-3).

$$N^* = 2^n - N = (2^n - 1 - N) + 1 = \overline{N} + 1$$

so the 2's complement can be formed by complementing $N$ bit-by-bit and then adding 1. An easier way to form the 2's complement of $N$ is to start at the right and complement all bits to the left of the first 1. For example, if

$$N = 0101100, \text{ then } N^* = 1010100$$

From Equations (1-2) and (1-3),

$$N = 2^n - N^* \qquad \text{and} \qquad N = (2^n - 1) - \overline{N}$$

Therefore, given a negative integer represented by its 2's complement ($N^*$), we can obtain the magnitude of the integer by taking the 2's complement of $N^*$. Similarly, to get the magnitude of a negative integer represented by its 1's complement ($\overline{N}$), we can take the 1's complement of $\overline{N}$.

In the 2's complement system the number of negative integers which can be represented is one more than the number of positive integers (not including 0). For example, in Table 1-1, 1000 represents $-8$, because a sign bit of 1 indicates a negative number, and if $N = 8$, $N^* = 10000 - 1000 = 1000$. In general, in a 2's complement system with a word length of $n$ bits, the number 100 . . . 000 (1 followed by $n - 1$ 0's) represents a negative number with a magnitude of

$$2^n - 2^{n-1} = 2^{n-1}$$

This special case occurs only for 2's complement. However, $-0$ has no representation in 2's complement, but $-0$ is a special case for 1's complement as well as for the sign and magnitude system.

## Addition of 2's Complement Numbers

The addition of $n$-bit signed binary numbers is straightforward using the 2's complement system. The addition is carried out just as if all the numbers were positive, and any carry from the sign position is ignored. This will always yield the correct result except when an overflow occurs. When the word length is $n$ bits, we say that an *overflow* has occurred if the correct representation of the sum (including sign) requires more than $n$ bits. The different cases which can occur are illustrated below for $n = 4$.

**1.** Addition of two positive numbers, sum $< 2^{n-1}$

$$\begin{array}{rr l} +3 & 0011 & \\ \underline{+4} & \underline{0100} & \\ +7 & 0111 & \text{(correct answer)} \end{array}$$

**2.** Addition of two positive numbers, sum $\geq 2^{n-1}$

$$\begin{array}{rr l} +5 & 0101 & \\ \underline{+6} & \underline{0110} & \\ & 1011 & \end{array}$$

$\longleftarrow$ wrong answer because of overflow (+11 requires 5 bits including sign)

**3.** Addition of positive and negative numbers (negative number has greater magnitude)

$$\begin{array}{rr l} +5 & 0101 & \\ \underline{-6} & \underline{1010} & \\ -1 & 1111 & \text{(correct answer)} \end{array}$$

**4.** Same as case 3 except positive number has greater magnitude

$$\begin{array}{rr} -5 & 1011 \\ \underline{+6} & \underline{0110} \\ +1 & (1)0001 \end{array}$$

$\longleftarrow$ correct answer when the carry from the sign bit is ignored (this is *not* an overflow)

**5.** Addition of two negative numbers, $|\text{sum}| \leq 2^{n-1}$

$$\begin{array}{rr} -3 & 1101 \\ \underline{-4} & \underline{1100} \\ -7 & (1)1001 \end{array}$$

$\longleftarrow$ correct answer when the last carry is ignored (this is *not* an overflow)

**6.** Addition of two negative numbers, $|\text{sum}| > 2^{n-1}$

$$\begin{array}{rr} -5 & 1011 \\ \underline{-6} & \underline{1010} \\ & (1)0101 \end{array}$$

$\longleftarrow$ wrong answer because of overflow (−11 requires 5 bits including sign)

Note that an overflow condition (cases 2 and 6) is easy to detect because in case 2 the addition of two positive numbers yields a negative result, and in case 6 the addition of two negative numbers yields a positive answer (for four bits).

The proof that throwing away the carry from the sign bit always gives the correct answer follows for cases 4 and 5:

Case 4: $-A + B$ (where $B > A$)

$$A^* + B = (2^n - A) + B = 2^n + (B - A) > 2^n$$

Throwing away the last carry is equivalent to subtracting $2^n$, so the result is $(B - A)$, which is correct.

Case 5: $-A - B$ (where $A + B \le 2^{n-1}$)

$$A^* + B^* = (2^n - A) + (2^n - B) = 2^n + 2^n - (A + B)$$

Discarding the last carry yields $2^n - (A + B) = (A + B)^*$, which is the correct representation of $-(A + B)$.

## Addition of 1's Complement Numbers

The addition of 1's complement numbers is similar to 2's complement except that instead of discarding the last carry, it is added to the $n$-bit sum in the position furthest to the right. This is referred to as an *end-around* carry. The addition of positive numbers is the same as illustrated for cases 1 and 2 under 2's complement. The remaining cases are illustrated below ($n = 4$).

**3.** Addition of positive and negative numbers (negative number with greater magnitude)

```
 +5      0101
 -6      1001
 --      ----
 -1      1110      (correct answer)
```

**4.** Same as case 3 except positive number has greater magnitude

```
 -5      1010
 +6      0110
 --      ----
     (1) 0000
      └─→   1      (end-around carry)
         ----
         0001      (correct answer, no overflow)
```

**5.** Addition of two negative numbers, $|\text{sum}| < 2^{n-1}$

```
 -3      1100
 -4      1011
 --      ----
     (1) 0111
      └─→   1      (end-around carry)
         ----
         1000      (correct answer, no overflow)
```

**6.** Addition of two negative numbers, $|\text{sum}| \ge 2^{n-1}$

```
 -5      1010
 -6      1001
 --      ----
     (1) 0011
      └─→   1      (end-around carry)
         ----
         0100      (wrong answer because of overflow)
```

Again, note that the overflow in case 6 is easy to detect because the addition of two negative numbers yields a positive result.

The proof that the end-round carry method gives the correct result follows for cases 4 and 5:

Case 4: $-A + B$ (where $B > A$)

$$\overline{A} + B = (2^n - 1 - A) + B = 2^n + (B - A) - 1$$

The end-around carry is equivalent to subtracting $2^n$ and adding 1, so the result is $(B - A)$, which is correct.

Case 5: $-A - B$ $(A + B < 2^{n-1})$

$$\overline{A} + \overline{B} = (2^n - 1 - A) + (2^n - 1 - B) = 2^n + [2^n - 1 - (A + B)] - 1$$

After the end-around carry, the result is $2^n - 1 - (A + B) = (\overline{A + B})$ which is the correct representation for $-(A + B)$.

The following examples illustrate the addition of 1's and 2's complement numbers for a word length of $n = 8$:

**1.** Add −11 and −20 in 1's complement.

+11 = 00001011 +20 = 00010100

taking the bit-by-bit complement,

−11 is represented by 11110100 and −20 by 11101011

```
    11110100    (−11)
    11101011  +(−20)
 (1) 11011111
  └──────→ 1    (end-around carry)
    11100000 = −31
```

**2.** Add −8 and +19 in 2's complement

+ 8 = 00001000

complementing all bits to the left of the first 1, −8, is represented by 11111000

```
    11111000   (−8)
    00010011   +19
 (1)00001011 = +11
  ↑ (discard last carry)
```

Note that in both cases, the addition produced a carry out of the furthest left bit position, but there is no overflow because the answer can be correctly represented by eight bits (including sign). A general rule for detecting overflow when adding two $n$-bit signed binary numbers (1's or 2's complement) to get an $n$-bit sum is:

> An overflow occurs if adding two positive numbers gives a negative answer or if adding two negative numbers gives a positive answer.

## 1.5 Binary Codes

Although most large computers work internally with binary numbers, the input-output equipment generally uses decimal numbers. Because most logic circuits only accept two-valued signals, the decimal numbers must be coded in terms of binary signals. In the simplest form of binary code, each decimal digit is replaced by its binary equivalent. For example, 937.25 is represented by

$$\begin{array}{cccccc} 9 & 3 & 7 & . & 2 & 5 \\ 1001 & 0011 & 0111 & . & 0010 & 0101 \end{array}$$

This representation is referred to as binary-coded-decimal (BCD) or more explicitly as 8-4-2-1 BCD. Note that the result is quite different than that obtained by converting the number as a whole into binary. Because there are only ten decimal digits, 1010 through 1111 are not valid BCD codes.

Table 1-2 shows several possible sets of binary codes for the ten decimal digits. Many other possibilities exist because the only requirement for a

**TABLE 1-2** Binary Codes for Decimal Digits

| Decimal Digit | 8-4-2-1 Code (BCD) | 6-3-1-1 Code | Excess-3 Code | 2-out-of-5 Code | Gray Code |
|---|---|---|---|---|---|
| 0 | 0000 | 0000 | 0011 | 00011 | 0000 |
| 1 | 0001 | 0001 | 0100 | 00101 | 0001 |
| 2 | 0010 | 0011 | 0101 | 00110 | 0011 |
| 3 | 0011 | 0100 | 0110 | 01001 | 0010 |
| 4 | 0100 | 0101 | 0111 | 01010 | 0110 |
| 5 | 0101 | 0111 | 1000 | 01100 | 1110 |
| 6 | 0110 | 1000 | 1001 | 10001 | 1010 |
| 7 | 0111 | 1001 | 1010 | 10010 | 1011 |
| 8 | 1000 | 1011 | 1011 | 10100 | 1001 |
| 9 | 1001 | 1100 | 1100 | 11000 | 1000 |

valid code is that each decimal digit be represented by a distinct combination of binary digits. To translate a decimal number to coded form, each decimal digit is replaced by its corresponding code. Thus 937 expressed in excess-3 code is 1100 0110 1010. The 8-4-2-1 (BCD) code and the 6-3-1-1 code are examples of weighted codes. A 4-bit weighted code has the property that if the weights are $w_3$, $w_2$, $w_1$, and $w_0$, the code $a_3a_2a_1a_0$ represents a decimal number $N$, where

$$N = w_3a_3 + w_2a_2 + w_1a_1 + w_0a_0$$

For example, the weights for the 6-3-1-1 code are $w_3 = 6$, $w_2 = 3$, $w_1 = 1$, and $w_0 = 1$. The binary code 1011 thus represents the decimal digit

$$N = 6 \cdot 1 + 3 \cdot 0 + 1 \cdot 1 + 1 \cdot 1 = 8$$

The excess-3 code is obtained from the 8-4-2-1 code by adding 3 (0011) to each of the codes. The 2-out-of-5 code has the property that exactly 2 out of the 5 bits are 1 for every valid code combination. This code has useful error-checking properties because if any one of the bits in a code combination is changed due to a malfunction of the logic circuitry, the number of 1 bits is no longer exactly two. The table shows one example of a Gray code. A Gray code has the property that the codes for successive decimal digits differ in exactly one bit. For example, the codes for 6 and 7 differ only in the fourth bit, and the codes for 9 and 0 differ only in the first bit. A Gray code is often used when translating an analog quantity, such as a shaft position, into digital form. In this case, a small change in the analog quantity will change only one bit in the code, which gives more reliable operation than if two or more bits changed at a time. The Gray and 2-out-of-5 codes are *not* weighted codes. In general, the decimal value of a coded digit *cannot* be computed by a simple formula when a non-weighted code is used.

Many applications of computers require the processing of data which contains numbers, letters, and other symbols such as punctuation marks. In order to transmit such alphanumeric data to or from a computer or store it internally in a computer, each symbol must be represented by a binary code. One common alphanumeric code is the ASCII code (American Standard Code for Information Interchange). This is a 7-bit code, so $2^7$ (128) different code combinations are available to represent letters, numbers, and other symbols. Table 1-3 shows a portion of the ASCII code; the code combinations not listed are used for special control functions such as "form feed" or "end of transmission." The word "*Start*" is represented in ASCII code as follows:

| 1010011 | 1110100 | 1100001 | 1110010 | 1110100 |
|---|---|---|---|---|
| S | t | a | r | t |

**TABLE 1-3** ASCII Code

| | ASCII Code | | | | | | | | ASCII Code | | | | | | | | ASCII Code | | | | | | |
|---|---|---|---|---|---|---|---|---|---|---|---|---|---|---|---|---|---|---|---|---|---|---|---|
| Character | $A_6$ | $A_5$ | $A_4$ | $A_3$ | $A_2$ | $A_1$ | $A_0$ | Character | $A_6$ | $A_5$ | $A_4$ | $A_3$ | $A_2$ | $A_1$ | $A_0$ | Character | $A_6$ | $A_5$ | $A_4$ | $A_3$ | $A_2$ | $A_1$ | $A_0$ |
| space | 0 | 1 | 0 | 0 | 0 | 0 | 0 | @ | 1 | 0 | 0 | 0 | 0 | 0 | 0 | ' | 1 | 1 | 0 | 0 | 0 | 0 | 0 |
| ! | 0 | 1 | 0 | 0 | 0 | 0 | 1 | A | 1 | 0 | 0 | 0 | 0 | 0 | 1 | a | 1 | 1 | 0 | 0 | 0 | 0 | 1 |
| " | 0 | 1 | 0 | 0 | 0 | 1 | 0 | B | 1 | 0 | 0 | 0 | 0 | 1 | 0 | b | 1 | 1 | 0 | 0 | 0 | 1 | 0 |
| # | 0 | 1 | 0 | 0 | 0 | 1 | 1 | C | 1 | 0 | 0 | 0 | 0 | 1 | 1 | c | 1 | 1 | 0 | 0 | 0 | 1 | 1 |
| $ | 0 | 1 | 0 | 0 | 1 | 0 | 0 | D | 1 | 0 | 0 | 0 | 1 | 0 | 0 | d | 1 | 1 | 0 | 0 | 1 | 0 | 0 |
| % | 0 | 1 | 0 | 0 | 1 | 0 | 1 | E | 1 | 0 | 0 | 0 | 1 | 0 | 1 | e | 1 | 1 | 0 | 0 | 1 | 0 | 1 |
| & | 0 | 1 | 0 | 0 | 1 | 1 | 0 | F | 1 | 0 | 0 | 0 | 1 | 1 | 0 | f | 1 | 1 | 0 | 0 | 1 | 1 | 0 |
| ' | 0 | 1 | 0 | 0 | 1 | 1 | 1 | G | 1 | 0 | 0 | 0 | 1 | 1 | 1 | g | 1 | 1 | 0 | 0 | 1 | 1 | 1 |
| ( | 0 | 1 | 0 | 1 | 0 | 0 | 0 | H | 1 | 0 | 0 | 1 | 0 | 0 | 0 | h | 1 | 1 | 0 | 1 | 0 | 0 | 0 |
| ) | 0 | 1 | 0 | 1 | 0 | 0 | 1 | I | 1 | 0 | 0 | 1 | 0 | 0 | 1 | i | 1 | 1 | 0 | 1 | 0 | 0 | 1 |
| * | 0 | 1 | 0 | 1 | 0 | 1 | 0 | J | 1 | 0 | 0 | 1 | 0 | 1 | 0 | j | 1 | 1 | 0 | 1 | 0 | 1 | 0 |
| + | 0 | 1 | 0 | 1 | 0 | 1 | 1 | K | 1 | 0 | 0 | 1 | 0 | 1 | 1 | k | 1 | 1 | 0 | 1 | 0 | 1 | 1 |
| , | 0 | 1 | 0 | 1 | 1 | 0 | 0 | L | 1 | 0 | 0 | 1 | 1 | 0 | 0 | l | 1 | 1 | 0 | 1 | 1 | 0 | 0 |
| – | 0 | 1 | 0 | 1 | 1 | 0 | 1 | M | 1 | 0 | 0 | 1 | 1 | 0 | 1 | m | 1 | 1 | 0 | 1 | 1 | 0 | 1 |
| . | 0 | 1 | 0 | 1 | 1 | 1 | 0 | N | 1 | 0 | 0 | 1 | 1 | 1 | 0 | n | 1 | 1 | 0 | 1 | 1 | 1 | 0 |
| / | 0 | 1 | 0 | 1 | 1 | 1 | 1 | O | 1 | 0 | 0 | 1 | 1 | 1 | 1 | o | 1 | 1 | 0 | 1 | 1 | 1 | 1 |
| 0 | 0 | 1 | 1 | 0 | 0 | 0 | 0 | P | 1 | 0 | 1 | 0 | 0 | 0 | 0 | p | 1 | 1 | 1 | 0 | 0 | 0 | 0 |
| 1 | 0 | 1 | 1 | 0 | 0 | 0 | 1 | Q | 1 | 0 | 1 | 0 | 0 | 0 | 1 | q | 1 | 1 | 1 | 0 | 0 | 0 | 1 |
| 2 | 0 | 1 | 1 | 0 | 0 | 1 | 0 | R | 1 | 0 | 1 | 0 | 0 | 1 | 0 | r | 1 | 1 | 1 | 0 | 0 | 1 | 0 |
| 3 | 0 | 1 | 1 | 0 | 0 | 1 | 1 | S | 1 | 0 | 1 | 0 | 0 | 1 | 1 | s | 1 | 1 | 1 | 0 | 0 | 1 | 1 |
| 4 | 0 | 1 | 1 | 0 | 1 | 0 | 0 | T | 1 | 0 | 1 | 0 | 1 | 0 | 0 | t | 1 | 1 | 1 | 0 | 1 | 0 | 0 |
| 5 | 0 | 1 | 1 | 0 | 1 | 0 | 1 | U | 1 | 0 | 1 | 0 | 1 | 0 | 1 | u | 1 | 1 | 1 | 0 | 1 | 0 | 1 |
| 6 | 0 | 1 | 1 | 0 | 1 | 1 | 0 | V | 1 | 0 | 1 | 0 | 1 | 1 | 0 | v | 1 | 1 | 1 | 0 | 1 | 1 | 0 |
| 7 | 0 | 1 | 1 | 0 | 1 | 1 | 1 | W | 1 | 0 | 1 | 0 | 1 | 1 | 1 | w | 1 | 1 | 1 | 0 | 1 | 1 | 1 |
| 8 | 0 | 1 | 1 | 1 | 0 | 0 | 0 | X | 1 | 0 | 1 | 1 | 0 | 0 | 0 | x | 1 | 1 | 1 | 1 | 0 | 0 | 0 |
| 9 | 0 | 1 | 1 | 1 | 0 | 0 | 1 | Y | 1 | 0 | 1 | 1 | 0 | 0 | 1 | y | 1 | 1 | 1 | 1 | 0 | 0 | 1 |
| : | 0 | 1 | 1 | 1 | 0 | 1 | 0 | Z | 1 | 0 | 1 | 1 | 0 | 1 | 0 | z | 1 | 1 | 1 | 1 | 0 | 1 | 0 |
| ; | 0 | 1 | 1 | 1 | 0 | 1 | 1 | [ | 1 | 0 | 1 | 1 | 0 | 1 | 1 | { | 1 | 1 | 1 | 1 | 0 | 1 | 1 |
| < | 0 | 1 | 1 | 1 | 1 | 0 | 0 | \ | 1 | 0 | 1 | 1 | 1 | 0 | 0 | \| | 1 | 1 | 1 | 1 | 1 | 0 | 0 |
| = | 0 | 1 | 1 | 1 | 1 | 0 | 1 | ] | 1 | 0 | 1 | 1 | 1 | 0 | 1 | } | 1 | 1 | 1 | 1 | 1 | 0 | 1 |
| > | 0 | 1 | 1 | 1 | 1 | 1 | 0 | ^ | 1 | 0 | 1 | 1 | 1 | 1 | 0 | ~ | 1 | 1 | 1 | 1 | 1 | 1 | 0 |
| ? | 0 | 1 | 1 | 1 | 1 | 1 | 1 | — | 1 | 0 | 1 | 1 | 1 | 1 | 1 | delete | 1 | 1 | 1 | 1 | 1 | 1 | 1 |

# Problems

**1.1** Convert to hexadecimal and then to binary:
(a) $757.25_{10}$ (b) $123.17_{10}$ (c) $356.89_{10}$ (d) $1063.5_{10}$

**1.2** Convert to octal. Convert to hexadecimal. Then convert both of your answers to decimal, and verify that they are the same.
(a) $111010110001.011_2$ (b) $10110011101.11_2$

1.3 Convert to base 6: $3BA.25_{14}$ (do all of the arithmetic in decimal).

1.4 (a) Convert to hexadecimal: $1457.11_{10}$. Round to two digits past the hexadecimal point.
(b) Convert your answer to binary, and then to octal.
(c) Devise a scheme for converting hexadecimal directly to base 4 and convert your answer to base 4.
(d) Convert to decimal: $DEC.A_{16}$.

1.5 Add, subtract, and multiply in binary:
(a) 1111 and 1010 (b) 110110 and 11101 (c) 100100 and 10110

1.6 Subtract in binary. Place a 1 over each column from which it was necessary to borrow.
(a) 11110100 − 1000111 (b) 1110110 − 111101 (c) 10110010 − 111101

1.7 Add the following numbers in binary using 2's complement to represent negative numbers. Use a word length of 6 bits (including sign) and indicate if an overflow occurs.
(a) $21 + 11$ (b) $(-14) + (-32)$ (c) $(-25) + 18$
(d) $(-12) + 13$ (e) $(-11) + (-21)$
Repeat (a), (c), (d), and (e) using 1's complement to represent negative numbers.

1.8 A computer has a word length of 8 bits (including sign). If 2's complement is used to represent negative numbers, what range of integers can be stored in the computer? If 1's complement is used? (Express your answers in decimal.)

1.9 Construct a table for 7-3-2-1 weighted code and write 3659 using this code.

1.10 Convert to hexadecimal and then to binary.
(a) $1305.375_{10}$ (b) $111.33_{10}$ (c) $301.12_{10}$ (d) $1644.875_{10}$

1.11 Convert to octal. Convert to hexadecimal. Then convert both of your answers to decimal, and verify that they are the same.
(a) $101111010100.101_2$ (b) $100001101111.01_2$

1.12 (a) Convert to base 3: $375.54_8$ (do all of the arithmetic in decimal).
(b) Convert to base 4: $384.74_{10}$.
(c) Convert to base 9: $A52.A4_{11}$ (do all of the arithmetic in decimal).

1.13 Convert to hexadecimal and then to binary: $544.1_9$.

1.14 Convert the decimal number $97.7_{10}$ into a number with exactly the same value represented in the following bases. The exact value requires an infinite repeating part in the fractional part of the number. Show the steps of your derivation.
(a) binary (b) octal (c) hexadecimal (d) base 3 (e) base 5

1.15 Devise a scheme for converting base 3 numbers directly to base 9. Use your method to convert the following number to base 9: $1110212.20211_3$

**1.16** Convert the following decimal numbers to octal and then to binary:
(a) $2983^{63}/_{64}$ (b) 93.70 (c) $1900^{31}/_{32}$ (d) 109.30

**1.17** Add, subtract, and multiply in binary:
(a) 1111 and 1001 (b) 1101001 and 110110 (c) 110010 and 11101

**1.18** Subtract in binary. Place a 1 over each column from which it was necessary to borrow.
(a) 10100100 – 01110011 (b) 10010011 – 01011001
(c) 11110011 – 10011110

**1.19** Divide in binary:
(a) 11101001 ÷ 101 (b) 110000001 ÷ 1110 (c) 1110010 ÷ 1001
Check your answers by multiplying out in binary and adding the remainder.

**1.20** Divide in binary:
(a) 10001101 ÷ 110 (b) 110000011 ÷ 1011 (c) 1110100 ÷ 1010

**1.21** Assume three digits are used to represent positive integers and also assume the following operations are correct. Determine the base of the numbers. Did any of the additions overflow?
(a) 654 + 013 = 000
(b) 024 + 043 + 013 + 033 = 223
(c) 024 + 043 + 013 + 033 = 201

**1.22** What is the lowest number of bits (digits) required in the binary number approximately equal to the decimal number $0.6117_{10}$ so that the binary number has the same or better precision?

**1.23** Convert $0.363636\ldots_{10}$ to its exact equivalent base 8 number.

**1.24** (a) Verify that a number in base $b$ can be converted to base $b^3$ by partitioning the digits of the base $b$ number into groups of three consecutive digits starting at the radix point and proceeding both left and right and converting each group into a base $b^3$ digit. (*Hint:* Represent the base $b$ number using the power series expansion.)
(b) Verify that a number in base $b^3$ can be converted to base $b$ by expanding each digit of the base $b^3$ number into three consecutive digits starting at the radix point and proceeding both left and right.

**1.25** Construct a table for 4-3-2-1 weighted code and write 9154 using this code.

**1.26** Is it possible to construct a 5-3-1-1 weighted code? A 6-4-1-1 weighted code? Justify your answers.

**1.27** Is it possible to construct a 5-4-1-1 weighted code? A 6-3-2-1 weighte code? Justify your answers.

**1.28** Construct a 6-2-2-1 weighted code for decimal digits. What number does 1100 0011 represent in this code?

**1.29** Construct a 5-2-2-1 weighted code for decimal digits. What numbers does 1110 0110 represent in this code?

**1.30** Construct a 7-3-2-1 code for base 12 digits. Write B4A9 using this code.

**1.31** (a) It is possible to have negative weights in a weighted code for the decimal digits, e.g., 8, 4, −2, and −1 can be used. Construct a table for this weighted code.
(b) If *d* is a decimal digit in this code, how can the code for 9 – *d* be obtained?

**1.32** Convert to hexadecimal, and then give the ASCII code for the resulting hexadecimal number (including the code for the hexadecimal point):
(a) $222.22_{10}$ (b) $183.81_{10}$

**1.33** Repeat 1.7 for the following numbers:
(a) $(-10) + (-11)$ (b) $(-10) + (-6)$ (c) $(-8) + (-11)$
(d) $11 + 9$ (e) $(-11) + (-4)$

**1.34** Because $A - B = A + (-B)$, the subtraction of signed numbers can be accomplished by adding the complement. Subtract each of the following pairs of 5-bit binary numbers by adding the complement of the subtrahend to the minuend. Indicate when an overflow occurs. Assume that negative numbers are represented in 1's complement. Then repeat using 2's complement.

| (a) | (b) | (c) | (d) | (e) |
|---|---|---|---|---|
| 01001 | 11010 | 10110 | 11011 | 11100 |
| −11010 | −11001 | −01101 | −00111 | −10101 |

**1.35** Work Problem 1.34 for the following pairs of numbers:

| (a) | (b) | (c) | (d) |
|---|---|---|---|
| 11010 | 01011 | 10001 | 10101 |
| −10100 | −11000 | −01010 | −11010 |

**1.36** (a) A = 101010 and B = 011101 are 1's complement numbers. Perform the following operations and indicate whether overflow occurs.
(i) A + B (ii) A − B
(b) Repeat Part (a) assuming the numbers are 2's complement numbers.

**1.37** (a) Assume the integers below are 1's complement integers. Find the 1's complement of each number, and give the decimal values of the original number and of its complement.
(i) 0000000 (ii) 1111111 (iii) 00110011 (iv) 1000000
(b) Repeat, assuming the numbers are 2's complement numbers and finding the 2's complement of them.

# Boolean Algebra

## Objectives

A list of 15 laws and theorems of Boolean algebra is given on page 55 of this unit. When you complete this unit, you should be familiar with and be able to use any of the first 12 of these. Specifically, you should be able to:

1. Understand the basic operations and laws of Boolean algebra.
2. Relate these operations and laws to circuits composed of AND gates, OR gates, and INVERTERS. Also relate these operations and laws to circuits composed of switches.
3. Prove any of these laws using a truth table.
4. Apply these laws to the manipulation of algebraic expressions including:
   a. Multiplying out an expression to obtain a sum of products (SOP).
   b. Factoring an expression to obtain a product of sums (POS).
   c. Simplifying an expression by applying one of the laws.
   d. Finding the complement of an expression.

# Study Guide

1. In this unit you will study Boolean algebra, the basic mathematics needed for the logic design of digital systems. Just as when you first learned ordinary algebra, you will need a fair amount of practice before you can use Boolean algebra effectively. However, by the end of the course, you should be just as comfortable with Boolean algebra as with ordinary algebra. Fortunately, many of the rules of Boolean algebra are the same as for ordinary algebra, but watch out for some surprises!

2. Study Sections 2.1 and 2.2, *Introduction* and *Basic Operations.*

   (a) How does the meaning of the symbols 0 and 1 as used in this unit differ from the meaning as used in Unit 1?

   (b) Two commonly used notations for the inverse or complement of $A$ are $\overline{A}$ and $A'$. The latter has the advantage that it is much easier for typists, printers, and computers. (Have you ever tried to get a computer to print a bar over a letter?) We will use $A'$ for the complement of $A$. You may use either notation in your work, but please do not mix notations in the same equation. Most engineers use + for OR and • (or no symbol) for AND, and we will follow this practice. An alternative notation, often used by mathematicians, is $\vee$ for OR and $\wedge$ for AND.

   (c) Many different symbols are used for AND, OR, and INVERTER logic blocks. Initially we will use

   for AND    for OR    for INVERTER

   The shapes of these symbols conform to those commonly used in industrial practice. We have added the + and • for clarity. These symbols point in the direction of signal flow. This makes it easier to read the circuit diagrams in comparison with the square or round symbols used in some books.

   (d) Determine the output of each of the following gates:

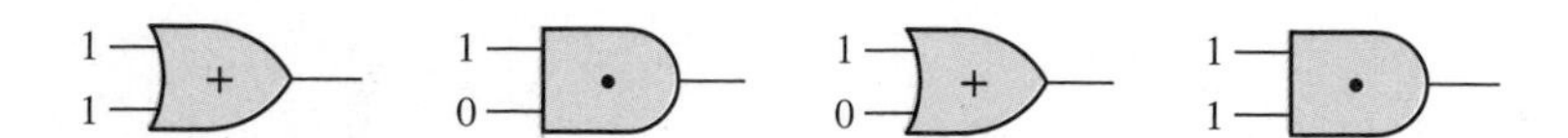


   (e) Determine the unspecified inputs to each of the following gates if the outputs are as shown:

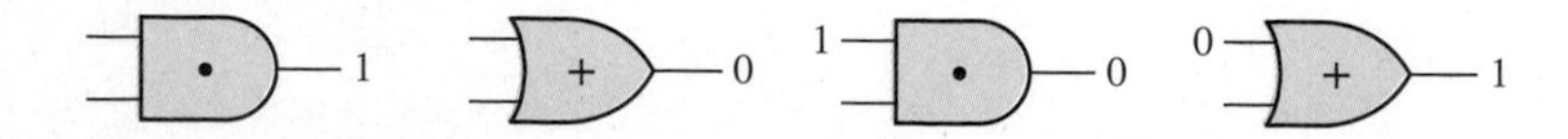

**3.** Study Section 2.3, *Boolean Expressions and Truth Tables.*

(a) How many *variables* does the following expression contain?
How many literals?

$$A'BC'D + AB + B'CD + D'$$

(b) For the following circuit, if $A = B = 0$ and $C = D = E = 1$, indicate the output of each gate (0 or 1) on the circuit diagram:

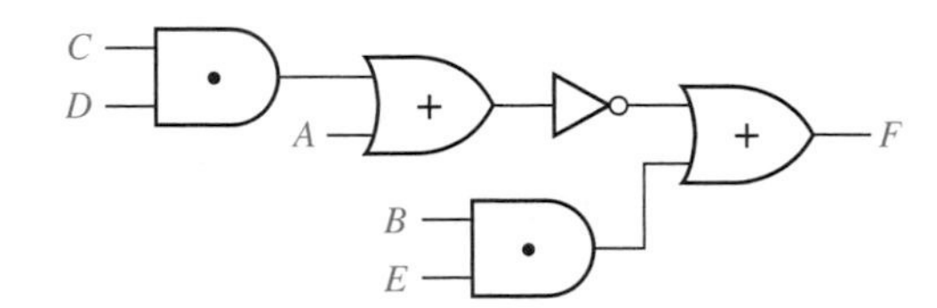


(c) Derive a Boolean expression for the circuit output. Then substitute $A = B = 0$ and $C = D = E = 1$ into your expression and verify that the value of $F$ obtained in this way is the same as that obtained on the circuit diagram in (b).

(d) Write an expression for the output of the following circuit and complete the truth table:

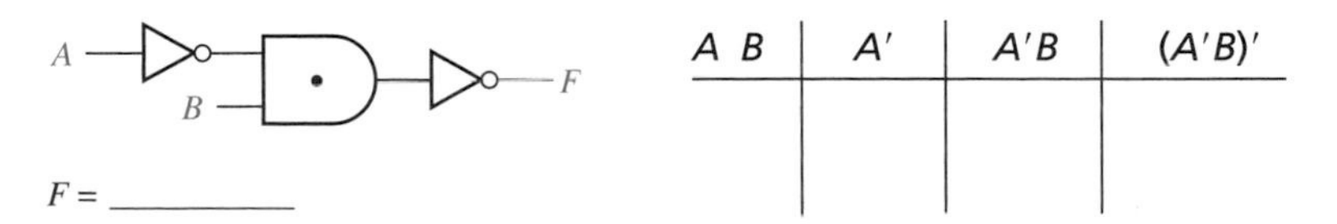


| A B | A' | A'B | (A'B)' |
| --- | --- | --- | --- |
| | | | |

(e) When filling in the combinations of values for the variables on the left side of a truth table, always list the combinations of 0's and 1's in binary order. For example, for a three-variable truth table, the first row should be 000, the next row 001, then 010, 011, 100, 101, 110, and 111. Write an expression for the output of the following circuit and complete the truth table:

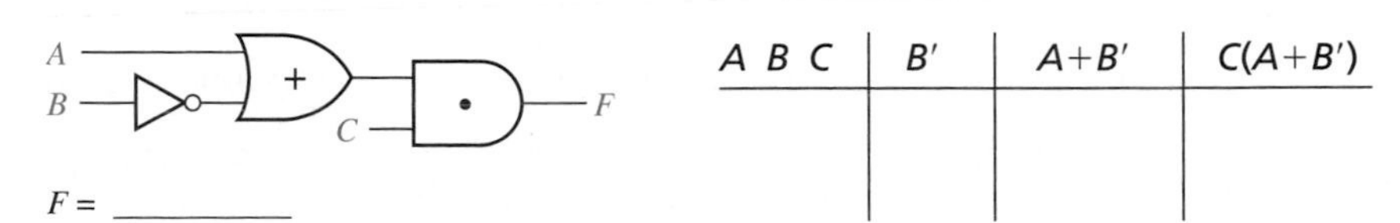


| A B C | B' | A+B' | C(A+B') |
| --- | --- | --- | --- |
| | | | |

(f) Draw a gate circuit which has an output

$$Z = [BC' + F(E + AD')]'$$

(*Hint:* Start with the innermost parentheses and draw the circuit for $AD'$ first.)

**4.** Study Section 2.4, *Basic Theorems.*

(a) Prove each of the Theorems (2-4) through (2-8D) by showing that it is valid for both $X = 0$ and $X = 1$.

(b) Determine the output of each of these gates:

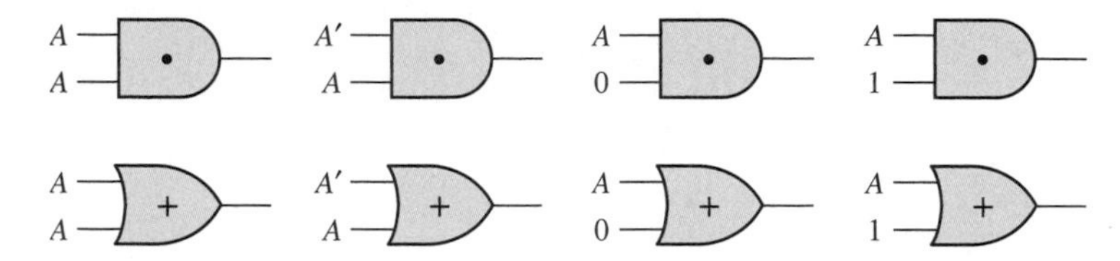


(c) State which of the basic theorems was used in simplifying each of the following expressions:

$(AB' + C) \cdot 0 = 0$ $\qquad$ $A(B + C') + 1 = 1$

$(BC' + A)(BC' + A) = BC' + A$ $\qquad$ $X(Y' + Z) + [X(Y' + Z)]' = 1$

$(X' + YZ)(X' + YZ)' = 0$ $\qquad$ $D'(E' + F) + D'(E' + F) = D'(E' + F)$

**5.** Study Section 2.5, *Commutative, Associative, and Distributive Laws.*

(a) State the associative law for OR.

(b) State the commutative law for AND.

(c) Simplify the following circuit by using the associative laws. Your answer should require only two gates.

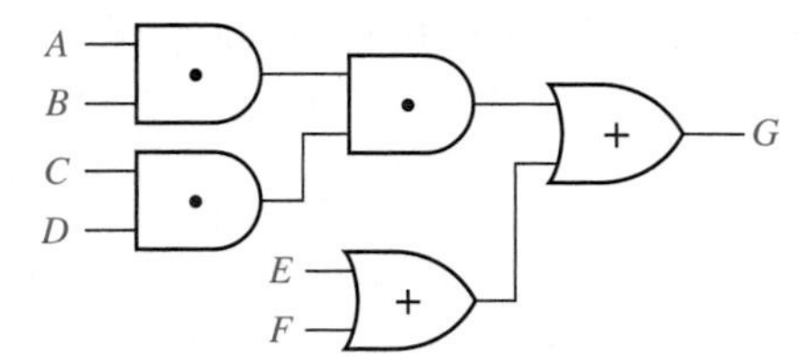


(d) For each gate determine the value of the unspecified input(s):

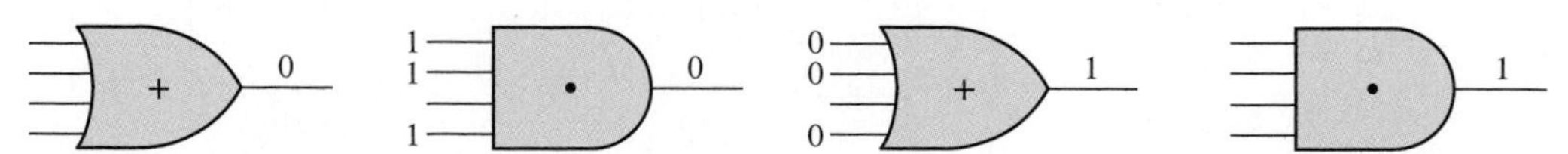


(e) Using a truth table, verify the distributive law, Equation (2-11).

(f) Illustrate the distributive laws, Equations (2-11) and (2-11D), using AND and OR gates.

(g) Verify Equation (2-3) using the second distributive law.

(h) Show how the second distributive law can be used to factor $RS + T'$.

**6.** Study Section 2.6, *Simplification Theorems.*

(a) By completing the truth table, prove that $XY' + Y = X + Y$.

| X Y | $XY'$ | $XY' + Y$ | $X + Y$ |
|---|---|---|---|
| 0 0 | | | |
| 0 1 | | | |
| 1 0 | | | |
| 1 1 | | | |

(b) Which one of Theorems (2-12) through (2-14D) was applied to simplify each of the following expressions? Identify $X$ and $Y$ in each case.

$$(A + B)(DE)' + DE = A + B + DE$$

$$AB' + AB'C'D = AB'$$

$$(A' + B)(CD + E') + (A' + B)(CD + E')' = A' + B$$

$$(A + BC' + D'E)(A + D'E) = A + D'E$$

(c) Simplify the following circuit to a single gate:

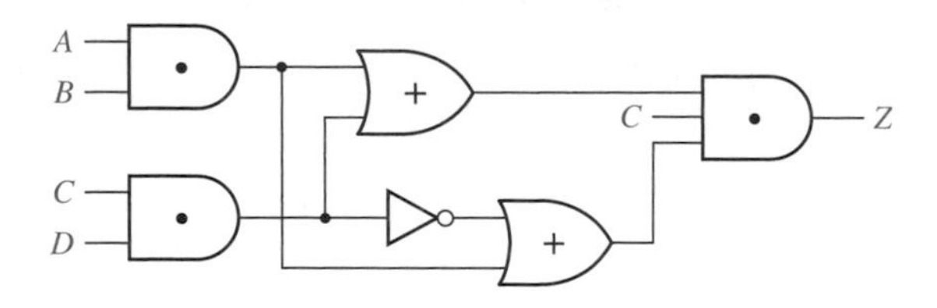


(d) Work Problems 2.1, 2.2, 2.3, and 2.4.

**7.** Study Section 2.7, *Multiplying Out and Factoring.*

(a) Indicate which of the following expressions are in the product-of-sums form, sum-of-products form, or neither:

$$AB' + D'EF' + G$$
$$(A + B'C')(A' + BC)$$
$$AB'(C' + D + E')(F' + G)$$
$$X'Y + WX(X' + Z) + A'B'C'$$

Your answer to this question should include one product-of-sums, one sum-of-products, and two neither, not necessarily in that order.

(b) When multiplying out an expression, why should the second distributive law be applied before the ordinary distributive law when possible?

(c) Factor as much as possible using the ordinary distributive law:

$$AD + B'CD + B'DE$$

Now factor your result using the second distributive law to obtain a product of sums.

(d) Work Problems 2.5, 2.6, and 2.7.

**8.** Probably the most difficult part of the unit is using the second distributive law for factoring or multiplying out an expression. If you have difficulty with Problems 2.5 or 2.6, or you cannot work them *quickly,* study the examples in Section 2.7 again, and then work the following problems.

Multiply out:

(a) $(B' + D + E)(B' + D + A)(AE + C')$

(b) $(A + C')(B' + D)(C' + D')(C + D)E$

As usual, when we say multiply out, we do not mean to multiply out by brute force, but rather to use the second distributive law whenever you can to cut down on the amount of work required.

The answer to (a) should be of the following form: $XX + XX + XX$ and (b) of the form: $XXX + XXXXX$, where each $X$ represents a single variable or its complement.

Now factor your answer to (a) to see that you can get back the original expression.

**9.** Study Section 2.8, *DeMorgan's Laws.*

**10.** Find the complement of each of the following expressions as indicated. In your answer, the complement operation should be applied only to single variables.

(a) $(ab'c')' =$

(b) $(a' + b + c + d')' =$

(c) $(a' + bc)' =$

(d) $(a'b' + cd)' =$

(e) $[a(b' + c'd)]' =$

**11.** Because $(X')' = X$, if you complement each of your answers to 10, you should get back the original expression. Verify that this is true.

(a)

(b)

(c)

(d)

(e)

**12.** Given that $F = a'b + b'c$, $F' =$

Complete the following truth table and verify that your answer is correct:

| a b c | a′b | b′c | a′b + b′c | (a + b′) | (b + c′) | F′ |
|---|---|---|---|---|---|---|
| 0 0 0 | | | | | | |
| 0 0 1 | | | | | | |
| 0 1 0 | | | | | | |
| 0 1 1 | | | | | | |
| 1 0 0 | | | | | | |
| 1 0 1 | | | | | | |
| 1 1 0 | | | | | | |
| 1 1 1 | | | | | | |

13. A fully simplified expression should have nothing complemented except the individual variables. For example, $F = (X + Y)'(W + Z)$ is *not* a minimum product of sums. Find the minimum product of sums for $F$.

14. Work Problems 2.8 and 2.9.

15. Find the dual of $(M + N')P'$.

16. Review the first 12 laws and theorems on page 55. Make sure that you can recognize when to apply them even if an expression has been substituted for a variable.

17. Reread the objectives of this unit. If you are satisfied that you can meet these objectives, take the readiness test.

    [*Note:* You will be provided with a copy of the theorem sheet (page 55) when you take the readiness test this time. However, by the end of Unit 3, you should know all the theorems by memory.]

# Boolean Algebra

## 2.1 Introduction

The basic mathematics needed for the study of the logic design of digital systems is Boolean algebra. Boolean algebra has many other applications including set theory and mathematical logic, but we will restrict ourselves to its application to switching circuits in this text. Because all of the switching devices which we will use are essentially two-state devices (such as a transistor with high or low output voltage), we will study the special case of Boolean algebra in which all of the variables assume only one of two values. This two-valued Boolean algebra is often referred to as switching algebra. George Boole developed Boolean algebra in 1847 and used it to solve problems in mathematical logic. Claude Shannon first applied Boolean algebra to the design of switching circuits in 1939.

We will use a Boolean variable, such as $X$ or $Y$, to represent the input or output of a switching circuit. We will assume that each of these variables can take on only two different values. The symbols "0" and "1" are used to represent these two different values. Thus, if $X$ is a Boolean (switching) variable, then either $X = 0$ or $X = 1$.

The symbols "0" and "1" used in Boolean algebra do not have a numeric value; instead they represent two different states in a logic circuit and are the two values of a switching variable. In a logic gate circuit, 0 (usually) represents a range of low voltages, and 1 represents a range of high voltages. In a switch circuit, 0 (usually) represents an open switch, and 1 represents a closed circuit. In general, 0 and 1 can be used to represent the two states in any binary-valued system.

## 2.2 Basic Operations

The basic operations of Boolean algebra are AND, OR, and complement (or inverse). The complement of 0 is 1, and the complement of 1 is 0. Symbolically, we write

$$0' = 1 \qquad \text{and} \qquad 1' = 0$$

where the prime (′) denotes complementation. If $X$ is a switching variable,

$$X' = 1 \text{ if } X = 0 \qquad \text{and} \qquad X' = 0 \text{ if } X = 1$$

An alternate name for complementation is inversion, and the electronic circuit which forms the inverse of $X$ is referred to as an inverter. Symbolically, we represent an inverter by

$X$ —▷o— $X'$

where the circle at the output indicates inversion. If a logic 0 corresponds to a low voltage and a logic 1 corresponds to a high voltage, a low voltage at the inverter input produces a high voltage at the output and vice versa. Complementation is sometimes referred to as the NOT operation because $X = 1$ if $X$ *is not* equal to 0.

The AND operation can be defined as follows:

$$0 \cdot 0 = 0 \qquad 0 \cdot 1 = 0 \qquad 1 \cdot 0 = 0 \qquad 1 \cdot 1 = 1$$

where "·" denotes AND. (Although this looks like binary multiplication, it is not, because 0 and 1 here are Boolean constants rather than binary numbers.) If we write the Boolean expression $C = A \cdot B$, then given the values of $A$ and $B$, we can determine $C$ from the following table:

| $A$ $B$ | $C = A \cdot B$ |
|---|---|
| 0 0 | 0 |
| 0 1 | 0 |
| 1 0 | 0 |
| 1 1 | 1 |

Note that $C = 1$ iff (if and only if) $A$ *and* $B$ are both 1, hence, the name AND operation. A logic gate which performs the AND operation is represented by

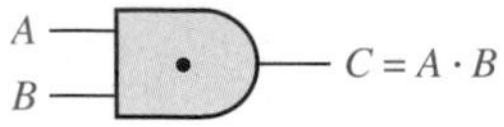


The dot symbol (·) is frequently omitted in a Boolean expression, and we will usually write $AB$ instead of $A \cdot B$. The AND operation is also referred to as logical (or Boolean) multiplication.

The OR operation can be defined as follows:

$$0 + 0 = 0 \qquad 0 + 1 = 1 \qquad 1 + 0 = 1 \qquad 1 + 1 = 1$$

where " + " denotes OR. If we write $C = A + B$, then given the values of $A$ and $B$, we can determine $C$ from the following table:

| $A$ $B$ | $C = A + B$ |
|---|---|
| 0 0 | 0 |
| 0 1 | 1 |
| 1 0 | 1 |
| 1 1 | 1 |

Note that $C = 1$ iff $A$ *or* $B$ (or both) is 1, hence, the name OR operation. This type of OR operation is sometimes referred to as inclusive-OR. A logic gate which performs the OR operation is represented by

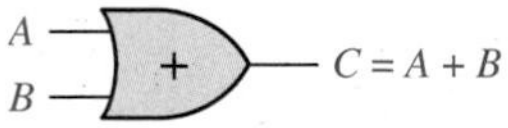


The OR operation is also referred to as logical (or Boolean) addition. Electronic circuits which realize inverters and AND and OR gates are described in Appendix A.

Next, we will apply switching algebra to describe circuits containing switches. We will label each switch with a variable. If switch $X$ is open, then we will define the value of $X$ to be 0; if switch $X$ is closed, then we will define the value of $X$ to be 1.

$X = 0 \rightarrow$ switch open
$X = 1 \rightarrow$ switch closed

Now consider a circuit composed of two switches in a series. We will define the transmission between the terminals as $T = 0$ if there is an open circuit between the terminals and $T = 1$ if there is a closed circuit between the terminals.

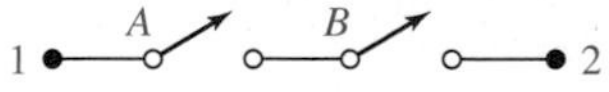


$T = 0 \rightarrow$ open circuit between terminals 1 and 2
$T = 1 \rightarrow$ closed circuit between terminals 1 and 2

Now we have a closed circuit between terminals 1 and 2 ($T = 1$) iff (if and only if) switch $A$ is closed *and* switch $B$ is closed. Stating this algebraically,

$$\mathrm{T} = A \cdot B$$

Next consider a circuit composed of two switches in parallel.

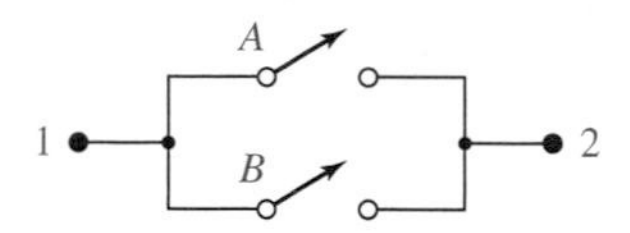


In this case, we have a closed circuit between terminals 1 and 2 iff switch $A$ is closed *or* switch $B$ is closed. Using the same convention for defining variables as above, an equation which describes the behavior of this circuit is

$$T = A + B$$

Thus, switches in a series perform the AND operation and switches in parallel perform the OR operation.

## 2.3 Boolean Expressions and Truth Tables

Boolean expressions are formed by application of the basic operations to one or more variables or constants. The simplest expressions consist of a single constant or variable, such as 0, $X$, or $Y'$. More complicated expressions are formed by combining two or more other expressions using AND or OR, or by complementing another expression. Examples of expressions are

$$AB' + C \quad (2\text{-}1)$$

$$[A(C + D)]' + BE \quad (2\text{-}2)$$

Parentheses are added as needed to specify the order in which the operations are performed. When parentheses are omitted, complementation is performed first followed by AND and then OR. Thus in Expression (2-1), $B'$ is formed first, then $AB'$, and finally $AB' + C$.

Each expression corresponds directly to a circuit of logic gates. Figure 2-1 gives the circuits for Expressions (2-1) and (2-2).

**FIGURE 2-1** Circuits for Expressions (2-1) and (2-2)

(a)

(b)

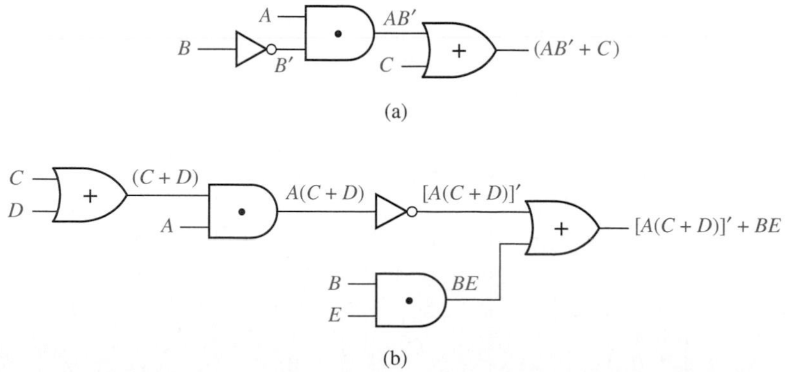

An expression is evaluated by substituting a value of 0 or 1 for each variable. If $A = B = C = 1$ and $D = E = 0$, the value of Expression (2-2) is

$$[A(C + D)]' + BE = [1(1 + 0)]' + 1 \cdot 0 = [1(1)]' + 0 = 0 + 0 = 0$$

Each appearance of a variable or its complement in an expression will be referred to as a *literal.* Thus, the following expression, which has three variables, has 10 literals:

$$ab'c + a'b + a'bc' + b'c'$$

When an expression is realized using logic gates, each literal in the expression corresponds to a gate input.

*A truth table* (also called a table of combinations) specifies the values of a Boolean expression for every possible combination of values of the variables in the expression. The name truth table comes from a similar table which is used in symbolic logic to list the truth or falsity of a statement under all possible conditions. We can use a truth table to specify the output values for a circuit of logic gates in terms of the values of the input variables. The output of the circuit in Figure 2-2(a) is $F = A' + B$. Figure 2-2(b) shows a truth table which specifies the output of the circuit for all possible combinations of values of the inputs $A$ and $B$. The first two columns list the four combinations of values of $A$ and $B$, and the next column gives the corresponding values of $A'$. The last column, which gives the values of $A' + B$, is formed by ORing together corresponding values of $A'$ and $B$ in each row.

FIGURE 2-2 Two-Input Circuit and Truth Table



| A B | A' | F = A' + B |
|---|---|---|
| 0 0 | 1 | 1 |
| 0 1 | 1 | 1 |
| 1 0 | 0 | 0 |
| 1 1 | 0 | 1 |

(b)

Next, we will use a truth table to specify the value of Expression (2-1) for all possible combinations of values of the variables $A$, $B$, and $C$. On the left side of Table 2-1, we list the values of the variables $A$, $B$, and $C$. Because each of the three variables can assume the value 0 or 1, there are $2 \times 2 \times 2 = 8$ combinations of values of the variables. These combinations are easily obtained by listing the binary numbers 000, 001, . . . , 111. In the next three columns of the truth table, we compute $B'$, $AB'$, and $AB' + C$, respectively.

Two expressions are equal if they have the same value for every possible combination of the variables. The expression $(A + C)(B' + C)$ is evaluated using the last three columns of Table 2-1. Because it has the same value as $AB' + C$ for all eight combinations of values of the variables $A$, $B$, and C, we conclude

TABLE 2-1

| A B C | B' | AB' | AB' + C | A + C | B' + C | (A + C)(B' + C) |
|---|---|---|---|---|---|---|
| 0 0 0 | 1 | 0 | 0 | 0 | 1 | 0 |
| 0 0 1 | 1 | 0 | 1 | 1 | 1 | 1 |
| 0 1 0 | 0 | 0 | 0 | 0 | 0 | 0 |
| 0 1 1 | 0 | 0 | 1 | 1 | 1 | 1 |
| 1 0 0 | 1 | 1 | 1 | 1 | 1 | 1 |
| 1 0 1 | 1 | 1 | 1 | 1 | 1 | 1 |
| 1 1 0 | 0 | 0 | 0 | 1 | 0 | 0 |
| 1 1 1 | 0 | 0 | 1 | 1 | 1 | 1 |

$$AB' + C = (A + C)(B' + C) \tag{2-3}$$

If an expression has $n$ variables, and each variable can have the value 0 or 1, the number of different combinations of values of the variables is

$$\underbrace{2 \times 2 \times 2 \times \ldots}_{n \text{ times}} = 2^n$$

Therefore, a truth table for an $n$-variable expression will have $2^n$ rows.

## 2.4 Basic Theorems

The following basic laws and theorems of Boolean algebra involve only a single variable:

Operations with 0 and 1:

$$X + 0 = X \quad (2\text{-}4) \qquad\qquad X \cdot 1 = X \quad (2\text{-}4\text{D})$$

$$X + 1 = 1 \quad (2\text{-}5) \qquad\qquad X \cdot 0 = 0 \quad (2\text{-}5\text{D})$$

Idempotent laws

$$X + X = X \quad (2\text{-}6) \qquad\qquad X \cdot X = X \quad (2\text{-}6\text{D})$$

Involution law

$$(X')' = X \quad (2\text{-}7)$$

Laws of complementarity

$$X + X' = 1 \quad (2\text{-}8) \qquad\qquad X \cdot X' = 0 \quad (2\text{-}8\text{D})$$

Each of these theorems is easily proved by showing that it is valid for both of the possible values of $X$. For example, to prove $X + X' = 1$, we observe that if

$$X = 0, \qquad 0 + 0' = 0 + 1 = 1, \qquad \text{and if } X = 1, \qquad 1 + 1' = 1 + 0 = 1$$

Any expression can be substituted for the variable $X$ in these theorems. Thus, by Theorem (2-5),

$$(AB' + D)E + 1 = 1$$

and by Theorem (2-8D),

$$(AB' + D)(AB' + D)' = 0$$

We will illustrate some of the basic theorems with circuits of switches. As before, 0 will represent an open circuit or open switch, and 1 will represent a closed circuit or closed switch. If two switches are both labeled with the variable $A$, this means that both switches are open when $A = 0$ and both are closed when $A = 1$. Thus the circuit

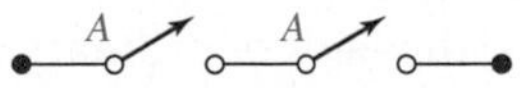

A A

can be replaced with a single switch:

A

This illustrates the theorem $A \cdot A = A$. Similarly,

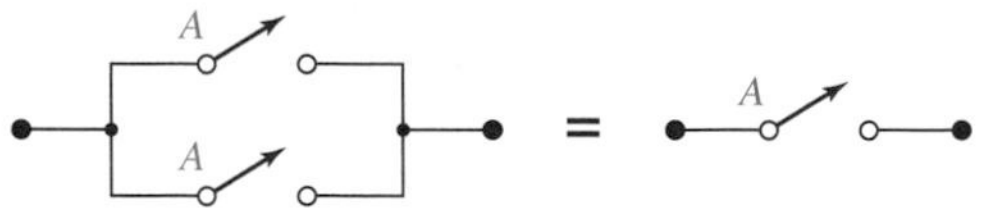

which illustrates the theorem $A + A = A$. A switch in parallel with an open circuit is equivalent to the switch alone

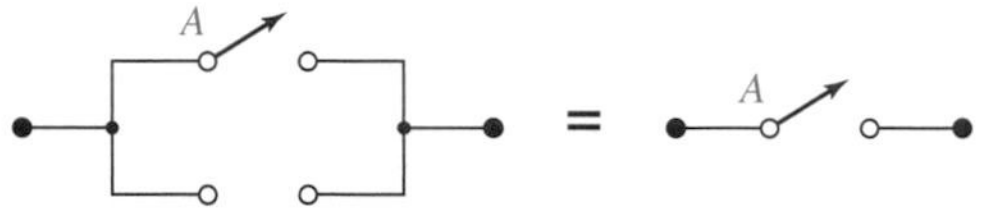

$(A + 0 = A)$

while a switch in parallel with a short circuit is equivalent to a short circuit.

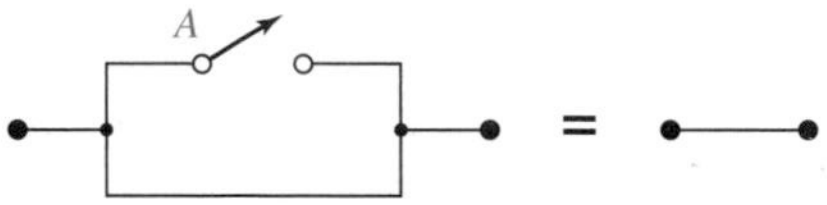

$(A + 1 = 1)$

If a switch is labeled $A'$, then it is open when $A$ is closed and conversely. Hence, $A$ in parallel with $A'$ can be replaced with a closed circuit because one or the other of the two switches is always closed.

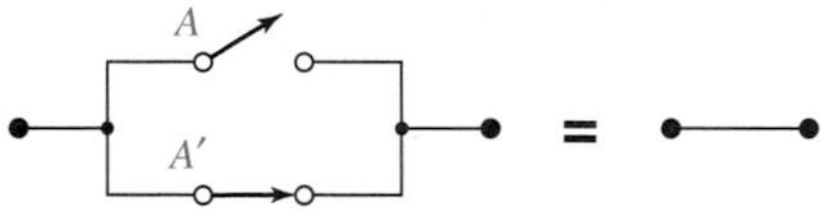

$(A + A' = 1)$

Similarly, switch $A$ in series with $A'$ can be replaced with an open circuit (why?).

$(A \cdot A' = 0)$

# 2.5 Commutative, Associative, and Distributive Laws

Many of the laws of ordinary algebra, such as the commutative and associative laws, also apply to Boolean algebra. The commutative laws for AND and OR, which follow directly from the definitions of the AND and OR operations, are

$$XY = YX \qquad (2\text{-}9) \qquad\qquad X + Y = Y + X \qquad (2\text{-}9D)$$

This means that the order in which the variables are written will not affect the result of applying the AND and OR operations.

The associative laws also apply to AND and OR:

$$(XY)Z = X(YZ) = XYZ \tag{2-10}$$

$$(X + Y) + Z = X + (Y + Z) = X + Y + Z \tag{2-10D}$$

When forming the AND (or OR) of three variables, the result is independent of which pair of variables we associate together first, so parentheses can be omitted as indicated in Equations (2-10) and (2-10D).

We will prove the associative law for AND by using a truth table (Table 2-2). On the left side of the table, we list all combinations of values of the variables $X$, $Y$, and $Z$. In the next two columns of the truth table, we compute $XY$ and $YZ$ for each combination of values of $X$, $Y$, and $Z$. Finally, we compute $(XY)Z$ and $X(YZ)$. Because $(XY)Z$ and $X(YZ)$ are equal for all possible combinations of values of the variables, we conclude that Equation (2-10) is valid.

**TABLE 2-2** Proof of Associative Law for AND

| X Y Z | XY | YZ | (XY)Z | X(YZ) |
|---|---|---|---|---|
| 0 0 0 | 0 | 0 | 0 | 0 |
| 0 0 1 | 0 | 0 | 0 | 0 |
| 0 1 0 | 0 | 0 | 0 | 0 |
| 0 1 1 | 0 | 1 | 0 | 0 |
| 1 0 0 | 0 | 0 | 0 | 0 |
| 1 0 1 | 0 | 0 | 0 | 0 |
| 1 1 0 | 1 | 0 | 0 | 0 |
| 1 1 1 | 1 | 1 | 1 | 1 |

Figure 2-3 illustrates the associative laws using AND and OR gates. In Figure 2-3(a) two two-input AND gates are replaced with a single three-input AND gate. Similarly, in Figure 2-3(b) two two-input OR gates are replaced with a single three-input OR gate.

**FIGURE 2-3** Associative Laws for AND and OR

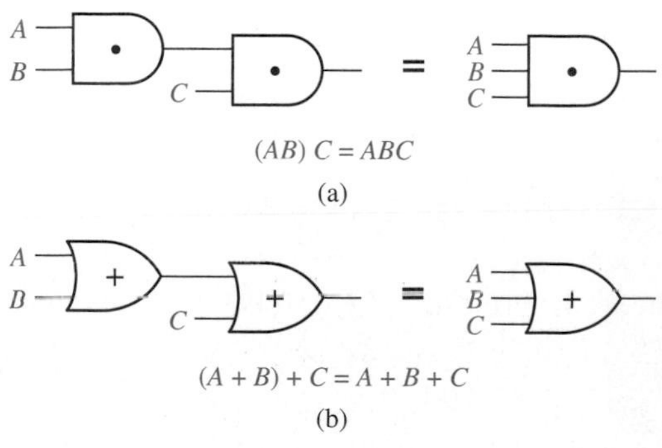

A B C = A B C

$(AB)\,C = ABC$

(a)

A B C = A B C

$(A + B) + C = A + B + C$

(b)

When two or more variables are ANDed together, the value of the result will be 1 iff all of the variables have the value 1. If any of the variables have the value 0, the result of the AND operation will be 0. For example,

$$XYZ = 1 \text{ iff } X = Y = Z = 1$$

When two or more variables are ORed together, the value of the result will be 1 if any of the variables have the value 1. The result of the OR operation will be 0 iff all of the variables have the value 0. For example,

$$X + Y + Z = 0 \text{ iff } X = Y = Z = 0$$

Using a truth table, it is easy to show that the distributive law is valid:

$$X(Y + Z) = XY + XZ \tag{2-11}$$

In addition to the ordinary distributive law, a second distributive law is valid for Boolean algebra but not for ordinary algebra:

$$X + YZ = (X + Y)(X + Z) \tag{2-11D}$$

Proof of the second distributive law follows:

$$\begin{aligned}(X + Y)(X + Z) &= X(X + Z) + Y(X + Z) = XX + XZ + YX + YZ \\ &\qquad \text{(by (2-11))} \\ &= X + XZ + XY + YZ = X \cdot 1 + XZ + XY + YZ \\ &\qquad \text{(by (2-6D) and (2-4D))} \\ &= X(1 + Z + Y) + YZ = X \cdot 1 + YZ = X + YZ \\ &\qquad \text{(by (2-11), (2-5), and (2-4D))}\end{aligned}$$

The ordinary distributive law states that the AND operation distributes over OR, while the second distributive law states that OR distributes over AND. This second law is very useful in manipulating Boolean expressions. In particular, an expression like $A + BC$, which cannot be factored in ordinary algebra, is easily factored using the second distributive law:

$$A + BC = (A + B)(A + C)$$

## 2.6 Simplification Theorems

The following theorems are useful in simplifying Boolean expressions:

$$XY + XY' = X \quad (2\text{-}12) \qquad (X + Y)(X + Y') = X \quad (2\text{-}12D)$$

$$X + XY = X \quad (2\text{-}13) \qquad X(X + Y) = X \quad (2\text{-}13D)$$

$$(X + Y')Y = XY \quad (2\text{-}14) \qquad XY' + Y = X + Y \quad (2\text{-}14D)$$

In each case, one expression can be replaced by a simpler one. Because each expression corresponds to a circuit of logic gates, simplifying an expression leads to simplifying the corresponding logic circuit.

Each of the preceding theorems can be proved by using a truth table, or they can be proved algebraically starting with the basic theorems.

Proof of (2-13): $X + XY = X \cdot 1 + XY = X(1 + Y) = X \cdot 1 = X$

Proof of (2-13D): $X(X + Y) = XX + XY = X + XY = X$

(by (2-6D) and (2-13))

Proof of (2-14D): $Y + XY' = (Y + X)(Y + Y') = (Y + X)1 = Y + X$

(by (2-11 D) and (2-8))

The proof of the remaining theorems is left as an exercise.

We will illustrate Theorem (2-14D), using switches. Consider the following circuit:

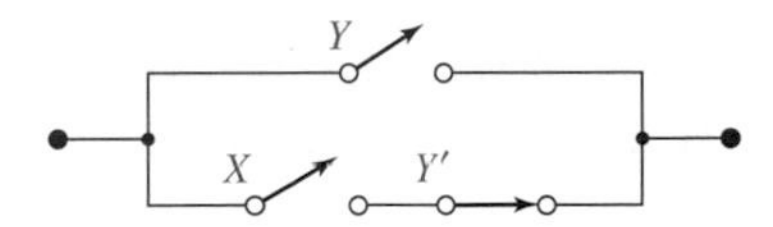


Its transmission is $T = Y + XY'$ because there is a closed circuit between the terminals if switch $Y$ is closed *or* switch $X$ is closed and switch $Y'$ is closed. The following circuit is equivalent because if $Y$ is closed ($Y = 1$) both circuits have a transmission of 1; if $Y$ is open ($Y' = 1$) both circuits have a transmission of $X$.

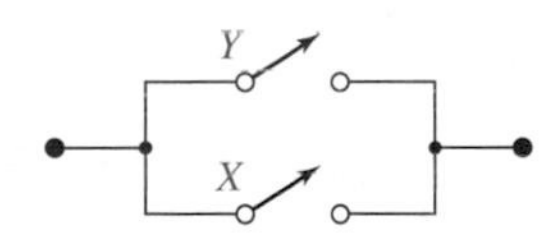


The following example illustrates simplification of a logic gate circuit using one of the theorems. In Figure 2-4, the output of circuit (a) is

$$F = A(A' + B)$$

By Theorem (2-14), the expression for $F$ simplifies to $AB$. Therefore, circuit (a) can be replaced with the equivalent circuit (b).

FIGURE 2-4 Equivalent Gate Circuits

(a) (b)

Any expressions can be substituted for $X$ and $Y$ in the theorems.

*Example 1*

Simplify $Z = A'BC + A'$

This expression has the same form as (2-13) if we let $X = A'$ and $Y = BC$. Therefore, the expression simplifies to $Z = X + XY = X = A'$.

*Example 2*

Simplify $Z = [\underbrace{A + B'C}_{} + \underbrace{D + EF}_{}]\,[\underbrace{A + B'C}_{} + \underbrace{(D + EF)}_{}']$

Substituting: $Z = [\quad X \quad + \quad Y \quad][\quad X \quad + \quad Y' \quad]$

Then, by (2-12D), the expression reduces to

$$Z = X = A + B'C$$

*Example 3*

Simplify $Z = \underbrace{(AB + C)}_{}\,\underbrace{(B'D + C'E')}_{} + \underbrace{(AB + C)'}_{}$

Substituting: $Z = \quad Y' \quad\quad X \quad + \quad Y$

By, (2-14D): $Z = X + Y = B'D + C'E' + (AB + C)'$

Note that in this example we let $Y = (AB + C)'$ rather than $(AB + C)$ in order to match the form of (2-14D).

## 2.7 Multiplying Out and Factoring

The two distributive laws are used to multiply out an expression to obtain a sum-of-products (SOP) form. An expression is said to be in *sum-of-products* form when all products are the products of single variables. This form is the end result when an expression is fully multiplied out. It is usually easy to recognize a sum-of-products expression because it consists of a sum of product terms:

$$AB' + CD'E + AC'E' \tag{2-15}$$

However, in degenerate cases, one or more of the product terms may consist of a single variable. For example,

$$ABC' + DEFG + H \tag{2-16}$$

and

$$A + B' + C + D'E \tag{2-17}$$

are still considered to be in sum-of-products form. The expression

$$(A + B)CD + EF$$

is not in sum-of-products form because the $A + B$ term enters into a product but is not a single variable.

When multiplying out an expression, apply the second distributive law first when possible. For example, to multiply out $(A + BC)(A + D + E)$ let

$$X = A, \qquad Y = BC, \qquad Z = D + E$$

Then

$$(X + Y)(X + Z) = X + YZ = A + BC(D + E) = A + BCD + BCE$$

Of course, the same result could be obtained the hard way by multiplying out the original expression completely and then eliminating redundant terms:

$$\begin{aligned}(A + BC)(A + D + E) &= A + AD + AE + ABC + BCD + BCE \\ &= A(1 + D + E + BC) + BCD + BCE \\ &= A + BCD + BCE\end{aligned}$$

You will save yourself a lot of time if you learn to apply the second distributive law instead of doing the problem the hard way.

Both distributive laws can be used to factor an expression to obtain a product-of-sums form. An expression is in *product-of-sums* (POS) form when all sums are the sums of single variables. It is usually easy to recognize a product-of-sums expression since it consists of a product of sum terms:

$$(A + B')(C + D' + E)(A + C' + E') \tag{2-18}$$

However, in degenerate cases, one or more of the sum terms may consist of a single variable. For example,

$$(A + B)(C + D + E)F \tag{2-19}$$

and

$$AB'C(D' + E) \tag{2-20}$$

are still considered to be in product-of-sums form, but $(A + B)(C + D) + EF$ is not. An expression is fully factored iff it is in product-of-sums form. Any expression not in this form can be factored further.

The following examples illustrate how to factor using the second distributive law:

**Example 1**

Factor $A + B'CD$. This is of the form $X + YZ$ where $X = A$, $Y = B'$, and $Z = CD$, so

$$A + B'CD = (X + Y)(X + Z) = (A + B')(A + CD)$$

$A + CD$ can be factored again using the second distributive law, so

$$A + B'CD = (A + B')(A + C)(A + D)$$

---

**Example 2**

Factor $AB' + C'D$.

$AB' + C'D = (AB' + C')(AB' + D)$ ← note how $X + YZ = (X + Y)(X + Z)$ was applied here

$= (A + C')(B' + C')(A + D)(B' + D)$ ← the second distributive law was applied again to each term

---

**Example 3**

Factor $C'D + C'E' + G'H$.

$$C'D + C'E' + G'H = C'(D + E') + G'H$$ ← first apply the ordinary distributive law, $XY + XZ = X(Y + Z)$

$$= (C' + G'H)(D + E' + G'H)$$ ← then apply the second distributive law

$$= (C' + G')(C' + H)(D + E' + G')(D + E' + H)$$ ← now identify $X$, $Y$, and $Z$ in each expression and complete the factoring

As in Example 3, the ordinary distributive law should be applied before the second law when factoring an expression.

A sum-of-products expression can always be realized directly by one or more AND gates feeding a single OR gate at the circuit output. Figure 2-5 shows the circuits for Equations (2-15) and (2-17). Inverters required to generate the complemented variables have been omitted.

A product-of-sums expression can always be realized directly by one or more OR gates feeding a single AND gate at the circuit output. Figure 2-6 shows the circuits for Equations (2-18) and (2-20). Inverters required to generate the complements have been omitted.

The circuits shown in Figures 2-5 and 2-6 are often referred to as two-level circuits because they have a maximum of two gates in series between an input and the circuit output.

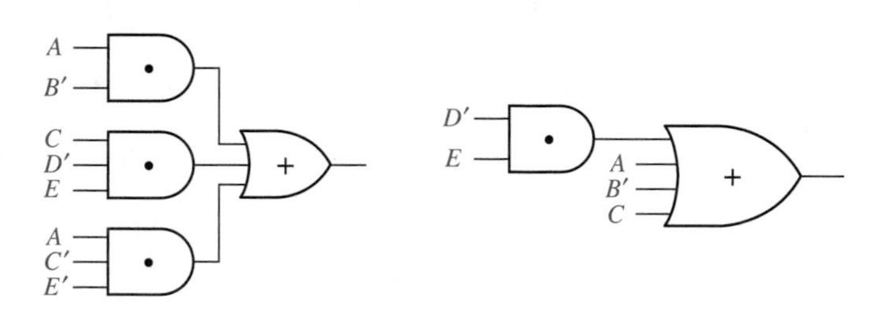


**FIGURE 2-5** Circuits for Equations (2-15) and (2-17)

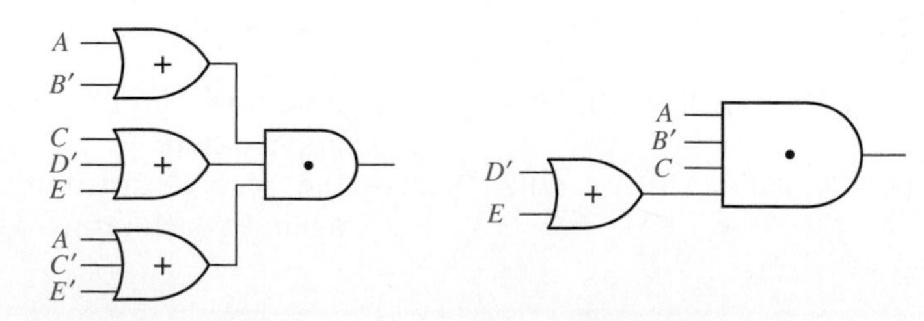


**FIGURE 2-6** Circuits for Equations (2-18) and (2-20)

## 2.8 DeMorgan's Laws

The inverse or complement of any Boolean expression can easily be found by successively applying the following theorems, which are frequently referred to as DeMorgan's laws:

$$(X + Y)' = X'\,Y' \tag{2-21}$$
$$(XY)' = X' + Y' \tag{2-22}$$

We will verify these laws using a truth table:

| X Y | X' Y' | X + Y | (X + Y)' | X' Y' | XY | (XY)' | X' + Y' |
|---|---|---|---|---|---|---|---|
| 0 0 | 1 1 | 0 | 1 | 1 | 0 | 1 | 1 |
| 0 1 | 1 0 | 1 | 0 | 0 | 0 | 1 | 1 |
| 1 0 | 0 1 | 1 | 0 | 0 | 0 | 1 | 1 |
| 1 1 | 0 0 | 1 | 0 | 0 | 1 | 0 | 0 |

DeMorgan's laws are easily generalized to $n$ variables:

$$(X_1 + X_2 + X_3 + \ldots + X_n)' = X_1'\,X_2'\,X_3' \ldots X_n' \tag{2-23}$$
$$(X_1X_2X_3 \ldots X_n)' = X_1' + X_2' + X_3' + \ldots + X_n' \tag{2-24}$$

For example, for $n = 3$,

$$(X_1 + X_2 + X_3)' = (X_1 + X_2)'X_3' = X_1'X_2'X_3'$$

Referring to the OR operation as the logical sum and the AND operation as logical product, DeMorgan's laws can be stated as

The complement of the product is the sum of the complements.

The complement of the sum is the product of the complements.

To form the complement of an expression containing both OR and AND operations, DeMorgan's laws are applied alternately.

**Example 1**

To find the complement of $(A' + B)C'$, first apply (2-22) and then (2-21).

$$[(A' + B)C']' = (A' + B)' + (C')' = AB' + C$$

**Example 2**

$$\begin{aligned} [(AB' + C)D' + E]' &= [(AB' + C)D']'E' && \text{(by (2-21))} \\ &= [(AB' + C)' + D]E' && \text{(by (2-22))} \\ &= [(AB')'C' + D]E' && \text{(by (2-21))} \\ &= [(A' + B)C' + D]E' && \text{(by (2-22))} \end{aligned} \tag{2-25}$$

Note that in the final expressions, the complement operation is applied only to single variables.

The inverse of $F = A'B + AB'$ is

$$F' = (A'B + AB')' = (A'B)'(AB')' = (A + B')(A' + B)$$
$$= AA' + AB + B'A' + BB' = A'B' + AB$$

We will verify that this result is correct by constructing a truth table for $F$ and $F'$:

| A B | A'B | AB' | F = A'B + AB' | A'B' | AB | F' = A'B' + AB |
|---|---|---|---|---|---|---|
| 0 0 | 0 | 0 | 0 | 1 | 0 | 1 |
| 0 1 | 1 | 0 | 1 | 0 | 0 | 0 |
| 1 0 | 0 | 1 | 1 | 0 | 0 | 0 |
| 1 1 | 0 | 0 | 0 | 0 | 1 | 1 |

In the table, note that for every combination of values of $A$ and $B$ for which $F = 0$, $F' = 1$; and whenever $F = 1, F' = 0$.

Given a Boolean expression, the *dual* is formed by replacing AND with OR, OR with AND, 0 with 1, and 1 with 0. Variables and complements are left unchanged. The dual of AND is OR and the dual of OR is AND:

$$(XYZ\ldots)^D = X + Y + Z + \ldots \qquad (X + Y + Z + \ldots)^D = XYZ\ldots \tag{2-26}$$

The dual of an expression may be found by complementing the entire expression and then complementing each individual variable. For example, to find the dual of $AB' + C$,

$$(AB' + C)' = (AB')'C' = (A' + B)C', \quad \text{so} \quad (AB' + C)^D = (A + B')C$$

The laws and theorems of Boolean algebra on page 55 are listed in dual pairs. For example, Theorem 11 is $(X + Y')Y = XY$ and its dual is $XY' + Y = X + Y$ (Theorem 11D).

# Problems

**2.1** Prove the following theorems algebraically:
(a) $X(X' + Y) = XY$ (b) $X + XY = X$
(c) $XY + XY' = X$ (d) $(A + B)(A + B') = A$

**2.2** Illustrate the following theorems using circuits of switches:
(a) $X + XY = X$ (b) $X + YZ = (X + Y)(X + Z)$
In each case, explain why the circuits are equivalent.

**2.3** Simplify each of the following expressions by applying *one* of the theorems. State the theorem used (see page 55).
(a) $X'Y'Z + (X'Y'Z)'$ (b) $(AB' + CD)(B'E + CD)$
(c) $ACF + AC'F$ (d) $A(C + D'B) + A'$
(e) $(A'B + C + D)(A'B + D)$ (f) $(A + BC) + (DE + F)(A + BC)'$

**2.4** For each of the following circuits, find the output and design a simpler circuit having the same output. (*Hint:* Find the circuit output by first finding the output of each gate, going from left to right, and simplifying as you go.)

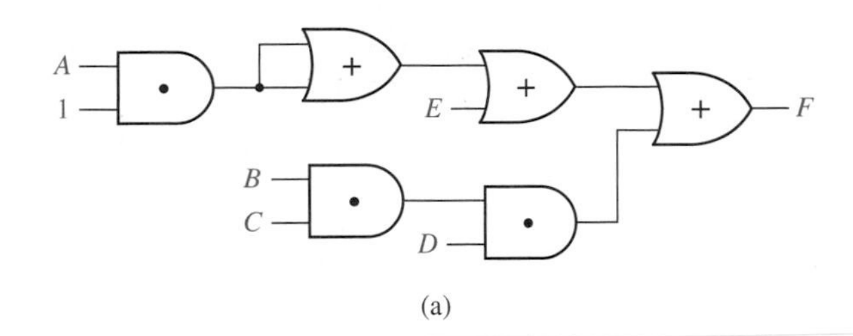


(a)

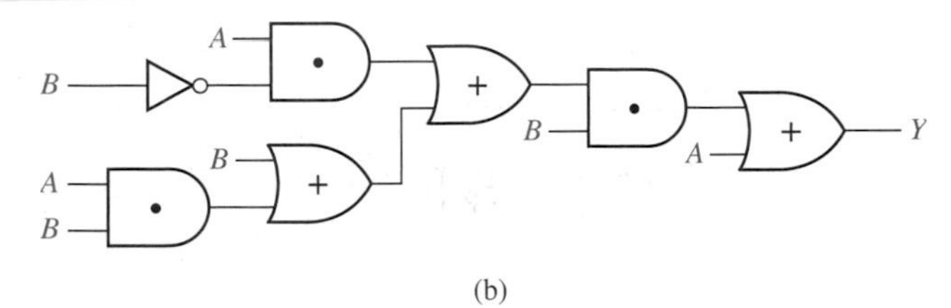


(b)

**2.5** Multiply out and simplify to obtain a sum of products:
(a) $(A + B)(C + B)(D' + B)(ACD' + E)$
(b) $(A' + B + C')(A' + C' + D)(B' + D')$

**2.6** Factor each of the following expressions to obtain a product of sums:
(a) $AB + C'D'$ (b) $WX + WY'X + ZYX$
(c) $A'BC + EF + DEF'$ (d) $XYZ + W'Z + XQ'Z$
(e) $ACD' + C'D' + A'C$ (f) $A + BC + DE$
(The answer to (f) should be the product of four terms, each a sum of three variables.)

**2.7** Draw a circuit that uses only one AND gate and one OR gate to realize each of the following functions:
(a) $(A + B + C + D)(A + B + C + E)(A + B + C + F)$
(b) $WXYZ + VXYZ + UXYZ$

**2.8** Simplify the following expressions to a minimum sum of products.
(a) $[(AB)' + C'D]'$ (b) $[A + B(C' + D)]'$ (c) $((A + B')C)'(A + B)(C + A)'$

**2.9** Find $F$ and $G$ and simplify:

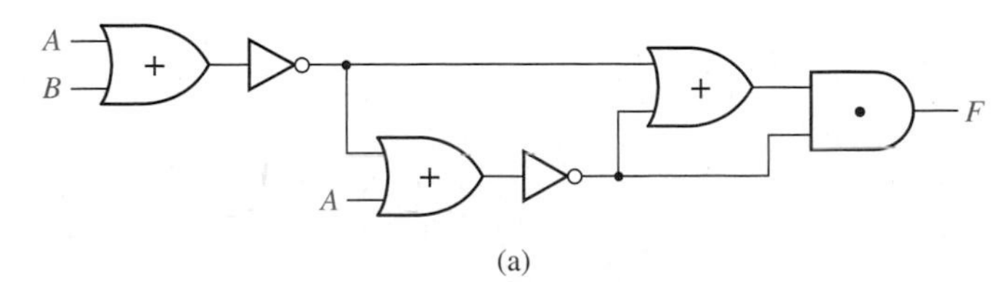


(a)

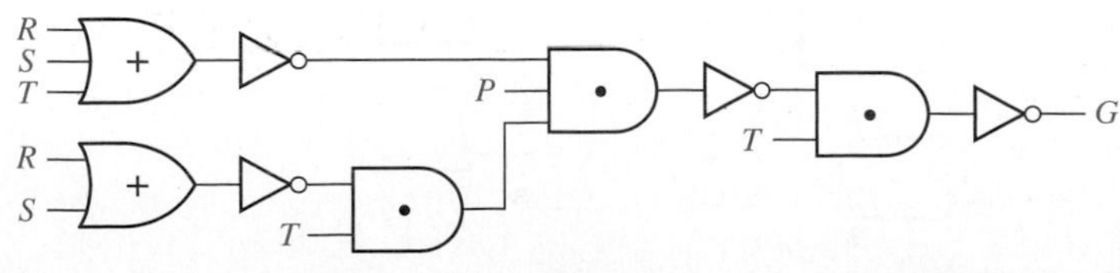


(b)

**2.10** Illustrate the following equations using circuits of switches:

(a) $XY + XY' = X$ (b) $(X + Y')Y = XY$
(c) $X + X'ZY = X + YZ$ (d) $(A + B)C + (A + B)C' = A + B$
(e) $(X + Y)(X + Z) = X + YZ$ (f) $X(X + Y) = X$

**2.11** Simplify each of the following expressions by applying *one* of the theorems. State the theorem used.

(a) $(A' + B' + C)(A' + B' + C)'$ (b) $AB(C' + D) + B(C' + D)$
(c) $AB + (C' + D)(AB)'$ (d) $(A'BF + CD')(A'BF + CEG)$
(e) $[AB' + (C + D)' + E'F](C + D)$ (f) $A'(B + C)(D'E + F)' + (D'E + F)$

**2.12** Simplify each of the following expressions by applying *one* of the theorems. State the theorem used.

(a) $(X + Y'Z) + (X + Y'Z)'$
(b) $[W + X'(Y + Z)][W' + X'(Y + Z)]$
(c) $(V'W + UX)'(UX + Y + Z + V'W)$
(d) $(UV' + W'X)(UV' + W'X + Y'Z)$
(e) $(W' + X)(Y + Z') + (W' + X)'(Y + Z')$
(f) $(V' + U + W)[(W + X) + Y + UZ'] + [(W + X) + UZ' + Y]$

**2.13** For each of the following circuits, find the output and design a simpler circuit that has the same output. (*Hint:* Find the circuit output by first finding the output of each gate, going from left to right, and simplifying as you go).

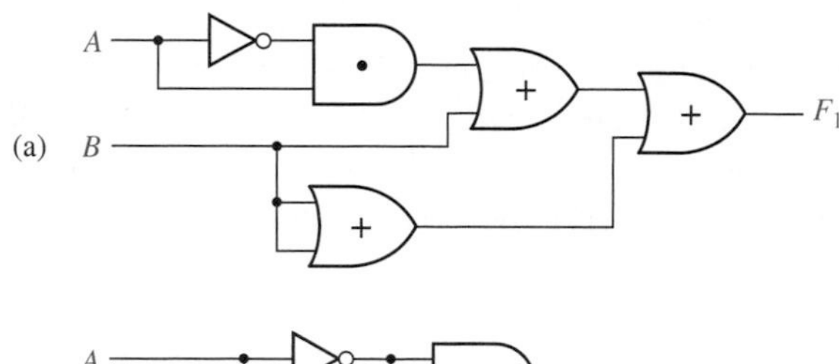


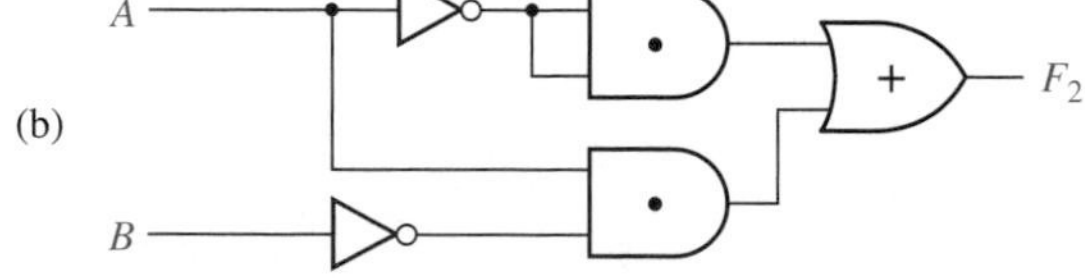


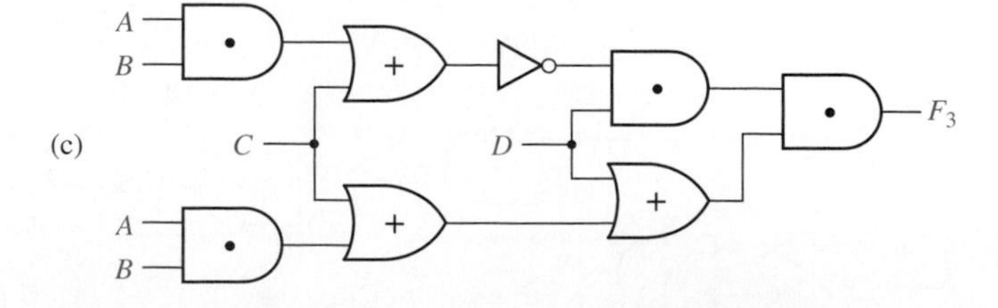

(d)

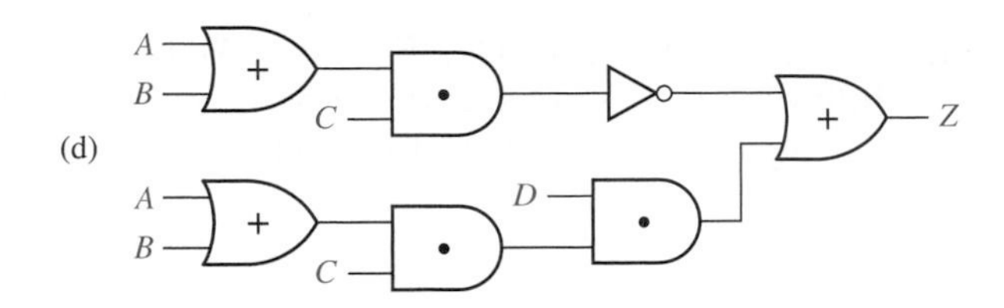


**2.14** Draw a circuit that uses only one AND gate and one OR gate to realize each of the following functions:

(a) $ABCF + ACEF + ACDF$

(b) $(V + W + Y + Z)(U + W + Y + Z)(W + X + Y + Z)$

**2.15** Use *only* DeMorgan's relationships and Involution to find the complements of the following functions:

(a) $f(A, B, C, D) = [A + (BCD)'][(AD)' + B(C' + A)]$

(b) $f(A, B, C, D) = AB'C + (A' + B + D)(ABD' + B')$

**2.16** Using *just* the definition of the dual of a Boolean algebra expression, find the duals of the following expressions:

(a) $f(A, B, C, D) = [A + (BCD)'][(AD)' + B(C' + A)]$

(b) $f(A, B, C, D) = AB'C + (A' + B + D)(ABD' + B')$

**2.17** For the following switching circuit, find the logic function expression describing the circuit by the three methods indicated, simplify each expression, and show they are equal.

(a) subdividing it into series and parallel connections of subcircuits until single switches are obtained

(b) finding all paths through the circuit (sometimes called *tie sets*), forming an AND term for each path and ORing the AND terms together

(c) finding all ways of breaking all paths through the circuit (sometimes called *cut sets*), forming an OR term for each cut set and ANDing the OR terms together.

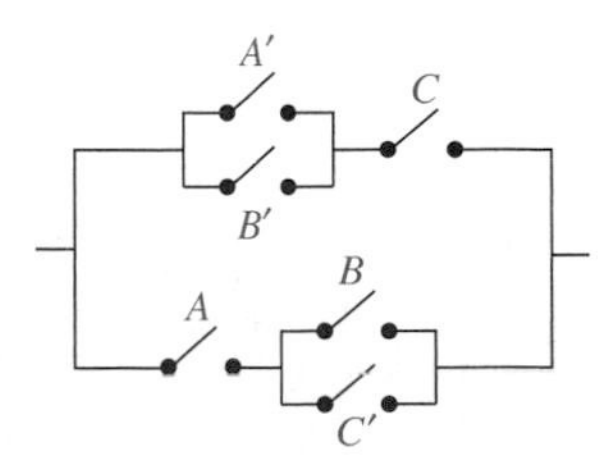


**2.18** For each of the following Boolean (or switching) algebra expressions, indicate which, if any, of the following terms describe the expression: product term, sum-of-products, sum term, and product-of-sums. (More than one may apply.)

(a) $X'Y$

(b) $XY' + YZ$

(c) $(X' + Y)(WX + Z)$

(d) $X + Z$

(e) $(X' + Y)(W + Z)(X + Y' + Z')$

**2.19** Construct a gate circuit using AND, OR, and NOT gates that corresponds one to one with the following switching algebra expression. Assume that inputs are available only in uncomplemented form. (Do not change the expression.)

$$(WX' + Y)[(W + Z)' + XYZ')]$$

**2.20** For the following switch circuit:

(a) derive the switching algebra expression that corresponds one to one with the switch circuit.

(b) derive an equivalent switch circuit with a structure consisting of a parallel connection of groups of switches connected in series. (Use 9 switches.)

(c) derive an equivalent switch circuit with a structure consisting of a series connection of groups of switches connected in parallel. (Use 6 switches.)

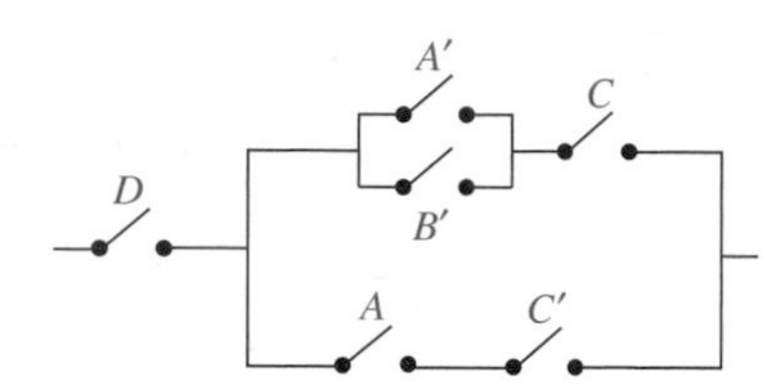


**2.21** In the following circuit, $F = (A' + B)C$. Give a truth table for $G$ so that $H$ is as specified in its truth table. If $G$ can be either 0 or 1 for some input combination, leave its value unspecified.

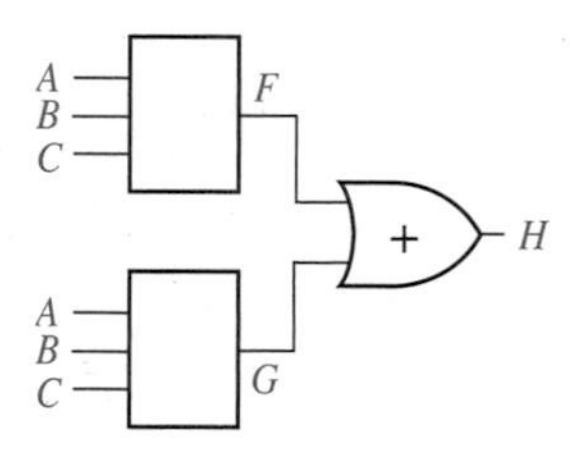


| A | B | C | H |
|---|---|---|---|
| 0 | 0 | 0 | 0 |
| 0 | 0 | 1 | 1 |
| 0 | 1 | 0 | 1 |
| 0 | 1 | 1 | 1 |
| 1 | 0 | 0 | 0 |
| 1 | 0 | 1 | 1 |
| 1 | 1 | 0 | 0 |
| 1 | 1 | 1 | 1 |

**2.22** Factor each of the following expressions to obtain a product of sums:

(a) $A'B' + A'CD + A'DE'$ (b) $H'I' + JK$

(c) $A'BC + A'B'C + CD'$ (d) $A'B' + (CD' + E)$

(e) $A'B'C + B'CD' + EF'$ (f) $WX'Y + W'X' + W'Y'$

**2.23** Factor each of the following expressions to obtain a product of sums:

(a) $W + U'YV$ (b) $TW + UY' + V$

(c) $A'B'C + B'CD' + B'E'$ (d) $ABC + ADE' + ABF'$

**2.24** Simplify the following expressions to a minimum sum of products. Only individual variables should be complemented.

(a) $[(XY')' + (X' + Y)'Z]$ (b) $(X + (Y'(Z + W)')')'$

(c) $[(A' + B')' + (A'B'C)' + C'D]'$ (d) $(A + B)CD + (A + B)'$

**2.25** For each of the following functions find a sum-of-products expression for $F'$.
(a) $F(P, Q, R, S) = (R' + PQ)S$
(b) $F(W, X, Y, Z) = X + YZ(W + X')$
(c) $F(A, B, C, D) = A' + B' + ACD$

**2.26** Find $F$, $G$, and $H$, and simplify:

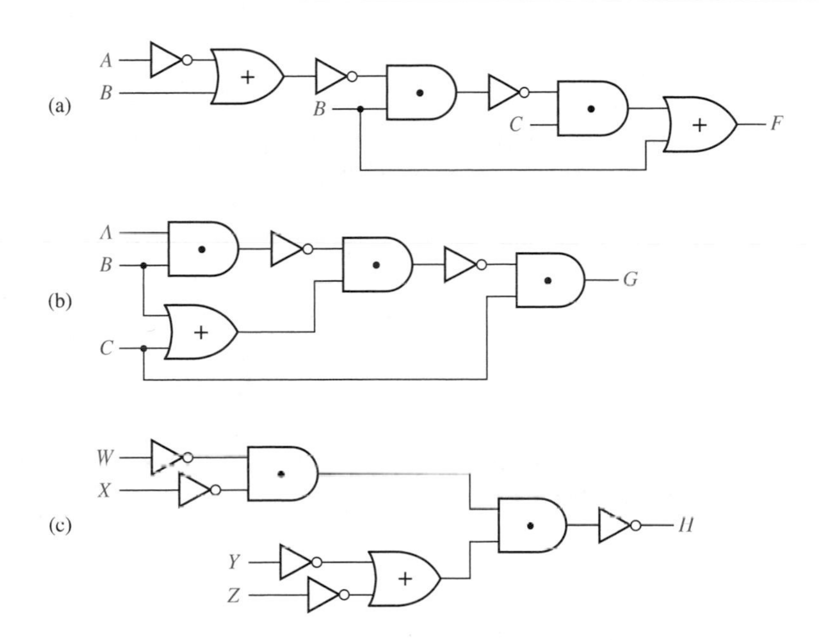


**2.27** Draw a circuit that uses two OR gates and two AND gates to realize the following function:

$$F = (V + W + X)(V + X + Y)(V + Z)$$

**2.28** Draw a circuit to realize the function:

$$F = ABC + A'BC + AB'C + ABC'$$

(a) using one OR gate and three AND gates. The AND gates should have two inputs.
(b) using two OR gates and two AND gates. All of the gates should have two inputs.

**2.29** Prove the following equations using truth tables:
(a) $(X + Y)(X' + Z) = XZ + X'Y$
(b) $(X + Y)(Y + Z)(X' + Z) = (X + Y)(X' + Z)$
(c) $XY + YZ + X'Z = XY + X'Z$

(d) $(A + C)(AB + C') = AB + AC'$
(e) $W'XY + WZ = (W' + Z)(W + XY)$
(*Note:* Parts (a), (b), and (c) are theorems that will be introduced in Unit 3.)

**2.30** Show that the following two gate circuits realize the same function.

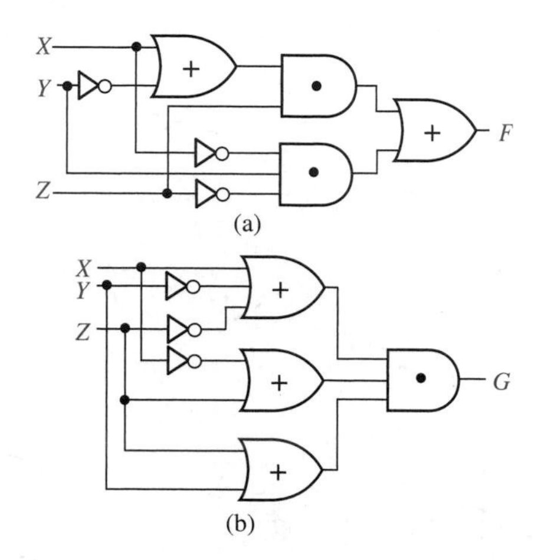

# Laws and Theorems of Boolean Algebra

Operations with 0 and 1:

1. $X + 0 = X$ — 1D. $X \cdot 1 = X$
2. $X + 1 = 1$ — 2D. $X \cdot 0 = 0$

Idempotent laws:

3. $X + X = X$ — 3D. $X \cdot X = X$

Involution law:

4. $(X')' = X$

Laws of complementarity:

5. $X + X' = 1$ — 5D. $X \cdot X' = 0$

Commutative laws:

6. $X + Y = Y + X$ — 6D. $XY = YX$

Associative laws:

7. $(X + Y) + Z = X + (Y + Z) = X + Y + Z$ — 7D. $(XY)Z = X(YZ) = XYZ$

Distributive laws:

8. $X(Y + Z) = XY + XZ$ — 8D. $X + YZ = (X + Y)(X + Z)$

Simplification theorems:

9. $XY + XY' = X$ — 9D. $(X + Y)(X + Y') = X$
10. $X + XY = X$ — 10D. $X(X + Y) = X$
11. $(X + Y')Y = XY$ — 11D. $XY' + Y = X + Y$

DeMorgan's laws:

12. $(X + Y + Z + \ldots)' = X'Y'Z' \ldots$ — 12D. $(XYZ \ldots)' = X' + Y' + Z' + \ldots$

Duality:

13. $(X + Y + Z + \ldots)^D = XYZ \ldots$ — 13D. $(XYZ \ldots)^D = X + Y + Z + \ldots$

Theorem for multiplying out and factoring:

14. $(X + Y)(X' + Z) = XZ + X'Y$ — 14D. $XY + X'Z = (X + Z)(X' + Y)$

Consensus theorem:

15. $XY + YZ + X'Z = XY + X'Z$ — 15D. $(X + Y)(Y + Z)(X' + Z) = (X + Y)(X' + Z)$

# Boolean Algebra (Continued)

## Objectives

When you complete this unit, you should know from memory and be able to use any of the laws and theorems of Boolean algebra listed at the end of Unit 2. Specifically, you should be able to

1. Apply these laws and theorems to the manipulation of algebraic expressions including:
   a. Simplifying an expression.
   b. Finding the complement of an expression.
   c. Multiplying out and factoring an expression.
2. Prove any of the theorems using a truth table or give an algebraic proof if appropriate.
3. Define the exclusive-OR and equivalence operations. State, prove, and use the basic theorems that concern these operations.
4. Use the consensus theorem to delete terms from and add terms to a switching expression.
5. Given an equation, prove algebraically that it is valid or show that it is not valid.

# Study Guide

**1.** Study Section 3.1, *Multiplying Out and Factoring Expressions.*

(a) List three laws or theorems which are useful when multiplying out or factoring expressions.

(b) Use Equation (3-3) to factor each of the following:

$ab'c + bd =$

$abc + (ab)'d =$

(c) In the following example, first group the terms so that (3-2) can be applied two times.

$$F_1 = (x + y' + z)(w' + x' + y)(w + x + y')(w' + y + z')$$

After applying (3-2), apply (3-3) and then finish multiplying out by using (3-1).

If we did not use (3-2) and (3-3) and used only (3-1) on the original $F_1$ expression, we would generate many more terms:

$$\begin{aligned} F_1 &= (w'x + w'y' + w'z + \cancel{xx'} + x'y' + x'z + xy + \cancel{yy'} + yz) \\ &\qquad (\cancel{ww'} + w'x + w'y' + wy + xy + \cancel{yy'} + wz' + xz' + y'z') \\ &= \underbrace{(w'x + w'xy' + w'xz + \cdots + yzy'z')}_{\text{49 terms in all}} \end{aligned}$$

This is obviously a *very inefficient* way to proceed! The moral to this story is to first group the terms and apply (3-2) and (3-3) where possible.

(d) Work Programmed Exercise 3.1. Then work Problem 3.6, being careful not to introduce any unnecessary terms in the process.

(e) In Unit 2 you learned how to factor a Boolean expression, using the two distributive laws. In addition, this unit introduced use of the theorem

$$XY + X'Z = (X + Z)(X' + Y)$$

in the factoring process. Careful choice of the order in which these laws and theorems are applied may cut down the amount of work required to factor an expression. When factoring, it is best to apply Equation (3-1) first, using as $X$ the variable or variables which appear most frequently. Then Equations (3-2) and (3-3) can be applied in either order, depending on circumstances.

(f) Work Programmed Exercise 3.2. Then work Problem 3.7.

**2.** Checking your answers:

A good way to partially check your answers for correctness is to substitute 0's or 1's for some of the variables. For example, if we substitute $A = 1$ in the first and last expression in Equation (3-5), we get

$$1 \cdot C + 0 \cdot BD' + 0 \cdot BE + 0 \cdot C'DE = (1 + B + C')(1 + B + D) \cdot (1 + B + E)(1 + D' + E)(0 + C)$$
$$C = 1 \cdot 1 \cdot 1 \cdot 1 \cdot C \checkmark$$

Similarly, substituting $A = 0, B = 0$ we get

$$0 + 0 + 0 + C'DE = (0 + C')(0 + D)(0 + E)(D' + E)(1 + C) = C'DE \checkmark$$

Verify that the result is also correct when $A = 0$ and $B = 1$.

**3.** The *method* which you use to get your answer is very important in this unit. If it takes you two pages of algebra and one hour of time to work a problem that can be solved in 10 minutes with three lines of work, you have not learned the material in this unit! Even if you get the correct answer, your work is not satisfactory if you worked the problem by an excessively long and time-consuming method. It is important that you learn to solve simple problems in a simple manner—otherwise, when you are asked to solve a complex problem, you will get bogged down and never get the answer. When you are given a problem to solve, do not just plunge in, but first ask yourself, "What is the easiest way to work this problem?" For example, when you are asked to multiply out an expression, do not just multiply it out by brute force, term by term. Instead, ask yourself, "How can I group the terms and which theorems should I apply first in order to reduce the amount of work?" (See Study Guide Part 1.) After you have worked out Problems 3.6 and 3.7, compare your solutions with those in the solution book. If your solution required substantially more work than the one in the solution book, rework the problem and try to get the answer in a more straightforward manner.

**4.** Study Section 3.2, *Exclusive-OR and Equivalence Operations.*

(a) Prove Theorems (3-8) through (3-13). You should be able to prove these both algebraically and by using a truth table.

(b) Show that $(xy' + x'y)' = xy + x'y'$. Memorize this result.

(c) Prove Theorem (3-15).

(d) Show that $(x \equiv 0) = x'$, $(x \equiv x) = 1$, and $(x \equiv y)' = (x \equiv y')$.

(e) Express $(x \equiv y)'$ in terms of exclusive OR.

(f) Work Problems 3.8 and 3.9.

**5.** Study Section 3.3, *The Consensus Theorem.* The consensus theorem is an important method for simplifying switching functions.

(a) In each of the following expressions, find the consensus term and eliminate it:

$$abc'd + a'be + bc'de$$
$$(a' + b + c)(a + d)(b + c + d)$$
$$ab'c + a'bd + bcd' + a'bc$$

(b) Eliminate two terms from the following expression by applying the consensus theorem:

$$A'B'C + BC'D' + A'CD + AB'D' + BCD + AC'D'$$

(*Hint:* First, compare the first term with each of the remaining terms to see if a consensus exists, then compare the second term with each of the remaining terms, etc.)

(c) Study the example given in Equations (3-22) and (3-23) carefully. Now let us start with the four-term form of the expression (Equation 3-22):

$$A'C'D + A'BD + ABC + ACD'$$

Can this be reduced directly to three terms by the application of the consensus theorem? Before we can reduce this expression, we must add another term. Which term can be added by applying the consensus theorem?

Add this term, and then reduce the expression to three terms. After this reduction, can the term which was added be removed? Why not?

(d) Eliminate two terms from the following expression by applying the dual consensus theorem:

$$(a' + c' + d)(a' + b + c)(a + b + d)(a' + b + d)(b + c' + d)$$

Use brackets to indicate how you formed the consensus terms. (*Hint:* First, find the consensus of the first two terms and eliminate it.)

(e) Derive Theorem (3-3) by using the consensus theorem.

(f) Work Programmed Exercise 3.3. Then work Problem 3.10.

**6.** Study Section 3.4, *Algebraic Simplification of Switching Expressions.*

(a) What theorems are used for:
Combining terms?

Eliminating terms?

Eliminating literals?

Adding redundant terms?

Factoring or multiplying out?

(b) Note that in the example of Equation (3-27), the redundant term $WZ'$ was added and then was eliminated later after it had been used to eliminate another term. Why was it possible to eliminate $WZ'$ in this example?

If a term has been added by the consensus theorem, it may not always be possible to eliminate the term later by the consensus theorem. Why?

(c) You will need considerable practice to develop skill in simplifying switching expressions. Work through Programmed Exercises 3.4 and 3.5.
(d) Work Problem 3.11.
(e) When simplifying an expression using Boolean algebra, two frequently asked questions are
(1) Where do I begin?
(2) How do I know when I am finished?

In answer to (1), it is generally best to try simple techniques such as combining terms or eliminating terms and literals before trying more complicated things such as using the consensus theorem or adding redundant terms. Question (2) is generally difficult to answer because it may be impossible to simplify some expressions without first adding redundant terms. We will usually tell you how many terms to expect in the minimum solution so that you will not have to waste time trying to simplify an expression which is already minimized. In Units 5 and 6, you will learn systematic techniques which will guarantee finding the minimum solution.

**7.** Study Section 3.5, *Proving Validity of an Equation.*

(a) When attempting to prove that an equation is valid, is it permissible to add the same expression to both sides? Explain.
(b) Work Problem 3.12.
(c) Show that (3-33) and (3-34) are true by considering both $x = 0$ and $x = 1$.

(d) Given that $a'(b + d') = a'(b + e')$, the following "proof" shows that $d = e$:

$$a'(b + d') = a'(b + e')$$
$$a + b'd = a + b'e$$
$$b'd = b'e$$
$$d = e$$

State two things that are wrong with the "proof." Give a set of values for $a$, $b$, $d$, and $e$ that demonstrates that the result is incorrect.

**8.** Reread the objectives of this unit. When you take the readiness test, you will be expected to know from memory the laws and theorems listed at the end of Unit 2. Where appropriate, you should know them "forward and backward"; that is, given either side of the equation, you should be able to supply the other. Test yourself to see if you can do this. When you are satisfied that you can meet the objectives, take the readiness test.

# Boolean Algebra (Continued)

In this unit we continue our study of Boolean algebra to learn additional methods for manipulating Boolean expressions. We introduce another theorem for multiplying out and factoring that facilitates conversion between sum-of-products and product-of-sums expressions. These algebraic manipulations allow us to realize a switching function in a variety of forms. The exclusive-OR and equivalence operations are introduced along with examples of their use. The consensus theorem provides a useful method for simplifying an expression. Then methods for algebraic simplification are reviewed and summarized. The unit concludes with methods for proving the validity of an equation.

## 3.1 Multiplying Out and Factoring Expressions

Given an expression in product-of-sums form, the corresponding sum-of-products expression can be obtained by multiplying out, using the two distributive laws:

$$X(Y + Z) = XY + XZ \tag{3-1}$$

$$(X + Y)(X + Z) = X + YZ \tag{3-2}$$

In addition, the following theorem is very useful for factoring and multiplying out:

$$(X + Y)(X' + Z) = XZ + X'Y \tag{3-3}$$

Note that the variable that is paired with $X$ on one side of the equation is paired with $X'$ on the other side, and vice versa.

Proof:

If $X = 0$, (3-3) reduces to $Y(1 + Z) = 0 + 1 \cdot Y$ or $Y = Y$.

If $X = 1$, (3-3) reduces to $(1 + Y)Z = Z + 0 \cdot Y$ or $Z = Z$.

Because the equation is valid for both $X = 0$ and $X = 1$, it is always valid.

The following example illustrates the use of Theorem (3-3) for factoring:

$$AB + A'C = (A + C)(A' + B)$$

Note that the theorem can be applied when we have two terms, one which contains a variable and another which contains its complement.

Theorem (3-3) is very useful for multiplying out expressions. In the following example, we can apply (3-3) because one factor contains the variable $Q$, and the other factor contains $Q'$.

$$(Q + AB')(C'D + Q') = QC'D + Q'AB'$$

If we simply multiplied out by using the distributive law, we would get four terms instead of two:

$$(Q + AB')(C'D + Q') = QC'D + QQ' + AB'C'D + AB'Q'$$

Because the term $AB'C'D$ is difficult to eliminate, it is much better to use (3-3) instead of the distributive law.

In general, when we multiply out an expression, we should use (3-3) along with (3-1) and (3-2). To avoid generating unnecessary terms when multiplying out, (3-2) and (3-3) should generally be applied before (3-1), and terms should be grouped to expedite their application.

***Example***

$$\begin{aligned}
&(A + B + C')(A + B + D)(A + B + E)(A + D' + E)(A' + C) \\
&= (A + B + C'D)(A + B + E)[AC + A'(D' + E)] \\
&= (A + B + C'DE)(AC + A'D' + A'E) \\
&= AC + ABC + A'BD' + A'BE + A'C'DE \qquad (3\text{-}4)
\end{aligned}$$

What theorem was used to eliminate $ABC$? (*Hint:* let $X = AC$.)

In this example, if the ordinary distributive law (3-1) had been used to multiply out the expression by brute force, 162 terms would have been generated, and 158 of these terms would then have to be eliminated.

---

The same theorems that are useful for multiplying out expressions are useful for factoring. By repeatedly applying (3-1), (3-2), and (3-3), any expression can be converted to a product-of-sums form.

***Example of Factoring***

$$\begin{aligned}
&AC + A'BD' + A'BE + A'C'DE \\
&= \underbrace{AC}_{XZ} + \underbrace{A'}_{X'}(\underbrace{BD' + BE + C'DE}_{Y}) \\
&= (A + BD' + BE + C'DE)(A' + C) \\
&= [\underbrace{A + C'DE}_{X} + \underbrace{B}_{Y}(\underbrace{D' + E}_{Z})](A' + C)
\end{aligned}$$

$$= (A + B + C'DE)(A + \cancel{C'DE} + D' + E)(A' + C)$$
$$= (A + B + C')(A + B + D)(A + B + E)(A + D' + E)(A' + C) \quad (3\text{-}5)$$

This is the same expression we started with in (3-4).

## 3.2 Exclusive-OR and Equivalence Operations

The *exclusive-OR* operation ($\oplus$) is defined as follows:

$$0 \oplus 0 = 0 \qquad 0 \oplus 1 = 1$$
$$1 \oplus 0 = 1 \qquad 1 \oplus 1 = 0$$

The truth table for $X \oplus Y$ is

| X Y | $X \oplus Y$ |
|---|---|
| 0 0 | 0 |
| 0 1 | 1 |
| 1 0 | 1 |
| 1 1 | 0 |

From this table, we can see that $X \oplus Y = 1$ iff $X = 1$ or $Y = 1$, but *not* both. The ordinary OR operation, which we have previously defined, is sometimes called inclusive OR because $X + Y = 1$ iff $X = 1$ or $Y = 1$, or both.

Exclusive OR can be expressed in terms of AND and OR. Because $X \oplus Y = 1$ iff $X$ is 0 and $Y$ is 1 or $X$ is 1 and $Y$ is 0, we can write

$$X \oplus Y = X'Y + XY' \quad (3\text{-}6)$$

The first term in (3-6) is 1 if $X = 0$ and $Y = 1$; the second term is 1 if $X = 1$ and $Y = 0$. Alternatively, we can derive Equation (3-6) by observing that $X \oplus Y = 1$ iff $X = 1$ or $Y = 1$ *and* $X$ and $Y$ are not both 1. Thus,

$$X \oplus Y = (X + Y)(XY)' = (X + Y)(X' + Y') = X'Y + XY' \quad (3\text{-}7)$$

In (3-7), note that $(X\,Y)' = 1$ if $X$ and $Y$ are not both 1.

We will use the following symbol for an exclusive-OR gate:

X, Y → $X \oplus Y$

The following theorems apply to exclusive OR:

$$X \oplus 0 = X \tag{3-8}$$

$$X \oplus 1 = X' \tag{3-9}$$

$$X \oplus X = 0 \tag{3-10}$$

$$X \oplus X' = 1 \tag{3-11}$$

$$X \oplus Y = Y \oplus X \text{ (commutative law)} \tag{3-12}$$

$$(X \oplus Y) \oplus Z = X \oplus (Y \oplus Z) = X \oplus Y \oplus Z \text{ (associative law)} \tag{3-13}$$

$$X(Y \oplus Z) = XY \oplus XZ \text{ (distributive law)} \tag{3-14}$$

$$(X \oplus Y)' = X \oplus Y' = X' \oplus Y = XY + X'Y' \tag{3-15}$$

Any of these theorems can be proved by using a truth table or by replacing $X \oplus Y$ with one of the equivalent expressions from Equation (3-7). Proof of the distributive law follows:

$$\begin{aligned} XY \oplus XZ &= XY(XZ)' + (XY)'XZ = XY(X' + Z') + (X' + Y')XZ \\ &= XYZ' + XY'Z \\ &= X(YZ' + Y'Z) = X(Y \oplus Z) \end{aligned}$$

The *equivalence* operation (≡) is defined by

$$\begin{array}{ll} (0 \equiv 0) = 1 & (0 \equiv 1) = 0 \\ (1 \equiv 0) = 0 & (1 \equiv 1) = 1 \end{array} \tag{3-16}$$

The truth table for $X \equiv Y$ is

| X Y | X ≡ Y |
|---|---|
| 0 0 | 1 |
| 0 1 | 0 |
| 1 0 | 0 |
| 1 1 | 1 |

From the definition of equivalence, we see that $(X \equiv Y) = 1$ iff $X = Y$. Because $(X \equiv Y) = 1$ iff $X = Y = 1$ or $X = Y = 0$, we can write

$$(X \equiv Y) = XY + X'Y' \tag{3-17}$$

Equivalence is the complement of exclusive-OR:

$$\begin{aligned} (X \oplus Y)' &= (X'Y + XY')' = (X + Y')(X' + Y) \\ &= XY + X'Y' = (X \equiv Y) \end{aligned} \tag{3-18}$$

Just as for exclusive-OR, the equivalence operation is commutative and associative.

We will use the following symbol for an equivalence gate:

X
Y
≡
$X \equiv Y$

Because equivalence is the complement of exclusive-OR, an alternate symbol for the equivalence gate is an exclusive-OR gate with a complemented output:

$X$, $Y$ → $(X \oplus Y)' = (X \equiv Y)$

The equivalence gate is also called an exclusive-NOR gate.

In order to simplify an expression which contains AND and OR as well as exclusive OR and equivalence, it is usually desirable to first apply (3-6) and (3-17) to eliminate the $\oplus$ and $\equiv$ operations. As an example, we will simplify

$$F = (A'B \equiv C) + (B \oplus AC')$$

By (3-6) and (3-17),

$$\begin{aligned} F &= [(A'B)C + (A'B)'C'] + [B'(AC') + B(AC')'] \\ &= A'BC + (A + B')C' + AB'C' + B(A' + C) \\ &= B(A'C + A' + C) + C'(A + B' + AB') = B(A' + C) + C'(A + B') \end{aligned}$$

When manipulating an expression that contains several exclusive-OR or equivalence operations, it is useful to note that

$$(XY' + X'Y)' = XY + X'Y' \tag{3-19}$$

For example,

$$\begin{aligned} A' \oplus B \oplus C &= [A'B' + (A')'B] \oplus C \\ &= (A'B' + AB)C' + (A'B' + AB)'C && \text{(by (3-6))} \\ &= (A'B' + AB)C' + (A'B + AB')C && \text{(by (3-19))} \\ &= A'B'C' + ABC' + A'BC + AB'C \end{aligned}$$

# 3.3 The Consensus Theorem

The consensus theorem is very useful in simplifying Boolean expressions. Given an expression of the form $XY + X'Z + YZ$, the term $YZ$ is redundant and can be eliminated to form the equivalent expression $XY + X'Z$.

The term that was eliminated is referred to as the *consensus term*. Given a pair of terms for which a variable appears in one term and the complement of that variable in another, the consensus term is formed by multiplying the two original terms together, leaving out the selected variable and its complement. For example, the consensus of $ab$ and $a'c$ is $bc$; the consensus of $abd$ and $b'de'$ is $(ad)(de') = ade'$. The consensus of terms $ab'd$ and $a'bd'$ is 0.

The consensus theorem can be stated as follows:

$$XY + X'Z + YZ = XY + X'Z \tag{3-20}$$

*Proof:*

$$\begin{aligned} XY + X'Z + YZ &= XY + X'Z + (X + X')YZ \\ &= (XY + XYZ) + (X'Z + X'YZ) \\ &= XY(1 + Z) + X'Z(1 + Y) = XY + X'Z \end{aligned}$$

The consensus theorem can be used to eliminate redundant terms from Boolean expressions. For example, in the following expression, $b'c$ is the consensus of $a'b'$ and $ac$, and $ab$ is the consensus of $ac$ and $bc'$, so both consensus terms can be eliminated:

$$a'b' + ac + bc' + b'c + ab = a'b' + ac + bc'$$

The brackets indicate how the consensus terms are formed.

The dual form of the consensus theorem is

$$(X + Y)(X' + Z)(Y + Z) = (X + Y)(X' + Z) \tag{3-21}$$

Note again that the key to recognizing the consensus term is to first find a pair of terms, one of which contains a variable and the other its complement. In this case, the consensus is formed by adding this pair of terms together leaving out the selected variable and its complement. In the following expression, $(a + b + d')$ is a consensus term and can be eliminated by using the dual consensus theorem:

$$(a + b + c')(a + b + d')(b + c + d') = (a + b + c')(b + c + d')$$

The final result obtained by application of the consensus theorem may depend on the order in which terms are eliminated.

**Example**

$$A'C'D + A'BD + \cancel{BCD} + ABC + ACD' \tag{3-22}$$

First, we eliminate $BCD$ as shown. (Why can it be eliminated?)

Now that $BCD$ has been eliminated, it is no longer there, and it *cannot* be used to eliminate another term. Checking all pairs of terms shows that no additional terms can be eliminated by the consensus theorem.

Now we start over again:

$$A'C'D + \cancel{A'BD} + BCD + \cancel{ABC} + ACD' \tag{3-23}$$

This time, we do not eliminate $BCD$; instead we eliminate two other terms by the consensus theorem. After doing this, observe that $BCD$ can no longer be eliminated. Note that the expression reduces to four terms if $BCD$ is eliminated first, but that it can be reduced to three terms if $BCD$ is not eliminated.

Sometimes it is impossible to directly reduce an expression to a minimum number of terms by simply eliminating terms. It may be necessary to first add a term using the consensus theorem and *then* use the added term to eliminate other terms. For example, consider the expression

$$F = ABCD + B'CDE + A'B' + BCE'$$

If we compare every pair of terms to see if a consensus term can be formed, we find that the only consensus terms are $ACDE$ (from $ABCD$ and $B'CDE$) and $A'CE'$ (from $A'B'$ and $BCE'$). Because neither of these consensus terms appears in the original expression, we cannot directly eliminate any terms using the consensus theorem. However, if we first add the consensus term $ACDE$ to $F$, we get

$$F = ABCD + B'CDE + A'B' + BCE' + ACDE$$

Then, we can eliminate $ABCD$ and $B'CDE$ using the consensus theorem, and $F$ reduces to

$$F = A'B' + BCE' + ACDE$$

The term $ACDE$ is no longer redundant and cannot be eliminated from the final expression.

## 3.4 Algebraic Simplification of Switching Expressions

In this section we review and summarize methods for simplifying switching expressions, using the laws and theorems of Boolean algebra. This is important because simplifying an expression reduces the cost of realizing the expression using gates. Later, we will learn graphical methods for simplifying switching functions, but we will learn algebraic methods first. In addition to multiplying out and factoring, three basic ways of simplifying switching functions are combining terms, eliminating terms, and eliminating literals.

1. *Combining terms.* Use the theorem $XY + XY' = X$ to combine two terms. For example,

$$abc'd' + abcd' = abd' \qquad [X = abd', Y = c] \tag{3-24}$$

When combining terms by this theorem, the two terms to be combined should contain exactly the same variables, and exactly one of the variables should appear complemented in one term and not in the other. Because $X + X = X$, a given term may be duplicated and combined with two or more other terms. For example,

$$ab'c + abc + a'bc = ab'c + abc + abc + a'bc = ac + bc$$

The theorem still can be used, of course, when $X$ and $Y$ are replaced with more complicated expressions. For example,

$$(a + bc)(d + e') + a'(b' + c')(d + e') = d + e'$$
$$[X = d + e', Y = a + bc, Y' = a'(b' + c')]$$

2. *Eliminating terms.* Use the theorem $X + XY = X$ to eliminate redundant terms if possible; then try to apply the consensus theorem $(XY + X'Z + YZ = XY + X'Z)$ to eliminate any consensus terms. For example,

$$a'b + a'bc = a'b \qquad [X = a'b]$$
$$a'bc' + bcd + a'bd = a'bc' + bcd \qquad [X = c, Y = bd, Z = a'b] \qquad (3\text{-}25)$$

3. *Eliminating literals.* Use the theorem $X + X'Y = X + Y$ to eliminate redundant literals. Simple factoring may be necessary before the theorem is applied.

**Example**

$$\begin{aligned} A'B + A'B'C'D' + ABCD' &= A'(B + B'C'D') + ABCD' \\ &= A'(B + C'D') + ABCD' \\ &= B(A' + ACD') + A'C'D' \\ &= B(A' + CD') + A'C'D' \\ &= A'B + BCD' + A'C'D' \end{aligned} \qquad (3\text{-}26)$$

---

The expression obtained after applying steps 1, 2, and 3 will not necessarily have a minimum number of terms or a minimum number of literals. If it does not and no further simplification can be made using steps 1, 2, and 3, the deliberate introduction of redundant terms may be necessary before further simplification can be made.

4. *Adding redundant terms.* Redundant terms can be introduced in several ways such as adding $xx'$, multiplying by $(x + x')$, adding $yz$ to $xy + x'z$, or adding $xy$ to $x$. When possible, the added terms should be chosen so that they will combine with or eliminate other terms.

**Example**

$$\begin{aligned} & WX + XY + X'Z' + WY'Z' && \text{(add } WZ' \text{ by consensus theorem)} \\ &= WX + XY + X'Z' + WY'Z' + WZ' && \text{(eliminate } WY'Z'\text{)} \\ &= WX + XY + X'Z' + WZ' && \text{(eliminate } WZ'\text{)} \\ &= WX + XY + X'Z' && (3\text{-}27) \end{aligned}$$

---

The following comprehensive example illustrates the use of all four methods:

**Example**

$$\underbrace{A'B'C'D' + A'BC'D'}_{①\ A'C'D'} + A'BD + \underbrace{\cancel{A'BC'D}}_{②} + ABCD + ACD' + B'CD'$$

$$= A'C'D' + BD(A' \underset{③}{+} AC) + ACD' + B'CD'$$

$$= A'C'D' + A'BD + \underbrace{BCD + ACD'}_{+\ ABC\ ④} + B'CD'$$

$$= A'C'D' + \underbrace{A'BD + \cancel{BCD} + \overbrace{\cancel{ACD'} + B'CD' + ABC}^{\text{consensus } ACD'}}_{\text{consensus } BCD}$$

$$= A'C'D' + A'BD + B'CD' + ABC \tag{3-28}$$

What theorems were used in steps 1, 2, 3, and 4?

---

If the simplified expression is to be left in a product-of-sums form instead of a sum-of-products form, the duals of the preceding theorems should be applied.

**Example**

$$\underbrace{(A' + B' + C')(A' + B' + C)}_{① \ (A' + B')}(B' + C)(A + C)\underset{②}{\cancel{(A + B + C)}}$$

$$= (A' + B')\underset{③}{\cancel{(B' + C)}}(A + C) = (A' + B')(A + C) \tag{3-29}$$

What theorems were used in steps 1, 2, and 3?

---

In general, there is no easy way of determining when a Boolean expression has a minimum number of terms or a minimum number of literals. Systematic methods for finding minimum sum-of-products and minimum product-of-sums expressions will be discussed in Units 5 and 6.

## 3.5 Proving Validity of an Equation

Often we will need to determine if an equation is valid for all combinations of values of the variables. Several methods can be used to determine if an equation is valid:

1. Construct a truth table and evaluate both sides of the equation for all combinations of values of the variables. (This method is rather tedious if the number of variables is large, and it certainly is not very elegant.)
2. Manipulate one side of the equation by applying various theorems until it is identical with the other side.
3. Reduce both sides of the equation independently to the same expression.
4. It is permissible to perform the same operation on both sides of the equation provided that the operation is reversible. For example, it is all right to complement both sides of the equation, but it is *not* permissible to multiply both sides of the equation by the same expression. (Multiplication is not reversible because division is not defined for Boolean algebra.) Similarly, it is *not* permissible to add the same term to both sides of the equation because subtraction is not defined for Boolean algebra.

To prove that an equation is *not* valid, it is sufficient to show one combination of values of the variables for which the two sides of the equation have different values. When using method 2 or 3 above to prove that an equation is valid, a useful strategy is to

1. First reduce both sides to a sum of products (or a product of sums).
2. Compare the two sides of the equation to see how they differ.
3. Then try to add terms to one side of the equation that are present on the other side.
4. Finally try to eliminate terms from one side that are not present on the other.

Whatever method is used, frequently compare both sides of the equation and let the difference between them serve as a guide for what steps to take next.

**Example 1**

Show that

$$A'BD' + BCD + ABC' + AB'D = BC'D' + AD + A'BC$$

Starting with the left side, we first add consensus terms, then combine terms, and finally eliminate terms by the consensus theorem.

$A'BD' + BCD + ABC' + AB'D$

$= A'BD' + BCD + ABC' + AB'D + BC'D' + A'BC + ABD$

(add consensus of $A'BD'$ and $ABC'$) → $BC'D'$

(add consensus of $A'BD'$ and $BCD$) → $A'BC$

(add consensus of $BCD$ and $ABC'$) → $ABD$

$= AD + A'BD' + BCD + ABC' + BC'D' + A'BC = BC'D' + AD + A'BC$

$ABC'$ ← (eliminate consensus of $BC'D'$ and $AD$)

$BCD$ ← (eliminate consensus of $AD$ and $A'BC$)

$A'BD'$ ← (eliminate consensus of $BC'D'$ and $A'BC$) (3-30)

---

**Example 2**

Show that the following equation is valid:

$$A'BC'D + (A' + BC)(A + C'D') + BC'D + A'BC'$$
$$= ABCD + A'C'D' + ABD + ABCD' + BC'D$$

First, we will reduce the left side:

$A'BC'D + (A' + BC)(A + C'D') + BC'D + A'BC'$

(eliminate $A'BC'D$ using (2-13))

$= (A' + BC)(A + C'D') + BC'D + A'BC'$

(multiply out using (3-3))

$= ABC + A'C'D' + BC'D + A'BC'$

(eliminate $A'BC'$ by consensus)

$= ABC + A'C'D' + BC'D$

Now we will reduce the right side:

$$= ABCD + A'C'D' + ABD + ABCD' + BC'D$$

(combine $ABCD$ and $ABCD'$)

$$= ABC + A'C'D' + ABD + BC'D$$

(eliminate $ABD$ by consensus)

$$= ABC + A'C'D' + BC'D$$

Because both sides of the original equation were independently reduced to the same expression, the original equation is valid.

---

As we have previously observed, some of the theorems of Boolean algebra are not true for ordinary algebra. Similarly, some of the theorems of ordinary algebra are *not* true for Boolean algebra. Consider, for example, the cancellation law for ordinary algebra:

$$\text{If } x + y = x + z, \quad \text{then} \quad y = z \tag{3-31}$$

The cancellation law is *not* true for Boolean algebra. We will demonstrate this by constructing a counterexample in which $x + y = x + z$ but $y \neq z$. Let $x = 1$, $y = 0$, $z = 1$. Then,

$$1 + 0 = 1 + 1 \text{ but } 0 \neq 1$$

In ordinary algebra, the cancellation law for multiplication is

$$\text{If } xy = xz, \quad \text{then} \quad y = z \tag{3-32}$$

This law is valid provided $x \neq 0$.

In Boolean algebra, the cancellation law for multiplication is also *not* valid when $x = 0$. (Let $x = 0$, $y = 0$, $z = 1$; then $0 \cdot 0 = 0 \cdot 1$, but $0 \neq 1$). Because $x = 0$ about half of the time in switching algebra, the cancellation law for multiplication cannot be used.

Even though Statements (3-31) and (3-32) are generally false for Boolean algebra, the converses

$$\text{If } y = z, \quad \text{then} \quad x + y = x + z \tag{3-33}$$

$$\text{If } y = z, \quad \text{then} \quad xy = xz \tag{3-34}$$

are true. Thus, we see that although adding the same term to both sides of a Boolean equation leads to a valid equation, the reverse operation of canceling or subtracting a term from both sides generally does not lead to a valid equation. Similarly, multiplying both sides of a Boolean equation by the same term leads to a valid equation, but not conversely. When we are attempting to prove that an equation is valid, it is *not* permissible to add the same expression to both sides of the equation or to multiply both sides by the same expression, because these operations are not reversible.

# Programmed Exercise 3.1

Cover the answers to this exercise with a sheet of paper and slide it down as you check your answers. Write your answer in the space provided before looking at the correct answer.

The following expression is to be multiplied out to form a sum of products:

$$(A + B + C')(A' + B' + D)(A' + C + D')(A + C' + D)$$

First, find a pair of sum terms which have two literals in common and apply the second distributive law. Also, apply the same law to the other pair of terms.

---

**Answer** $(A + C' + BD)[A' + (B' + D)(C + D')]$

(*Note:* This answer was obtained by using $(X + Y)(X + Z) = X + YZ$.)

---

Next, find a pair of sum terms which have a variable in one and its complement in the other. Use the appropriate theorem to multiply these sum terms together without introducing any redundant terms. Apply the same theorem a second time.

---

**Answer** $(A + C' + BD)(A' + B'D' + CD) = A(B'D' + CD) + A'(C' + BD)$ or
$A(B' + D)(C + D') + A'(C' + BD) = A(B'D' + CD) + A'(C' + BD)$

(*Note:* This answer was obtained using $(X + Y)(X' + Z) = XZ + X'Y$.)

---

Complete the problem by multiplying out using the ordinary distributive law.

---

**Final Answer** $AB'D' + ACD + A'C' + A'BD$

---

# Programmed Exercise 3.2

Cover the answers to this exercise with a sheet of paper and slide it down as you check your answers. Write your answer in the space provided before looking at the correct answer.

The following expression is to be factored to form a product of sums:

$$WXY' + W'X'Z + WY'Z + W'YZ'$$

First, factor as far as you can using the ordinary distributive law.

**Answer** $WY'(X + Z) + W'(X'Z + YZ')$

Next, factor further by using a theorem which involves a variable and its complement. Apply this theorem twice.

**Answer**

$$(W + X'Z + YZ')[W' + Y'(X + Z)] = [W + (X' + Z')(Y + Z)][W' + Y'(X + Z)]$$

or

$$WY'(X + Z) + W'(X' + Z')(Y + Z) = [W + (X' + Z')(Y + Z)][W' + Y'(X + Z)]$$

[*Note:* This answer was obtained by using $AB + A'C = (A + C)(A' + B)$.]

Now, complete the factoring by using the second distributive law.

**Final answer** $(W + X' + Z')(W + Y + Z)(W' + Y')(W' + X + Z)$

# Programmed Exercise 3.3

Cover the answers to this exercise with a sheet of paper and slide it down as you check your answers. Write your answer in the space provided before looking at the correct answer.

The following expression is to be simplified using the consensus theorem:

$$AC' + AB'D + A'B'C + A'CD' + B'C'D'$$

First, find all of the consensus terms by checking all pairs of terms.

**Answer** The consensus terms are indicated.

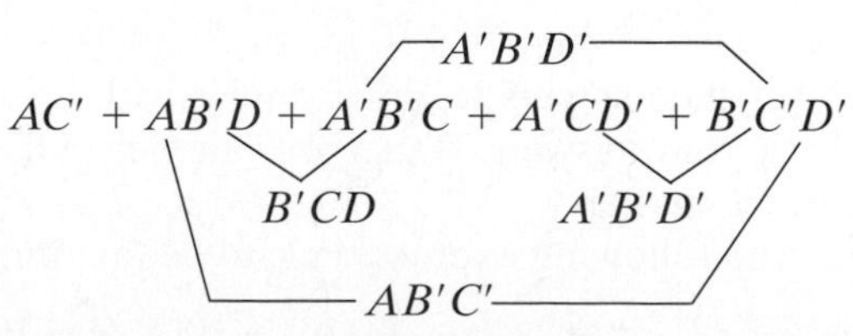

$$AC' + AB'D + A'B'C + A'CD' + B'C'D'$$

Consensus terms: $A'B'D'$ (from $A'B'C$ and $B'C'D'$), $B'CD$ (from $AB'D$ and $A'B'C$), $A'B'D'$ (from $A'CD'$ and $B'C'D'$), $AB'C'$ (from $AC'$ and $B'C'D'$)

Can the original expression be simplified by the direct application of the consensus theorem?

---

**Answer** No, because none of the consensus terms appears in the original expression.

---

Now add the consensus term $B'CD$ to the original expression. Compare the added term with each of the original terms to see if any consensus exists. Eliminate as many of the original terms as you can.

---

**Answer**

$$\overbrace{AC' + \cancel{AB'D} + \cancel{A'B'C} + \underbrace{A'CD' + B'C'D' + B'CD}_{(A'B'C)}}^{(AB'D)}$$

---

Now that we have eliminated two terms, can $B'CD$ also be eliminated? What is the final reduced expression?

---

**Answer** No, because the terms used to form $B'CD$ are gone. Final answer is

$$AC' + A'CD' + B'C'D' + B'CD$$

---

# Programmed Exercise 3.4

Keep the answers to this exercise covered with a sheet of paper and slide it down as you check your answers.

*Problem:* The following expression is to be simplified

$$ab'cd'e + acd + acf'gh' + abcd'e + acde' + e'h'$$

State a theorem which can be used to combine a pair of terms and apply it to combine two of the terms in the above expression.

---

**Answer** Apply $XY + XY' = X$ to the terms $ab'cd'e$ and $abcd'e$, which reduces the expression to

$$acd'e + acd + acf'gh' + acde' + e'h'$$

---

Now state a theorem (other than the consensus theorem) which can be used to eliminate terms and apply it to eliminate a term in this expression.

---

**Answer** Apply $X + XY = X$ to eliminate $acde'$. (What term corresponds to $X$?) The result is

$$acd'e + acd + acf'gh' + e'h'$$

---

Now state a theorem that can be used to eliminate literals and apply it to eliminate a literal from one of the terms in this expression. (*Hint*: It may be necessary to factor out some common variables from a pair of terms before the theorem can be applied.)

---

**Answer** Use $X + X'Y = X + Y$ to eliminate a literal from $acd'e$. To do this, first factor $ac$ out of the first two terms: $acd'e + acd = ac(d + d'e)$. After eliminating $d'$, the resulting expression is

$$ace + acd + acf'gh' + e'h'$$

---

(a) Can any term be eliminated from this expression by the direct application of the consensus theorem?
(b) If not, add a redundant term using the consensus theorem, and use this redundant term to eliminate one of the other terms.
(c) Finally, reduce your expression to three terms.

---

**Answer**
(a) No
(b) Add the consensus of $ace$ and $e'h'$:
$$ace + acd + acf'gh' + e'h' + ach'$$
Now eliminate $acf'gh'$ (by $X + XY = X$)
$$ace + acd + e'h' + ach'$$
(c) Now eliminate $ach'$ by the consensus theorem. The final answer is
$$ace + acd + e'h'$$

---

# Programmed Exercise 3.5

Keep the answers to this exercise covered with a sheet of paper and slide it down as you check your answers.

$$Z = (A + C' + F' + G)(A + C' + F + G)(A + B + C' + D' + G)$$
$$(A + C + E + G)(A' + B + G)(B + C' + F + G)$$

This is to be simplified to the form

$$(X + X + X)(X + X + X)(X + X + X)$$

where each $X$ represents a literal.

State a theorem which can be used to combine the first two sum terms of $Z$ and apply it. (*Hint:* The two sum terms differ in only one variable.)

---

**Answer**

$$(X + Y)(X + Y') = X$$
$$Z = (A + C' + G)(A + B + C' + D' + G)(A + C + E + G)(A' + B + G)$$
$$(B + C' + F + G)$$

---

Now state a theorem (other than the consensus theorem) which can be used to eliminate a sum term and apply it to this expression.

---

**Answer**

$$X(X + Y) = X$$
$$Z = (A + C' + G)(A + C + E + G)(A' + B + G)(B + C' + F + G)$$

---

Next, eliminate one literal from the second term, leaving the expression otherwise unchanged. (*Hint*: This cannot be done by the direct application of one theorem; it will be necessary to partially multiply out the first two sum terms before eliminating the literal.)

---

**Answer**

$$(A + C' + G)(A + C + E + G) = A + G + C'(C + E) = A + G + C'E$$

Therefore,

$$Z = (A + C' + G)(A + E + G)(A' + B + G)(B + C' + F + G)$$

---

(a) Can any term be eliminated from this expression by the direct application of the consensus theorem?
(b) If not, add a redundant sum term using the consensus theorem, and use this redundant term to eliminate one of the other terms.
(c) Finally, reduce your expression to a product of three sum terms.

---

**Answer**
(a) No
(b) Add $B + C' + G$ (consensus of $A + C' + G$ and $A' + B + G$).
Use $X(X + Y) = X$, where $X = B + C' + G$, to eliminate $B + C' + F + G$.
(c) Now eliminate $B + C' + G$ by consensus. The final answer is

$$Z = (A + C' + G)\,(A + E + G)\,(A' + B + G)$$

---

# Problems

**3.6** In each case, multiply out to obtain a sum of products: (Simplify where possible.)
(a) $(W + X' + Z')\,(W' + Y')\,(W' + X + Z')\,(W + X')\,(W + Y + Z)$
(b) $(A + B + C + D)\,(A' + B' + C + D')\,(A' + C)\,(A + D)\,(B + C + D)$

**3.7** Factor to obtain a product of sums. (Simplify where possible.)
(a) $BCD + C'D' + B'C'D + CD$
(b) $A'C'D' + ABD' + A'CD + B'D$

**3.8** Write an expression for $F$ and simplify.

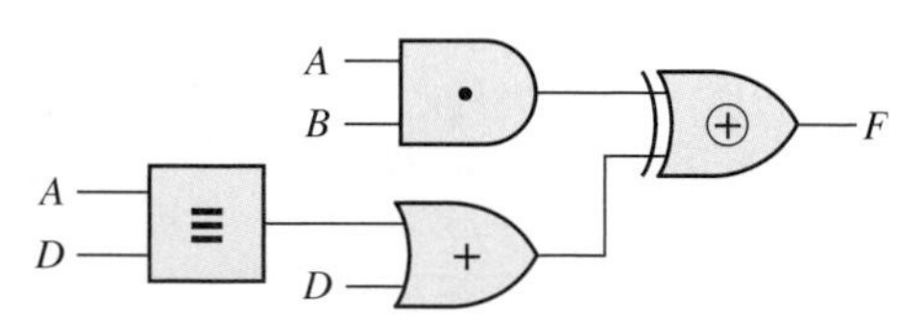


**3.9** Is the following distributive law valid? $A \oplus BC = (A \oplus B)(A \oplus C)$ Prove your answer.

**3.10** (a) Reduce to a minimum sum of products (three terms):
$(X + W)\,(Y \oplus Z) + XW'$
(b) Reduce to a minimum sum of products (four terms):
$(A \oplus BC) + BD + ACD$
(c) Reduce to a minimum product of sums (three terms):
$(A' + C' + D')\,(A' + B + C')\,(A + B + D)\,(A + C + D)$

**3.11** Simplify algebraically to a minimum sum of products (five terms):
$(A + B' + C + E')(A + B' + D' + E)(B' + C' + D' + E')$

**3.12** Prove algebraically that the following equation is valid:
$A'CD'E + A'B'D' + ABCE + ABD = A'B'D' + ABD + BCD'E$

**3.13** Simplify each of the following expressions:
(a) $KLMN' + K'L'MN + MN'$
(b) $KL'M' + MN' + LM'N'$
(c) $(K + L')(K' + L' + N)(L' + M + N')$
(d) $(K' + L + M' + N)(K' + M' + N + R)(K' + M' + N + R')KM$

**3.14** Factor to obtain a product of sums:
(a) $K'L'M + KM'N + KLM + LM'N'$ (four terms)
(b) $KL + K'L' + L'M'N' + LMN'$ (four terms)
(c) $KL + K'L'M + L'M'N + LM'N'$ (four terms)
(d) $K'M'N + KL'N' + K'MN' + LN$ (four terms)
(e) $WXY + WX'Y + WYZ + XYZ'$ (three terms)

**3.15** Multiply out to obtain a sum of products:
(a) $(K' + M' + N)(K' + M)(L + M' + N')(K' + L + M)(M + N)$ (three terms)
(b) $(K' + L' + M')(K + M + N')(K + L)(K' + N)(K' + M + N)$
(c) $(K' + L' + M)(K + N')(K' + L + N')(K + L)(K + M + N')$
(d) $(K + L + M)(K' + L' + N')(K' + L' + M')(K + L + N)$
(e) $(K + L + M)(K + M + N)(K' + L' + M')(K' + M' + N')$

**3.16** Eliminate the exclusive-OR, and then factor to obtain a minimum product of sums:
(a) $(KL \oplus M) + M'N'$
(b) $M'(K \oplus N') + MN + K'N$

**3.17** Algebraically prove identities involving the equivalence (exclusive-NOR) operation:
(a) $x \equiv 0 = x'$
(b) $x \equiv 1 = x$
(c) $x \equiv x = 1$
(d) $x \equiv x' = 0$
(e) $x \equiv y = y \equiv x$
(f) $(x \equiv y) \equiv z = x \equiv (y \equiv z)$
(g) $(x \equiv y)' = x' \equiv y = x \equiv y'$

**3.18** Algebraically prove identities involving the exclusive-OR operation:
(a) $x \oplus 0 = x$
(b) $x \oplus 1 = x'$
(c) $x \oplus x = 0$
(d) $x \oplus x' = 1$
(e) $x \oplus y = y \oplus x$
(f) $(x \oplus y) \oplus z = x \oplus (y \oplus z)$
(g) $(x \oplus y)' = x' \oplus y = x \oplus y'$

**3.19** Algebraically prove the following identities:
(a) $x + y = x \oplus y \oplus xy$
(b) $x + y = x \equiv y \equiv xy$

**3.20** Algebraically prove or disprove the following distributive identities:
(a) $x(y \oplus z) = xy \oplus xz$
(b) $x + (y \oplus z) = (x + y) \oplus (x + z)$
(c) $x(y \equiv z) = xy \equiv xz$
(d) $x + (y \equiv z) = (x + y) \equiv (x + z)$

**3.21** Simplify each of the following expressions using only the consensus theorem (or its dual):
(a) $BC'D' + ABC' + AC'D + AB'D + A'BD'$ (reduce to three terms)
(b) $W'Y' + WYZ + XY'Z + WX'Y$ (reduce to three terms)
(c) $(B + C + D)(A + B + C)(A' + C + D)(B' + C' + D')$
(d) $W'XY + WXZ + WY'Z + W'Z'$
(e) $A'BC' + BC'D' + A'CD + B'CD + A'BD$
(f) $(A + B + C)(B + C' + D)(A + B + D)(A' + B' + D')$

**3.22** Factor $Z = ABC + DE + ACF + AD' + AB'E'$ and simplify it to the form $(X + X)(X + X)(X + X + X + X)$ (where each $X$ represents a literal). Now express $Z$ as a minimum sum of products in the form:

$$XX + XX + XX + XX$$

**3.23** Repeat Problem 3.22 for $F = A'B + AC + BC'D' + BEF + BDF$.

**3.24** Factor to obtain a product of four terms and then reduce to three terms by applying the consensus theorem: $X'Y'Z' + XYZ$

**3.25** Simplify each of the following expressions:
(a) $xy + x'yz' + yz$
(b) $(xy' + z)(x + y')z$
(c) $xy' + z + (x' + y)z'$
(d) $a'd(b' + c) + a'd'(b + c') + (b' + c)(b + c')$
(e) $w'x' + x'y' + yz + w'z'$
(f) $A'BCD + A'BC'D + B'EF + CDE'G + A'DEF + A'B'EF$ (reduce to a sum of three terms)
(g) $[(a' + d' + b'c)(b + d + ac')]' + b'c'd' + a'c'd$ (reduce to three terms)

**3.26** Simplify to a sum of three terms:
(a) $A'C'D' + AC' + BCD + A'CD' + A'BC + AB'C'$
(b) $A'B'C' + ABD + A'C + A'CD' + AC'D + AB'C'$

**3.27** Reduce to a minimum sum of products:

$$F = WXY' + (W'Y' \equiv X) + (Y \oplus WZ).$$

**3.28** Determine which of the following equations are always valid (give an algebraic proof):
(a) $a'b + b'c + c'a = ab' + bc' + ca'$
(b) $(a + b)(b + c)(c + a) = (a' + b')(b' + c')(c' + a')$
(c) $abc + ab'c' + b'cd + bc'd + ad = abc + ab'c' + b'cd + bc'd$
(d) $xy' + x'z + yz' = x'y + xz' + y'z$
(e) $(x + y)(y + z)(x + z) = (x' + y')(y' + z')(x' + z')$
(f) $abc' + ab'c + b'c'd + bcd = ab'c + abc' + ad + bcd + b'c'd$

**3.29** The following circuit is implemented using two half-adder circuits. The expressions for the half-adder outputs are $S = A \oplus B$ where $\oplus$ represents the exclusive-OR function, and $C = AB$. Derive simplified sum-of-products expressions for the circuit outputs SUM and $C_o$. Give the truth table for the outputs.

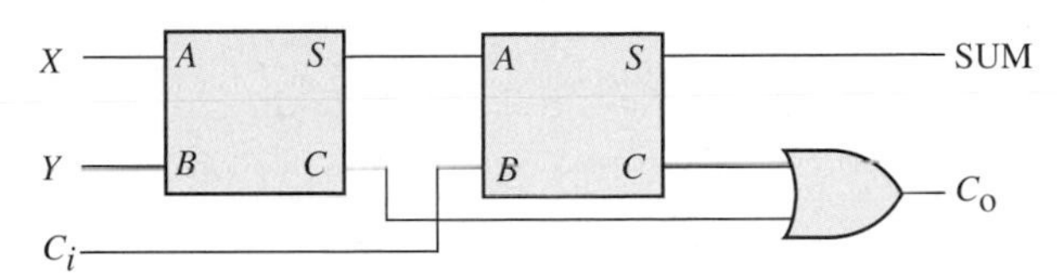


**3.30** The output of a majority circuit is 1 if a majority (more than half) of its inputs are equal to 1, and the output is 0 otherwise. Construct a truth table for a three-input majority circuit and derive a simplified sum-of-products expression for its output.

**3.31** Prove algebraically:
(a) $(X' + Y')(X \equiv Z) + (X + Y)(X \oplus Z) = (X \oplus Y) + Z'$
(b) $(W' + X + Y')(W + X' + Y)(W + Y' + Z) = X'Y' + WX + XYZ + W'YZ$
(c) $ABC + A'C'D' + A'BD' + ACD = (A' + C)(A + D')(B + C' + D)$

**3.32** Which of the following statements are always true? Justify your answers.
(a) If $A + B = C$, then $AD' + BD' = CD'$
(b) If $A'B + A'C = A'D$, then $B + C = D$
(c) If $A + B = C$, then $A + B + D = C + D$
(d) If $A + B + C = C + D$, then $A + B = D$

**3.33** Find all possible terms that could be added to each expression using the consensus theorem. Then reduce to a minimum sum of products.
(a) $A'C' + BC + AB' + A'BD + B'C'D' + ACD'$
(b) $A'C'D' + BC'D + AB'C' + A'BC$

**3.34** Simplify the following expression to a sum of two terms and then factor the result to obtain a product of sums: $abd'f' + b'cegh' + abd'f + acd'e + b'ce$

**3.35** Multiply out the following expression and simplify to obtain a sum-of-products expression with three terms: $(a + c)(b' + d)(a + c' + d')(b' + c' + d')$

**3.36** Factor and simplify to obtain a product-of-sums expression with four terms: $abc' + d'e + ace + b'c'd'$

**3.37** (a) Show that $x \oplus y = (x \equiv y)'$
(b) Realize $a'b'c' + a'bc + ab'c + abc'$ using only two-input equivalence gates.

# Applications of Boolean Algebra Minterm and Maxterm Expansions

## Objectives

1. Given a word description of the desired behavior of a logic circuit, write the output of the circuit as a function of the input variables. Specify this function as an algebraic expression or by means of a truth table, as is appropriate.
2. Given a truth table, write the function (or its complement) as both a minterm expansion (standard sum of products) and a maxterm expansion (standard product of sums). Be able to use both alphabetic and decimal notation.
3. Given an algebraic expression for a function, expand it algebraically to obtain the minterm or maxterm form.
4. Given one of the following: minterm expansion for $F$, minterm expansion for $F'$, maxterm expansion for $F$, or maxterm expansion for $F'$, find any of the other three forms.
5. Write the general form of the minterm and maxterm expansion of a function of $n$ variables.
6. Explain why some functions contain don't-care terms.
7. Explain the operation of a full adder and a full subtracter and derive logic equations for these modules. Draw a block diagram for a parallel adder or subtracter and trace signals on the block diagram.

# Study Guide

In the previous units, we placed a dot (•) inside the AND-gate symbol, a plus sign (+) inside the OR-gate symbol, and a ⊕ inside the Exclusive-OR. Because you are now familiar with the relationship between the shape of the gate symbol and the logic function performed, we will omit the •, +, and ⊕ and use the standard gate symbols for AND, OR, and Exclusive-OR in the rest of the book.

**1.** Study Section 4.1, *Conversion of English Sentences to Boolean Equations.*

(a) Use braces to identify the phrases in each of the following sentences:

(1) The tape reader should stop if the manual stop button is pressed,

if an error occurs, or if an end-of-tape signal is present.

(2) He eats eggs for breakfast if it is not Sunday and

he has eggs in the refrigerator.

(3) Addition should occur iff an add instruction is given and

the signs are the same, or if a subtract instruction is given and

the signs are not the same.

(b) Write a Boolean expression which represents each of the sentences in (a). Assign a variable to each phrase, and use a complemented variable to represent a phrase which contains "not".

(Your answers should be in the form $F = S'E$, $F = AB + SB'$, and $F = A + B + C$, but not necessarily in that order.)

(c) If $X$ represents the phrase "$N$ is greater than 3", how can you represent the phrase "$N$ is less than or equal to 3"?

(d) Work Problems 4.1 and 4.2.

**2.** Study Section 4.2, *Combinational Logic Design Using a Truth Table.* Previously, you have learned how to go from an algebraic expression for a function to a truth table; in this section you will learn how to go from a truth table to an algebraic expression.

(a) Write a product term which is 1 iff $a = 0, b = 0$, and $c = 1$.

(b) Write a sum term which is 0 iff $a = 0, b = 0$, and $c = 1$.

(c) Verify that your answers to (a) and (b) are complements.

(d) Write a product term which is 1 iff $a = 1, b = 0, c = 0$, and $d = 1$.

(e) Write a sum term which is 0 iff $a = 0, b = 0, c = 1$, and $d = 1$.

(f) For the given truth table, write $F$ as a sum of four product terms which correspond to the four 1's of $F$.

(g) From the truth table write $F$ as a product of four sum terms which correspond to the four 0's of $F$.

| a b c | F |
|---|---|
| 0 0 0 | 1 |
| 0 0 1 | 1 |
| 0 1 0 | 0 |
| 0 1 1 | 1 |
| 1 0 0 | 1 |
| 1 0 1 | 0 |
| 1 1 0 | 0 |
| 1 1 1 | 0 |

(h) Verify that your answers to both (f) and (g) reduce to $F = b'c' + a'c$.

**3.** Study Section 4.3, *Minterm and Maxterm Expansions.*

(a) Define the following terms:
minterm (for $n$ variables)

maxterm (for $n$ variables)

(b) Study Table 4-1 and observe the relation between the values of $A$, $B$, and $C$ and the corresponding minterms and maxterms.
If $A = 0$, then does $A$ or $A'$ appear in the minterm?
In the maxterm?
If $A = 1$, then does $A$ or $A'$ appear in the minterm?
In the maxterm?
What is the relation between minterm, $m_i$, and the corresponding maxterm, $M_i$?

(c) For the table given in Study Guide Question 2(f), write the minterm expansion for $F$ in $m$-notation and in decimal notation.

For the same table, write the maxterm expansion for $F$ in $M$-notation and in decimal notation.

Check your answers by converting your answer to 2(f) to $m$-notation and your answer to 2(g) to $M$-notation.

(d) Given a sum-of-products expression, how do you expand it to a standard sum of products (minterm expansion)?

(e) Given a product-of-sums expression, how do you expand it to a standard product of sums (maxterm expansion)?

(f) In Equation (4-11), what theorems were used to factor $f$ to obtain the maxterm expansion?

(g) Why is the following expression not a maxterm expansion?

$$f(A, B, C, D) = (A + B' + C + D)(A' + B + C')(A' + B + C + D')$$

(h) Assuming that there are three variables ($A$, $B$, $C$), identify each of the following as a minterm expansion, maxterm expansion, or neither:

(1) $AB + B'C'$ (2) $(A' + B + C')(A + B' + C)$
(3) $A + B + C$ (4) $(A' + B)(B' + C)(A' + C)$
(5) $A'BC' + AB'C + ABC$ (6) $AB'C'$

Note that it is possible for a minterm or maxterm expansion to have only one term.

**4.** (a) Given a minterm in terms of its variables, the procedure for conversion to decimal notation is
(1) Replace each complemented variable with a _____ and replace each uncomplemented variable with a _____.
(2) Convert the resulting binary number to decimal.

(b) Convert the minterm $AB'C'DE$ to decimal notation.

(c) Given that $m_{13}$ is a minterm of the variables $A$, $B$, $C$, $D$, and $E$, write the minterm in terms of these variables.

(d) Given a maxterm in terms of its variables, the procedure for conversion to decimal notation is
(1) Replace each complemented variable with a _____ and replace each uncomplemented variable with a _____.
(2) Group these 0's and 1's to form a binary number and convert to decimal.

(e) Convert the maxterm $A' + B + C + D' + E'$ to decimal notation.

(f) Given that $M_{13}$ is a maxterm of the variables $A$, $B$, $C$, $D$, and $E$, write the maxterm in terms of these variables.

(g) Check your answers to (b), (c), (e), and (f) by using the relation $M_i = m_i'$.

(h) Given $f(a, b, c, d, e) = \Pi\, M(0, 10, 28)$, express $f$ in terms of $a$, $b$, $c$, $d$, and $e$. (Your answer should contain only five complemented variables.)

**5.** Study Section 4.4, *General Minterm and Maxterm Expansions.* Make sure that you understand the notation here and can follow the algebra in all of the equations. If you have difficulty with this section, ask for help *before* you take the readiness test.

(a) How many different functions of four variables are possible?

(b) Explain why there are $2^{2^n}$ functions of $n$ variables.

(c) Write the function of Figure 4-1 in the form of Equation (4-13) and show that it reduces to Equation (4-3).

(d) For Equation (4-19), write out the indicated summations in full for the case $n = 2$.

(e) Study Tables 4-3 and 4-4 carefully and make sure you understand why each table entry is valid. Use the truth table for $f$ and $f'$ (Figure 4-1) to verify the entries in Table 4-4. If you understand the relationship between Table 4-3 and the truth table for $f$ and $f'$, you should be able to perform the conversions without having to memorize the table.

(f) Given that $f(A, B, C) = \Sigma m(0, 1, 3, 4, 7)$

The maxterm expansion for $f$ is ______________________

The minterm expansion for $f'$ is ______________________

The maxterm expansion for $f'$ is ______________________

(g) Work Problem 4.3 and 4.4.

**6.** Study Section 4.5, *Incompletely Specified Functions.*

(a) State two reasons why some functions have don't-care terms.

(b) Given the following table, write the minterm expansion for $Z$ in decimal form.

(c) Write the maxterm expansion in decimal form.

| *A B C* | *Z* |
|---|---|
| 0 0 0 | 1 |
| 0 0 1 | X |
| 0 1 0 | 0 |
| 0 1 1 | X |
| 1 0 0 | X |
| 1 0 1 | 1 |
| 1 1 0 | 0 |
| 1 1 1 | 0 |

(d) Work Problems 4.5 and 4.6.

7. Study Section 4.6, *Examples of Truth Table Construction.* Finding the truth table from the problem statement is probably the most difficult part of the process of designing a switching circuit. Make sure that you understand how to do this.

8. Work Problems 4.7 through 4.10.

9. Study Section 4.7, *Design of Binary Adders.*

   (a) For the given parallel adder, show the 0's and 1's at the full adder (FA) inputs and outputs when the following unsigned numbers are added: 11 + 14 = 25. Verify that the result is correct if $C_4S_3S_2S_1S_0$ is taken as a 5-bit sum. If the sum is limited to 4 bits, explain why this is an overflow condition.

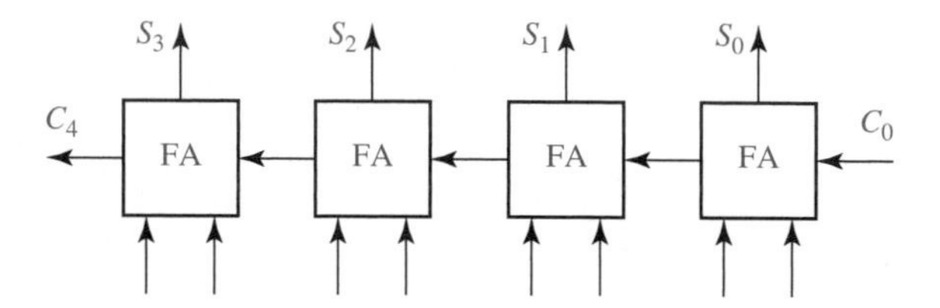


   (b) Review Section 1.4, *Representation of Negative Numbers.* If we use the 2's complement number system to add (−5) + (−2), verify that the FA inputs and outputs are exactly the same as in Part (a). However, for 2's complement, the interpretation of the results is quite different. After discarding $C_4$, verify that the resultant 4-bit sum is correct, and therefore no overflow has occurred.

   (c) If we use the 1's complement number system to add (−5) + (−2), show the FA inputs and outputs on the diagram below before the end-around carry is added in. Assume that $C_0$ is initially 0. Then add the end-around carry ($C_4$) to the rightmost FA, add the new carry ($C_1$) into the next cell, and continue until no further changes occur. Verify that the resulting sum is the correct 1's complement representation of −7.

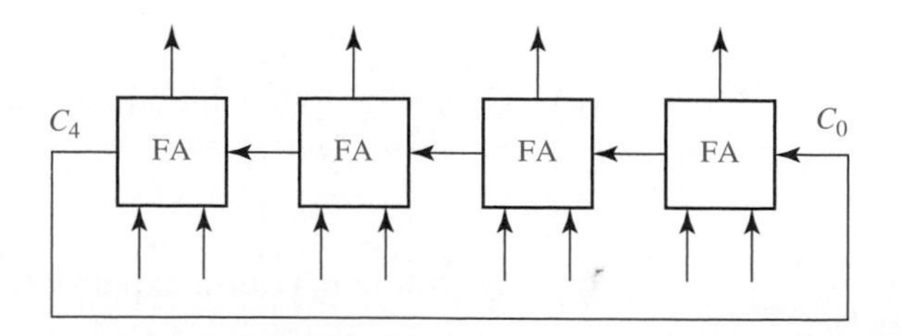

**10.** (a) Work the following subtraction example. As you subtract each column, place a 1 over the next column if you have to borrow, otherwise place a 0. For each column, as you compute $x_i - y_i - b_i$, fill in the corresponding values of $b_{i+1}$ and $d_i$ in the truth table. If you have done this correctly, the resulting table should match the full subtracter truth table (Table 4-6).

| | |
|---|---|
| | ← borrows |
| 1 1 0 0 0 1 1 0 | ← $X$ |
| −0 1 0 1 1 0 1 0 | ← $Y$ |
| | ← difference |

| $x_i$ $y_i$ $b_i$ | $b_{i+1}$ $d_i$ |
|---|---|
| 0 0 0 | |
| 0 0 1 | |
| 0 1 0 | |
| 0 1 1 | |
| 1 0 0 | |
| 1 0 1 | |
| 1 1 0 | |
| 1 1 1 | |

(b) Work Problems 4.11 and 4.12.

**11.** Read the following and then work Problem 4.13 or 4.14 as assigned:
When looking at an expression to determine the required number of gates, keep in mind that the number of required gates is generally *not* equal to the number of AND and OR operations which appear in the expression. For example,

$$AB + CD + EF(G + H)$$

contains four AND operations and three OR operations, but it only requires three AND gates and two OR gates:

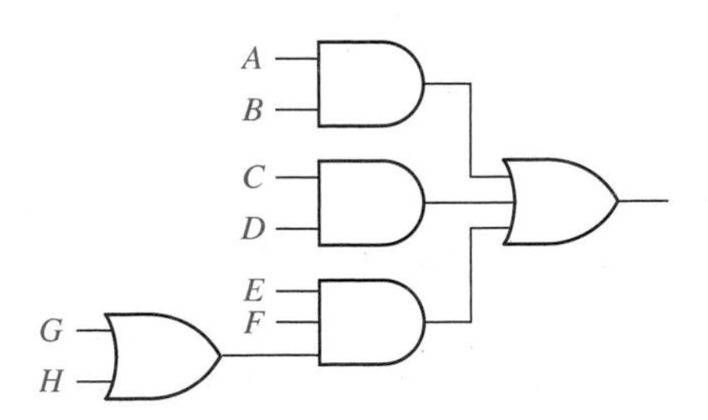


**12.** *Simulation Exercise.* (Must be completed before you take the readiness test.) One purpose of this exercise is to acquaint you with the simulator that you will be using later in more complex design problems. Follow the instructions on the Unit 4 lab assignment sheet.

**13.** Reread the objectives of this unit. Make sure that you understand the difference in the procedures for converting maxterms and minterms from decimal to algebraic notation. When you are satisfied that you can meet the objectives, take the readiness test. When you come to take the readiness test, turn in a copy of your solution to assigned simulation exercise.

# Applications of Boolean Algebra Minterm and Maxterm Expansions

In this unit you will learn how to design a combinational logic circuit starting with a word description of the desired circuit behavior. The first step is usually to translate the word description into a truth table or into an algebraic expression. Given the truth table for a Boolean function, two standard algebraic forms of the function can be derived—the standard sum of products (minterm expansion) and the standard product of sums (maxterm expansion). Simplification of either of these standard forms leads directly to a realization of the circuit using AND and OR gates.

## 4.1 Conversion of English Sentences to Boolean Equations

The three main steps in designing a single-output combinational switching circuit are

1. Find a switching function that specifies the desired behavior of the circuit.
2. Find a simplified algebraic expression for the function.
3. Realize the simplified function using available logic elements.

For simple problems, it may be possible to go directly from a word description of the desired behavior of the circuit to an algebraic expression for the output function. In other cases, it is better to first specify the function by means of a truth table and then derive an algebraic expression from the truth table.

Logic design problems are often stated in terms of one or more English sentences. The first step in designing a logic circuit is to translate these sentences into Boolean equations. In order to do this, we must break down each sentence into phrases and associate a Boolean variable with each phrase. If a phrase can have a value of true or false, then we can represent that phrase by a Boolean variable. Phrases such as "she goes to the store" or "today is Monday" can be either true or false, but a command like "go to the store" has no truth value. If a sentence has several phrases, we will mark each phrase with a brace. The following sentence has three phrases:

$\underbrace{\text{Mary watches TV}}$ if $\underbrace{\text{it is Monday night}}$ and $\underbrace{\text{she has finished her homework.}}$

The "if" and "and" are not included in any phrase; they show the relationships among the phrases.

We will define a two-valued variable to indicate the truth or falsity of each phrase:

$F = 1$ if "Mary watches TV" is true; otherwise, $F = 0$.

$A = 1$ if "it is Monday night" is true; otherwise, $A = 0$.

$B = 1$ if "she has finished her homework" is true; otherwise $B = 0$.

Because $F$ is "true" if $A$ and $B$ are both "true", we can represent the sentence by $F = A \cdot B$

The following example illustrates how to go from a word statement of a problem directly to an algebraic expression which represents the desired circuit behavior. An alarm circuit is to be designed which operates as follows:

> The alarm will ring iff the alarm switch is turned on and the door is not closed, or it is after 6 P.M. and the window is not closed.

The first step in writing an algebraic expression which corresponds to the above sentence is to associate a Boolean variable with each phrase in the sentence. This variable will have a value of 1 when the phrase is true and 0 when it is false. We will use the following assignment of variables:

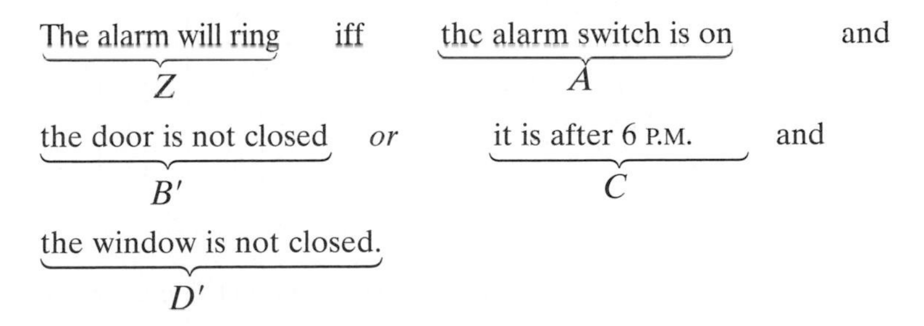


This assignment implies that if $Z = 1$, the alarm will ring. If the alarm switch is turned on, $A = 1$, and if it is after 6 P.M., $C = 1$. If we use the variable $B$ to represent the phrase "the door is closed", then $B'$ represents "the door is not closed". Thus, $B = 1$ if the door is closed, and $B' = 1$ $(B = 0)$ if the door is not closed. Similarly, $D = 1$ if the window is closed, and $D' = 1$ if the window is not closed. Using this assignment of variables, the above sentence can be translated into the following Boolean equation:

$$Z = AB' + CD'$$

This equation corresponds to the following circuit:

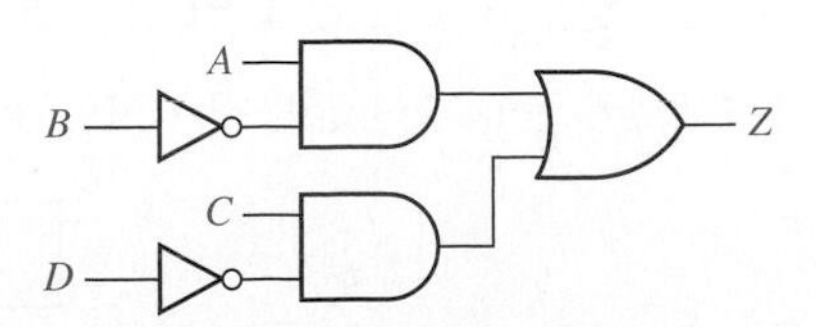

In this circuit, $A$ is a signal which is 1 when the alarm switch is on, $C$ is a signal from a time clock which is 1 when it is after 6 P.M., $B$ is a signal from a switch on the door which is 1 when the door is closed, and similarly $D$ is 1 when the window is closed. The output $Z$ is connected to the alarm so that it will ring when $Z = 1$.

## 4.2 Combinational Logic Design Using a Truth Table

The next example illustrates logic design using a truth table. A switching circuit has three inputs and one output, as shown in Figure 4-1(a). The inputs $A$, $B$, and $C$ represent the first, second, and third bits, respectively, of a binary number $N$. The output of the circuit is to be $f = 1$ if $N \geq 011_2$ and $f = 0$ if $N < 011_2$. The truth table for $f$ is shown in Figure 4-1(b).

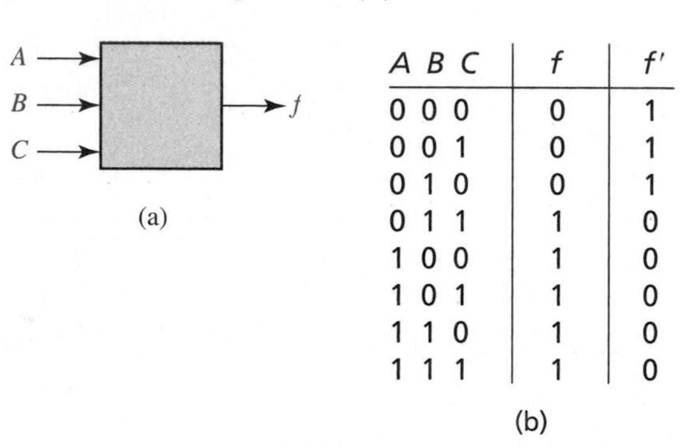


| A B C | f | f' |
|---|---|---|
| 0 0 0 | 0 | 1 |
| 0 0 1 | 0 | 1 |
| 0 1 0 | 0 | 1 |
| 0 1 1 | 1 | 0 |
| 1 0 0 | 1 | 0 |
| 1 0 1 | 1 | 0 |
| 1 1 0 | 1 | 0 |
| 1 1 1 | 1 | 0 |

(b)

**FIGURE 4-1** Combinational Circuit with Truth Table

Next, we will derive an algebraic expression for $f$ from the truth table by using the combinations of values of $A$, $B$, and $C$ for which $f = 1$. The term $A'BC$ is 1 only if $A = 0$, $B = 1$, and $C = 1$. Similarly, the term $AB'C'$ is 1 only for the combination 100, $AB'C$ is 1 only for 101, $ABC'$ is 1 only for 110, and $ABC$ is 1 only for 111. ORing these terms together yields

$$f = A'BC + AB'C' + AB'C + ABC' + ABC \tag{4-1}$$

This expression equals 1 if $A$, $B$, and $C$ take on any of the five combinations of values 011, 100, 101, 110, or 111. If any other combination of values occurs, $f$ is 0 because all five terms are 0.

Equation (4-1) can be simplified by first combining terms and then eliminating $A'$:

$$f = A'BC + AB' + AB = A'BC + A = A + BC \tag{4-2}$$

Equation (4-2) leads directly to the following circuit:

B
C
A
f

Instead of writing $f$ in terms of the 1's of the function, we may also write $f$ in terms of the 0's of the function. The function defined by Figure 4-1 is 0 for three combinations of input values. Observe that the term $A + B + C$ is 0 only if $A = B = C = 0$. Similarly, $A + B + C'$ is 0 only for the input combination 001, and $A + B' + C$ is 0 only for the combination 010. ANDing these terms together yields

$$f = (A + B + C)(A + B + C')(A + B' + C) \tag{4-3}$$

This expression equals 0 if $A$, $B$, and $C$ take on any of the combinations of values 000, 001, or 010. For any other combination of values, $f$ is 1 because all three terms are 1. Because Equation (4-3) represents the same function as Equation (4-1) they must both reduce to the same expression. Combining terms and using the second distributive law, Equation (4-3) simplifies to

$$f = (A + B)(A + B' + C) = A + B(B' + C) = A + BC \tag{4-4}$$

which is the same as Equation (4-2).

Another way to derive Equation (4-3) is to first write $f'$ as a sum of products, and then complement the result. From Figure 4-1, $f'$ *is* 1 for input combinations $ABC = 000, 001,$ and 010, so

$$f' = A'B'C' + A'B'C + A'BC'$$

Taking the complement of $f'$ yields Equation (4-3).

# 4.3 Minterm and Maxterm Expansions

Each of the terms in Equation (4-1) is referred to as a minterm. In general, a *minterm* of $n$ variables is a product of $n$ literals in which each variable appears exactly once in either true or complemented form, but not both. (A *literal* is a variable or its complement.) Table 4-1 lists all of the minterms of the three variables $A$, $B$, and $C$. Each minterm has a value of 1 for exactly one combination of values of the variables $A$, $B$, and $C$. Thus if $A = B = C = 0$, $A'B'C' = 1$; if $A = B = 0$ and $C = 1$, $A'B'C = 1$; and so forth. Minterms are often written in abbreviated form—$A'B'C'$ is designated $m_0$, $A'B'C$ is designated $m_1$, etc. In general, the minterm which corresponds to row $i$ of the truth table is designated $m_i$ ($i$ is usually written in decimal).

**TABLE 4-1** Minterms and Maxterms for Three Variables

| Row No. | A B C | Minterms | Maxterms |
|---|---|---|---|
| 0 | 0 0 0 | $A'B'C' = m_0$ | $A + B + C = M_0$ |
| 1 | 0 0 1 | $A'B'C = m_1$ | $A + B + C' = M_1$ |
| 2 | 0 1 0 | $A'BC' = m_2$ | $A + B' + C = M_2$ |
| 3 | 0 1 1 | $A'BC = m_3$ | $A + B' + C' = M_3$ |
| 4 | 1 0 0 | $AB'C' = m_4$ | $A' + B + C = M_4$ |
| 5 | 1 0 1 | $AB'C = m_5$ | $A' + B + C' = M_5$ |
| 6 | 1 1 0 | $ABC' = m_6$ | $A' + B' + C = M_6$ |
| 7 | 1 1 1 | $ABC = m_7$ | $A' + B' + C' = M_7$ |

When a function $f$ is written as a sum of minterms as in Equation (4-1), this is referred to as a *minterm expansion* or a *standard sum of products*.[1] If $f = 1$ for row $i$ of the truth table, then $m_i$ must be present in the minterm expansion because $m_i = 1$ only for the combination of values of the variables corresponding to row i of the table. Because the minterms present in $f$ are in one-to-one correspondence with the 1's of $f$ in the truth table, the minterm expansion for a function $f$ *is* unique. Equation (4-1) can be rewritten in terms of $m$-notation as

$$f(A, B, C) = m_3 + m_4 + m_5 + m_6 + m_7 \tag{4-5}$$

This can be further abbreviated by listing only the decimal subscripts in the form

$$f(A, B, C) = \Sigma\, m(3, 4, 5, 6, 7) \tag{4-5a}$$

Each of the sum terms (or factors) in Equation (4-3) is referred to as a *maxterm*. In general, a maxterm of $n$ variables is a sum of $n$ literals in which each variable appears exactly once in either true or complemented form, but not both. Table 4-1 lists all of the maxterms of the three variables $A$, $B$, and $C$. Each maxterm has a value of 0 for exactly one combination of values for $A$, $B$, and $C$. Thus, if $A = B = C = 0$, $A + B + C = 0$; if $A = B = 0$ and $C = 1$, $A + B + C' = 0$; and so forth. Maxterms are often written in abbreviated form using $M$-notation. The maxterm which corresponds to row $i$ of the truth table is designated $M_i$. Note that each maxterm is the complement of the corresponding minterm, that is, $M_i = m'_i$.

When a function $f$ is written as a product of maxterms, as in Equation (4-3), this is referred to as a *maxterm expansion* or *standard product of sums*. If $f = 0$ for row $i$ of the truth table, then $M_i$ must be present in the maxterm expansion because $M_i = 0$ only for the combination of values of the variables corresponding to row $i$ of the table. Note that the maxterms are multiplied together so that if any one of them is 0, $f$ will be 0. Because the maxterms are in one-to-one correspondence with the 0's of $f$ in the truth table, the maxterm expansion for a function $f$ *is* unique. Equation (4-3) can be rewritten in $M$-notation as

$$f(A, B, C) = M_0 M_1 M_2 \tag{4-6}$$

This can be further abbreviated by listing only the decimal subscripts in the form

$$f(A, B, C) = \Pi\, M(0, 1, 2) \tag{4-6a}$$

where $\Pi$ means a product.

Because if $f \neq 1$ then $f = 0$, it follows that if $m_i$ is *not* present in the minterm expansion of $f$, then $M_i$ is present in the maxterm expansion. Thus, given a minterm expansion of an $n$-variable function $f$ in decimal notation, the maxterm expansion is obtained by listing those decimal integers ($0 \leq i \leq 2^n - 1$) not in the minterm list. Using this method, Equation (4-6a) can be obtained directly from Equation (4-5a).

[1]Other names used in the literature for standard sum of products are canonical sum of products and disjunctive normal form. Similarly, a standard product of sums may be called a canonical product of sums or a conjunctive normal form.

Given the minterm or maxterm expansions for $f$, the minterm or maxterm expansions for the complement of $f$ are easy to obtain. Because $f'$ is 1 when $f$ is 0, the minterm expansion for $f'$ contains those minterms not present in $f$. Thus, from Equation (4-5),

$$f' = m_0 + m_1 + m_2 = \Sigma\, m(0, 1, 2) \tag{4-7}$$

Similarly, the maxterm expansion for $f'$ contains those maxterms not present in $f$. From Equation (4-6),

$$f' = \Pi\, M(3, 4, 5, 6, 7) = M_3 M_4 M_5 M_6 M_7 \tag{4-8}$$

Because the complement of a minterm is the corresponding maxterm, Equation (4-8) can be obtained by complementing Equation (4-5):

$$f' = (m_3 + m_4 + m_5 + m_6 + m_7)' = m_3' m_4' m_5' m_6' m_7' = M_3 M_4 M_5 M_6 M_7$$

Similarly, Equation (4-7) can be obtained by complementing Equation (4-6):

$$f' = (M_0 M_1 M_2)' = M_0' + M_1' + M_2' = m_0 + m_1 + m_2$$

A general switching expression can be converted to a minterm or maxterm expansion either using a truth table or algebraically. If a truth table is constructed by evaluating the expression for all different combinations of the values of the variables, the minterm and maxterm expansions can be obtained from the truth table by the methods just discussed. Another way to obtain the minterm expansion is to first write the expression as a sum of products and then introduce the missing variables in each term by applying the theorem $X + X' = 1$.

**Example**

Find the minterm expansion of $f(a,b,c,d) = a'(b' + d) + acd'$.

$$\begin{aligned} f &= a'b' + a'd + acd' \\ &= a'b'(c + c')(d + d') + a'd(b + b')(c + c') + acd'(b + b') \\ &= a'b'c'd' + a'b'c'd + a'b'cd' + a'b'cd + \cancel{a'b'c'd} + \cancel{a'b'cd} \\ &\qquad + a'bc'd + a'bcd + abcd' + ab'cd' \end{aligned} \tag{4-9}$$

Duplicate terms have been crossed out, because $X + X = X$. This expression can then be converted to decimal notation:

$$\begin{array}{cccccccc} f = a'b'c'd' + & a'b'c'd + & a'b'cd' + & a'b'cd + & a'bc'd + & a'bcd + & abcd' + & ab'cd' \\ 0000 & 0001 & 0010 & 0011 & 0101 & 0111 & 1110 & 1010 \end{array}$$

$$f = \Sigma\, m(0, 1, 2, 3, 5, 7, 10, 14) \tag{4-10}$$

The maxterm expansion for $f$ can then be obtained by listing the decimal integers (in the range 0 to 15) which do not correspond to minterms of $f$:

$$f = \Pi\, M(4, 6, 8, 9, 11, 12, 13, 15)$$

An alternate way of finding the maxterm expansion is to factor $f$ to obtain a product of sums, introduce the missing variables in each sum term by using $XX' = 0$, and then factor again to obtain the maxterms. For Equation (4-9),

$$\begin{aligned}
f &= a'(b' + d) + acd' \\
&= (a' + cd')(a + b' + d) = (a' + c)(a' + d')(a + b' + d) \\
&= (a' + bb' + c + dd')(a' + bb' + cc' + d')(a + b' + cc' + d) \\
&= (a' + bb' + c + d)(a' + bb' + c + d')\cancel{(a' + bb' + c + d')} \\
&\quad (a' + bb' + c' + d')(a + b' + cc' + d) \\
&= \underset{1000}{(a' + b + c + d)}\underset{1100}{(a' + b' + c + d)}\underset{1001}{(a' + b + c + d')}\underset{1101}{(a' + b' + c + d')} \\
&\quad \underset{1011}{(a' + b + c' + d')}\underset{1111}{(a' + b' + c' + d')}\underset{0100}{(a + b' + c + d)}\underset{0110}{(a + b' + c' + d)} \\
&= \Pi\, M(4, 6, 8, 9, 11, 12, 13, 15) \qquad (4\text{-}11)
\end{aligned}$$

Note that when translating the maxterms to decimal notation, a primed variable is first replaced with a 1 and an unprimed variable with a 0.

Because the terms in the minterm expansion of a function $F$ correspond one-to-one with the rows of the truth table for which $F = 1$, the minterm expansion of $F$ is unique. Thus, we can prove that an equation is valid by finding the minterm expansion of each side and showing that these expansions are the same.

*Example*

Show that $a'c + b'c' + ab = a'b' + bc + ac'$.

We will find the minterm expansion of each side by supplying the missing variables. For the left side,

$$\begin{aligned}
&a'c(b + b') + b'c'(a + a') + ab(c + c') \\
&\qquad = a'bc + a'b'c + ab'c' + a'b'c' + abc + abc' \\
&\qquad = m_3 + m_1 + m_4 + m_0 + m_7 + m_6
\end{aligned}$$

For the right side,

$$\begin{aligned}
&a'b'(c + c') + bc(a + a') + ac'(b + b') \\
&\qquad = a'b'c + a'b'c' + abc + a'bc + abc' + ab'c' \\
&\qquad = m_1 + m_0 + m_7 + m_3 + m_6 + m_4
\end{aligned}$$

Because the two minterm expansions are the same, the equation is valid.

## 4.4 General Minterm and Maxterm Expansions

Table 4-2 represents a truth table for a general function of three variables. Each $a_i$ is a constant with a value of 0 or 1. To completely specify a function, we must assign values to all of the $a_i$'s. Because each $a_i$ can be specified in two ways, there are $2^8$

**TABLE 4-2**
General Truth Table for Three Variables

| A B C | F |
|---|---|
| 0 0 0 | $a_0$ |
| 0 0 1 | $a_1$ |
| 0 1 0 | $a_2$ |
| 0 1 1 | $a_3$ |
| 1 0 0 | $a_4$ |
| 1 0 1 | $a_5$ |
| 1 1 0 | $a_6$ |
| 1 1 1 | $a_7$ |

ways of filling the $F$ column of the truth table; therefore, there are 256 different functions of three variables (this includes the degenerate cases, $F$ identically equal to 0 and $F$ identically equal to 1). For a function of $n$ variables, there are $2^n$ rows in the truth table, and because the value of $F$ can be 0 or 1 for each row, there are $2^{2^n}$ possible functions of $n$ variables.

From Table 4-2, we can write the minterm expansion for a general function of three variables as follows:

$$F = a_0m_0 + a_1m_1 + a_2m_2 + \cdots + a_7m_7 = \sum_{i=0}^{7} a_im_i \tag{4-12}$$

Note that if $a_i = 1$, minterm $m_i$ is present in the expansion; if $a_i = 0$, the corresponding minterm is not present. The maxterm expansion for a general function of three variables is

$$F = (a_0 + M_0)(a_1 + M_1)(a_2 + M_2) \cdots (a_7 + M_7) = \prod_{i=0}^{7}(a_i + M_i) \tag{4-13}$$

Note that if $a_i = 1$, $a_i + M_i = 1$, and $M_i$ drops out of the expansion; however, $M_i$ is present if $a_i = 0$.

From Equation (4-13), the minterm expansion of $F'$ is

$$F' = \left[\prod_{i=0}^{7}(a_i + M_i)\right]' = \sum_{i=0}^{7} a_i'M_i' = \sum_{i=0}^{7} a_i'm_i \tag{4-14}$$

Note that all minterms which are not present in $F$ are present in $F'$.

From Equation (4-12), the maxterm expansion of $F'$ is

$$F' = \left[\sum_{i=0}^{7} a_im_i\right]' = \prod_{i=0}^{7}(a_i' + m_i') = \prod_{i=0}^{7}(a_i' + M_i) \tag{4-15}$$

Note that all maxterms which are not present in $F$ are present in $F'$. Generalizing Equations (4-12), (4-13), (4-14), and (4-15) to $n$ variables, we have

$$F = \sum_{i=0}^{2^n-1} a_im_i = \prod_{i=0}^{2^n-1} (a_i + M_i) \tag{4-16}$$

$$F' = \sum_{i=0}^{2^n-1} a_i' m_i = \prod_{i=0}^{2^n-1} (a_i' + M_i) \tag{4-17}$$

Given two different minterms of $n$ variables, $m_i$ and $m_j$, at least one variable appears complemented in one of the minterms and uncomplemented in the other. Therefore, if $i \neq j$, $m_i m_j = 0$. For example, for $n = 3$, $m_1 m_3 = (A'B'C)(A'BC) = 0$. Given minterm expansions for two functions

$$f_1 = \sum_{i=0}^{2^n-1} a_i m_i \qquad f_2 = \sum_{j=0}^{2^n-1} b_j m_j \tag{4-18}$$

the product is

$$f_1 f_2 = \left(\sum_{i=0}^{2^n-1} a_i m_i\right)\left(\sum_{j=0}^{2^n-1} b_j m_j\right) = \sum_{i=0}^{2^n-1}\sum_{j=0}^{2^n-1} a_i b_j m_i m_j$$

$$= \sum_{i=0}^{2^n-1} a_i b_i m_i \quad \text{(because } m_i m_j = 0 \text{ unless } i = j\text{)} \tag{4-19}$$

Note that all of the cross-product terms ($i \neq j$) drop out so that $f_1 f_2$ contains only those minterms which are present in both $f_1$ and $f_2$. For example, if

$$f_1 = \Sigma\, m(0, 2, 3, 5, 9, 11) \qquad \text{and} \qquad f_2 = \Sigma\, m(0, 3, 9, 11, 13, 14)$$
$$f_1 f_2 = \Sigma\, m(0, 3, 9, 11)$$

Table 4-3 summarizes the procedures for conversion between minterm and maxterm expansions of $F$ and $F'$, assuming that all expansions are written as lists of decimal numbers. When using this table, keep in mind that the truth table for an $n$-variable function has $2^n$ rows so that the minterm (or maxterm) numbers range from 0 to $2^n - 1$. Table 4-4 illustrates the application of Table 4-3 to the three-variable function given in Figure 4-1.

**TABLE 4-3** Conversion of Forms

| GIVEN FORM | DESIRED FORM: Minterm Expansion of $F$ | Maxterm Expansion of $F$ | Minterm Expansion of $F'$ | Maxterm Expansion of $F'$ |
|---|---|---|---|---|
| Minterm Expansion of $F$ | ——— | maxterm nos. are those nos. not on the minterm list for $F$ | list minterms not present in $F$ | maxterm nos. are the same as minterm nos. of $F$ |
| Maxterm Expansion of $F$ | minterm nos. are those nos. not on the maxterm list for $F$ | ——— | minterm nos. are the same as maxterm nos. of $F$ | list maxterms not present in $F$ |

TABLE 4-4
Application of Table 4.3

| GIVEN FORM | DESIRED FORM: Minterm Expansion of $f$ | Maxterm Expansion of $f$ | Minterm Expansion of $f'$ | Maxterm Expansion of $f'$ |
|---|---|---|---|---|
| $f = \Sigma\, m(3, 4, 5, 6, 7)$ | ______ | $\Pi\, M(0, 1, 2)$ | $\Sigma\, m(0, 1, 2)$ | $\Pi\, M(3, 4, 5, 6, 7)$ |
| $f = \Pi\, M(0, 1, 2)$ | $\Sigma\, m(3, 4, 5, 6, 7)$ | ______ | $\Sigma\, m(0, 1, 2)$ | $\Pi\, M(3, 4, 5, 6, 7)$ |

## 4.5 Incompletely Specified Functions

A large digital system is usually divided into many subcircuits. Consider the following example in which the output of circuit $N_1$ drives the input of circuit $N_2$.

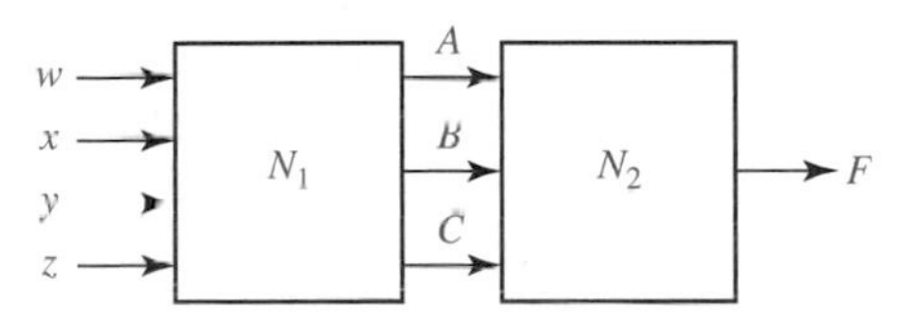


Let us assume that the output of $N_1$ does not generate all possible combinations of values for $A$, $B$, and $C$. In particular, we will assume that there are no combinations of values for $w$, $x$, $y$, and $z$ which cause $A$, $B$, and $C$ to assume values of 001 or 110. Hence, when we design $N_2$, it is not necessary to specify values of $F$ for $ABC = 001$ or 110 because these combinations of values can never occur as inputs to $N_2$. For example, $F$ might be specified by Table 4-5.

The X's in the table indicate that we don't care whether the value of 0 or 1 is assigned to $F$ for the combinations $ABC = 001$ or 110. In this example, we don't care what the value of $F$ is because these input combinations never occur anyway. The function $F$ is then *incompletely specified.* The minterms $A'B'C$ and $ABC'$ are referred to as don't-care minterms, since we don't care whether they are present in the function or not.

TABLE 4-5
Truth Table with Don't-Cares

| $A\ B\ C$ | $F$ |
|---|---|
| 0 0 0 | 1 |
| 0 0 1 | X |
| 0 1 0 | 0 |
| 0 1 1 | 1 |
| 1 0 0 | 0 |
| 1 0 1 | 0 |
| 1 1 0 | X |
| 1 1 1 | 1 |

When we realize the function, we must specify values for the don't-cares. It is desirable to choose values which will help simplify the function. If we assign the value 0 to both X's, then

$$F = A'B'C' + A'BC + ABC = A'B'C' + BC$$

If we assign 1 to the first X and 0 to the second, then

$$F = A'B'C' + A'B'C + A'BC + ABC = A'B' + BC$$

If we assign 1 to both X's, then

$$F = A'B'C' + A'B'C + A'BC + ABC' + ABC = A'B' + BC + AB$$

The second choice of values leads to the simplest solution.

We have seen one way in which incompletely specified functions can arise, and there are many other ways. In the preceding example, don't-cares were present because certain combinations of circuit inputs did not occur. In other cases, all input combinations may occur, but the circuit output is used in such a way that we do not care whether it is 0 or 1 for certain input combinations.

When writing the minterm expansion for an incompletely specified function, we will use $m$ to denote the required minterms and $d$ to denote the don't-care minterms. Using this notation, the minterm expansion for Table 4-5 is

$$F = \Sigma\, m(0, 3, 7) + \Sigma d(1, 6)$$

For each don't-care minterm there is a corresponding don't-care maxterm. For example, if $F = \text{X}$ (don't-care) for input combination 001, $m_1$ is a don't-care minterm and $M_1$ is a don't-care maxterm. We will use $D$ to represent a don't-care maxterm, and we write the maxterm expansion of the function in Table 4-5 as

$$F = \Pi\, M(2, 4, 5) \bullet \Pi\, D\,(1, 6)$$

which implies that maxterms $M_2$, $M_4$, and $M_5$ are present in $F$ and don't-care maxterms $M_1$ and $M_6$ are optional.

## 4.6 Examples of Truth Table Construction

### *Example 1*

We will design a simple binary adder that adds two 1-bit binary numbers, $a$ and $b$, to give a 2-bit sum. The numeric values for the adder inputs and output are as follows:

| $a$ | $b$ | Sum | |
|---|---|---|---|
| 0 | 0 | 00 | (0 + 0 = 0) |
| 0 | 1 | 01 | (0 + 1 = 1) |
| 1 | 0 | 01 | (1 + 0 = 1) |
| 1 | 1 | 10 | (1 + 1 = 2) |

We will represent inputs to the adder by the logic variables $A$ and $B$ and the 2-bit sum by the logic variables $X$ and $Y$, and we construct a truth table:

| A B | X Y |
|---|---|
| 0 0 | 0 0 |
| 0 1 | 0 1 |
| 1 0 | 0 1 |
| 1 1 | 1 0 |

Because a numeric value of 0 is represented by a logic 0 and a numeric value of 1 by a logic 1, the 0's and 1's in the truth table are exactly the same as in the previous table. From the truth table,

$$X = AB \text{ and } Y = A'B + AB' = A \oplus B$$

---

**Example 2**

An adder is to be designed which adds two 2-bit binary numbers to give a 3-bit binary sum. Find the truth table for the circuit. The circuit has four inputs and three outputs as shown:

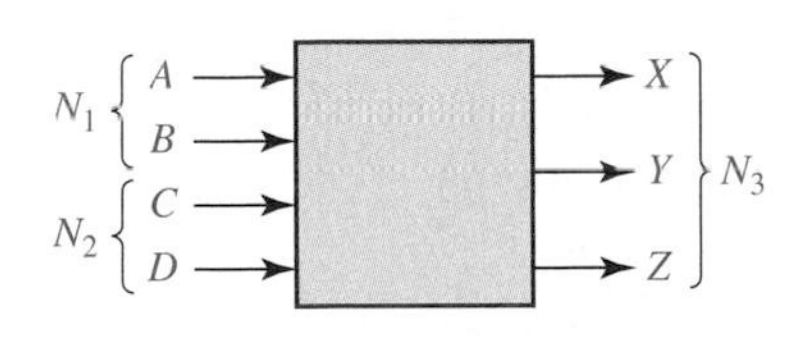


TRUTH TABLE:

| $N_1$ | $N_2$ | $N_3$ |
|---|---|---|
| A B | C D | X Y Z |
| 0 0 | 0 0 | 0 0 0 |
| 0 0 | 0 1 | 0 0 1 |
| 0 0 | 1 0 | 0 1 0 |
| 0 0 | 1 1 | 0 1 1 |
| 0 1 | 0 0 | 0 0 1 |
| 0 1 | 0 1 | 0 1 0 |
| 0 1 | 1 0 | 0 1 1 |
| 0 1 | 1 1 | 1 0 0 |
| 1 0 | 0 0 | 0 1 0 |
| 1 0 | 0 1 | 0 1 1 |
| 1 0 | 1 0 | 1 0 0 |
| 1 0 | 1 1 | 1 0 1 |
| 1 1 | 0 0 | 0 1 1 |
| 1 1 | 0 1 | 1 0 0 |
| 1 1 | 1 0 | 1 0 1 |
| 1 1 | 1 1 | 1 1 0 |

Inputs $A$ and $B$ taken together represent a binary number $N_1$. Inputs $C$ and $D$ taken together represent a binary number $N_2$. Outputs $X$, $Y$, and $Z$ taken together represent a binary number $N_3$, where $N_3 = N_1 + N_2$ (+ of course represents ordinary addition here).

In this example we have used $A$, $B$, $C$, and $D$ to represent both numeric values and logic values, but this should not cause any confusion because the numeric and logic values are the same. In forming the truth table, the variables were treated like binary numbers having *numeric* values. Now we wish to derive the switching functions for the output variables. In doing so, we will treat $A$, $B$, $C$, $D$, $X$, $Y$, and $Z$ as switching variables having nonnumeric values 0 and 1. (Remember that in this case the 0 and 1 may represent low and high voltages, open and closed switches, etc.)

From inspection of the table, the output functions are

$$X(A, B, C, D) = \Sigma\, m(7, 10, 11, 13, 14, 15)$$
$$Y(A, B, C, D) = \Sigma\, m(2, 3, 5, 6, 8, 9, 12, 15)$$
$$Z(A, B, C, D) = \Sigma\, m(1, 3, 4, 6, 9, 11, 12, 14)$$

---

**Example 3**

Design an error detector for 6-3-1-1 binary-coded-decimal digits. The output ($F$) is to be 1 iff the four inputs ($A$, $B$, $C$, $D$) represent an invalid code combination.

The valid 6-3-1-1 code combinations are listed in Table 1-2. If any other combination occurs, this is not a valid 6-3-1-1 binary-coded-decimal digit, and the circuit output should be $F = 1$ to indicate that an error has occurred. This leads to the following truth table:

| A B C D | F |
|---|---|
| 0 0 0 0 | 0 |
| 0 0 0 1 | 0 |
| 0 0 1 0 | 1 |
| 0 0 1 1 | 0 |
| 0 1 0 0 | 0 |
| 0 1 0 1 | 0 |
| 0 1 1 0 | 1 |
| 0 1 1 1 | 0 |
| 1 0 0 0 | 0 |
| 1 0 0 1 | 0 |
| 1 0 1 0 | 1 |
| 1 0 1 1 | 0 |
| 1 1 0 0 | 0 |
| 1 1 0 1 | 1 |
| 1 1 1 0 | 1 |
| 1 1 1 1 | 1 |

The corresponding output function is

$$\begin{aligned} F &= \Sigma\, m(2, 6, 10, 13, 14, 15) \\ &= A'B'CD' + A'BCD' + AB'CD' + ABCD' + ABC'D + ABCD \\ &= A'CD' + ACD' + ABD = CD' + ABD \end{aligned}$$

The realization using AND and OR gates is

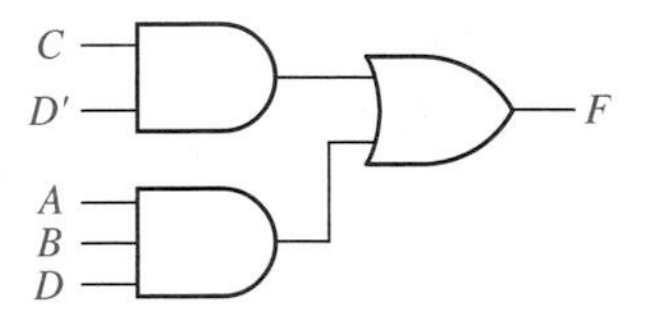


**Example 4**

The four inputs to a circuit ($A$, $B$, $C$, $D$) represent an 8-4-2-1 binary-coded-decimal digit. Design the circuit so that the output ($Z$) is 1 iff the decimal number represented by the inputs is exactly divisible by 3. Assume that only valid BCD digits occur as inputs.

The digits 0, 3, 6, and 9 are exactly divisible by 3, so $Z = 1$ for the input combinations $ABCD$ = 0000, 0011, 0110, and 1001. The input combinations 1010, 1011, 1100, 1101, 1110, and 1111 do not represent valid $BCD$ digits and will never occur, so $Z$ is a don't-care for these combinations. This leads to the following truth table:

| A B C D | Z |
|---|---|
| 0 0 0 0 | 1 |
| 0 0 0 1 | 0 |
| 0 0 1 0 | 0 |
| 0 0 1 1 | 1 |
| 0 1 0 0 | 0 |
| 0 1 0 1 | 0 |
| 0 1 1 0 | 1 |
| 0 1 1 1 | 0 |
| 1 0 0 0 | 0 |
| 1 0 0 1 | 1 |
| 1 0 1 0 | X |
| 1 0 1 1 | X |
| 1 1 0 0 | X |
| 1 1 0 1 | X |
| 1 1 1 0 | X |
| 1 1 1 1 | X |

The corresponding output function is

$$Z = \Sigma\, m(0, 3, 6, 9) + \Sigma\, d(10, 11, 12, 13, 14, 15)$$

In order to find the simplest circuit which will realize $Z$, we must choose some of the don't-cares (X's) to be 0 and some to be 1. The easiest way to do this is to use a Karnaugh map as described in Unit 5.

## 4.7 Design of Binary Adders and Subtracters

In this section, we will design a parallel adder that adds two 4-bit unsigned binary numbers and a carry input to give a 4-bit sum and a carry output (see Figure 4-2). One approach would be to construct a truth table with nine inputs and five outputs and then derive and simplify the five output equations. Because each equation would be a function of nine variables before simplification, this approach would be very difficult, and the resulting logic circuit would be very complex. A better method is to design a logic module that adds two bits and a carry, and then connect four of these modules together to form a 4-bit adder as shown in Figure 4-3. Each of the modules is called a *full adder.* The carry output from the first full adder serves as the carry input to the second full adder, etc.

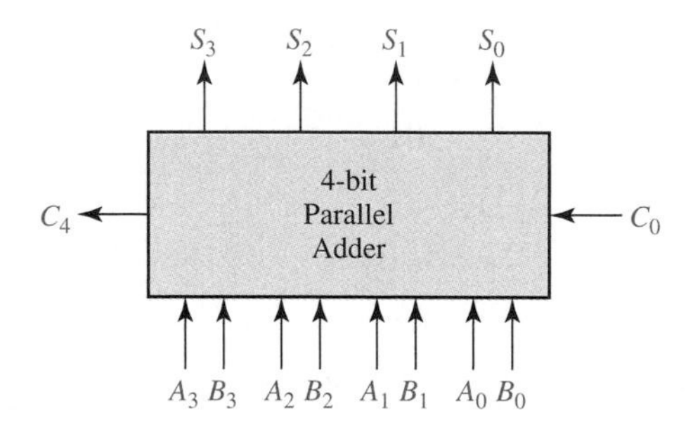


**FIGURE 4-2** Parallel Adder for 4-Bit Binary Numbers

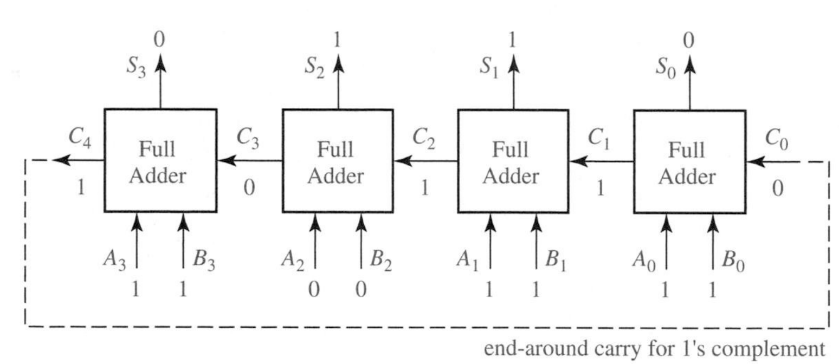


**FIGURE 4-3** Parallel Adder Composed of Four Full Adders

In the example of Figure 4-3, we perform the following addition:

$$\begin{array}{r l} 10110 & \text{(carries)} \\ 1011 & \\ +\ 1011 & \\ \hline 10110 & \end{array}$$

The full adder to the far right adds $A_0 + B_0 + C_0 = 1 + 1 + 0$ to give a sum of $10_2$, which gives a sum $S_0 = 0$ and a carry out of $C_1 = 1$. The next full adder adds $A_1 + B_1 + C_1 = 1 + 1 + 1 = 11_2$, which gives a sum $S_1 = 1$ and a carry $C_2 = 1$. The carry continues to propagate from right to left until the left cell produces a final carry of $C_4 = 1$.

**FIGURE 4-4**
Truth Table for a Full Adder

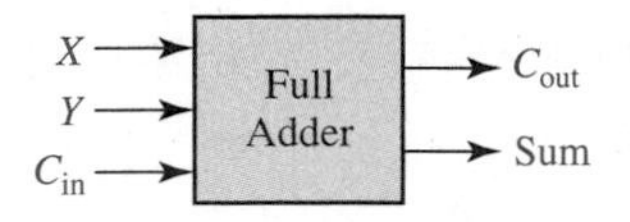


| $X$ | $Y$ | $C_{in}$ | $C_{out}$ | $Sum$ |
|---|---|---|---|---|
| 0 | 0 | 0 | 0 | 0 |
| 0 | 0 | 1 | 0 | 1 |
| 0 | 1 | 0 | 0 | 1 |
| 0 | 1 | 1 | 1 | 0 |
| 1 | 0 | 0 | 0 | 1 |
| 1 | 0 | 1 | 1 | 0 |
| 1 | 1 | 0 | 1 | 0 |
| 1 | 1 | 1 | 1 | 1 |

Figure 4-4 gives the truth table for a full adder with inputs $X$, $Y$, and $C_{in}$. The outputs for each row of the table are found by adding up the input bits ($X + Y + C_{in}$) and splitting the result into a carry out ($C_{i+1}$) and a sum bit ($S_i$). For example, in the 101 row $1 + 0 + 1 = 10_2$, so $C_{i+1} = 1$ and $S_i = 0$. Figure 4-5 shows the implementation of the full adder using gates. The logic equations for the full adder derived from the truth table are

$$\begin{aligned} Sum &= X'Y'C_{\text{in}} + X'YC'_{\text{in}} + XY'C'_{\text{in}} + XYC_{\text{in}} \\ &= X'(Y'C_{\text{in}} + YC'_{\text{in}}) + X(Y'C'_{\text{in}} + YC_{\text{in}}) \\ &= X'(Y \oplus C_{\text{in}}) + X(Y \oplus C_{\text{in}})' = X \oplus Y \oplus C_{\text{in}} \end{aligned} \tag{4-20}$$

$$\begin{aligned} C_{out} &= X'YC_{\text{in}} + XY'C_{\text{in}} + XYC'_{\text{in}} + XYC_{\text{in}} \\ &= (X'YC_{\text{in}} + XYC_{\text{in}}) + (XY'C_{\text{in}} + XYC_{\text{in}}) + (XYC'_{\text{in}} + XYC_{\text{in}}) \\ &= YC_{\text{in}} + XC_{\text{in}} + XY \end{aligned} \tag{4-21}$$

Note that the term $XYC_{\text{in}}$ was used three times in simplifying $C_{\text{out}}$. Figure 4-5 shows the logic circuit for Equations (4-20) and (4-21).

**FIGURE 4-5**
Implementation of Full Adder

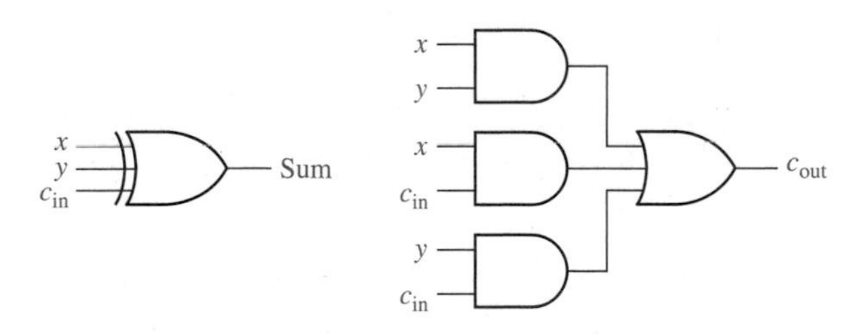


Although designed for unsigned binary numbers, the parallel adder of Figure 4-3 can also be used for signed binary numbers with negative numbers expressed in complement form. When 2's complement is used, the last carry ($C_4$) is discarded, and there is no carry into the first cell. Because $C_0 = 0$, the equations for the first cell may be simplified to

$$S_0 = A_0 \oplus B_0 \text{ and } C_1 = A_0\, B_0$$

When 1's complement is used, the end-around carry is accomplished by connecting $C_4$ to the $C_0$ input, as shown by the dashed line in Figure 4-3.

When adding signed binary numbers with negative numbers expressed in complement form, the sign bit of the sum is wrong when an overflow occurs. That is, an overflow has occurred if adding two positive numbers gives a negative result, or adding two negative numbers gives a positive result. We will define a signal $V$ that is

1 when an overflow occurs. For Figure 4-3, we can use the sign bits of $A$, $B$, and $S$ (the sum) to determine the value of $V$:

$$V = A_3'B_3'S_3 + A_3B_3S_3' \tag{4-22}$$

If the number of bits is large, a parallel binary adder of the type shown in Figure 4-4 may be rather slow because the carry generated in the first cell might have to propagate all of the way to the last cell. Other types of adders, such as a carry-look-ahead adder,[2] may be used to speed up the carry propagation.

Subtraction of binary numbers is most easily accomplished by adding the complement of the number to be subtracted. To compute $A - B$, add the complement of $B$ to $A$. This gives the correct answer because $A + (-B) = A - B$. Either 1's or 2's complement is used depending on the type of adder employed.

The circuit of Figure 4-6 may be used to form $A - B$ using the 2's complement representation for negative numbers. The 2's complement of $B$ can be formed by first finding the 1's complement and then adding 1. The 1's complement is formed by inverting each bit of $B$, and the addition of 1 is effectively accomplished by putting a 1 into the carry input of the first full adder.

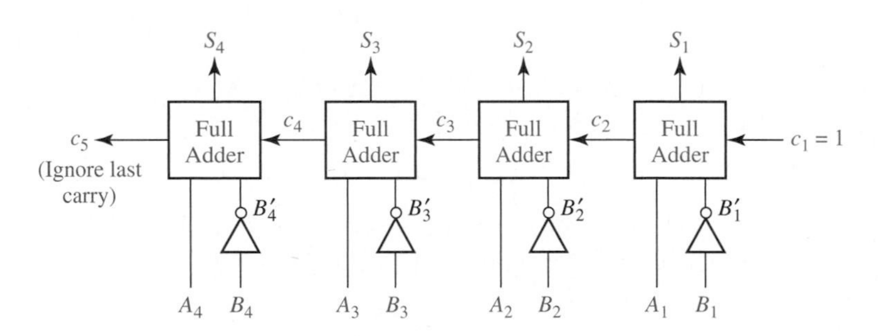


**FIGURE 4-6**
Binary Subtracter Using Full Adders

***Example***

$A = 0110 \quad (+6)$
$B = 0011 \quad (+3)$

The adder output is

$$\begin{array}{rll} & 0110 & (+6) \\ & +1100 & \text{(1's complement of 3)} \\ & +\quad 1 & \text{(first carry input)} \\ \hline (1) & 0011 = 3 = 6 - 3 & \end{array}$$

Alternatively, direct subtraction can be accomplished by employing a full subtracter in a manner analogous to a full adder. A block diagram for a parallel subtracter which subtracts $Y$ from $X$ is shown in Figure 4-7. The first two bits are subtracted in the rightmost cell to give a difference $d_1$, and a borrow signal ($b_2 = 1$) is generated if it is necessary to borrow from the next column. A typical cell (cell $i$) has inputs $x_i$, $y_i$, and $b_i$, and outputs $b_{i+1}$ and $d_i$. An input $b_i = 1$ indicates that we must borrow 1 from $x_i$ in that cell, and borrowing 1 from $x_i$ is equivalent to subtracting 1 from $x_i$. In cell $i$, bits $b_i$ and

[2]See, for example, J. F., Wakerly, *Digital Design Principles and Practices,* 4th ed (Prentice Hall, 2006).

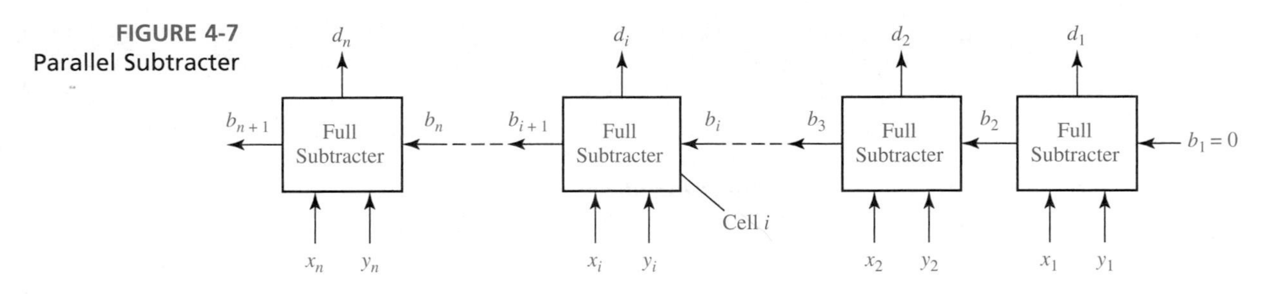


FIGURE 4-7
Parallel Subtracter

TABLE 4-6
Truth Table for Binary Full Subtracter

| $x_i\ y_i\ b_i$ | $b_{i+1}\ d_i$ |
|---|---|
| 0 0 0 | 0 0 |
| 0 0 1 | 1 1 |
| 0 1 0 | 1 1 |
| 0 1 1 | 1 0 |
| 1 0 0 | 0 1 |
| 1 0 1 | 0 0 |
| 1 1 0 | 0 0 |
| 1 1 1 | 1 1 |

$y_i$ are subtracted from $x_i$ to form the difference $d_i$, and a borrow signal ($b_{i+1} = 1$) is generated if it is necessary to borrow from the next column.

Table 4-6 gives the truth table for a binary full subtracter. Consider the following case, where $x_i = 0$, $y_i = 1$ and $b_i = 1$:

| | Column $i$ Before Borrow | Column $i$ After Borrow | |
|---|---|---|---|
| $x_i$ | 0 | 10 | |
| $-b_i$ | −1 | −1 | |
| $-y_i$ | −1 | −1 | |
| $d_i$ | | 0 | ($b_{i+1} = 1$) |

Note that in column $i$, we cannot immediately subtract $y_i$ and $b_i$ from $x_i$. Hence, we must borrow from column $i + 1$. Borrowing 1 from column $i + 1$ is equivalent to setting $b_{i+1}$ to 1 and adding 10 ($2_{10}$) to $x_i$. We then have $d_i = 10 - 1 - 1 = 0$. Verify that Table 4-6 is correct for the other input combinations and use it to work out several examples of binary subtraction.

# Problems

**4.1** Represent each of the following sentences by a Boolean equation.

(a) The company safe should be unlocked only when Mr. Jones is in the office or Mr. Evans is in the office, and only when the company is open for business, and only when the security guard is present.

(b) You should wear your overshoes if you are outside in a heavy rain and you are wearing your new suede shoes, or if your mother tells you to.
(c) You should laugh at a joke if it is funny, it is in good taste, and it is not offensive to others, or if it is told in class by your professor (regardless of whether it is funny and in good taste) and it is not offensive to others.
(d) The elevator door should open if the elevator is stopped, it is level with the floor, and the timer has not expired, or if the elevator is stopped, it is level with the floor, and a button is pressed.

**4.2** A flow rate sensing device used on a liquid transport pipeline functions as follows. The device provides a 5-bit output where all five bits are zero if the flow rate is less than 10 gallons per minute. The first bit is 1 if the flow rate is at least 10 gallons per minute; the first and second bits are 1 if the flow rate is at least 20 gallons per minute; the first, second, and third bits are 1 if the flow rate is at least 30 gallons per minute; and so on. The five bits, represented by the logical variables $A$, $B$, $C$, $D$, and $E$, are used as inputs to a device that provides two outputs $Y$ and $Z$.
(a) Write an equation for the output $Y$ if we want $Y$ to be 1 iff the flow rate is less than 30 gallons per minute.
(b) Write an equation for the output $Z$ if we want $Z$ to be 1 iff the flow rate is at least 20 gallons per minute but less than 50 gallons per minute.

**4.3** Given $F_1 = \Sigma\, m(0, 4, 5, 6)$ and $F_2 = \Sigma\, m(0, 3, 6, 7)$ find the minterm expression for $F_1 + F_2$. State a general rule for finding the expression for $F_1 + F_2$ given the minterm expansions for $F_1$ and $F_2$. Prove your answer by using the general form of the minterm expansion.

**4.4** (a) How many switching functions of two variables ($x$ and $y$) are there?
(b) Give each function in truth table form and in reduced algebraic form.

**4.5** A combinational circuit is divided into two subcircuits $N_1$ and $N_2$ as shown. The truth table for $N_1$ is given. Assume that the input combinations $ABC = 110$ and $ABC = 010$ never occur. Change as many of the values of $D$, $E$, and $F$ to don't-cares as you can without changing the value of the output $Z$.

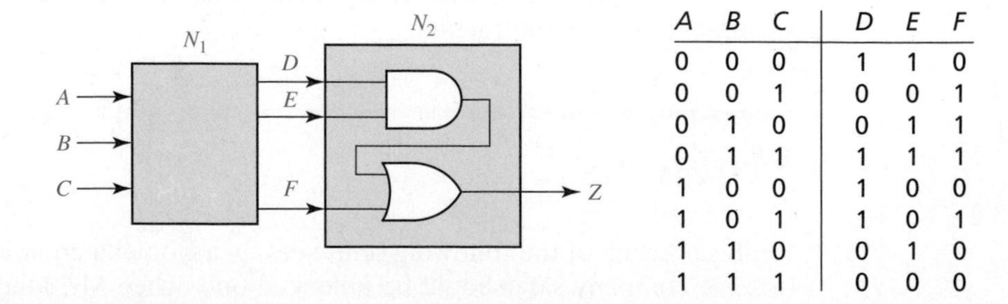


| A | B | C | D | E | F |
|---|---|---|---|---|---|
| 0 | 0 | 0 | 1 | 1 | 0 |
| 0 | 0 | 1 | 0 | 0 | 1 |
| 0 | 1 | 0 | 0 | 1 | 1 |
| 0 | 1 | 1 | 1 | 1 | 1 |
| 1 | 0 | 0 | 1 | 0 | 0 |
| 1 | 0 | 1 | 1 | 0 | 1 |
| 1 | 1 | 0 | 0 | 1 | 0 |
| 1 | 1 | 1 | 0 | 0 | 0 |

**4.6** Work (a) and (b) with the following truth table:

| A | B | C | F | G |
|---|---|---|---|---|
| 0 | 0 | 0 | 1 | 0 |
| 0 | 0 | 1 | X | 1 |
| 0 | 1 | 0 | 0 | X |
| 0 | 1 | 1 | 0 | 1 |
| 1 | 0 | 0 | 0 | 0 |
| 1 | 0 | 1 | X | 1 |
| 1 | 1 | 0 | 1 | X |
| 1 | 1 | 1 | 1 | 1 |

(a) Find the simplest expression for $F$, and specify the values of the don't-cares that lead to this expression.
(b) Repeat (a) for $G$. (*Hint:* Can you make $G$ the same as one of the inputs by properly choosing the values for the don't-care?)

**4.7** Each of three coins has two sides, heads and tails. Represent the heads or tails status of each coin by a logical variable ($A$ for the first coin, $B$ for the second coin, and $C$ for the third) where the logical variable is 1 for heads and 0 for tails. Write a logic function $F(A, B, C)$ which is 1 iff exactly one of the coins is heads after a toss of the coins. Express $F$
(a) as a minterm expansion.
(b) as a maxterm expansion.

**4.8** A switching circuit has four inputs as shown. $A$ and $B$ represent the first and second bits of a binary number $N_1$. $C$ and $D$ represent the first and second bits of a binary number $N_2$. The output is to be 1 only if the product $N_1 \times N_2$ is less than or equal to 2.
(a) Find the minterm expansion for $F$.
(b) Find the maxterm expansion for $F$.
Express your answers in both decimal notation and algebraic form.

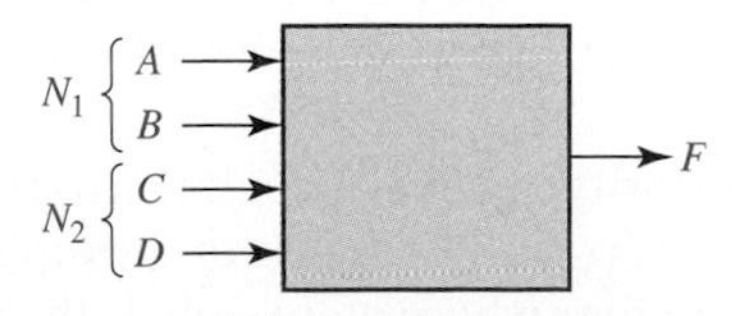


**4.9** Given: $F(a, b, c) = abc' + b'$.
(a) Express $F$ as a minterm expansion. (Use $m$-notation.)
(b) Express $F$ as a maxterm expansion. (Use $M$-notation.)
(c) Express $F'$ as a minterm expansion. (Use $m$-notation.)
(d) Express $F'$ as a maxterm expansion. (Use $M$-notation.)

**4.10** Work Problem 4.9 using:
$F(a, b, c, d) = (a + b + d)(a' + c)(a' + b' + c')(a + b + c' + d')$

**4.11** (a) Implement a full subtracter using a minimum number of gates.
(b) Compare the logic equations for the full adder and full subtracter. What is the relation between $s_i$ and $d_i$? Between $c_{i+1}$ and $b_{i+1}$?

**4.12** Design a circuit which will perform the following function on three 4-bit numbers:

$$(X_3X_2X_1X_0 + Y_3Y_2Y_1Y_0) - Z_3Z_2Z_1Z_0$$

It will give a result $S_3S_2S_1S_0$, a carry, and a borrow. Use eight full adders and any other type of gates. Assume that negative numbers are represented in 2's complement.

**4.13** A combinational logic circuit has four inputs ($A$, $B$, $C$, and $D$) and one output $Z$. The output is 1 iff the input has three consecutive 0's or three consecutive 1's. For example, if $A = 1, B = 0, C = 0$, and $D = 0$, then $Z = 1$, but if $A = 0, B = 1, C = 0$, and $D = 0$, then $Z = 0$. Design the circuit using one four-input OR gate and four three-input AND gates.

**4.14** Design a combinational logic circuit which has one output $Z$ and a 4-bit input $ABCD$ representing a binary number. $Z$ should be 1 iff the input is at least 5, but is no greater than 11. Use one OR gate (three inputs) and three AND gates (with no more than three inputs each).

**4.15** A logic circuit realizing the function $f$ has four inputs $A$, $B$, $C$, and $D$. The three inputs $A$, $B$, and $C$ are the binary representation of the digits 0 through 7 with $A$ being the most-significant bit. The input $D$ is an odd-parity bit, i.e., the value of $D$ is such that $A$, $B$, $C$, and $D$ always contain an odd number of 1's. (For example, the digit 1 is represented by $ABC = 001$ and $D = 0$, and the digit 3 is represented by $ABCD = 0111$.) The function $f$ has value 1 if the input digit is a prime number. (A number is prime if it is divisible only by itself and 1; 1 is considered to be prime and 0 is not.)
(a) List the minterms and don't-care minterms of $f$ in algebraic form.
(b) List the maxterms and don't-care maxterms of $f$ in algebraic form.

**4.16** A priority encoder circuit has four inputs, $x_3, x_2, x_1$, and $x_0$. The circuit has three outputs: $z$, $y_1$, and $y_0$. If one of the inputs is 1, $z$ is 1 and $y_1$ and $y_0$ represent a 2-bit, binary number whose value equals the index of the highest numbered input that is 1. For example, if $x_2$ is 1 and $x_3$ is 0, then the outputs are $z = 1$ and $y_1 = 1$ and $y_0 = 0$. If all inputs are 0, $z = 0$ and $y_1$ and $y_0$ are don't-cares.
(a) List in decimal form the minterms and don't-care minterms of each output.
(b) List in decimal form the maxterms and don't-care maxterms of each output.

**4.17** The 9's complement of a decimal digit $d$ (0 to 9) is defined to be $9 - d$. A logic circuit produces the 9's complement of an input digit where the input and output

digits are represented in BCD. Label the inputs $A$, $B$, $C$, and $D$, and label the outputs $W$, $X$, $Y$ and $Z$.
(a) Determine the minterms and don't-care minterms for each of the outputs.
(b) Determine the maxterms and don't-care maxterms for each of the outputs.

**4.18** Repeat Problem 4.17 for the case where the input and output digits are represented using the 4-2-2-1 weighted code. (If only one weight of 2 is required for decimal digits less than 5, select the rightmost 2. In addition, select the codes so that $W = A'$, $X = B'$, $Y = C'$, and $Z = D'$. (There are two possible codes with these restrictions.)

**4.19** Each of the following sentences has two possible interpretations depending on whether the AND or OR is done first. Write an equation for each interpretation.
(a) The buzzer will sound if the key is in the ignition switch, and the car door is open, or the seat belts are not fastened.
(b) You will gain weight if you eat too much, or you do not exercise enough, and your metabolism rate is too low.
(c) The speaker will be damaged if the volume is set too high, and loud music is played, or the stereo is too powerful.
(d) The roads will be very slippery if it snows, or it rains, and there is oil on the road.

**4.20** A bank vault has three locks with a different key for each lock. Each key is owned by a different person. To open the door, at least two people must insert their keys into the assigned locks. The signal lines $A$, $B$, and $C$ are 1 if there is a key inserted into lock 1, 2, or 3, respectively. Write an equation for the variable $Z$ which is 1 iff the door should open.

**4.21** A paper tape reader used as an input device to a computer has five rows of holes as shown. A hole punched in the tape indicates a logic 1, and no hole indicates a logic 0. As each hole pattern passes under the photocells, the pattern is translated into logic signals on lines $A$, $B$, $C$, $D$, and $E$. All patterns of holes indicate a valid character with two exceptions. A pattern consisting of none of the possible holes punched is not used because it is impossible to distinguish between this pattern and the unpunched space between patterns. An incorrect pattern punched on the tape is erased by punching all five holes in that position. Therefore, a valid character punched on the tape will have at least one hole but will not have all five holes punched.
(a) Write an equation for a variable $Z$ which is 1 iff a valid character is being read.
(b) Write an equation for a variable $Y$ which is 1 iff the hole pattern being read has holes punched only in rows $C$ and $E$.

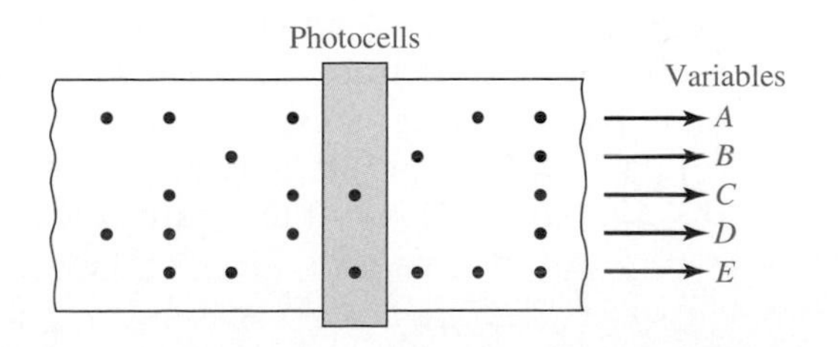

4.22 A computer interface to a line printer has seven data lines that control the movement of the paper and the print head and determine which character to print. The data lines are labeled $A$, $B$, $C$, $D$, $E$, $F$, and $G$, and each represents a binary 0 or 1. When the data lines are interpreted as a 7-bit binary number with line $A$ being the most significant bit, the data lines can represent the numbers 0 to $127_{10}$. The number $13_{10}$ is the command to return the print head to the beginning of a line, the number $10_{10}$ means to advance the paper by one line, and the numbers $32_{10}$ to $127_{10}$ represent printing characters.

(a) Write an equation for the variable $X$ which is 1 iff the data lines indicate a command to return the print head to the beginning of the line.

(b) Write an equation for the variable $Y$ which is 1 iff there is an advance paper command on the data lines.

(c) Write an equation for the variable $Z$ which is 1 iff the data lines indicate a printable character. (*Hint:* Consider the binary representations of the numbers 0–31 and 32–127 and write the equation for $Z$ with only two terms.)

4.23 Given $F_1 = \Pi\, M(0,4,5,6)$ and $F_2 = \Pi\, M(0,4,7)$, find the maxterm expansion for $F_1F_2$. State a general rule for finding the maxterm expansion of $F_1F_2$ given the maxterm expansions of $F_1$ and $F_2$.
Prove your answer by using the general form of the maxterm expansion.

4.24 Given $F_1 = \Pi\, M(0, 4, 5, 6)$ and $F_2 = \Pi\, M(0, 4, 7)$, find the maxterm expansion for $F_1 + F_2$.
State a general rule for finding the maxterm expansion of $F_1 + F_2$, given the maxterm expansions of $F_1$ and $F_2$.
Prove your answer by using the general form of the maxterm expansion.

4.25 Four chairs are placed in a row:

| A | B | C | D |
|---|---|---|---|

Each chair may be occupied (1) or empty (0). Give the minterm and maxterm expansion for each logic function described.

(a) $F(A, B, C, D)$ is 1 iff there are no adjacent empty chairs.

(b) $G(A, B, C, D)$ is 1 iff the chairs on the ends are both empty.

(c) $H(A, B, C, D)$ is 1 iff at least three chairs are full.

(d) $J(A, B, C, D)$ is 1 iff there are more people sitting in the left two chairs than in the right two chairs.

4.26 Four chairs ($A$, $B$, $C$, and $D$) are placed in a circle: $A$ next to $B$, $B$ next to $C$, $C$ next to $D$, and $D$ next to $A$. Each chair may be occupied (1) or empty (0). Give the minterm and maxterm expansion for each of the following logic functions:

(a) $F(A, B, C, D)$ is 1 iff there are no adjacent empty chairs.

(b) $G(A, B, C, D)$ is 1 iff there are at least three adjacent empty chairs.

(c) $H(A, B, C, D)$ is 1 iff at least three chairs are full.
(d) $J(A, B, C, D)$ is 1 iff there are more people sitting in chairs $A$ and $B$ than chairs $C$ and $D$.

**4.27** Given $f(a, b, c) = a(b + c')$.
(a) Express $f$ as a minterm expansion (use $m$-notation).
(b) Express $f$ as maxterm expansion (use $M$-notation).
(c) Express $f'$ as a minterm expansion (use $m$-notation).
(d) Express $f'$ as a maxterm expansion (use $M$-notation).

**4.28** Work Problem 4.27 using $f(a, b, c, d) = acd + bd' + a'c'd + ab'cd + a'b'cd'$.

**4.29** Find both the minterm expansion and maxterm expansion for the following functions, using *algebraic manipulations*:
(a) $f(A, B, C, D) = AB + A'CD$
(b) $f(A, B, C, D) = (A + B + D')(A' + C)(C + D)$

**4.30** Given $F'(A, B, C, D) = \Sigma\, m(0, 1, 2, 6, 7, 13, 15)$.
(a) Find the minterm expansion for $F$ (both decimal and algebraic form).
(b) Find the maxterm expansion for $F$ (both decimal and algebraic form).

**4.31** Repeat Problem 4.30 for $F'(A, B, C, D) = \Sigma\, m(1, 2, 5, 6, 10, 15)$.

**4.32** Work parts (a) through (d) with the given truth table.

| $A$ | $B$ | $C$ | $F_1$ | $F_2$ | $F_3$ | $F_4$ |
|---|---|---|---|---|---|---|
| 0 | 0 | 0 | 1 | 1 | 0 | 1 |
| 0 | 0 | 1 | X | 0 | 0 | 0 |
| 0 | 1 | 0 | 0 | 1 | X | 0 |
| 0 | 1 | 1 | 0 | 0 | 1 | 1 |
| 1 | 0 | 0 | 0 | 1 | 1 | 1 |
| 1 | 0 | 1 | X | 0 | 1 | 0 |
| 1 | 1 | 0 | 0 | X | X | X |
| 1 | 1 | 1 | 1 | X | 1 | X |

(a) Find the simplest expression for $F_1$, and specify the values for the don't-cares that lead to this expression.
(b) Repeat for $F_2$.
(c) Repeat for $F_3$.
(d) Repeat for $F_4$.

**4.33** Work Problem 4.5 using the following circuits and truth table. Assume that the input combinations of $ABC = 011$ and $ABC = 110$ will never occur.

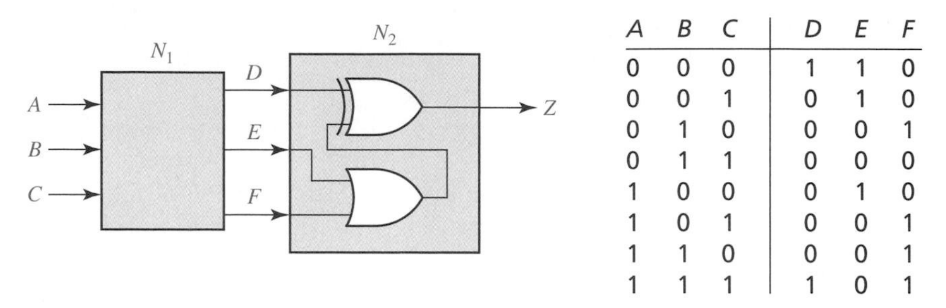


| A | B | C | D | E | F |
|---|---|---|---|---|---|
| 0 | 0 | 0 | 1 | 1 | 0 |
| 0 | 0 | 1 | 0 | 1 | 0 |
| 0 | 1 | 0 | 0 | 0 | 1 |
| 0 | 1 | 1 | 0 | 0 | 0 |
| 1 | 0 | 0 | 0 | 1 | 0 |
| 1 | 0 | 1 | 0 | 0 | 1 |
| 1 | 1 | 0 | 0 | 0 | 1 |
| 1 | 1 | 1 | 1 | 0 | 1 |

**4.34** Work Problem 4.7 for the following logic functions:

(a) $G_1(A, B, C)$ is 1 iff all the coins landed on the same side (heads or tails).

(b) $G_2(A, B, C)$ is 1 iff the second coin landed on the same side as the first coin.

**4.35** A combinational circuit has four inputs $(A, B, C, D)$ and three outputs $(X, Y, Z)$. $XYZ$ represents a binary number whose value equals the number of 1's at the input. For example if $ABCD = 1011$, $XYZ = 011$.

(a) Find the minterm expansions for $X$, $Y$, and $Z$.

(b) Find the maxterm expansions for $Y$ and $Z$.

**4.36** A combinational circuit has four inputs $(A, B, C, D)$ and four outputs $(W, X, Y, Z)$. $WXYZ$ represents an excess-3 coded number whose value equals the number of 1's at the input. For example, if $ABCD = 1101$, $WXYZ = 0110$.

(a) Find the minterm expansions for $X$, $Y$, and $Z$.

(b) Find the maxterm expansions for $Y$ and $Z$.

**4.37** A combinational circuit has four inputs $(A, B, C, D)$, which represent a binary-coded-decimal digit. The circuit has two groups of four outputs—$S$, $T$, $U$, $V$, and $W$, $X$, $Y$, $Z$. Each group represents a BCD digit. The output digits represent a decimal number which is five times the input number. For example, if $ABCD = 0111$, the outputs are 0011 0101. Assume that invalid BCD digits do not occur as inputs.

(a) Construct the truth table.

(b) Write down the minimum expressions for the outputs by inspection of the truth table. (*Hint:* Try to match output columns in the table with input columns.)

**4.38** Work Problem 4.37 where the BCD outputs represent a decimal number that is 1 more than four times the input number. For example, if $ABCD = 0011$, the outputs are 0001 0011.

**4.39** Design a circuit which will add a 4-bit binary number to a 5-bit binary number. Use five full adders. Assume negative numbers are represented in 2's complement. (*Hint:* How do you make a 4-bit binary number into a 5-bit binary number, without making a negative number positive or a positive number negative? Try writing down the representation for −3 as a 3-bit 2's complement number, a 4-bit 2's complement number, and a 5-bit 2's complement number. Recall that one way to find the 2's complement of a binary number is to complement *all* bits to the left of the first 1.)

**4.40** A half adder is a circuit that adds two bits to give a sum and a carry. Give the truth table for a half adder, and design the circuit using only two gates. Then design a circuit which will find the 2's complement of a 4-bit binary number. Use four half adders and any additional gates. (*Hint:* Recall that one way to find the 2's complement of a binary number is to complement *all* bits, and then add 1.)

**4.41** (a) Write the switching function $f(x, y) = x + y$ as a sum of minterms and as a product of maxterms.

(b) Consider the Boolean algebra of four elements $\{0, 1, a, b\}$ specified by the following operation tables and the Boolean function $f(x, y) = ax + by$ where $a$ and $b$ are two of the elements in the Boolean algebra. Write $f(x, y)$ in a sum-of-minterms form.

(c) Write the Boolean function of part (b) in a product-of-maxterms form.

(d) Give a table of combinations for the Boolean function of Part (b). (*Note*: The table of combinations has 16 rows, not just 4.)

(e) Which four rows of the table of combinations completely specify the function of Part (b)? Verify your answer.

| | ′ |
|---|---|
| 0 | 1 |
| 1 | 0 |
| a | b |
| b | a |

| + | 0 | 1 | a | b |
|---|---|---|---|---|
| 0 | 0 | 1 | a | b |
| 1 | 1 | 1 | 1 | 1 |
| a | a | 1 | a | 1 |
| b | b | 1 | 1 | b |

| • | 0 | 1 | a | b |
|---|---|---|---|---|
| 0 | 0 | 0 | 0 | 0 |
| 1 | 0 | 1 | a | b |
| a | 0 | a | a | 0 |
| b | 0 | b | 0 | b |

**4.42** (a) If $m_1$ and $m_2$ are minterms of $n$ variables, prove that $m_1 + m_2 = m_1 \oplus m_2$.

(b) Prove that any switching function can be written as the exclusive-OR sum of products where each product does not contain a complemented literal. [*Hint*: Start with the function written as a sum of minterms and use Part (a).]

# Karnaugh Maps

## Objectives

1. Given a function (completely or incompletely specified) of three to five variables, plot it on a Karnaugh map. The function may be given in minterm, maxterm, or algebraic form.
2. Determine the essential prime implicants of a function from a map.
3. Obtain the minimum sum-of-products or minimum product-of-sums form of a function from the map.
4. Determine all of the prime implicants of a function from a map.
5. Understand the relation between operations performed using the map and the corresponding algebraic operations.

# Study Guide

In this unit we will study the Karnaugh (pronounced "car-no") map. Just about any type of algebraic manipulation we have done so far can be facilitated by using the map, provided the number of variables is small.

1. Study Section 5.1, *Minimum Forms of Switching Functions.*

   (a) Define a minimum sum of products.

   (b) Define a minimum product of sums.

2. Study Section 5.2, *Two- and Three-Variable Karnaugh Maps.*

   (a) Plot the given truth table on the map. Then, loop two pairs of 1's on the map and write the simplified form of $F$.

| P Q | F |
|---|---|
| 0 0 | 1 |
| 0 1 | 1 |
| 1 0 | 0 |
| 1 1 | 1 |

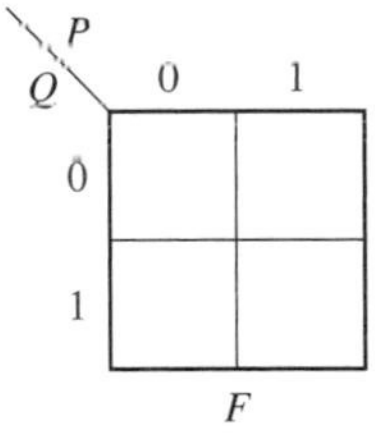

| Q \ P | 0 | 1 |
|---|---|---|
| 0 | | |
| 1 | | |

$F$

$F =$ __________

   Now simplify $F$ algebraically and verify that your answer is correct.

   (b) $F(a, b, c)$ is plotted below. Find the truth table for $F$.

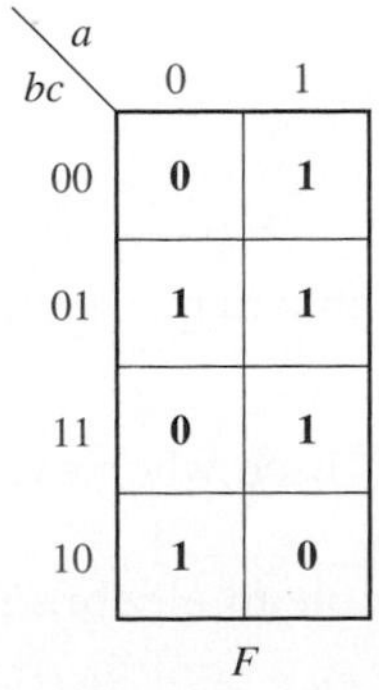

| bc \ a | 0 | 1 |
|---|---|---|
| 00 | 0 | 1 |
| 01 | 1 | 1 |
| 11 | 0 | 1 |
| 10 | 1 | 0 |

$F$

| a b c | F |
|---|---|
| 0 0 0 | |
| 0 0 1 | |
| 0 1 0 | |
| 0 1 1 | |
| 1 0 0 | |
| 1 0 1 | |
| 1 1 0 | |
| 1 1 1 | |

(c) Plot the following functions on the given Karnaugh maps:

$$F_1(R, S, T) = \Sigma\, m(0, 1, 5, 6) \qquad F_2(R, S, T) = \Pi\, M(2, 3, 4, 7)$$

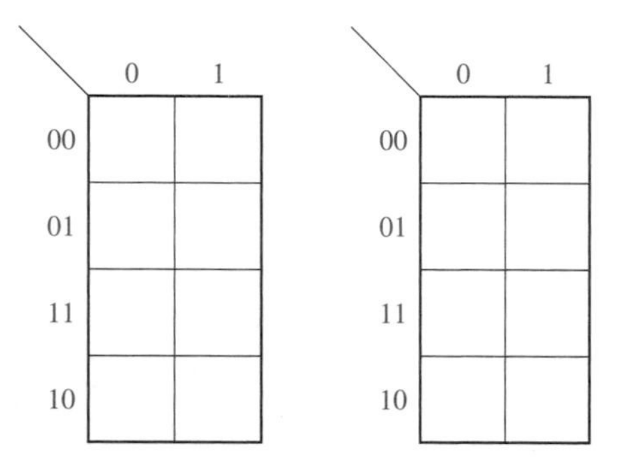


Why are the two maps the same?

(d) Plot the following function on the given map:

$$f(x, y, z) = z' + x'z + yz$$

Do *not* make a minterm expansion or a truth table before plotting.

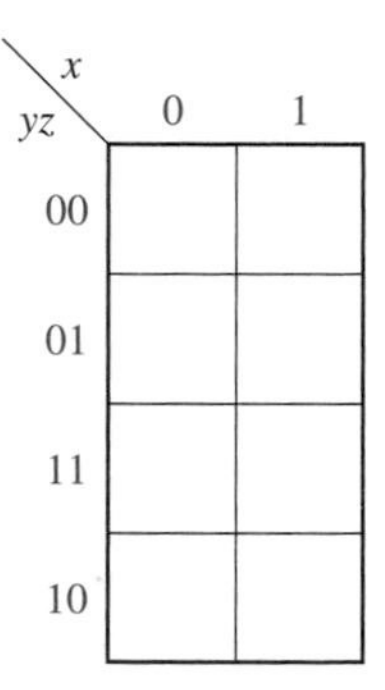


(e) For a three-variable map, which squares are "adjacent" to square 2?

__________

(f) What theorem is used when two terms in adjacent squares are combined?

(g) What law of Boolean algebra justifies using a given 1 on a map in two or more loops?

(h) Each of the following solutions *is not* minimum.

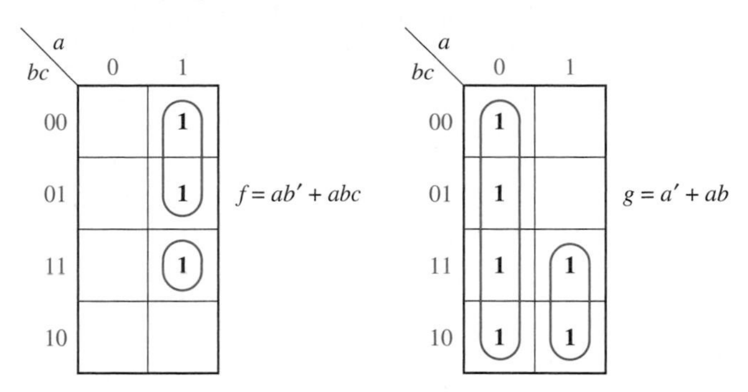


In each case, change the looping on the map so that the minimum solution is obtained.

(i) Work Problem 5.3.

(j) Find two different minimum sum-of-products expressions for the function $G$, which is plotted below.

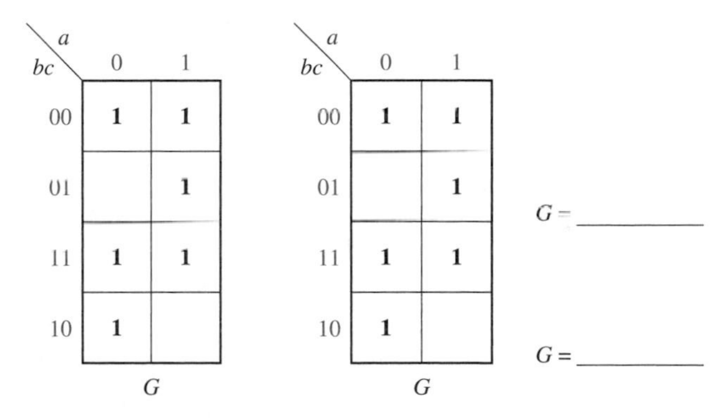


**3.** Study Section 5.3, *Four-Variable Karnaugh Maps.*

(a) Note the locations of the minterms on three- and four-variable maps [Figures 5-3(b) and 5-10]. Memorize this ordering. This will save you a lot of time when you are plotting Karnaugh maps.

This ordering is valid only for the order of the variables given. If we label the maps as shown below, fill in the locations of the minterms:

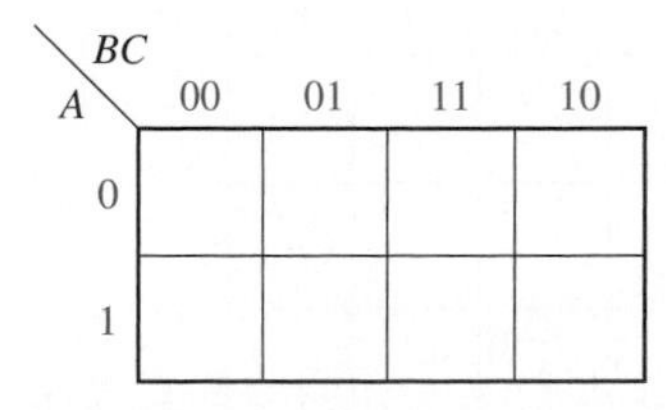


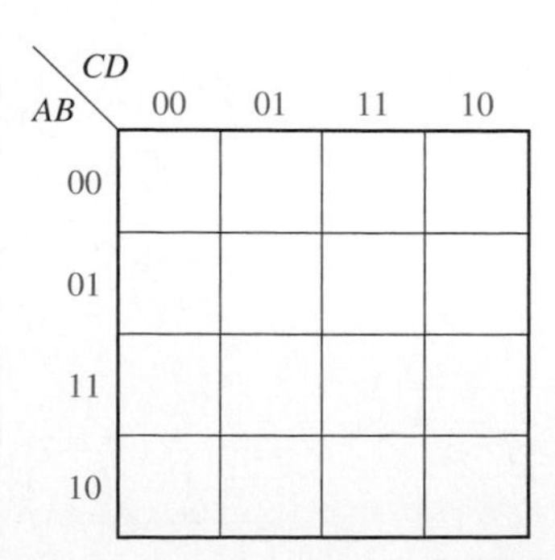

(b) Given the following map, write the minterm and maxterm expansions for $F$ in decimal form:

| $cd \backslash ab$ | 00 | 01 | 11 | 10 |
|---|---|---|---|---|
| 00 | | | 1 | 1 |
| 01 | 1 | 1 | | 1 |
| 11 | | 1 | 1 | |
| 10 | | | 1 | |

$F =$ ________________

$F =$ ________________

(c) Plot the following functions on the given maps:

(1) $f(w, x, y, z) = \Sigma\, m(0, 1, 2, 5, 7, 8, 9, 10, 13, 14)$

(2) $f(w, x, y, z) = x'z' + y'z + w'xz + wyz'$

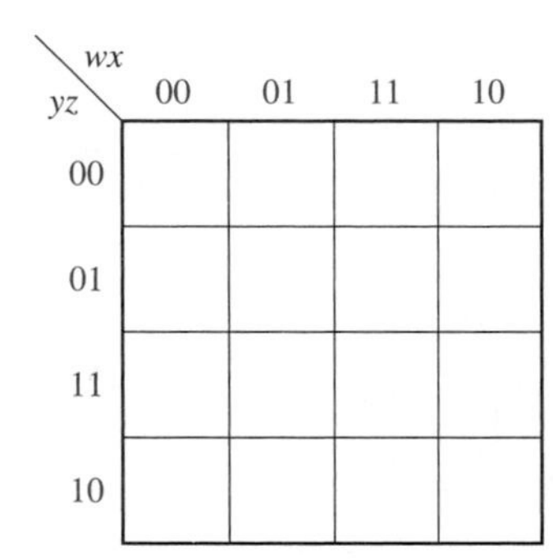


| $yz \backslash wx$ | 00 | 01 | 11 | 10 |
|---|---|---|---|---|
| 00 | | | | |
| 01 | | | | |
| 11 | | | | |
| 10 | | | | |

Your answers to (1) and (2) should be the same.

(d) For a four-variable map, which squares are adjacent to square 14? _______

To square 8? __________

(e) When we combine two adjacent 1's on a map, this corresponds to applying the theorem $xy' + xy = x$ to eliminate the variable in which the two terms differ. Thus, looping the two 1's as indicated on the following map is equivalent to combining the corresponding minterms algebraically:

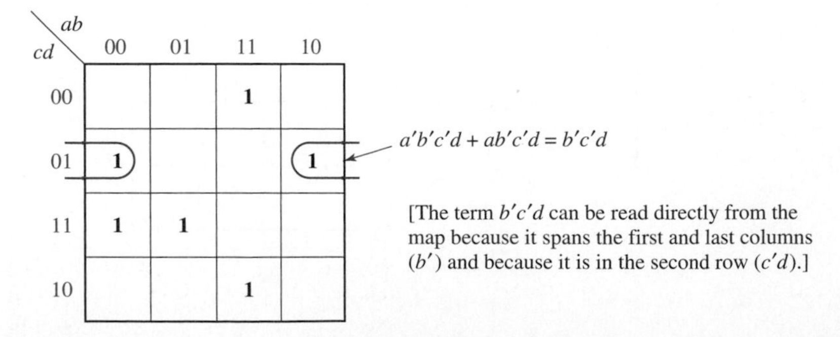

Loop two other pairs of adjacent 1's on this map and state the algebraic equivalent of looping these terms. Now read the loops directly off the map and check your algebra.

(f) When we combine four adjacent 1's on a map (either four in a line or four in a square) this is equivalent to applying $xy + xy' = x$ three times:

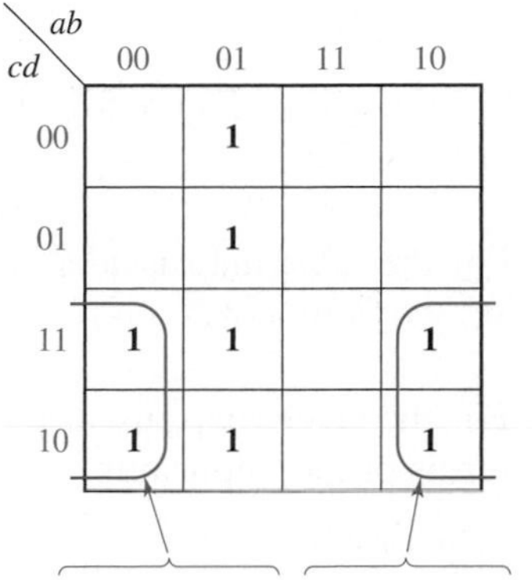


$$a'b'cd + a'b'cd' + ab'cd + ab'cd' = a'b'c + ab'c = b'c$$

Loop the other four 1's on the map and state the algebraic equivalent.

(g) For each of the following maps, loop a minimum number of terms which will cover all of the 1's.

| cd \ ab | 00 | 01 | 11 | 10 |
|---|---|---|---|---|
| 00 | | 1 | 1 | |
| 01 | | 1 | 1 | 1 |
| 11 | 1 | 1 | | |
| 10 | | 1 | | |

$f_1$

| cd \ ab | 00 | 01 | 11 | 10 |
|---|---|---|---|---|
| 00 | | | | 1 |
| 01 | | 1 | | |
| 11 | 1 | 1 | 1 | 1 |
| 10 | 1 | | | 1 |

$f_2$

(For each part you should have looped two groups of four 1's and two groups of two 1's).

Write down the minimum sum-of-products expression for $f_1$ and $f_2$ from these maps.

$f_1$ = ______________________________________

$f_2$ = ______________________________________

(h) Why is it not possible to combine three or six minterms together rather than just two, four, eight, etc.?

(i) Note the procedure for deriving the minimum *product of sums* from the map. You will probably make fewer mistakes if you write down $f'$ as a sum of products first and then complement it, as illustrated by the example in Figure 5-14.

(j) Work Problems 5.4 and 5.5.

**4.** Study Section 5.4, *Determination of Minimum Expressions Using Essential Prime Implicants.*

(a) For the map of Figure 5-15, list three implicants of $F$ other than those which are labeled.

For the same map, is $ac'd'$ a prime implicant of $F$?
Why or why not?

(b) For the given map, are any of the circled terms prime implicants?

Why or why not?

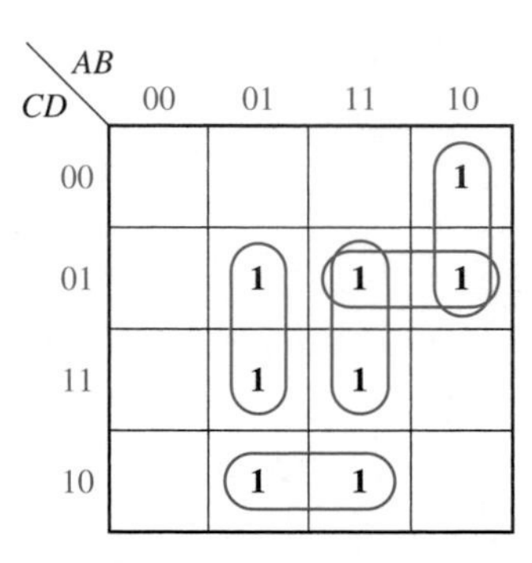


**5.** Study Figure 5-18 carefully and then answer the following questions for the given map:

(a) How many 1's are adjacent to $m_0$?

(b) Are all these 1's covered by a single prime implicant?

(c) From your answer to (b), can you determine whether $B'C'$ is essential?

(d) How many 1's are adjacent to $m_9$?

(e) Are all of these 1's covered by a single prime implicant?

(f) From your answer to (e), is $B'C'$ essential?

(g) How many 1's are adjacent to $m_7$?

(h) Why is $A'C$ essential?

| CD \ AB | 00 | 01 | 11 | 10 |
|---|---|---|---|---|
| 00 | 1 (0) | 1 (4) | | 1 (8) |
| 01 | 1 (1) | | | 1 (9) |
| 11 | 1 (3) | 1 (7) | | |
| 10 | 1 (2) | 1 (6) | | 1 (10) |

(i) Find two other essential prime implicants and tell which minterm makes them essential.

**6.** (a) How do you determine if a prime implicant is essential using a Karnaugh map?

(b) For the following map, why is $A'B'$ *not* essential?

Why is $BD'$ essential?

Is $A'D'$ essential? Why?

Is $BC'$ essential? Why?

Is $B'CD$ essential? Why?

Find the minimum sum of products.

| CD \ AB | 00 | 01 | 11 | 10 |
|---|---|---|---|---|
| 00 | 1 | 1 | 1 | |
| 01 | 1 | 1 | 1 | |
| 11 | 1 | | | 1 |
| 10 | 1 | 1 | 1 | |

(c) Work Programmed Exercise 5.1.

(d) List all 1's and X's that are adjacent to $1_0$.

| CD \ AB | 00 | 01 | 11 | 10 |
|---|---|---|---|---|
| 00 | $1_0$ | $1_4$ | $1_{12}$ | 8 |
| 01 | $X_1$ | $1_5$ | $X_{13}$ | 9 |
| 11 | 3 | $X_7$ | $1_{15}$ | $1_{11}$ |
| 10 | 2 | 6 | $X_{14}$ | 10 |

Why is $A'C'$ an essential prime implicant?

List all 1's and X's adjacent to $1_{15}$.

Based on this list, why can you not find an essential prime implicant that covers $1_{15}$?

Does this mean that there is no essential prime implicant that covers $1_{15}$?

What essential prime implicant covers $1_{11}$?

Can you find an essential prime implicant that covers $1_{12}$? Explain.

Find two prime implicants that cover $1_{12}$.

Give two minimum expressions for $F$.

(e) Work Problem 5.6.

(f) If you have a copy of the *LogicAid* program available, use the Karnaugh map tutorial mode to help you learn to find minimum solutions from Karnaugh maps. This program will check your work at each step to make sure that you loop the terms in the correct order. It also will check your final answer. Work Problem 5.7 using the Karnaugh map tutor.

**7.** (a) In Example 4, page 103, we derived the following function:

$$Z = \Sigma\, m(0, 3, 6, 9) + \Sigma\, d(10, 11, 12, 13, 14, 15)$$

Plot $Z$ on the given map using X's to represent don't-care terms.

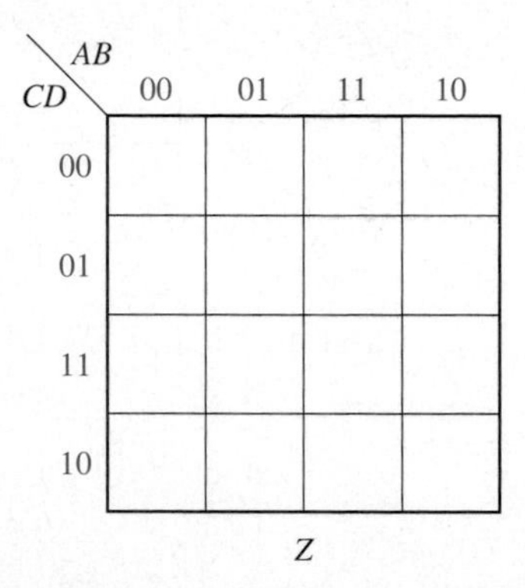


(b) Show that the minimum sum of products is

$$Z = A'B'C'D' + B'CD + AD + BCD'$$

Which *four* don't-care minterms were assigned the value 1 when forming your solution?

(c) Show that the minimum product of sums for $Z$ is

$$Z = (B' + C)(B' + D')(A' + D)(A + C + D')(B + C' + D)$$

Which *one* don't-care term of $Z$ was assigned the value 1 when forming your solution?

(d) Work Problem 5.8.

**8.** Study Section 5.5, *Five-Variable Karnaugh Maps.*

(a) The figure below shows a three-dimensional five-variable map. Plot the 1's and loops on the corresponding two-dimensional map, and give the minimum sum-of-products expression for the function.

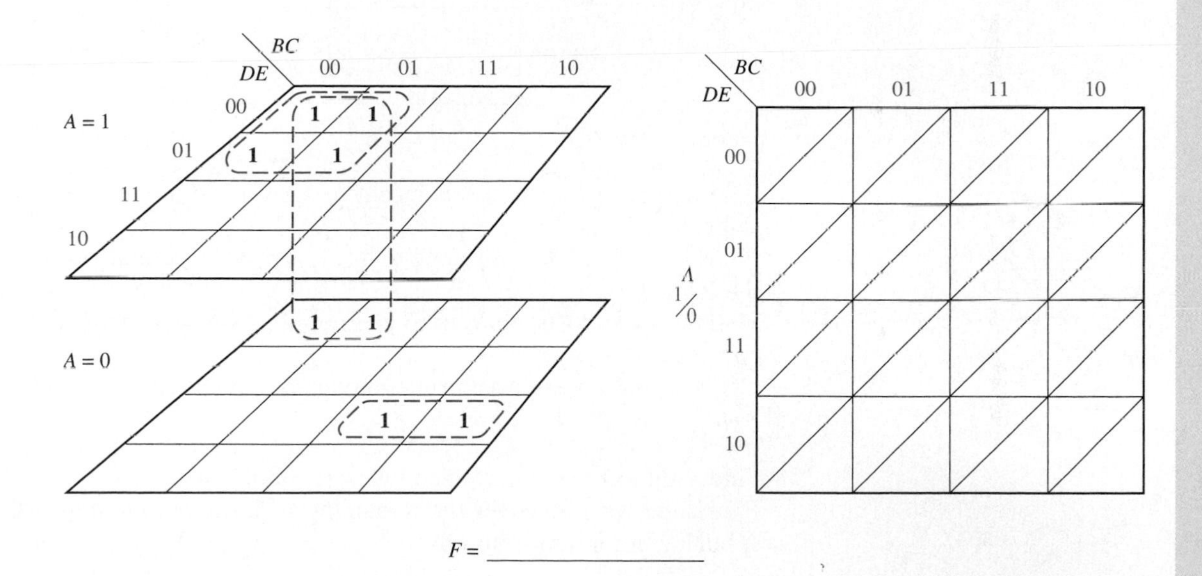


$F =$ ____________________

(b) On a five-variable map (Figure 5-21), what are the five minterms adjacent to minterm 24?

(c) Work through all of the examples in this section carefully and make sure that you understand all of the steps.

(d) Two minimum solutions are given for Figure 5-24. There is a third minimum sum-of-products solution. What is it?

(e) Work Programmed Exercise 5.2.

(f)

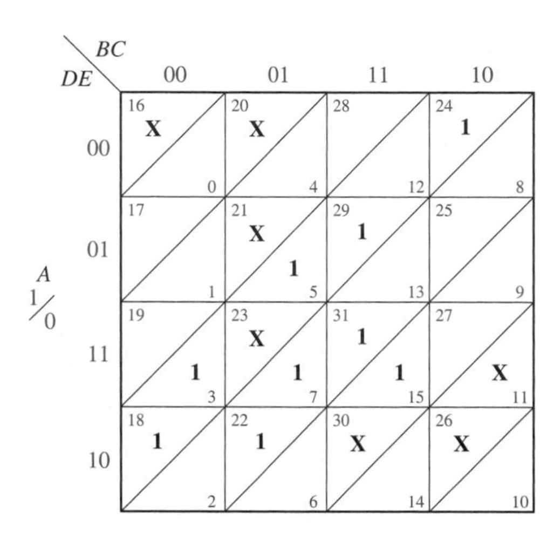


Find the three 1's and X's adjacent to $1_{18}$. Can these all be looped with a single loop?
Find the 1's and X's adjacent to $1_{24}$. Loop the essential prime implicant that covers $1_{24}$.
Find the 1's and X's adjacent to $1_3$. Loop the essential prime implicant that covers $1_3$.
Can you find an essential prime implicant that covers $1_{22}$? Explain.

Find and loop two more essential prime implicants.
Find three ways to cover the remaining 1 on the map and give the corresponding minimum solutions.

(g) If you have the *LogicAid* program available, work Problem 5.9, using the Karnaugh map tutor.

**9.** Study Section 5.6, *Other Uses of Karnaugh Maps*. Refer to Figure 5-8 and note that a consensus term exists if there are two adjacent, but nonoverlapping prime implicants. Observe how this principle is applied in Figure 5-26.

**10.** Work Problems 5.10, 5.11, 5.12, and 5.13 When deriving the minimum solution from the map, always write down the *essential* prime implicants first. If you do not, it is quite likely that you will not get the minimum solution. In addition, make sure you can find *all* of the prime implicants from the map [see Problem 5.10(b)].

**11.** Review the objectives and take the readiness test.

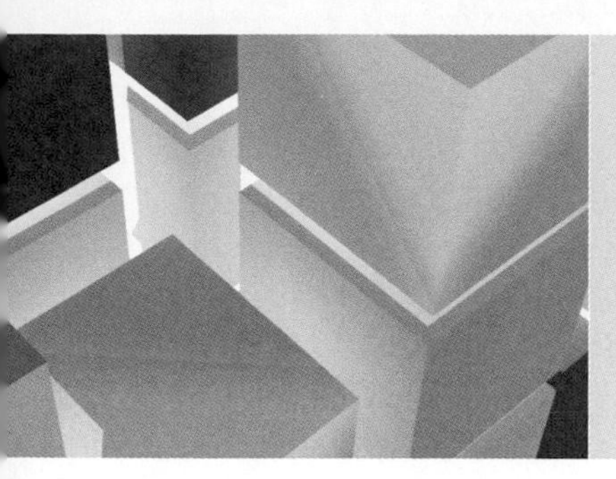

# Karnaugh Maps

Switching functions can generally be simplified by using the algebraic techniques described in Unit 3. However, two problems arise when algebraic procedures are used:

1. The procedures are difficult to apply in a systematic way.
2. It is difficult to tell when you have arrived at a minimum solution.

The Karnaugh map method studied in this unit and the Quine-McCluskey procedure studied in Unit 6 overcome these difficulties by providing systematic methods for simplifying switching functions. The Karnaugh map is an especially useful tool for simplifying and manipulating switching functions of three or four variables, but it can be extended to functions of five or more variables. Generally, you will find the Karnaugh map method is faster and easier to apply than other simplification methods.

## 5.1 Minimum Forms of Switching Functions

When a function is realized using AND and OR gates, the cost of realizing the function is directly related to the number of gates and gate inputs used. The Karnaugh map techniques developed in this unit lead directly to minimum cost two-level circuits composed of AND and OR gates. An expression consisting of a sum of product terms corresponds directly to a two-level circuit composed of a group of AND gates feeding a single OR gate (see Figure 2-5). Similarly, a product-of-sums expression corresponds to a two-level circuit composed of OR gates feeding a single AND gate (see Figure 2-6). Therefore, to find minimum cost two-level AND-OR gate circuits, we must find minimum expressions in sum-of-products or product-of-sums form.

A *minimum sum-of-products* expression for a function is defined as a sum of product terms which (a) has a minimum number of terms and (b) of all those expressions which have the same minimum number of terms, has a minimum number of literals. The minimum sum of products corresponds directly to a minimum two-level gate circuit which has (a) a minimum number of gates and (b) a minimum number of gate inputs. Unlike the minterm expansion for a function, the minimum sum of products is not necessarily unique; that is, a given function may have two different minimum sum-of-products forms, each with the same number of terms and the same number of literals. Given a minterm expansion, the minimum sum-of-products form can often be obtained by the following procedure:

1. Combine terms by using $XY' + XY = X$. Do this repeatedly to eliminate as many literals as possible. A given term may be used more than once because $X + X = X$.
2. Eliminate redundant terms by using the consensus theorem or other theorems.

Unfortunately, the result of this procedure may depend on the order in which terms are combined or eliminated so that the final expression obtained is not necessarily minimum.

**Example**

Find a minimum sum-of-products expression for

$$F(a, b, c) = \Sigma m(0, 1, 2, 5, 6, 7)$$

$$F = a'b'c' + a'b'c + a'bc' + ab'c + abc' + abc$$

$$= a'b' + b'c + bc' + ab \tag{5-1}$$

None of the terms in the above expression can be eliminated by consensus. However, combining terms in a different way leads directly to a minimum sum of products:

$$F = a'b'c' + a'b'c + a'bc' + ab'c + abc' + abc$$

$$= a'b' + bc' + ac \tag{5-2}$$

A *minimum product-of-sums* expression for a function is defined as a product of sum terms which (a) has a minimum number of factors, and (b) of all those expressions which have the same number of factors, has a minimum number of literals. Unlike the maxterm expansion, the minimum product-of-sums form of a function is not necessarily unique. Given a maxterm expansion, the minimum product of sums can often be obtained by a procedure similar to that used in the minimum sum-of-products case, except that the theorem $(X + Y)(X + Y') = X$ is used to combine terms.

**Example**

$$(A + B' + C + D')(A + B' + C' + D')(A + B' + C' + D)(A' + B' + C' + D)(A + B + C' + D)(A' + B + C' + D)$$

$$= (A + B' + D') \quad (A + B' + C') \quad (B' + C' + D) \quad (B + C' + D)$$

$$= (A + B' + D') \quad \underbrace{(A + B' + C')} \quad (C' + D)$$

← eliminate by consensus

$$= (A + B' + D')(C' + D) \tag{5-3}$$

## 5.2 Two- and Three-Variable Karnaugh Maps

Just like a truth table, the Karnaugh map of a function specifies the value of the function for every combination of values of the independent variables. A two-variable Karnaugh map is shown. The values of one variable are listed across the top of the map, and the values of the other variable are listed on the left side. Each square of the map corresponds to a pair of values for $A$ and $B$ as indicated.

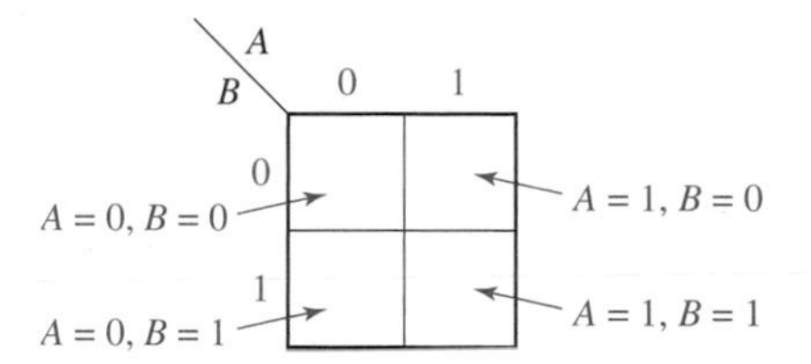


Figure 5-1 shows the truth table for a function $F$ and the corresponding Karnaugh map. Note that the value of $F$ for $A = B = 0$ is plotted in the upper left square, and the other map entries are plotted in a similar way in Figure 5-1(b). Each 1 on the map corresponds to a minterm of $F$. We can read the minterms from the map just like we can read them from the truth table. A 1 in square 00 of Figure 5-1(c) indicates that $A'B'$ is a minterm of $F$. Similarly, a 1 in square 01 indicates that $A'B$ is a minterm. Minterms in adjacent squares of the map can be combined since they differ in only one variable. Thus, $A'B'$ and $A'B$ combine to form $A'$, and this is indicated by looping the corresponding 1's on the map in Figure 5-1(d).

FIGURE 5-1

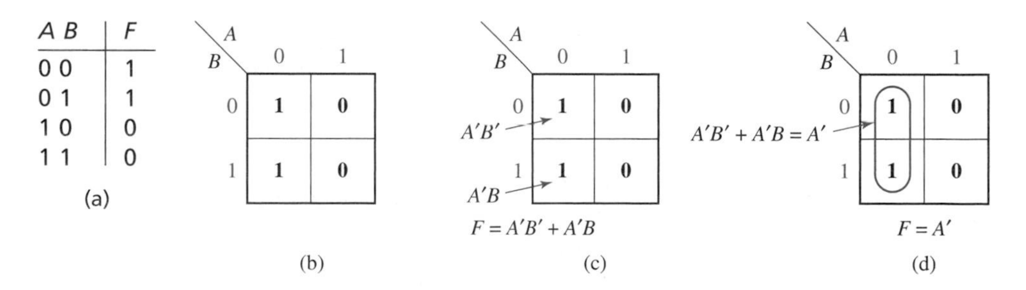


Figure 5-2 shows a three-variable truth table and the corresponding Karnaugh map (see Figure 5-27 for an alternative way of labeling maps). The value of one variable ($A$) is listed across the top of the map, and the values of the other two variables ($B$, $C$) are listed along the side of the map. The rows are labeled in the sequence 00, 01, 11, 10 so that values in adjacent rows differ in only one variable. For each combination of values of the variables, the value of $F$ is read from the truth table and plotted in the appropriate map square. For example, for the input combination $ABC = 001$, the value $F = 0$ is plotted in the square for which $A = 0$ and $BC = 01$. For the combination $ABC = 110$, $F = 1$ is plotted in the $A = 1$, $BC = 10$ square.

FIGURE 5-2
Truth Table and Karnaugh Map for Three-Variable Function

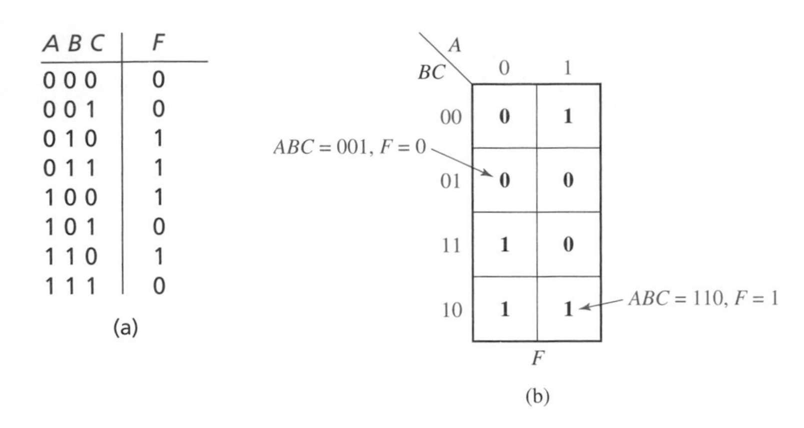

| A B C | F |
|---|---|
| 0 0 0 | 0 |
| 0 0 1 | 0 |
| 0 1 0 | 1 |
| 0 1 1 | 1 |
| 1 0 0 | 1 |
| 1 0 1 | 0 |
| 1 1 0 | 1 |
| 1 1 1 | 0 |

(a)



(b)

Figure 5-3 shows the location of the minterms on a three-variable map. Minterms in adjacent squares of the map differ in only one variable and therefore can be combined using the theorem $XY' + XY = X$. For example, minterm 011 ($a'bc$) is adjacent to the three minterms with which it can be combined—001 ($a'b'c$), 010 ($a'bc'$), and 111 ($abc$). In addition to squares which are physically adjacent, the top and bottom rows of the map are defined to be adjacent because the corresponding minterms in these rows differ in only one variable. Thus 000 and 010 are adjacent, and so are 100 and 110.

FIGURE 5-3
Location of Minterms on a Three-Variable Karnaugh Map

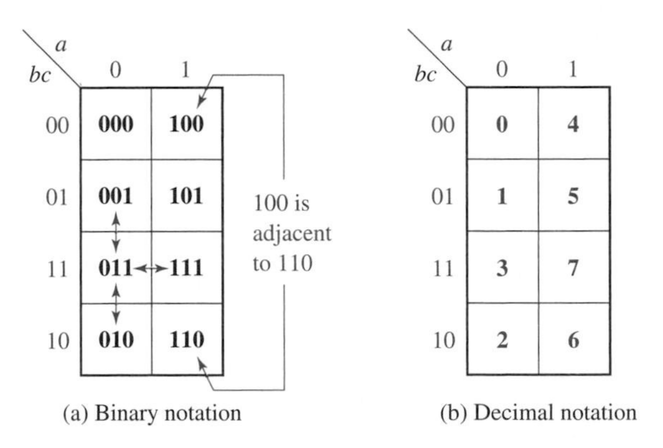


(a) Binary notation (b) Decimal notation

Given the minterm expansion of a function, it can be plotted on a map by placing 1's in the squares which correspond to minterms of the function and 0's in the remaining squares (the 0's may be omitted if desired). Figure 5-4 shows the plot of $F(a, b, c) = m_1 + m_3 + m_5$. If $F$ is given as a maxterm expansion, the map is plotted by placing 0's in the squares which correspond to the maxterms and then by filling in the remaining squares with 1's. Thus, $F(a, b, c) = M_0 M_2 M_4 M_6 M_7$ gives the same map as Figure 5-4.

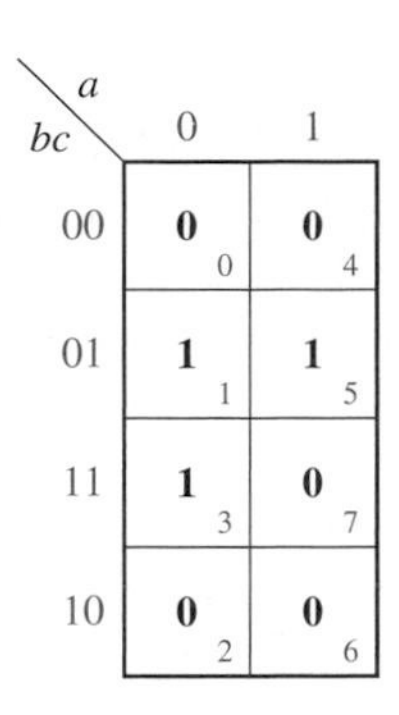


**FIGURE 5-4** Karnaugh Map of $F(a, b, c) = \Sigma\, m(1, 3, 5) = \Pi\, M(0, 2, 4, 6, 7)$

Figure 5-5 illustrates how product terms can be plotted on Karnaugh maps. To plot the term $b$, 1's are entered in the four squares of the map where $b = 1$. The term $bc'$ is 1 when $b = 1$ and $c = 0$, so 1's are entered in the two squares in the $bc = 10$ row. The term $ac'$ is 1 when $a = 1$ and $c = 0$, so 1's are entered in the $a = 1$ column in the rows where $c = 0$.

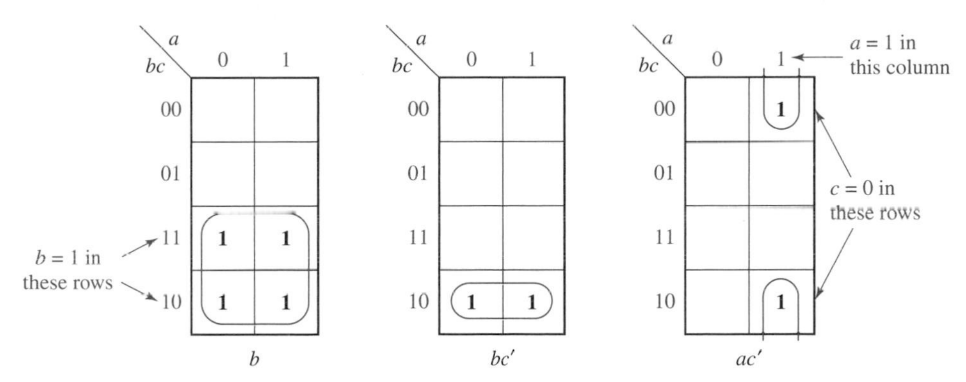


**FIGURE 5-5** Karnaugh Maps for Product Terms

If a function is given in algebraic form, it is unnecessary to expand it to minterm form before plotting it on a map. If the algebraic expression is converted to sum-of-products form, then each product term can be plotted directly as a group of 1's on the map. For example, given that

$$f(a, b, c) = abc' + b'c + a'$$

we would plot the map as follows:

1. The term $abc'$ is 1 when $a = 1$ and $bc = 10$, so we place a 1 in the square which corresponds to the $a = 1$ column and the $bc = 10$ row of the map.
2. The term $b'c$ is 1 when $bc = 01$, so we place 1's in both squares of the $bc = 01$ row of the map.
3. The term $a'$ is 1 when $a = 0$, so we place 1's in all the squares of the $a = 0$ column of the map. (Note: Since there already is a 1 in the $abc = 001$ square, we do not have to place a second 1 there because $x + x = x$.)

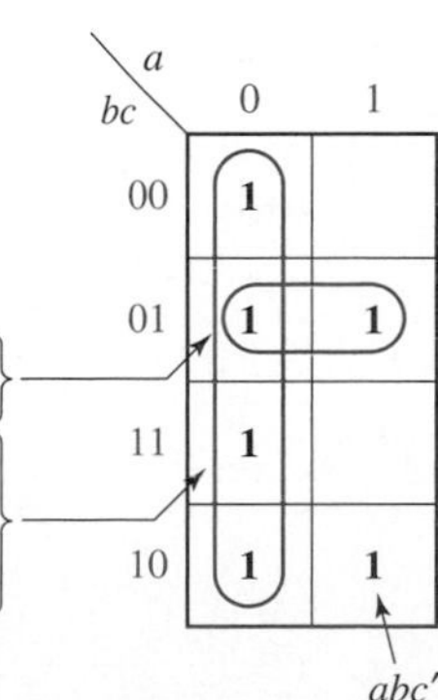

Figure 5-6 illustrates how a simplified expression for a function can be derived using a Karnaugh map. The function to be simplified is first plotted on a Karnaugh map in Figure 5-6(a). Terms in adjacent squares on the map differ in only one variable and can be combined using the theorem $XY' + XY = X$. Thus $a'b'c$ and $a'bc$ combine to form $a'c$, and $a'b'c$ and $ab'c$ combine to form $b'c$, as shown in Figure 5-6(b). A loop around a group of minterms indicates that these terms have been combined. The looped terms can be read directly off the map. Thus, for Figure 5-6(b), term $T_1$ is in the $a = 0$ $(a')$ column, and it spans the rows where $c = 1$, so $T_1 = a'c$. Note that $b$ has been eliminated because the two minterms in $T_1$ differ in the variable $b$. Similarly, the term $T_2$ is in the $bc = 01$ row so $T_2 = b'c$, and $a$ has been eliminated because $T_2$ spans the $a = 0$ and $a = 1$ columns. Thus, the minimum sum-of-products form for $F$ is $a'c + b'c$.

**FIGURE 5-6** Simplification of a Three-Variable Function

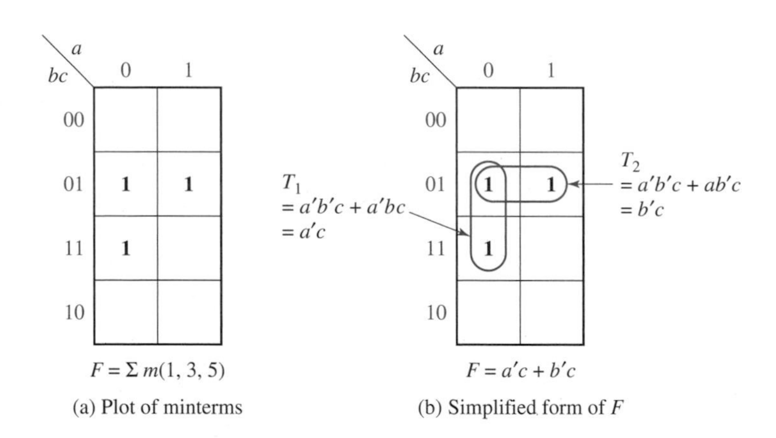


The map for the complement of $F$ (Figure 5-7) is formed by replacing 0's with 1's and 1's with 0's on the map of $F$. To simplify $F'$, note that the terms in the top row combine to form $b'c'$, and the terms in the bottom row combine to form $bc'$. Because $b'c'$ and $bc'$ differ in only one variable, the top and bottom rows can then be combined to form a group of four 1's, thus eliminating two variables and leaving $T_1 = c'$. The remaining 1 combines, as shown, to form $T_2 = ab$, so the minimum sum-of-products form for $F'$ is $c' + ab$.

**FIGURE 5-7** Complement of Map in Figure 5.6(a)

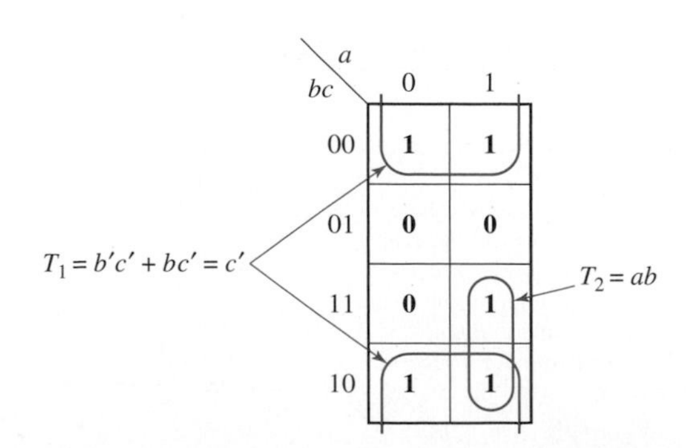

The Karnaugh map can also illustrate the basic theorems of Boolean algebra. Figure 5-8 illustrates the consensus theorem, $XY + X'Z + YZ = XY + X'Z$. Note that the consensus term ($YZ$) is redundant because its 1's are covered by the other two terms.

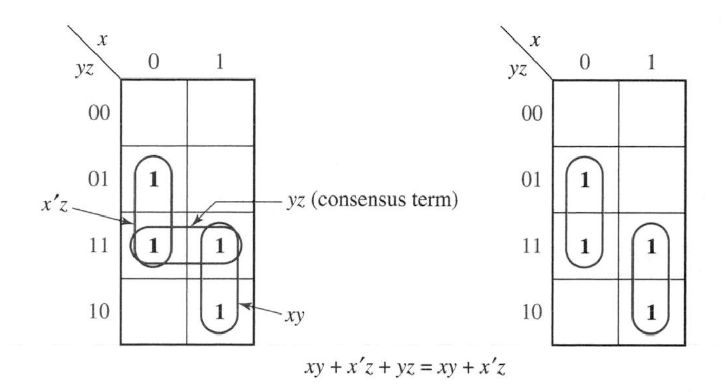


**FIGURE 5-8** Karnaugh Maps that Illustrate the Consensus Theorem

If a function has two or more minimum sum-of-products forms, all of these forms can be determined from a map. Figure 5-9 shows the two minimum solutions for $F = \Sigma\, m(0, 1, 2, 5, 6, 7)$.

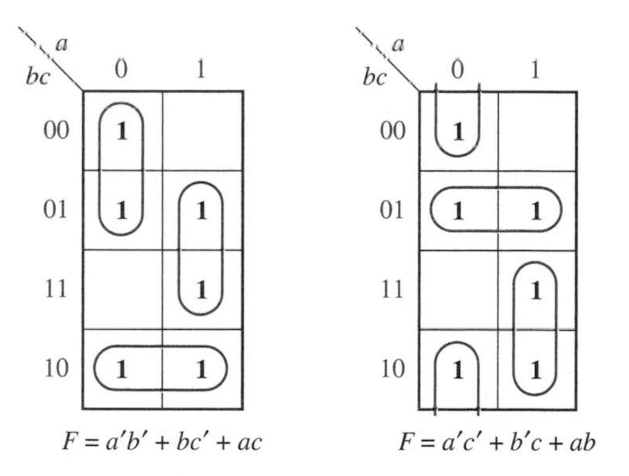


**FIGURE 5-9** Function with Two Minimum Forms

# 5.3 Four-Variable Karnaugh Maps

Figure 5-10 shows the location of minterms on a four-variable map. Each minterm is located adjacent to the four terms with which it can combine. For example, $m_5$ (0101) could combine with $m_1$ (0001), $m_4$ (0100), $m_7$ (0111), or $m_{13}$ (1101) because it differs in only one variable from each of the other minterms. The definition of adjacent squares must be extended so that not only are top and bottom rows adjacent as in the three-variable map, but the first and last columns are also adjacent. This requires numbering the columns in the sequence 00, 01, 11, 10 so that minterms 0 and 8, 1 and 9, etc., are in adjacent squares.

**FIGURE 5-10** Location of Minterms on Four-Variable Karnaugh Map

| CD \ AB | 00 | 01 | 11 | 10 |
|---|---|---|---|---|
| 00 | 0 | 4 | 12 | 8 |
| 01 | 1 | 5 | 13 | 9 |
| 11 | 3 | 7 | 15 | 11 |
| 10 | 2 | 6 | 14 | 10 |

We will now plot the following four-variable expression on a Karnaugh map (Figure 5-11):

$$f(a, b, c, d) = acd + a'b + d'$$

The first term is 1 when $a = c = d = 1$, so we place 1's in the two squares which are in the $a = 1$ column and $cd = 11$ row. The term $a'b$ is 1 when $ab = 01$, so we place four 1's in the $ab = 01$ column. Finally, $d'$ is 1 when $d = 0$, so we place eight 1's in the two rows for which $d = 0$. (Duplicate 1's are not plotted because $1 + 1 = 1$.)

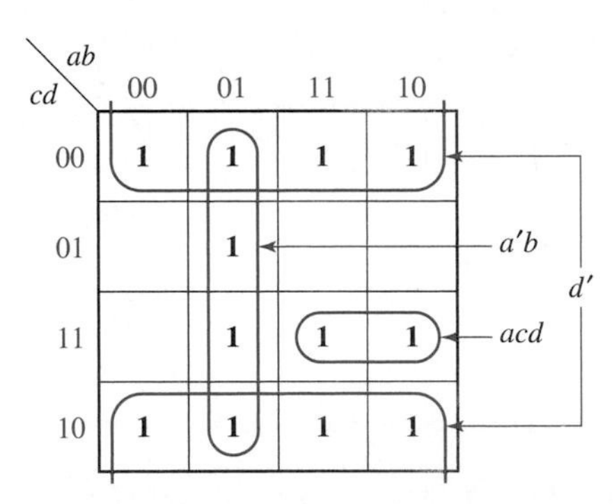


**FIGURE 5-11** Plot of $acd + a'b + d'$

Next, we will simplify the functions $f_1$ and $f_2$ given in Figure 5-12. Because the functions are specified in minterm form, we can determine the locations of the 1's on the map by referring to Figure 5-10. After plotting the maps, we can then combine adjacent groups of 1's. Minterms can be combined in groups of two, four, or eight to eliminate one, two, or three variables, respectively. In Figure 5-12(a), the pair of 1's in the $ab = 00$ column and also in the $d = 1$ rows represents $a'b'd$. The group of four 1's in the $b = 1$ columns and $c = 0$ rows represents $bc'$.

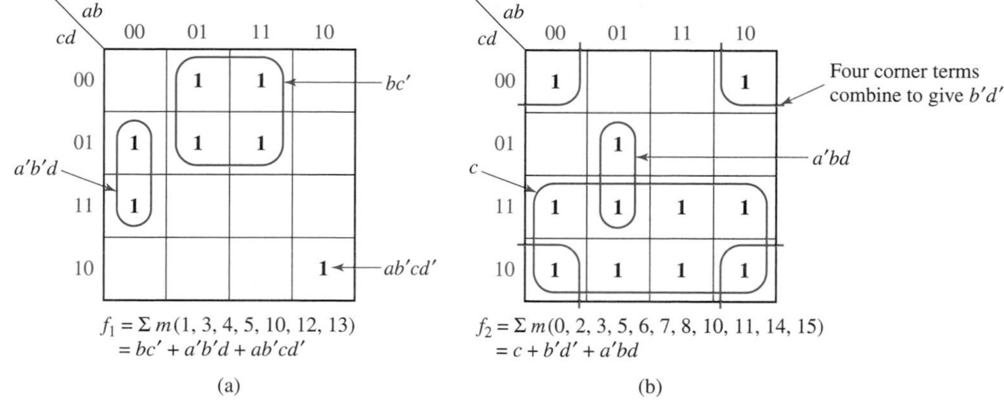


**FIGURE 5-12** Simplification of Four-Variable Functions

In Figure 5-12(b), note that the four corner 1's span the $b = 0$ columns and $d = 0$ rows and, therefore, can be combined to form the term $b'd'$. The group of eight 1's covers both rows where $c = 1$ and, therefore, represents the term $c$. The pair of 1's which is looped on the map represents the term $a'bd$ because it is in the $ab = 01$ column and spans the $d = 1$ rows.

The Karnaugh map method is easily extended to functions with don't-care terms. The required minterms are indicated by 1's on the map, and the don't-care minterms are indicated by X's. When choosing terms to form the minimum sum of products, all the 1's must be covered, but the X's are only used if they will simplify the resulting expression. In Figure 5-13, the only don't-care term used in forming the simplified expression is 13.

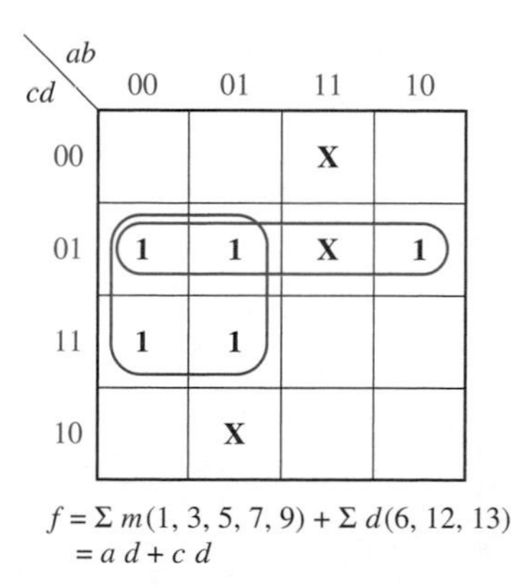


**FIGURE 5-13** Simplification of an Incompletely Specified Function

The use of Karnaugh maps to find a minimum sum-of-products form for a function has been illustrated in Figures 5-1, 5-6, and 5-12. A minimum product of sums can also be obtained from the map. Because the 0's of $f$ are 1's of $f'$, the minimum sum of products for $f'$ can be determined by looping the 0's on a map of $f$. The complement of the minimum sum of products for $f'$ is then the minimum product of sums for $f$. The following example illustrates this procedure for

$$f = x'z' + wyz + w'y'z' + x'y$$

First, the 1's of $f$ are plotted in Figure 5-14. Then, from the 0's,

$$f' = y'z + wxz' + w'xy$$

and the minimum product of sums for $f$ is

$$f = (y + z')(w' + x' + z)(w + x' + y')$$

FIGURE 5-14

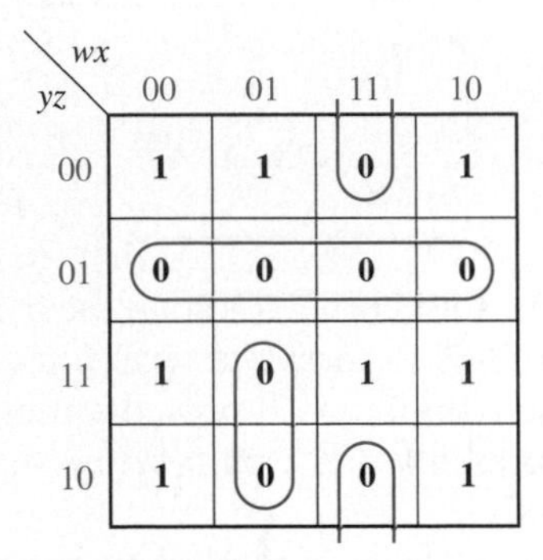


# 5.4 Determination of Minimum Expressions Using Essential Prime Implicants

Any single 1 or any group of 1's which can be combined together on a map of the function $F$ represents a product term which is called an *implicant* of $F$ (see Section 6.1 for a formal definition of implicant and prime implicant). Several implicants of $F$ are indicated in Figure 5-15. A product term implicant is called a *prime implicant* if it cannot be combined with another term to eliminate a variable. In Figure 5-15,

FIGURE 5-15

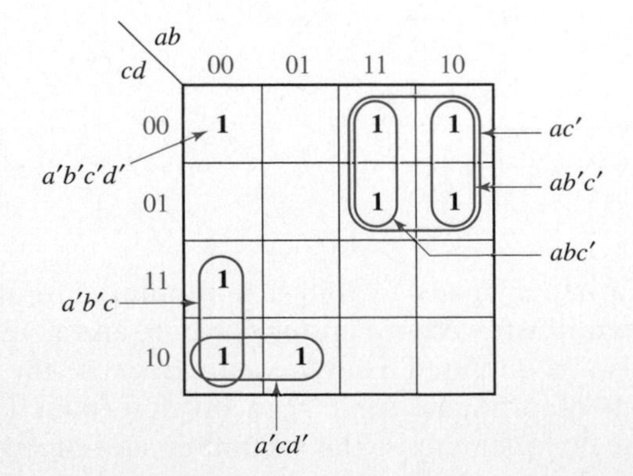

$a'b'c$, $a'cd'$, and $ac'$ are prime implicants because they cannot be combined with other terms to eliminate a variable. On the other hand, $a'b'c'd'$ is not a prime implicant because it can be combined with $a'b'cd'$ or $ab'c'd'$. Neither $abc'$, nor $ab'c'$ *is* a prime implicant because these terms can be combined together to form $ac'$.

All of the prime implicants of a function can be obtained from a Karnaugh map. A single 1 on a map represents a prime implicant if it is not adjacent to any other 1's. Two adjacent 1's on a map form a prime implicant if they are not contained in a group of four 1's; four adjacent 1's form a prime implicant if they are not contained in a group of eight 1's, etc.

The minimum sum-of-products expression for a function consists of some (but not necessarily all) of the prime implicants of a function. In other words, a sum-of-products expression containing a term which is not a prime implicant cannot be minimum. This is true because if a nonprime term were present, the expression could be simplified by combining the nonprime term with additional minterms. In order to find the minimum sum of products from a map, we must find a minimum number of prime implicants which cover all of the 1's on the map. The function plotted in Figure 5-16 has six prime implicants. Three of these prime implicants cover all of the 1's on the map, and the minimum solution is the sum of these three prime implicants. The shaded loops represent prime implicants which are not part of the minimum solution.

**FIGURE 5-16**
Determination of All Prime Implicants

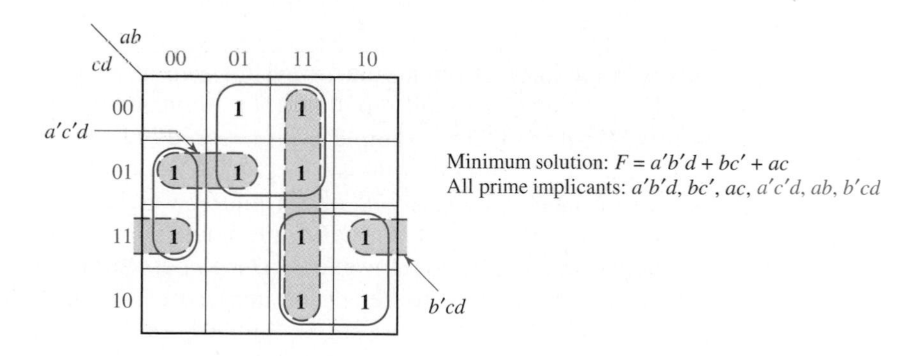


When writing down a list of *all* of the prime implicants from the map, note that there are often prime implicants which are not included in the minimum sum of products. Even though all of the 1's in a term have already been covered by prime implicants, that term may still be a prime implicant provided that it is not included in a larger group of 1's. For example, in Figure 5-16, $a'c'd$ is a prime implicant because it cannot be combined with other 1's to eliminate another variable. However, $abd$ is not a prime implicant because it can be combined with two other 1's to form $ab$. The term $b'cd$ is also a prime implicant even though both of its 1's are already covered by other prime implicants. In the process of finding prime implicants, don't-cares are treated just like 1's. However, a prime implicant composed entirely of don't-cares can never be part of the minimum solution.

Because all of the prime implicants of a function are generally not needed in forming the minimum sum of products, a systematic procedure for selecting prime implicants is needed. If prime implicants are selected from the map in the wrong order, a nonminimum solution may result. For example, in Figure 5-17, if $CD$ is chosen first, then $BD$, $B'C$, and $AC$ are needed to cover the remaining 1's, and the solution contains four terms. However, if the prime implicants indicated in Figure 5-17(b) are chosen first, all 1's are covered and $CD$ is not needed.

FIGURE 5-17

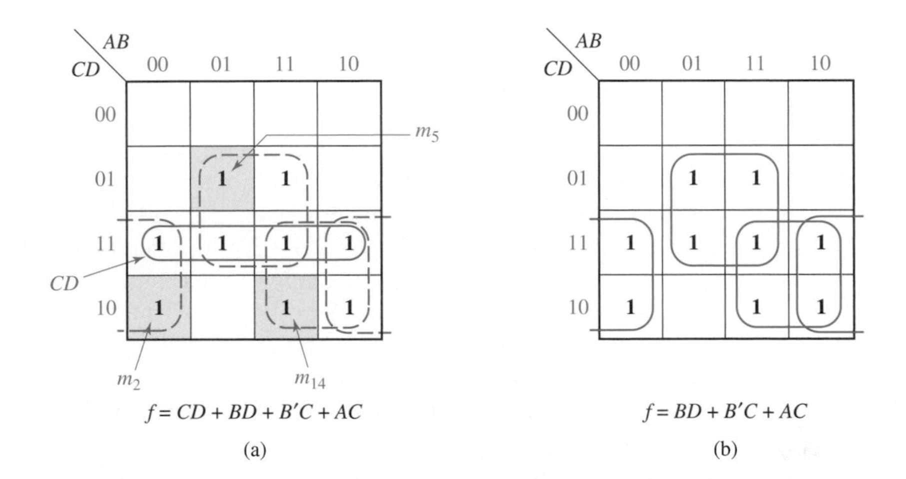


Note that some of the minterms on the map of Figure 5-17(a) can be covered by only a single prime implicant, but other minterms can be covered by two different prime implicants. For example, $m_2$ is covered only by $B'C$, but $m_3$ is covered by both $B'C$ and $CD$. If a minterm is covered by only one prime implicant, that prime implicant is said to be *essential*, and it must be included in the minimum sum of products. Thus, $B'C$ *is* an essential prime implicant because $m_2$ is not covered by any other prime implicant. However, $CD$ is *not* essential because each of the 1's in $CD$ can be covered by another prime implicant. The only prime implicant which covers $m_5$ is $BD$, so $BD$ is essential. Similarly, $AC$ is essential because no other prime implicant covers $m_{14}$. In this example, if we choose all of the essential prime implicants, all of the 1's on the map are covered and the nonessential prime implicant $CD$ is not needed.

In general, in order to find a minimum sum of products from a map, we should first loop all of the essential prime implicants. One way of finding essential prime implicants on a map is simply to look at each 1 on the map that has not already been covered, and check to see how many prime implicants cover that 1. If there is only one prime implicant which covers the 1, that prime implicant is essential. If there are two or more prime implicants which cover the 1, we cannot say whether these prime implicants are essential or not without checking the other minterms. For simple problems, we can locate the essential prime implicants in this way by inspection of each 1 on the map. For example, in Figure 5-16, $m_4$ is covered only by the prime implicant $bc'$, and $m_{10}$ is covered only by the prime implicant $ac$. All other 1's on the map are covered by two prime implicants; therefore, the only essential prime implicants are $bc'$ and $ac$.

For more complicated maps, and especially for maps with five or more variables, we need a more systematic approach for finding the essential prime implicants. When checking a minterm to see if it is covered by only one prime implicant, we must look at all squares adjacent to that minterm. If the given minterm and all of the 1's adjacent to it are covered by a single term, then that term is an *essential* prime implicant.[1] If all of the 1's adjacent to a given minterm are *not* covered by a single term, then there are two or more prime implicants which cover that minterm, and we cannot say whether these prime implicants are essential or not without checking the other minterms. Figure 5-18 illustrates this principle.

**FIGURE 5-18**

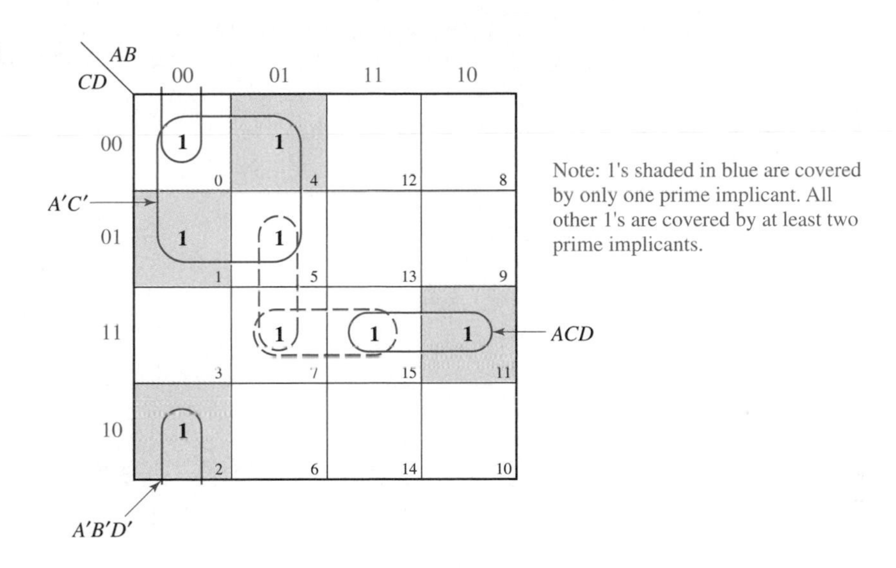
AB
CD
00
01
11
10
A′C′
ACD
A′B′D′
Note: 1's shaded in blue are covered by only one prime implicant. All other 1's are covered by at least two prime implicants.

The adjacent 1's for minterm $m_0$ ($1_0$) are $1_1$, $1_2$, and $1_4$. Because no single term covers these four 1's, no essential prime implicant is yet apparent. The adjacent 1's for $1_1$ are $1_0$ and $1_5$, so the term which covers these three 1's ($A'C'$) is an essential prime implicant. Because the only 1 adjacent to $1_2$ is $1_0$, $A'B'D'$ is also essential. Because the 1's adjacent to $1_7$ ($1_5$ and $1_{15}$) are not covered by a single term, neither $A'BD$ nor $BCD$ is essential at this point. However, because the only 1 adjacent to $1_{11}$ is $1_{15}$, $ACD$ is essential. To complete the minimum solution, one of the nonessential prime implicants is needed. Either $A'BD$ or $BCD$ may be selected. The final solution is

$$A'C' + A'B'D' + ACD + \left\{\begin{matrix} A'BD \\ \text{or} \\ BCD \end{matrix}\right\}$$

[1] This statement is proved in Appendix D.

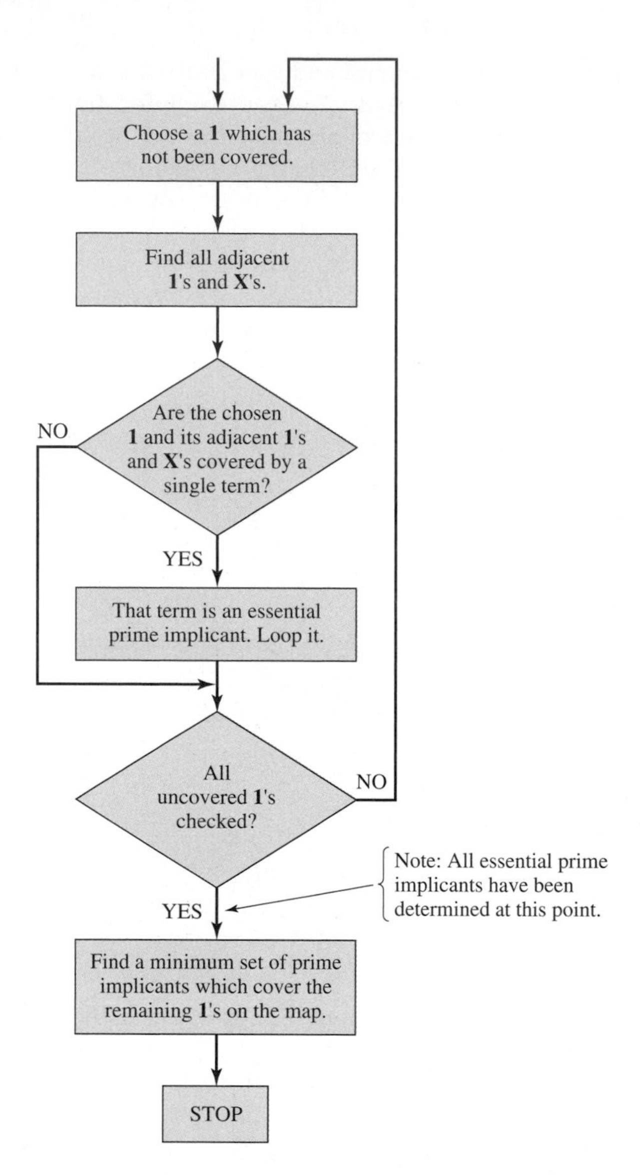


**FIGURE 5-19** Flowchart for Determining a Minimum Sum of Products Using a Karnaugh Map

If a don't-care minterm is present on the map, we do not have to check it to see if it is covered by one or more prime implicants. However, when checking a 1 for adjacent 1's, we treat the adjacent don't-cares as if they were 1's because don't-cares may be combined with 1's in the process of forming prime implicants. The following procedure can then be used to obtain a minimum sum of products from a Karnaugh map:

1. Choose a minterm (a 1) which has not yet been covered.
2. Find all 1's and X's adjacent to that minterm. (Check the $n$ adjacent squares on an $n$-variable map.)
3. If a single term covers the minterm and all of the adjacent 1's and X's, then that term is an essential prime implicant, so select that term. (Note that don't-care terms are treated like 1's in steps 2 and 3 but not in step 1.)

4. Repeat steps 1, 2, and 3 until all essential prime implicants have been chosen.
5. Find a minimum set of prime implicants which cover the remaining 1's on the map. (If there is more than one such set, choose a set with a minimum number of literals.)

Figure 5-19 gives a flowchart for this procedure. The following example (Figure 5-20) illustrates the procedure. Starting with $1_4$, we see that the adjacent 1's and X's ($X_0$, $1_5$, and $1_6$) are *not* covered by a single term, so no essential prime implicant is apparent. However, $1_6$ and its adjacent 1's and X's ($1_4$ and $X_7$) are covered by $A'B$, so $A'B$ is an essential prime implicant. Next, looking at $1_{13}$, we see that its adjacent 1's and X's ($1_5$, $1_9$, and $X_{15}$) are *not* covered by a single term, so no essential prime implicant is apparent. Similarly, an examination of the terms adjacent to $1_8$ and $1_9$ reveals no essential prime implicants. However, $1_{10}$ has only $1_8$ adjacent to it, so $AB'D'$ *is* an essential prime implicant because it covers both $1_{10}$ and $1_8$. Having first selected the essential prime implicants, we now choose $AC'D$ because it covers both of the remaining 1's on the map.

Judicious selection of the order in which the minterms are selected (step 1) reduces the amount of work required in applying this procedure. As will be seen in the next section, this procedure is especially helpful in obtaining minimum solutions for five- and six-variable problems.

**FIGURE 5-20**

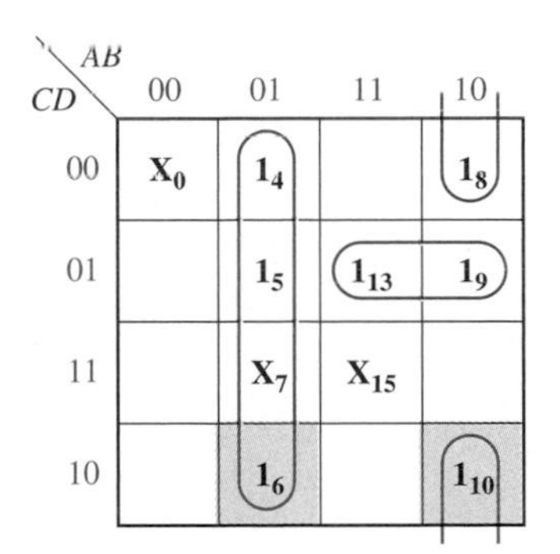


Shaded 1's are covered by only one prime implicant.

# 5.5 Five-Variable Karnaugh Maps

A five-variable map can be constructed in three dimensions by placing one four-variable map on top of a second one. Terms in the bottom layer are numbered 0 through 15 and corresponding terms in the top layer are numbered 16 through 31, so that terms in the bottom layer contain $A'$ and those in the top layer contain $A$. To represent the map in two dimensions, we will divide each square in a four-variable map by a diagonal line and place terms in the bottom layer below the line and terms in the top layer above the line (Figure 5-21). Terms in the top or bottom layer combine just like terms on a four-variable map. In addition, two terms in the same square which are separated by a diagonal line differ in only one variable and can be combined.

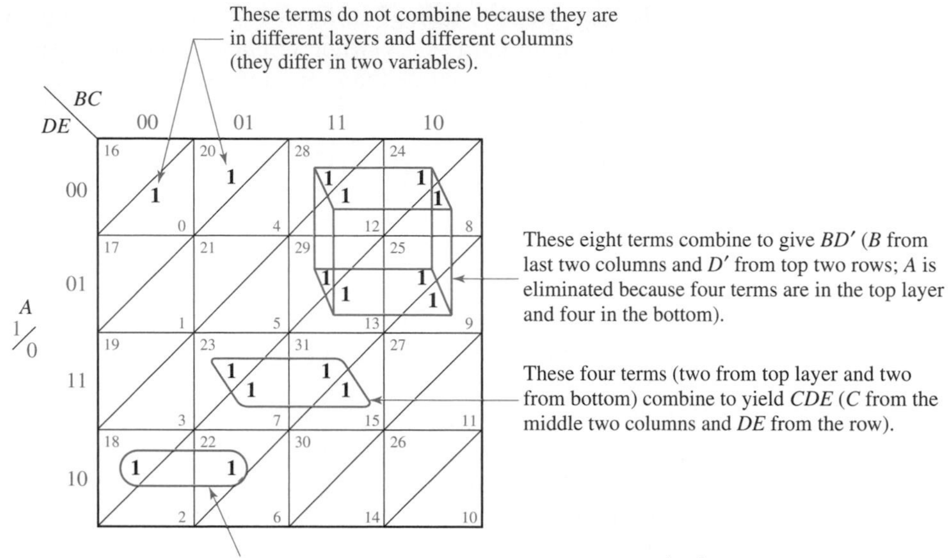
These terms do not combine because they are in different layers and different columns (they differ in two variables).
BC
DE
00
01
11
10
A
1/0
These eight terms combine to give BD′ (B from last two columns and D′ from top two rows; A is eliminated because four terms are in the top layer and four in the bottom).
These four terms (two from top layer and two from bottom) combine to yield CDE (C from the middle two columns and DE from the row).
These two terms in the top layer combine to give AB′DE′.

**FIGURE 5-21** A Five-Variable Karnaugh Map

However, some terms which appear to be physically adjacent are not. For example, terms 0 and 20 are not adjacent because they appear in a different column and a different layer. Each term can be adjacent to exactly five other terms, four in the same layer and one in the other layer (Figure 5-22). An alternate representation for five-variable maps is to draw the two layers side-by-side, as in Figure 5-28, but most individuals find adjacencies more difficult to see when this form is used.

When checking for adjacencies, each term should be checked against the five possible adjacent squares. (In general, the number of adjacent squares is equal to the number of variables.) Two examples of five-variable minimization using maps follow. Figure 5-23 is a map of

$$F(A, B, C, D, E) = \Sigma\, m(0, 1, 4, 5, 13, 15, 20, 21, 22, 23, 24, 26, 28, 30, 31)$$

**FIGURE 5-22**

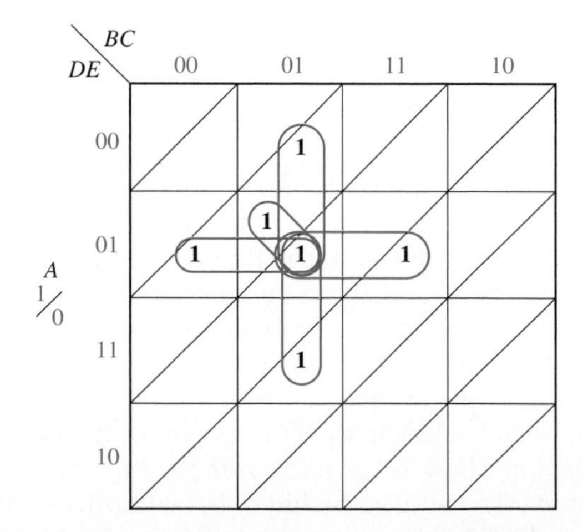
BC
DE
00
01
11
10
A
1/0

FIGURE 5-23

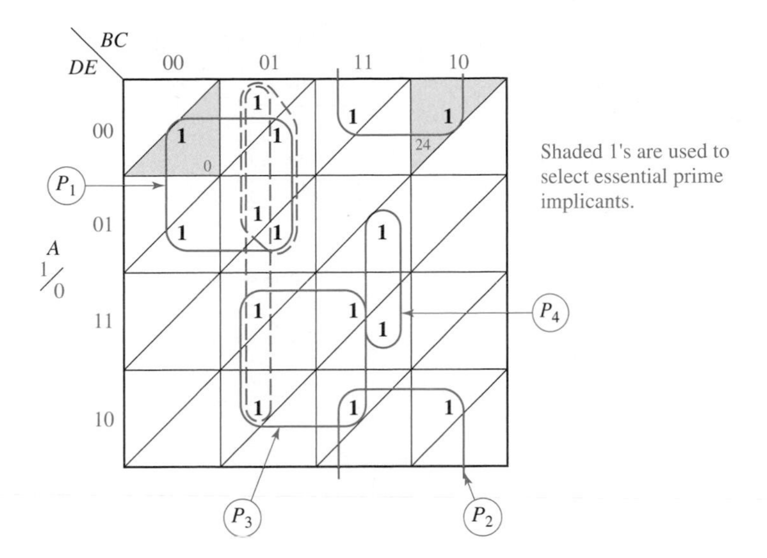


Prime implicant $P_1$ is chosen first because all of the 1's adjacent to minterm 0 are covered by $P_1$. Prime implicant $P_2$ is chosen next because all of the 1's adjacent to minterm 24 are covered by $P_2$. All of the remaining 1's on the map can be covered by at least two different prime implicants, so we proceed by trial and error. After a few tries, it becomes apparent that the remaining 1's can be covered by three prime implicants. If we choose prime implicants $P_3$ and $P_4$ next, the remaining two 1's can be covered by two different groups of four. The resulting minimum solution is

$$F = \underset{P_1}{A'B'D'} + \underset{P_2}{ABE'} + \underset{P_3}{ACD} + \underset{P_4}{A'BCE} + \left\{\begin{matrix} AB'C \\ \text{or} \\ B'CD' \end{matrix}\right\}$$

Figure 5-24 is a map of

$$F(A, B, C, D, E) = \Sigma\, m(0, 1, 3, 8, 9, 14, 15, 16, 17, 19, 25, 27, 31)$$

All 1's adjacent to $m_{16}$ are covered by $P_1$, so choose $P_1$ first. All 1's adjacent to $m_3$ are covered by $P_2$, so $P_2$ is chosen next. All 1's adjacent to $m_8$ are covered by $P_3$, so $P_3$ is chosen. Because $m_{14}$ is only adjacent to $m_{15}$, $P_4$ is also essential. There are no more essential prime implicants, and the remaining 1's can be covered by two terms, $P_5$ and (1-9-17-25) or (17-19-25-27). The final solution is

$$F = \underset{P_1}{B'C'D'} + \underset{P_2}{B'C'E} + \underset{P_3}{A'C'D'} + \underset{P_4}{A'BCD} + \underset{P_5}{ABDE} + \left\{\begin{matrix} C'D'E \\ \text{or} \\ AC'E \end{matrix}\right\}$$

FIGURE 5-24

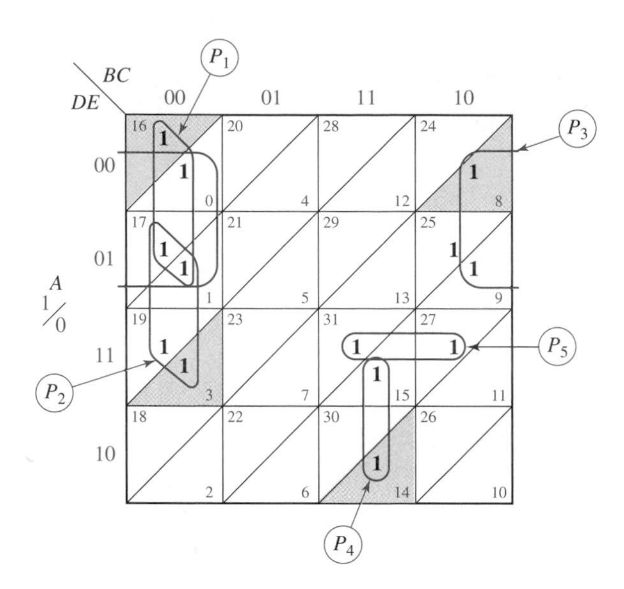


## 5.6 Other Uses of Karnaugh Maps

Many operations that can be performed using a truth table or algebraically can be done using a Karnaugh map. A map conveys the same information as a truth table—it is just arranged in a different format. If we plot an expression for $F$ on a map, we can read off the minterm and maxterm expansions for $F$ and for $F'$. From the map of Figure 5-14, the minterm expansion of $f$ is

$$f = \Sigma\, m(0, 2, 3, 4, 8, 10, 11, 15)$$

and because each 0 corresponds to a maxterm, the maxterm expansion of $f$ is

$$f = \Pi\, M(1, 5, 6, 7, 9, 12, 13, 14)$$

We can prove that two functions are equal by plotting them on maps and showing that they have the same Karnaugh map. We can perform the AND operation (or the OR operation) on two functions by ANDing (or ORing) the 1's and 0's which appear in corresponding positions on their maps. This procedure is valid because it is equivalent to doing the same operations on the truth tables for the functions.

A Karnaugh map can facilitate factoring an expression. Inspection of the map reveals terms which have one or more variables in common. For the map of Figure 5-25, the two terms in the first column have $A'B'$ in common; the two terms in the lower right corner have $AC$ in common.

FIGURE 5-25

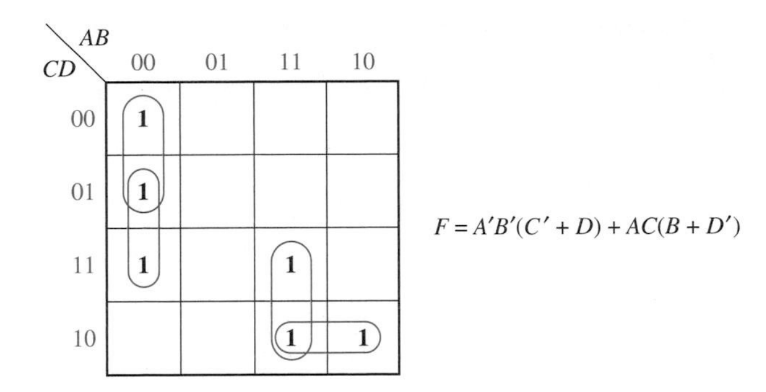


When simplifying a function algebraically, the Karnaugh map can be used as a guide in determining what steps to take. For example, consider the function

$$F = ABCD + B'CDE + A'B' + BCE'$$

From the map (Figure 5-26), we see that in order to get the minimum solution, we must add the term $ACDE$. We can do this using the consensus theorem:

$$F = ABCD + B'CDE + A'B' + BCE' + ACDE$$

As can be seen from the map, this expression now contains two redundant terms, $ABCD$ and $B'CDE$. These can be eliminated using the consensus theorem, which gives the minimum solution:

$$F = A'B' + BCE' + ACDE$$

FIGURE 5-26

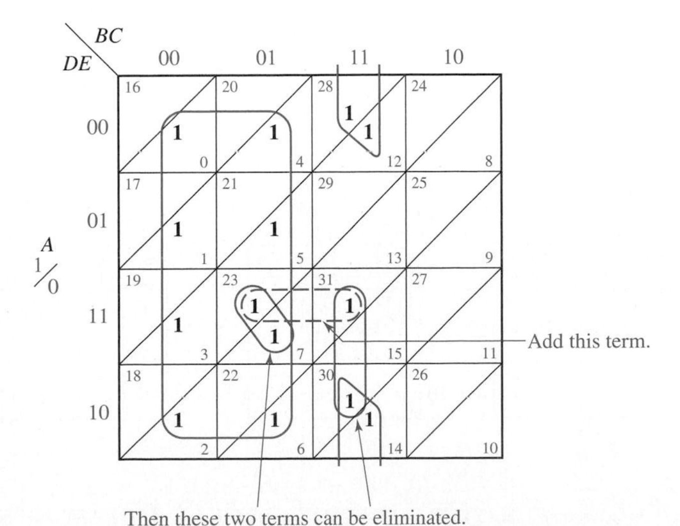

## 5.7 Other Forms of Karnaugh Maps

Instead of labeling the sides of a Karnaugh map with 0's and 1's, some people prefer to use the labeling shown in Figure 5-27. For the half of the map labeled $A$, $A = 1$; and for the other half, $A = 0$. The other variables have a similar interpretation. A map labeled this way is sometimes referred to as a Veitch diagram. It is particularly useful for plotting functions given in algebraic form rather than in minterm or maxterm form. However, when utilizing Karnaugh maps to solve sequential circuit problems (Units 12 through 16), the use of 0's and 1's to label the maps is more convenient.

**FIGURE 5-27**
Veitch Diagrams

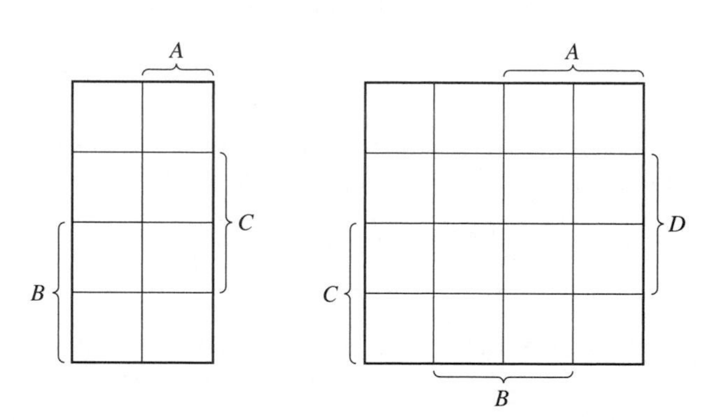


Two alternative forms for five-variable maps are used. One form simply consists of two four-variable maps side-by-side as in Figure 5-28(a). A modification of this uses a *mirror image* map as in Figure 5-28(b). In this map, first and eighth columns are "adjacent" as are second and seventh columns, third and sixth columns, and fourth and fifth columns. The same function is plotted on both these maps.

**FIGURE 5-28**
Other Forms of Five-Variable Karnaugh Maps

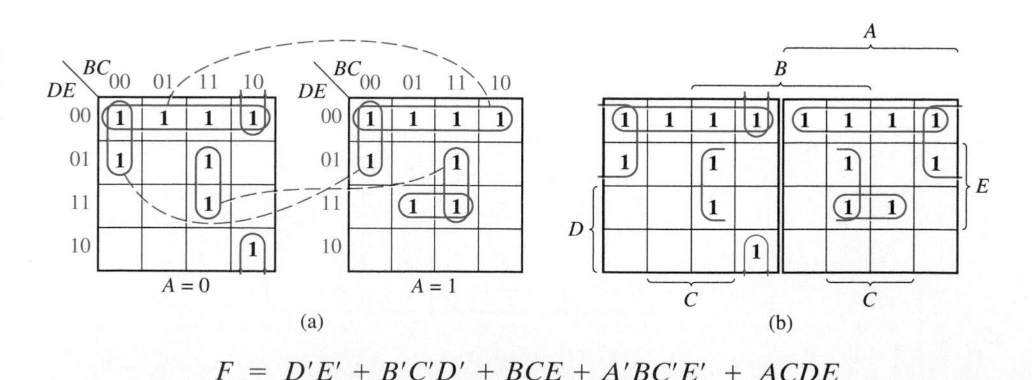


$$F = D'E' + B'C'D' + BCE + A'BC'E' + ACDE$$

# Programmed Exercise 5.1

Cover the answers to this exercise with a sheet of paper and slide it down as you check your answers. Write your answers in the space provided before looking at the correct answer.

***Problem:*** Determine the minimum sum of products and minimum product of sums for

$$f = b'c'd' + bcd + acd' + a'b'c + a'bc'd$$

First, plot the map for $f$.

| | 00 | 01 | 11 | 10 |
|---|---|---|---|---|
| 00 | | | | |
| 01 | | | | |
| 11 | | | | |
| 10 | | | | |

---

**Answer:**

| cd \ ab | 00 | 01 | 11 | 10 |
|---|---|---|---|---|
| 00 | **1** | | | **1** |
| 01 | | **1** | | |
| 11 | **1** | **1** | **1** | |
| 10 | **1** | | **1** | **1** |

---

(a) The minterms adjacent to $m_0$ on the preceding map are __________ and __________.

(b) Find an essential prime implicant containing $m_0$ and loop it.

(c) The minterms adjacent to $m_3$ are __________ and __________.

(d) Is there an essential prime implicant which contains $m_3$?

(e) Find the remaining essential prime implicant(s) and loop it (them).

---

**Answers:**

(a) $m_2$ and $m_8$
(b)
(c) $m_2$ and $m_7$
(e)
(d) No

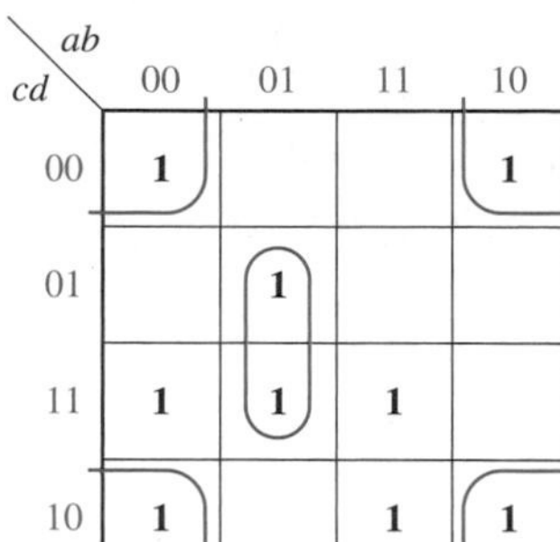

---

Loop the remaining 1's using a minimum number of loops.
The two possible minimum sum-of-products forms for $f$ are

$f =$ ______________________________ and

$f =$ ______________________________

---

**Answer:**

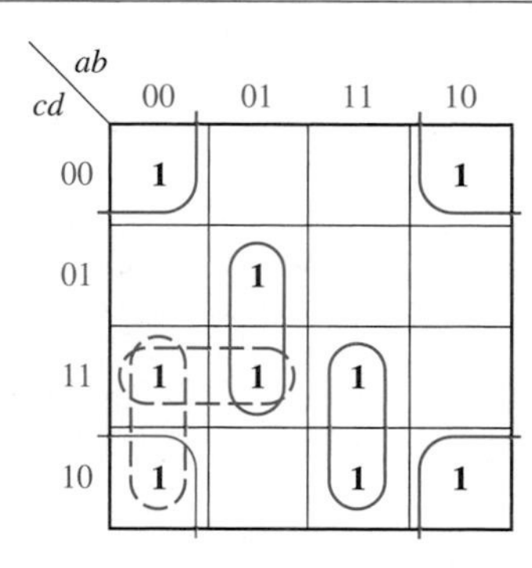

$$f = b'd' + a'bd + abc + \left\{ \begin{matrix} a'cd \\ or \\ a'b'c \end{matrix} \right\}$$

---

Next, we will find the minimum product of sums for $f$. Start by plotting the map for $f'$.

Loop all essential prime implicants of $f'$ and indicate which minterm makes each one essential.

|  | 00 | 01 | 11 | 10 |
|---|---|---|---|---|
| 00 |  |  |  |  |
| 01 |  |  |  |  |
| 11 |  |  |  |  |
| 10 |  |  |  |  |

$f'$

**Answer:**

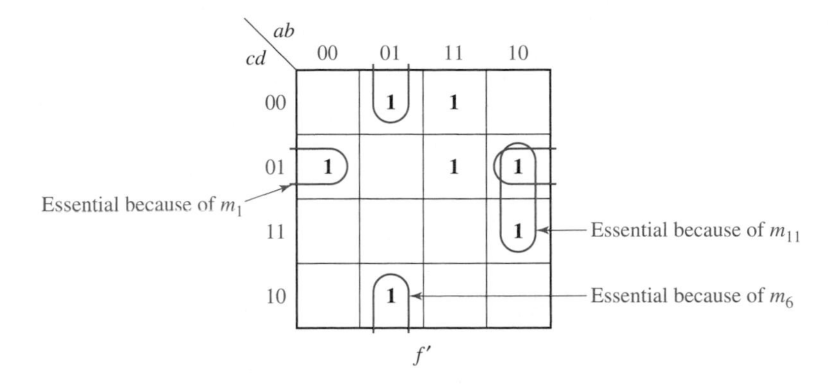


Loop the remaining 1's and write the minimum sum of products for $f'$.

$f'$ = ______________________

The minimum product of sums for $f$ is therefore

$f$ – ______________________

**Final Answer:** $f' = b'c'd + a'bd' + ab'd + abc'$
$f = (b + c + d')(a + b' + d)(a' + b + d')(a' + b' + c)$

# Programmed Exercise 5.2

***Problem:*** Determine a minimum sum-of-products expression for

$$f(a, b, c, d, e) = (a' + c + d)(a' + b + e)(a + c' + e')(c + d + e')$$
$$(b + c + d' + e)(a' + b' + c + e')$$

The first step in the solution is to plot a map for $f$. Because $f$ is given in product-of-sums form, it is easier to first plot the map for $f'$ and then complement the map. Write $f'$ as a sum of products:
$f'$ = ______________________
Now plot the map for $f'$. (Note that there are three terms in the upper layer, one term in the lower layer, and two terms which span the two layers.)

Next, convert your map for $f'$ to a map for $f$.

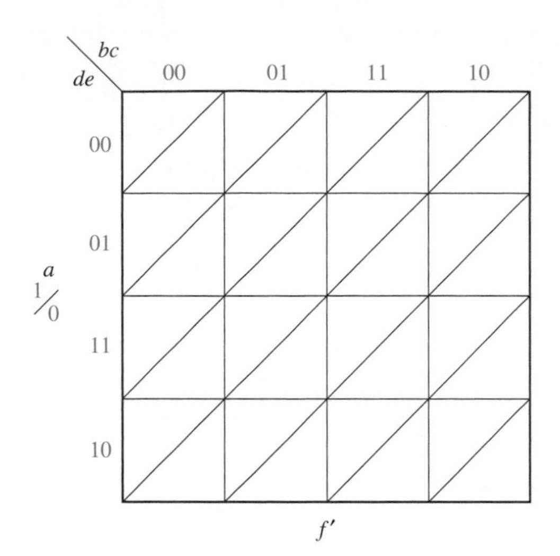


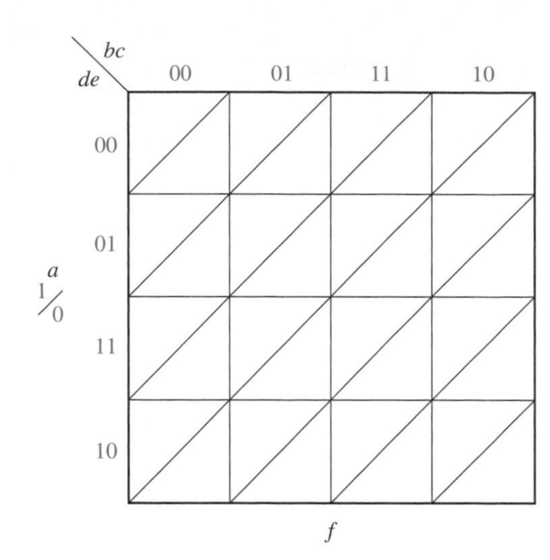


**Answer:**

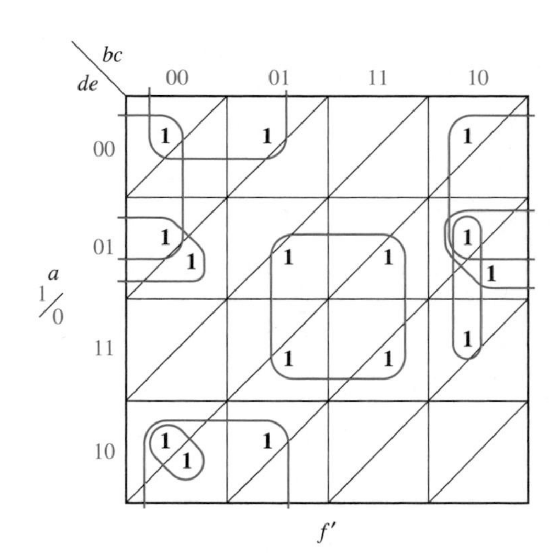


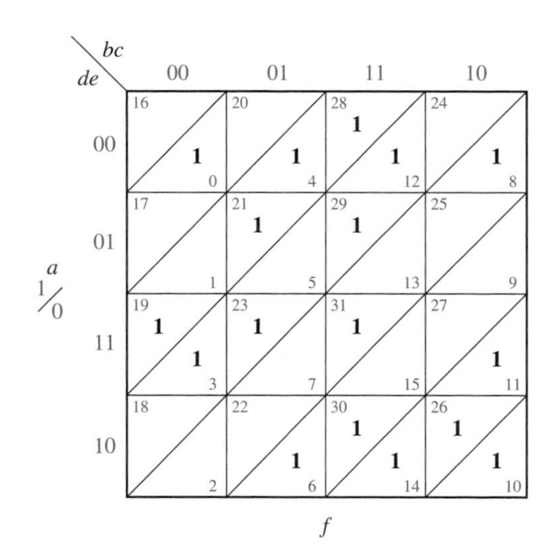


The next step is to determine the essential prime implicants of $f$.

(a) Why is $a'd'e'$ an essential prime implicant?

(b) Which minterms are adjacent to $m_3$? __________ To $m_{19}$? __________

(c) Is there an essential prime implicant which covers $m_3$ and $m_{19}$?

(d) Is there an essential prime implicant which covers $m_{21}$?

(e) Loop the essential prime implicants which you have found. Then, find two more essential prime implicants and loop them.

**Answers:**

(a) It covers $m_0$ and both adjacent minterms.
(b) $m_{19}$ and $m_{11}$; $m_3$ and $m_{23}$
(c) No
(d) Yes
(e)

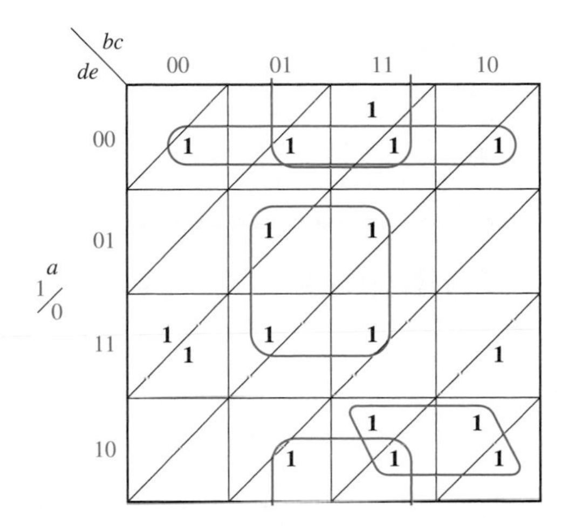


(a) Why is there no essential prime implicant which covers $m_{11}$?
(b) Why is there no essential prime implicant which covers $m_{28}$?

Because there are no more essential prime implicants, loop a minimum number of terms which cover the remaining 1's.

**Answers:**

(a) All adjacent 1's of $m_{11}$ ($m_3, m_{10}$) cannot be covered by one grouping.
(b) All adjacent 1's of $m_{28}$ ($m_{12}, m_{30}, m_{29}$) cannot be covered by one grouping.

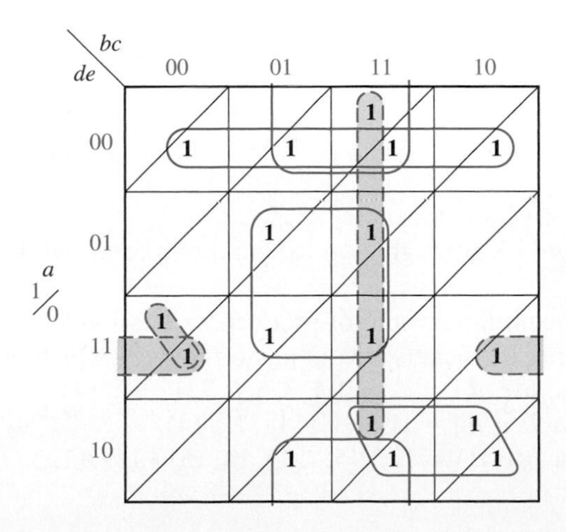


Note: There are five other possible ways to loop the four remaining 1's.

Write down two different minimum sum-of-products expressions for $f$.

$f =$ ________________________________

$f =$ ________________________________

**Answer:**

$$f = a'd'e' + ace + a'ce' + bde' + \begin{Bmatrix} abc \\ or \\ bce' \end{Bmatrix} + \begin{Bmatrix} b'c'de + a'c'de \\ b'c'de + a'bc'd \\ ab'de + a'c'de \end{Bmatrix}$$

# Problems

5.3 Find the minimum sum of products for each function using a Karnaugh map.
(a) $f_1(a, b, c) = m_0 + m_2 + m_5 + m_6$ (b) $f_2(d, e, f) = \Sigma\, m(0,1,2,4)$
(c) $f_3(r, s, t) = rt' + r's' + r's$ (d) $f_4(x, y, z) = M_0 \bullet M_5$

5.4 (a) Plot the following function on a Karnaugh map. (Do not expand to minterm form before plotting.)

$$F(A,B,C,D) = BD' + B'CD + ABC + ABC'D + B'D'$$

(b) Find the minimum sum of products.
(c) Find the minimum product of sums.

5.5 A switching circuit has two control inputs ($C_1$ and $C_2$), two data inputs ($X_1$ and $X_2$), and one output ($Z$). The circuit performs one of the logic operations AND, OR, EQU (equivalence), or XOR (exclusive OR) on the two data inputs. The function performed depends on the control inputs:

| $C_1$ | $C_2$ | Function Performed by Circuit |
|---|---|---|
| 0 | 0 | OR |
| 0 | 1 | XOR |
| 1 | 0 | AND |
| 1 | 1 | EQU |

(a) Derive a truth table for $Z$.
(b) Use a Karnaugh map to find a minimum AND-OR gate circuit to realize $Z$.

5.6 Find the minimum sum-of-products expression for each function. Underline the essential prime implicants in your answer and tell which minterm makes each one essential.
(a) $f(a, b, c, d) = \Sigma\, m(0, 1, 3, 5, 6, 7, 11, 12, 14)$
(b) $f(a, b, c, d) = \Pi\, M(1, 9, 11, 12, 14)$
(c) $f(a, b, c, d) = \Pi\, M(5, 7, 13, 14, 15) \bullet \Pi\, D(1, 2, 3, 9)$

**5.7** Find the minimum sum-of-products expression for each function.
(a) $f(a, b, c, d) = \Sigma\, m(0, 2, 3, 4, 7, 8, 14)$
(b) $f(a, b, c, d) = \Sigma\, m(1, 2, 4, 15) + \Sigma\, d(0, 3, 14)$
(c) $f(a, b, c, d) = \Pi\, M(1, 2, 3, 4, 9, 15)$
(d) $f(a, b, c, d) = \Pi\, M(0, 2, 4, 6, 8) \bullet \Pi\, D(1, 12, 9, 15)$

**5.8** Find the minimum sum of products and the minimum product of sums for each function:
(a) $f(a, b, c, d) = \Pi\, M(0, 1, 6, 8, 11, 12) \bullet \Pi\, D(3, 7, 14, 15)$
(b) $f(a, b, c, d) = \Sigma\, m(1, 3, 4, 11) + \Sigma\, d(2, 7, 8, 12, 14, 15)$

**5.9** Find the minimum sum of products and the minimum product of sums for each function:
(a) $F(A, B, C, D, E) = \Sigma\, m(0, 1, 2, 6, 7, 9, 10, 15, 16, 18, 20, 21, 27, 30) + \Sigma\, d(3, 4, 11, 12, 19)$
(b) $F(A, B, C, D, E) = \Pi\, M(0, 3, 6, 9, 11, 19, 20, 24, 25, 26, 27, 28, 29, 30) \bullet \Pi\, D(1, 2, 12, 13)$

**5.10** $F(a, b, c, d, e) = \Sigma\, m(0, 3, 4, 5, 6, 7, 8, 12, 13, 14, 16, 21, 23, 24, 29, 31)$
(a) Find the essential prime implicants using a Karnaugh map, and indicate why each one of the chosen prime implicants is essential (there are four essential prime implicants).
(b) Find all of the prime implicants by using the Karnaugh map. (There are nine in all.)

**5.11** Find a minimum product-of-sums solution for $f$. Underline the essential prime implicants.

$$f(a, b, c, d, e) = \Sigma\, m(2, 4, 5, 6, 7, 8, 10, 12, 14, 16, 19, 27, 28, 29, 31) + \Sigma\, d(1, 30)$$

**5.12** Given $F = AB'D' + A'B + A'C + CD$.
(a) Use a Karnaugh map to find the maxterm expression for $F$ (express your answer in both decimal and algebric notation).
(b) Use a Karnaugh map to find the minimum sum-of-products form for $F'$.
(c) Find the minimum product of sums for $F$.

**5.13** Find the minimum sum of products for the given expression. Then, make minterm 5 a don't-care term and verify that the minimum sum of products is unchanged. Now, start again with the original expression and find each minterm which could *individually* be made a don't-care without changing the minimum sum of products.

$$F(A, B, C, D) = A'C' + B'C + ACD' + BC'D$$

**5.14** Find the minimum sum-of-products expressions for each of these functions.
(a) $f_1(A, B, C) = m_1 + m_2 + m_5 + m_7$
(b) $f_2(d, e, f) = \Sigma\, m(1, 5, 6, 7)$
(c) $f_3(r, s, t) = rs' + r's' + st'$
(d) $f_4(a, b, c) = m_0 + m_2 + m_3 + m_7$
(e) $f_5(n, p, q) = \Sigma\, m(1, 3, 4, 5)$
(f) $f_6(x, y, z) = M_1 M_7$

**5.15** Find the minimum product-of-sums expression for each of the functions in Problem 5.14.

**5.16** Find the minimum sum of products for each of these functions.
(a) $f_1(A, B, C) = m_1 + m_3 + m_4 + m_6$
(b) $f_2(d, e, f) = \Sigma\, m(1, 4, 5, 7)$
(c) $f_3(r, s, t) = r't' + rs' + rs$
(d) $f_1(a, b, c) = m_3 + m_4 + m_6 + m_7$
(e) $f_2(n, p, q) = \Sigma\, m(2, 3, 5, 7)$
(f) $f_4(x, y, z) = M_3 M_6$

**5.17** (a) Plot the following function on a Karnaugh map. (Do not expand to minterm form before plotting.)

$$F(A,B,C,D) = A'B' + CD' + ABC + A'B'CD' + ABCD'$$

(b) Find the minimum sum of products.
(c) Find the minimum product of sums.

**5.18** Work Problem 5.17 for the following:

$$f(A,B,C,D) = A'B' + A'B'C' + A'BD' + AC'D + A'BD + AB'CD'$$

**5.19** A switching circuit has two control inputs ($C_1$ and $C_2$), two data inputs ($X_1$ and $X_2$), and one output ($Z$). The circuit performs logic operations on the two data inputs, as shown in this table:

| $C_1$ | $C_2$ | Function Performed by Circuit |
|---|---|---|
| 0 | 0 | $X_1X_2$ |
| 0 | 1 | $X_1 \oplus X_2$ |
| 1 | 0 | $X_1' + X_2$ |
| 1 | 1 | $X_1 \equiv X_2$ |

(a) Derive a truth table for $Z$.
(b) Use a Karnaugh map to find a minimum OR-AND gate circuit to realize $Z$.

**5.20** Use Karnaugh maps to find all possible minimum sum-of-products expressions for each function.
(a) $F(a, b, c) = \Pi\, M(3, 4)$
(b) $g(d, e, f) = \Sigma\, m(1, 4, 6) + \Sigma\, d(0, 2, 7)$
(c) $F(p, q, r) = (p + q' + r)(p' + q + r')$
(d) $F(s, t, u) = \Sigma\, m(1, 2, 3) + \Sigma\, d(0, 5, 7)$
(e) $f(a, b, c) = \Pi\, M(2, 3, 4)$
(f) $G(D, E, F) = \Sigma\, m(1, 6) + \Sigma\, d(0, 3, 5)$

**5.21** Simplify the following expression first by using a map and then by using Boolean algebra. Use the map as a guide to determine which theorems to apply to which terms for the algebraic simplification.

$$F = a'b'c' + a'c'd + bcd + abc + ab'$$

**5.22** Find all prime implicants and all minimum sum-of-products expressions for each of the following functions.
(a) $f(A,B,C,D) = \Sigma\, m(4, 11, 12, 13, 14) + \Sigma\, d(5, 6, 7, 8, 9, 10)$
(b) $f(A,B,C,D) = \Sigma\, m(3, 11, 12, 13, 14) + \Sigma\, d(5, 6, 7, 8, 9, 10)$
(c) $f(A,B,C,D) = \Sigma\, m(1, 2, 4, 13, 14) + \Sigma\, d(5, 6, 7, 8, 9, 10)$
(d) $f(A,B,C,D) = \Sigma\, m(4, 15) + \Sigma\, d(5, 6, 7, 8, 9, 10)$
(e) $f(A,B,C,D) = \Sigma\, m(3, 4, 11, 15) + \Sigma\, d(5, 6, 7, 8, 9, 10)$
(f) $f(A,B,C,D) = \Sigma\, m(4) + \Sigma\, d(5, 6, 7, 8, 9, 10, 11, 12, 13, 14)$
(g) $f(A,B,C,D) = \Sigma\, m(4, 15) + \Sigma\, d(0, 1, 2, 5, 6, 7, 8, 9, 10)$

**5.23** For each function in Problem 5.22, find all minimum product-of-sums expressions.

**5.24** Find the minimum sum-of-products expression for
(a) $\Sigma\, m(0, 2, 3, 5, 6, 7, 11, 12, 13)$
(b) $\Sigma\, m(2, 4, 8) + \Sigma\, d(0, 3, 7)$
(c) $\Sigma\, m(1, 5, 6, 7, 13) + \Sigma\, d(4, 8)$
(d) $f(w, x, y, z) = \Sigma\, m(0, 3, 5, 7, 8, 9, 10, 12, 13) + \Sigma\, d(1, 6, 11, 14)$
(e) $\Pi\, M(0, 1, 2, 5, 7, 9, 11) \bullet \Pi\, D(4, 10, 13)$

**5.25** Work Problem 5.24 for the following:
(a) $f(a, b, c, d) = \Sigma\, m(1, 3, 4, 5, 7, 9, 13, 15)$
(b) $f(a, b, c, d) = \Pi\, M(0, 3, 5, 8, 11)$
(c) $f(a, b, c, d) = \Sigma\, m(0, 2, 6, 9, 13, 14) + \Sigma\, d(3, 8, 10)$
(d) $f(a, b, c, d) = \Pi\, M(0, 2, 6, 7, 9, 12, 13) \bullet \Pi\, D(1, 3, 5)$

**5.26** Find the minimum product of sums for the following. Underline the essential prime implicants in your answer.
(a) $\Pi\, M(0, 2, 4, 5, 6, 9, 14) \bullet \Pi\, D(10, 11)$
(b) $\Sigma\, m(1, 3, 8, 9, 15) + \Sigma\, d(6, 7, 12)$

**5.27** Find a minimum sum-of-products and a minimum product-of-sums expression for each function:
(a) $f(A, B, C, D) = \Pi\, M(0, 2, 10, 11, 12, 14, 15) \bullet \Pi\, D(5, 7)$
(b) $f(w, x, y, z) = \Sigma\, m(0, 3, 5, 7, 8, 9, 10, 12, 13) + \Sigma\, d(1, 6, 11, 14)$

**5.28** A logic *circuit* realizes the function $F(a, b, c, d) = a'b' + a'cd + ac'd + ab'd'$. Assuming that $a = c$ never occurs when $b = d = 1$, find a simplified expression for $F$.

**5.29** Given $F = AB'D' + A'B + A'C + CD$.
(a) Use a Karnaugh map to find the maxterm expression for $F$ (express your answer in both decimal and algebric notation).

(b) Use a Karnaugh map to find the minimum sum-of-products form for $F'$.
(c) Find the minimum product of sums for $F$.

**5.30** Assuming that the inputs $ABCD = 0101$, $ABCD = 1001$, $ABCD = 1011$ never occur, find a simplified expression for

$$F = A'BC'D + A'B'D + A'CD + ABD + ABC$$

**5.31** Find all of the prime implicants for each of the functions plotted on page 150.

**5.32** Find all of the prime implicants for each of the plotted functions:

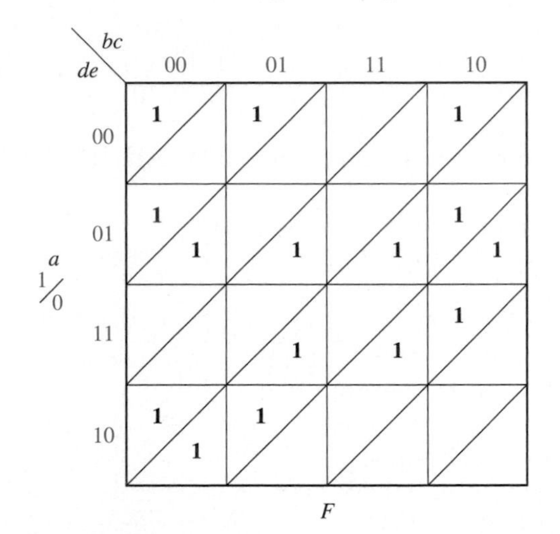


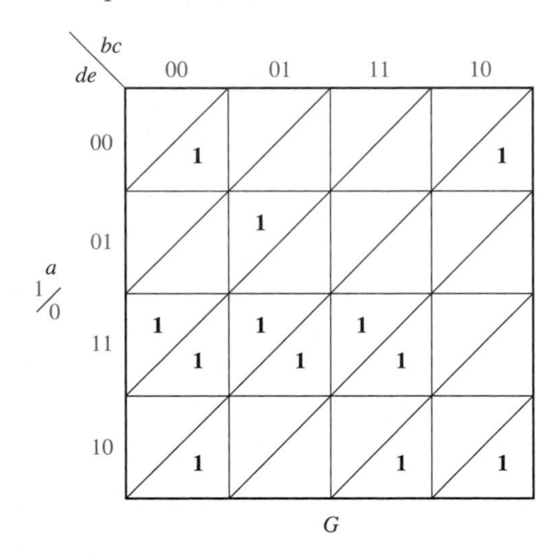


**5.33** Given that $f(a, b, c, d, e) = \Sigma\, m(6, 7, 9, 11, 12, 13, 16, 17, 18, 20, 21, 23, 25, 28)$, using a Karnaugh map,
(a) Find the essential prime implicants (three).
(b) Find the minimum sum of products (7 terms).
(c) Find all of the prime implicants (twelve).

**5.34** A logic circuit realizing the function $f$ has four inputs $a, b, c, d$. The three inputs $a$, $b$, and $c$ are the binary representation of the digits 0 through 7 with $a$ being the most significant bit. The input $d$ is an odd-parity bit; that is, the value of $d$ is such that $a, b, c$, and $d$ always contains an odd number of 1's. (For example, the digit 1 is represented by $abc = 001$ and $d = 0$, and the digit 3 is represented by $abcd = 0111$.) The function $f$ has value 1 if the input digit is a prime number. (A number is prime if it is divisible only by itself and 1; 1 is considered to be prime, and 0 is not.)
(a) Draw a Karnaugh map for $f$.
(b) Find all prime implicants of $f$.
(c) Find all minimum sum of products for $f$.
(d) Find all prime implicants of $f'$.
(e) Find all minimum product of sums for $f$.

**5.35** The decimal digits 0 though 9 are represented using five bits $A, B, C, D$, and $E$. The bits $A, B, C$, and $D$ are the BCD representation of the decimal digit, and bit $E$ is a parity bit that makes the five bits have odd parity. The function $F(A, B, C, D, E)$ has value 1 if the decimal digit represented by $A, B, C, D$, and $E$ is divisible by either 3 or 4. (Zero is divisible by 3 and 4.)

(a) Draw a Karnaugh map for $f$.
(b) Find all prime implicants of $f$. (Prime implicants containing only don't-cares need not be included.)
(c) Find all minimum sum of products for $f$.
(d) Find all prime implicants of $f'$.
(e) Find all minimum product of sums for $f$.

**5.36** Rework Problem 5.35 assuming the decimal digits are represented in excess-3 rather than BCD.

**5.37** The function $F(A, B, C, D, E) = \Sigma\, m(1, 7, 8, 13, 16, 19) + \Sigma\, d(0, 3, 5, 6, 9, 10, 12, 15, 17, 18, 20, 23, 24, 27, 29, 30)$.

(a) Draw a Karnaugh map for $f$.
(b) Find all prime implicants of $f$. (Prime implicants containing only don't-cares need not be included.)
(c) Find all minimum sum of products for $f$.
(d) Find all prime implicants of $f'$.
(e) Find all minimum product of sums for $f$.

**5.38** $F(a, b, c, d, e) = \Sigma\, m(0, 1, 4, 5, 9, 10, 11, 12, 14, 18, 20, 21, 22, 25, 26, 28)$

(a) Find the essential prime implicants using a Karnaugh map, and indicate why each one of the chosen prime implicants is essential (there are four essential prime implicants).
(b) Find all of the prime implicants by using the Karnaugh map (there are 13 in all).

**5.39** Find the minimum sum-of-products expression for $F$. Underline the essential prime implicants in this expression.

(a) $f(a, b, c, d, e) = \Sigma\, m(0, 1, 3, 4, 6, 7, 8, 10, 11, 15, 16, 18, 19, 24, 25, 28, 29, 31) + \Sigma\, d(5, 9, 30)$
(b) $f(a, b, c, d, e) = \Sigma\, m(1, 3, 5, 8, 9, 15, 16, 20, 21, 23, 27, 28, 31)$

**5.40** Work Problem 5.39 with

$$F(A, B, C, D, E) = \Pi\, M(2, 3, 4, 8, 9, 10, 14, 15, 16, 18, 19, 20, 23, 24, 30, 31)$$

**5.41** Find the minimum sum-of-products expression for $F$. Underline the essential prime implicants in your expression.

$$F(A, B, C, D, E) = \Sigma\, m(0, 2, 3, 5, 8, 11, 13, 20, 25, 26, 30) + \Sigma\, d(6, 7, 9, 24)$$

**5.42** $F(V, W, X, Y, Z) = \Pi\, M(0, 3, 5, 6, 7, 8, 11, 13, 14, 15, 18, 20, 22, 24) \bullet \Pi\, D(1, 2, 16, 17)$

(a) Find a minimum sum-of-products expression for $F$. Underline the essential prime implicants.

(b) Find a minimum product-of-sums expression for $F$. Underline the essential prime implicants.

**5.43** Find the minimum product of sums for
(a) $F(a, b, c, d, e) = \Sigma\, m(1, 2, 3, 4, 5, 6, 25, 26, 27, 28, 29, 30, 31)$
(b) $F(a, b, c, d, e) = \Sigma\, m(1, 5, 12, 13, 14, 16, 17, 21, 23, 24, 30, 31) + \Sigma\, d(0, 2, 3, 4)$

**5.44** Find a minimum product-of-sums expression for each of the following functions:
(a) $F(v, w, x, y, z) = \Sigma\, m(4, 5, 8, 9, 12, 13, 18, 20, 21, 22, 25, 28, 30, 31)$
(b) $F(a, b, c, d, e) = \Pi\, M(2, 4, 5, 6, 8, 10, 12, 13, 16, 17, 18, 22, 23, 24) \cdot \Pi\, D(0, 11, 30, 31)$

**5.45** Find the minimum sum of products for each function. Then, make the specified minterm a don't-care and verify that the minimum sum of products is unchanged. Now, start again with the original expression and find each minterm which could individually be made a don't-care, without changing the minimum sum of products.
(a) $F(A, B, C, D) = A'C' + A'B' + ACD' + BC'D$, minterm 2
(b) $F(A, B, C, D) = A'BD + AC'D + AB' + BCD + A'C'D$, minterm 7

**5.46** $F(V, W, X, Y, Z) = \Pi\, M(0, 3, 6, 9, 11, 19, 20, 24, 25, 26, 27, 28, 29, 30) \cdot \Pi\, D(1, 2, 12, 13)$
(a) Find two minimum sum-of-products expressions for $F$.
(b) Underline the essential prime implicants in your answer and tell why each one is essential.

# Quine-McCluskey Method

## Objectives

1. Find the prime implicants of a function by using the Quine-McCluskey method. Explain the reasons for the procedures used.
2. Define *prime implicant* and *essential prime implicant*.
3. Given the prime implicants, find the essential prime implicants and a minimum sum-of-products expression for a function, using a prime implicant chart and using Petrick's method.
4. Minimize an incompletely specified function, using the Quine-McCluskey method.
5. Find a minimum sum-of-products expression for a function, using the method of map-entered variables.

# Study Guide

1. Review Section 5.1, *Minimum Forms of Switching Functions*.

2. Read the introduction to this unit and, then, study Section 6.1. *Determination of Prime Implicants*.

   (a) Using variables $A, B, C, D$, and $E$, give the algebraic equivalent of
   10110 + 10010 = 10–10
   10–10 + 10–11 = 10–1–

   (b) Why will the following pairs of terms not combine?
   01101 + 00111
   10–10 + 001–0

   (c) When using the Quine-McCluskey method for finding prime implicants, why is it necessary to compare terms only from adjacent groups?

   (d) How can you determine if two minterms from adjacent groups will combine by looking at their decimal representations?

   (e) When combining terms, why is it permissible to use a term which has already been checked off?

   (f) In forming Column II of Table 6-1, note that terms 10 and 14 were combined to form 10, 14 even though both 10 and 14 had already been checked off. If this had not been done, which term in Column II could not be eliminated (checked off)?

   (g) In forming Column III of Table 6-1, note that minterms 0, 1, 8, and 9 were combined in two different ways to form –00–. This is equivalent to looping the minterms in two different ways on the Karnaugh map, as shown.

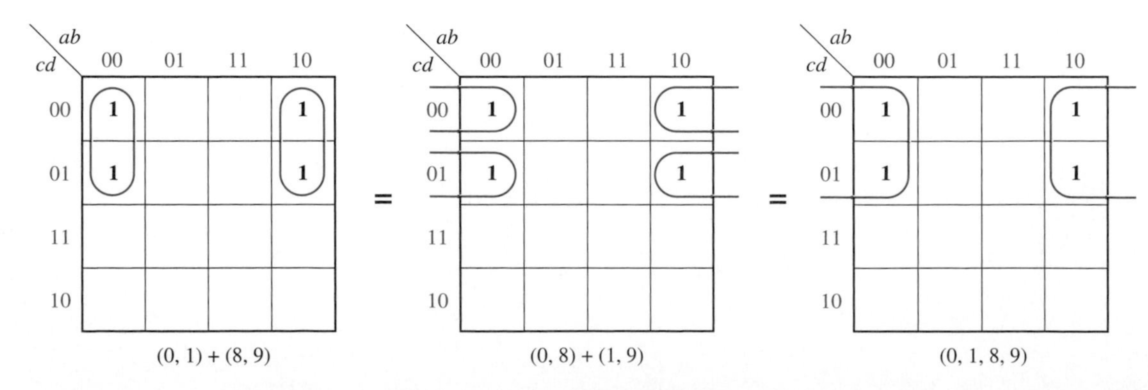

(h) Using a map, find *all* of the prime implicants of Equation (6-2) and compare your answer with Equation (6-3).

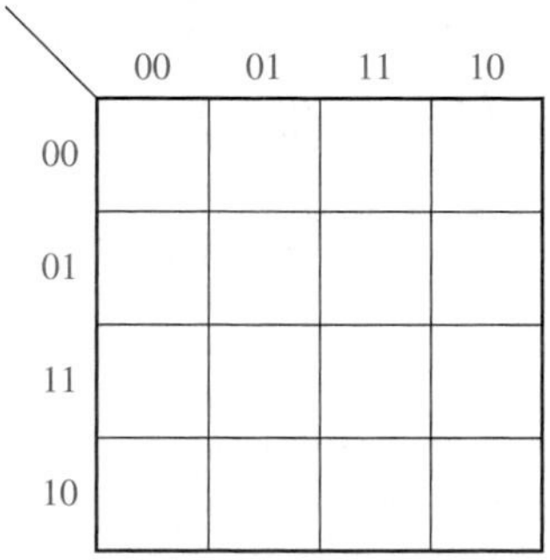

| | 00 | 01 | 11 | 10 |
|---|---|---|---|---|
| 00 | | | | |
| 01 | | | | |
| 11 | | | | |
| 10 | | | | |

(i) The prime implicants of $f(a, b, c, d) = \Sigma\, m(4, 5, 6, 7, 12, 13, 14, 15)$ are to be found using the Quine-McCluskey method. Column III is given; find Column IV and check off the appropriate terms in Column III.

| | Column III | Column IV |
|---|---|---|
| (4, 5, 6, 7) | 01-- | |
| (4, 5, 12, 13) | –10– | |
| (4, 6, 12, 14) | –1–0 | |
| (5, 7, 13, 15) | –1–1 | |
| (6, 7, 14,15) | –11– | |
| (12, 13, 14, 15) | 11-- | |

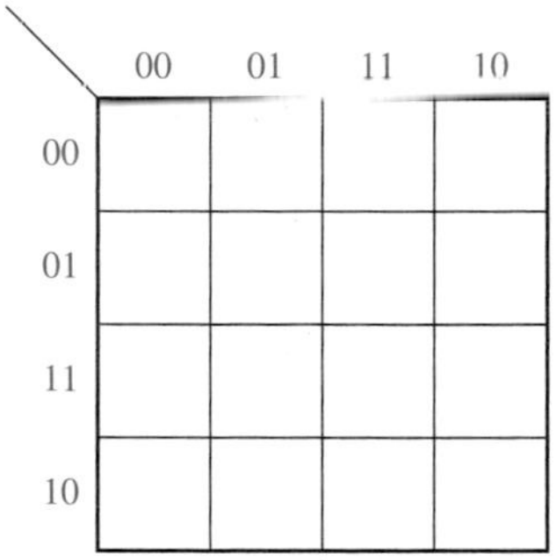

| | 00 | 01 | 11 | 10 |
|---|---|---|---|---|
| 00 | | | | |
| 01 | | | | |
| 11 | | | | |
| 10 | | | | |

Check your answer using a Karnaugh map.

**3.** (a) List all seven product term implicants of $F(a, b, c) = \Sigma\, m(0, 1, 5, 7)$

Which of these implicants are prime?

Why is $a'c$ not an implicant?

(b) Define a prime implicant.

(c) Why must every term in a minimum sum-of-products expression be a prime implicant?

(d) Given that $F(A, B, C, D) = \Sigma\, m(0, 1, 4, 5, 7, 10, 15)$, which of the following terms are *not* prime implicants and why?

$A'B'C'$ $\quad$ $A'C'$ $\quad$ $BCD$ $\quad$ $ABC$ $\quad$ $AB'CD'$

**4.** Study Section 6.2, *The Prime Implicant Chart.*

(a) Define an *essential* prime implicant.

(b) Find all of the essential prime implicants from the following chart.

| | a b c d | 0 | 4 | 5 | 10 | 11 | 12 | 13 | 15 |
|---|---|---|---|---|---|---|---|---|---|
| (0, 4) | 0 – 0 0 | × | × | | | | | | |
| (4, 5, 12, 13) | – 1 0 – | | × | × | | | × | × | |
| (13, 15) | 1 1 – 1 | | | | | | | × | × |
| (11, 15) | 1 – 1 1 | | | | | × | | | × |
| (10, 11) | 1 0 1 – | | | | × | × | | | |

Check your answer using a Karnaugh map.

(c) Why must all essential prime implicants of a function be included in the minimum sum of products?

(d) Complete the solution of Table 6-5.
(e) Work Programmed Exercise 6.1.
(f) Work Problems 6.2 and 6.3.

**5.** Study Section 6.3, *Petrick's Method* (optional).

(a) Consider the following reduced prime implicant chart for a function $F$:

| | | $m_4$ | $m_5$ | $m_7$ | $m_{13}$ |
|---|---|---|---|---|---|
| $P_1$ | $bd$ | | × | × | × |
| $P_2$ | $bc'$ | × | × | | × |
| $P_3$ | $a'b$ | × | × | × | |
| $P_4$ | $c'd$ | | × | | × |

We will find all minimum solutions using Petrick's method. Let $P_i = 1$ mean the prime implicant in row $P_i$ is included in the solution.
Which minterm is covered iff $(P_1 + P_3) = 1$?__________
Write a sum term which is 1 iff $m_4$ is covered.__________

Write a product-of-sum terms which is 1 iff all $m_4$, $m_5$, $m_7$, and $m_{13}$ are all covered:

$P =$ ______________________________

(b) Reduce $P$ to a minimum sum of products. (Your answer should have four terms, each one of the form $P_iP_j$.)

$P =$ ______________________________

If $P_1P_2 = 1$, which prime implicants are included in the solution? __________

How many minimum solutions are there? __________

Write out each solution in terms of $a$, $b$, $c$, and $d$.

(1) $F =$ (2) $F =$

(3) $F =$ (4) $F =$

**6.** Study Section 6.4, *Simplification of Incompletely Specified Functions*.

(a) Why are don't-care terms treated like required minterms when finding the prime implicants?

(b) Why are the don't-care terms not listed at the top of the prime implicant chart when finding the minimum solution?

(c) Work Problem 6.4.

(d) Work Problem 6.5, and check your solution using a Karnaugh map.

**7.** If you have *LogicAid* or a similar computer program available, use it to check your answers to some of the problems in this unit. *LogicAid* accepts Boolean functions in the form of equations, minterms or maxterms, and truth tables. It finds simplified sum-of-products and product-of-sums expressions for the functions using a modified version of the Quine-McCluskey method or Espresso-II. It can also find one or all of the minimum solutions using Petrick's method.

**8.** Study Section 6.5, *Simplification Using Map-Entered Variables*.

(a) For the following map, find $MS_0$, $MS_1$, and $F$. Verify that your solution for $F$ is minimum by using a four-variable map.

| $BC$ \ $A$ | 0 | 1 |
|---|---|---|
| 00 | $D$ | 1 |
| 01 | | |
| 11 | 1 | $D$ |
| 10 | 1 | X |

(b) Use the method of map-entered variables to find an expression for $F$ from the following map. Treat $C$ and $C'$ as if they were independent variables. Is the result a correct representation of $F$? Is it minimum?

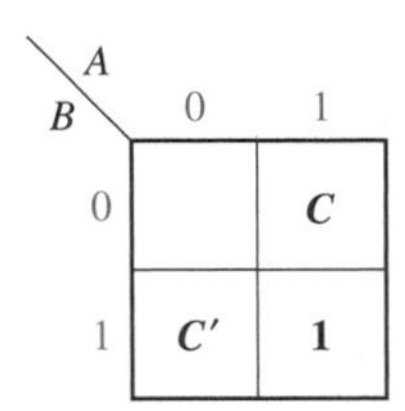


(c) Work Problem 6.6.

**9.** In this unit you have learned a "turn-the-crank" type procedure for finding minimum sum-of-products forms for switching functions. In addition to learning how to "turn the crank" and grind out minimum solutions, you should have learned several very important concepts in this unit. In particular, make sure you know:

(a) What a prime implicant is
(b) What an essential prime implicant is
(c) Why the minimum sum-of-products form is a sum of prime implicants
(d) How don't-cares are handled when using the Quine-McCluskey method and the prime implicant chart

**10.** Reread the objectives of the unit. If you are satisfied that you can meet the objectives, take the readiness test.

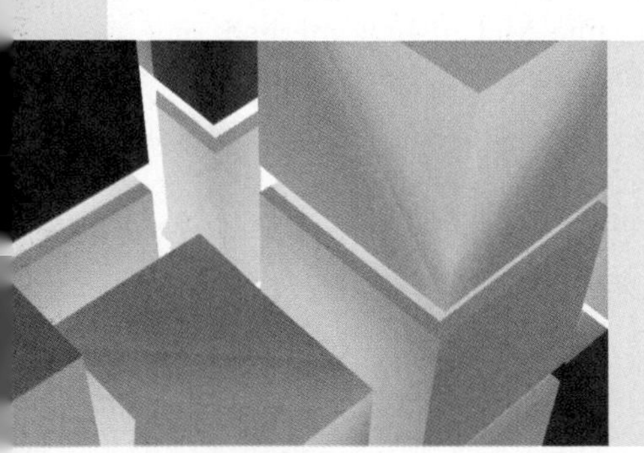

# Quine-McCluskey Method

The Karnaugh map method described in Unit 5 is an effective way to simplify switching functions which have a small number of variables. When the number of variables is large or if several functions must be simplified, the use of a digital computer is desirable. The Quine-McCluskey method presented in this unit provides a systematic simplification procedure which can be readily programmed for a digital computer.

The Quine-McCluskey method reduces the minterm expansion (standard sum-of-products form) of a function to obtain a minimum sum of products. The procedure consists of two main steps:

1. Eliminate as many literals as possible from each term by systematically applying the theorem $XY + XY' = X$. The resulting terms are called prime implicants.
2. Use a prime implicant chart to select a minimum set of prime implicants which, when ORed together, are equal to the function being simplified and which contain a minimum number of literals.

# 6.1 Determination of Prime Implicants

In order to apply the Quine-McCluskey method to determine a minimum sum-of-products expression for a function, the function must be given as a sum of minterms. (If the function is not in minterm form, the minterm expansion can be found by using one of the techniques given in Section 5.3.) In the first part of the Quine-McCluskey method, all of the prime implicants of a function are systematically formed by combining minterms. The minterms are represented in binary notation and combined using

$$XY + XY' = X \tag{6-1}$$

where $X$ represents a product of literals and $Y$ is a single variable. Two minterms will combine if they differ in exactly one variable. The examples given below show both the binary notation and its algebraic equivalent.

$$AB'CD' + AB'CD = AB'C$$

$$\underbrace{1\ 0\ 1}_{X}\ \underbrace{0}_{Y} + \underbrace{1\ 0\ 1}_{X}\ \underbrace{1}_{Y'} = \underbrace{1\ 0\ 1}_{X}\ - \quad \text{(the dash indicates a missing variable)}$$

$$A'BC'D + A'BCD' \quad \text{(will not combine)}$$

$$0\ 1\ 0\ 1 + 0\ 1\ 1\ 0 \quad \text{(will not combine)}$$

In order to find all of the prime implicants, all possible pairs of minterms should be compared and combined whenever possible. To reduce the required number of comparisons, the binary minterms are sorted into groups according to the number of 1's in each term. Thus,

$$f(a, b, c, d) = \Sigma\, m(0, 1, 2, 5, 6, 7, 8, 9, 10, 14) \tag{6-2}$$

is represented by the following list of minterms:

| | | |
|---|---|---|
| group 0 | 0 | 0000 |
| group 1 | 1 | 0001 |
| | 2 | 0010 |
| | 8 | 1000 |
| group 2 | 5 | 0101 |
| | 6 | 0110 |
| | 9 | 1001 |
| | 10 | 1010 |
| group 3 | 7 | 0111 |
| | 14 | 1110 |

In this list, the term in group 0 has zero 1's, the terms in group 1 have one 1, those in group 2 have two 1's, and those in group 3 have three 1's.

Two terms can be combined if they differ in exactly one variable. Comparison of terms in nonadjacent groups is unnecessary because such terms will always differ in at least two variables and cannot be combined using $XY + XY' = X$. Similarly, the comparison of terms within a group is unnecessary because two terms with the same number of 1's must differ in at least two variables. Thus, only terms in adjacent groups must be compared.

First, we will compare the term in group 0 with all of the terms in group 1. Terms 0000 and 0001 can be combined to eliminate the fourth variable, which yields 000–. Similarly, 0 and 2 combine to form 00–0 ($a'b'd'$), and 0 and 8 combine to form –000 ($b'c'd'$). The resulting terms are listed in Column II of Table 6-1.

Whenever two terms combine, the corresponding decimal numbers differ by a power of 2 (1, 2, 4, 8, etc.). This is true because when the binary representations differ in exactly one column and if we subtract these binary representations, we

**TABLE 6-1** Determination of Prime Implicants

| | Column I | | Column II | | Column III | |
|---|---|---|---|---|---|---|
| group 0 | 0 | 0000 ✓ | 0, 1 | 000– ✓ | 0, 1, 8, 9 | –00– |
| group 1 | 1 | 0001 ✓ | 0, 2 | 00–0 ✓ | 0, 2, 8, 10 | –0–0 |
| | 2 | 0010 ✓ | 0, 8 | –000 ✓ | ~~0, 8, 1, 9~~ | ~~–00–~~ |
| | 8 | 1000 ✓ | 1, 5 | 0–01 | ~~0, 8, 2, 10~~ | ~~–0–0~~ |
| group 2 | 5 | 0101 ✓ | 1, 9 | –001 ✓ | 2, 6, 10, 14 | --10 |
| | 6 | 0110 ✓ | 2, 6 | 0–10 ✓ | ~~2, 10, 6, 14~~ | ~~--10~~ |
| | 9 | 1001 ✓ | 2, 10 | –010 ✓ | | |
| | 10 | 1010 ✓ | 8, 9 | 100– ✓ | | |
| group 3 | 7 | 0111 ✓ | 8, 10 | 10–0 ✓ | | |
| | 14 | 1110 ✓ | 5, 7 | 01–1 | | |
| | | | 6, 7 | 011– | | |
| | | | 6, 14 | –110 ✓ | | |
| | | | 10, 14 | 1–10 ✓ | | |

get a 1 only in the column in which the difference exists. A binary number with a 1 in exactly one column is a power of 2.

Because the comparison of group 0 with groups 2 and 3 is unnecessary, we proceed to compare terms in groups 1 and 2. Comparing term 1 with all terms in group 2, we find that it combines with 5 and 9 but not with 6 or 10. Similarly, term 2 combines only with 6 and 10, and term 8 only with 9 and 10. The resulting terms are listed in Column II. Each time a term is combined with another term, it is checked off. A term may be used more than once because $X + X = X$. Even though two terms have already been combined with other terms, they still must be compared and combined if possible. This is necessary because the resultant term may be needed to form the minimum sum solution. At this stage, we may generate redundant terms, but these redundant terms will be eliminated later. We finish with Column I by comparing terms in groups 2 and 3. New terms are formed by combining terms 5 and 7, 6 and 7, 6 and 14, and 10 and 14.

Note that the terms in Column II have been divided into groups, according to the number of 1's in each term. Again, we apply $XY + XY' = X$ to combine pairs of terms in Column II. In order to combine two terms, the terms must have the same variables, and the terms must differ in exactly one of these variables. Thus, it is necessary only to compare terms which have dashes (missing variables) in corresponding places and which differ by exactly one in the number of 1's.

Terms in the first group in Column II need only be compared with terms in the second group which have dashes in the same places. Term 000– (0, 1) combines only with term 100– (8, 9) to yield –00–. This is algebraically equivalent to $a'b'c' + ab'c' = b'c'$. The resulting term is listed in Column III along with the designation 0, 1, 8, 9 to indicate that it was formed by combining minterms 0, 1, 8, and 9. Term (0, 2) combines only with (8, 10), and term (0, 8) combines with both (1, 9) and (2, 10). Again, the terms which have been combined are checked off. Comparing terms from the second and third groups in Column II, we find that (2,6) combines with (10, 14), and (2, 10) combines with (6,14).

Note that there are three pairs of duplicate terms in Column III. These duplicate terms were formed in each case by combining the same set of four minterms in a different order. After deleting the duplicate terms, we compare terms from the two groups in Column III. Because no further combination is possible, the process terminates. In general, we would keep comparing terms and forming new groups of terms and new columns until no more terms could be combined.

The terms which have not been checked off because they cannot be combined with other terms are called prime implicants. Because every minterm has been included in at least one of the prime implicants, the function is equal to the sum of its prime implicants. In this example we have

$$\begin{array}{cccccc} f = a'c'd + & a'bd + & a'bc + & b'c' + & b'd' + & cd' \\ (1,5) & (5,7) & (6,7) & (0,1,8,9) & (0,2,8,10) & (2,6,10,14) \end{array} \tag{6-3}$$

In this expression, each term has a minimum number of literals, but the number of terms is not minimum. Using the consensus theorem to eliminate redundant terms yields

$$f = a'bd + b'c' + cd' \tag{6-4}$$

which is the minimum sum-of-products expression for $f$. Section 6.2 discusses a better method of eliminating redundant prime implicants using a prime implicant chart.

Next, we will define implicant and prime implicant and relate these terms to the Quine-McCluskey method.

---

**Definition** Given a function $F$ of $n$ variables, a product term $P$ is an *implicant* of $F$ iff for every combination of values of the $n$ variables for which $P = 1$, $F$ is also equal to 1.

---

In other words, if for some combination of values of the variables, $P = 1$ and $F = 0$, then $P$ is *not* an implicant of $F$. For example, consider the function

$$F(a, b, c) = a'b'c' + ab'c' + ab'c + abc = b'c' + ac \quad (6\text{-}5)$$

*If* $a'b'c' = 1$, then $F = 1$; if $ac = 1$, then $F = 1$; etc. Hence, the terms $a'b'c'$, $ac$, etc., are implicants of $F$. In this example, $bc$ is *not* an implicant of $F$ because when $a = 0$ and $b = c = 1$, $bc = 1$ and $F = 0$. In general, if $F$ is written in sum-of-products form, every product term is an implicant. Every minterm of $F$ is also an implicant of $F$, and so is any term formed by combining two or more minterms. For example, in Table 6-1, all of the terms listed in any of the columns are implicants of the function given in Equation (6-2).

---

**Definition** A *prime implicant* of a function $F$ is a product term implicant which is no longer an implicant if any literal is deleted from it.

---

In Equation (6-5), the implicant $a'b'c'$ *is not* a *prime* implicant because $a'$ can be eliminated, and the resulting term ($b'c'$) is still an implicant of $F$. The implicants $b'c'$ and $ac$ are *prime implicants* because *if* we delete a literal from either term, the term will no longer be an implicant of $F$. Each prime implicant of a function has a minimum number of literals in the sense that no more literals can be eliminated from it by combining it with other terms.

The Quine-McCluskey method, as previously illustrated, finds all of the product term implicants of a function. The implicants which are nonprime are checked off in the process of combining terms so that the remaining terms are prime implicants.

A minimum sum-of-products expression for a function consists of a sum of some (but not necessarily all) of the prime implicants of that function. In other words, a sum-of-products expression which contains a term which is not a prime implicant cannot be minimum. This is true because the nonprime term does not contain a minimum number of literals—it can be combined with additional minterms to form a prime implicant which has fewer literals than the nonprime term. Any nonprime term in a sum-of-products expression can thus be replaced with a prime implicant, which reduces the number of literals and simplifies the expression.

## 6.2 The Prime Implicant Chart

The second part of the Quine-McCluskey method employs a prime implicant chart to select a minimum set of prime implicants. The minterms of the function are listed across the top of the chart, and the prime implicants are listed down the side. A prime implicant is equal to a sum of minterms, and the prime implicant is said to cover these minterms. If a prime implicant covers a given minterm, an X is placed at the intersection of the corresponding row and column. Table 6-2 shows the prime implicant chart derived from Table 6-1. All of the prime implicants (terms which have not been checked off in Table 6-1) are listed on the left.

In the first row, X's are placed in columns 0, 1, 8, and 9, because prime implicant $b'c'$ was formed from the sum of minterms 0, 1, 8, and 9. Similarly, X's are placed in columns 0, 2, 8, and 10 opposite the prime implicant $b'd'$ and so forth.

**TABLE 6-2** Prime Implicant Chart

| | | 0 | 1 | 2 | 5 | 6 | 7 | 8 | 9 | 10 | 14 |
|---|---|---|---|---|---|---|---|---|---|---|---|
| (0, 1, 8, 9) | $b'c'$ | × | × | | | | | × | ⊗ | | |
| (0, 2, 8, 10) | $b'd'$ | × | | × | | | | × | | × | |
| (2, 6, 10, 14) | $cd'$ | | | × | | × | | | | × | ⊗ |
| (1, 5) | $a'c'd$ | | × | | × | | | | | | |
| (5, 7) | $a'bd$ | | | | × | | × | | | | |
| (6, 7) | $a'bc$ | | | | | × | × | | | | |

If a minterm is covered by only one prime implicant, then that prime implicant is called an *essential* prime implicant and must be included in the minimum sum of products. Essential prime implicants are easy to find using the prime implicant chart. If a given column contains only one X, then the corresponding row is an essential prime implicant. In Table 6-2, columns 9 and 14 each contain one X, so prime implicants $b'c'$ and $cd'$ are essential.

Each time a prime implicant is selected for inclusion in the minimum sum, the corresponding row should be crossed out. After doing this, the columns which correspond to all minterms covered by that prime implicant should also be crossed out. Table 6-3 shows the resulting chart when the essential prime implicants and the corresponding rows and columns of Table 6-2 are crossed out. A minimum set of prime implicants must now be chosen to cover the remaining columns. In this example, $a'bd$ covers the remaining two columns, so it is chosen. The resulting minimum sum of products is

$$f = b'c' + cd' + a'bd$$

which is the same as Equation (6-4). Note that even though the term $a'bd$ is included in the minimum sum of products, $a'bd$ is *not* an *essential* prime implicant. It is the sum of minterms $m_5$ and $m_7$; $m_5$ is also covered by $a'c'd$, and $m_7$ is also covered by $a'bc$.

**TABLE 6-3**

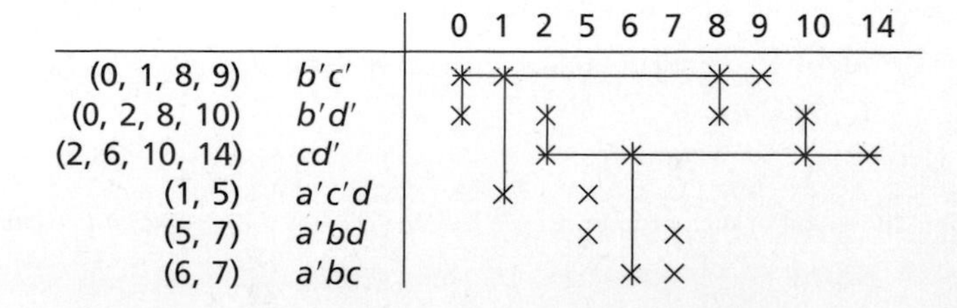

| | | 0 | 1 | 2 | 5 | 6 | 7 | 8 | 9 | 10 | 14 |
|---|---|---|---|---|---|---|---|---|---|---|---|
| (0, 1, 8, 9) | $b'c'$ | × | × | | | | | × | × | | |
| (0, 2, 8, 10) | $b'd'$ | × | | × | | | | × | | × | |
| (2, 6, 10, 14) | $cd'$ | | | × | | × | | | | × | × |
| (1, 5) | $a'c'd$ | | × | | × | | | | | | |
| (5, 7) | $a'bd$ | | | | × | | × | | | | |
| (6, 7) | $a'bc$ | | | | | × | × | | | | |

When the prime implicant chart is constructed, some minterms may be covered by only a single prime implicant, although other minterms may be covered by two or more prime implicants. A prime implicant is *essential* (or necessary) to a function $f$ iff the prime implicant contains a minterm which is not covered by any other prime implicant of $f$. The essential prime implicants are chosen first because all essential prime implicants must be included in every minimum sum. After the essential prime implicants have been chosen, the minterms which they cover can be eliminated from the prime implicant chart by crossing out the corresponding columns. If the essential prime implicants do not cover all of the minterms, then additional nonessential prime implicants are needed. In simple cases, the nonessential prime implicants needed to form the minimum solution may be selected by trial and error. For larger prime implicant charts, additional procedures for chart reduction can be employed.[1] Some functions have two or more minimum sum-of-products expressions, each having the same number of terms and literals. The next example shows such a function.

**Example**

A prime implicant chart which has two or more X's in every column is called a *cyclic* prime implicant chart. The following function has such a chart:

$$F = \Sigma\, m(0, 1, 2, 5, 6, 7) \tag{6-6}$$

Derivation of prime implicants:

| | | | | |
|---|---|---|---|---|
| 0 | 000 ✓ | | 0, 1 | 00– |
| 1 | 001 ✓ | | 0, 2 | 0–0 |
| 2 | 010 ✓ | | 1, 5 | –01 |
| 5 | 101 ✓ | | 2, 6 | –10 |
| 6 | 110 ✓ | | 5, 7 | 1–1 |
| 7 | 111 ✓ | | 6, 7 | 11– |

Table 6-4 shows the resulting prime implicant chart. All columns have two X's, so we will proceed by trial and error. Both (0, 1) and (0, 2) cover column 0, so we will try (0, 1). After crossing out row (0, 1) and columns 0 and 1, we examine column 2, which is covered by (0, 2) and (2, 6). The best choice is (2, 6) because it covers two of the remaining columns while (0, 2) covers only one of the remaining columns. After crossing out row (2, 6) and columns 2 and 6, we see that (5, 7) covers the remaining columns and completes the solution. Therefore, one solution is $F = a'b' + bc' + ac$.

TABLE 6-4

| | | | 0 | 1 | 2 | 5 | 6 | 7 |
|---|---|---|---|---|---|---|---|---|
| ① → | (0, 1) | $a'b'$ | X | X | | | | |
| | (0, 2) | $a'c'$ | X | | X | | | |
| | (1, 5) | $b'c$ | | X | | X | | |
| ② → | (2, 6) | $bc'$ | | | X | | X | |
| ③ → | (5, 7) | $ac$ | | | | X | | X |
| | (6, 7) | $ab$ | | | | | X | X |

[1]For a discussion of such procedures, see E. J. McCluskey, *Logic Design Principles.* (Prentice-Hall, 1986).

However, we are not guaranteed that this solution is minimum. We must go back and solve the problem over again starting with the other prime implicant that covers column 0. The resulting table (Table 6-5) is

TABLE 6-5

| | | | 0 | 1 | 2 | 5 | 6 | 7 |
|---|---|---|---|---|---|---|---|---|
| $P_1$ | (0, 1) | $a'b'$ | X | X | | | | |
| $P_2$ | (0, 2) | $a'c'$ | X | | X | | | |
| $P_3$ | (1, 5) | $b'c$ | | X | | X | | |
| $P_4$ | (2, 6) | $bc'$ | | | X | | X | |
| $P_5$ | (5, 7) | $ac$ | | | | X | | X |
| $P_6$ | (6, 7) | $ab$ | | | | | X | X |

Finish the solution and show that $F = a'c' + b'c + ab$. Because this has the same number of terms and same number of literals as the expression for $F$ derived in Table 6-4, there are two minimum sum-of-products solutions to this problem. Compare these two minimum solutions for Equation (6-6) with the solutions obtained in Figure 5-9 using Karnaugh maps. Note that each minterm on the map can be covered by two different loops. Similarly, each column of the prime implicant chart (Table 6-4) has two X's, indicating that each minterm can be covered by two different prime implicants.

## 6.3 Petrick's Method

Petrick's method is a technique for determining all minimum sum-of-products solutions from a prime implicant chart. The example shown in Tables 6-4 and 6-5 has two minimum solutions. As the number of variables increases, the number of prime implicants and the complexity of the prime implicant chart may increase significantly. In such cases, a large amount of trial and error may be required to find the minimum solution(s). Petrick's method is a more systematic way of finding all minimum solutions from a prime implicant chart than the method used previously. Before applying Petrick's method, all essential prime implicants and the minterms they cover should be removed from the chart.

We will illustrate Petrick's method using Table 6-5. First, we will label the rows of the table $P_1$, $P_2$, $P_3$, etc. We will form a logic function, $P$, which is true when all of the minterms in the chart have been covered. Let $P_1$ be a logic variable which is true when the prime implicant in row $P_1$ is included in the solution, $P_2$ be a logic variable which is true when the prime implicant in row $P_2$ is included in the solution, etc. Because column 0 has X's in rows $P_1$ and $P_2$, we must choose row $P_1$ or $P_2$ in order to cover minterm 0. Therefore, the expression $(P_1 + P_2)$ must be true. In order to cover minterm 1, we must choose row $P_1$ or $P_3$; therefore, $(P_1 + P_3)$ must be true. In order to cover minterm 2, $(P_2 + P_4)$ must be true. Similarly, in order to cover minterms 5, 6, and 7, the expressions $(P_3 + P_5)$, $(P_4 + P_6)$ and $(P_5 + P_6)$ must be true. Because we must cover all of the minterms, the following function must be true:

$$P = (P_1 + P_2)(P_1 + P_3)(P_2 + P_4)(P_3 + P_5)(P_4 + P_6)(P_5 + P_6) = 1$$

The expression for $P$ in effect means that we must choose row $P_1$ or $P_2$, *and* row $P_1$ or $P_3$, *and* row $P_2$ or $P_4$, etc.

The next step is to reduce $P$ to a minimum sum of products. This is easy because there are no complements. First, we multiply out, using $(X + Y)(X + Z) = X + YZ$ and the ordinary distributive law:

$$\begin{aligned} P &= (P_1 + P_2P_3)(P_4 + P_2P_6)(P_5 + P_3P_6) \\ &= (P_1P_4 + P_1P_2P_6 + P_2P_3P_4 + P_2P_3P_6)(P_5 + P_3P_6) \\ &= P_1P_4P_5 + P_1P_2P_5P_6 + P_2P_3P_4P_5 + P_2P_3P_5P_6 + P_1P_3P_4P_6 \\ &\quad + P_1P_2P_3P_6 + P_2P_3P_4P_6 + P_2P_3P_6 \end{aligned}$$

Next, we use $X + XY = X$ to eliminate redundant terms from $P$, which yields

$$P = P_1P_4P_5 + P_1P_2P_5P_6 + P_2P_3P_4P_5 + P_1P_3P_4P_6 + P_2P_3P_6$$

Because $P$ must be true $(P = 1)$ in order to cover all of the minterms, we can translate the equation back into words as follows. In order to cover all of the minterms, we must choose rows $P_1$ and $P_4$ and $P_5$, or rows $P_1$ and $P_2$ and $P_5$ and $P_6$, or . . . or rows $P_2$ and $P_3$ and $P_6$. Although there are five possible solutions, only two of these have the minimum number of rows. Thus, the two solutions with the minimum number of prime implicants are obtained by choosing rows $P_1$, $P_4$, and $P_5$ or rows $P_2$, $P_3$, and $P_6$. The first choice leads to $F = a'b' + bc' + ac$, and the second choice to $F = a'c' + b'c + ab$, which are the two minimum solutions derived in Section 6.2.

In summary, Petrick's method is as follows:

1. Reduce the prime implicant chart by eliminating the essential prime implicant rows and the corresponding columns.
2. Label the rows of the reduced prime implicant chart $P_1$, $P_2$, $P_3$, etc.
3. Form a logic function $P$ which is true when all columns are covered. $P$ consists of a product of sum terms, each sum term having the form $(P_{i0} + P_{i1} + \dots)$, where $P_{i0}, P_{i1} \dots$ represent the rows which cover column $i$.
4. Reduce $P$ to a minimum sum of products by multiplying out and applying $X + XY = X$.
5. Each term in the result represents a solution, that is, a set of rows which covers all of the minterms in the table. To determine the minimum solutions (as defined in Section 5.1), find those terms which contain a minimum number of variables. Each of these terms represents a solution with a minimum number of prime implicants.
6. For each of the terms found in step 5, count the number of literals in each prime implicant and find the total number of literals. Choose the term or terms which correspond to the minimum total number of literals, and write out the corresponding sums of prime implicants.

The application of Petrick's method is very tedious for large charts, but it is easy to implement on a computer.

## 6.4 Simplification of Incompletely Specified Functions

Given an incompletely specified function, the proper assignment of values to the don't-care terms is necessary in order to obtain a minimum form for the function. In this section, we will show how to modify the Quine-McCluskey method in order to obtain a minimum solution when don't-care terms are present. In the process of finding the prime implicants, we will treat the don't-care terms as if they were required minterms. In this way, they can be combined with other minterms to eliminate as many literals as possible. If extra prime implicants are generated because of the don't-cares, this is correct because the extra prime implicants will be eliminated in the next step anyway. When forming the prime implicant chart, the don't-cares are *not* listed at the top. This way, when the prime implicant chart is solved, all of the required minterms will be covered by one of the selected prime implicants. However, the don't-care terms are not included in the final solution unless they have been used in the process of forming one of the selected prime implicants. The following example of simplifying an incompletely specified function should clarify the procedure.

$$F(A, B, C, D) = \Sigma\, m(2, 3, 7, 9, 11, 13) + \Sigma\, d(1, 10, 15)$$

(the terms following $d$ are don't-care terms)

The don't-care terms are treated like required minterms when finding the prime implicants:

| | | | | | |
|---|---|---|---|---|---|
| 1 | 0001 ✓ | (1, 3) | 00–1 ✓ | (1, 3, 9, 11) | –0–1 |
| 2 | 0010 ✓ | (1, 9) | –001 ✓ | (2, 3, 10, 11) | –01– |
| 3 | 0011 ✓ | (2, 3) | 001– ✓ | (3, 7, 11, 15) | --11 |
| 9 | 1001 ✓ | (2, 10) | –010 ✓ | (9, 11, 13, 15) | 1--1 |
| 10 | 1010 ✓ | (3, 7) | 0–11 ✓ | | |
| 7 | 0111 ✓ | (3, 11) | –011 ✓ | | |
| 11 | 1011 ✓ | (9, 11) | 10–1 ✓ | | |
| 13 | 1101 ✓ | (9, 13) | 1–01 ✓ | | |
| 15 | 1111 ✓ | (10, 11) | 101– ✓ | | |
| | | (7, 15) | –111 ✓ | | |
| | | (11, 15) | 1–11 ✓ | | |
| | | (13, 15) | 11–1 ✓ | | |

The don't-care columns are omitted when forming the prime implicant chart:

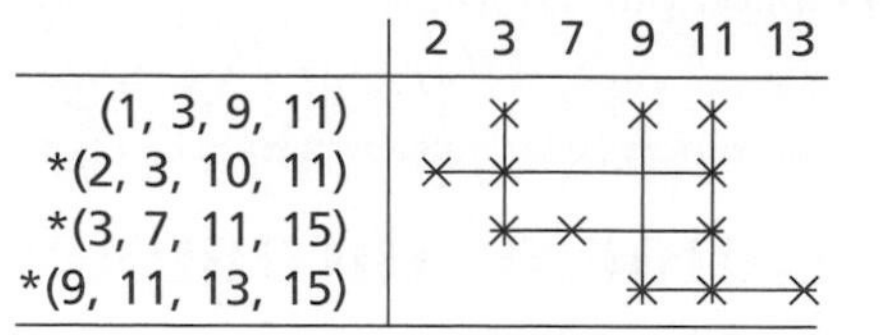

| | 2 | 3 | 7 | 9 | 11 | 13 |
|---|---|---|---|---|---|---|
| (1, 3, 9, 11) | | X | | X | X | |
| *(2, 3, 10, 11) | X | X | | | X | |
| *(3, 7, 11, 15) | | X | X | | X | |
| *(9, 11, 13, 15) | | | | X | X | X |

*indicates an essential prime implicant.

$$F = B'C + CD + AD$$

Note that although the original function was incompletely specified, the final simplified expression for $F$ is defined for all combinations of values for $A$, $B$, $C$, and $D$ and is therefore completely specified. In the process of simplification, we have automatically assigned values to the don't-cares in the original truth table for $F$. If we replace each term in the final expression for $F$ by its corresponding sum of minterms, the result is

$$F = (m_2 + m_3 + m_{10} + m_{11}) + (\cancel{m_3} + m_7 + \cancel{m_{11}} + m_{15}) + (m_9 + \cancel{m_{11}} + m_{13} + \cancel{m_{15}})$$

Because $m_{10}$ and $m_{15}$ appear in this expression and $m_1$ does not, this implies that the don't-care terms in the original truth table for $F$ have been assigned as follows:

for $ABCD = 0001, F = 0$;    for $1010, F = 1$;    for $1111, F = 1$

## 6.5 Simplification Using Map-Entered Variables

Although the Quine-McCluskey method can be used with functions with a fairly large number of variables, it is not very efficient for functions that have many variables and relatively few terms. Some of these functions can be simplified by using a modification of the Karnaugh map method. By using map-entered variables, Karnaugh map techniques can be extended to simplify functions with more than four or five variables. Figure 6-1(a) shows a four-variable map with two additional variables entered in the squares in the map. When $E$ appears in a square, this means that

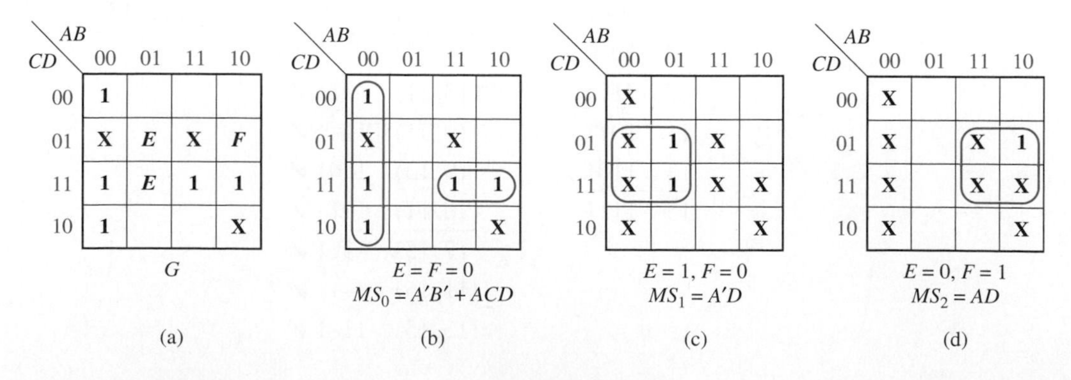


FIGURE 6-1 Use of Map-Entered Variables

if $E = 1$, the corresponding minterm is present in the function $G$, and if $E = 0$, the minterm is absent. Thus, the map represents the six-variable function

$$G(A, B, C, D, E, F) = m_0 + m_2 + m_3 + Em_5 + Em_7 + Fm_9 + m_{11} + m_{15}$$
$$(+ \text{ don't-care terms})$$

where the minterms are minterms of the variables $A$, $B$, $C$, and $D$. Note that $m_9$ is present in $G$ only when $F = 1$.

We will now use a three-variable map to simplify the function:

$$F(A, B, C, D) = A'B'C + A'BC + A'BC'D + ABCD + (AB'C)$$

where the $AB'C$ is a don't-care term. Because $D$ appears in only two terms, we will choose it as a map-entered variable, which leads to Figure 6-2(a). We will simplify $F$ by first considering $D = 0$ and then $D = 1$. First set $D = 0$ on the map, and $F$ reduces to $A'C$. Setting $D = 1$ leads to the map of Figure 6-2(b). The two 1's on the original map have already been covered by the term $A'C$, so they are changed to X's because we do not care whether they are covered again or not. From Figure 6-2(b), when $D = 1$. Thus, the expression

$$F = A'C + D(C + A'B) = A'C + CD + A'BD$$

gives the correct value of $F$ both when $D = 0$ and when $D = 1$. This is a minimum expression for $F$, as can be verified by plotting the original function on a four-variable map; see Figure 6-2(c).

Next, we will discuss a general method of simplifying functions using map-entered variables. In general, if a variable $P_i$ is placed in square $m_j$ of a map of function $F$, this means that $F = 1$ when $P_i = 1$, and the variables are chosen so that $m_j = 1$. Given a map with variables $P_1, P_2, \ldots$ entered into some of the squares, the minimum sum-of-products form of $F$ can be found as follows:

Find a sum-of-products expression for $F$ of the form

$$F = MS_0 + P_1MS_1 + P_2MS_2 + \cdots$$

where

$MS_0$ is the minimum sum obtained by setting $P_1 = P_2 = \cdots = 0$.

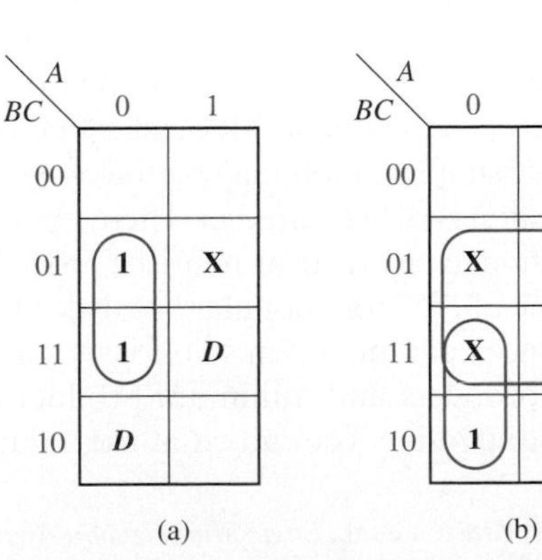


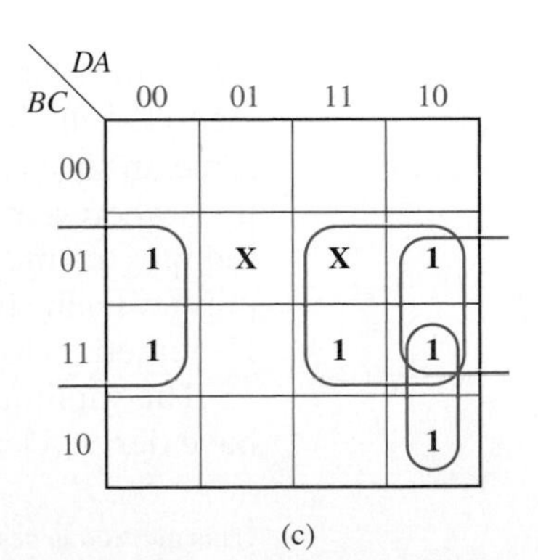


**FIGURE 6-2** Simplification Using a Map-Entered Variable

$MS_1$ is the minimum sum obtained by setting $P_1 = 1, P_j = 0\ (j \neq 1)$, and replacing all 1's on the map with don't-cares.
$MS_2$ is the minimum sum obtained by setting $P_2 = 1, P_j = 0\ (j \neq 2)$ and replacing all 1's on the map with don't-cares.

(Corresponding minimum sums can be found in a similar way for any remaining map-entered variables.)

The resulting expression for $F$ will always be a correct representation of $F$. This expression will be minimum provided that the values of the map-entered variables can be assigned independently. On the other hand, the expression will not generally be minimum if the variables are not independent (for example, if $P_1 = P_2'$).

For the example of Figure 6-1(a), maps for finding $MS_0$, $MS_1$ and $MS_2$ are shown in Figures 6-1(b), (c), and (d), where $E$ corresponds to $P_1$ and $F$ corresponds to $P_2$. The resulting expression is a minimum sum of products for $G$:

$$G = A'B' + ACD + EA'D + FAD$$

After some practice, it should be possible to write the minimum expression directly from the original map without first plotting individual maps for each of the minimum sums.

## 6.6 Conclusion

We have discussed four methods for reducing a switching expression to a minimum sum-of-products or a minimum product-of-sums form: algebraic simplification, Karnaugh maps, Quine-McCluskey method, and Petrick's method. Many other methods of simplification are discussed in the literature, but most of these methods are based on variations or extensions of the Karnaugh map or Quine-McCluskey techniques. Karnaugh maps are most useful for functions with three to five variables. The Quine-McCluskey technique can be used with a high-speed digital computer to simplify functions with up to 15 or more variables. Such computer programs are of greatest value when used as part of a computer-aided design (CAD) package that assists with deriving the equations as well as implementing them. Algebraic simplification is still valuable in many cases, especially when different forms of the expressions are required. For problems with a large number of variables and a small number of terms, it may be impossible to use the Karnaugh map, and the Quine-McCluskey method may be very cumbersome. In such cases, algebraic simplification may be the easiest method to use. In situations where a minimum solution is not required or where obtaining a minimum solution requires too much computation to be practical, heuristic procedures may be used to simplify switching functions. One of the more popular heuristic procedures is the Espresso-II method,[2] which can produce near minimum solutions for a large class of problems.

The minimum sum-of-products and minimum product-of-sums expressions we have derived lead directly to two-level circuits that use a minimum number of AND

[2]This method is described in R. K. Brayton et al., *Logic Minimization Algorithms for VLSI Synthesis* (Kluwer Academic Publishers, 1984).

and OR gates and have a minimum number of gate inputs. As discussed in Unit 7, these circuits are easily transformed into circuits that contain NAND or NOR gates. These minimum expressions may also be useful when designing with some types of array logic, as discussed in Unit 9. However, many situations exist where minimum expressions do not lead to the best design. For practical designs, many other factors must be considered, such as the following:

What is the maximum number of inputs a gate can have?
What is the maximum number of outputs a gate can drive?
Is the speed with which signals propagate through the circuit fast enough?
How can the number of interconnections in the circuit be reduced?
Does the design lead to a satisfactory circuit layout on a printed circuit board or on a silicon chip?

Until now, we have considered realizing only one switching function at a time. Unit 7 describes design techniques and Unit 9 describes components that can be used when several functions must be realized by a single circuit.

# Programmed Exercise 6.1

Cover the answers to this exercise with a sheet of paper and slide it down as you check your answers.

Find a minimum sum-of-products expression for the following function:

$$f(A, B, C, D, E) = \Sigma\, m(0, 2, 3, 5, 7, 9, 11, 13, 14, 16, 18, 24, 26, 28, 30)$$

Translate each decimal minterm into binary and sort the binary terms into groups according to the number of 1's in each term.

---

**Answer:**

| | | | | |
|---|---|---|---|---|
| 0 | 00000 | ✓ | 0, 2 | 000-0 |
| 2 | 00010 | ✓ | | |
| 16 | 10000 | | | |
| 3 | 00011 | | | |
| 5 | 00101 | | | |
| 9 | 01001 | | | |
| 18 | 10010 | | | |
| 24 | 11000 | | | |
| 7 | 00111 | | | |
| 11 | 01011 | | | |
| 13 | 01101 | | | |
| 14 | 01110 | | | |
| 26 | 11010 | | | |
| 28 | 11100 | | | |
| 30 | 11110 | | | |

---

Compare pairs of terms in adjacent groups and combine terms where possible. (Check off terms which have been combined.)

Answer:

| | | | | | |
|---|---|---|---|---|---|
| 0 | 00000 ✓ | 0, 2 | 000–0 ✓ | 0, 2, 16, 18 | -00-0 |
| 2 | 00010 ✓ | 0, 16 | -0000 | | |
| 16 | 10000 ✓ | 2, 3 | 0001– | | |
| 3 | 00011 ✓ | 2, 18 | -0010 | | |
| 5 | 00101 ✓ | 16, 18 | 100–0 ✓ | | |
| 9 | 01001 ✓ | 16, 24 | 1–000 | | |
| 18 | 10010 ✓ | 3, 7 | 00–11 | | |
| 24 | 11000 ✓ | 3, 11 | 0–011 | | |
| 7 | 00111 ✓ | 5, 7 | 001–1 | | |
| 11 | 01011 ✓ | 5, 13 | 0–101 | | |
| 13 | 01101 ✓ | 9, 11 | 010–1 | | |
| 14 | 01110 ✓ | 9, 13 | 01–01 | | |
| 26 | 11010 ✓ | 18, 26 | 1–010 | | |
| 28 | 11100 ✓ | 24, 26 | 110–0 | | |
| 30 | 11110 ✓ | 24, 28 | 11–00 | | |
| | | 14, 30 | -1110 | | |
| | | 26, 30 | 11–10 | | |
| | | 28, 30 | 111–0 | | |

Now, compare pairs of terms in adjacent groups in the second column and combine terms where possible. (Check off terms which have been combined.) Check your work by noting that each new term can be formed in two ways. (Cross out duplicate terms.)

Answer: (third column)

| | | |
|---|---|---|
| 0, 2, 16, 18 | -00–0 | (check off (0, 2), (16, 18), (0, 16), and (2, 18)) |
| 16, 18, 24, 26 | 1–0–0 | (check off (16, 18), (24, 26), (16, 24), and (18, 26)) |
| 24, 26, 28, 30 | 11--0 | (check off (24, 26), (28, 30), (24, 28), and (26, 30)) |

Can any pair of terms in the third column be combined?
Complete the given prime implicant chart.

| | 0 | 2 |
|---|---|---|
| (0, 2, 16, 18) | | |

**Answer:** No pair of terms in the third column combine.

| | 0 | 2 | 3 | 5 | 7 | 9 | 11 | 13 | 14 | 16 | 18 | 24 | 26 | 28 | 30 |
|---|---|---|---|---|---|---|---|---|---|---|---|---|---|---|---|
| (0, 2, 16, 18) | X | X | | | | | | | | X | X | | | | |
| (16, 18, 24, 26) | | | | | | | | | | X | X | X | X | | |
| (24, 26, 28, 30) | | | | | | | | | | | | X | X | X | X |
| (2, 3) | | X | X | | | | | | | | | | | | |
| (3, 7) | | | X | | X | | | | | | | | | | |
| (3, 11) | | | X | | | | X | | | | | | | | |
| (5, 7) | | | | X | X | | | | | | | | | | |
| (5, 13) | | | | X | | | | X | | | | | | | |
| (9, 11) | | | | | | X | X | | | | | | | | |
| (9, 13) | | | | | | X | | X | | | | | | | |
| (14, 30) | | | | | | | | | X | | | | | | X |

Determine the essential prime implicants, and cross out the corresponding rows and columns.

**Answer:**

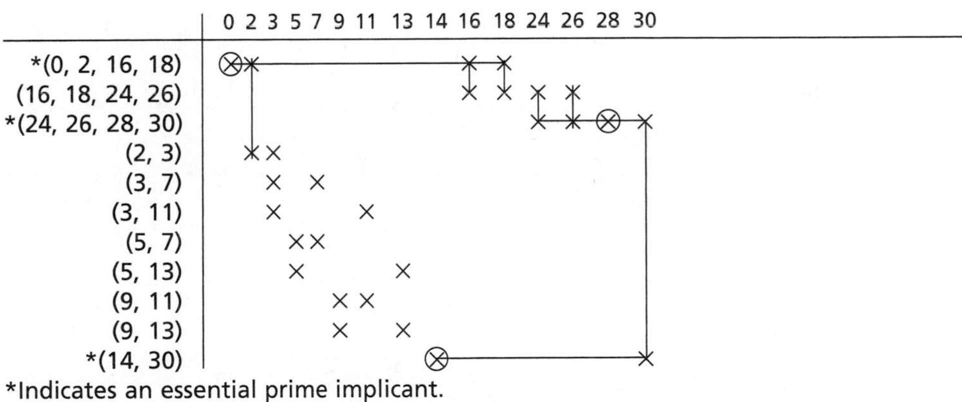

| | 0 | 2 | 3 | 5 | 7 | 9 | 11 | 13 | 14 | 16 | 18 | 24 | 26 | 28 | 30 |
|---|---|---|---|---|---|---|---|---|---|---|---|---|---|---|---|
| *(0, 2, 16, 18) | ⊗ | X | | | | | | | | X | X | | | | |
| (16, 18, 24, 26) | | | | | | | | | | X | X | X | X | | |
| *(24, 26, 28, 30) | | | | | | | | | | | | X | X | ⊗ | X |
| (2, 3) | | X | X | | | | | | | | | | | | |
| (3, 7) | | | X | | X | | | | | | | | | | |
| (3, 11) | | | X | | | | X | | | | | | | | |
| (5, 7) | | | | X | X | | | | | | | | | | |
| (5, 13) | | | | X | | | | X | | | | | | | |
| (9, 11) | | | | | | X | X | | | | | | | | |
| (9, 13) | | | | | | X | | X | | | | | | | |
| *(14, 30) | | | | | | | | | ⊗ | | | | | | X |

*Indicates an essential prime implicant.

Note that all remaining columns contain two or more X's. Choose the first column which has two X's and then select the prime implicant which covers the first X in that column. Then, choose a minimum number of prime implicants which cover the remaining columns in the chart.

**Answer:**

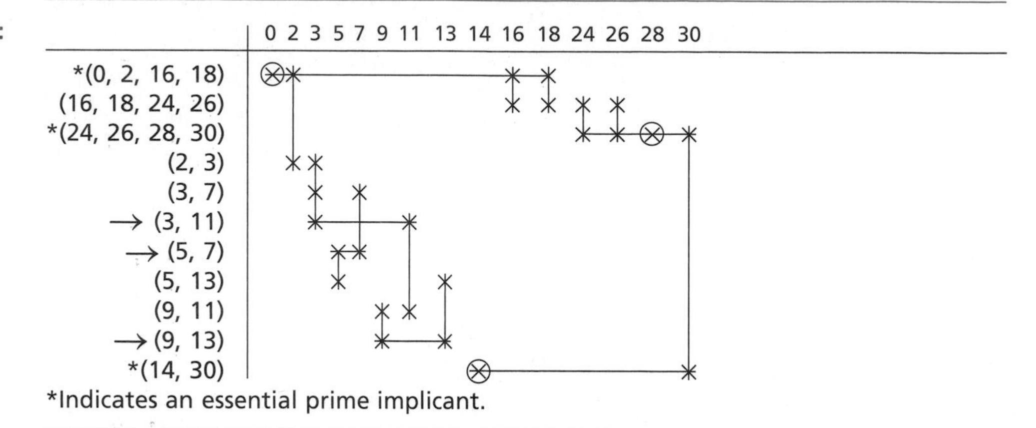


From this chart, write down the chosen prime implicants in 0, 1, and – notation.

Then, write the minimum sum of products in algebraic form.

**Answer:** –00–0, 11--0, 0–011, 001–1, 01–01, and –1110

$f = B'C'E' + ABE' + A'C'DE + A'B'CE + A'BD'E + BCDE'$

The prime implicant chart with the essential prime implicants crossed out is repeated here.
Find a second minimum sum-of-products solution.

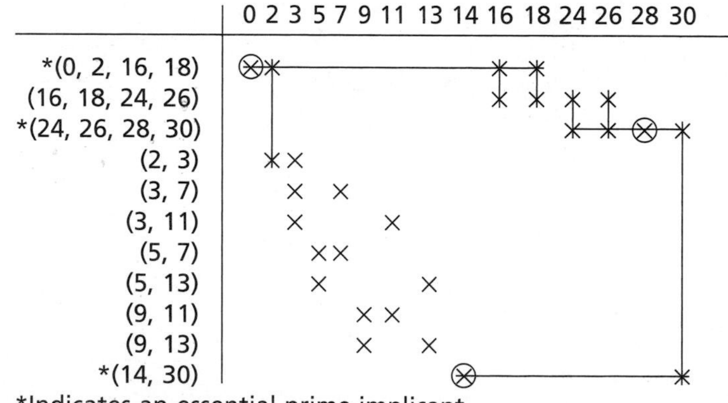


**Answer:** Start by choosing prime implicant (5, 13).

$f = BCDE' + B'C'E' + ABE' + A'B'DE + A'CD'E + A'BC'E$

# Problems

**6.2** For each of the following functions, find all of the prime implicants, using the Quine-McCluskey method.
(a) $f(a, b, c, d) = \Sigma\, m(1, 5, 7, 9, 11, 12, 14, 15)$
(b) $f(a, b, c, d) = \Sigma\, m(0, 1, 3, 5, 6, 7, 8, 10, 14, 15)$

**6.3** Using a prime implicant chart, find *all* minimum sum-of-products solutions for each of the functions given in Problem 6.2.

**6.4** For this function, find a minimum sum-of-products solution, using the Quine-McCluskey method.
$f(a, b, c, d) = \Sigma\, m(1, 3, 4, 5, 6, 7, 10, 12, 13) + \Sigma\, d(2, 9, 15)$

**6.5** Find all prime implicants of the following function and then find all minimum solutions using Petrick's method:
$F(A, B, C, D) = \Sigma\, m(9, 12, 13, 15) + \Sigma\, d(1, 4, 5, 7, 8, 11, 14)$

**6.6** Using the method of map-entered variables, use four-variable maps to find a minimum sum-of-products expression for
(a) $F(A, B, C, D, E) = \Sigma\, m(0, 4, 5, 7, 9) + \Sigma\, d(6, 11) + E(m_1 + m_{15})$, where the $m$'s represent minterms of the variables $A$, $B$, $C$, and $D$.
(b) $Z(A, B, C, D, E, F, G) = \Sigma\, m(0, 3, 13, 15) + \Sigma\, d(1, 2, 7, 9, 14) + E(m_6 + m_8) + Fm_{12} + Gm_5$

**6.7** For each of the following functions, find all of the prime implicants using the Quine-McCluskey method.
(a) $f(a, b, c, d) = \Sigma\, m(0, 3, 4, 5, 7, 9, 11, 13)$
(b) $f(a, b, c, d) = \Sigma\, m(2, 4, 5, 6, 9, 10, 11, 12, 13, 15)$

**6.8** Using a prime implicant chart, find *all* minimum sum-of-products solutions for each of the functions given in Problem 6.7.

**6.9** For each function, find a minimum sum-of-products solution using the Quine-McCluskey method.
(a) $f(a, b, c, d) = \Sigma\, m(2, 3, 4, 7, 9, 11, 12, 13, 14) + \Sigma\, d(1, 10, 15)$
(b) $f(a, b, c, d) = \Sigma\, m(0, 1, 5, 6, 8, 9, 11, 13) + \Sigma\, d(7, 10, 12)$
(c) $f(a, b, c, d) = \Sigma\, m(3, 4, 6, 7, 8, 9, 11, 13, 14) + \Sigma\, d(2, 5, 15)$

**6.10** Work Problem 5.24(a) using the Quine-McCluskey method.

**6.11** $F(A, B, C, D, E) = \Sigma\, m(0, 2, 6, 7, 8, 10, 11, 12, 13, 14, 16, 18, 19, 29, 30) + \Sigma\, d(4, 9, 21)$

Find the minimum sum-of-products expression for $F$, using the Quine-McCluskey method. Underline the essential prime implicants in this expression.

**6.12** Using the Quine-McCluskey method, find all minimum sum-of-products expressions for

(a) $f(A, B, C, D, E) = \Sigma\, m(0, 1, 2, 3, 4, 8, 9, 10, 11, 19, 21, 22, 23, 27, 28, 29, 30)$

(b) $f(A, B, C, D, E) = \Sigma\, m(0, 1, 2, 4, 8, 11, 13, 14, 15, 17, 18, 20, 21, 26, 27, 30, 31)$

**6.13** Using the Quine-McCluskey method, find all minimum product-of-sums expressions for the functions of Problem 6.12.

**6.14** (a) Using the Quine-McCluskey, method find all prime implicants of $f(A, B, C, D) = \Sigma\, m(1, 3, 5, 6, 8, 9, 12, 14, 15) + \Sigma\, d(4, 10, 13)$. Identify all essential prime implicants and find all minimum sum-of-products expressions.

(b) Repeat Part (a) for $f'$.

**6.15** (a) Use the Quine-McCluskey method to find all prime implicants of $f(a, b, c, d, e) = \Sigma\, m(1, 2, 4, 5, 6, 7, 9, 12, 13, 15, 17, 20, 22, 25, 28, 30)$. Find all essential prime implicants, and find all minimum sum-of-products expressions.

(b) Repeat Part (a) for $f'$.

**6.16** $G(A, B, C, D, E, F) = \Sigma\, m(1, 2, 3, 16, 17, 18, 19, 26, 32, 39, 48, 63) + \Sigma\, d(15, 28, 29, 30)$

(a) Find all minimum sum-of-products expressions for $G$.

(b) Circle the *essential* prime implicants in your answer.

(c) If there were no don't-care terms present in the original function, how would your answer to part (a) change? (Do this by inspection of the prime implicant chart; do *not* rework the problem.)

**6.17** (a) Use the Quine-McCluskey procedure to find *all* prime implicants of the function $G(A, B, C, D, E, F) = \Sigma\, m(1, 7, 11, 12, 15, 33, 35, 43, 47, 59, 60) + \Sigma\, d(30, 50, 54, 58)$. Identify all essential prime implicants and find all minimum sum-of-products expressions.

(b) Repeat Part (a) for $G'$.

**6.18** The following prime implicant table (chart) is for a four-variable function $f(A, B, C, D)$.

(a) Give the decimal representation for each of the prime implicants.

(b) List the maxterms of $f$.

(c) List the don't-cares of $f$, if any.

(d) Give the algebraic expression for each of the essential prime implicants.

| | 2 | 3 | 7 | 9 | 11 | 13 |
|---|---|---|---|---|---|---|
| –0–1 | | × | | × | × | |
| –01– | × | × | | | × | |
| --11 | | × | × | | × | |
| 1--1 | | | | × | × | × |

**6.19** Packages arrive at the stockroom and are delivered on carts to offices and laboratories by student employees. The carts and packages are various sizes and shapes. The students are paid according to the carts used. There are five carts and the pay for their use is
Cart C1: $2
Cart C2: $1
Cart C3: $4
Cart C4: $2
Cart C5: $2
On a particular day, seven packages arrive, and they can be delivered using the five carts as follows:
C1 can be used for packages P1, P3, and P4.
C2 can be used for packages P2, P5, and P6.
C3 can be used for packages P1, P2, P5, P6, and P7.
C4 can be used for packages P3, P6, and P7.
C5 can be used for packages P2 and P4.

The stockroom manager wants the packages delivered at minimum cost. Using minimization techniques described in this unit, present a systematic procedure for finding the minimum cost solution.

**6.20** Use the Quine-McCluskey procedure to find all prime implicants of the function $h(A, B, C, D, E, F, G) = \Sigma\, m(24, 28, 39, 47, 70, 86, 88, 92, 102, 105, 118)$.
Express the prime implicants *algebraically*.

**6.21** Find all prime implicants of the following function, and then find all minimum solutions using Petrick's method:

$$F(A, B, C, D) = \Sigma\, m(7, 12, 14, 15) + \Sigma\, d(1, 3, 5, 8, 10, 11, 13)$$

**6.22** Using the method of map-entered variables, use four-variable maps to find a minimum sum-of-products expression for
(a) $F(A, B, C, D, E) = \Sigma\, m(0, 4, 6, 13, 14) + \Sigma\, d(2, 9) + E(m_1 + m_{12})$
(b) $Z(A, B, C, D, E, F, G) = \Sigma\, m(2, 5, 6, 9) + \Sigma\, d(1, 3, 4, 13, 14) + E(m_{11} + m_{12}) + F(m_{10}) + G(m_0)$

**6.23** (a) Rework Problem 6.6(a), using a five-variable map.
(b) Rework Problem 6.6(a), using the Quine-McCluskey method. Note that you must express $F$ in terms of minterms of all five variables; the original four-variable minterms cannot be used.

**6.24** Using map-entered variables, find the minimum sum-of-products expressions for the following function:
$G = C'E'F + DEF + AD'E'F' + BDE'F + AD'EF'$

# Multi-Level Gate Circuits NAND and NOR Gates

## Objectives

1. Design a minimal two-level or multi-level circuit of AND and OR gates to realize a given function. (Consider *both* circuits with an OR gate at the output and circuits with an AND gate at the output.)
2. Design or analyze a two-level gate circuit using any one of the eight basic forms (AND-OR, NAND-NAND, OR-NAND, NOR-OR, OR-AND, NOR-NOR, AND-NOR, and NAND-AND).
3. Design or analyze a multi-level NAND-gate or NOR-gate circuit.
4. Convert circuits of AND and OR gates to circuits of NAND gates or NOR gates, and conversely, by adding or deleting inversion bubbles.
5. Design a minimal two-level, multiple-output AND-OR, OR-AND, NAND-NAND, or NOR-NOR circuit using Karnaugh maps.

# Study Guide

**1.** Study Section 7.1, *Multi-Level Gate Circuits.*

(a) What are two ways of changing the number of levels in a gate circuit?

(b) By constructing a tree diagram, determine the number of gates, gate inputs, and levels of gates required to realize $Z_1$ and $Z_2$:

$$Z_1 = [(A + B)C + DE(F + G)]H \qquad Z_2 = A + B[C + DE(F + G)]$$

Check your answers by drawing the corresponding gate circuits.

(c) In order to find a minimum two-level solution, why is it necessary to consider both a sum-of-products form and a product-of-sums form for the function?

(d) One realization of $Z = ABC(D + E) + FG$ is

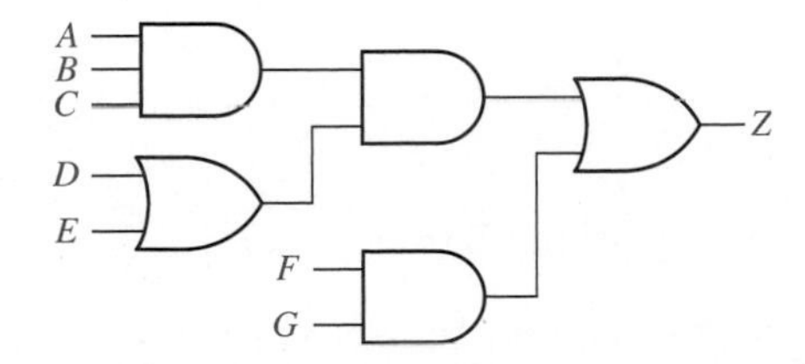


Redraw the circuit so that it uses one less gate and so that the output of an AND gate never goes directly to the input of another AND gate.

(e) Work Problems 7.1 and 7.2. Unless otherwise specified, you may always assume that both the variables and their complements are available as circuit inputs.

2. Study Section 7.2, *NAND and NOR Gates*

(a) For each gate, specify the missing inputs:

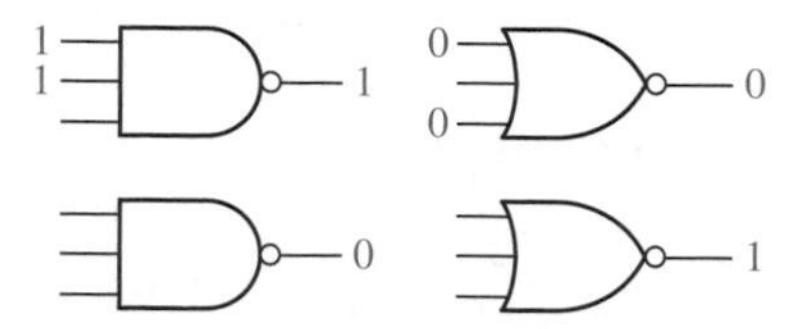


(b) What is meant by functionally complete set of logic gates?

(c) How can you show that a set of logic gates is functionally complete?

(d) Show that the NOR gate itself is functionally complete.

(e) Using NAND gates, draw a circuit for $F = (A'(BC)')'$.

(f) Using NOR gates, draw a circuit for $F = ((X + Y)' + (X' + Z)')'$

3. Study Section 7.3, *Design of Two-Level NAND- and NOR-Gate Circuits.*

(a) Draw the circuit corresponding to Equation (7-17).

(b) Derive Equation (7-18).

(c) Make sure that you understand the relation between Equations (7-13) through (7-21) and the diagrams of Figure 7-11.

(d) Why is the NOR-NAND form degenerate?

(e) What assumption is made about the types of inputs available when the procedures for designing two-level NAND-NAND and NOR-NOR circuits are used?

(f) For these procedures the literal inputs to the output gate are complemented but not the literal inputs to the other gates. Explain why. Use an equation to illustrate.

(g) A general OR-AND circuit follows. Transform this to a NOR-NOR circuit and prove that your transformation is valid.

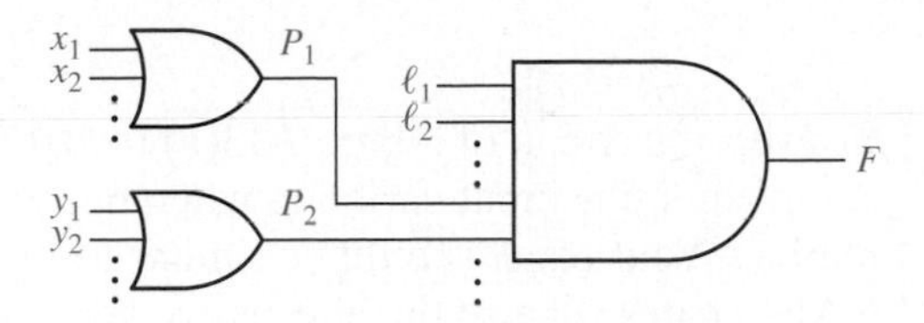


(h) Work Problem 7.3.

**4.** Study Section 7.4, *Design of Multi-Level NAND- and NOR-Gate Circuits.*

(a) Verify that the NAND circuit of Figure 7-13 is correct by dividing the corresponding circuit of AND and OR gates into two-level subcircuits and transforming each subcircuit.

(b) If you wish to design a two-level circuit using only NOR gates, should you start with a minimum sum of products or a minimum product of sums?

(c) Note that direct conversion of a circuit of AND and OR gates to a NAND gate circuit requires starting with an OR gate at the output, but the direct conversion to a NOR gate circuit requires starting with an AND gate at the output. This is easy to remember because a NAND is equivalent to an OR with the inputs inverted:

and a NOR is equivalent to an AND with the inputs inverted:

(d) Convert the circuit of Figure 7-1(b) to all NAND gates.
(e) Work Problems 7.4, 7.5, 7.6, and 7.7.

**5.** Study Section 7.5, *Circuit Conversion Using Alternative Gate Symbols*.

(a) Determine the logic function realized by each of the following circuits:

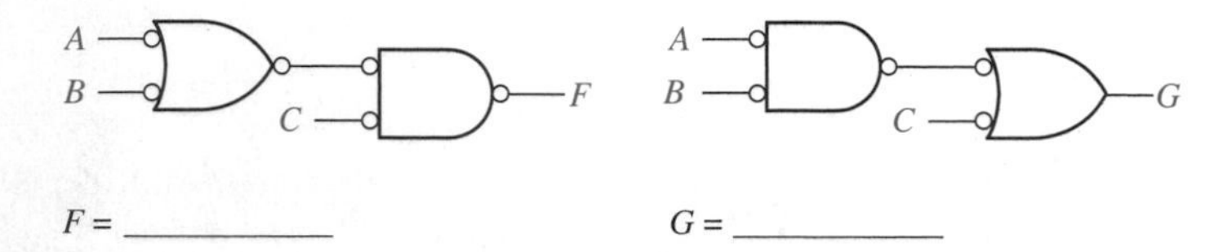


$F =$ __________ $G =$ __________

(b) Convert the circuit of Figure 7-13(a) to NAND gates by adding bubbles and complementing input variables when necessary. (You should have added 12 bubbles. Your result should be similar to Figure 7-13(b), except some of the NAND gates will use the alternative symbol.)
(c) Draw a circuit of AND and OR gates for the following equation:

$$Z = A[BC + D + E(F + GH)]$$

Then convert to NOR gates by adding bubbles and complementing inputs when necessary. (You should have added 10 bubbles and complemented six input variables.)

(d) Work Problem 7.8.

**6.** Study Section 7.6, *Design of Two-Level, Multiple-Output Circuits*.

(a) In which of the following cases would you replace a term $xy'$ with $xy'z + xy'z'$?
(1) Neither $xy'z$ or $xy'z'$ is used in another function.
(2) Both $xy'z$ and $xy'z'$ are used in other functions.
(3) Term $xy'z$ is used in another function, but $xy'z'$ is not.
(b) In the second example (Figure 7-21), in $f_2$, $c$ could have been replaced by $bc + b'c$ because $bc$ and $b'c$ were available "free" from $f_1$ and $f_3$. Why was this replacement not made?

(c) In the following example, compute the cost of realizing $f_1$ and $f_2$ separately; then compute the cost using the term $a'b'c$ in common between the two functions. Use a two-level AND-OR circuit in both cases.

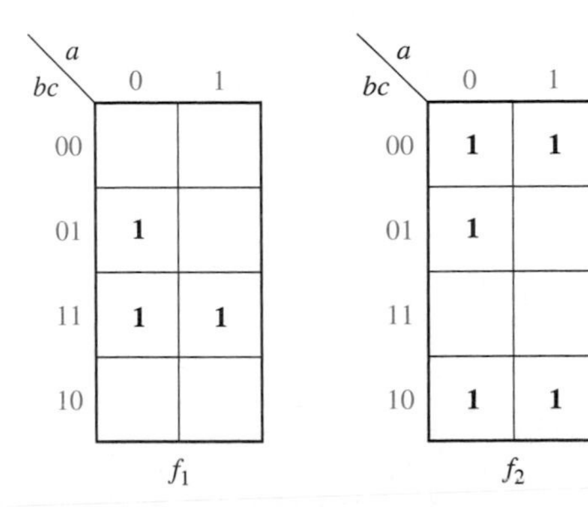


(d) Find expressions which correspond to a two-level, minimum multiple-output, AND-OR realization of $F_1$, $F_2$, and $F_3$. Why should the term *cd not* be included in $F_1$?

$F_1$:

| cd \ ab | 00 | 01 | 11 | 10 |
|---|---|---|---|---|
| 00 | | | | 1 |
| 01 | | | | 1 |
| 11 | 1 | 1 | 1 | 1 |
| 10 | | | | |

$F_2$:

| cd \ ab | 00 | 01 | 11 | 10 |
|---|---|---|---|---|
| 00 | | 1 | | |
| 01 | | 1 | | |
| 11 | 1 | 1 | 1 | |
| 10 | | | 1 | 1 |

$F_3$:

| cd \ ab | 00 | 01 | 11 | 10 |
|---|---|---|---|---|
| 00 | | | | |
| 01 | 1 | 1 | | |
| 11 | | 1 | 1 | 1 |
| 10 | | | | |

$F_1 =$

$F_2 =$

$F_3 =$

(e) Work Problems 7.9, 7.10, and 7.11.

(f) Work Problem 7.12. (*Hint*: Work with the 0's on the maps and first find a minimum solution for $f_1'$, $f_2'$, and $f_3'$.)

**7.** Study Section 7.7, *Multiple-Output NAND- and NOR-Gate Circuits*.

(a) Derive expressions for the $F_1$ and $F_2$ outputs of the NOR circuits of Figure 7-24(b) by finding the equation for each gate output, and show that these expressions reduce to the original expressions for $F_1$ and $F_2$.

(b) Convert Figure 7-24(a) to 7-24(b) by using the bubble method.

(c) Work Problem 7.13.

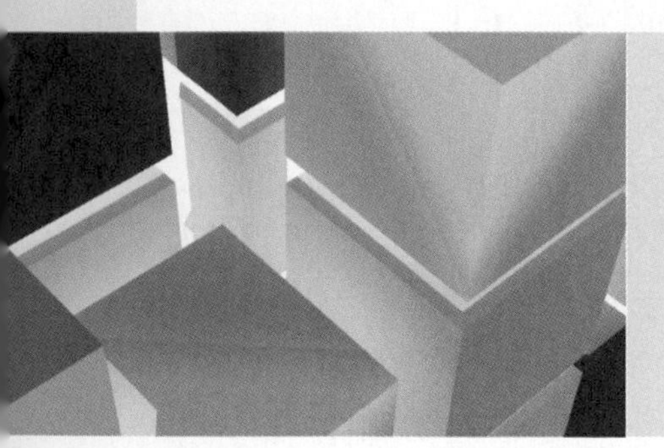

# Multi-Level Gate Circuits NAND and NOR Gates

In the first part of this unit, you will learn how to design circuits which have more than two levels of AND and OR gates. In the second part you will learn techniques for designing with NAND and NOR gates. These techniques generally consist of first designing a circuit of AND and OR gates and then converting it to the desired type of gates. These techniques are easy to apply *provided* that you start with the proper form of circuit.

## 7.1 Multi-Level Gate Circuits

The maximum number of gates cascaded in series between a circuit input and the output is referred to as the number of *levels* of gates (not to be confused with voltage levels). Thus, a function written in sum-of-products form or in product-of-sums form corresponds directly to a two-level gate circuit. As is usually the case in digital circuits where the gates are driven from flip-flop outputs (as discussed in Unit 11), we will assume that all variables and their complements are available as circuit inputs. For this reason, we will not normally count inverters which are connected directly to input variables when determining the number of levels in a circuit. In this unit we will use the following terminology:

1. *AND-OR circuit* means a two-level circuit composed of a level of AND gates followed by an OR gate at the output.
2. *OR-AND circuit* means a two-level circuit composed of a level of OR gates followed by an AND gate at the output.
3. *OR-AND-OR circuit* means a three-level circuit composed of a level of OR gates followed by a level of AND gates followed by an OR gate at the output.
4. Circuit of AND and OR gates implies no particular ordering of the gates; the output gate may be either AND or OR.

The number of levels in an AND-OR circuit can usually be increased by factoring the sum-of-products expression from which it was derived. Similarly, the number of levels in an OR-AND circuit can usually be increased by multiplying out some of the terms in the product-of-sums expression from which it was derived. Logic designers are concerned with the number of levels in a circuit for several reasons. Sometimes factoring (or multiplying out) to increase the number of levels of gates will reduce the required number of gates and gate inputs and, thus, reduce the cost of building the circuit, but in other cases increasing the number of levels will increase the cost. In many applications, the number of gates which can be cascaded is limited by gate delays. When the input of a gate is switched, there is a finite time before the output changes. When several gates are cascaded, the time between an input change and the corresponding change in the circuit output may become excessive and slow down the operation of the digital system.

The number of gates, gate inputs, and levels in a circuit can be determined by inspection of the corresponding expression. In the example of Figure 7-1(a), the tree diagram drawn below the expression for $Z$ indicates that the corresponding circuit will have four levels, six gates, and 13 gate inputs, as verified in Figure 7-1(b). Each

**FIGURE 7-1** Four-Level Realization of $Z$

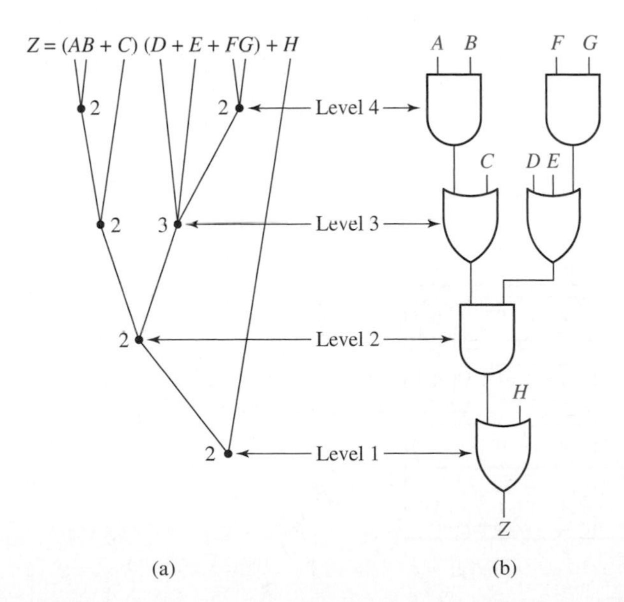

FIGURE 7-2 Three-Level Realization of Z

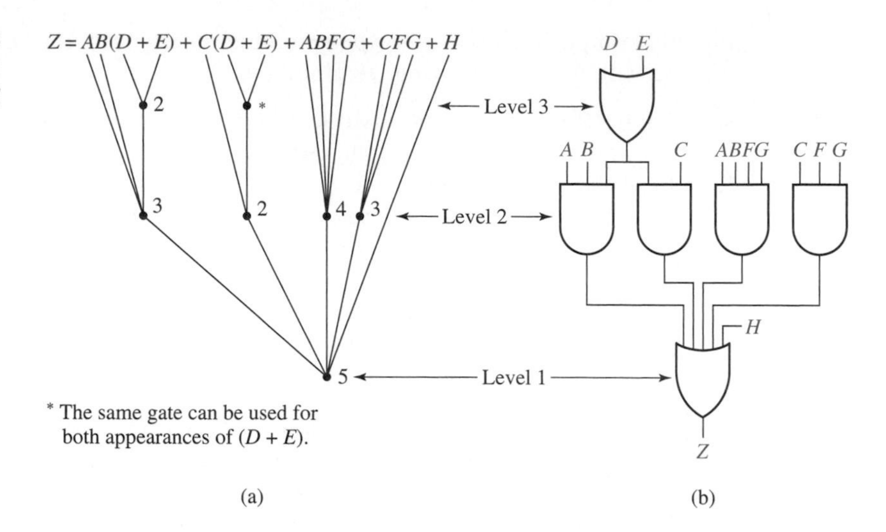


node on the tree diagram represents a gate, and the number of gate inputs is written beside each node.

We can change the expression for $Z$ to three levels by partially multiplying it out:

$$Z = (AB + C)[(D + E) + FG] + H$$
$$= AB(D + E) + C(D + E) + ABFG + CFG + H$$

As shown in Figure 7-2, the resulting circuit requires three levels, six gates, and 19 gate inputs.

### *Example of Multi-Level Design Using AND and OR Gates*

*Problem:* Find a circuit of AND and OR gates to realize

$$f(a, b, c, d) = \Sigma\, m(1, 5, 6, 10, 13, 14)$$

Consider solutions with two levels of gates and three levels of gates. Try to minimize the number of gates and the total number of gate inputs. Assume that all variables and their complements are available as inputs.

***Solution:*** First, simplify $f$ by using a Karnaugh map (Figure 7-3):

FIGURE 7-3

| cd \ ab | 00 | 01 | 11 | 10 |
|---|---|---|---|---|
| 00 | 0 | 0 | 0 | 0 |
| 01 | 1 | 1 | 1 | 0 |
| 11 | 0 | 0 | 0 | 0 |
| 10 | 0 | 1 | 1 | 1 |

$$f = a'c'd + bc'd + bcd' + acd' \tag{7-1}$$

This leads directly to a two-level AND-OR gate circuit (Figure 7-4):

FIGURE 7-4

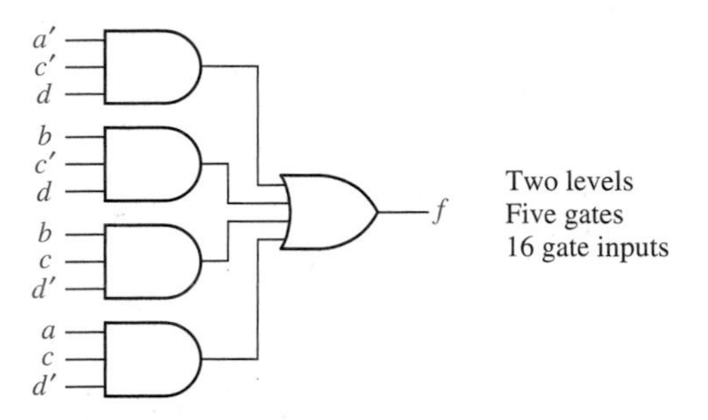


Factoring Equation (7-1) yields

$$f = c'd(a' + b) + cd'(a + b) \tag{7-2}$$

which leads to the following three-level OR-AND-OR gate circuit (Figure 7-5):

FIGURE 7-5

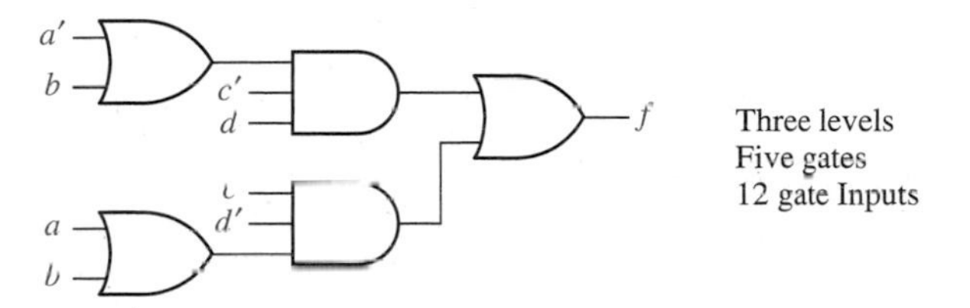


Both of these solutions have an OR gate at the output. A solution with an AND gate at the output might have fewer gates or gate inputs. A two-level OR-AND circuit corresponds to a product-of-sums expression for the function. This can be obtained from the 0's on the Karnaugh map as follows:

$$f' = c'd' + ab'c' + cd + a'b'c \tag{7-3}$$

$$f = (c + d)(a' + b + c)(c' + d')(a + b + c') \tag{7-4}$$

Equation (7-4) leads directly to a two-level OR-AND circuit (Figure 7-6):

FIGURE 7-6

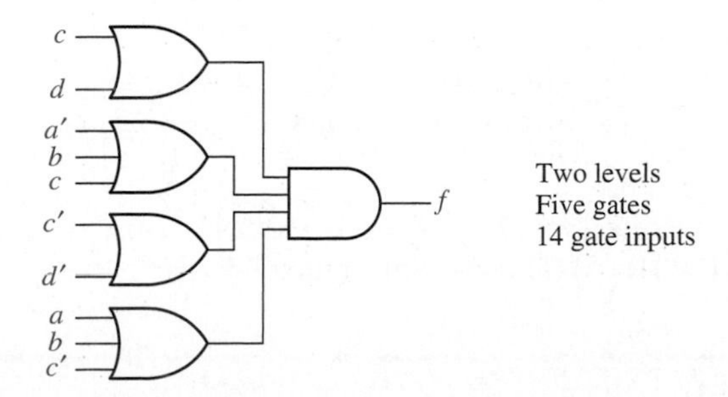

To get a three-level circuit with an AND gate output, we partially multiply out Equation (7-4) using $(X + Y)(X + Z) = X + YZ$:

$$f = [c + d(a' + b)][c' + d'(a + b)] \tag{7-5}$$

Equation (7-5) would require four levels of gates to realize; however, if we multiply out $d'(a + b)$ and $d(a' + b)$, we get

$$f = (c + a'd + bd)(c' + ad' + bd') \tag{7-6}$$

which leads directly to a three-level AND-OR-AND circuit (Figure 7-7):

**FIGURE 7-7**

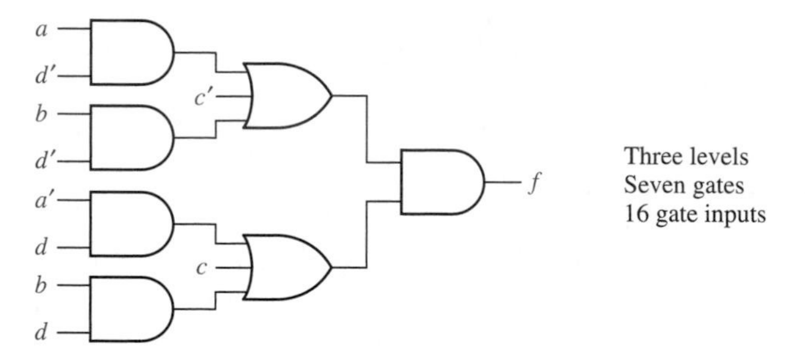


For this particular example, the best two-level solution had an AND gate at the output (Figure 7-6), and the best three-level solution had an OR gate at the output (Figure 7-5). In general, to be sure of obtaining a minimum solution, one must find *both* the circuit with the AND-gate output and the one with the OR-gate output.

If an expression for $f'$ has $n$ levels, the complement of that expression is an $n$-level expression for $f$. Therefore, to realize $f$ as an $n$-level circuit with an AND-gate output, one procedure is first to find an $n$-level expression for $f'$ with an OR operation at the output level and then complement the expression for $f'$. In the preceding example, factoring Equation (7-3) gives a three-level expression for $f'$:

$$\begin{aligned} f' &= c'(d' + ab') + c(d + a'b') \\ &= c'(d' + a)(d' + b') + c(d + a')(d + b') \end{aligned} \tag{7-7}$$

Complementing Equation (7-7) gives Equation (7-6), which corresponds to the three-level AND-OR-AND circuit of Figure 7-7.

## 7.2 NAND and NOR Gates

Until this point we have designed logic circuits using AND gates, OR gates, and inverters. Exclusive-OR and equivalence gates have also been introduced in Unit 3. In this section we will define NAND and NOR gates. Logic designers frequently use NAND and NOR gates because they are generally faster and use fewer components than AND or OR gates. As will be shown later, any logic function can be implemented using only NAND gates or only NOR gates.

Figure 7-8(a) shows a three-input NAND gate. The small circle (or "bubble") at the gate output indicates inversion, so the NAND gate is equivalent to an AND gate followed by an inverter, as shown in Figure 7-8(b). A more appropriate name would be an AND-NOT gate, but we will follow common usage and call it a NAND gate.
The gate output is

$$F = (ABC)' = A' + B' + C'$$

The output of the $n$-input NAND gate in Figure 7-8(c) is

$$F = (X_1X_2 \ldots X_n)' = X_1' + X_2' + \ldots + X_n' \tag{7-8}$$

The output of this gate is 1 iff one or more of its inputs are 0.

FIGURE 7-8
NAND Gates

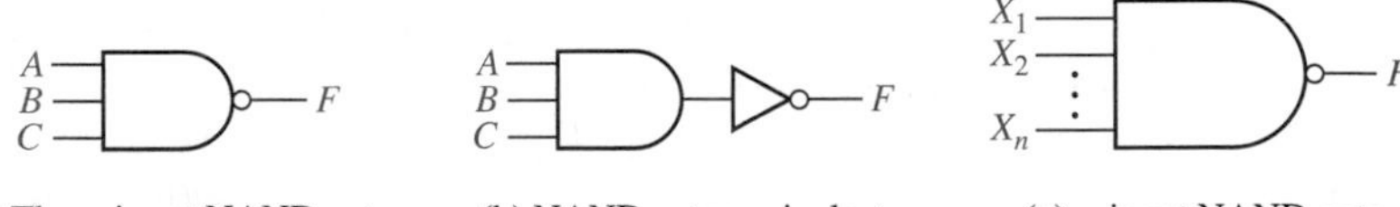

(a) Three-input NAND gate (b) NAND gate equivalent (c) $n$-input NAND gate

Figure 7-9(a) shows a three-input NOR gate. The small circle at the gate output indicates inversion, so the NOR gate is equivalent to an OR gate followed by an inverter. A more appropriate name would be an OR-NOT gate, but we will follow common usage and call it a NOR gate. The gate output is

$$F = (A + B + C)' = A'B'C'$$

FIGURE 7-9
NOR Gates

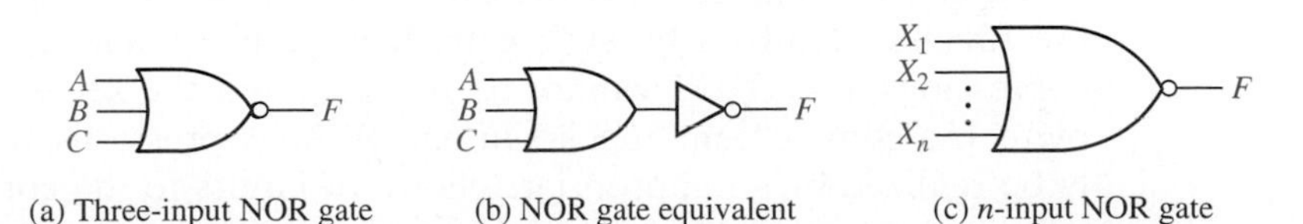

(a) Three-input NOR gate (b) NOR gate equivalent (c) $n$-input NOR gate

The output of an $n$-input NOR gate, shown in Figure 7-9(c), is

$$F = (X_1 + X_2 + \ldots + X_n)' = X_1' X_2' \ldots X_n' \tag{7-9}$$

A set of logic operations is said to be *functionally complete* if any Boolean function can be expressed in terms of this set of operations. The set AND, OR, and NOT is obviously functionally complete because any function can be expressed in sum-of-products form, and a sum-of-products expression uses only the AND, OR, and NOT operations. Similarly, a set of logic gates is functionally complete if all switching functions can be realized using this set of gates. Because the set of operations AND, OR, and NOT is functionally complete, any set of logic gates which can realize AND, OR, and NOT is also functionally complete. AND and NOT are a functionally complete set of gates because OR can also be realized using AND and NOT:

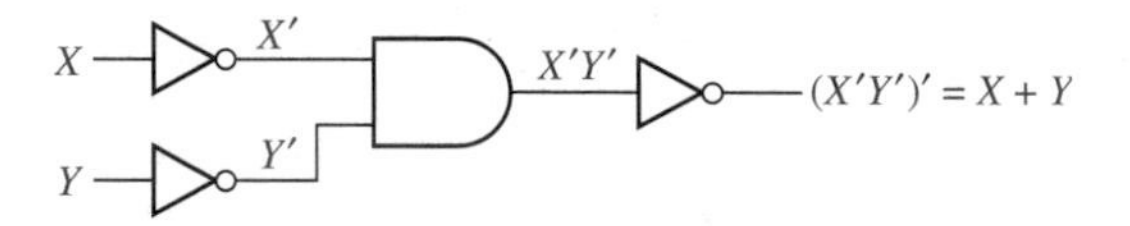


If a single gate forms a functionally complete set by itself, then any switching function can be realized using only gates of that type. The NAND gate is an example of such a gate. Because the NAND gate performs the AND operation followed by an inversion, NOT, AND, and OR can be realized using only NAND gates, as shown in Figure 7-10. Thus, any switching function can be realized using only NAND gates. An easy method for converting an AND-OR circuit to a NAND circuit is discussed in the next section. Similarly, any function can be realized using only NOR gates.

**FIGURE 7-10**
NAND Gate Realization of NOT, AND, and OR

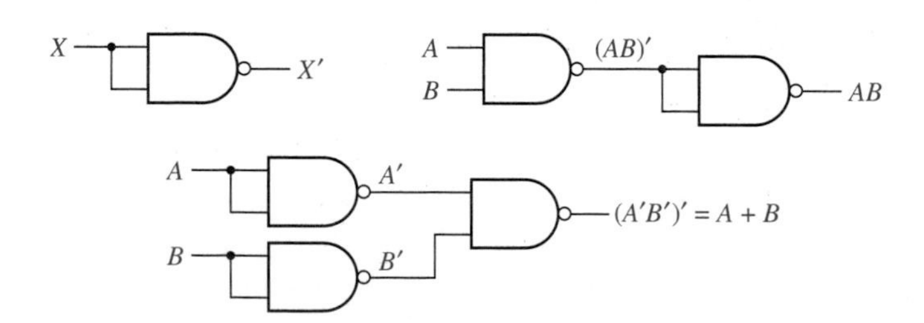


The following procedure can be used to determine if a given set of gates is functionally complete. First, write out a minimum sum-of-products expression for the function realized by each gate. If no complement appears in any of these expressions, then NOT cannot be realized, and the set is not functionally complete. If a complement appears in one of the expressions, then NOT can generally be realized by an appropriate choice of inputs to the corresponding gate. (We will always assume that 0 and 1 are available as gate inputs). Next, attempt to realize AND or OR, keeping in mind that NOT is now available. Once AND or

OR has been realized, the other one can always be realized using DeMorgan's laws if no more direct procedure is apparent. For example, if OR and NOT are available, AND can be realized by

$$XY = (X' + Y')'$$

## 7.3 Design of Two-Level NAND- and NOR-Gate Circuits

A two-level circuit composed of AND and OR gates is easily converted to a circuit composed of NAND gates or NOR gates. This conversion is carried out by using $F = (F')'$ and then applying DeMorgan's laws:

$$(X_1 + X_2 + \ldots + X_n)' = X_1' X_2' \ldots X_n' \quad (7\text{-}11)$$

$$(X_1 X_2 \ldots X_n)' = X_1' + X_2' + \ldots + X_n' \quad (7\text{-}12)$$

The following example illustrates conversion of a minimum sum-of-products form to several other two-level forms:

$$F = A + BC' + B'CD = [(A + BC' + B'CD)']' \quad (7\text{-}13)$$

$$= [A' \cdot (BC')' \cdot (B'CD)']' \quad \text{(by 7-11)} \quad (7\text{-}14)$$

$$= [A' \cdot (B' + C) \cdot (B + C' + D')]' \quad \text{(by 7-12)} \quad (7\text{-}15)$$

$$= A + (B' + C)' + (B + C' + D')' \quad \text{(by 7-12)} \quad (7\text{-}16)$$

Equations (7-13), (7-14), (7-15), and (7-16) represent the AND-OR, NAND-NAND, OR-NAND, and NOR-OR forms, respectively, as shown in Figure 7-11.

Rewriting Equation (7-16) in the form

$$F = \{[A + (B' + C)' + (B + C' + D')' ]'\}' \quad (7\text{-}17)$$

leads to a three-level NOR-NOR-INVERT circuit. However, if we want a two-level circuit containing only NOR gates, we should start with the minimum product-of-sums form for $F$ instead of the minimum sum of products. After obtaining the minimum product of sums from a Karnaugh map, $F$ can be written in the following two-level forms:

$$F = (A + B + C)(A + B' + C')(A + C' + D) \quad (7\text{-}18)$$

$$= \{[(A + B + C)(A + B' + C')(A + C' + D)]'\}'$$

$$= [(A + B + C)' + (A + B' + C')' + (A + C' + D)']' \quad \text{(by 7-12)} \quad (7\text{-}19)$$

$$= (A'B'C' + A'BC + A'CD')' \quad \text{(by 7-11)} \quad (7\text{-}20)$$

$$= (A'B'C')' \cdot (A'BC)' \cdot (A'CD')' \quad \text{(by 7-11)} \quad (7\text{-}21)$$

**FIGURE 7-11**
Eight Basic Forms for Two-Level Circuits

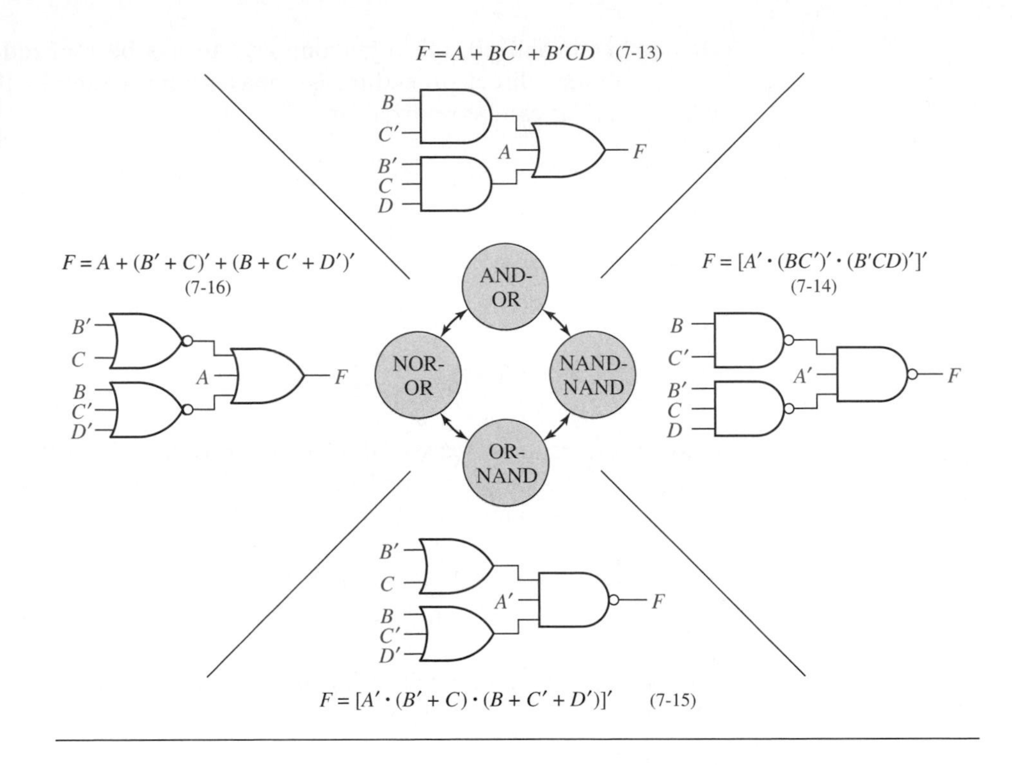


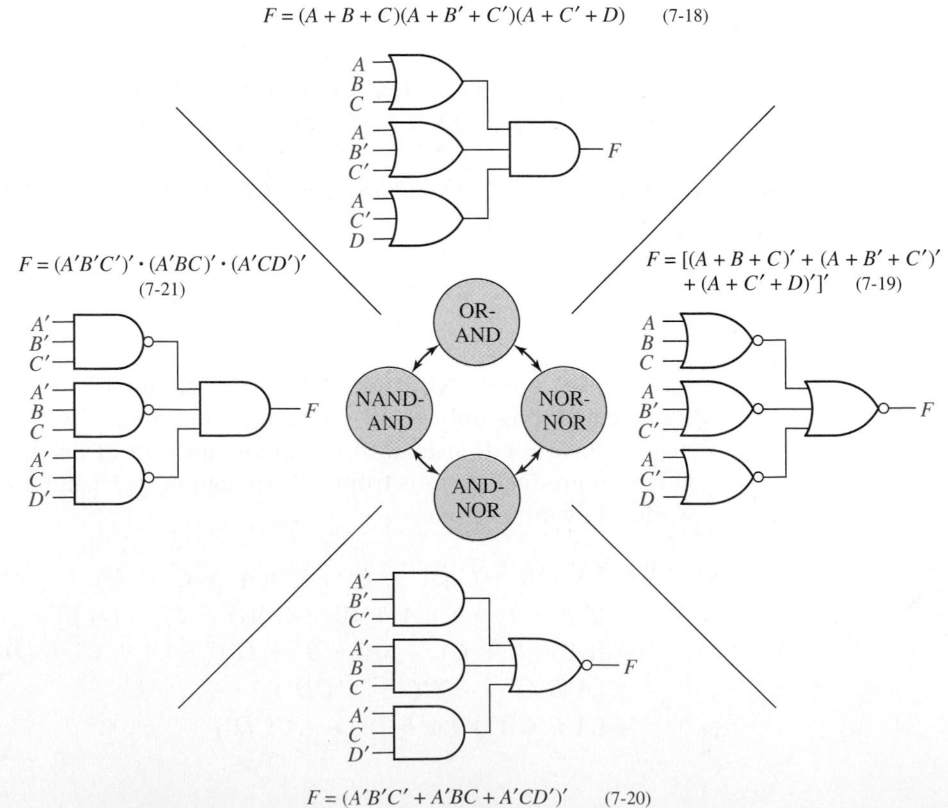

Equations (7-18),(7-19),(7-20), and (7-21) represent the OR-AND, NOR-NOR, AND-NOR, and NAND-AND forms, respectively, as shown in Figure 7-11. Two-level AND-NOR (AND-OR-INVERT) circuits are available in integrated-circuit form. Some types of NAND gates can also realize AND-NOR circuits when the so-called *wired OR* connection is used.

The other eight possible two-level forms (AND-AND, OR-OR, OR-NOR, AND-NAND, NAND-NOR, NOR-NAND, etc.) are degenerate in the sense that they cannot realize all switching functions. Consider, for example, the following NAND-NOR circuit:

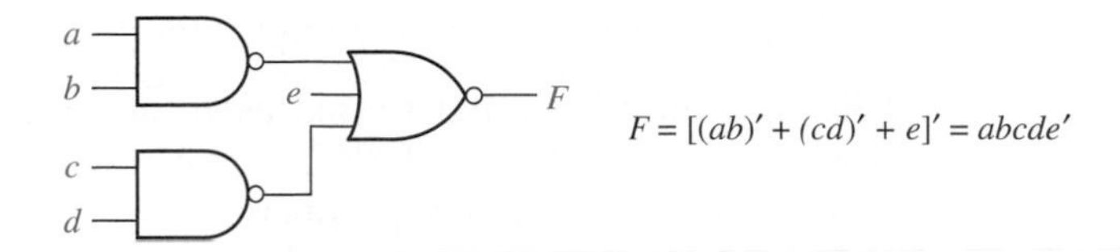


From this example, it is clear that the NAND-NOR form can realize only a product of literals and not a sum of products.

Because NAND and NOR gates are readily available in integrated circuit form, two of the most commonly used circuit forms are the NAND-NAND and the NOR-NOR. Assuming that all variables and their complements are available as inputs, the following method can be used to realize $F$ with NAND gates:

*Procedure for designing a minimum two-level NAND-NAND circuit:*

1. Find a minimum *sum-of-products* expression for $F$.
2. Draw the corresponding two-level AND-OR circuit.
3. Replace all gates with NAND gates leaving the gate interconnections unchanged. If the output gate has any single literals as inputs, complement these literals.

Figure 7-12 illustrates the transformation of step 3. Verification that this transformation leaves the circuit output unchanged follows. In general, $F$ is a sum of literals ($\ell_1, \ell_2, \ldots$) and product terms ($P_1, P_2, \ldots$):

$$F = \ell_1 + \ell_2 + \cdots + P_1 + P_2 + \cdots$$

After applying DeMorgan's law,

$$F = (\ell_1' \ell_2' \cdots P_1' P_2' \cdots)'$$

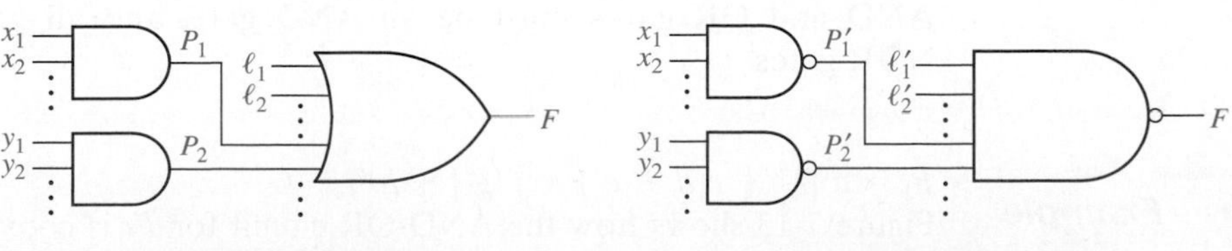


**FIGURE 7-12** AND-OR to NAND-NAND Transformation

So the output OR gate is replaced with a NAND gate with inputs $\ell'_1, \ell'_2, \ldots, P'_1, P'_2, \ldots$ Because product terms $P_1, P_2, \ldots$ are each realized with an AND gate, $P'_1, P'_2, \ldots$ are each realized with a NAND gate in the transformed circuit.

Assuming that all variables and their complements are available as inputs, the following method can be used to realize $F$ with NOR gates:

*Procedure for designing a minimum two-level NOR-NOR circuit:*

1. Find a minimum *product-of-sums* expression for $F$.
2. Draw the corresponding two-level OR-AND circuit.
3. Replace all gates with NOR gates leaving the gate interconnections unchanged. If the output gate has any single literals as inputs, complement these literals.

This procedure is similar to that used for designing NAND-NAND circuits. Note, however, that for the NOR-NOR circuit, the starting point is a minimum *product of sums* rather than a sum of products.

## 7.4 Design of Multi-Level NAND- and NOR-Gate Circuits

The following procedure may be used to design multi-level NAND-gate circuits:

1. Simplify the switching function to be realized.
2. Design a multi-level circuit of AND and OR gates. The output gate must be OR. AND gate outputs cannot be used as AND-gate inputs; OR-gate outputs cannot be used as OR-gate inputs.
3. Number the levels starting with the output gate as level 1. Replace all gates with NAND gates, leaving all interconnections between gates unchanged. Leave the inputs to levels 2, 4, 6, . . . unchanged. Invert any literals which appear as inputs to levels 1, 3, 5, . . . .

The validity of this procedure is easily proven by dividing the multi-level circuit into two-level subcircuits and applying the previous results for two-level circuits to each of the two-level subcircuits. The example of Figure 7-13 illustrates the procedure. Note that if step 2 is performed correctly, each level of the circuit will contain only AND gates or only OR gates.

The procedure for the design of multi-level NOR-gate circuits is exactly the same as for NAND-gate circuits except the output gate of the circuit of AND and OR gates must be an AND gate, and all gates are replaced with NOR gates.

**Example**

$F_1 = a'[b' + c(d + e') + f'g'] + hi'j + k$

Figure 7-13 shows how the AND-OR circuit for $F_1$ is converted to the corresponding NAND circuit.

**FIGURE 7-13** Multi-Level Circuit Conversion to NAND Gates

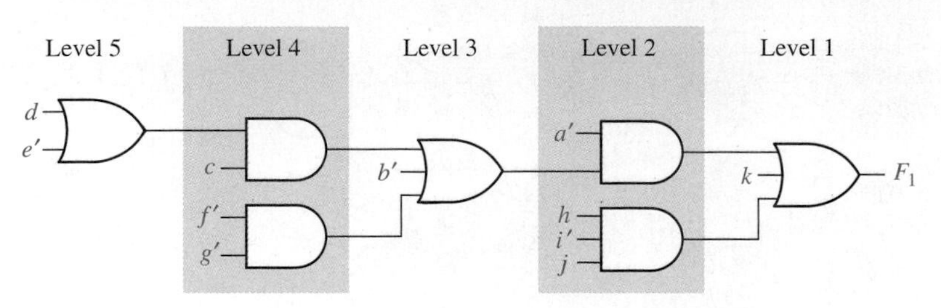


(a) AND-OR network

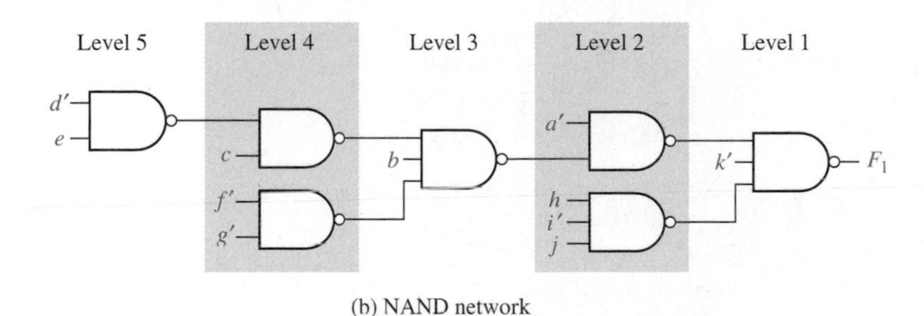


(b) NAND network

## 7.5 Circuit Conversion Using Alternative Gate Symbols

Logic designers who design complex digital systems often find it convenient to use more than one representation for a given type of gate. For example, an inverter can be represented by

$A$ — $A'$ or $A$ — $A'$

In the second case, the inversion "bubble" is at the input instead of the output. Figure 7-14 shows some alternative representations for AND, OR, NAND, and NOR gates. These equivalent gate symbols are based on DeMorgan's Laws.

**FIGURE 7-14** Alternative Gate Symbols

| $AB = (A' + B')'$ | $A + B = (A'B')'$ | $(AB)' = A' + B'$ | $(A + B)' = A'B'$ |
|---|---|---|---|
| (a) AND | (b) OR | (c) NAND | (d) NOR |

These alternative symbols can be used to facilitate the analysis and design of NAND and NOR gate circuits. Figure 7-15(a) shows a simple NAND-gate circuit. To analyze the circuit, we will replace the NAND gates at the first and third levels with the alternative NAND gate symbol. This eliminates the inversion bubble at the circuit output.

**FIGURE 7-15**
NAND Gate Circuit Conversion

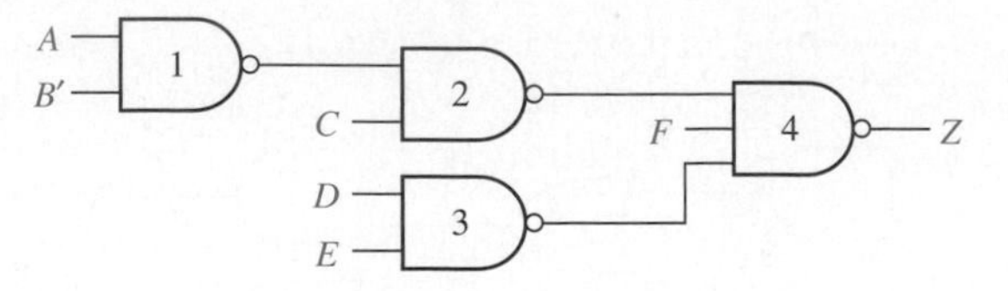


(a) NAND gate network

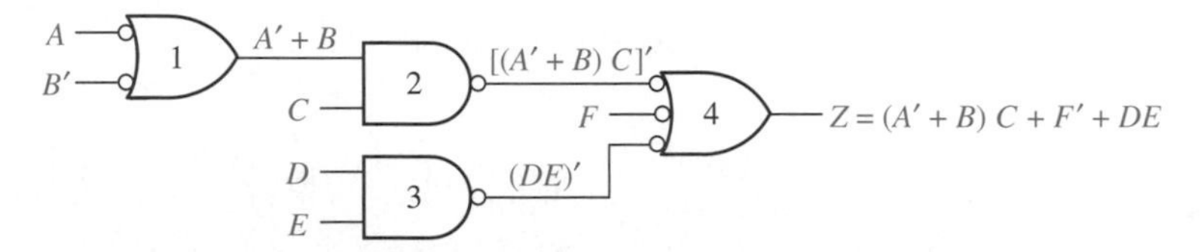


(b) Alternate form for NAND gate network

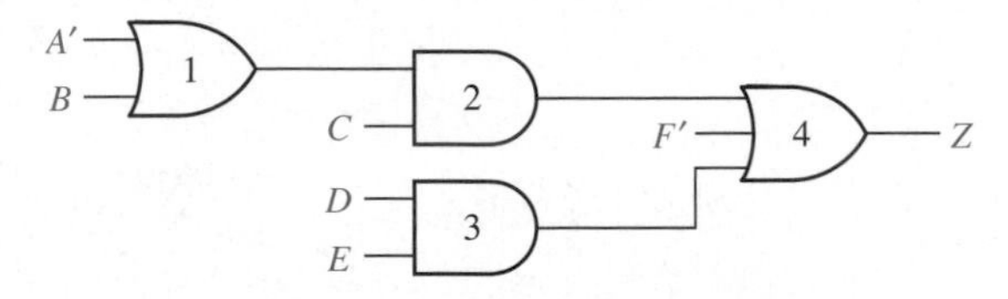


(c) Equivalent AND-OR network

In the resulting circuit [Figure 7-15(b)], inverted outputs (those with a bubble) are always connected to inverted inputs, and noninverted outputs are connected to noninverted inputs. Because two inversions in a row cancel each other out, we can easily analyze the circuit without algebraically applying DeMorgan's laws. Note, for example, that the output of gate 2 is $[(A' + B)C]'$, but the term $(A' + B)C$ appears in the output function. We can also convert the circuit to an AND-OR circuit by simply removing the double inversions [see Figure 7-15(c)]. When a single input variable is connected to an inverted input, we must also complement that variable when we remove the inversion from the gate input. For example, $A$ in Figure 7-15(b) becomes $A'$ in Figure 7-15(c).

The circuit of AND and OR gates shown in Figure 7-16(a) can easily be converted to a NOR-gate circuit because the output gate is an AND gate, and AND and OR gates alternate throughout the circuit. That is, AND gate outputs connect only to OR gate inputs, and OR gate outputs connect only to AND gate inputs. To carry out conversion to NOR gates, we first replace all of the OR and AND gates with NOR gates, as shown in Figure 7-16(b). Because each inverted gate output drives an inverted gate input, the pairs of inversions cancel. However, when an input variable drives an inverted input, we have added a single inversion, so we must complement the variable to compensate. Therefore, we have complemented $C$ and $G$. The resulting NOR-gate circuit is equivalent to the original AND-OR circuit.

Even if AND and OR gates do not alternate, we can still convert an AND-OR circuit to a NAND or NOR circuit, but it may be necessary to add extra inverters so that each added inversion is cancelled by another inversion. The following procedure may be used to convert to a NAND (or NOR) circuit:

1. Convert all AND gates to NAND gates by adding an inversion bubble at the output. Convert all OR gates to NAND gates by adding inversion bubbles at the

**FIGURE 7-16**
Conversion to NOR Gates

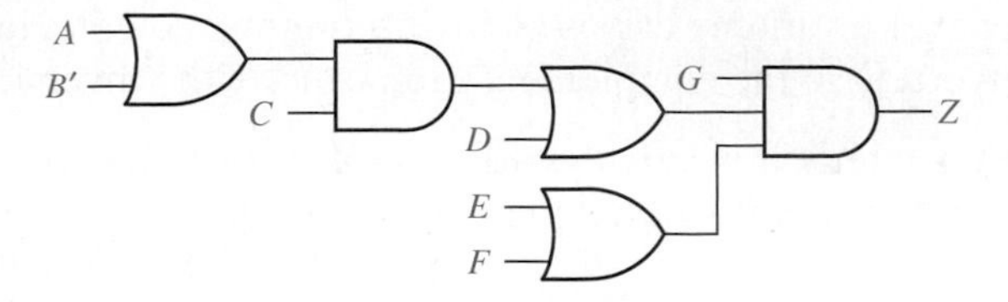


(a) Circuit with OR and AND gates

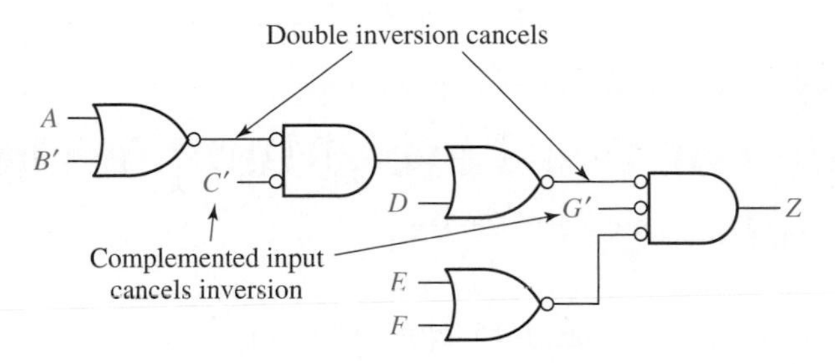


(b) Equivalent circuit with NOR gates

inputs. (To convert to NOR, add inversion bubbles at all OR gate outputs and all AND gate inputs.)

2. Whenever an inverted output drives an inverted input, no further action is needed because the two inversions cancel.
3. Whenever a noninverted gate output drives an inverted gate input or vice versa, insert an inverter so that the bubbles will cancel. (Choose an inverter with the bubble at the input or output as required.)

**FIGURE 7-17**
Conversion of AND-OR Circuit to NAND Gates

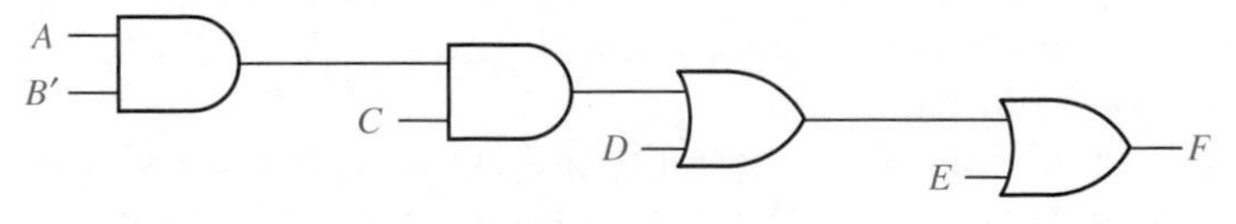


(a) AND-OR network

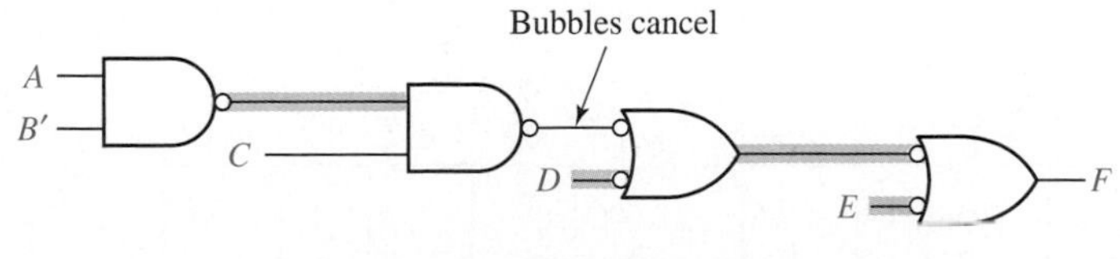


(b) First step in NAND conversion

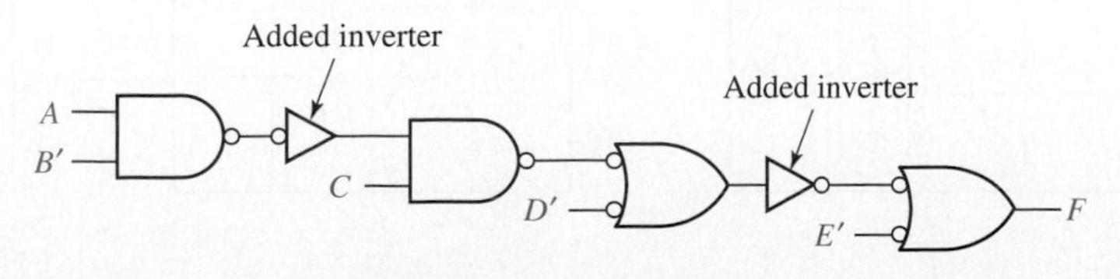


(c) Completed conversion

4. Whenever a variable drives an inverted input, complement the variable (or add an inverter) so the complementation cancels the inversion at the input.

In other words, if we always add bubbles (or inversions) in pairs, the function realized by the circuit will be unchanged. To illustrate the procedure we will convert Figure 7-17(a) to NANDs. First, we add bubbles to change all gates to NAND gates (Figure 7-17(b)). In four places (highlighted in blue), we have added only a single inversion. This is corrected in Figure 7-17(c) by adding two inverters and complementing two variables.

## 7.6 Design of Two-Level, Multiple-Output Circuits

Solution of digital design problems often requires the realization of several functions of the same variables. Although each function could be realized separately, the use of some gates in common between two or more functions sometimes leads to a more economical realization. The following example illustrates this:

Design a circuit with four inputs and three outputs which realizes the functions

$$
\begin{aligned}
F_1(A, B, C, D) &= \Sigma\, m(11, 12, 13, 14, 15) \\
F_2(A, B, C, D) &= \Sigma\, m(3, 7, 11, 12, 13, 15) \\
F_3(A, B, C, D) &= \Sigma\, m(3, 7, 12, 13, 14, 15) \qquad (7\text{-}22)
\end{aligned}
$$

First, each function will be realized individually. The Karnaugh maps, functions, and resulting circuit are given in Figures 7-18 and 7-19. The cost of this circuit is 9 gates and 21 gate inputs.

An obvious way to simplify this circuit is to use the same gate for $AB$ in both $F_1$ and $F_3$. This reduces the cost to eight gates and 19 gate inputs. (Another, but less obvious, way to simplify the circuit is possible.) Observing that the term $ACD$ is

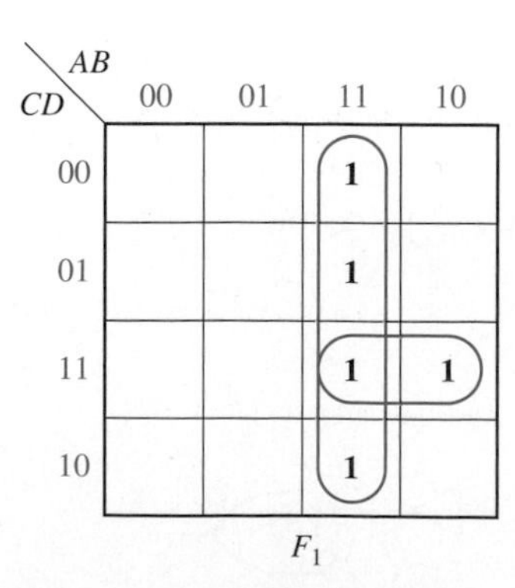


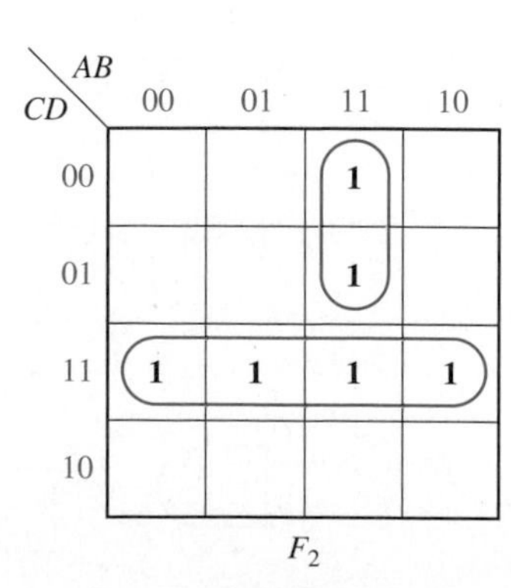


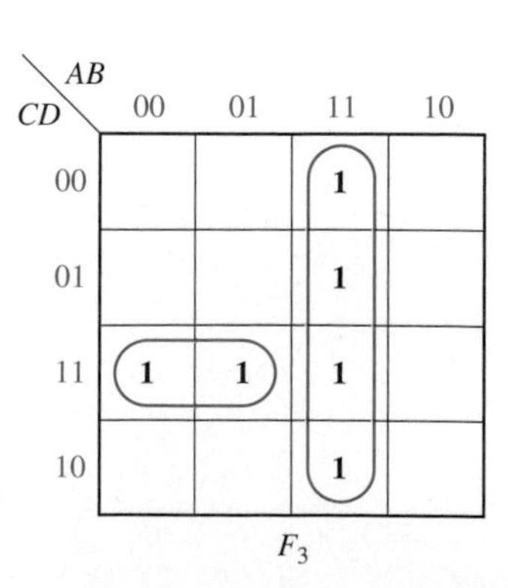


**FIGURE 7-18** Karnaugh Maps for Equations (7-22)

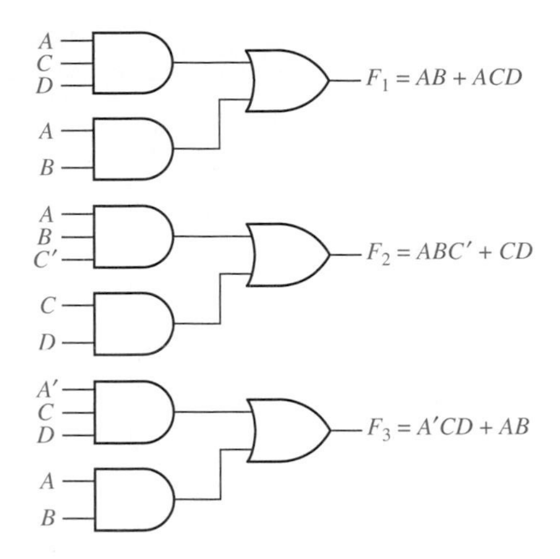


**FIGURE 7-19** Realization of Equations (7-22)

necessary for the realization of $F_1$ and $A'CD$ is necessary for $F_3$, if we replace $CD$ in $F_2$ by $A'CD + ACD$, the realization of $CD$ is unnecessary and one gate is saved. Figure 7-20 shows the reduced circuit, which requires seven gates and 18 gate inputs. Note that $F_2$ is realized by the expression $ABC' + A'CD + ACD$ which is not a minimum sum of products, and two of the terms are not prime implicants of $F_2$. Thus in realizing multiple-output circuits, the use of a minimum sum of prime implicants for each function does not necessarily lead to a minimum cost solution for the circuit as a whole.

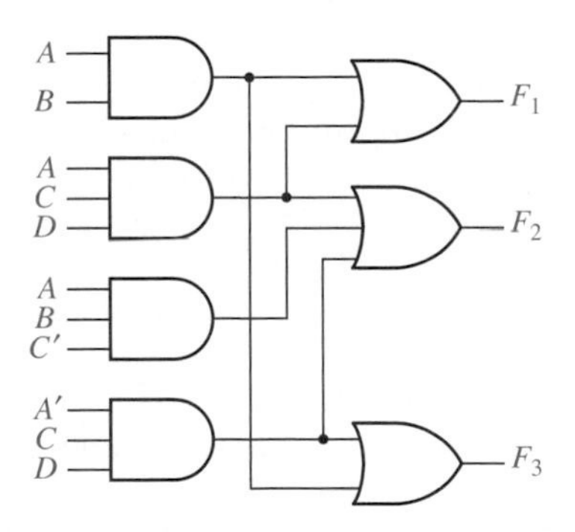


**FIGURE 7-20** Multiple-Output Realization of Equations (7-22)

When designing multiple-output circuits, you should try to minimize the total number of gates required. If several solutions require the same number of gates, the one with the minimum number of gate inputs should be chosen. The next example further illustrates the use of common terms to save gates. A four-input, three-output circuit is to be designed to realize

$$
\begin{aligned}
f_1 &= \Sigma\, m(2, 3, 5, 7, 8, 9, 10, 11, 13, 15) \\
f_2 &= \Sigma\, m(2, 3, 5, 6, 7, 10, 11, 14, 15) \\
f_3 &= \Sigma\, m(6, 7, 8, 9, 13, 14, 15)
\end{aligned}
\qquad (7\text{-}23)
$$

FIGURE 7-21

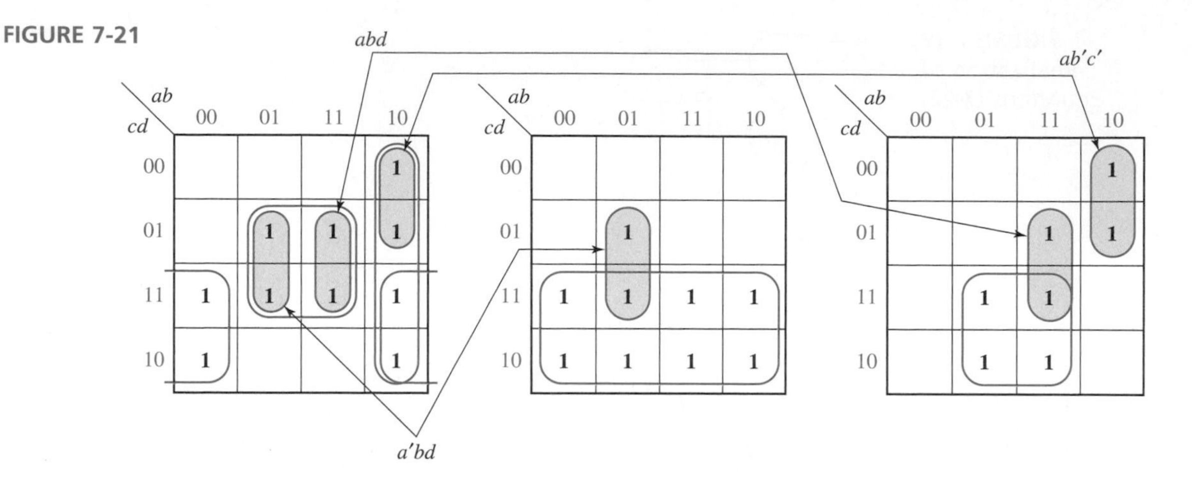


First, we plot maps for $f_1$, $f_2$, and $f_3$ (Figure 7-21). If each function is minimized separately, the result is

$$f_1 = bd + b'c + ab'$$
$$f_2 = c + a'bd$$
$$f_3 = bc + ab'c' + \left\{\begin{matrix} abd \\ \text{or} \\ ac'd \end{matrix}\right\} \begin{matrix} \text{10 gates,} \\ \text{25 gate inputs} \end{matrix} \tag{7-23(a)}$$

By inspecting the maps, we can see that terms $a'bd$ (from $f_2$), $abd$ (from $f_3$), and $ab'c'$ (from $f_3$) can be used in $f_1$. If $bd$ is replaced with $a'bd + abd$, then the gate needed to realize $bd$ can be eliminated. Because $m_{10}$ and $m_{11}$ in $f_1$ are already covered by $b'c$, $ab'c'$ (from $f_3$) can be used to cover $m_8$ and $m_9$, and the gate needed to realize $ab'$ can be eliminated. The minimal solution is therefore

$$\begin{aligned} f_1 &= \underline{a'bd} + \underline{abd} + \underline{ab'c'} + b'c \\ f_2 &= c + \underline{a'bd} && \text{eight gates} \\ f_3 &= bc + \underline{ab'c'} + \underline{abd} && \text{22 gate inputs} \end{aligned} \tag{7-23(b)}$$

(Terms which are used in common between two functions are underlined.)

When designing multiple-output circuits, it is sometimes best not to combine a 1 with its adjacent 1's, as illustrated in the example of Figure 7-22.

The solution with the maximum number of common terms is not necessarily best, as illustrated in the example of Figure 7-23.

## Determination of Essential Prime Implicants for Multiple-Output Realization

As a first step in determining a minimum two-level, multiple-output realization, it is often desirable to determine essential prime implicants. However, we must be careful because some of the prime implicants essential to an individual function may not be essential to the multiple-output realization. For example, in Figure 7-21, $bd$ is an

**FIGURE 7-22**

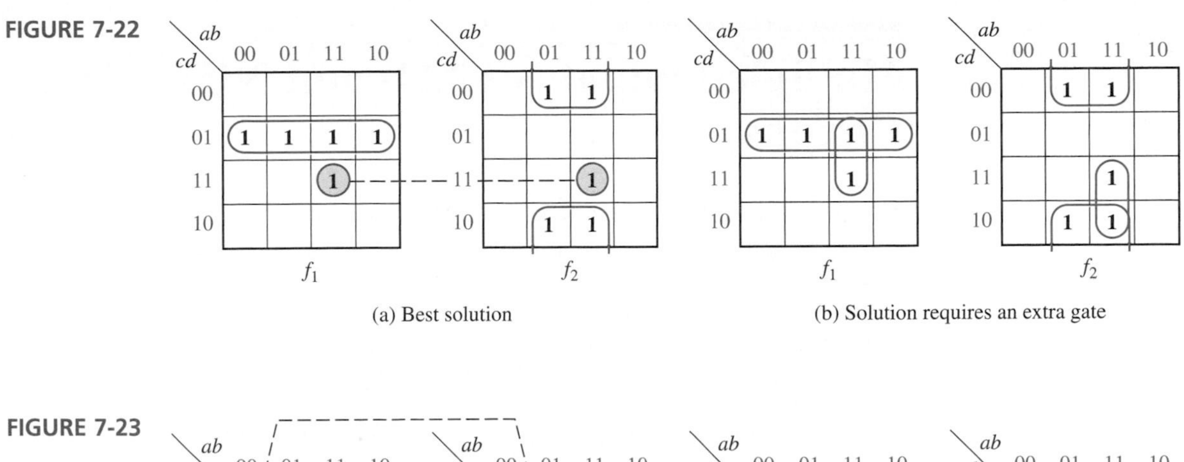

(a) Best solution

(b) Solution requires an extra gate

**FIGURE 7-23**

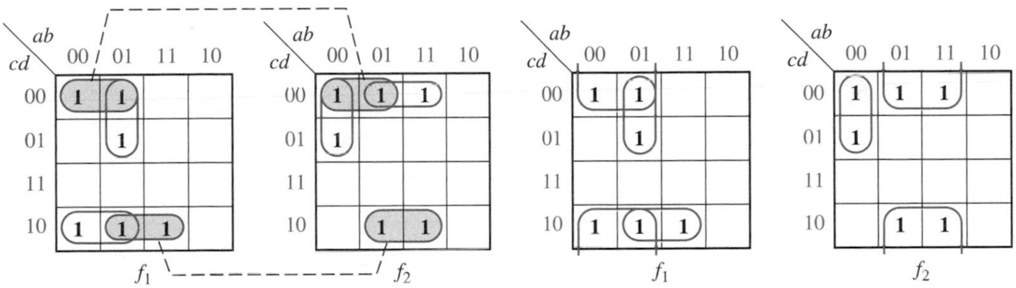

(a) Solution with maximum number of common terms requires 8 gates, 26 inputs

(b) Best solution requires 7 gates, 18 inputs and has no common terms

essential prime implicant of $f_1$ (only prime implicant which covers $m_5$), but it is not essential to the multiple-output realization. The reason that *bd* is *not* essential is that $m_5$ also appears on the $f_2$ map and, hence, might be covered by a term which is shared by $f_1$ and $f_2$.

We can find prime implicants which are essential to one of the functions *and* to the multiple-output realization by a modification of the procedure used for the single-output case. In particular, when we check each 1 on the map to see if it is covered by only one prime implicant, we will only check those 1's which do not appear on the other function maps. Thus, in Figure 7-22 we find that $c'd$ *is* essential to $f_1$ for the multiple-output realization (because of $m_1$), but *abd* is not essential because $m_{15}$ also appears on the $f_2$ map. In Figure 7-23, the only minterms of $f_1$ which do not appear on the $f_2$ map are $m_2$ and $m_5$. The only prime implicant which covers $m_2$ is $a'd'$; hence, $a'd'$ *is* essential to $f_1$ in the multiple-output realization. Similarly, the only prime implicant which covers $m_5$ is $a'bc'$, and $a'bc'$ is essential. On the $f_2$ map, $bd'$ *is* essential. Why?

Once the essential prime implicants for $f_1$ and $f_2$ have been looped, selection of the remaining terms to form the minimum solution is obvious in this example. The techniques for finding essential prime implicants outlined above cannot be applied in a problem such as Figure 7-21 where every minterm of $f_1$ also appears on the $f_2$ or $f_3$ map. More sophisticated techniques are available for finding essential multiple-output terms for such problems, but these techniques are beyond the scope of this text.

# 7.7 Multiple-Output NAND- and NOR-Gate Circuits

The procedure given in Section 7.4 for design of single-output, multi-level NAND- and NOR-gate circuits also applies to multiple-output circuits. If all of the output gates are OR gates, direct conversion to a NAND-gate circuit is possible. If all of the output gates are AND, direct conversion to a NOR-gate circuit is possible. Figure 7-24 gives an example of converting a 2-output circuit to NOR gates. Note that the inputs to the first and third levels of NOR gates are inverted.

$$F_1 = [(a + b')c + d](e' + f) \qquad F_2 = [(a + b')c + g'](e' + f)h$$

**FIGURE 7-24** Multi-level Circuit Conversion to NOR Gates

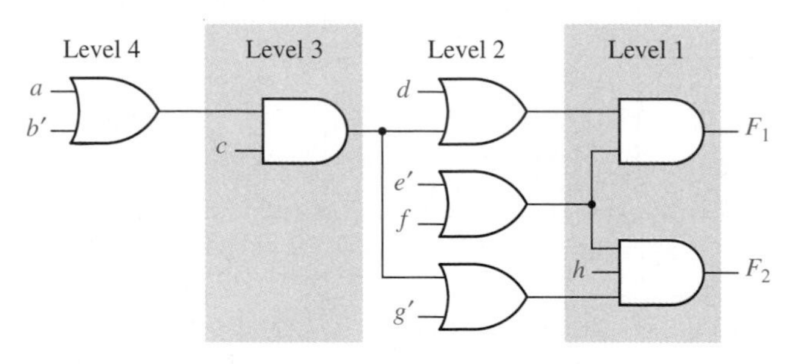


(a) Network of AND and OR gates

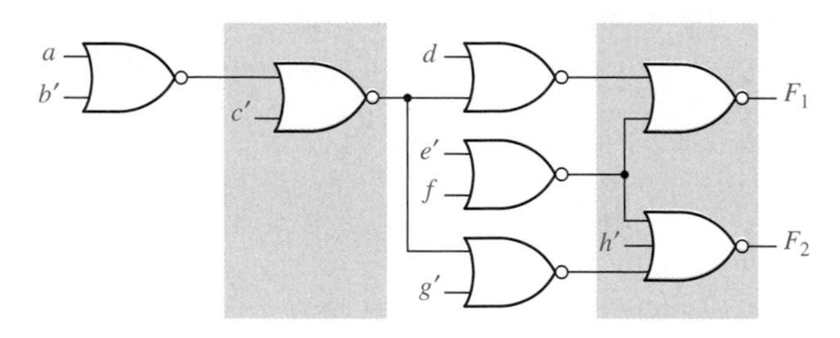


(b) NOR network

# Problems

**7.1** Using AND and OR gates, find a minimum circuit to realize

$$f(a, b, c, d) = m_4 + m_6 + m_7 + m_8 + m_9 + m_{10}$$

(a) using two-level logic

(b) using three-level logic (12 gate inputs minimum)

**7.2** Realize the following functions using AND and OR gates. Assume that there are no restrictions on the number of gates which can be cascaded and minimize the number of gate inputs.

(a) $AC'D + ADE' + BE' + BC' + A'D'E'$

(b) $AE + BDE + BCE + BCFG + BDFG + AFG$

**7.3** Find eight different simplified two-level gate circuits to realize

$$F(a, b, c, d) = a'bd + ac'd$$

**7.4** Find a minimum three-level NAND gate circuit to realize

$$F(A, B, C, D) = \Sigma\, m(5, 10, 11, 12, 13) \quad \text{(four gates)}$$

**7.5** Realize $Z = A'D + A'C + AB'C'D'$ using four NOR gates.

**7.6** Realize $Z = ABC + AD + C'D'$ using only two-input NAND gates. Use as few gates as possible.

**7.7** Realize $Z = AE + BDE + BCEF$ using only two-input NOR gates. Use as few gates as possible.

**7.8** (a) Convert the following circuit to all NAND gates, by adding bubbles and inverters where necessary.

(b) Convert to all NOR gates (an inverter at the output is allowed).

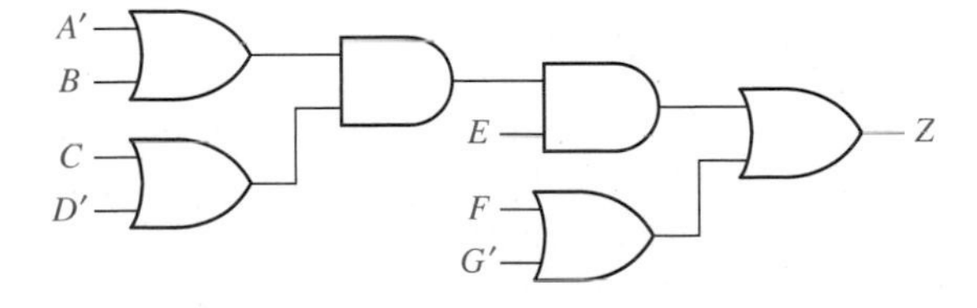


**7.9** Find a two-level, multiple-output AND-OR gate circuit to realize the following functions. Minimize the required number of gates (six gates minimum).

$$f_1 = ac + ad + b'd \quad \text{and} \quad f_2 = a'b' + a'd' + cd'$$

**7.10** Find a minimum two-level, multiple-output AND-OR gate circuit to realize these functions.

$$f_1(a, b, c, d) = \Sigma\, m(3, 4, 6, 9, 11)$$
$$f_2(a, b, c, d) = \Sigma\, m(2, 4, 8, 10, 11, 12)$$
$$f_3(a, b, c, d) = \Sigma\, m(3, 6, 7, 10, 11) \quad \text{(11 gates minimum)}$$

**7.11** Find a minimum two-level OR-AND circuit to simultaneously realize

$$F_1(a, b, c, d) = \Sigma\, m(2, 3, 8, 9, 14, 15)$$
$$F_2(a, b, c, d) = \Sigma\, m(0, 1, 5, 8, 9, 14, 15)$$

(minimum solution has eight gates)

**7.12** Find a minimum two-level OR-AND circuit to realize the functions given in Equations (7-23) on page 205 (nine gates minimum)

**7.13** (a) Find a minimum two-level NAND-NAND circuit to realize the functions given in Equations (7-23) on page 205.
(b) Find a minimum two-level NOR-NOR circuit to realize the functions given in Equations (7-23).

**7.14** Using AND and OR gates, find a minimum circuit to realize

$$f(a, b, c, d) = M_0 M_1 M_3 M_{13} M_{14} M_{15}$$

(a) using two-level logic
(b) using three-level logic (12 gate inputs minimum)

**7.15** Using AND and OR gates, find a minimum two-level circuit to realize
(a) $F = a'c + bc'd + ac'd$
(b) $F = (b' + c)(a + b' + d)(a + b + c' + d)$
(c) $F = a'cd' + a'bc + ad$
(d) $F = a'b + ac + bc + bd'$

**7.16** Realize the following functions using AND and OR gates. Assume that there are no restrictions on the number of gates which can be cascaded and minimize the number of gate inputs.
(a) $ABC' + ACD + A'BC + A'C'D$
(b) $ABCE + ABEF + ACD' + ABEG + ACDE$

**7.17** A combinational switching circuit has four inputs ($A$, $B$, $C$, $D$) and one output ($F$). $F = 0$ iff three or four of the inputs are 0.
(a) Write the maxterm expansion for $F$.
(b) Using AND and OR gates, find a minimum three-level circuit to realize $F$ (five gates, 12 inputs).

**7.18** Find eight different simplified two-level gate circuits to realize
(a) $F(w, x, y, z) = (x + y' + z)(x' + y + z)w$
(b) $F(a, b, c, d) = \Sigma\, m(4, 5, 8, 9, 13)$

**7.19** Implement $f(x, y, z) = \Sigma\, m(0, 1, 3, 4, 7)$ as a two-level gate circuit, using a minimum number of gates.
(a) Use AND gates and NAND gates.
(b) Use NAND gates only.

**7.20** Implement $f(a, b, c, d) = \Sigma\, m(3, 4, 5, 6, 7, 11, 15)$ as a two-level gate circuit, using a minimum number of gates.
(a) Use OR gates and NOR gates.
(b) Use NOR gates only.

**7.21** Realize each of the following functions as a minimum two-level NAND-gate circuit and as a minimum two-level NOR-gate circuit.

(a) $F(A, B, C, D) = BD' + B'CD + A'BC + A'BC'D + B'D'$

(b) $f(a, b, c, d) = \Pi\, M(0, 1, 7, 9, 10, 13) \bullet \Pi\, D(2, 6, 14, 15)$

(c) $f(a, b, c, d) = \Sigma\, m(0, 2, 5, 10) + \Sigma\, d(3, 6, 9, 13, 14, 15)$

(d) $F(A, B, C, D, E) = \Sigma\, m(0, 2, 4, 5, 11, 14, 16, 17, 18, 22, 23, 25, 26, 31) + \Sigma\, d(3, 19, 20, 27, 28)$

(e) $F(A, B, C, D, E) = \Pi\, M(3, 4, 8, 9, 10, 11, 12, 13, 14, 16, 19, 22, 25, 27) \bullet \Pi\, D(16, 18, 28, 29)$

(f) $f(a, b, c, d) = \Pi\, M(1, 3, 10, 11, 13, 14, 15) \bullet \Pi\, D(4, 6)$

(g) $f(w, x, y, z) = \Sigma\, m(1, 2, 4, 6, 8, 9, 11, 12, 13) + \Sigma\, d(0, 7, 10, 15)$

**7.22** A combinational switching circuit has four inputs and one output as shown. $F = 0$ iff three or four of the inputs are 1.

(a) Write the maxterm expansion for $F$.

(b) Using AND and OR gates, find a minimum three-level circuit to realize $F$ (5 gates, 12 inputs).

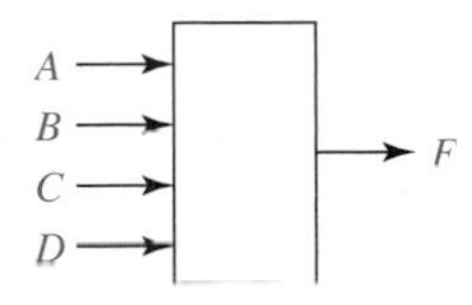


**7.23** Implement $f(a, b, c, d) = \Sigma\, m(3, 4, 5, 6, 7, 11, 15)$ as a two-level gate circuit, using a minimum number of gates.

(a) Use AND gates and NAND gates.

(b) Use OR gates and NAND gates.

(c) Use NAND gates only.

**7.24** (a) Use gate equivalences to convert the circuit into a four-level circuit containing only NAND gates and a minimum number of inverters. (Assume the inputs are available only in uncomplemented form.)

(b) Derive a minimum SOP expression for $f$.

(c) By manipulating the expression for $f$, find a three-level circuit containing only five NAND gates and inverters.

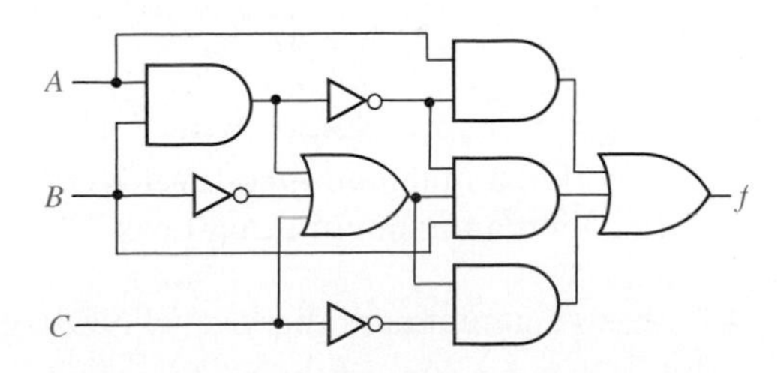

**7.25** (a) Use gate equivalences to convert the circuit of Problem 7.24 into a five-level circuit containing only NOR gates and a minimum number of inverters. (Assume the inputs are available only in uncomplemented form.)
(b) Derive a minimum POS expression *for f*.
(c) By manipulating the expression for *f*, find a four-level circuit containing only six NOR gates and inverters.

**7.26** In the circuit, replace each NOR gate by an AND or OR gate so that the resulting circuit contains the fewest inverters possible. Assume the inputs are available in both true and complemented form. Do *not* replace the exclusive-OR gates.

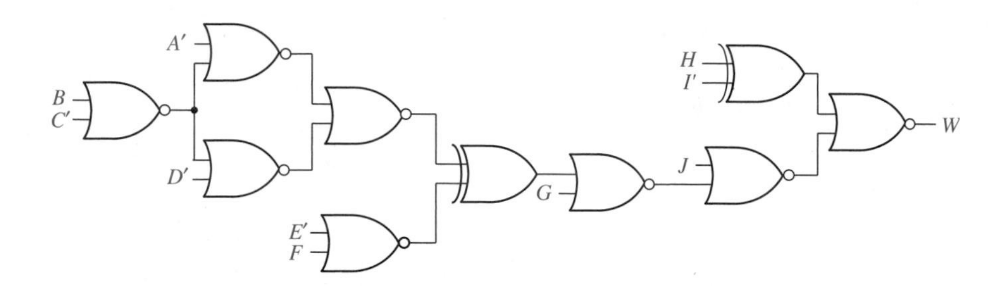


**7.27** (a) Convert the circuit shown into a four-level circuit only containing AND and OR gates and a minimum number of inverters.
(b) Derive a sum-of-products expression for *f*.
(c) Find a circuit that realizes $f'$ containing only NOR gates (no internal inverters). (*Hint:* Use gate conversions to convert the NAND gates in the given circuit to NOR gates.)

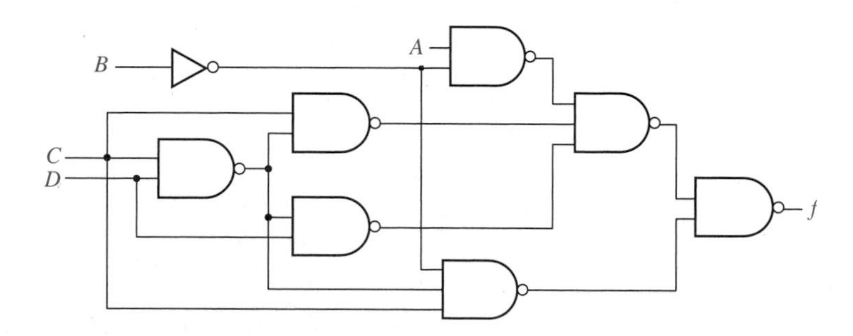


**7.28** $f(a, b, c, d, e) = \Sigma\, m(2, 3, 6, 12, 13, 16, 17, 18, 19, 22, 24, 25, 27, 28, 29, 31)$
(a) Find a minimum two-level NOR-gate circuit to realize *f*.
(b) Find a minimum three-level NOR-gate circuit to realize *f*.

**7.29** Design a minimum three-level NOR-gate circuit to realize

$$f = a'b' + abd + acd$$

**7.30** Find a minimum four-level NAND- or NOR-gate circuit to realize
(a) $Z = abe'f + c'e'f + d'e'f + gh$
(b) $Z = (a' + b + e + f)(c' + a' + b)(d' + a' + b)(g + h)$

**7.31** Implement $abde' + a'b' + c$ using four NOR gates.

**7.32** Implement $x'yz + xvy'w' + xvy'z'$ using a three-level NAND-gate circuit.

**7.33** Design a logic circuit that has a 4-bit binary number as an input and one output. The output should be 1 iff the input is a prime number (greater than 1) or zero.
(a) Use a two-level NAND-gate circuit.
(b) Use a two-level NOR-gate circuit.
(c) Use only two-input NAND gates.

**7.34** Work Problem 7.33 for a circuit that has an output 1 iff the input is evenly divisible by 3 (0 is divisible by 3).

**7.35** Realize the following functions, using only two-input NAND gates. Repeat using only two-input NOR gates.
(a) $F = A'BC' + BD + AC + B'CD'$
(b) $F = A'CD + AB'C'D + ABD' + BC$

**7.36** (a) Find a minimum circuit of two-input AND and two-input OR gates to realize $F(A, B, C, D) = \Sigma\, m(0, 1, 2, 3, 4, 5, 7, 9, 11, 13, 14, 15)$
(b) Convert your circuit to two-input NAND gates. Add inverters where necessary.
(c) Repeat (b), except convert to two-input NOR gates.

**7.37** Realize $Z = A[BC' + D + E(F' + GH)]$ using NOR gates. Add inverters if necessary.

**7.38** In which of the following two-level circuit forms can an arbitrary switching function be realized? Verify your answers. (Assume the inputs are available in both complemented and uncomplemented form.)
(a) NOR-AND
(b) NOR-OR
(c) NOR-NAND
(d) NOR-XOR
(e) NAND-AND
(f) NAND-OR
(g) NAND-NOR
(h) NAND-XOR

**7.39** Find a minimum two-level, multiple-output AND-OR gate circuit to realize these functions (eight gates minimum).

$$f_1(a, b, c, d) = \Sigma\, m(10, 11, 12, 15) + \Sigma\, d(4, 8, 14)$$
$$f_2(a, b, c, d) = \Sigma\, m(0, 4, 8, 9) + \Sigma\, d(1, 10, 12)$$
$$f_3(a, b, c, d) = \Sigma\, m(4, 11, 13, 14, 15) + \Sigma\, d(5, 9, 12)$$

**7.40** Repeat 7.39 for the following functions (six gates).

$$f_1(a, b, c, d) = \Sigma\, m(2, 3, 5, 6, 7, 8, 10)$$
$$f_2(a, b, c, d) = \Sigma\, m(0, 1, 2, 3, 5, 7, 8, 10)$$

**7.41** Repeat 7.39 for the following functions (eight gates).

$$f_1(x, y, z) = \Sigma\, m(2, 3, 4, 5)$$
$$f_2(x, y, z) = \Sigma\, m(1, 3, 5, 6)$$
$$f_3(x, y, z) = \Sigma\, m(1, 2, 4, 5, 6)$$

**7.42** (a) Find a minimum two-level, multiple-output OR-AND circuit to realize $f_1 = b'd + a'b' + c'd$ and $f_2 = a'd' + bc' + bd'$.
(b) Realize the same functions with a minimum two-level NAND-NAND circuit.

**7.43** Repeat Problem 7.42 for $f_1 = ac' + b'd + c'd$ and $f_2 = b'c + a'd + cd'$.

**7.44** (a) Find a minimum two-level, multiple-output NAND-NAND circuit to realize $f_1 = \Sigma\, m(3, 6, 7, 11, 13, 14, 15)$ and $f_2 = \Sigma\, m(3, 4, 6, 11, 12, 13, 14)$.
(b) Repeat for a minimum two-level, NOR-NOR circuit.

**7.45** (a) Find a minimum two-level, multiple-output NAND-NAND circuit to realize $f_1 = \Sigma\, m(0, 2, 4, 6, 7, 10, 14)$ and $f_2 = \Sigma\, m(0, 1, 4, 5, 7, 10, 14)$.
(b) Repeat for a minimum two-level, multiple-output NOR-NOR circuit.

**7.46** Draw a multi-level, multiple-output, circuit equivalent to Figure 7-24(a) using:
(a) NAND and AND gates.
(b) NAND gates only (a direct conversion is not possible).

# Combinational Circuit Design and Simulation Using Gates

## Objectives

1. Draw a timing diagram for a combinational circuit with gate delays.
2. Define static 0- and 1-hazards and dynamic hazards. Given a combinational circuit, find all of the static 0- and 1-hazards. For each hazard, specify the order in which the gate outputs must switch in order for the hazard to actually produce a false output.
3. Given a switching function, realize it using a two-level circuit which is free of static and dynamic hazards (for single input variable changes).
4. Design a multiple-output NAND or NOR circuit using gates with limited fan-in.
5. Explain the operation of a logic simulator that uses four-valued logic.
6. Test and debug a logic circuit design using a simulator.

# Study Guide

1. Obtain your design problem assignment from your instructor.

2. Study Section 8.1, *Review of Combinational Circuit Design.*

3. Generally, it is possible to redesign a circuit which has two AND gates cascaded or two OR gates cascaded so that AND and OR gates alternate. If this is not practical, the conversion to a NAND or NOR circuit by the techniques of Section 7.4 is still possible by introducing a dummy one-input OR (AND) gate between the two AND (OR) gates. When the conversion is carried out, the dummy gate becomes an inverter. Try this technique and convert the following circuit to all NAND gates. Alternatively, you may use the procedures given in Section 7.5 to do the conversion.

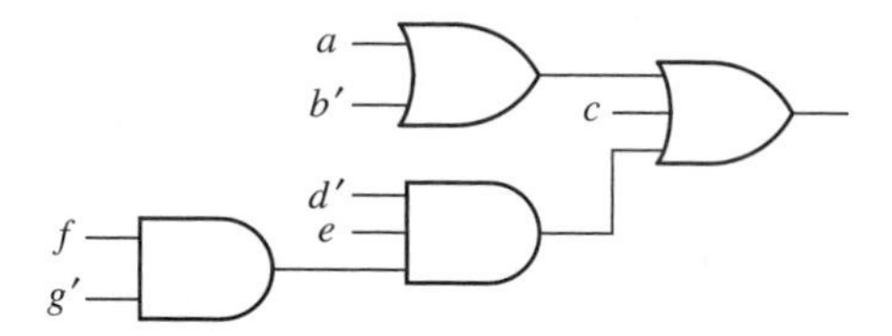


4. Study Section 8.2, *Design of Circuits with Limited Gate Fan-In.*

   (a) If a realization of a switching expression requires too many inputs on one or more gates, what should be done?

   (b) Assuming that all variables and their complements are available as inputs and that both AND and OR gates are available, does realizing the complement of an expression take the same number of gates and gate inputs as realizing the original expression?

   (c) When designing multiple-output circuits with limited gate fan-in, why is the procedure of Section 7.6 of little help?

5. (a) Study Section 8.3, *Gate Delays and Timing Diagrams.* Complete the timing diagram for the given circuit. Assume that the AND gate has a 30-nanosecond (ns) propagation delay and the inverter has a 20-ns delay.

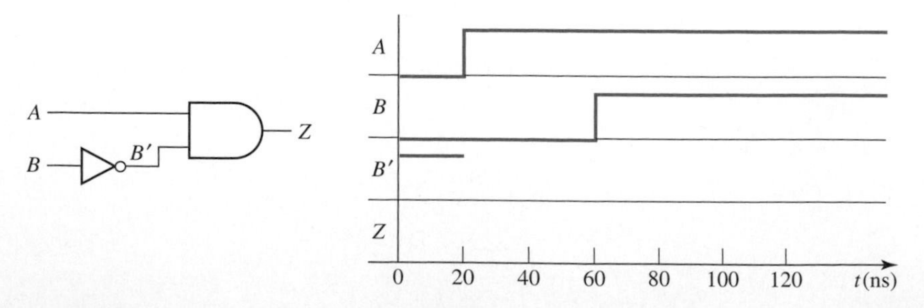

(b) Work Problem 8.1.

**6.** Study Section 8.4, *Hazards in Combinational Logic.*

(a) Even though all of the gates in a circuit are of the same type, each individual gate may have a different propagation delay. For example, for one type of TTL NAND gate the manufacturer specifies a minimum propagation delay of 5 ns and a maximum delay of 30 ns. Sketch the gate outputs for the following circuit when the $x$ input changes from 1 to 0, assuming the following gate delays:

(a) gate 1–5 ns (b) gate 2–20 ns (c) gate 3–10 ns.

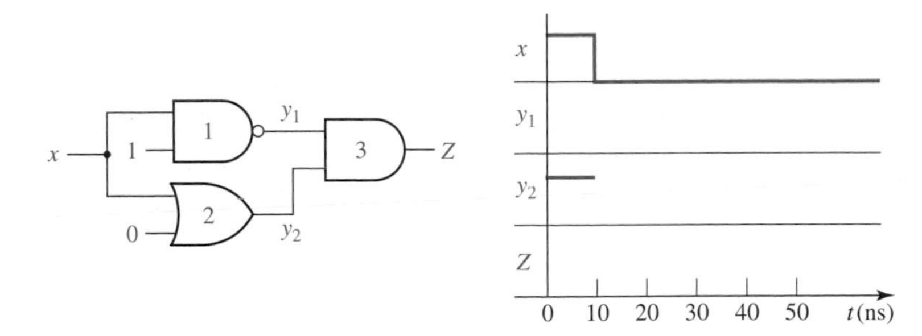


(b) Define static 0-hazard, static 1-hazard, and dynamic hazard.

(c) Using a Karnaugh map, explain why $F = a'b + ac$ has a 1-hazard for the input change $abc$ = 011 to 111, but not for 011 to 010. Then explain it without using the map.

(d) Explain why $F = (a' + b')(b + c)$ has a 0-hazard for the input change $abc$ = 100 to 110, but not for 100 to 000.

(e) Under what condition does a sum-of-products expression represent a hazard-free, two-level AND-OR circuit?

(f) Under what condition does a product-of-sums expression represent a hazard-free, two-level OR-AND circuit?

(g) If a hazard-free circuit of AND and OR gates is transformed to NAND or NOR gates using the procedure given in Unit 7, why will the results be hazard-free?

(h) Work Problems 8.2 and 8.3.

7. Study Section 8.5, *Simulation and Testing of Logic Circuits.*

   (a) Verify that Table 8-1 is correct. Consider both the case where the unknown value, X, is 0 and the case where it is 1.

   (b) The following circuit was designed to realize the function

$$F = [A' + B + C'D]\,[A + B' + (C' + D')(C + D)]$$

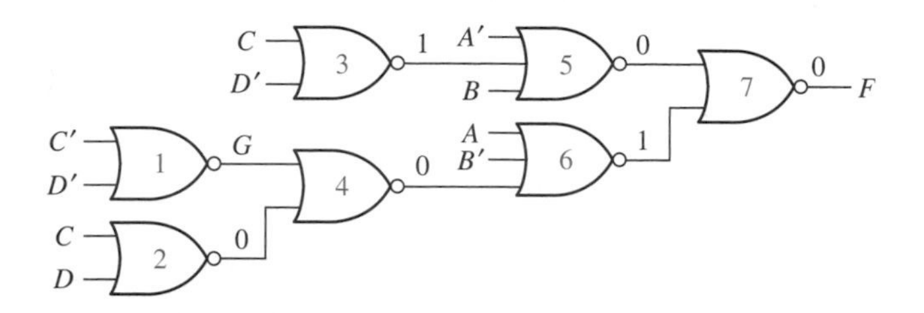


When a student builds the circuit in lab, he finds that when $A = C = 0$ and $B = D = 1$, the output $F$ has the wrong value and that the gate outputs are as shown. Determine some possible causes of the incorrect output if $G = 0$ and if $G = 1$.

   (c) Work Problems 8.4 and 8.5.

8. Study your assigned design problem and prepare a design which meets specifications. Note that only two-, three-, and four-input NAND gates (or NOR gates as specified) and inverters are available for this project; therefore, factoring some of the equations will be necessary. Try to make an economical design by using common terms; however, do not waste time trying to get an absolute minimum solution. When counting gates, count both NAND (or NOR) gates and inverters, but do not count the inverters needed for the input variables.

9. Check your design carefully before simulating it. Test it *on paper* by applying some input combinations of 0's and 1's and tracing the signals through to make sure that the outputs are correct. If you have a CAD program such as *LogicAid* available, enter the truth table for your design into the computer, derive the minimum two-level equations, and compare them with your solution.

10. In designing multi-level, multiple-output circuits of the type used in the design problems in this unit, it is very difficult and time-consuming to find a minimum solution. You are not expected to find the best possible solution to these problems. All of these solutions involve some "tricks," and it is unlikely that you could find them without trying a large number of different ways of factoring your equations. Therefore, if you already have an acceptable solution, do not waste time trying to find the minimum solution. Because integrated circuit gates are quite inexpensive, it is not good engineering practice to spend a large amount of time finding the absolute minimum solution unless a very large number of units of the same type are to be manufactured.

11. Obtain a Unit 8 supplement from your instructor and follow the instructions therein regarding simulating and testing your design.

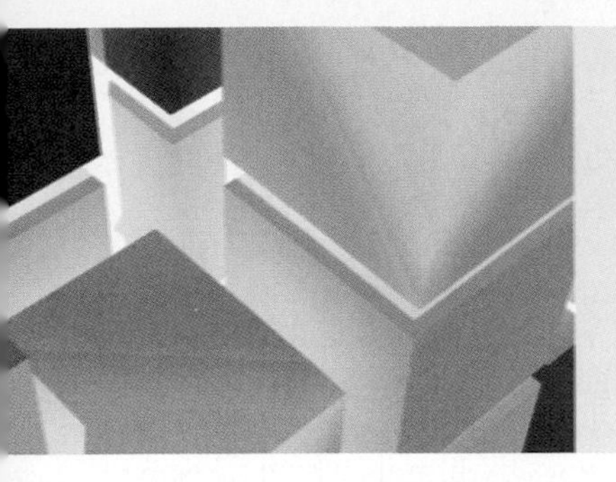

# Combinational Circuit Design and Simulation Using Gates

## 8.1 Review of Combinational Circuit Design

The first step in the design of a combinational switching circuit is usually to set up a truth table which specifies the output(s) as a function of the input variables. For $n$ input variables this table will have $2^n$ rows. If a given combination of values for the input variables can never occur at the circuit inputs, the corresponding output values are don't-cares. The next step is to derive simplified algebraic expressions for the output functions using Karnaugh maps, the Quine-McCluskey method, or a similar procedure. In some cases, particularly if the number of variables is large and the number of terms is small, it may be desirable to go directly from the problem statement to algebraic equations, without writing down a truth table. The resulting equations can then be simplified algebraically. The simplified algebraic expressions are then manipulated into the proper form, depending on the type of gates to be used in realizing the circuit.

The number of levels in a gate circuit is equal to the maximum number of gates through which a signal must pass when going between the input and output terminals. The minimum sum of products (or product of sums) leads directly to a minimum two-level gate circuit. However, in some applications it is desirable to increase the number of levels by factoring (or multiplying out) because this may lead to a reduction in the number of gates or gate inputs.

When a circuit has two or more outputs, common terms in the output functions can often be used to reduce the total number of gates or gate inputs. If each function is minimized separately, this does not always lead to a minimum multiple-output circuit. For a two-level circuit, Karnaugh maps of the output functions can be used to find the common terms. All of the terms in the minimum multiple-output circuit will not necessarily be prime implicants of the individual functions. When designing circuits with three or more levels, looking for common terms on the Karnaugh maps may be of little value. In this case, the designer will often minimize the functions separately and, then, use ingenuity to factor the expressions in such a way to create common terms.

Minimum two-level AND-OR, NAND-NAND, OR-NAND, and NOR-OR circuits can be realized using the minimum sum of products as a starting point. Minimum two-level OR-AND, NOR-NOR, AND-NOR, and NAND-AND circuits can be realized using the minimum product of sums as a starting point. Design of multi-level, multiple-output NAND-gate circuits is most easily accomplished by first designing a circuit of AND and OR gates. Usually, the best starting point is the minimum sum-of-products expressions for the output functions. These expressions are then factored in various ways until an economical circuit of the desired form can be found. If this circuit has an OR gate at each output and is arranged so that an AND gate (or OR gate) output is never connected to the same type of gate, a direct conversion to a NAND-gate circuit is possible. Conversion is accomplished by replacing all of the AND and OR gates with NAND gates and then inverting any literals which appear as inputs to the first, third, fifth, . . . levels (output gates are the first level).

If the AND-OR circuit has an AND gate (or OR gate) output connected to the same type of gate, then extra inverters must be added in the conversion process (see Section 7.5, *Circuit Conversion Using Alternative Gate Symbols*.)

Similarly, design of multi-level, multiple-output NOR-gate circuits is most easily accomplished by first designing a circuit of AND and OR gates. In this case the best starting point is usually the minimum sum-of-products expressions for the *complements* of the output functions. *After* factoring these expressions to the desired form, they are then complemented to get expressions for the output functions, and the corresponding circuit of AND and OR gates is drawn. If this circuit has an AND gate at each output, and an AND gate (or OR gate) output is never connected to the same type of gate, a direct conversion to a NOR-gate circuit is possible. Otherwise, extra inverters must be added in the conversion process.

## 8.2 Design of Circuits with Limited Gate Fan-In

In practical logic design problems, the maximum number of inputs on each gate (or the fan-in) is limited. Depending on the type of gates used, this limit may be two, three, four, eight, or some other number. If a two-level realization of a circuit requires more gate inputs than allowed, factoring the logic expression to obtain a multi-level realization is necessary.

**Example**

Realize $f(a, b, c, d) = \Sigma\, m(0, 3, 4, 5, 8, 9, 10, 14, 15)$ using three-input NOR gates.

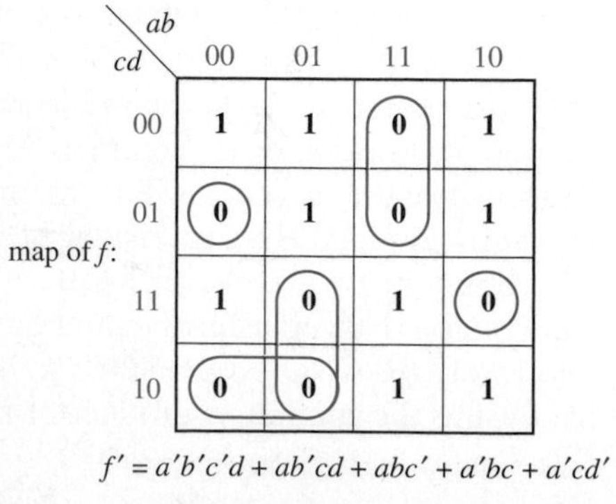

map of $f$:

| cd \ ab | 00 | 01 | 11 | 10 |
|---|---|---|---|---|
| 00 | 1 | 1 | 0 | 1 |
| 01 | 0 | 1 | 0 | 1 |
| 11 | 1 | 0 | 1 | 0 |
| 10 | 0 | 0 | 1 | 1 |

$f' = a'b'c'd + ab'cd + abc' + a'bc + a'cd'$

As can be seen from the preceding expression, a two-level realization requires two four-input gates and one five-input gate. The expression for $f'$ is factored to reduce the maximum number of gate inputs to three and, then, it is complemented:

$$f' = b'd(a'c' + ac) + a'c(b + d') + abc'$$
$$f = [b + d' + (a + c)(a' + c')][a + c' + b'd][a' + b' + c]$$

The resulting NOR-gate circuit is shown in Figure 8-1.

FIGURE 8-1

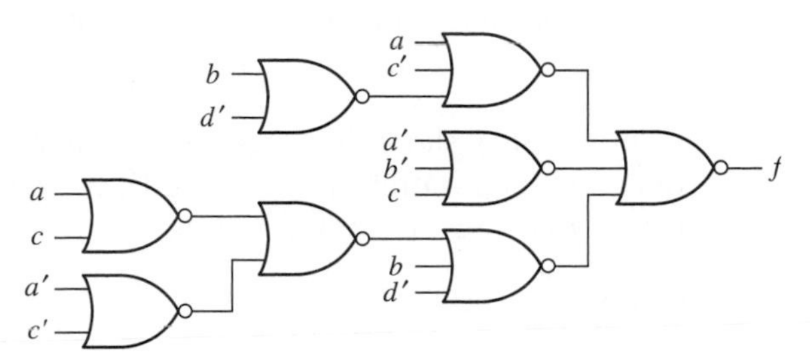


The techniques for designing two-level, multiple-output circuits given in Section 7.6 are not very effective for designing multiple-output circuits with more than two levels. Even if the two-level expressions had common terms, most of these common terms would be lost when the expressions were factored. Therefore, when designing multiple-output circuits with more than two levels, it is usually best to minimize each function separately. The resulting two-level expressions must then be factored to increase the number of levels. This factoring should be done in such a way as to introduce common terms wherever possible.

**Example**

Realize the functions given in Figure 8-2, using only two-input NAND gates and inverters. If we minimize each function separately, the result is

$$f_1 = b'c' + ab' + a'b$$
$$f_2 = b'c' + bc + a'b$$
$$f_3 = a'b'c + ab + bc'$$

FIGURE 8-2

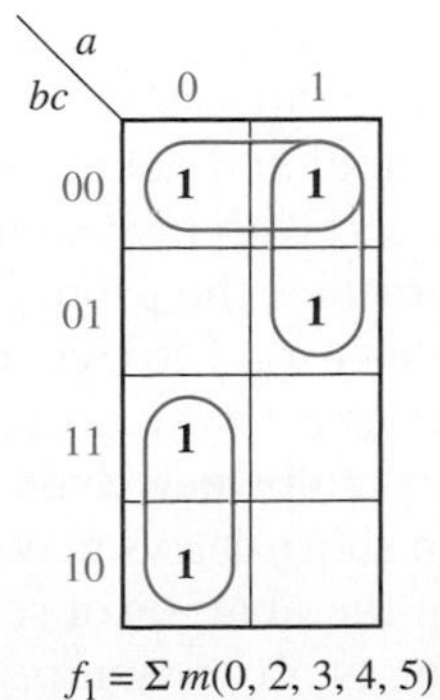


$f_1 = \Sigma\, m(0, 2, 3, 4, 5)$

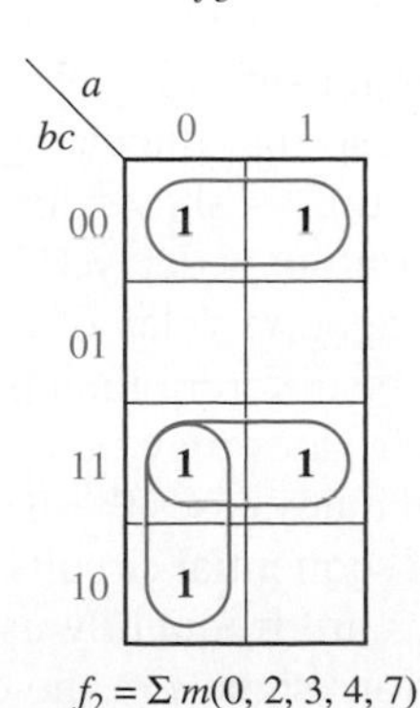


$f_2 = \Sigma\, m(0, 2, 3, 4, 7)$

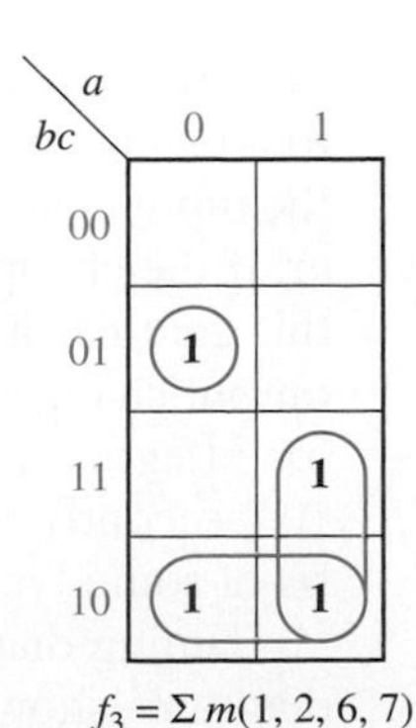


$f_3 = \Sigma\, m(1, 2, 6, 7)$

Each function requires a three-input OR gate, so we will factor to reduce the number of gate inputs:

$$f_1 = b'(\underline{a + c'}) + \underline{a'b}$$
$$f_2 = b(a' + c) + b'c' \qquad \text{or} \qquad f_2 = (b' + c)(b + c') + \underline{a'b}$$
$$f_3 = a'b'c + b(\underline{a + c'})$$

The second expression for $f_2$ has a term common to $f_1$, so we will choose the second expression. We can eliminate the remaining three-input gate from $f_3$ by noting that

$$a'b'c = a'(b'c) = a'(b + c')'$$

Figure 8-3(a) shows the resulting circuit, using common terms $a'b$ and $a + c'$. Because each output gate is an OR, the conversion to NAND gates, as shown in Figure 8-3(b), is strainghtforward.

FIGURE 8-3 Realization of Figure 8-2

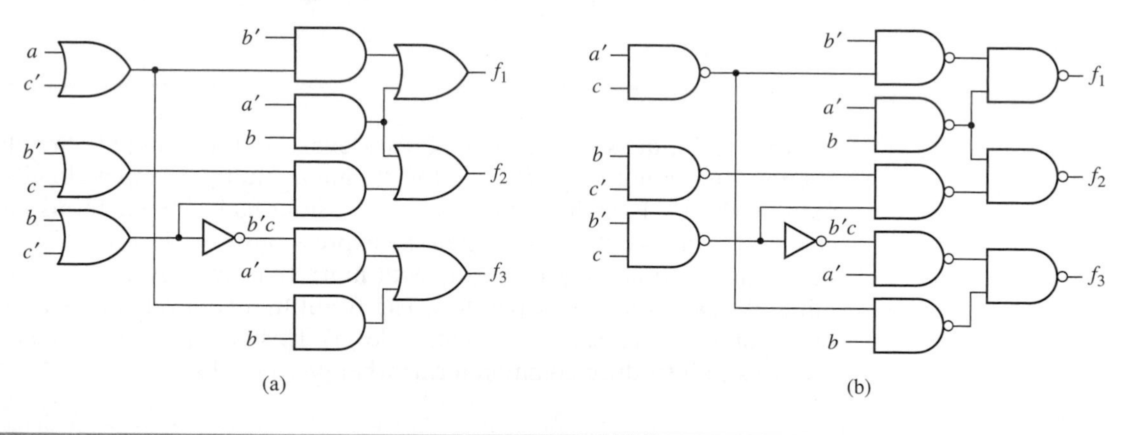


## 8.3 Gate Delays and Timing Diagrams

When the input to a logic gate is changed, the output will not change instantaneously. The transistors or other switching elements within the gate take a finite time to react to a change in input, so that the change in the gate output is delayed with respect to the input change. Figure 8-4 shows possible input and output waveforms for an inverter. If the change in output is delayed by time, $\epsilon$, with respect to the input, we say that this gate has a propagation delay of $\epsilon$. In practice, the propagation delay for a 0 to 1 output change may be different than the delay for a 1 to 0 change. Propagation delays for integrated circuit gates may be as short as a few nanoseconds (1 nanosecond = $10^{-9}$ second), and in many cases these delays can be neglected. However, in the analysis of some types of sequential circuits, even short delays may be important.

Timing diagrams are frequently used in the analysis of sequential circuits. These diagrams show various signals in the circuit as a function of time. Several variables are usually plotted with the same time scale so that the times at which these variables change with respect to each other can easily be observed.

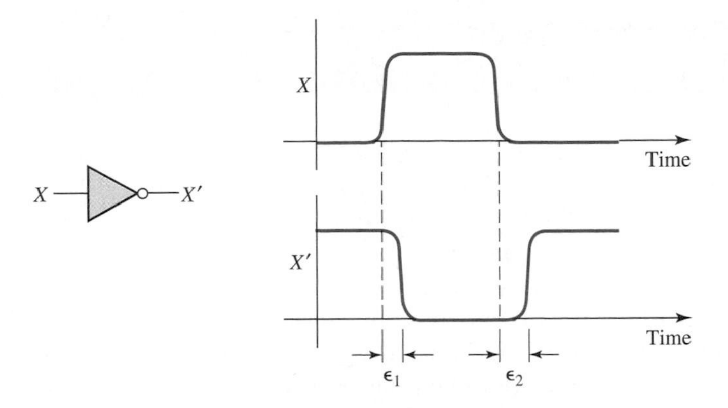


FIGURE 8-4 Propagation Delay in an Inverter

Figure 8-5 shows the timing diagram for a circuit with two gates. We will assume that each gate has a propagation delay of 20 ns (nanoseconds). This timing diagram indicates what happens when gate inputs $B$ and $C$ are held at constant values 1 and 0, respectively, and input $A$ is changed to 1 at $t = 40$ ns and then changed back to 0 at $t = 100$ ns. The output of gate $G_1$ changes 20 ns after $A$ changes, and the output of gate $G_2$ changes 20 ns after $G_1$ changes.

Figure 8-6 shows a timing diagram for a circuit with an added delay element. The input $X$ consists of two pulses, the first of which is 2 microseconds ($2 \times 10^{-6}$ second) wide and the second is 3 microseconds wide. The delay element has an output $Y$ which

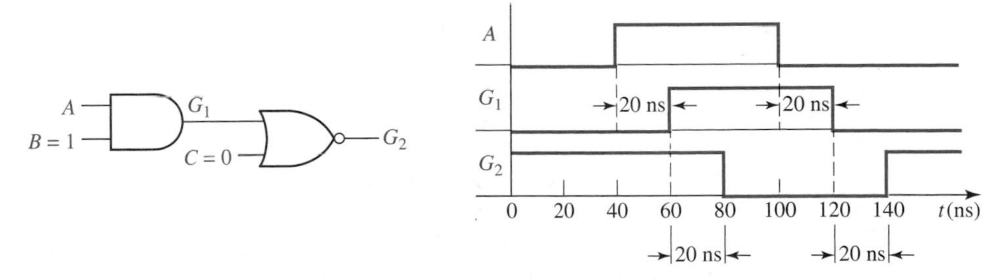


FIGURE 8-5 Timing Diagram for AND-NOR Circuit

is the same as the input except that it is delayed by 1 microsecond. That is, $Y$ changes to a 1 value 1 microsecond after the rising edge of the $X$ pulse and returns to 0 1 microsecond after the falling edge of the $X$ pulse. The output $(Z)$ of the AND gate should be 1 during the time interval in which both $X$ and $Y$ are 1. If we assume a small propagation delay in the AND gate ($\epsilon$), then $Z$ will be as shown in Figure 8-6.

FIGURE 8-6 Timing Diagram for Circuit with Delay

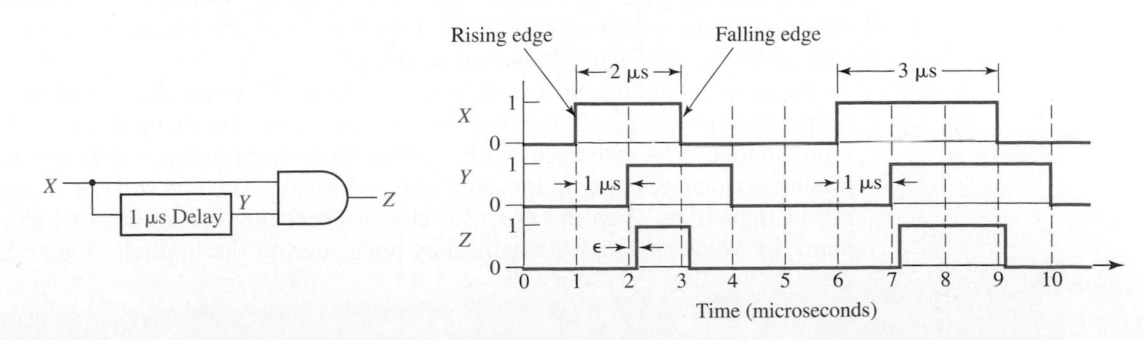

## 8.4 Hazards in Combinational Logic

When the input to a combinational circuit changes, unwanted switching transients may appear in the output. These transients occur when different paths from input to output have different propagation delays. If, in response to any single input change and for some combination of propagation delays, a circuit output may momentarily go to 0 when it should remain a constant 1, we say that the circuit has a static 1-hazard. Similarly, if the output may momentarily go to 1 when it should remain a 0, we say that the circuit has a static 0-hazard. If, when the output is supposed to change from 0 to 1 (or 1 to 0), the output may change three or more times, we say that the circuit has a dynamic hazard. Figure 8-7 shows possible outputs from a circuit with hazards. In each case the steady-state output of the circuit is correct, but a switching transient appears at the circuit output when the input is changed.

**FIGURE 8-7** Types of Hazards

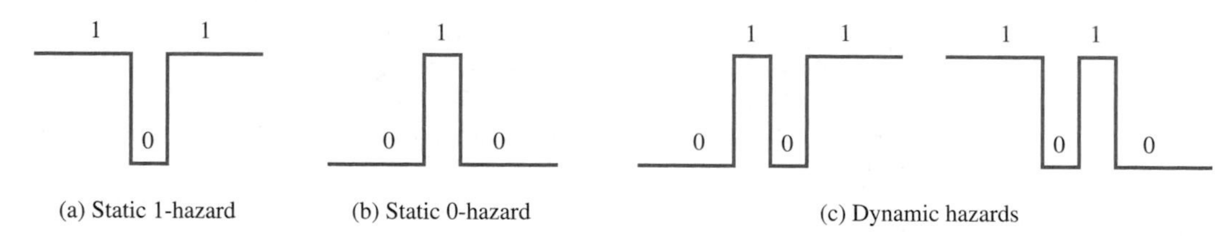


Figure 8-8(a) illustrates a circuit with a static 1-hazard. If $A = C = 1$, then $F = B + B' = 1$, so the $F$ output should remain a constant 1 when $B$ changes from 1 to 0. However, as shown in Figure 8-8(b), if each gate has a propagation delay of 10 ns, $E$ will go to 0 before $D$ goes to 1, resulting in a momentary 0 (a glitch caused by the 1-hazard) appearing at the output $F$. Note that right after $B$ changes to 0, both the inverter input ($B$) and output ($B'$) are 0 until the propagation delay has elapsed. During this period, both terms in the equation for $F$ are 0, so $F$ momentarily goes to 0.

Note that hazards are properties of the circuit and are independent of the delays existing in the circuit. If the circuit is free of hazards, then for any combination of delays that might exist in the circuit and for any single input change, the output will not contain a transient. On the other hand, if a circuit contains a hazard, then there is some combination of delays and some input change for which the circuit output contains a transient. The combination of delays that produces the transient may or may not be likely to occur in an implementation of the circuit; in some cases it is very unlikely that such delays would occur.

Besides depending on the delays existing in a circuit, the occurrence of transients depends on how gates respond to input changes. In some cases, if multiple input changes to a gate occur within a short time period, a gate may not respond to the input changes. For example, in Figure 8-8 assume the inverter has a delay of 2 ns rather than 10 ns. Then the $D$ and $E$ changes reaching the output OR gate are 2 ns apart, in which case the OR gate may not generate the 0 glitch. A gate exhibiting

**FIGURE 8-8** Detection of a 1-Hazard

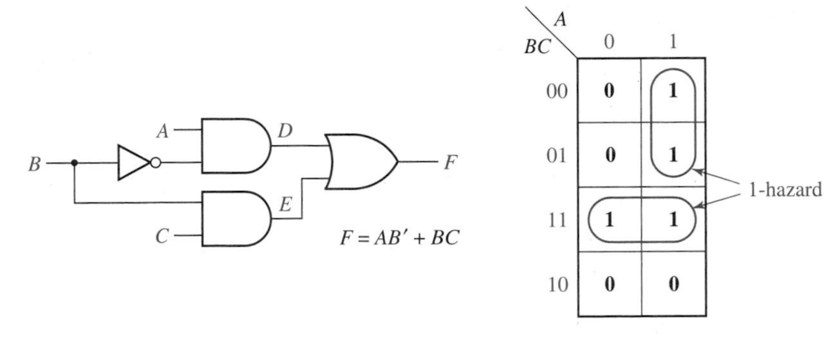


(a) Circuit with a static 1-hazard

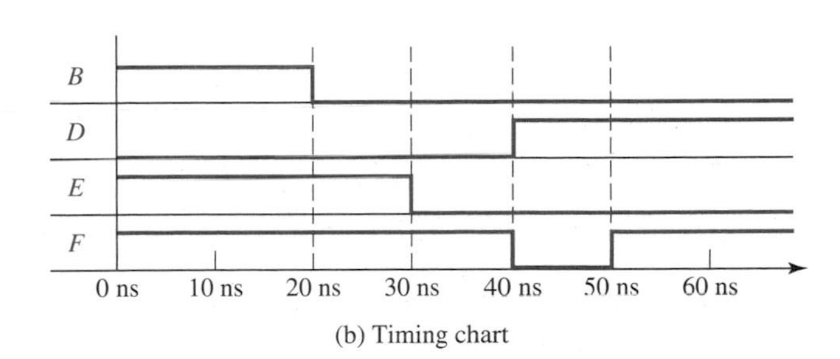


(b) Timing chart

this behavior is said to have an *inertial* delay. Quite often the inertial delay value is assumed to be the same as the propagation delay of the gate; if this is the case, the circuit of Figure 8-8 will generate the 0 glitch only for inverter delays greater than 10 ns. In contrast, if a gate always responds to input changes (with a propagation delay), no matter how closely spaced the input changes may be, the gate is said to have an *ideal* or *transport* delay. If the OR gate in Figure 8-8 has an ideal delay, then the 0 glitch would be generated for any nonzero value of the inverter delay. (Inertial and transport delay models are discussed more in Unit 10.) Unless otherwise noted, the examples and problems in this unit assume that gates have an ideal delay.

Hazards can be detected using a Karnaugh map [Figure 8-8(a)]. As seen on the map, no loop covers both minterms $ABC$ and $AB'C$. So if $A = C = 1$ and $B$ changes, both terms can momentarily go to 0, resulting in a glitch in $F$. We can detect hazards in a two-level AND-OR circuit, using the following procedure:

1. Write down the sum-of-products expression for the circuit.
2. Plot each term on the map and loop it.
3. If any two adjacent 1's are not covered by the same loop, a 1-hazard exists for the transition between the two 1's. For an $n$-variable map, this transition occurs when one variable changes and the other $n-1$ variables are held constant.

If we add a loop to the map of Figure 8-8(a) and, then, add the corresponding gate to the circuit (Figure 8-9), this eliminates the hazard. The term $AC$ remains 1 while $B$ is changing, so no glitch can appear in the output. Note that $F$ is no longer a minimum sum of products.

FIGURE 8-9
Circuit with Hazard Removed

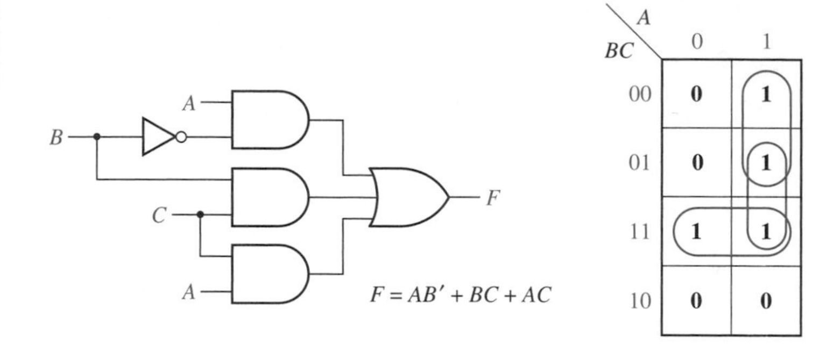


Figure 8-10(a) shows a circuit with several 0-hazards. The product-of-sums representation for the circuit output is

$$F = (A + C)(A' + D')(B' + C' + D)$$

The Karnaugh map for this function (Figure 8-10(b)) shows four pairs of adjacent 0's that are not covered by a common loop as indicated by the arrows. Each of these pairs corresponds to a 0-hazard. For example, when $A = 0, B = 1, D = 0$, and $C$ changes from 0 to 1, a spike may appear at the $Z$ output for some combination of gate delays. The timing diagram of Figure 8-10(c) illustrates this

FIGURE 8-10
Detection of a Static 0-Hazard

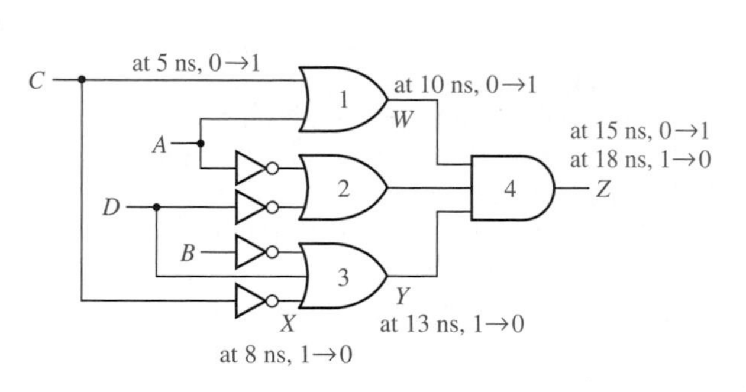


(a) Circuit with a static 0-hazard

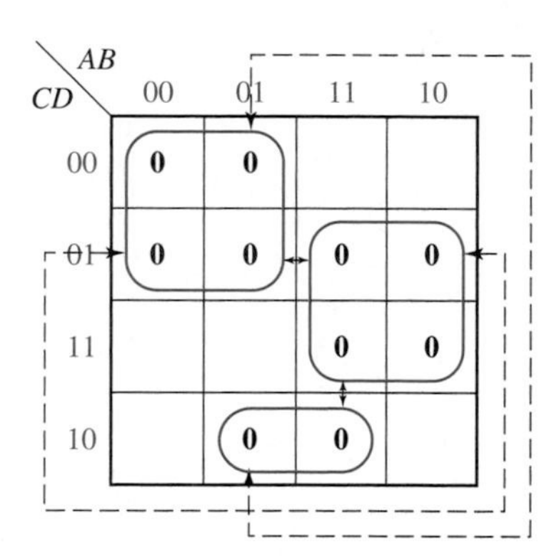


(b) Karnaugh map for circuit of (a)

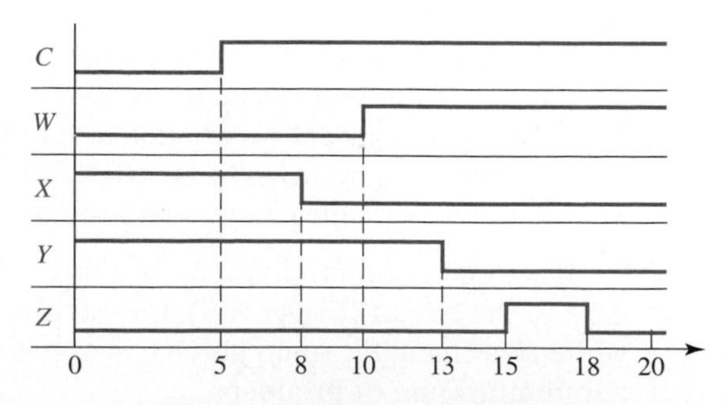


(c) Timing diagram illustrating 0-hazard of (a)

assuming gate delays of 3 ns for each inverter, and of 5 ns for each AND gate and each OR gate.

We can eliminate the 0-hazards by looping additional prime implicants that cover the adjacent 0's that are not already covered by a common loop. This requires three additional loops as shown in Figure 8-11. The resulting equation is

$$F = (A + C)(A' + D')(B' + C' + D)(C + D')(A + B' + D)(A' + B' + C')$$

and the resulting circuit requires seven gates in addition to the inverters.

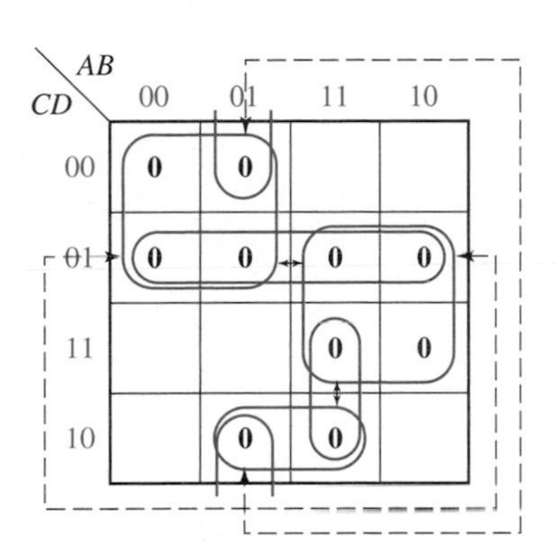


**FIGURE 8-11** Karnaugh Map Removing Hazards of Figure 8-10.

Hazards in circuits with more than two levels can be determined by deriving either a SOP or POS expression for the circuit that represents a two-level circuit containing the same hazards as the original circuit. The SOP or POS expression is derived in the normal manner except that the complementation laws are *not* used, i.e., $xx' = 0$ and $x + x' = 1$ are not used. Consequently, the resulting SOP (POS) expression may contain products (sums) of the form $xx'\alpha$ $(x + x' + \beta)$. ($\alpha$ is a product of literals or it may be null; $\beta$ is a sum of literals or it may be empty.) The complementation laws are not used because we are analyzing the circuit behavior resulting from an input change. As that input change propagates through the circuit, at a given point in time a line tending toward the value $x$ may not have the value that is the complement of a line tending toward the value $x'$. In the SOP expression, a product of the form $xx'\alpha$ represents a pseudo gate that may temporarily have the output value 1 as $x$ changes and if $\alpha = 1$.

Given the SOP expression, the circuit is analyzed for static 1-hazards the same as for a two-level AND-OR circuit, i.e., the products are mapped on a Karnaugh map and if two 1's are adjacent on the map and not included in one of the products, they correspond to a static 1-hazard. The circuit can have a static 0-hazard or a dynamic hazard only if the SOP expression contains a term of the form $xx'\alpha$. A static 0-hazard exists if there are two adjacent 0's on the Karnaugh map for which $\alpha = 1$ and the two input combinations differ just in the value of $x$. A dynamic hazard exists if there is a term of the form $xx'\alpha$ and two conditions are satisfied: (1) There are adjacent input combinations on the Karnaugh map differing in the value of $x$, with $\alpha = 1$ and with opposite function values, and (2) for these input combinations the change in $x$ propagates over at least three paths through the circuit.

As an example consider the circuit of Figure 7-7 (page 194). The expression for the circuit output is

$$
\begin{aligned}
f &= (c' + ad' + bd')(c + a'd + bd) \\
&= cc' + acd' + bcd' + a'c'd + aa'dd' + a'bdd' + bc'd + abdd' + bdd' \\
&= cc' + acd' + bcd' + a'c'd + aa'dd' + bc'd + bdd'
\end{aligned}
$$

The Karnaugh map for this function is shown as the circled 1's in Figure 7-3 (page 192). It is derived in the normal way ignoring the product terms containing both a variable and its complement. The circuit does not contain any static 1-hazards because each pair of adjacent 1's are covered by one of the product terms. Potentially, the terms $cc'$ and $bdd'$ may cause either static 0- or dynamic hazards or both; the first for $c$ changing and the second for $d$ changing. (The term $aa'dd'$ cannot cause either hazard because, for example, if $a$ changes the $dd'$ part of the product forces it to 0.) With $a = 0, b = 0$, and $d = 0$ and $c$ changing, the circuit output is 0 before and after the change, and because the $cc'$ term can cause the output to temporarily become 1, this transition is a static 0-hazard. Similarly, a change in $c$, with $a = 1, b = 0$ and $d = 1$, is a static 0-hazard. The $cc'$ term cannot cause a dynamic hazard because there are only two physical paths from input $c$ to the circuit output.

The term $bdd'$ can cause a static 0- or dynamic hazard only if $b = 1$. From the Karnaugh map, it is seen that, with $b = 1$ and $d$ changing, the circuit output changes for any combination of $a$ and $c$, so the only possibility is that of a dynamic hazard. There are four physical paths from $d$ to the circuit output, so a dynamic hazard exists if a $d$ change can propagate over at least three of those paths. However, this cannot happen because, with $c = 0$, propagation over the upper two paths is blocked at the upper OR gate because $c' = 1$ forces the OR gate output to be 1, and with $c = 1$ propagation over the lower two paths is blocked at the lower OR gate. The circuit does not contain a dynamic hazard.

Another approach to finding the hazards is as follows: If we factor the original expression for the circuit output (without using the complementation laws), we get

$$f = (c' + a + b)(c' + d')(c + a' + b)(c + d)$$

Plotting the 0's of $f$ from this expression on a Karnaugh map reveals that there are 0-hazards when $a = b = d = 0$ and $c$ changes, and also when $b = 0, a = d = 1$, and $c$ changes. An expression of the form $x + x'$ does not appear in any sum term of $f$, and this indicates that there are no 1-hazards or dynamic hazards.

To design a circuit which is free of static and dynamic hazards, the following procedure may be used:

1. Find a sum-of-products expression ($F^t$) for the output in which every pair of adjacent 1's is covered by a 1-term. (The sum of all prime implicants will always satisfy this condition.) A two-level AND-OR circuit based on this $F^t$ will be free of 1-, 0-, and dynamic hazards.
2. If a different form of the circuit is desired, manipulate $F^t$ to the desired form by simple factoring, DeMorgan's laws, etc. Treat each $x_i$ and $x_i'$ as independent variables to prevent introduction of hazards.

Alternatively, you can start with a product-of-sums expression in which every pair of adjacent 0's is covered by a 0-term, and follow the dual procedure to design a hazard-free two-level OR-AND circuit.

It should be emphasized that the discussion of hazards and the possibility of resulting glitches in this section has assumed that only a single input can change at a time and that no other input will change until the circuit has stabilized. If more than one input can change at one time, then nearly all circuits will contain hazards, and they cannot be eliminated by modifying the circuit implementation. The circuit corresponding to the Karnaugh map of Figure 8-11 illustrates this. Consider the input change $(A, B, C, D) = (0,1,0,1)$ to $(0,1,1,0)$ with both $C$ and $D$ changing. The output is 0 before the change and will be 0 after the circuit has stabilized; however, if the $C$ change propagates through the circuit before the $D$ change, then the circuit will output a transient 1. Effectively, the input combination to the circuit can temporarily become $(A, B, C, D) = (0, 1, 1, 1)$, and the circuit output will temporarily become 1 no matter how it is implemented.

Glitches are of most importance in asynchronous sequential circuits. The latches and flip-flops discussed in Unit 11 are the most important examples of asynchronous sequential circuits. Although more than one input can change at the same time for some of these circuits, restrictions are placed on the changes so that it is necessary to analyze the circuits for hazards only when a single input changes. Consequently, the discussion in this section is relevant to this important class of circuits.

## 8.5 Simulation and Testing of Logic Circuits

An important part of the logic design process is verifying that the final design is correct and debugging the design if necessary. Logic circuits may be tested either by actually building them or by simulating them on a computer. Simulation is generally easier, faster, and more economical. As logic circuits become more and more complex, it is very important to simulate a design before actually building it. This is particularly true when the design is built in integrated circuit form, because fabricating an integrated circuit may take a long time and correcting errors may be very expensive. Simulation is done for several reasons, including (1) verification that the design is logically correct, (2) verification that the timing of the logic signals is correct, and (3) simulation of faulty components in the circuit as an aid to finding tests for the circuit.

To use a computer program for simulating logic circuits, you must first specify the circuit components and connections; then, specify the circuit inputs; and, finally, observe the circuit outputs. The circuit description may be input into a simulator in the form of a list of connections between the gates and other logic elements in the circuit, or the description may be in the form of a logic diagram drawn on a computer screen. Most modern logic simulators use the latter approach. A typical simulator which runs on a personal computer uses switches or input boxes to specify the inputs and probes to read the logic outputs. Alternatively, the inputs and outputs may be specified as sequences of 0's and 1's or in the form of timing diagrams.

A simple simulator for combinational logic works as follows:

1. The circuit inputs are applied to the first set of gates in the circuit, and the outputs of those gates are calculated.
2. The outputs of the gates which changed in the previous step are fed into the next level of gate inputs. If the input to any gate has changed, then the output of that gate is calculated.
3. Step 2 is repeated until no more changes in gate inputs occur. The circuit is then in a steady-state condition, and the outputs may be read.
4. Steps 1 through 3 are repeated every time a circuit input changes.

The two logic values, 0 and 1, are not sufficient for simulating logic circuits. At times, the value of a gate input or output may be unknown, and we will represent this unknown value by X. At other times we may have no logic signal at an input, as in the case of an open circuit when an input is not connected to any output. We use the logic value Z to represent an open circuit, or *high impedance* (hi-Z) connection. The discussion that follows assumes we are using a four-valued logic simulator with logic values 0, 1, X (unknown), and Z (hi-Z).

Figure 8-12(a) shows a typical simulation screen on a personal computer. The switches are set to 0 or 1 for each input. The probes indicate the value of each gate output. In Figure 8-12(b), one gate has no connection to one of its inputs. Because that gate has a 1 input and a hi-Z input, we do not know what the hardware will do, and the gate output is unknown. This is indicated by an X in the probe.

**FIGURE 8-12**

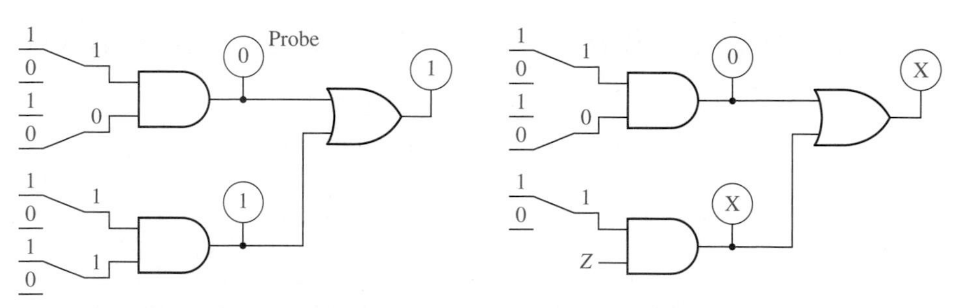


(a) Simulation screen showing switches

(b) Simulation screen with missing gate input

Table 8-1 shows AND and OR functions for four-valued logic simulation. These functions are defined in a manner similar to the way real gates work. For an AND gate, if one of the inputs is 0, the output is always 0 regardless of the other input. If one input is 1 and the other input is X (we do not know what the other input is), then the output is X (we do not know what the output is). If one input is 1 and the other input is Z (it has no logic signal), then the output is X (we do not know what the hardware will do).

**TABLE 8-1** AND and OR Functions for Four-Valued Simulation

| · | 0 | 1 | X | Z |
|---|---|---|---|---|
| 0 | 0 | 0 | 0 | 0 |
| 1 | 0 | 1 | X | X |
| X | 0 | X | X | X |
| Z | 0 | X | X | X |

| + | 0 | 1 | X | Z |
|---|---|---|---|---|
| 0 | 0 | 1 | X | X |
| 1 | 1 | 1 | 1 | 1 |
| X | X | 1 | X | X |
| Z | X | 1 | X | X |

For an OR gate, if one of the inputs is 1, the output is 1 regardless of the other input. If one input is 0 and the other input is X or Z, the output is unknown. For gates with more than two inputs, the operations may be applied several times.

A combinational logic circuit with a small number of inputs may easily be tested with a simulator or in lab by checking the circuit outputs for all possible combinations of the input values. When the number of inputs is large, it is usually possible to find a relatively small set of input test patterns which will test for all possible faulty gates in the circuit.[1]

If a circuit output is wrong for some set of input values, this may be due to several possible causes:

1. Incorrect design
2. Gates connected wrong
3. Wrong input signals to the circuit

If the circuit is built in lab, other possible causes include

4. Defective gates
5. Defective connecting wires

Fortunately, if the output of a combinational logic circuit is wrong, it is very easy to locate the problem systematically by starting at the output and working back through the circuit until the trouble is located. For example, if the output gate has the wrong output and its inputs are correct, this indicates that the gate is defective. On the other hand, if one of the inputs is wrong, then either the gate is connected wrong, the gate driving this input has the wrong output, or the input connection is defective.

***Example***

The function $F = AB(C'D + CD') + A'B'(C + D)$ is realized by the circuit of Figure 8-13:

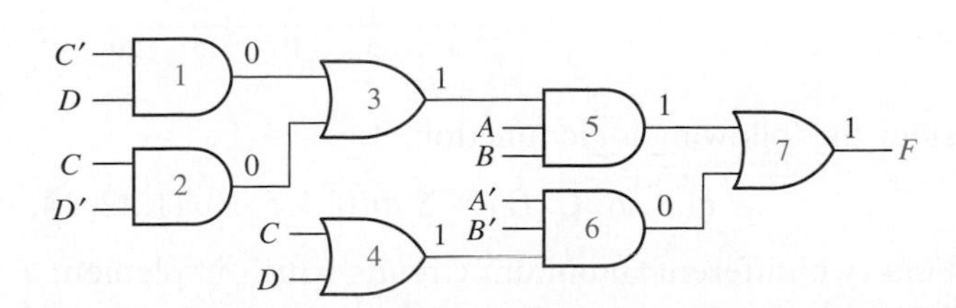


**FIGURE 8-13** Logic Circuit with Incorrect Output

[1] Methods for test pattern generation are described in Alexander Miczo, *Digital Logic Testing and Simulation*, 2nd ed (John Wiley & Sons, 2003).

When a student builds the circuit in a lab, he finds that when $A = B = C = D = 1$, the output $F$ has the wrong value, and that the gate outputs are as shown in Figure 8-13. The reason for the incorrect value of $F$ can be determined as follows:

1. The output of gate 7 ($F$) is wrong, but this wrong output is consistent with the inputs to gate 7, that is, $1 + 0 = 1$. Therefore, one of the inputs to gate 7 must be wrong.
2. In order for gate 7 to have the correct output ($F = 0$), both inputs must be 0. Therefore, the output of gate 5 is wrong. However, the output of gate 5 is consistent with its inputs because $1 \cdot 1 \cdot 1 = 1$. Therefore, one of the inputs to gate 5 must be wrong.
3. Either the output of gate 3 is wrong, or the $A$ or $B$ input to gate 5 is wrong. Because $C'D + CD' = 0$, the output of gate 3 is wrong.
4. The output of gate 3 is not consistent with the outputs of gates 1 and 2 because $0 + 0 \neq 1$. Therefore, either one of the inputs to gate 3 is connected wrong, gate 3 is defective, or one of the input connections to gate 3 is defective.

This example illustrates how to troubleshoot a logic circuit by starting at the output gate and working back until the wrong connection or defective gate is located.

# Problems

**8.1** Complete the timing diagram for the given circuit. Assume that both gates have a propagation delay of 5 ns.

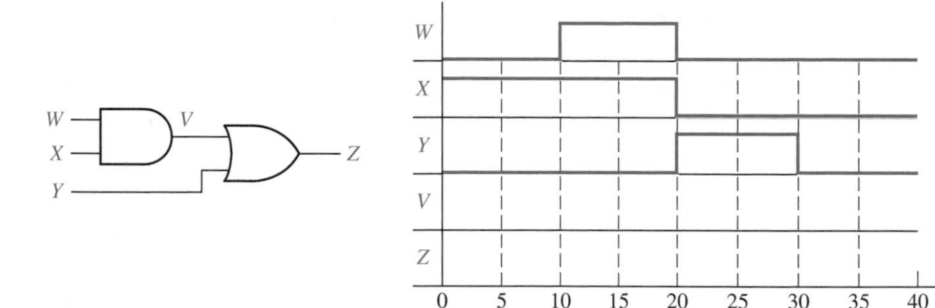


**8.2** Consider the following logic function.

$$F(A, B, C, D) = \Sigma\, m(0, 4, 5, 10, 11, 13, 14, 15)$$

(a) Find two different minimum circuits which implement $F$ using AND and OR gates. Identify two hazards in each circuit.
(b) Find an AND-OR circuit for $F$ which has no hazards.
(c) Find an OR-AND circuit for $F$ which has no hazards.

**8.3** For the following circuit:

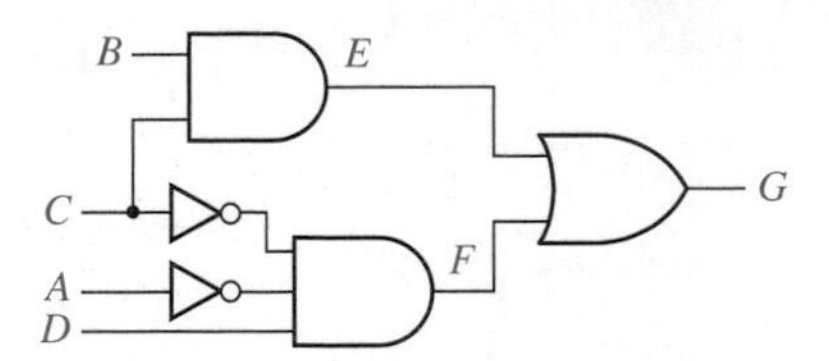


(a) Assume that the inverters have a delay of 1 ns and the other gates have a delay of 2 ns. Initially $A = 0$ and $B = C = D = 1$, and $C$ changes to 0 at time = 2 ns. Draw a timing diagram and identify the transient that occurs.

(b) Modify the circuit to eliminate the hazard.

**8.4** Using four-valued logic, find $A$, $B$, $C$, $D$, $E$, $F$, $G$, and $H$.

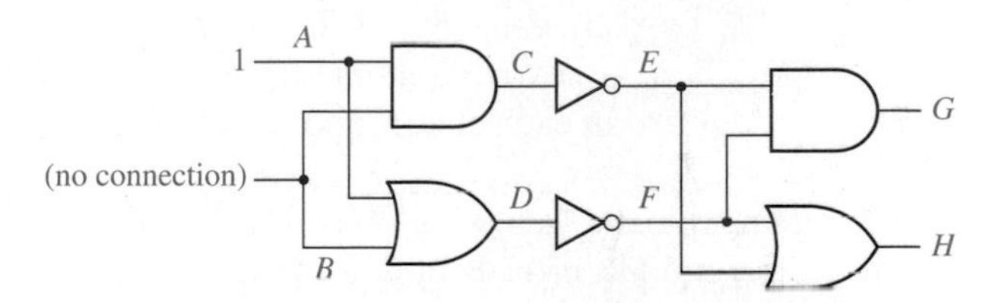


**8.5** The circuit below was designed to implement the logic equation $F = AB'D + BC'D' + BCD$, but it is not working properly. The input wires to gates 1, 2, and 3 are so tightly packed, it would take you a while to trace them all back to see whether the inputs are correct. It would be nice to only have to trace whichever one is incorrectly wired. When $A = B = 0$ and $C = D = 1$, the inputs and outputs of gate 4 are as shown. Is gate 4 working properly? If so, which of the other gates either is connected incorrectly or is malfunctioning?

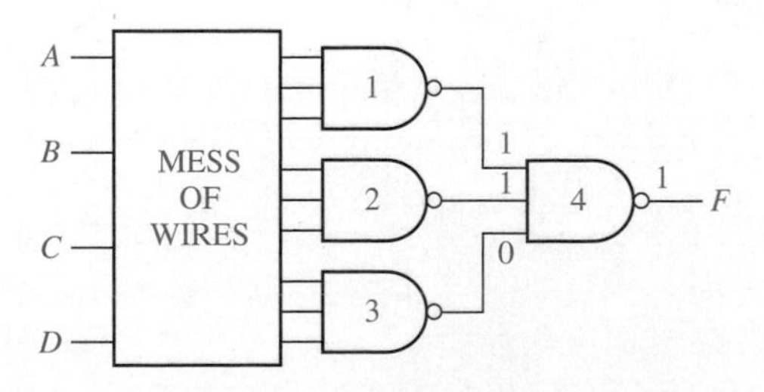


**8.6** (a) Assume the inverters have a delay of 1 ns and the other gates have a delay of 2 ns. Initially $A = B = 0$ and $C = D = 1$; $C$ changes to 0 at time 2 ns. Draw a timing diagram showing the glitch corresponding to the hazard.

(b) Modify the circuit so that it is hazard free. (Leave the circuit as a two-level, OR-AND circuit.)

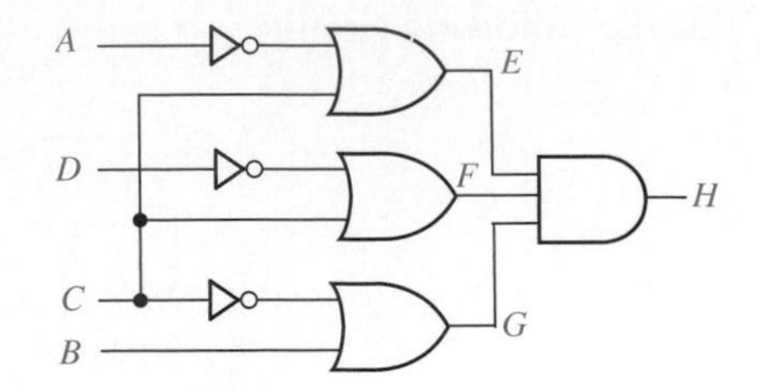


**8.7** A two-level, NOR-NOR circuit implements the function
$f(a, b, c, d) = (a + d')(b' + c + d)(a' + c' + d')(b' + c' + d)$.
(a) Find all hazards in the circuit.
(b) Redesign the circuit as a two-level, NOR-NOR circuit free of all hazards and using a minimum number of gates.

**8.8** $F(A, B, C, D) = \Sigma\, m(0, 2, 3, 5, 6, 7, 8, 9, 13, 15)$
(a) Find three different minimum AND-OR circuits that implement F. Identify two hazards in each circuit. Then find an AND-OR circuit for $F$ that has no hazards.
(b) There are two minimum OR-AND circuits for $F$; each has one hazard. Identify the hazard in each circuit, and then find an OR-AND circuit for $F$ that has no hazards.

**8.9** Consider the following three-level NOR circuit:
(a) Find all hazards in this circuit.
(b) Redesign the circuit as a three-level NOR circuit that is free of all hazards.

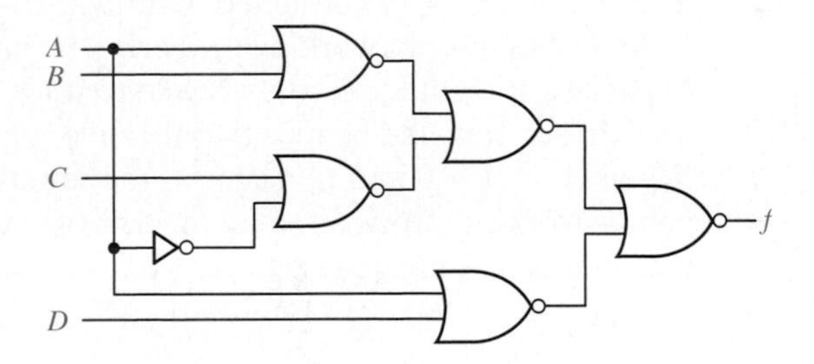


**8.10** Draw the timing diagram for $V$ and $Z$ for the circuit. Assume that the AND gate has a delay of 10 ns and the OR gate has a delay of 5 ns.

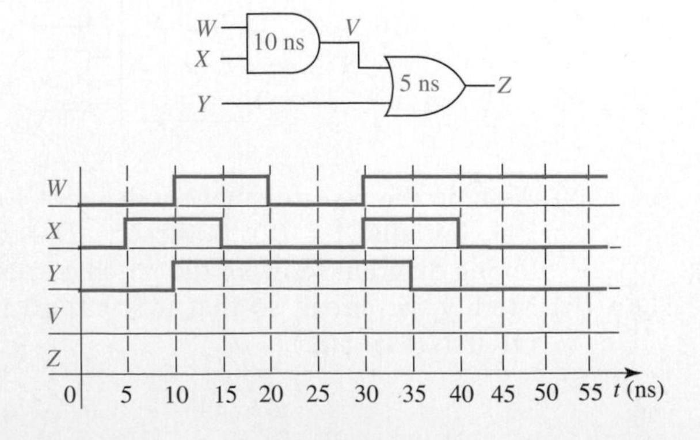

**8.11** Consider the three-level circuit corresponding to the expression f(A, B, C, D) = (A + B)(B′C′ + BD′).
(a) Find all hazards in this circuit.
(b) Redesign the circuit as a three-level NOR circuit that is free of all hazards.

**8.12** Complete the timing diagram for the given circuit. Assume that both gates have a propagation delay of 5 ns.

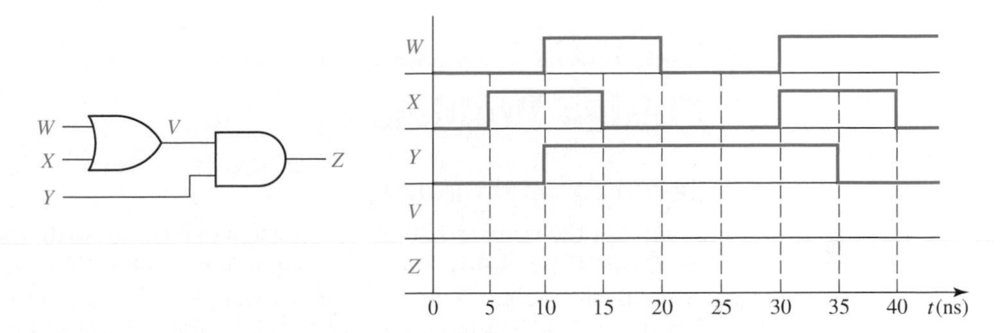


**8.13** Implement the logic function from Figure 8.10(b) as a minimum sum of products. Find the static hazards and tell what minterms they are between. Implement the same logic function as a sum of products without any hazards.

**8.14** Using four-valued logic, find $A$, $B$, $C$, $D$, $E$, $F$, $G$, and $H$.

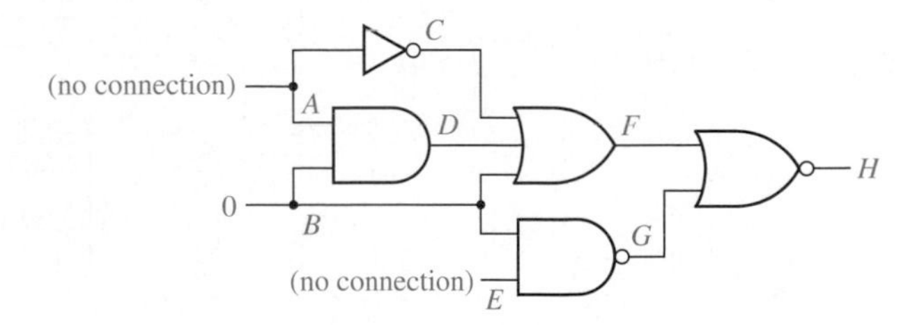


**8.15** The following circuit was designed to implement the logic equation $F = (A + B' + C')(A' + B + C')(A' + B' + C)$, but it is not working properly. The input wires to gates 1, 2, and 3 are so tightly packed, it would take you a while to trace them all back to see whether the inputs are correct. It would be nice to only have to trace whichever one is incorrectly wired. When $A = B = C = 1$, the inputs and outputs of gate 4 are as shown. Is gate 4 working properly? If so, which of the other gates either is connected incorrectly or is malfunctioning?

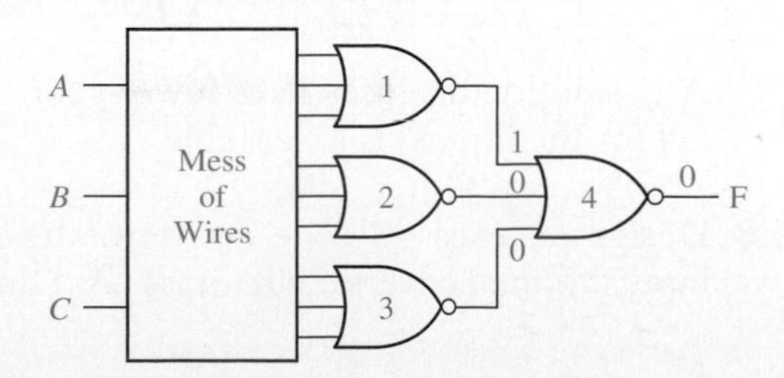

**8.16** Consider the following logic function.

$$F(A, B, C, D) = \Sigma\, m(0, 2, 5, 6, 7, 8, 9, 12, 13, 15)$$

(a) Find two different minimum AND-OR circuits which implement $F$. Identify two hazards in each circuit. Then find an AND-OR circuit for $F$ that has no hazards.

(b) The minimum OR-AND circuit for $F$ has one hazard. Identify it, and then find an OR-AND circuit for $F$ that has no hazards.

# Design Problems

## Seven-Segment Indicator

Several of the problems involve the design of a circuit to drive a seven-segment indicator (see Figure 8-14). The seven-segment indicator can be used to display any one of the decimal digits 0 through 9. For example, "1" is displayed by lighting segments 2 and 3, "2" by lighting segments 1, 2, 7, 5, and 4, and "8" by lighting all seven segments. A segment is lighted when a logic 1 is applied to the corresponding input on the display module.

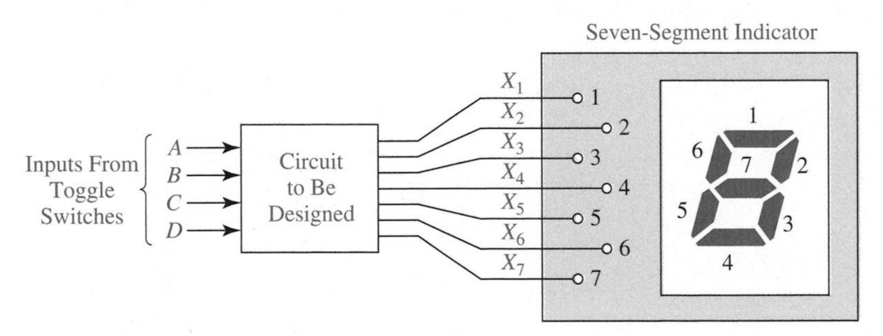


**FIGURE 8-14** Circuit Driving Seven-Segment Module

**8.A** Design an 8-4-2-1 BCD code converter to drive a seven-segment indicator. The four inputs to the converter circuit ($A$, $B$, $C$, and $D$ in Figure 8-14) represent an 8-4-2-1 binary-coded-decimal digit. Assume that only input combinations representing the digits 0 through 9 can occur as inputs, so that the combinations 1010 through 1111 are don't-cares. Design your circuit using only two-, three-, and four-input NAND gates and inverters. Try to minimize the number of gates required. The variables $A$, $B$, $C$, and $D$ will be available from toggle switches.

Use b (not 6) for 6. Use 9 (not 9) for 9.

Any solution that uses 18 or fewer gates and inverters (not counting the four inverters for the inputs) is acceptable.

**8.B** Design an excess-3 code converter to drive a seven-segment indicator. The four inputs to the converter circuit ($A$, $B$, $C$, and $D$ in Figure 8-14) represent an excess-3 coded decimal digit. Assume that only input combinations representing the digits 0 through 9 can occur as inputs, so that the six unused combinations are don't-cares. Design your circuit using only two-, three-, and four-input NAND gates and inverters. Try to minimize the number of gates and inverters required. The variables $A$, $B$, $C$, and $D$ will be available from toggle switches.

Use 6 (not 6) for 6. Use 9 (not 9) for 9.

Any solution with 16 or fewer gates and inverters (not counting the four inverters for the inputs) is acceptable.

**8.C** Design a circuit which will yield the product of two binary numbers, $n_2$ and $m_2$, where $00_2 \le n_2 \le 11_2$ and $000_2 \le m_2 \le 101_2$. For example, if $n_2 = 10_2$ and $m_2 = 001_2$, then the product is $n_2 \times m_2 = 10_2 \times 001_2 = 0010_2$. Let the variables $A$ and $B$ represent the first and second digits of $n_2$, respectively (i.e., in this example $A = 1$ and $B = 0$). Let the variables $C$, $D$, and $E$ represent the first, second, and third digits of $m_2$, respectively (in this example $C = 0$, $D = 0$, and $E = 1$). Also let the variables $W$, $X$, $Y$, and $Z$ represent the first, second, third, and fourth digits of the product. (In this example $W = 0$, $X = 0$, $Y = 1$, and $Z = 0$.) Assume that $m_2 > 101_2$ never occurs as a circuit input.

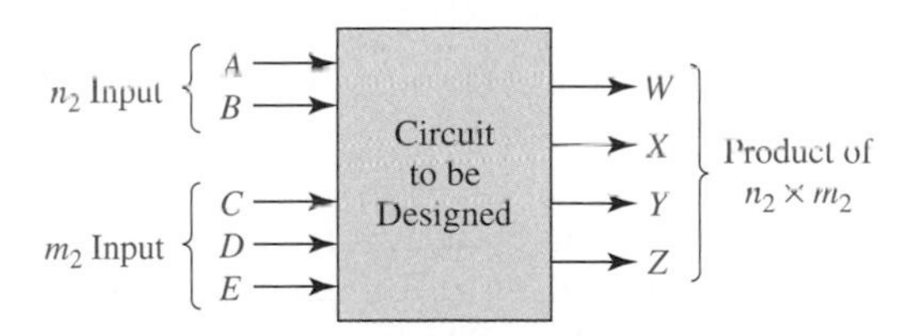


Design the circuit using only two-, three-, and four-input NOR gates and inverters. Try to minimize the total number of gates and inverters required. The variables $A$, $B$, $C$, $D$, and $E$ will be available from toggle switches. Any solution that uses 15 or fewer gates and inverters (not counting the five inverters for the inputs) is acceptable.

**8.D** Work Design Problem 8.C using two-, three-, and four-input NAND gates and inverters instead of NOR gates and inverters. Any solution that uses 14 gates and inverters or less (not counting the five inverters for the inputs) is acceptable.

**8.E** Design a circuit which multiplies two 2-bit binary numbers and displays the answer in decimal on a seven-segment indicator. In Figure 8-14, $A$ and $B$ are two bits of a binary number $N_1$, and $C$ and $D$ are two bits of a binary number $N_2$. The product $(N_1 \times N_2)$ is to be displayed in decimal by lighting appropriate segments of the seven-segment indicator. For example, if $A = 1$, $B = 0$, $C = 1$, and $D = 0$, the number "4" is displayed by lighting segments 2, 3, 6, and 7.

Use 6 (not 6) for 6. Use 9 (not 9) for 9.

Design your circuit using only two-, three-, and four-input NAND gates and inverters. Try to minimize the number of gates required. The variables $A$, $B$, $C$, and $D$ will be available from toggle switches. Any solution that uses 18 or fewer gates and inverters (not counting the four inverters for the inputs) is acceptable.

**8.F** Design a Gray code converter to drive a seven-segment indicator. The four inputs to the converter circuit ($A$, $B$, $C$, and $D$ in Figure 8-14) represent a decimal digit coded using the Gray code of Table 1-2. Assume that only input combinations representing the digits 0 through 9 can occur as inputs, so that the six unused combinations are don't-care terms. Design your circuit using only two-, three-, and four-input NAND gates and inverters. Try to minimize the numbers of gates and inverters required. The variables $A$, $B$, $C$, and $D$ will be available from toggle switches.

Use 6 (not b) for 6. Use 9 (not 9) for 9.

Any solution with 20 or fewer gates and inverters (not counting the four inverters for the inputs) is acceptable.

**8.G** Design a circuit that will add either 1 or 2 to a 4-bit binary number $N$. Let the inputs $N_3$, $N_2$, $N_1$, $N_0$ represent $N$. The input $K$ is a control signal. The circuit should have outputs $M_3$, $M_2$, $M_1$, $M_0$, which represent the 4-bit binary number $M$. When $K = 0$, $M = N + 1$. When $K = 1$, $M = N + 2$. Assume that the inputs for which $M > 1111_2$ will never occur.

Design the circuit using only two-, three-, and four-input NAND gates and inverters. Try to minimize the total number of gates and inverters required. The input variables $K$, $N_3$, $N_2$, $N_1$, and $N_0$ will be available from toggle switches. Any solution that uses 13 or fewer gates and inverters (not counting the five inverters for the inputs) is acceptable.

**8.H** Work Problem 8.A, except use 4-2-1-8 code instead of 8-4-2-1 code. For example, in 4-2-1-8 code, 9 is represented by 0011. Also change the representations of digits 6 and 9 to the opposite form given in Problem 8.A. Any solution with 20 or fewer gates and inverters (not counting the four inverters for the inputs) is acceptable.

**8.I** Work Problem 8.B, except use excess-2 code instead of excess-3 code. (In excess-2 code, 0 is represented by 0010, 1 by 0011, 2 by 0100, etc.). Any solution with 17 or fewer gates and inverters (not counting the four inverters for the inputs) is acceptable.

**8.J** Design a circuit which will multiply a 3-bit binary number $CDE$ by 2, 3, or 5, depending on the value of a 2-bit code $AB$ (00, 01, or 10), to produce a 4-bit result $WXYZ$. If the result has a value greater than or equal to 15, $WXYZ$ should be 1111 to indicate an overflow. Assume that the code $AB = 11$ will never occur. Design your circuit using only two-, three-, and four-input NOR gates and inverters. Try to minimize the number of gates required. The inputs $A$, $B$, $C$, $D$, and $E$ will be available from toggle switches. Any solution which uses 19 or fewer gates and inverters (not counting the five inverters for the inputs) is acceptable.

**8.K** Design a circuit which will divide a 5-bit binary number by 3 to produce a 4-bit binary quotient. Assume that the input number is in the range 0 through 27 and that numbers in the range 28 through 31 will never occur as inputs. Design your circuit using only two-, three-, and four-input NAND gates and inverters. Try to minimize the number of gates required. The inputs $A$, $B$, $C$, $D$, and $E$ will be available from toggle switches. Any solution which uses 22 or fewer gates and inverters (not counting the five inverters for the inputs) is acceptable.

**8.L** Design an excess-3 code converter to drive a seven-segment indicator. The four inputs ($A$, $B$, $C$, $D$) to the converter circuit represent an excess-3 digit. Input combinations representing the numbers 0 through 9 should be displayed as decimal digits. The input combinations 0000, 0001, and 0010 should be interpreted as an error, and an "E" should be displayed. Assume that the input combinations 1101, 1110, and 1111 will never occur. Design your circuit using only two-, three-, and four-input NOR gates and inverters. Any solution with 18 or fewer gates and inverters (not counting the four inverters for the inputs) is acceptable.

Use 6 (not 6) for 6. Use 9 (not 9) for 9.

**8.M** Design a circuit which displays the letters A through J on a seven-segment indicator. The circuit has four inputs $W$, $X$, $Y$, $Z$ which represent the last 4 bits of the ASCII code for the letter to be displayed. For example, if $WXYZ$ = 0001, "A" will be displayed. The letters should be displayed in the following form:

A b C d E F g H I J

Design your circuit using only two-, three-, and four-input NOR gates and inverters. Any solution with 22 or fewer gates and inverters (not counting the four inverters for the inputs) is acceptable.

**8.N** A simple security system for two doors consists of a card reader and a keypad.

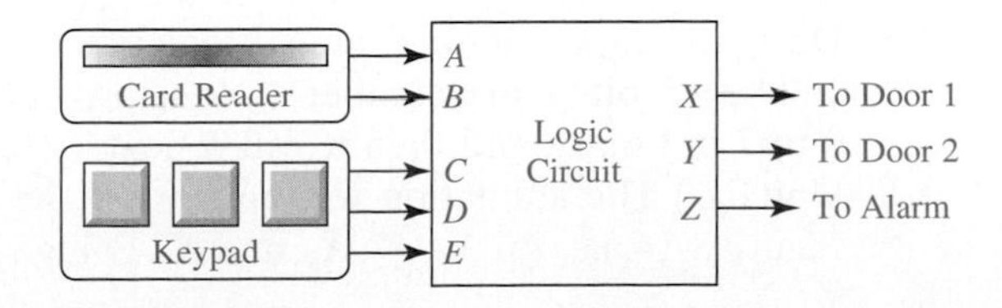

A person may open a particular door if he or she has a card containing the corresponding code and enters an authorized keypad code for that card. The outputs from the card reader are as follows:

| | *A* | *B* |
|---|---|---|
| No card inserted | 0 | 0 |
| Valid code for door 1 | 0 | 1 |
| Valid code for door 2 | 1 | 1 |
| Invalid card code | 1 | 0 |

To unlock a door, a person must hold down the proper keys on the keypad and, then, insert the card in the reader. The authorized keypad codes for door 1 are 101 and 110, and the authorized keypad codes for door 2 are 101 and 011. If the card has an invalid code or if the wrong keypad code is entered, the alarm will ring when the card is inserted. If the correct keypad code is entered, the corresponding door will be unlocked when the card is inserted.

Design the logic circuit for this simple security system. Your circuit's inputs will consist of a card code $AB$, and a keypad code $CDE$. The circuit will have three outputs $XYZ$ (if $X$ or $Y = 1$, door 1 or 2 will be opened; if $Z = 1$, the alarm will sound). Design your circuit using only two-, three-, and four-input NOR gates and inverters. Any solution with 19 or fewer gates and inverters (not counting the five inverters for the inputs) is acceptable. Use toggle switches for inputs $A$, $B$, $C$, $D$, and $E$ when you test your circuit.

**8.O** Work Design Problem 8.A using two-, three-, and four-input NOR gates and inverters instead of NAND gates and inverters. Any solution that uses 19 gates and inverters or fewer (not counting the four inverters for the inputs) is acceptable.

**8.P** Work Design Problem 8.F using two-, three-, and four-input NOR gates and inverters instead of NAND gates and inverters. Any solution that uses 21 gates and inverters or fewer (not counting the four inverters for the inputs) is acceptable.

**8.Q** Work Design Problem 8.H using two-, three-, and four-input NOR gates and inverters instead of NAND gates and inverters. Any solution that uses 17 gates and inverters or fewer (not counting the four inverters for the inputs) is acceptable.

**8.R** Work Design Problem 8.I using two-, three-, and four-input NOR gates and inverters instead of NAND gates and inverters. Any solution that uses 16 gates and inverters or fewer (not counting the four inverters for the inputs) is acceptable.

**8.S** Design a "disk spinning" animation circuit for a CD player. The input to the circuit will be a 3-bit binary number $A_1A_2A_3$ provided by another circuit. It will count from 0 to 7 in binary, and then it will repeat. (You will learn to design such counters in Unit 12.) The animation will appear on the top four lights of the LED display of Figure 8-14, i.e., on $X_1$, $X_2$, $X_7$, and $X_6$, going clockwise. The animation should consist of a blank spot on a disk spinning around once, beginning with $X_1$. Then, the entire disk should blink on and off twice. The pattern is shown.

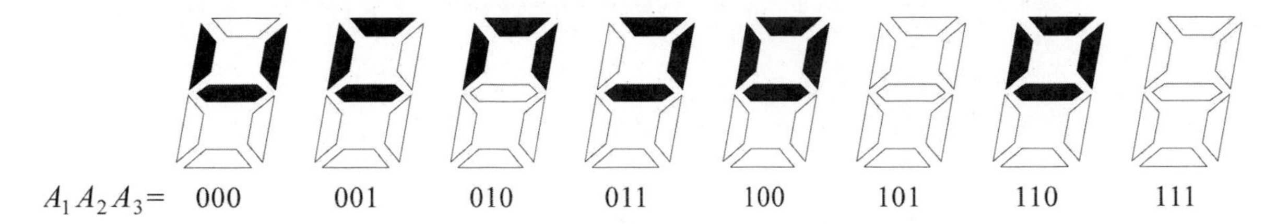


Design your circuit using only two-, three-, and four-input NOR gates and inverters. Try to minimize the number of gates required. Any solution which uses 11 or fewer gates (not counting the four inverters for the inputs) is acceptable.

# Multiplexers, Decoders, and Programmable Logic Devices

## Objectives

1. Explain the function of a multiplexer. Implement a multiplexer using gates.
2. Explain the operation of three-state buffers. Determine the resulting output when three-state buffer outputs are connected together. Use three-state buffers to multiplex signals onto a bus.
3. Explain the operation of a decoder and encoder. Use a decoder with added gates to implement a set of logic functions. Implement a decoder or priority encoder using gates.
4. Explain the operation of a read-only memory (ROM). Use a ROM to implement a set of logic functions.
5. Explain the operation of a programmable logic array (PLA). Use a PLA to implement a set of logic functions. Given a PLA table or an internal connection diagram for a PLA, determine the logic functions realized.
6. Explain the operation of a programmable array logic device (PAL). Determine the programming pattern required to realize a set of logic functions with a PAL.
7. Explain the operation of a complex programmable logic device (CPLD) and a field-programmable gate array (FPGA).
8. Use Shannon's expansion theorem to decompose a switching function.

# Study Guide

**1.** Read Section 9.1, *Introduction.*

**2.** Study Section 9.2, *Multiplexers.*

(a) Draw a logic circuit for a 2-to-1 multiplexer (MUX) using gates.

(b) Write the equation for a 4-to-1 MUX with control inputs $A$ and $C$.

$Z =$ ______________________________

(c) By tracing signals on Figure 9-3, determine what will happen to $Z$ if $A = 1$, $B = 0$ and $C$ changes from 0 to 1.

(d) Use three 2-to-1 MUXes to make a 4-to-1 MUX with control inputs $A$ and $B$. Draw the circuit. (*Hint:* One MUX should have $I_0$ and $I_1$ inputs, and another should have $I_2$ and $I_3$ inputs.)

(e) Observe that if $A = 0, A \oplus B = B$, and that if $A = 1, A \oplus B = B'$. Using this observation, construct an exclusive-OR gate using a 2-to-1 multiplexer and one inverter.

(f) Work Problems 9.1 and 9.2.

(g) This section introduces bus notation. The bus symbol $A$ —/— (4) represents a group of four wires:

$A_3$ ____________

$A_2$ ____________

$A_1$ ____________

$A_0$ ____________

Draw the bus symbol for

$B_2$ ——————
$B_1$ ——————
$B_0$ ——————

(h) Represent the circuit of Figure 4-3 by one 4-bit full adder with two bus inputs, one bus output, and terminals for carry input $C_0$ and output $C_4$. Note that the carries $C_3$, $C_2$, and $C_1$ will not appear on your circuit diagram because they are signals internal to the 4-bit adder.

**3.** Study Section 9.3, *Three-State Buffers*.

(a) Determine the output of each three-state buffer:

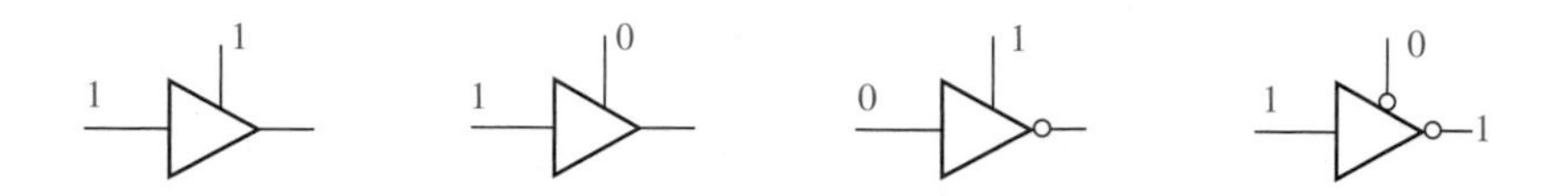


(b) Determine the inputs for each three-state buffer (use X if an input is a don't-care).

(c) Determine the output for each circuit. Use X to represent an unknown output.

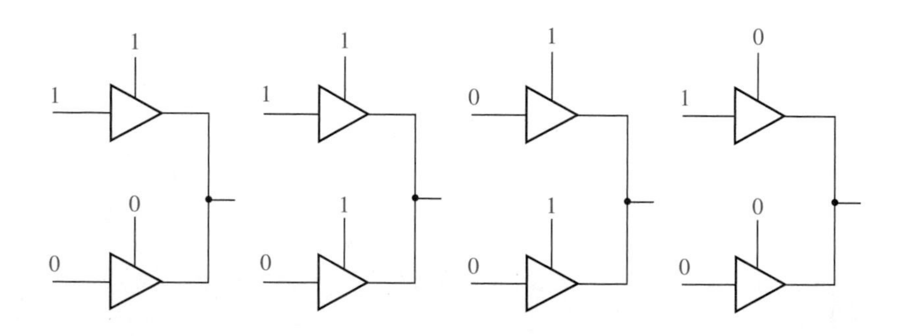


(d) The symbol $A \overset{2}{\not-} \triangleright \overset{2}{\not-} B$ (with control input $C$) represents 2 three-state buffers with a common control input:

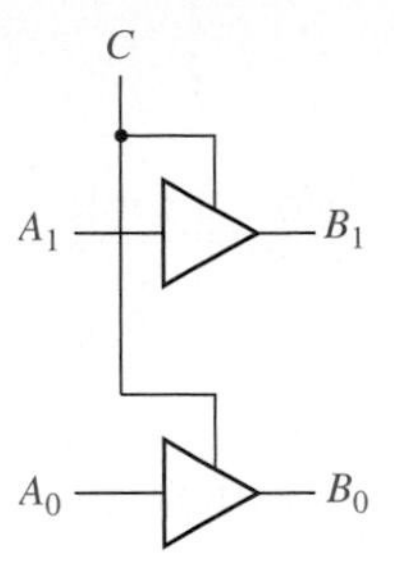


Using bus notation, draw an equivalent circuit for:

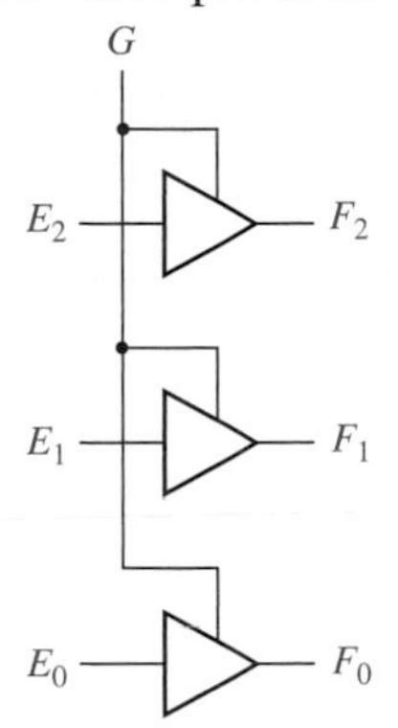


(e) For the following circuit, determine the 4-bit output ($P$) if $M = 0$. ______________ Repeat for $M = 1$. ______________

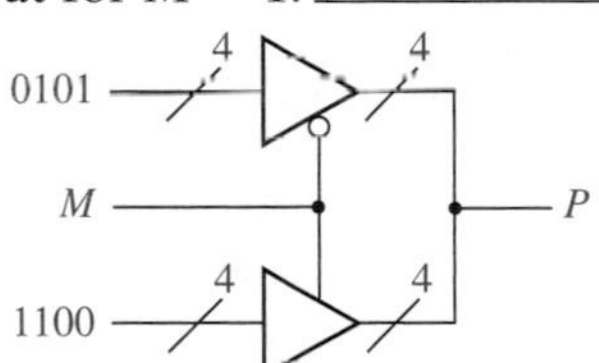


(f) Specify the AND-gate inputs so that the given circuit is equivalent to the 4-to-1 MUX in Figure 9-2. ($Z$ in the following figure represents an output terminal, not high impedance.)

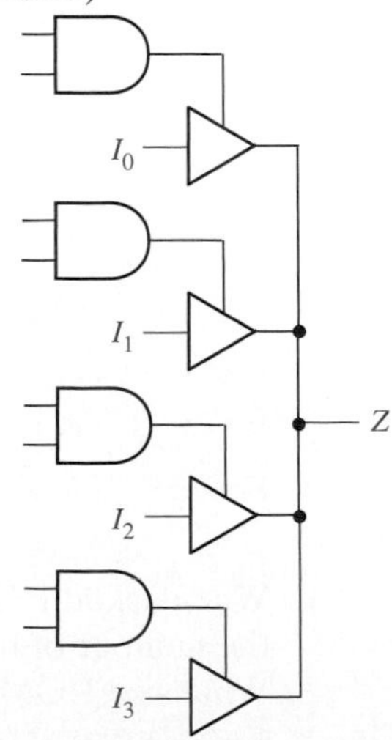

(g) Work Problem 9.3.

**4.** Study Section 9.4, *Decoders and Encoders.*

(a) The 7442 4-to-10 line decoder (Figure 9-14) can be used as a 3-to-8 line decoder. To do this, which three lines should be used as inputs?

________________________

The remaining input line should be set equal to ____________ .

(b) Complete the following table for a 4-to-2 priority encoder:

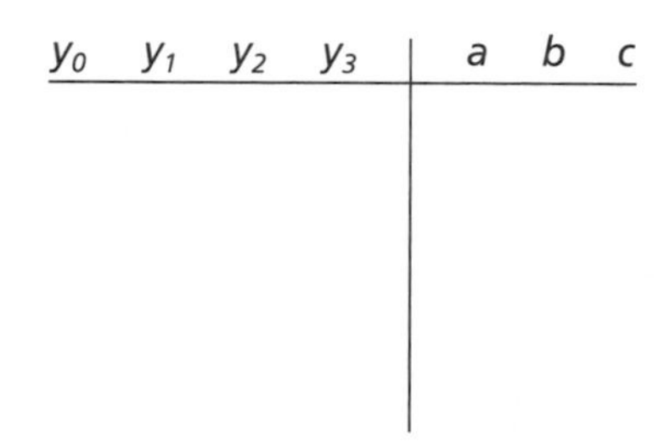

| $y_0$ | $y_1$ | $y_2$ | $y_3$ | $a$ | $b$ | $c$ |
|---|---|---|---|---|---|---|

What will $a$, $b$, and $c$ be if $y_0\ y_1\ y_2\ y_3$ is 0101?

(c) Work Problems 9.4, 9.5, and 9.6.

**5.** Study Section 9.5, *Read-Only Memories.*

(a) The following diagram shows the pattern of 0's and 1's stored in a ROM with eight words and four bits per word. What will be the values of $F_1$, $F_2$, $F_3$, and $F_4$ if $A = 0$ and $B = C = 1$?

Give the minterm expansions for $F_1$ and $F_2$:

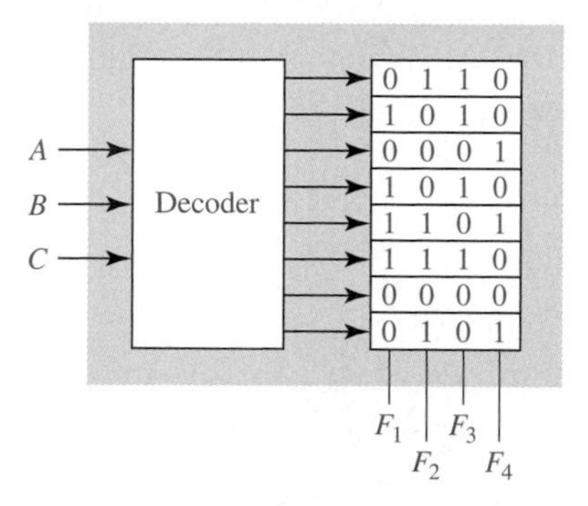


$F_1 =$

$F_2 =$

(b) When asked to specify the size of a ROM, give the number of words and the number of bits per word.

What size ROM is required to realize four functions of 5 variables?

What size ROM is required to realize eight functions of 10 variables?

(c) When specifying the size of a ROM, assume that you are specifying a standard size ROM with $2^n$ words. What size ROM is required to convert 8-4-2-1 BCD code to 2-out-of-5 code? (See Table 1-2, page 21.)

What size ROM would be required to realize the decoder given in Figure 9-14?

(d) Draw an internal connection diagram for a ROM which would perform the same function as the circuit of Figure 7-20. (Indicate the presence of switching elements by dots at the intersection of the word lines and output lines.)

(e) Explain the difference between a mask-programmable ROM and an EEPROM. Which would you use for a new design which had not yet been debugged?

(f) Work Problem 9.7.

**6.** Study Section 9.6, *Programmable Logic Devices.*

(a) When you are asked to specify the size of a PLA, give the number of inputs, the number of product terms, and the number of outputs.
What size PLA would be required to realize Equations (7-22) if no simplification of the minterm expansions were performed?

(b) If the realization of Equations (7-22) shown in Figure 7-20 were converted to a PLA realization, what size PLA would be required?

(c) Specify the contents of the PLA of question (b) in tabular form. Your table should have four rows. (You will only need seven 1's on the right side of your table. If you get eight 1's, you are probably doing more work than is necessary.)

(d) Draw an internal connection diagram for the PLA of (b). (Use X's to indicate the presence of switching elements in the AND and OR arrays.)

(e) Given the following PLA table, plot maps for $Z_1$, $Z_2$, and $Z_3$.

| A B C | $Z_1$ $Z_2$ $Z_3$ |
|---|---|
| – 0 0 | 1 1 0 |
| 0 1 – | 1 1 0 |
| 1 0 – | 1 0 0 |
| 1 1 1 | 0 1 1 |
| 0 – 1 | 1 0 1 |
| 0 0 0 | 0 0 1 |

$Z_1$:

| BC \ A | 0 | 1 |
|---|---|---|
| 00 | | |
| 01 | | |
| 11 | | |
| 10 | | |

$Z_2$:

| BC \ A | 0 | 1 |
|---|---|---|
| 00 | | |
| 01 | | |
| 11 | | |
| 10 | | |

$Z_3$:

| BC \ A | 0 | 1 |
|---|---|---|
| 00 | | |
| 01 | | |
| 11 | | |
| 10 | | |

(The $Z_1$ map should have six 1's, $Z_2$ should have five, and $Z_3$ should have four.)

(f) For a truth table, any combination of input values will select exactly one row. Is this statement true for a PLA table?

For any combination of input values, the output values from a PLA can be determined by inspection of the PLA table. Consider Table 9-1, which represents a PLA with three inputs and four outputs. If the inputs are $ABC = 110$, which three rows in the table are selected?

In a given output column, what is the output if some of the selected rows are 1's and some are 0's? (Remember that the output bits for the selected rows are ORed together.)

When $ABC = 110$, what are the values of $F_0F_1F_2F_3$ at the PLA output?

When $ABC = 010$, which rows are selected and what are the values of $F_0F_1F_2F_3$ at the PLA output?

(g) Which interconnection points in Figure 9-28(a) must be set in order to realize the function shown in Figure 9-28(b)?

(h) What size of PAL could be used to realize the 8-to-1 MUX of Figure 9-3? The quad MUX of Figure 9-5? Give the number of inputs, the number of OR gates, and the maximum number of inputs to an OR gate.

(i) Work Problems 9.8, 9.9, and 9.10.

**7.** Study Section 9.7, *Complex Programmable Logic Devices.* Work Problem 9.11.

**8.** Study Section 9.8, *Field-Programmable Gate Arrays.*

(a) For the CLB of Figure 9-33, write a logic equation for $H$ in terms of $F$, $G$, and $H_1$.

(b) How many 4-variable function generators are required to implement a four-input OR gate? A 4-variable function with 13 minterms?

(c) Expand the function of Equation 9-7 about the variable $c$ instead of $a$. Expand it algebraically and, then, expand it by using the Karnaugh map of Figure 9-35. (*Hint:* How should you split the map into two halves?)

(d) Draw a diagram showing how to implement Equation 9-10 using four function generators and a 4-to-1 MUX.

(e) In the worst case, how many 4-variable function generators are required to realize a 7-variable function (assume the necessary MUXes are available).

(f) Show how to realize $K = abcdefg$ using only two 4-variable function generators. (*Hint:* Use the output of one function generator as an input to the other.)

(g) Work Problems 9.12 and 9.13.

9. When you are satisfied that you can meet all of the objectives, take the readiness test.

# Multiplexers, Decoders, and Programmable Logic Devices

## 9.1 Introduction

Until this point we have mainly been concerned with basic principles of logic design. We have illustrated these principles using gates as our basic building blocks. In this unit we introduce the use of more complex integrated circuits (ICs) in logic design. Integrated circuits may be classified as small-scale integration (SSI), medium-scale integration (MSI), large-scale integration (LSI), or very-large-scale integration (VLSI), depending on the number of gates in each integrated circuit package and the type of function performed. SSI functions include NAND, NOR, AND, and OR gates, inverters, and flip-flops. SSI integrated circuit packages typically contain one to four gates, six inverters, or one or two flip-flops. MSI integrated circuits, such as adders, multiplexers, decoders, registers, and counters, perform more complex functions. Such integrated circuits typically contain the equivalent of 12 to 100 gates in one package. More complex functions such as memories and microprocessors are classified as LSI or VLSI integrated circuits. An LSI integrated circuit generally contains 100 to a few thousand gates in a single package, and a VLSI integrated circuit contains several thousand gates or more.

It is generally uneconomical to design digital systems using only SSI and MSI integrated circuits. By using LSI and VLSI functions, the required number of integrated circuit packages is greatly reduced. The cost of mounting and wiring the integrated circuits as well as the cost of designing and maintaining the digital system may be significantly lower when LSI and VLSI functions are used.

This unit introduces the use of multiplexers, decoders, encoders, and three-state buffers in logic design. Then read-only memories (ROMs) are described and used to implement multiple-output combinational logic circuits. Finally, other types of programmable logic devices (PLDs), including programmable logic arrays (PLAs), programmable array logic devices (PALs), complex programmable logic devices (CPLDs), and field-programmable gate arrays (FPGAs) are introduced and used in combinational logic design.

# 9.2 Multiplexers

A multiplexer (or data selector, abbreviated as MUX) has a group of data inputs and a group of control inputs. The control inputs are used to select one of the data inputs and connect it to the output terminal. Figure 9-1 shows a 2-to-1 multiplexer and its switch analog. When the control input $A$ is 0, the switch is in the upper position and the MUX output is $Z = I_0$; when $A$ is 1, the switch is in the lower position and the MUX output is $Z = I_1$. In other words, a MUX acts like a switch that selects one of the data inputs ($I_0$ or $I_1$) and transmits it to the output. The logic equation for the 2-to-1 MUX is therefore:

$$Z = A'I_0 + AI_1$$

**FIGURE 9-1**
2-to-1 Multiplexer and Switch Analog

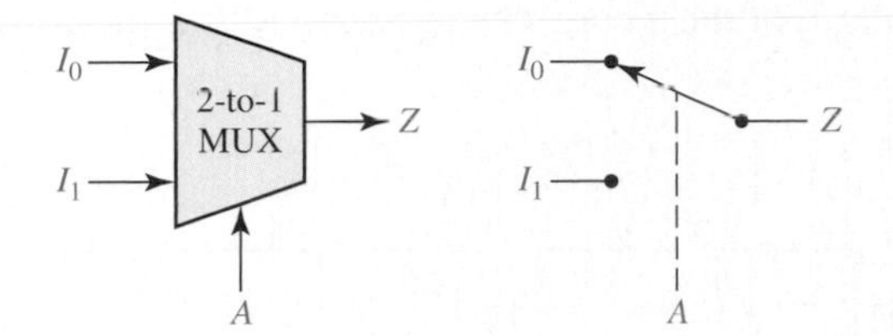


Figure 9-2 shows diagrams for a 4-to-1 multiplexer, 8-to-1 multiplexer, and $2^n$-to-1 multiplexer. The 4-to-1 MUX acts like a four-position switch that transmits one of the four inputs to the output. Two control inputs ($A$ and $B$) are needed to select one of the four inputs. If the control inputs are $AB = 00$, the output is $I_0$; similarly, the control inputs 01, 10, and 11 give outputs of $I_1$, $I_2$, and $I_3$, respectively. The 4-to-1 multiplexer is described by the equation

$$Z = A'B'I_0 + A'BI_1 + AB'I_2 + ABI_3 \tag{9-1}$$

**FIGURE 9-2**
Multiplexers

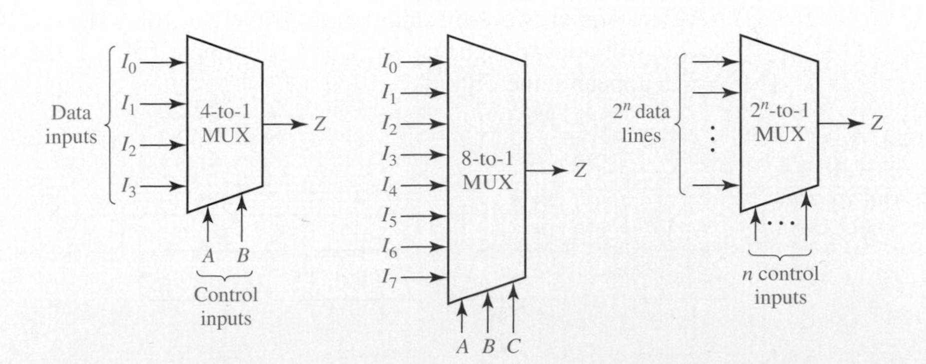

Similarly, the 8-to-1 MUX selects one of eight data inputs using three control inputs. It is described by the equation

$$Z = A'B'C'I_0 + A'B'CI_1 + A'BC'I_2 + A'BCI_3 + AB'C'I_4 + AB'CI_5 + ABC'I_6 + ABCI_7 \qquad (9\text{-}2)$$

When the control inputs are $ABC = 011$, the output is $I_3$, and the other outputs are selected in a similar manner. Figure 9-3 shows an internal logic diagram for the 8-to-1 MUX. In general, a multiplexer with $n$ control inputs can be used to select any one of $2^n$ data inputs. The general equation for the output of a MUX with $n$ control inputs and $2^n$ data inputs is

$$Z = \sum_{k=0}^{2^n-1} m_k I_k$$

where $m_k$ is a minterm of the $n$ control variables and $I_k$ is the corresponding data input.

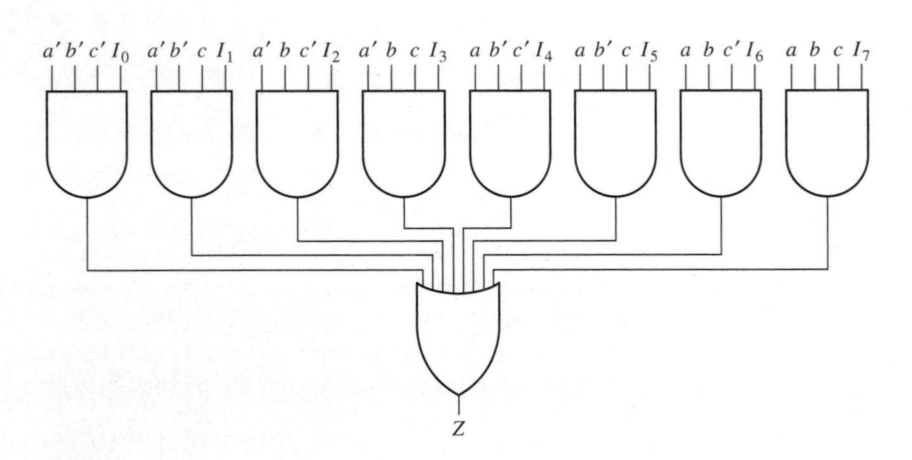


**FIGURE 9-3** Logic Diagram for 8-to-1 MUX

Multiplexers are frequently used in digital system design to select the data which is to be processed or stored. Figure 9-4 shows how a quadruple 2-to-1 MUX is used to select one of two 4-bit data words. If the control is $A = 0$, the values of $x_0$, $x_1$, $x_2$, and $x_3$ will appear at the $z_0$, $z_1$, $z_2$, and $z_3$ outputs; if $A = 1$, the values of $y_0$, $y_1$, $y_2$, and $y_3$ will appear at the outputs.

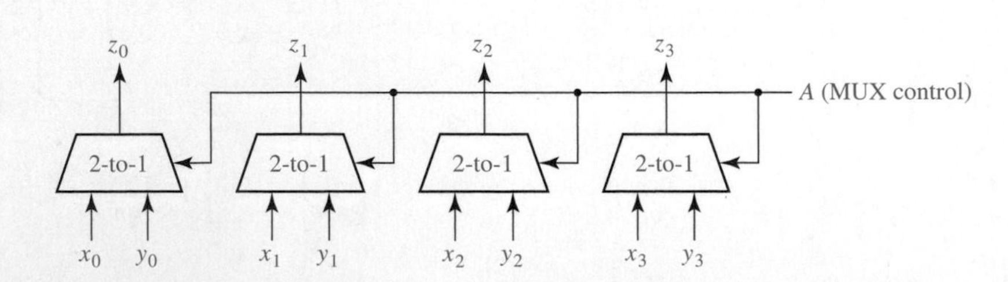


**FIGURE 9-4** Quad Multiplexer Used to Select Data

Several logic signals that perform a common function may be grouped together to form a bus. For example, the sum outputs of a 4-bit binary adder can be grouped together to form a 4-bit bus. Instead of drawing the individual wires that make up a bus, we often represent a bus by a single heavy line. The quad MUX of Figure 9-4 is redrawn in Figure 9-5, using bus inputs $X$ and $Y$, and bus output $Z$. The $X$ bus represents the four signals $x_0$, $x_1$, $x_2$, and $x_3$, and similarly for the $Y$ and $Z$ buses. When $A = 0$, the signals on bus $X$ appear on bus $Z$; otherwise, the signals on bus $Y$ appear. A diagonal slash through a bus with a number beside it specifies the number of bits in the bus.

**FIGURE 9-5** Quad Multiplexer with Bus Inputs and Output

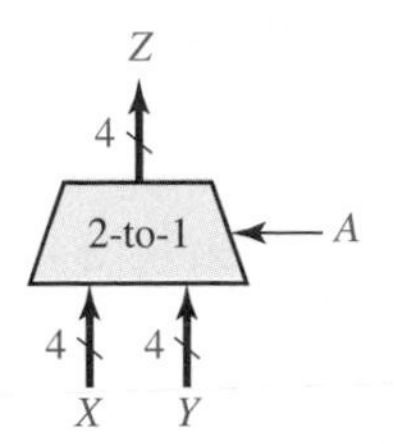


The preceding multiplexers do not invert the data inputs as they are routed to the output. Some multiplexers do invert the inputs, e.g., if the OR gate in Figure 9-3 is replaced by a NOR gate, then the 8-to-1 MUX inverts the selected input. To distinguish between these two types of multiplexers, we will say that the multiplexers without the inversion have active high outputs, and the multiplexers with the inversion have active low outputs.

Another type of multiplexer has an additional input called an *enable*. The 8-to-1 MUX in Figure 9-3 can be modified to include an enable by changing the AND gates to five-input gates. The enable signal $E$ is connected to the fifth input of each of the AND gates. Then, if $E = 0$, $Z = 0$ independent of the gate inputs $I_i$ and the select inputs $a$, $b$, and $c$. However, if $E = 1$, then the MUX functions as an ordinary 8-to-1 multiplexer. The terminology used for the MUX output, i.e., active high and active low, can be used for the enable as well. As described above, the enable is active high; $E$ must be 1 for the MUX to function as a multiplexer. If an inverter is inserted between $E$ and the AND gates, $E$ must be 0 for the MUX to function as a multiplexer; the enable is active low.

Four combinations of multiplexers with an enable are possible. The output can be active high or active low, whereas the enable can be active high or active low. In a block diagram for the MUX, an active low line is indicated by inserting a bubble on the line to indicate the inclusion of an inversion.

## 9.3 Three-State Buffers

A gate output can only be connected to a limited number of other device inputs without degrading the performance of a digital system. A simple buffer may be used to increase the driving capability of a gate output. Figure 9-6 shows a buffer connected

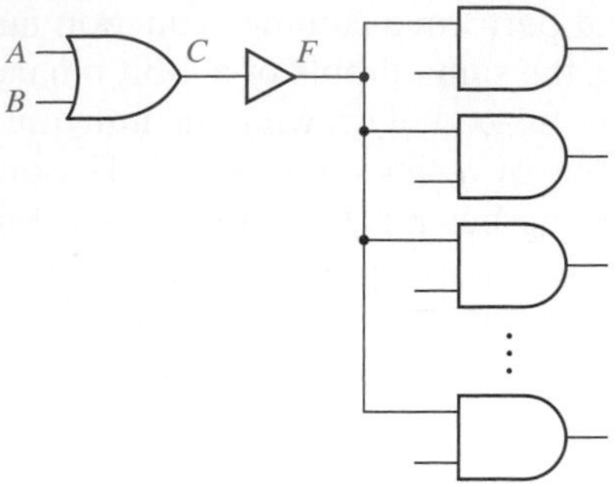

**FIGURE 9-6** Gate Circuit with Added Buffer



between a gate output and several gate inputs. Because no bubble is present at the buffer output, this is a noninverting buffer, and the logic values of the buffer input and output are the same, that is, $F = C$.

Normally, a logic circuit will not operate correctly if the outputs of two or more gates or other logic devices are directly connected to each other. For example, if one gate has a 0 output (a low voltage) and another has a 1 output (a high voltage), when the gate outputs are connected together the resulting output voltage may be some intermediate value that does not clearly represent either a 0 or a 1. In some cases, damage to the gates may result if the outputs are connected together.

Use of three-state logic permits the outputs of two or more gates or other logic devices to be connected together. Figure 9-7 shows a three-state buffer and its logical equivalent. When the enable input $B$ is 1, the output $C$ equals $A$; when $B$ is 0, the output $C$ acts like an open circuit. In other words, when $B$ is 0, the output $C$ is effectively disconnected from the buffer output so that no current can flow. This is often referred to as a Hi-Z (high-impedance) state of the output because the circuit offers a very high resistance or impedance to the flow of current. Three-state buffers are also called tri-state buffers.

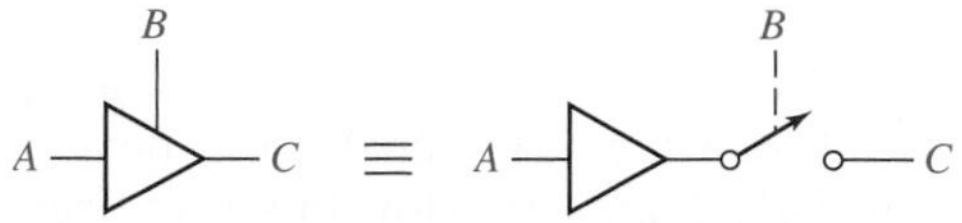

**FIGURE 9-7** Three-State Buffer



Figure 9-8 shows the truth tables for four types of three-state buffers. In Figures 9-8(a) and (b), the enable input $B$ is not inverted, so the buffer output is enabled when $B = 1$ and disabled when $B = 0$. That is, the buffer operates normally when $B = 1$, and the buffer output is effectively an open circuit when $B = 0$. We use the symbol **Z** to represent this high-impedance state. In Figure 9-8(b), the buffer output is inverted so that $C = A'$ when the buffer is enabled. The buffers in 9-8(c) and (d) operate the same as in (a) and (b) except that the enable input is inverted, so the buffer is enabled when $B = 0$.

FIGURE 9-8
Four Kinds of Three-State Buffers

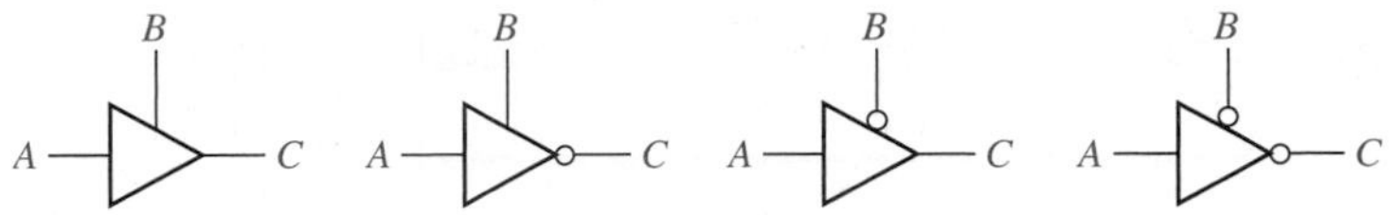

(a)

| B | A | C |
|---|---|---|
| 0 | 0 | Z |
| 0 | 1 | Z |
| 1 | 0 | 0 |
| 1 | 1 | 1 |

(b)

| B | A | C |
|---|---|---|
| 0 | 0 | Z |
| 0 | 1 | Z |
| 1 | 0 | 1 |
| 1 | 1 | 0 |

(c)

| B | A | C |
|---|---|---|
| 0 | 0 | 0 |
| 0 | 1 | 1 |
| 1 | 0 | Z |
| 1 | 1 | Z |

(d)

| B | A | C |
|---|---|---|
| 0 | 0 | 1 |
| 0 | 1 | 0 |
| 1 | 0 | Z |
| 1 | 1 | Z |

In Figure 9-9, the outputs of two three-state buffers are tied together. When $B = 0$, the top buffer is enabled, so that $D = A$; when $B = 1$, the lower buffer is enabled, so that $D = C$. Therefore, $D = B'A + BC$. This is logically equivalent to using a 2-to-1 multiplexer to select the $A$ input when $B = 0$ and the $C$ input when $B = 1$.

When we connect two three-state buffer outputs together, as shown in Figure 9-10, if one of the buffers is disabled (output = Z), the combined output $F$ is the same as the other buffer output. If both buffers are disabled, the output is Z. If both buffers are enabled, a conflict can occur. If $A = 0$ and $C = 1$, we do not know what the hardware will do, so the $F$ output is unknown (X). If one of the buffer inputs is unknown, the $F$ output will also be unknown. The table in Figure 9-10 summarizes the operation of the circuit. S1 and S2 represent the outputs the two buffers would have if they were not connected together. When a bus is driven by three-state buffers, we call it a three-state bus. The signals on this bus can have values of 0, 1, Z, and perhaps X.

A multiplexer may be used to select one of several sources to drive a device input. For example, if an adder input must come from four different sources, a 4-to-1 MUX may be used to select one of the four sources. An alternative is to

FIGURE 9-9
Data Selection Using Three-State Buffers

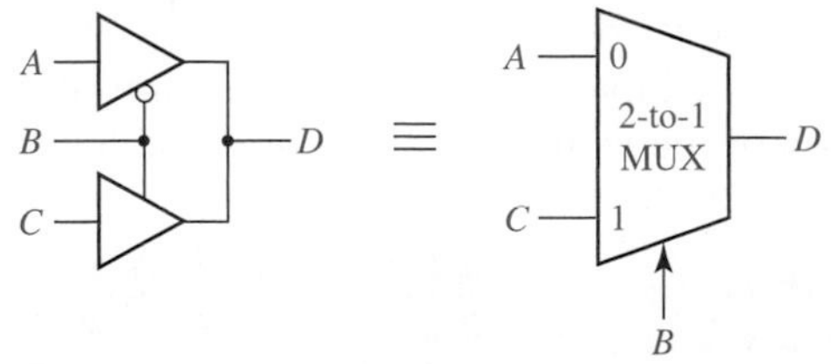

FIGURE 9-10
Circuit with Two Three-State Buffers

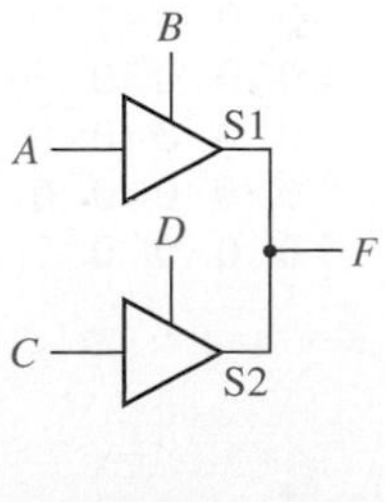

| S1 \ S2 | X | 0 | 1 | Z |
|---|---|---|---|---|
| X | X | X | X | X |
| 0 | X | 0 | X | 0 |
| 1 | X | X | 1 | 1 |
| Z | X | 0 | 1 | Z |

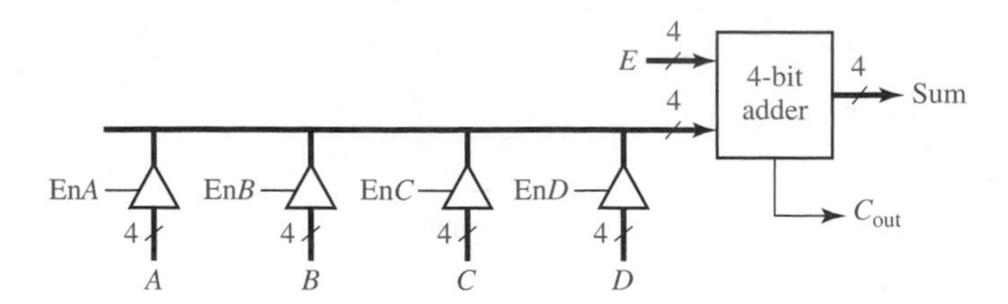


**FIGURE 9-11** 4-Bit Adder with Four Sources for One Operand

set up a three-state bus, using three-state buffers to select one of the sources (see Figure 9-11). In this circuit, each buffer symbol actually represents four three-state buffers that have a common enable signal.

Integrated circuits are often designed using bi-directional pins for input and output. Bi-directional means that the same pin can be used as an input pin and as an output pin, but not both at the same time. To accomplish this, the circuit output is connected to the pin through a three-state buffer, as shown in Figure 9-12. When the buffer is enabled, the pin is driven with the output signal. When the buffer is disabled, an external source can drive the input pin.

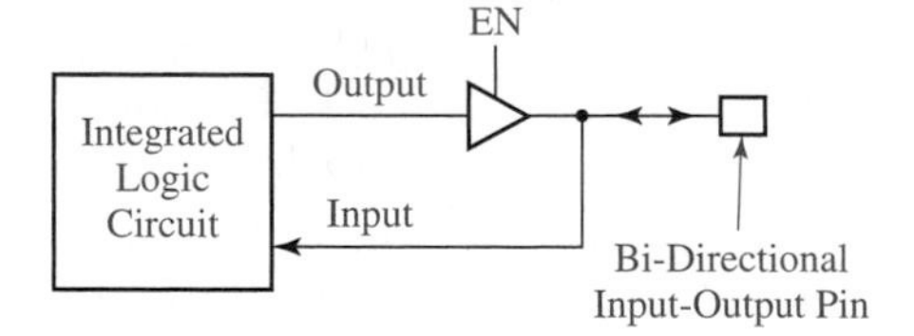


**FIGURE 9-12** Integrated Circuit with Bi-Directional Input-Output Pin

# 9.4 Decoders and Encoders

The decoder is another commonly used type of integrated circuit. Figure 9-13 shows the diagram and truth table for a 3-to-8 line decoder. This decoder generates all of the minterms of the three input variables. Exactly one of the output lines will be 1 for each combination of the values of the input variables.

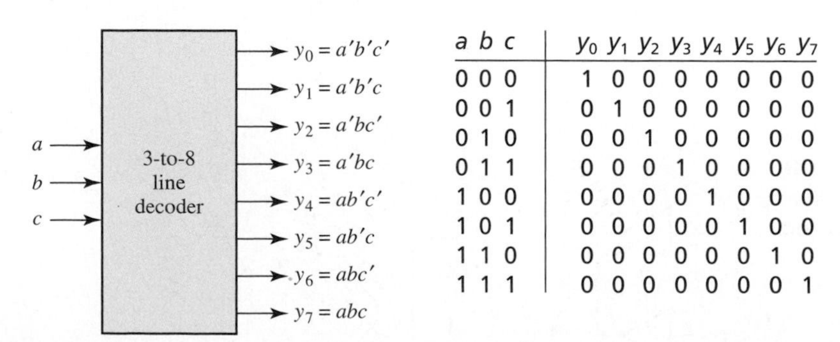


| a b c | $y_0$ | $y_1$ | $y_2$ | $y_3$ | $y_4$ | $y_5$ | $y_6$ | $y_7$ |
|---|---|---|---|---|---|---|---|---|
| 0 0 0 | 1 | 0 | 0 | 0 | 0 | 0 | 0 | 0 |
| 0 0 1 | 0 | 1 | 0 | 0 | 0 | 0 | 0 | 0 |
| 0 1 0 | 0 | 0 | 1 | 0 | 0 | 0 | 0 | 0 |
| 0 1 1 | 0 | 0 | 0 | 1 | 0 | 0 | 0 | 0 |
| 1 0 0 | 0 | 0 | 0 | 0 | 1 | 0 | 0 | 0 |
| 1 0 1 | 0 | 0 | 0 | 0 | 0 | 1 | 0 | 0 |
| 1 1 0 | 0 | 0 | 0 | 0 | 0 | 0 | 1 | 0 |
| 1 1 1 | 0 | 0 | 0 | 0 | 0 | 0 | 0 | 1 |

**FIGURE 9-13** A 3-to-8 Line Decoder

Figure 9-14 illustrates a 4-to-10 decoder. This decoder has inverted outputs (indicated by the small circles). For each combination of the values of the inputs, exactly one of the output lines will be 0. When a binary-coded-decimal digit is used as an input to this decoder, one of the output lines will go low to indicate which of the 10 decimal digits is present.

**FIGURE 9-14**
A 4-to-10 Line Decoder

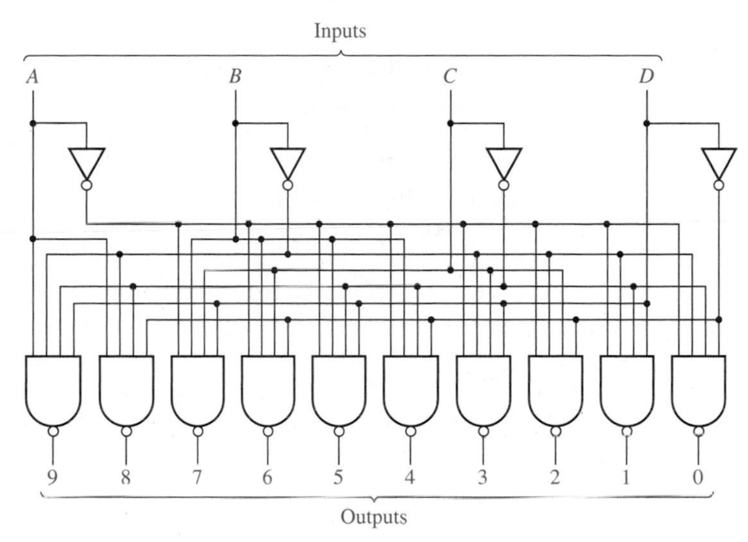


(a) Logic diagram

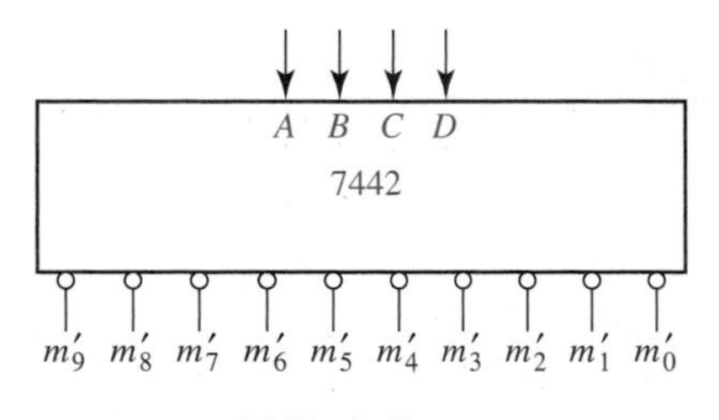


(b) Block diagram

| BCD Input | | | | Decimal Output | | | | | | | | | |
|---|---|---|---|---|---|---|---|---|---|---|---|---|---|
| *A* | *B* | *C* | *D* | 0 | 1 | 2 | 3 | 4 | 5 | 6 | 7 | 8 | 9 |
| 0 | 0 | 0 | 0 | 0 | 1 | 1 | 1 | 1 | 1 | 1 | 1 | 1 | 1 |
| 0 | 0 | 0 | 1 | 1 | 0 | 1 | 1 | 1 | 1 | 1 | 1 | 1 | 1 |
| 0 | 0 | 1 | 0 | 1 | 1 | 0 | 1 | 1 | 1 | 1 | 1 | 1 | 1 |
| 0 | 0 | 1 | 1 | 1 | 1 | 1 | 0 | 1 | 1 | 1 | 1 | 1 | 1 |
| 0 | 1 | 0 | 0 | 1 | 1 | 1 | 1 | 0 | 1 | 1 | 1 | 1 | 1 |
| 0 | 1 | 0 | 1 | 1 | 1 | 1 | 1 | 1 | 0 | 1 | 1 | 1 | 1 |
| 0 | 1 | 1 | 0 | 1 | 1 | 1 | 1 | 1 | 1 | 0 | 1 | 1 | 1 |
| 0 | 1 | 1 | 1 | 1 | 1 | 1 | 1 | 1 | 1 | 1 | 0 | 1 | 1 |
| 1 | 0 | 0 | 0 | 1 | 1 | 1 | 1 | 1 | 1 | 1 | 1 | 0 | 1 |
| 1 | 0 | 0 | 1 | 1 | 1 | 1 | 1 | 1 | 1 | 1 | 1 | 1 | 0 |
| 1 | 0 | 1 | 0 | 1 | 1 | 1 | 1 | 1 | 1 | 1 | 1 | 1 | 1 |
| 1 | 0 | 1 | 1 | 1 | 1 | 1 | 1 | 1 | 1 | 1 | 1 | 1 | 1 |
| 1 | 1 | 0 | 0 | 1 | 1 | 1 | 1 | 1 | 1 | 1 | 1 | 1 | 1 |
| 1 | 1 | 0 | 1 | 1 | 1 | 1 | 1 | 1 | 1 | 1 | 1 | 1 | 1 |
| 1 | 1 | 1 | 0 | 1 | 1 | 1 | 1 | 1 | 1 | 1 | 1 | 1 | 1 |
| 1 | 1 | 1 | 1 | 1 | 1 | 1 | 1 | 1 | 1 | 1 | 1 | 1 | 1 |

(c) Truth Table

In general, an $n$-to-$2^n$ line decoder generates all $2^n$ minterms (or maxterms) of the $n$ input variables. The outputs are defined by the equations

$$y_i = m_i, \qquad i = 0 \text{ to } 2^n - 1 \qquad \text{(noninverted outputs)} \tag{9-3}$$

or

$$y_i = m_i' = M_i, \qquad i = 0 \text{ to } 2^n - 1 \qquad \text{(inverted outputs)} \tag{9-4}$$

where $m_i$ is a minterm of the $n$ input variables and $M_i$ is a maxterm.

Because an $n$-input decoder generates all of the minterms of $n$ variables, $n$-variable functions can be realized by ORing together selected minterm outputs from a decoder. If the decoder outputs are inverted, then NAND gates can be used to generate the functions, as illustrated in the following example. Realize

$$f_1(a, b, c, d) = m_1 + m_2 + m_4 \text{ and } f_2(a, b, c, d) = m_4 + m_7 + m_9$$

using the decoder of Figure 9-14. Rewriting $f_1$ and $f_2$, we have

$$f_1 = (m_1' m_2' m_4')' \qquad f_2 = (m_4' m_7' m_9')'$$

Then $f_1$ and $f_2$ can be generated using NAND gates, as shown in Figure 9-15.

An encoder performs the inverse function of a decoder. Figure 9-16 shows an 8-to-3 priority encoder with inputs $y_0$ through $y_7$. If input $y_i$ is 1 and the other inputs are 0, then the $abc$ outputs represent a binary number equal to $i$. For example, if $y_3 = 1$, then $abc = 011$. If more than one input can be 1 at the same time, the output can be defined using a priority scheme. The truth table in Figure 9-16 uses the following

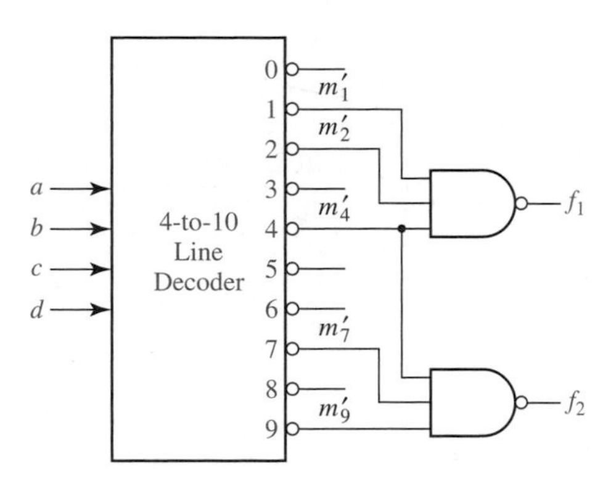


**FIGURE 9-15** Realization of a Multiple-Output Circuit Using a Decoder

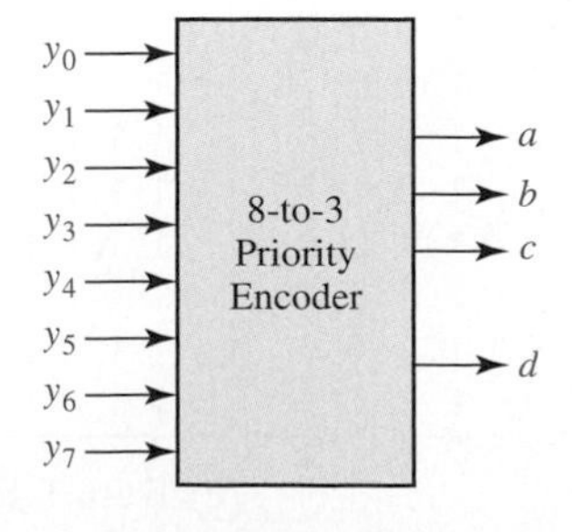


**FIGURE 9-16** An 8-to-3 Priority Encoder

| $y_0$ | $y_1$ | $y_2$ | $y_3$ | $y_4$ | $y_5$ | $y_6$ | $y_7$ | $a$ | $b$ | $c$ | $d$ |
|---|---|---|---|---|---|---|---|---|---|---|---|
| 0 | 0 | 0 | 0 | 0 | 0 | 0 | 0 | 0 | 0 | 0 | 0 |
| 1 | 0 | 0 | 0 | 0 | 0 | 0 | 0 | 0 | 0 | 0 | 1 |
| X | 1 | 0 | 0 | 0 | 0 | 0 | 0 | 0 | 0 | 1 | 1 |
| X | X | 1 | 0 | 0 | 0 | 0 | 0 | 0 | 1 | 0 | 1 |
| X | X | X | 1 | 0 | 0 | 0 | 0 | 0 | 1 | 1 | 1 |
| X | X | X | X | 1 | 0 | 0 | 0 | 1 | 0 | 0 | 1 |
| X | X | X | X | X | 1 | 0 | 0 | 1 | 0 | 1 | 1 |
| X | X | X | X | X | X | 1 | 0 | 1 | 1 | 0 | 1 |
| X | X | X | X | X | X | X | 1 | 1 | 1 | 1 | 1 |

scheme: If more than one input is 1, the highest numbered input determines the output. For example, if inputs $y_1$, $y_4$, and $y_5$ are 1, the output is $abc = 101$. The X's in the table are don't-cares; for example, if $y_5$ is 1, we do not care what inputs $y_0$ through $y_4$ are. Output $d$ is 1 if any input is 1, otherwise, $d$ is 0. This signal is needed to distinguish the case of all 0 inputs from the case where only $y_0$ is 1.

# 9.5 Read-Only Memories

A read-only memory (ROM) consists of an array of semiconductor devices that are interconnected to store an array of binary data. Once binary data is stored in the ROM, it can be read out whenever desired, but the data that is stored cannot be changed under normal operating conditions. Figure 9-17(a) shows a ROM which has three input lines and four output lines. Figure 9-17(b) shows a typical truth table which relates the ROM inputs and outputs. For each combination of input values on the three input lines, the corresponding pattern of 0's and 1's appears on the ROM output lines. For example, if the combination $ABC = 010$ is applied to the input lines, the pattern $F_0F_1F_2F_3 = 0111$ appears on the output lines. Each of the output patterns that is stored in the ROM is called a *word*. Because the ROM has three input lines, we have $2^3$ = eight different combinations of input values. Each input combination serves as an *address* which can select one of the eight words stored in the memory. Because there are four output lines, each word is four bits long, and the size of this ROM is 8 words $\times$ 4 bits.

A ROM which has $n$ input lines and $m$ output lines (Figure 9-18) contains an array of $2^n$ words, and each word is $m$ bits long. The input lines serve as an address to select one of the $2^n$ words. When an input combination is applied to the ROM, the pattern of 0's and 1's which is stored in the corresponding word in the memory appears at the output lines. For the example in Figure 9-18, if 00 . . . 11 is applied to the input (address lines) of the ROM, the word 110 . . . 010 will be selected and transferred to the output lines. A $2^n \times m$ ROM can realize $m$ functions of $n$ variables because it can store a truth table with $2^n$ rows and $m$ columns. Typical sizes for commercially available ROMs range from 32 words $\times$ 4 bits to 512K words $\times$ 8 bits, or larger.

FIGURE 9-17
An 8-Word $\times$ 4-Bit ROM

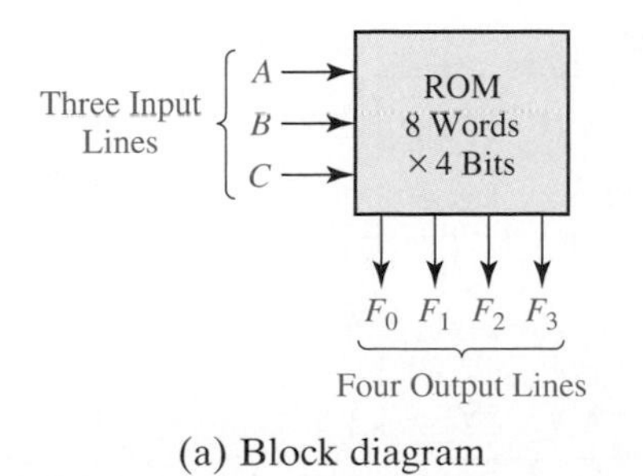


(a) Block diagram

| $A$ $B$ $C$ | $F_0$ $F_1$ $F_2$ $F_3$ |
|---|---|
| 0 0 0 | 1 0 1 0 |
| 0 0 1 | 1 0 1 0 |
| 0 1 0 | 0 1 1 1 |
| 0 1 1 | 0 1 0 1 |
| 1 0 0 | 1 1 0 0 |
| 1 0 1 | 0 0 0 1 |
| 1 1 0 | 1 1 1 1 |
| 1 1 1 | 0 1 0 1 |

Typical Data Stored in ROM ($2^3$ words of 4 bits each)

(b) Truth table for ROM

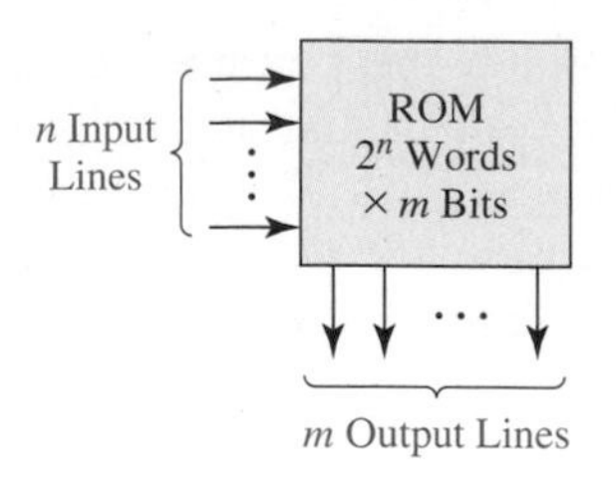


| n Input Variables | m Output Variables |
|---|---|
| 00 ··· 00 | 100 ··· 110 |
| 00 ··· 01 | 010 ··· 111 |
| 00 ··· 10 | 101 ··· 101 |
| 00 ··· 11 | 110 ··· 010 |
| ⋮ | ⋮ |
| 11 ··· 00 | 001 ··· 011 |
| 11 ··· 01 | 110 ··· 110 |
| 11 ··· 10 | 011 ··· 000 |
| 11 ··· 11 | 111 ··· 101 |

Typical Data Array Stored in ROM ($2^n$ words of $m$ bits each)

**FIGURE 9-18** Read-Only Memory with $n$ Inputs and $m$ Outputs

A ROM basically consists of a decoder and a memory array, as shown in Figure 9-19. When a pattern of $n$ 0's and 1's is applied to the decoder inputs, exactly one of the $2^n$ decoder outputs is 1. This decoder output line selects one of the words in the memory array, and the bit pattern stored in this word is transferred to the memory output lines.

Figure 9-20 illustrates one possible internal structure of the 8-word × 4-bit ROM shown in Figure 9-17. The decoder generates the eight minterms of the three input variables. The memory array forms the four output functions by ORing together selected minterms. A switching element is placed at the intersection of a *word line* and an *output line* if the corresponding minterm is to be included in the output function; otherwise, the switching element is omitted (or not connected). If a switching element connects an output line to a word line which is 1, the output line will be 1. Otherwise, the pull-down resistors at the top of Figure 9-20 cause the output line to be 0. So the switching elements which are connected in this way in the memory array effectively form an OR gate for each of the output functions. For example, $m_0$, $m_1$, $m_4$, and $m_6$ are ORed together to form $F_0$. Figure 9-21 shows the equivalent OR gate.

In general, those minterms which are connected to output line $F$ by switching elements are ORed together to form the output $F_i$. Thus, the ROM in Figure 9-20 generates the following functions:

$$\begin{aligned} F_0 &= \Sigma\, m(0, 1, 4, 6) = A'B' + AC' \\ F_1 &= \Sigma\, m(2, 3, 4, 6, 7) = B + AC' \\ F_2 &= \Sigma\, m(0, 1, 2, 6) = A'B' + BC' \\ F_3 &= \Sigma\, m(2, 3, 5, 6, 7) = AC + B \end{aligned} \tag{9-5}$$

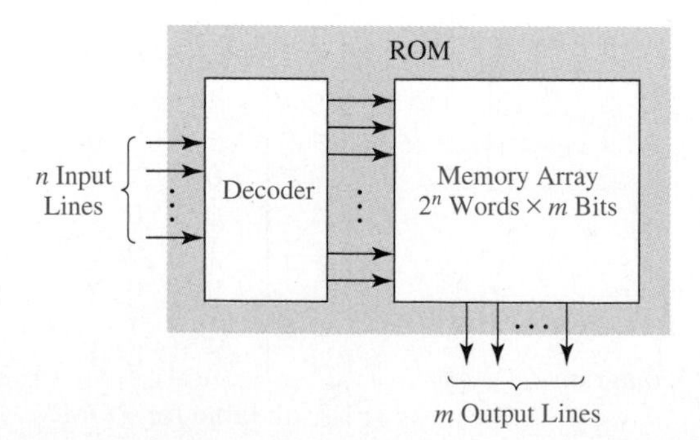


**FIGURE 9-19** Basic ROM Structure

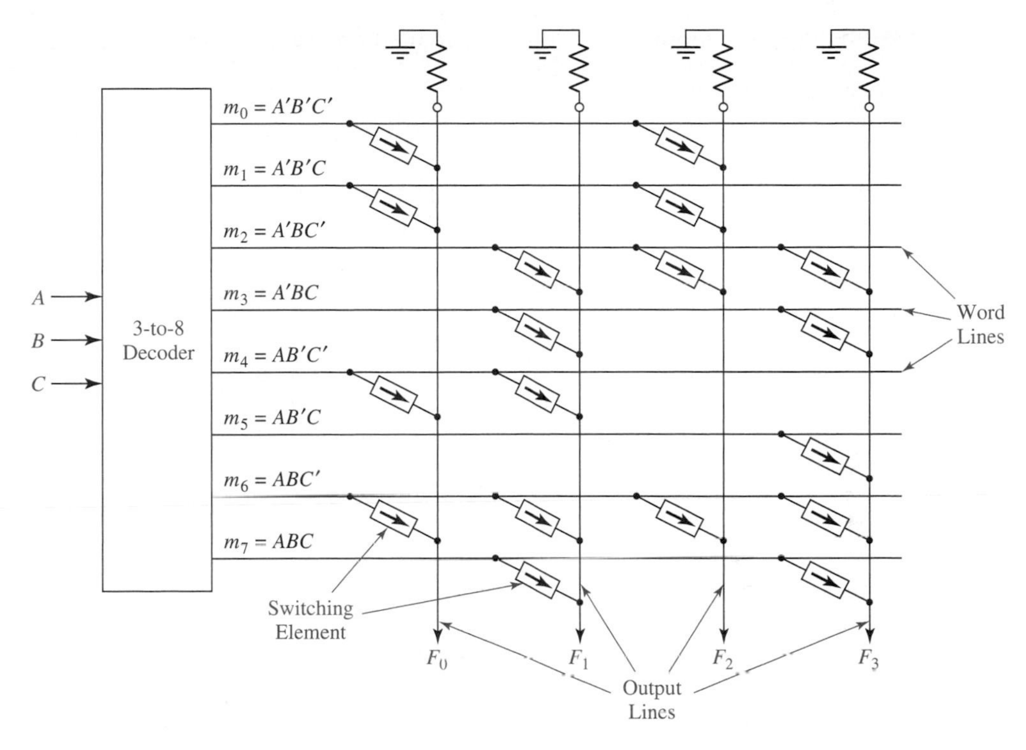


**FIGURE 9-20**
An 8-Word × 4-Bit ROM

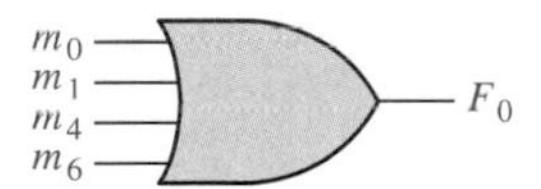


**FIGURE 9-21**
Equivalent OR Gate for $F_0$

The contents of a ROM are usually specified by a truth table. The truth table of Figure 9-17(b) specifies the ROM in Figure 9-20. Note that a 1 or 0 in the output part of the truth table corresponds to the presence or absence of a switching element in the memory array of the ROM.

Multiple-output combinational circuits can easily be realized using ROMs. As an example, we will realize a code converter that converts a 4-bit binary number to a hexadecimal digit and outputs the 7-bit ASCII code. Figure 9-22 shows the truth table and logic circuit for the converter. Because $A_5 = A_4$, and $A_6 = A_4'$, the ROM needs only five outputs. Because there are four address lines, the ROM size is 16 words by 5 bits. Columns $A_4A_3A_2A_1A_0$ of the truth table are stored in the ROM. Figure 9-23 shows an internal diagram of the ROM. The switching elements at the intersections of the rows and columns of the memory array are indicated using X's. An X indicates that the switching element is present and connected, and no X indicates that the corresponding element is absent or not connected.

Three common types of ROMs are mask-programmable ROMs, programmable ROMs (PROMs), and electrically erasable programmable ROMs (EEPROMs). At the time of manufacture, the data array is permanently stored in a mask-programmable ROM. This is accomplished by selectively including or omitting the switching elements at the row-column intersections of the memory array. This requires preparation

**FIGURE 9-22** Hexadecimal-to-ASCII Code Converter

| Input $W\ X\ Y\ Z$ | Hex Digit | ASCII Code for Hex Digit $A_6\ A_5\ A_4\ A_3\ A_2\ A_1\ A_0$ |
|---|---|---|
| 0 0 0 0 | 0 | 0 1 1 0 0 0 0 |
| 0 0 0 1 | 1 | 0 1 1 0 0 0 1 |
| 0 0 1 0 | 2 | 0 1 1 0 0 1 0 |
| 0 0 1 1 | 3 | 0 1 1 0 0 1 1 |
| 0 1 0 0 | 4 | 0 1 1 0 1 0 0 |
| 0 1 0 1 | 5 | 0 1 1 0 1 0 1 |
| 0 1 1 0 | 6 | 0 1 1 0 1 1 0 |
| 0 1 1 1 | 7 | 0 1 1 0 1 1 1 |
| 1 0 0 0 | 8 | 0 1 1 1 0 0 0 |
| 1 0 0 1 | 9 | 0 1 1 1 0 0 1 |
| 1 0 1 0 | A | 1 0 0 0 0 0 1 |
| 1 0 1 1 | B | 1 0 0 0 0 1 0 |
| 1 1 0 0 | C | 1 0 0 0 0 1 1 |
| 1 1 0 1 | D | 1 0 0 0 1 0 0 |
| 1 1 1 0 | E | 1 0 0 0 1 0 1 |
| 1 1 1 1 | F | 1 0 0 0 1 1 0 |

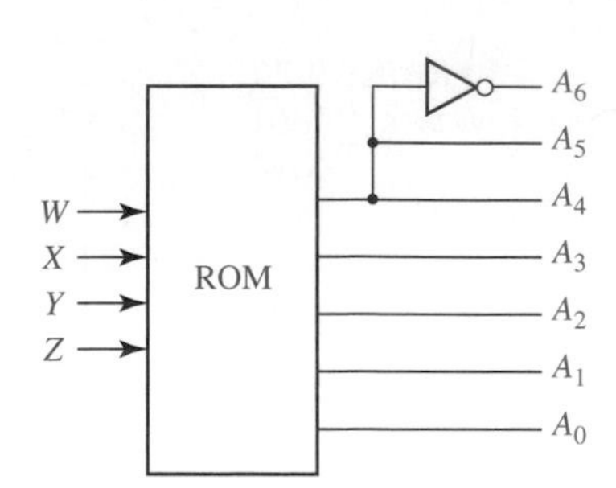


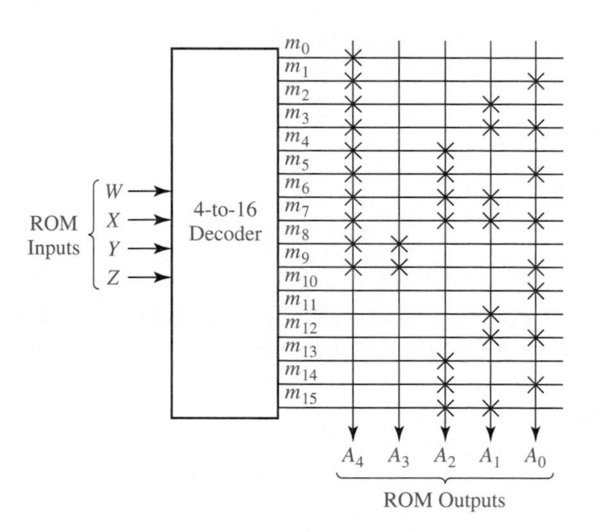


**FIGURE 9-23** ROM Realization of Code Converter

of a special *mask*, which is used during fabrication of the integrated circuit. Preparation of this mask is expensive, so the use of mask-programmable ROMs is economically feasible only if a large quantity (typically several thousand or more) is required with the same data array. If a small quantity of ROMs is required with a given data array, EEPROMs may be used.

Modification of the data stored in a ROM is often necessary during the developmental phases of a digital system, so EEPROMs are used instead of mask-programmable ROMs. EEPROMs use a special charge-storage mechanism to enable or disable the switching elements in the memory array. A PROM programmer is used to provide appropriate voltage pulses to store electronic charges in the memory array locations. Data stored in this manner is generally permanent until erased. After erasure, a new set of data can be stored in the EEPROM. An EEPROM can be erased and reprogrammed only a limited number of times, typically 100 to 1000 times. Flash memories are similar to EEPROMs, except that they use a different charge-storage mechanism. They usually have built-in programming and erase capability so that data can be written to the flash memory while it is in place in a circuit without the need for a separate programmer.

## 9.6 Programmable Logic Devices

A programmable logic device (or PLD) is a general name for a digital integrated circuit capable of being programmed to provide a variety of different logic functions. In this section we will discuss several types of combinational PLDs, and later we will discuss sequential PLDs. Simple combinational PLDs are capable of realizing from 2 to 10 functions of 4 to 16 variables with a single integrated circuit. More complex PLDs may contain thousands of gates and flip-flops. Thus, a single PLD can replace a large number of integrated circuits, and this leads to lower cost designs. When a digital system is designed using a PLD, changes in the design can easily be made by changing the programming of the PLD without having to change the wiring in the system.

### Programmable Logic Arrays

A programmable logic array (PLA) performs the same basic function as a ROM. A PLA with $n$ inputs and $m$ outputs (Figure 9-24) can realize $m$ functions of $n$ variables. The internal organization of the PLA is different from that of the ROM. The decoder is replaced with an AND array which realizes selected product terms of the input variables. The OR array ORs together the product terms needed to form the output functions, so a PLA implements a sum-of-products expression, while a ROM directly implements a truth table.

Figure 9-25 shows a PLA which realizes the same functions as the ROM of Figure 9-20. Product terms are formed in the AND array by connecting switching elements at appropriate points in the array. For example, to form $A'B'$, switching elements are used to connect the first word line with the $A'$ and $B'$ lines. Switching elements are

**FIGURE 9-24** Programmable Logic Array Structure

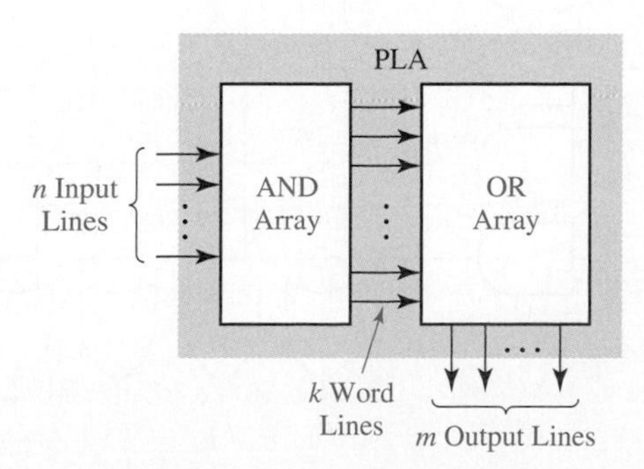

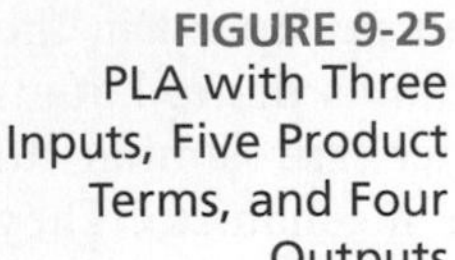

**FIGURE 9-25** PLA with Three Inputs, Five Product Terms, and Four Outputs

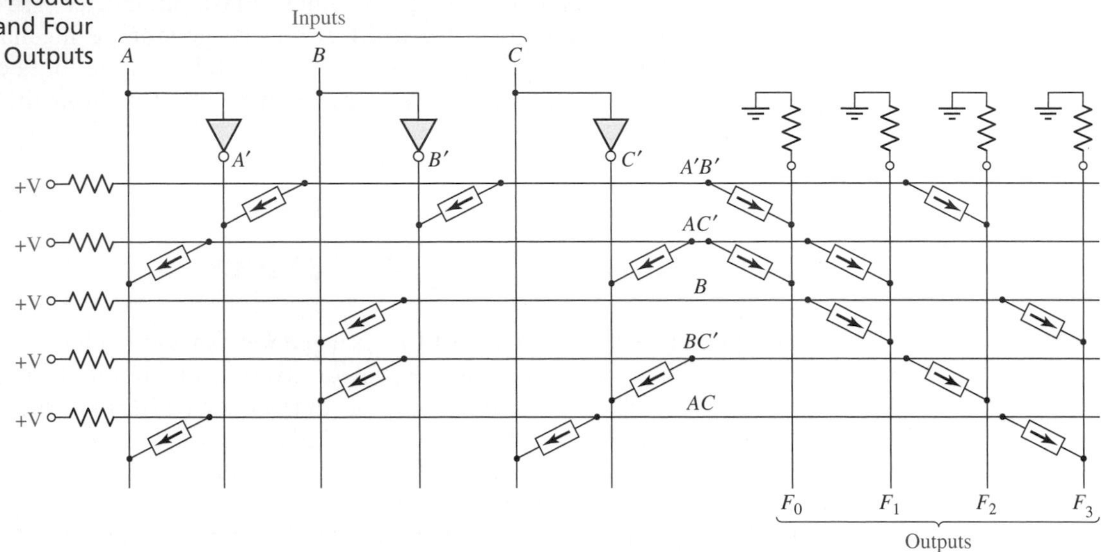


connected in the OR array to select the product terms needed for the output functions. For example, because $F_0 = A'B' + AC'$, switching elements are used to connect the $A'B'$ and $AC'$ lines to the $F_0$ line. The connections in the AND and OR arrays of this PLA make it equivalent to the AND-OR array of Figure 9-26.

The contents of a PLA can be specified by a PLA table. Table 9-1 specifies the PLA in Figure 9-25. The input side of the table specifies the product terms. The symbols 0, 1,

**FIGURE 9-26** AND-OR Array Equivalent to Figure 9-25

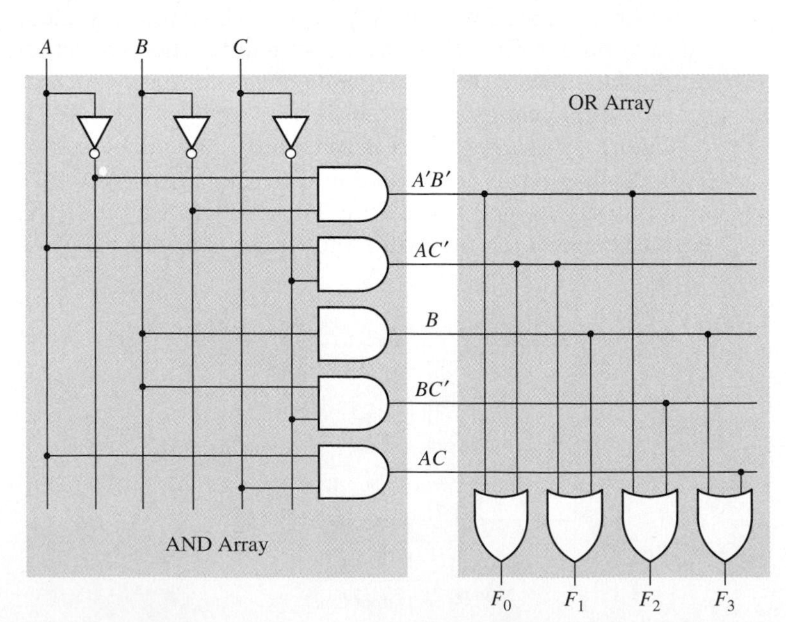

**TABLE 9-1**
PLA Table for Figure 9-25

| Product Term | Inputs $A\ B\ C$ | Outputs $F_0\ F_1\ F_2\ F_3$ |  |
|---|---|---|---|
| $A'B'$ | 0 0 – | 1 0 1 0 | $F_0 = A'B' + AC'$ |
| $AC'$ | 1 – 0 | 1 1 0 0 | $F_1 = AC' + B$ |
| $B$ | – 1 – | 0 1 0 1 | $F_2 = A'B' + BC'$ |
| $BC'$ | – 1 0 | 0 0 1 0 | $F_3 = B + AC$ |
| $AC$ | 1 – 1 | 0 0 0 1 |  |

and – indicate whether a variable is complemented, not complemented, or not present in the corresponding product term. The output side of the table specifies which product terms appear in each output function. A 1 or 0 indicates whether a given product term is present or not present in the corresponding output function. Thus, the first row of Table 9-1 indicates that the term $A'B'$ is present in output functions $F_0$ and $F_2$, and the second row indicates that $AC'$ is present in $F_0$ and $F_1$.

Next, we will realize Equations (7-23) using a PLA. Using the minimum multiple-output solution given in Equations (7-23b), we can construct a PLA table, Figure 9-27(a), with one row for each distinct product term. Figure 9-27(b) shows the corresponding PLA structure, which has four inputs, six product terms, and three outputs. A dot at the intersection of a word line and an input or output line indicates the presence of a switching element in the array.

**FIGURE 9-27**
PLA Realization of Equations (7-23b)

| $a\ b\ c\ d$ | $f_1\ f_2\ f_3$ |
|---|---|
| 0 1 – 1 | 1 1 0 |
| 1 1 – 1 | 1 0 1 |
| 1 0 0 – | 1 0 1 |
| – 0 1 – | 1 0 0 |
| – – 1 – | 0 1 0 |
| – 1 1 – | 0 0 1 |

(a) PLA table

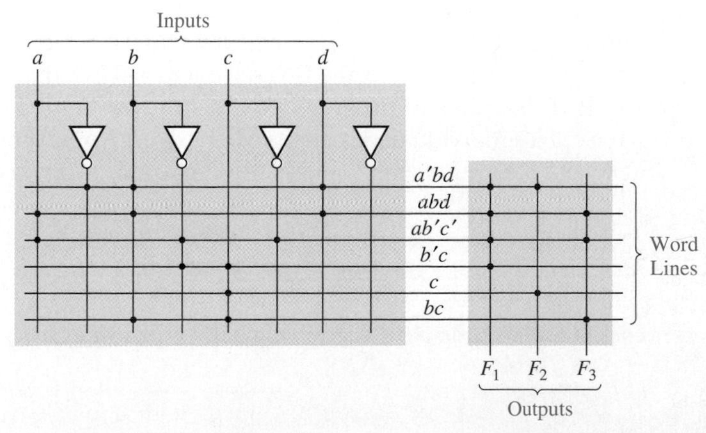


(b) PLA structure

A PLA table is significantly different than a truth table for a ROM. In a truth table each row represents a minterm; therefore, exactly one row will be selected by each combination of input values. The 0's and 1's of the output portion of the selected row determine the corresponding output values. On the other hand, each row in a PLA table represents a general product term. Therefore, zero, one, or more rows may be selected by each combination of input values. To determine the value of $f_i$ for a given input combination, the values of $f_i$ in the selected rows of the PLA table must be ORed together. The following examples refer to the PLA table of Figure 9-27(a). If $abcd = 0001$, no rows are selected, and all $f$'s are 0. If $abcd = 1001$, only the third row is selected, and $f_1 f_2 f_3 = 101$. If $abcd = 0111$, the first, fifth, and sixth rows are selected. Therefore, $f_1 = 1 + 0 + 0 = 1$, $f_2 = 1 + 1 + 0 = 1$, and $f_3 = 0 + 0 + 1 = 1$.

Both mask-programmable and field-programmable PLAs are available. The mask-programmable type is programmed at the time of manufacture in a manner similar to mask-programmable ROMs. The field-programmable logic array (FPLA) has programmable interconnection points that use electronic charges to store a pattern in the AND and OR arrays. An FPLA with 16 inputs, 48 product terms, and eight outputs can be programmed to implement eight functions of 16 variables, provided that the total number of product terms does not exceed 48.

When the number of input variables is small, a PROM may be more economical to use than a PLA. However, when the number of input variables is large, PLAs often provide a more economical solution than PROMs. For example, to realize eight functions of 24 variables would require a PROM with over 16 million 8-bit words. Because PROMs of this size are not readily available, the functions would have to be decomposed so that they could be realized using a number of smaller PROMs. The same eight functions of 24 variables could easily be realized using a single PLA, provided that the total number of product terms is small. If more terms are required, the outputs of several PLAs can be ORed together.

## Programmable Array Logic

The PAL (programmable array logic) is a special case of the programmable logic array in which the AND array is programmable and the OR array is fixed. The basic structure of the PAL is the same as the PLA shown in Figure 9-24. Because only the AND array is programmable, the PAL is less expensive than the more general PLA, and the PAL is easier to program. For this reason, logic designers frequently use PALs to replace individual logic gates when several logic functions must be realized.

Figure 9-28(a) represents a segment of an unprogrammed PAL. The symbol

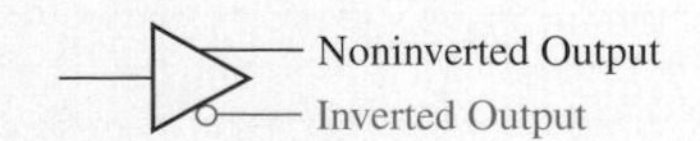


represents an input buffer which is logically equivalent to

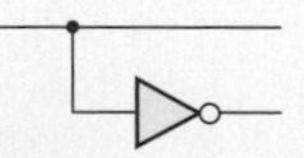

A buffer is used because each PAL input must drive many AND gate inputs. When the PAL is programmed, some of the interconnection points are programmed to make the desired connections to the AND gate inputs. Connections to the AND gate inputs in a PAL are represented by X's as shown:

As an example, we will use the PAL segment of Figure 9-28(a) to realize the function $I_1I_2' + I_1'I_2$. The X's in Figure 9-28(b) indicate that $I_1$ and $I_2'$ lines are connected to the first AND gate, and the $I_1'$ and $I_2$ lines are connected to the other gate.

When designing with PALs, we must simplify our logic equations and try to fit them into one (or more) of the available PALs. Unlike the more general PLA, the AND terms cannot be shared among two or more OR gates; therefore, each function to be realized can be simplified by itself without regard to common terms. For a given type of PAL, the number of AND terms that feed each output OR gate is fixed and limited. If the number of AND terms in a simplified function is too large, we may be forced to choose a PAL with more gate inputs and fewer outputs.

**FIGURE 9-28**
PAL Segment

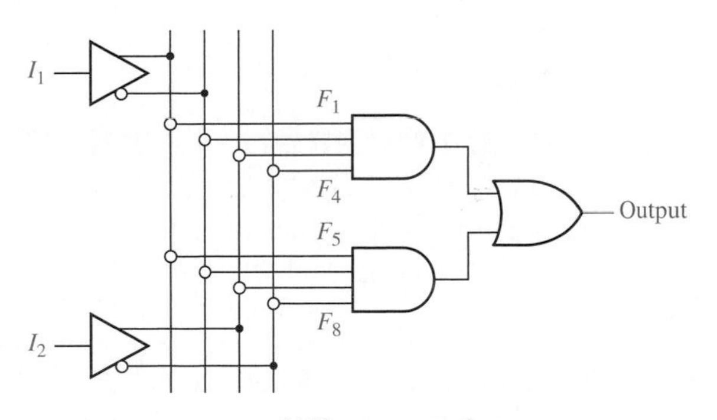


(a) Unprogrammed

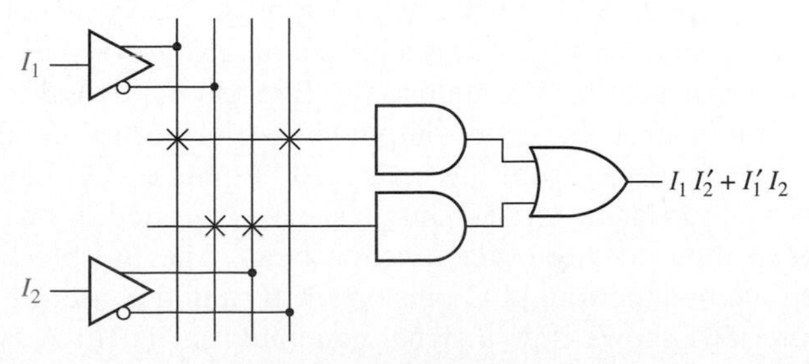


(b) Programmed

As an example of programming a PAL, we will implement a full adder. The logic equations for the full adder are

$$Sum = X'Y'C_{\text{in}} + X'YC'_{\text{in}} + XY'C'_{\text{in}} + XYC_{\text{in}}$$

$$C_{\text{out}} = XC_{\text{in}} + YC_{\text{in}} + XY$$

Figure 9-29 shows a section of a PAL where each OR gate is driven by four AND gates. The X's on the diagram show the connections that are programmed into the PAL to implement the full adder equations. For example, the first row of X's implements the product term $X'Y'C_{\text{in}}$.

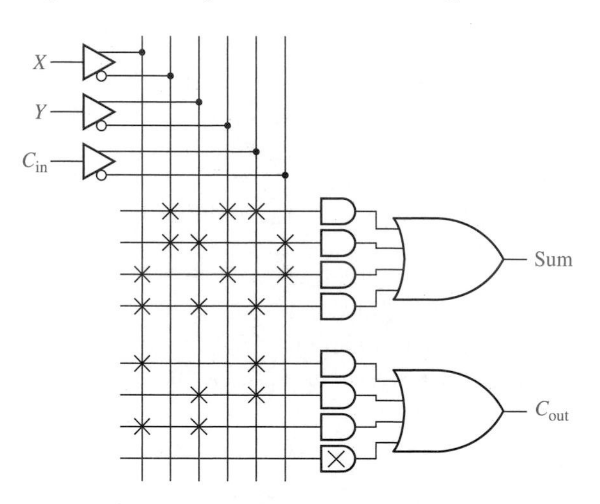


**FIGURE 9-29** Implementation of a Full Adder Using a PAL

## 9.7 Complex Programmable Logic Devices

As integrated circuit technology continues to improve, more and more gates can be placed on a single chip. This has allowed the development of complex programmable logic devices (CPLDs). Instead of a single PAL or PLA on a chip, many PALs or PLAs can be placed on a single CPLD chip and interconnected. When storage elements such as flip-flops are also included on the same IC, a small digital system can be implemented with a single CPLD.

Figure 9-30 shows the basic architecture of a Xilinx XCR3064XL CPLD. This CPLD has four function blocks, and each block has 16 associated macrocells (MC1, MC2, . . .). Each function block is a programmable AND-OR array that is configured as a PLA. Each macrocell contains a flip-flop and multiplexers that route signals from the function block to the input-output (I/O) block or to the interconnect array (IA). The IA selects signals from the macrocell outputs or I/O blocks and connects them back to function block inputs. Thus, a signal generated in one function block can be used as an input to any other function block. The I/O blocks provide an interface between the bi-directional I/O pins on the IC and the interior of the CPLD.

Figure 9-31 shows how a signal generated in the PLA is routed to an I/O pin through a macrocell. Any of the 36 outputs from the IA (or their complements) can

FIGURE 9-30 Architecture of Xilinx XCR3064XL CPLD (Figure based on figures and text owned by Xilinx, Inc., Courtesy of Xilinx, Inc. © Xilinx, Inc. 1999–2003. All rights reserved.)

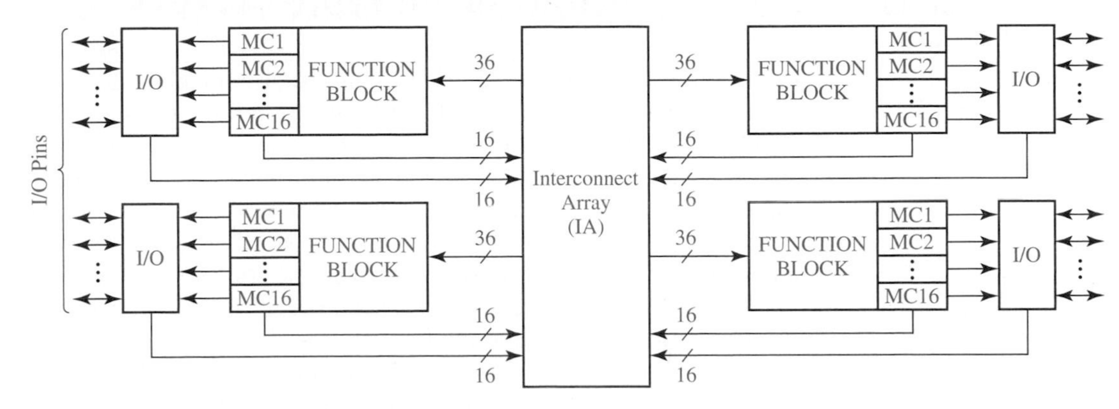


be connected to any inputs of the 48 AND gates. Each OR gate can accept up to 48 product term inputs from the AND array. The macrocell logic in this diagram is a simplified version of the actual logic. The first MUX (1) can be programmed to select the OR-gate output or its complement. Details of the flip-flop operation will be discussed in Unit 11. The MUX (2) at the output of the macrocell can be programmed to select either the combinational output ($G$) or the flip-flop output ($Q$). This output goes to the interconnect array and to the output cell. The output cell includes a three-state buffer (3) to drive the I/O pin. The buffer enable input can be programmed from several sources. When the I/O pin is used as an input, the buffer must be disabled.

Sophisticated CAD software is available for fitting logic circuits into a PLD and for programming the interconnections within the PLD. The input to this software can be in several forms such as a logic circuit diagram, a set of logic equations, or code written in a hardware description language (HDL). Unit 10 discusses the use of an HDL. The CAD software processes the input, determines the logic equations to be implemented, fits these equations into the PLD, determines the required interconnections within the PLD, and generates a bit pattern for programming the PLD.

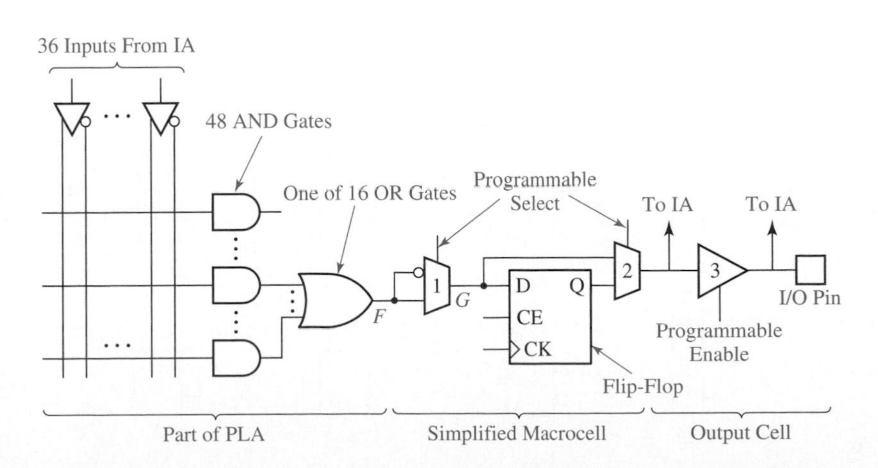


FIGURE 9-31 CPLD Function Block and Macrocell (A Simplified Version of XCR3064XL)

# 9.8 Field-Programmable Gate Arrays

In this section we introduce the use of field-programmable gate arrays (FPGAs) in combinational logic design. An FPGA is an IC that contains an array of identical logic cells with programmable interconnections. The user can program the functions realized by each logic cell and the connections between the cells. Figure 9-32 shows the layout of part of a typical FPGA. The interior of the FPGA consists of an array of logic cells, also called configurable logic blocks (CLBs). The array of CLBs is surrounded by a ring of input-output interface blocks. These I/O blocks connect the CLB signals to IC pins. The space between the CLBs is used to route connections between the CLB outputs and inputs.

Figure 9-33 shows a simplified version of a CLB. This CLB contains two function generators, two flip-flops, and various multiplexers for routing signals within the CLB. Each function generator has four inputs and can implement any function of up to four variables. The function generators are implemented as lookup tables (LUTs). A four-input LUT is essentially a reprogrammable ROM with 16 1-bit words. This ROM stores the truth table for the function being generated. The $H$ multiplexer selects either $F$ or $G$ depending on the value of $H_1$. The CLB has two combinational outputs

**FIGURE 9-32**
Layout of a Typical FPGA

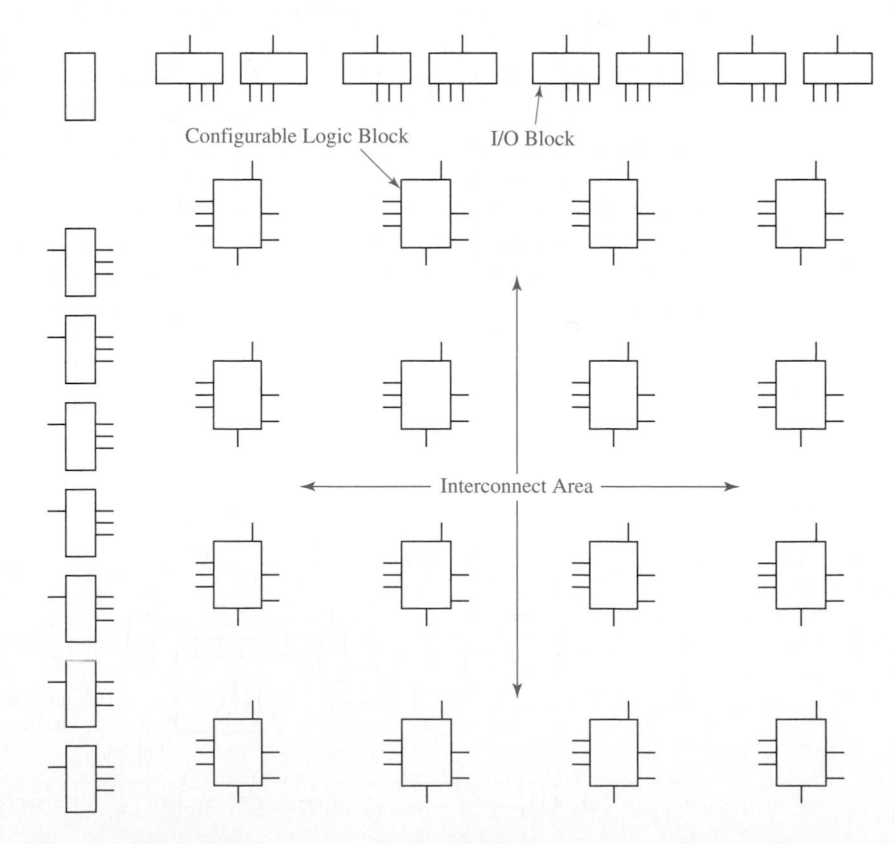

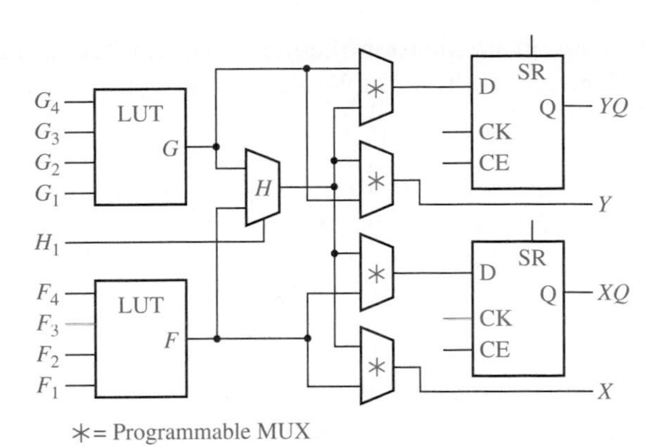


**FIGURE 9-33** Simplified Configurable Logic Block (CLB)

($X$ and $Y$) and two flip-flop outputs ($XQ$ and $YQ$). The $X$ and $Y$ outputs and the flip-flop inputs are selected by programmable multiplexers. The select inputs to these MUXes are programmed when the FPGA is configured. For example, the $X$ output can come from the $F$ function generator, and the $Y$ output from the $H$ multiplexer. Operation of the CLB flip-flops will be described in Unit 11.

Figure 9-34 shows one way to implement a function generator with inputs $a, b, c, d$. The numbers in the squares represent the bits stored in the LUT. These bits enable particular minterms. Because the function being implemented is stored as a truth table, a function with only one minterm or with as many as 15 minterms requires a single function generator. The functions

$$F = abc$$

and

$$F = a'b'c'd + a'b'cd + a'bc'd + a'bcd' + ab'c'd + ab'cd' + abc'd' + abcd$$

each require a single function generator.

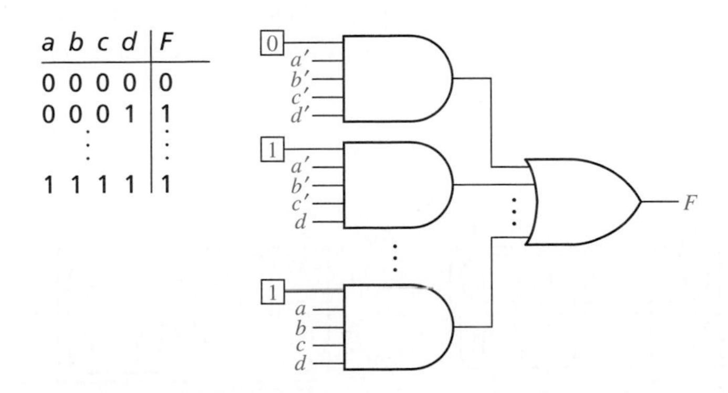


**FIGURE 9-34** Implementation of a Lookup Table (LUT)

## Decomposition of Switching Functions

In order to implement a switching function of more than four variables using 4-variable function generators, the function must be decomposed into subfunctions where each subfunction requires only four variables. One method of decomposition is based on Shannon's expansion theorem. We will first illustrate this theorem by expanding a function of the variables $a$, $b$, $c$, and $d$ about the variable $a$:

$$f(a, b, c, d) = a'f(0, b, c, d) + a\,f(1, b, c, d) = a'\,f_0 + a\,f_1 \tag{9-6}$$

The 3-variable function $f_0 = f(0, b, c, d)$ is formed by replacing $a$ with 0 in $f(a, b, c, d)$, and $f_1 = f(1, b, c, d)$ is formed by replacing $a$ with 1 in $f(a, b, c, d)$. To verify that Equation (9-6) is correct, first set $a$ to 0 on both sides, and then set $a$ to 1 on both sides. An example of applying Equation (9-6) is as follows:

$$\begin{aligned} f(a, b, c, d) &= c'd' + a'b'c + bcd + ac' \qquad (9\text{-}7)\\ &= a'(c'd' + b'c + bcd) + a(c'd' + bcd + c')\\ &= a'(c'd' + b'c + cd) + a(c' + bd) = a'\,f_0 + a\,f_1 \end{aligned}$$

Note that before simplification, the terms $c'd'$ and $bcd$ appear in both $f_0$ and $f_1$ because neither term contains $a'$ or $a$.

Expansion can also be accomplished using a truth table or a Karnaugh map. Figure 9-35 shows the map for Equation (9-7). The left half of the map where $a = 0$ is in effect a 3-variable map for $f_0(b, c, d)$. Looping terms on the left half gives $f_0 = c'd' + b'c + cd$, which is the same as the previous result. Similarly the right half where $a = 1$ is a 3-variable map for $f_1(b, c, d)$, and looping terms on the right half gives $f_1 = c' + bd$. The expressions for $f_0$ and $f_1$ obtained from the map are the same as those obtained algebraically in Equation (9-7).

The general form of Shannon's expansion theorem for expanding an $n$-variable function about the variable $x_i$ is

$$\begin{aligned} &f(x_1, x_2, \ldots, x_{i-1}, x_i, x_{i+1}, \ldots, x_n)\\ &\quad = x_i'\,f(x_1, x_2, \ldots, x_{i-1}, 0, x_{i+1}, \ldots, x_n) + x_i\,f(x_1, x_2, \ldots, x_{i-1}, 1, x_{i+1}, \ldots, x_n)\\ &\quad = x_i'\,f_0 + x_i\,f_1 \qquad (9\text{-}8) \end{aligned}$$

where $f_0$ is the $(n-1)$-variable function obtained by setting $x_i$ to 0 in the original function and $f_1$ is the $(n-1)$-variable function obtained by setting $x_i$ to 1 in the original function. The theorem is easily proved for switching algebra by first setting $x_i$

**FIGURE 9-35**
Function Expansion Using a Karnaugh Map

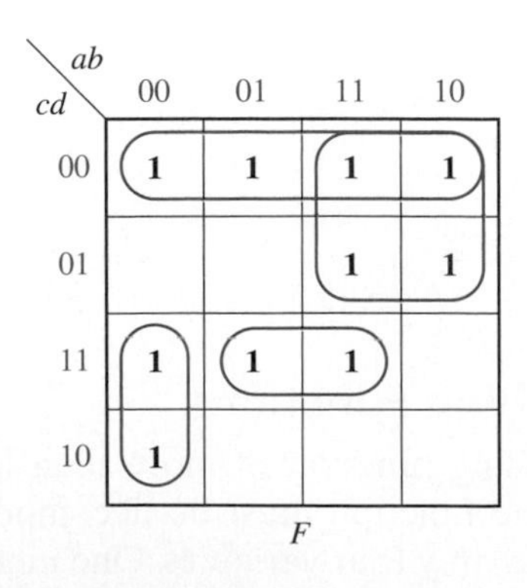


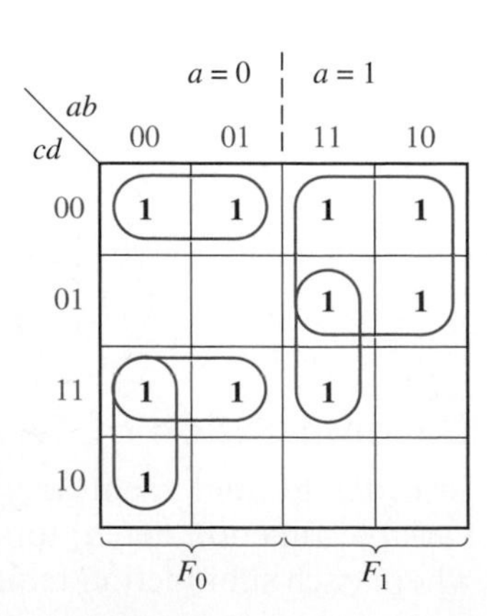

to 0 in Equation (9-8), and, then, setting $x_i$ to 1. Because both sides of the equation are equal for $x_i = 0$ and for $x_i = 1$, the theorem is true for switching algebra.

Applying the expansion theorem to a 5-variable function gives

$$f(a, b, c, d, e) = a'\, f(0, b, c, d, e) + a\, f(1, b, c, d, e) = a'\, f_0 + a\, f_1 \qquad \text{(9-9)}$$

This shows that any 5-variable function can be realized using two 4-variable function generators and a 2-to-1 MUX [Figure 9-36(a)]. This implies that any 5-variable function can be implemented using a CLB of the type shown in Figure 9-33.

To realize a 6-variable function using 4-variable function generators, we apply the expansion theorem twice:

$$G(a, b, c, d, e, f) = a'\, G(0, b, c, d, e, f) + a\, G(1, b, c, d, e, f) = a'\, G_0 + a\, G_1$$
$$G_0 = b' G(0, 0, c, d, e, f) + b\, G(0, 1, c, d, e, f) = b' G_{00} + b\, G_{01}$$
$$G_1 = b' G(1, 0, c, d, e, f) + b\, G(1, 1, c, d, e, f) = b' G_{10} + b G_{11}$$

Because $G_{00}$, $G_{01}$, $G_{10}$, and $G_{11}$ are all 4-variable functions, we can realize any 6-variable function using four 4-variable function generators and three 2-to-1 MUXes, as shown in Figure 9-36(b). Thus, we can realize any 6-variable function using two CLBs of the type shown in Figure 9-31. Alternatively, we can write

$$G(a, b, c, d, e, f) = a'b' G_{00} + a'b\, G_{01} + ab' G_{10} + ab\, G_{11} \qquad \text{(9-10)}$$

and realize $G$ using four function generators and a 4-to-1 MUX. In general, we can realize any $n$-variable function ($n > 4$) using $2^{n-4}$ 4-variable function generators and one $2^{n-4}$-to-1 MUX. This is a worst-case situation because many functions of $n$ variables can be realized with fewer function generators.

**FIGURE 9-36** Realization of 5- and 6-Variable Functions with Function Generators

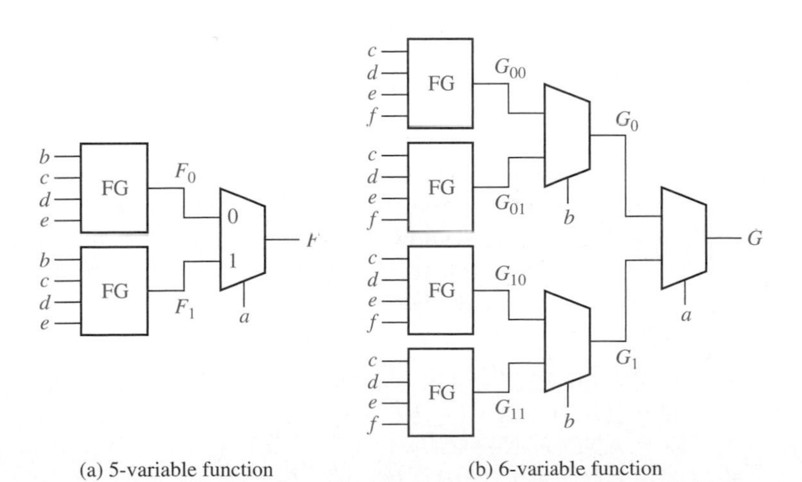


(a) 5-variable function

(b) 6-variable function

## Problems

**9.1** (a) Show how two 2-to-1 multiplexers (with no added gates) could be connected to form a 3-to-1 MUX. Input selection should be as follows:
If $AB = 00$, select $I_0$
If $AB = 01$, select $I_1$
If $AB = 1-$ ($B$ is a don't-care), select $I_2$
(b) Show how two 4-to-1 and one 2-to-1 multiplexers could be connected to form an 8-to-1 MUX with three control inputs.
(c) Show how four 2-to-1 and one 4-to-1 multiplexers could be connected to form an 8-to-1 MUX with three control inputs.

**9.2** Design a circuit which will either subtract $X$ from $Y$ or $Y$ from $X$, depending on the value of $A$. If $A = 1$, the output should be $X - Y$, and if $A = 0$, the output should be $Y - X$. Use a 4-bit subtracter and two 4-bit 2-to-1 multiplexers (with bus inputs and outputs as in Figure 9-5).

**9.3** Repeat 9.2 using a 4-bit subtracter, four 4-bit three-state buffers (with bus inputs and outputs), and one inverter.

**9.4** Realize a full adder using a 3-to-8 line decoder (as in Figure 9-13) and
(a) two OR gates.
(b) two NOR gates.

**9.5** Derive the logic equations for a 4-to-2 priority encoder. Refer to your table in the Study Guide, Part 4(b).

**9.6** Design a circuit equivalent to Figure 9-11 using a 4-to-1 MUX (with bus inputs as in Figure 9-5). Use a 4-to-2 line priority encoder to generate the control signals.

**9.7** An adder for Gray-coded-decimal digits (see Table 1-2) is to be designed using a ROM. The adder should add two Gray-coded digits and give the Gray-coded sum and a carry. For example, $1011 + 1010 = 0010$ with a carry of 1 ($7 + 6 = 13$). Draw a block diagram showing the required ROM inputs and outputs. What size ROM is required? Indicate how the truth table for the ROM would be specified by giving some typical rows.

**9.8** The following PLA will be used to implement the following equations:
$X = AB'D + A'C' + BC + C'D'$
$Y = A'C' + AC + C'D'$
$Z = CD + A'C' + AB'D$

(a) Indicate the connections that will be made to program the PLA to implement these equations.

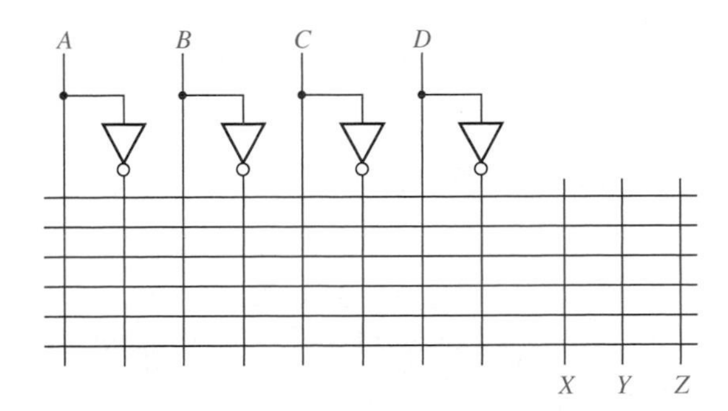


(b) Specify the truth table for a ROM which realizes these same equations.

**9.9** Show how to implement a full subtracter using a PAL. See Figure 9-29.

**9.10** (a) If the ROM in the hexadecimal to ASCII code converter of Figure 9-22 is replaced with a PAL, give the internal connection diagram.
(b) If the same ROM is replaced with a PLA, give the PLA table.

**9.11** (a) Sometimes the programmable MUX (1) in Figure 9-31 helps us to save AND gates. Consider the case in which $F = c'd' + bc' + a'c$. If programmable MUX (1) is not set to invert $F$ (i.e., $G = F$), how many AND gates are needed? If the MUX is set to invert $F$ (i.e., $G = F'$), how many AND gates are needed?
(b) Repeat (a) for $F = a'b' + c'd'$.

**9.12** (a) Implement a 3-variable function generator using a PAL with inputs $a$, $b$, $c$, and 1 (use the input inverter to get 0 also). Give the internal connection diagram. Leave the connections to 0 and 1 disconnected, so that any 3-variable function can be implemented by connecting only 0 and 1.
(b) Now connect 0 and 1 so that the function generator implements the sum function for a full adder. See Figure 9-34.

**9.13** Expand the following function about the variable $b$.
$F = ab'cde' + bc'd'e + a'cd'e + ac'de'$

**9.14** (a) Implement the following function using only 2-to-1 MUXes:
$R = ab'h' + bch' + eg'h + fgh$.
(b) Repeat using only tri-state buffers.

**9.15** Show how to make a 4-to-1 MUX, using an 8-to-1 MUX.

**9.16** Implement a 32-to-1 multiplexer using two 16-to-1 multiplexers and a 2-to-1 multiplexer in two ways: (a) Connect the most significant select line to the 2-to-1 multiplexer, and (b) connect the least significant select line to the 2-to-1 multiplexer.

**9.17** 2-to-1 multiplexers with an active high output and active high enable are to be used in the following implementations:

(a) Show how to implement a 4-to-1 multiplexer with an active high output and no enable using two of the 2-to-1 MUXes and a minimum number of additional gates.

(b) Repeat part (a) for a 4-to-1 multiplexer with an active low output.

(c) Repeat part (b) assuming the output of the 2-to-1 MUX is 1 (rather than 0) when the enable is 0.

**9.18** Realize a BCD to excess-3 code converter using a 4-to-10 decoder with active low outputs and a minimum number of gates.

**9.19** Use a 4-to-1 multiplexer and a minimum number of external gates to realize the function $F(w, x, y, z) = \Sigma\, m(3, 4, 5, 7, 10, 14) + \Sigma\, d(1, 6, 15)$.
The inputs are only available uncomplemented.

**9.20** Realize the function $f(a, b, c, d, e) = \Sigma\, m(6, 7, 9, 11, 12, 13, 16, 17, 18, 20, 21, 23, 25, 28)$ using a 16-to-1 MUX with control inputs $b$, $c$, $d$, and $e$. Each data input should be 0, 1, $a$, or $a'$. Hint: Start with a minterm expansion of $F$ and combine minterms to eliminate $a$ and $a'$ where possible.

**9.21** Implement a full adder

(a) using two 8-to-1 MUXes. Connect $X$, $Y$, and $C_{in}$ to the control inputs of the MUXes and connect 1 or 0 to each data input.

(b) using two 4-to-1 MUXes and one inverter. Connect $X$ and $Y$ to the control inputs of the MUXes, and connect 1's, 0's, $C_{in}$, or $C_{in}'$ to each data input.

(c) again using two 4-to-1 MUXes, but this time connect $C_{in}$ and $Y$ to the control inputs of the MUXes, and connect 1's, 0's, $X$, or $X'$ to each data input. Note that in this fashion, any $N$-variable logic function may be implemented using a $2^{(N-1)}$-to-1 MUX.

**9.22** Repeat Problem 9.21 for a full subtracter, except use $B_{in}$ instead of $C_{in}$.

**9.23** Make a circuit which gives the absolute value of a 4-bit binary number. Use four full adders, four multiplexers, and four inverters. Assume negative numbers are represented in 2's complement. Recall that one way to find the 2's complement of a binary number is to invert all of the bits and then add 1.

**9.24** Show how to make a 4-to-1 MUX using four three-state buffers and a decoder.

**9.25** Show how to make an 8-to-1 MUX using two 4-to-1 MUXes, two three-state buffers, and one inverter.

**9.26** Realize a full subtracter using a 3-to-8 line decoder with inverting outputs and

(a) two NAND gates.

(b) two AND gates.

**9.27** Show how to make the 8-to-3 priority encoder of Figure 9-16 using two 4-to-2 priority encoders and any additional necessary gates.

**9.28** Design an adder for excess-3 decimal digits (see Table 1-2) using a ROM. Add two excess-3 digits and give the excess-3 sum and a carry. For example, 1010 + 1001 = 0110 with a carry of 1 (7 + 6 = 13). Draw a block diagram showing the required ROM inputs and outputs. What size ROM is required? Indicate how the truth table for the ROM would be specified by giving some typical rows.

**9.29** A circuit has four inputs *RSTU* and four outputs *VWYZ*. *RSTU* represents a binary-coded-decimal digit. *VW* represents the quotient and *YZ* the remainder when *RSTU* is divided by 3 (*VW* and *YZ* represent 2-bit binary numbers). Assume that invalid inputs do not occur. Realize the circuit using
(a) a ROM.
(b) a minimum two-level NAND-gate circuit.
(c) a PLA (specify the PLA table).

**9.30** Repeat Problem 9.29 if the inputs *RSTU* represent a decimal digit in Gray code (see Table 1-2).

**9.31** (a) Find a minimum two-level NOR gate circuit to realize $F_1$ and $F_2$. Use as many common gates as possible.
$F_1(a, b, c, d) = \Sigma\, m(1, 2, 4, 5, 6, 8, 10, 12, 14)$
$F_2(a, b, c, d) = \Sigma\, m(2, 4, 6, 8, 10, 11, 12, 14, 15)$
(b) Realize $F_1$ and $F_2$ using a PLA. Give the PLA table and internal connection diagram for the PLA.

**9.32** Braille is a system which allows a blind person to read alphanumerics by feeling a pattern of raised dots. Design a circuit that converts BCD to Braille. The table shows the correspondence between BCD and Braille.
(a) Use a multiple-output NAND-gate circuit.

| | | | | W \| X |
|---|---|---|---|---|
| *A* | *B* | *C* | *D* | Z \| Y |
| 0 | 0 | 0 | 0 | |
| 0 | 0 | 0 | 1 | |
| 0 | 0 | 1 | 0 | |
| 0 | 0 | 1 | 1 | |
| 0 | 1 | 0 | 0 | |
| 0 | 1 | 0 | 1 | |
| 0 | 1 | 1 | 0 | |
| 0 | 1 | 1 | 1 | |
| 1 | 0 | 0 | 0 | |
| 1 | 0 | 0 | 1 | |

(b) Use a PLA. Give the PLA table.
(c) Specify the connection pattern for the PLA.

**9.33** (a) Implement your solution to Problem 7.10 using a PLA. Specify the PLA table and draw the internal connection diagram for the PLA using dots to indicate the presence of switching elements.
(b) Repeat (a) for Problem 7.41.
(c) Repeat (a) for Problem 7.43.

**9.34** Show how to make an 8-to-1 MUX using a PAL. Assume that PAL has 14 inputs and six outputs and assume that each output OR gate may have up to four AND terms as inputs, as in Figure 9-29. (*Hint*: Wire some outputs of the PAL around to the inputs, external to the PAL. Some PALs allow this inside the PAL to save inputs.)

**9.35** Work Problem 9.34 but make the 8-to-3 priority encoder of Figure 9-16 instead of a MUX.

**9.36** The function $F = CD'E + CDE + A'D'E + A'B'DE' + BCD$ is to be implemented in an FPGA which uses 3-variable lookup tables.
(a) Expand $F$ about the variables $A$ and $B$
(b) Expand $F$ about the variables $B$ and $C$.
(c) Expand $F$ about the variables $A$ and $C$.
(d) Any 5-variable function can be implemented using four 3-variable lookup tables and a 4-to-1 MUX, but this time we are lucky. Use your preceding answers to implement $F$ using only three 3-variable lookup tables and a 4-to-1 MUX. Give the truth tables for the lookup tables.

**9.37** Work Problem 9.36 for $F = B'D'E' + AB'C + C'DE' + A'BC'D$.

**9.38** Implement a 4-to-1 MUX using a CLB of the type shown in Figure 9-33. Specify the function realized by each function generator.

**9.39** Realize the function $f(A, B, C, D) = A'C' + A'B'D' + ACD + A'BD$.
(a) Use a single 8-to-1 multiplexer with an active low enable and an active high output. Use $A$, $C$, and $D$ as the select inputs where $A$ is the most significant and $D$ is the least significant.
(b) Repeat Part (a) assuming the multiplexer enable is active high and output is active low.
(c) Use a single 4-to-1 multiplexer with an active low enable and an active high output and a minimum of additional gates. Show the function expansion both algebraically and on a Karnaugh map.

**9.40** Repeat Problem 9.39 for the function
$f(A, B, C, D, E) = A'C'E' + A'B'D'E' + ACDE' + A'BDE'$.

**9.41** $F(a, b, c, d) = a' + ac'd' + b'cd' + ad$.

(a) Using Shannon's expansion theorem, expand $F$ about the variable $d$.

(b) Use the expansion in Part (a) to realize the function using two 4-variable LUTs and a 2-to-1 MUX. Specify the LUT inputs.

(c) Give the truth table for each LUT.

**9.42** Repeat 9.41 for $F(a, b, c, d) = cd' + ad' + a'b'cd + bc'$.

**9.43** Repeat 9.41 for $F(a, b, c, d) = bd + bc' + ac'd + a'd'$.

UNIT 10

# Introduction to VHDL

## Objectives

1. Represent gates and combinational logic by concurrent VHDL statements.
2. Given a set of concurrent VHDL statements, draw the corresponding combinational logic circuit.
3. Write a VHDL module for a combinational circuit
   **(a)** by using concurrent VHDL statements to represent logic equations.
   **(b)** by interconnecting VHDL components.
4. Compile and simulate a VHDL module.
5. Use the basic VHDL operators and understand their order of precedence.
6. Use the VHDL types: bit, bit_vector, Boolean, and integer.
   Define and use an array-type.
7. Use IEEE Standard Logic. Use std_logic_vectors, together with overloaded operators, to perform arithmetic operations.

# Study Guide

**1.** Study Section 10.1, *VHDL Description of Combinational Circuits.*

(a) Draw a circuit that corresponds to the following VHDL statements:

```
C <= not A;          D <= C and B;
```

(b) If A changes at time 5 ns, at what time do each of the following concurrent statements execute? At what times are C and D updated?

```
C <= A;
D <= A;
```

(c) Write a VHDL statement that corresponds to the following circuit. The inverter has a delay of 5 ns. Draw the waveform for $M$ assuming that $M$ is initially 0.

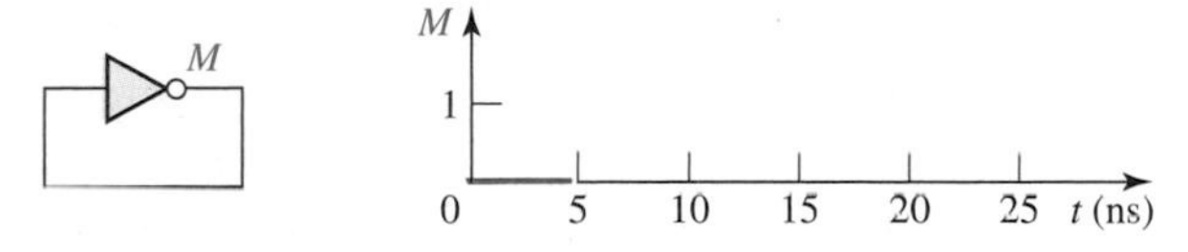


(d) Write a VHDL statement to implement $A = B \oplus C$ without using the xor or xnor operator. Do not include gate delays.

(e) Work Problems 10.1 and 10.2.

**2.** Study Section 10.2, *VHDL Models for Multiplexers.*

(a) Implement the following VHDL conditional assignment statement, using a 2-to-1 MUX:

```
F <= A when C = '1' else B;
```

(b) Write a VHDL conditional assignment statement that represents the 4-to-1 MUX of Figure 9-2. Assume $I_0 = 1$, $I_1 = 0$, and $I_2 = I_3 = C$.

(c) Write a VHDL selected signal assignment for the same circuit as in (b).

**3.** Study Section 10.3, *VHDL Modules*, and Section 10.4, *Signals and Constants.*

(a) Write an entity for the module MOD1. $A$, $B$, $C$, $D$, and $E$ are all of type bit.

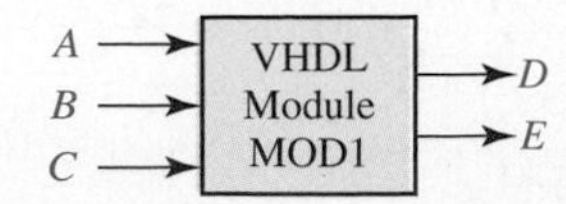

(b) Write the architecture for MOD1 if $D = ABC$ and $E = D'$.

(c) What changes must be made in the code of Figure 10-12 to implement a 5-bit adder?

(d) Given the concurrent VHDL statements

```
R <= A after 5 ns; -- statement 1
S <= R after 10 ns; -- statement 2
```

If A changes at time 3 ns, at what time will statement 1 be executed?
At what time will R be updated?
At what time will statement 2 be executed?
At what time will S be updated?

Answers: 3 ns, 8 ns, 8 ns, and 18 ns

(e) Write a statement that defines a bit_vector constant C1 equal to 10101011.

(f) The circuit of Figure 8-5 is implemented as a module without gate delays as follows.
(In the figure, *B* is set to 1 and *C* is set to 0, but here, assume they are inputs.)

```
entity fig8_5 is
    port (A, B, C: in bit; G2: out bit);
end fig8_5;
architecture circuit of fig8_5 is
begin
    G2 <= not(C or (A and B));
end circuit;
```

Each gate in Figure 8-5 has a delay of 20 ns. Modify the module to include gate delays. (*Hint*: You will need a **signal** declaration to introduce G1 as an internal signal.)

(g) Work Problems 10.3 and 10.4.

**4.** Study Section 10.5, *Arrays*.

(a) Write VHDL statements that define a ROM that is 16 words of 8 bits each. Leave the values stored in the ROM unspecified.

(b) Work Problem 10.5.

**5.** Study Section 10.6, *VHDL Operators.*

(a) For each of the following statements, eliminate one set of parentheses without changing the order of operation.

```
(i) not ((A & B) xor "10")
(ii) (not (A & B) xor "10")
```

(b) If A(0 **to** 7) = "11011011", what will be the result of executing the following concurrent statement?

```
B <= A(6 to 7)&A(0 to 5);
```

What problem will occur when the following concurrent statement is executed?

```
A <= A(6 to 7)&A(0 to 5);
```

(*Hint*: A concurrent statement executes every time the right-hand side changes.)

(c) Work Problem 10.6(a).

**6.** Study Section 10.7, *Packages and Libraries.*

Give the entity and architecture that describes a three-input AND gate with 2-ns delay. Assume that all signals are of type bit.

**7.** Study Section 10.8, *IEEE Standard Logic.*

(a) Suppose A, B, C, D, E, and F are of type std_logic. If the following concurrent statements are executed, what are the values of A, B, C, D, E, and F?

```
A <= '1'; A <= 'Z';
B <= '0'; B <= A;
C <= '0';
D <= A when C = '0' else 'Z';
D <= C when C = '1' else 'Z';
E <= '0' when A = '1' else C;
E <= A when C = '0' else '1';
F <= '1' when A = '1' and C = '1' else 'Z';
F <= '0' when A = '0' and C = '0' else 'Z';
```

(b) Given the concurrent statements

```
F <= '0';
F <= '1' after 2 ns;
```

What will happen if F is of type bit? What if F is of type std_logic?

(c) Suppose in Figure 10-19 that A is 1011, B is 0111, and Cin is 1. What is Addout? Sum? Cout?

(d) If A is a 6-bit std_logic_vector and B is a 4-bit std_logic_vector, write concurrent VHDL statements that will add A and B to result in a 6-bit sum and a carry.

(e) Draw a circuit that implements the following VHDL code:

```
signal A, B, C, D: std_logic_vector(1 to 3);
signal E, F, G: std_logic;
---------------------------------------------------
D <= A when E = '1' else "ZZZ";
D <= B when F = '1' else "ZZZ";
D <= C when G = '1' else "ZZZ";
```

(f) Work Problems 10.6(b), 10.7, and 10.8.

**8.** Before you take the test on Unit 10, pick up a lab assignment sheet and work the assigned lab problems. Turn in your VHDL code and simulation results.

# Introduction to VHDL

As integrated circuit technology has improved to allow more and more components on a chip, digital systems have continued to grow in complexity. As digital systems have become more complex, detailed design of the systems at the gate and flip-flop level has become very tedious and time consuming. For this reason, the use of hardware description languages in the digital design process continues to grow in importance. A hardware description language allows a digital system to be designed and debugged at a higher level before implementation at the gate and flip-flop level. The use of computer-aided design tools to do this conversion is becoming more widespread. This is analogous to writing software programs in a high-level language such as C and then using a compiler to convert the programs to machine language. The two most popular hardware description languages are VHDL and Verilog.

VHDL is a hardware description language that is used to describe the behavior and structure of digital systems. The acronym VHDL stands for VHSIC Hardware Description Language, and VHSIC in turn stands for Very High Speed Integrated Circuit. However, VHDL is a general-purpose hardware description language which can be used to describe and simulate the operation of a wide variety of digital systems, ranging in complexity from a few gates to an interconnection of many complex integrated circuits. VHDL was originally developed to allow a uniform method for specifying digital systems. The VHDL language became an IEEE standard in 1987, and it is widely used in industry. IEEE published a revised VHDL standard in 1993, and the examples in this text conform to that standard.

VHDL can describe a digital system at several different levels—behavioral, data flow, and structural. For example, a binary adder could be described at the behavioral level in terms of its function of adding two binary numbers, without giving any implementation details. The same adder could be described at the data flow level by giving the logic equations for the adder. Finally, the adder could be described at the structural level by specifying the interconnections of the gates which make up the adder.

VHDL leads naturally to a top-down design methodology in which the system is first specified at a high level and tested using a simulator. After the system is debugged at this level, the design can gradually be refined, eventually leading to a structural description which is closely related to the actual hardware implementation. VHDL was designed to be technology independent. If a design is described in VHDL and implemented in today's technology, the same VHDL description could be used as a starting point for a design in some future technology.

In this chapter, we introduce VHDL and illustrate how we can describe simple combinational circuits using VHDL. We will use VHDL in later units to design sequential circuits and more complex digital systems. In Unit 17, we introduce the use of CAD software tools for automatic synthesis from VHDL descriptions. These synthesis tools will derive a hardware implementation from the VHDL code.

## 10.1 VHDL Description of Combinational Circuits

We begin by describing a simple gate circuit using VHDL. A VHDL signal is used to describe a signal in a physical system. (Section 10.4 contains a summary of signals, constants, and types. The VHDL language also includes variables similar to variables in programming languages, but to obtain synthesizable code for hardware, signals should be used to represent hardware signals. VHDL variables are not used in this text.) The gate circuit of Figure 10-1 has five signals: A, B, C, D

FIGURE 10-1 Gate Circuit



```
C <= A and B after 5 ns;
E <= C or D after 5 ns;
```

and E. The symbol "<=" is the signal assignment operator which indicates that the value computed on the right-hand side is assigned to the signal on the left side. A *behavioral* description of the circuit in Figure 10-1 is

```
E <= D or (A and B);
```

Parentheses are used to specify the order of operator execution.

The two assignment statements in Figure 10-1 give a *dataflow* description of the circuit where it is assumed that each gate has a 5-ns propagation delay. When the statements in Figure 10-1 are simulated, the first statement will be evaluated any time A or B changes, and the second statement will be evaluated any time C or D changes. Suppose that initially A = 1, and B = C = D = E = 0. If B changes to 1 at time 0, C will change to 1 at time = 5 ns. Then, E will change to 1 at time = 10 ns.

The circuit of Figure 10-1 can also be described using *structural* VHDL code. To do so requires that a two-input AND-gate component and a two-input OR-gate component be declared and defined. Components may be declared and defined either in a library or within the architecture part of the VHDL code. (VHDL architectures are discussed in Section 10.3, and packages and libraries are discussed in Section 10.7.) Instantiation statements are used to specify how components are connected. Each copy of a component requires a separate instantiation statement to specify how it is connected to other components and to the port inputs and outputs. An instantiation statement is a concurrent statement that executes anytime one of the input signals in its port map changes. The circuit of Figure 10-1 is described by instantiating the AND gate and the OR gate as follows:

```
Gate1: AND2 port map (A, B, D);
Gate2: OR2 port map (C, D, E);
```

The port map for Gate1 connects A and B to the AND-gate inputs, and it connects D to the AND-gate output. Since an instantiation statement is concurrent, whenever A or B changes, these changes go to the Gate1 inputs, and then the component computes a new value of D. Similarly, the second statement passes changes in C or D to the Gate2 inputs, and then the component computes a new value of E. This is exactly how the real hardware works. (The order in which the instantiation statements appear is irrelevant.) Instantiating a component is different than calling a function in a computer program. A function returns a new value whenever it is called, but an instantiated component computes a new output value whenever its input changes.

VHDL signal assignment statements, such as the ones in Figure 10-1, are examples of concurrent statements. The VHDL simulator monitors the right side of each concurrent statement, and any time a signal changes, the expression on the right side is immediately re-evaluated. The new value is assigned to the signal on the left side after an appropriate delay. This is exactly the way the hardware works. Any time a gate input changes, the gate output is recomputed by the hardware, and the output changes after the gate delay.

When we initially describe a circuit, we may not be concerned about propagation delays. If we write

```
C <= A and B;
E <= C or D;
```

this implies that the propagation delays are 0 ns. In this case, the simulator will assume an infinitesimal delay referred to as $\Delta$ (delta). Assume that initially $A = 1$ and $B = C = D = E = 0$. If B is changed to 1 at time = 1 ns, then C will change at time $1 + \Delta$ and E will change at time $1 + 2\Delta$.

Unlike a sequential program, the order of the above concurrent statements is unimportant. If we write

```
E <= C or D;
C <= A and B;
```

the simulation results would be exactly the same as before.

In general, a signal assignment statement has the form

```
signal_name <= expression [after delay];
```

The expression is evaluated when the statement is executed, and the signal on the left side is scheduled to change after delay. The square brackets indicate that after delay is optional; they are not part of the statement. If after delay is omitted, then the signal is scheduled to be updated after a delta delay. Note that the time at which the statement executes and the time at which the signal is updated are not the same.

Even if a VHDL program has no explicit loops, concurrent statements may execute repeatedly as if they were in a loop. Figure 10-2 shows an inverter with the output connected back to the input. If the output is '0', then this '0' feeds back to the input and the inverter output changes to '1' after the inverter delay, assumed to be 10 ns. Then, the '1' feeds back to the input, and the output changes to '0' after the inverter delay. The signal CLK will continue to oscillate between '0' and '1', as shown in the waveform. The corresponding concurrent VHDL statement will produce the same result. If CLK is initialized to '0', the statement executes and CLK changes to '1' after 10 ns. Because CLK has changed, the statement executes again, and CLK will change back to '0' after another 10 ns. This process will continue indefinitely.

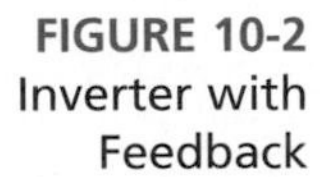

**FIGURE 10-2** Inverter with Feedback

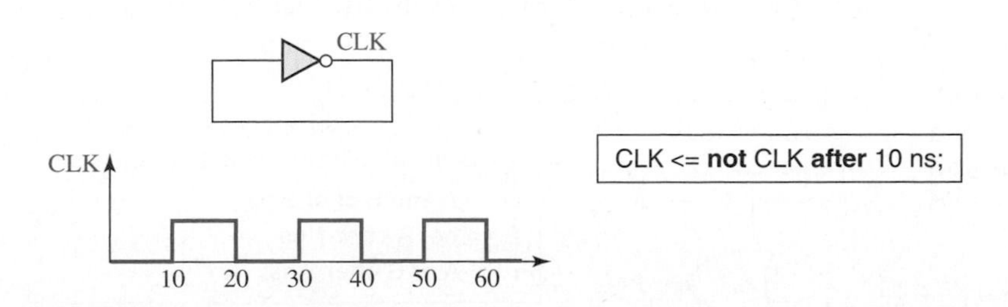

The statement in Figure 10-2 generates a clock waveform with a half period of 10 ns. On the other hand, the concurrent statement

```
CLK <= not CLK;
```

will cause a run-time error during simulation. Because there is 0 delay, the value of CLK will change at times $0 + \Delta$, $0 + 2\Delta$, $0 + 3\Delta$, etc. Because $\Delta$ is an infinitesimal time, time will never advance to 1 ns.

In general, VHDL is not case sensitive, that is, capital and lower case letters are treated the same by the compiler and the simulator. Thus, the statements

```
    Clk <= NOT clk After 10 NS;
and CLK <= not CLK after 10 ns;
```

would be treated exactly the same. Signal names and other VHDL identifiers may contain letters, numbers, and the underscore character (_). An identifier must start with a letter, and it cannot end with an underscore. Thus, C123 and ab_23 are legal identifiers, but 1ABC and ABC_ are not. Every VHDL statement must be terminated with a semicolon. Spaces, tabs, and carriage returns are treated in the same way. This means that a VHDL statement can be continued over several lines, or several statements can be placed on one line. In a line of VHDL code, anything following a double dash (--) is treated as a comment. Words such as **and**, **or**, and **after** are reserved words (or keywords) which have a special meaning to the VHDL compiler. In this text, we will put all reserved words in boldface type.

Figure 10-3 shows three gates that have the signal A as a common input and the corresponding VHDL code. The three concurrent statements execute simultaneously whenever A changes, just as the three gates start processing the signal change at the same time. However, if the gates have different delays, the gate outputs can change at different times. If the gates have delays of 2 ns, 1 ns, and 3 ns, respectively, and A changes at time 5 ns, then the gate outputs D, E, and F can change at times 7 ns, 6 ns, and 8 ns, respectively. The VHDL statements work in the same way. Even though the statements execute simultaneously, the signals D, E, and F are updated at times 7 ns, 6 ns, and 8 ns. However, if no delays were specified, then D, E, and F would all be updated at time $5 + \Delta$.

In these examples, every signal is of type bit, which means it can have a value of '0' or '1'. (Bit values in VHDL are enclosed in single quotes to distinguish them from integer values.) In digital design, we often need to perform the same operation on a group of signals. A one-dimensional array of bit signals is referred to as a bit-vector. If a 4-bit vector named B has an index range 0 through 3, then the four elements of the bit-vector are designated B(0), B(1), B(2), and B(3). The statement B <= "0110" assigns '0' to B(0), '1' to B(1), '1' to B(2), and '0' to B(3).

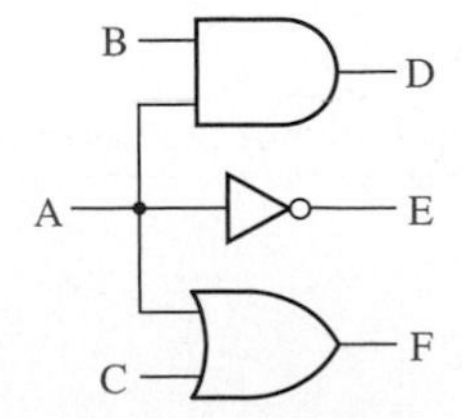


```
-- when A changes, these concurrent
-- statements all execute at the same time
D <= A and B after 2 ns;
E <= not A after 1 ns;
F <= A or C after 3 ns;
```

**FIGURE 10-3** Three Gates with a Common Input and Different Delays

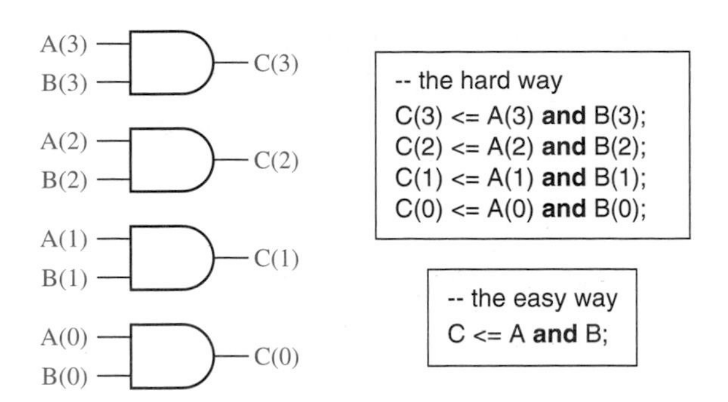


**FIGURE 10-4** Array of AND Gates

Figure 10-4 shows an array of four AND gates. The inputs are represented by bit-vectors A and B, and the outputs by bit-vector C. Although we can write four VHDL statements to represent the four gates, it is much more efficient to write a single VHDL statement that performs the **and** operation on the bit-vectors A and B. When applied to bit-vectors, the **and** operator performs the **and** operation on corresponding pairs of elements.

The preceding signal assignment statements containing "**after delay**" create what is called an **inertial** delay model. Consider a device with an inertial delay of D time units. If an input change to the device will cause its output to change, then the output changes D time units later. However, this is not what happens if the device receives two input changes within a period of D time units and both input changes should cause the output to change. In this case the device output does not change in response to either input change. As an example, consider the signal assignment

```
C <= A and B after 10 ns;
```

Assume A and B are initially 1, and A changes to 0 at 15 ns, to 1 at 30 ns, and to 0 at 35 ns. Then C changes to 1 at 10 ns and to 0 at 25 ns, but C does not change in response to the A changes at 30 ns and 35 ns because these two changes occurred less than 10 ns apart. A device with an inertial delay of D time units filters out output changes that would occur in less than or equal to D time units.

VHDL can also model devices with an **ideal** (**transport**) delay. Output changes caused by input changes to a device exhibiting an ideal (transport) delay of D time units are delayed by D time units, and the output changes occur even if they occur within D time units. The VHDL signal assignment statement that models ideal (transport) delay is

```
signal_name <= transport expression after delay
```

As an example, consider the signal assignment

```
C <= transport A and B after 10 ns;
```

Assume A and B are initially 1 and A changes to 0 at 15 ns, to 1 at 30 ns, and to 0 at 35 ns. Then C changes to 1 at 10 ns, to 0 at 25 ns, to 1 at 40 ns, and to 0 at 45 ns. Note that the last two changes are separated by just 5 ns.

# 10.2 VHDL Models for Multiplexers

Figure 10-5 shows a 2-to-1 multiplexer (MUX) with two data inputs and one control input. The MUX output is $F = A' \cdot I0 + A \cdot I1$. The corresponding VHDL statement is

```
F <= (not A and I0) or (A and I1);
```

Alternatively, we can represent the MUX by a conditional signal assignment statement, as shown in Figure 10-5. This statement executes whenever A, I0, or I1 changes. The MUX output is I0 when A = '0', and else it is I1. In the conditional statement, I0, I1, and F can either be bits or bit-vectors.

**FIGURE 10-5**
2-to-1 Multiplexer

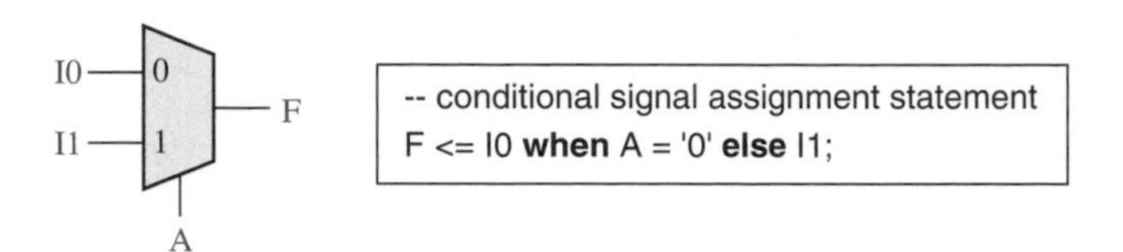
I0
I1
0
1
F
A
-- conditional signal assignment statement
F <= I0 when A = '0' else I1;

The general form of a conditional signal assignment statement is

```
signal_name <= expression1 when condition1
               else expression2 when condition2
               [else expressionN];
```

This concurrent statement is executed whenever a change occurs in a signal used in one of the expressions or conditions. If condition1 is true, signal_name is set equal to the value of expression1, or else if condition2 is true, signal_name is set equal to the value of expression2, etc. The line in square brackets is optional. Figure 10-6 shows how two cascaded MUXes can be represented by a conditional signal assignment statement. The output MUX selects A when E = '1'; or else it selects the output of the first MUX, which is B when D = '1', or else it is C.

**FIGURE 10-6**
Cascaded 2-to-1 MUXes

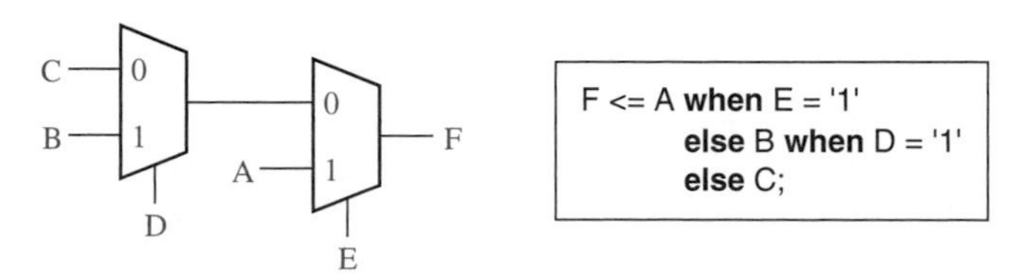
C
B
0
1
D
A
0
1
E
F
F <= A when E = '1'
else B when D = '1'
else C;

Figure 10-7 shows a 4-to-1 MUX with four data inputs and two control inputs, A and B. The control inputs select which one of the data inputs is transmitted to the output. The logic equation for the 4-to-1 MUX is

$$F = A'B'I_0 + A'BI_1 + AB'I_2 + ABI_3$$

Thus, one way to model the MUX is with the VHDL statement

```
F <= (not A and not B and I0) or (not A and B and I1) or
     (A and not B and I2) or (A and B and I3);
```

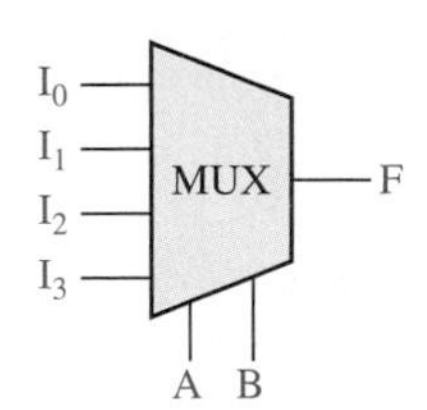


```
sel <= A&B;
-- selected signal assignment statement
with sel select
  F <= I0 when "00",
     I1 when "01",
     I2 when "10",
     I3 when "11";
```

FIGURE 10-7
4-to-1 Multiplexer

Another way to model the 4-to-1 MUX is to use a conditional assignment statement:

```
F <= I0 when A&B = "00"
else I1 when A&B = "01"
else I2 when A&B = "10"
else I3;
```

The expression A&B means A concatenated with B, that is, the two bits A and B are merged together to form a 2-bit vector. This bit vector is tested, and the appropriate MUX input is selected. For example, if A = '1' and B = '0', A&B = "10" and I2 is selected. Instead of concatenating A and B, we could use a more complex condition:

```
F <= I0 when A = '0' and B = '0'
  else I1 when A = '0' and B = '1'
  else I2 when A = '1' and B = '0'
  else I3;
```

A third way to model the MUX is to use a selected signal assignment statement, as shown in Figure 10-7. A&B cannot be used in this type of statement, so we first set Sel equal to A&B. The value of Sel then selects the MUX input that is assigned to F.

The general form of a selected signal assignment statement is

```
with expression_s select
   signal_s <= expression1 [after delay-time] when choice1,
               expression2 [after delay-time] when choice2,
               . . .
               [expression_n [after delay-time] when others];
```

This concurrent statement executes whenever a signal changes in any of the expressions. First, expression_s is evaluated. If it equals choice1, signal_s is set equal to expression1; if it equals choice2, signal_s is set equal to expression2; etc. If all possible choices for the value of expression_s are given, the last line should be omitted; otherwise, the last line is required. When it is present, if expression_s is not equal to any of the enumerated choices, signal_s is set equal to expression_n. The signal_s is updated after the specified delay-time, or after Δ if the "**after** delay-time" is omitted.

# 10.3 VHDL Modules

To write a complete VHDL module, we must declare all of the input and output signals using an **entity** declaration, and then specify the internal operation of the module using an **architecture** declaration. As an example, consider Figure 10-8. The entity declaration gives the name "two_gates" to the module. The port declaration specifies the inputs and outputs to the module. A, B, and D are input signals of type bit, and E is an output signal of type bit. The architecture is named "gates". The signal C is declared within the architecture because it is an internal signal. The two concurrent statements that describe the gates are placed between the keywords **begin** and **end**.

FIGURE 10-8
VHDL Module with Two Gates

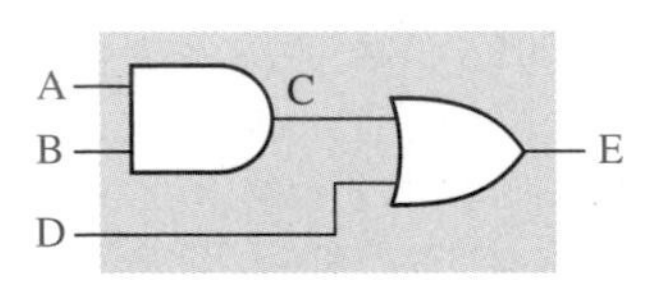


```
entity two_gates is
   port (A,B,D: in bit; E: out bit);
end two_gates;
architecture gates of two_gates is
   signal C: bit;
begin
   C <= A and B; -- concurrent
   E <= C or D; -- statements
end gates;
```

When we describe a system in VHDL, we must specify an entity and an architecture at the top level, and also specify an entity and architecture for each of the component modules that are part of the system (see Figure 10-9). Each entity declaration includes a list of interface signals that can be used to connect to other modules or to the outside world. We will use entity declarations of the form:

```
entity entity-name is
   [port(interface-signal-declaration);]
end [entity] [entity-name];
```

The items enclosed in square brackets are optional. The interface-signal-declaration normally has the following form:

```
list-of-interface-signals: mode type [: = initial-value]
{; list-of-interface-signals: mode type [: = initial-value]};
```

FIGURE 10-9
VHDL Program Structure

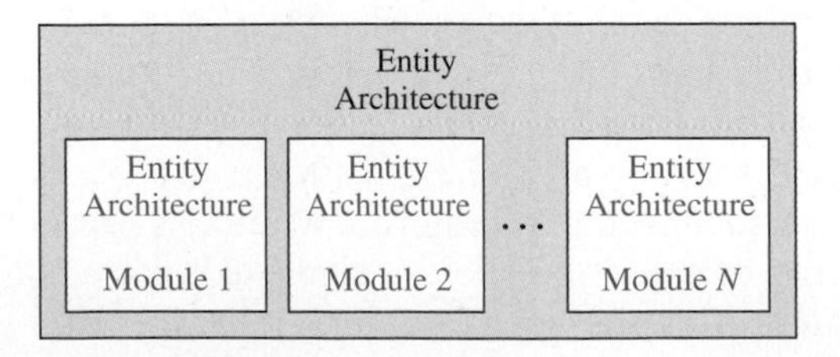

The curly brackets indicate zero or more repetitions of the enclosed clause. Input signals are of mode **in**, output signals are of mode **out**, and bi-directional signals (see Figure 9-12) are of mode **inout**.

So far, we have only used type bit and bit_vector; other types are described in Section 10.4. The optional initial-value is used to initialize the signals on the associated list; otherwise, the default initial value is used for the specified type. For example, the port declaration

```
port(A, B: in integer : = 2; C, D: out bit);
```

indicates that A and B are input signals of type integer that are initially set to 2, and C and D are output signals of type bit that are initialized by default to '0'.

Associated with each entity is one or more architecture declarations of the form

```
architecture architecture-name of entity-name is
    [declarations]
begin
    architecture body
end [architecture] [architecture-name];
```

In the declarations section, we can declare signals and components that are used within the architecture. The architecture body contains statements that describe the operation of the module.

Next, we will write the entity and architecture for a full adder module (refer to Section 4.7 for a description of a full adder). The entity specifies the inputs and outputs of the adder module, as shown in Figure 10-10. The port declaration specifies that X, Y and Cin are input signals of type bit, and that Cout and Sum are output signals of type bit.

**FIGURE 10-10** Entity Declaration for a Full Adder Module

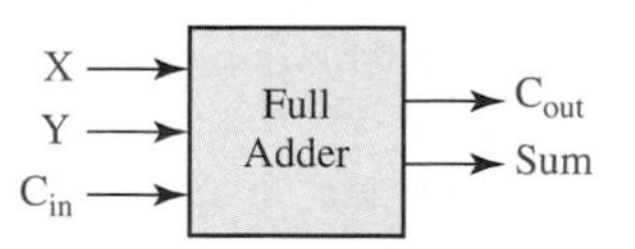


```
entity FullAdder is
   port (X,Y,Cin: in bit;          -- Inputs
            Cout, Sum: out bit);  -- Outputs
end FullAdder;
```

The operation of the full adder is specified by an architecture declaration:

```
architecture Equations of FullAdder is
begin           -- concurrent assignment statements
    Sum <= X xor Y xor Cin after 10 ns;
    Cout <= (X and Y) or (X and Cin) or (Y and Cin) after 10 ns;
end Equations;
```

In this example, the architecture name (Equations) is arbitrary, but the entity name (FullAdder) must match the name used in the associated entity declaration.

The VHDL assignment statements for Sum and Cout represent the logic equations for the full adder. Several other architectural descriptions such as a truth table or an interconnection of gates could have been used instead. In the Cout equation, parentheses are required around (X **and** Y) because VHDL does not specify an order of precedence for the logic operators.

## Four-Bit Full Adder

Next, we will show how to use the FullAdder module defined above as a component in a system which consists of four full adders connected to form a 4-bit binary adder (see Figure 10-11). We first declare the 4-bit adder as an entity (see Figure 10-12). Because the inputs and the sum output are four bits wide, we declare them as bit_vectors which are dimensioned 3 **downto** 0. (We could have used a range 1 **to** 4 instead.)

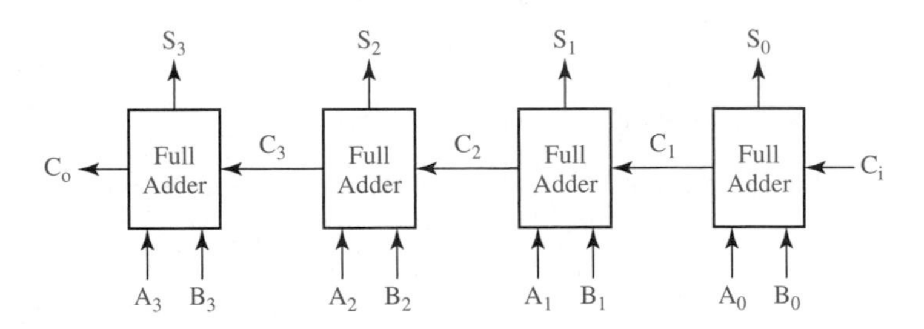


**FIGURE 10-11**
4-Bit Binary Adder

Next, we specify the FullAdder as a component within the architecture of Adder4 (Figure 10-12). The component specification is very similar to the entity declaration for the full adder, and the input and output port signals correspond to those declared for the full adder. Following the component statement, we declare a 3-bit internal carry signal C.

In the body of the architecture, we create several instances of the FullAdder component. (In CAD jargon, we *instantiate* four copies of the FullAdder.) Each copy of FullAdder has a name (such as FA0) and a port map. The signal names following the port map correspond one-to-one with the signals in the component port. Thus, A(0), B(0), and Ci correspond to the inputs X, Y, and Cin, respectively. C(1) and S(0) correspond to the Cout and Sum outputs. Note that the order of the signals in the port map must be the same as the order of the signals in the port of the component declaration.

In preparation for simulation, we can place the entity and architecture for the FullAdder and for Adder4 together in one file and compile. Alternatively, we could compile the FullAdder separately and place the resulting code in a library which is linked in when we compile Adder4.

All of the simulation examples in this text use the ModelSim simulator from Model Tech. Most other VHDL simulators use similar command files and can

FIGURE 10-12 Structural Description of 4-Bit Adder

```
entity Adder4 is
    port (A, B: in bit_vector(3 downto 0); Ci: in bit;    -- Inputs
        S: out bit_vector(3 downto 0); Co: out bit);   -- Outputs
end Adder4;
architecture Structure of Adder4 is
component FullAdder
    port (X, Y, Cin: in bit;   -- Inputs
        Cout, Sum: out bit);   -- Outputs
end component;
signal C: bit_vector(3 downto 1);
begin   -- instantiate four copies of the FullAdder
    FA0: FullAdder port map (A(0), B(0), Ci, C(1), S(0));
    FA1: FullAdder port map (A(1), B(1), C(1), C(2), S(1));
    FA2: FullAdder port map (A(2), B(2), C(2), C(3), S(2));
    FA3: FullAdder port map (A(3), B(3), C(3), Co, S(3));
end Structure;
```

produce output in a similar format. We will use the following simulator commands to test Adder4:

```
add list A B Co C Ci S     -- put these signals on the output list
force A 1111               -- set the A inputs to 1111
force B 0001               -- set the B inputs to 0001
force Ci 1                 -- set Ci to 1
run 50 ns                  -- run the simulation for 50 ns
force Ci 0
force A 0101
force B 1110
run 50 ns
```

We have chosen to run the simulation for 50 ns because this is more than enough time for the carry to propagate through all of the full adders. The simulation results for the above command list are:

| ns | delta | a | b | co | c | ci | s |
|---|---|---|---|---|---|---|---|
| 0 | +0 | 0000 | 0000 | 0 | 000 | 0 | 0000 |
| 0 | +1 | 1111 | 0001 | 0 | 000 | 1 | 0000 |
| 10 | +0 | 1111 | 0001 | 0 | 001 | 1 | 1111 |
| 20 | +0 | 1111 | 0001 | 0 | 011 | 1 | 1101 |
| 30 | +0 | 1111 | 0001 | 0 | 111 | 1 | 1001 |
| 40 | +0 | 1111 | 0001 | 1 | 111 | 1 | 0001 |
| 50 | +0 | 0101 | 1110 | 1 | 111 | 0 | 0001 |
| 60 | +0 | 0101 | 1110 | 1 | 110 | 0 | 0101 |
| 70 | +0 | 0101 | 1110 | 1 | 100 | 0 | 0111 |
| 80 | +0 | 0101 | 1110 | 1 | 100 | 0 | 0011 |

The listing shows how the carry propagates one position every 10 ns. The full adder inputs change at time = Δ:

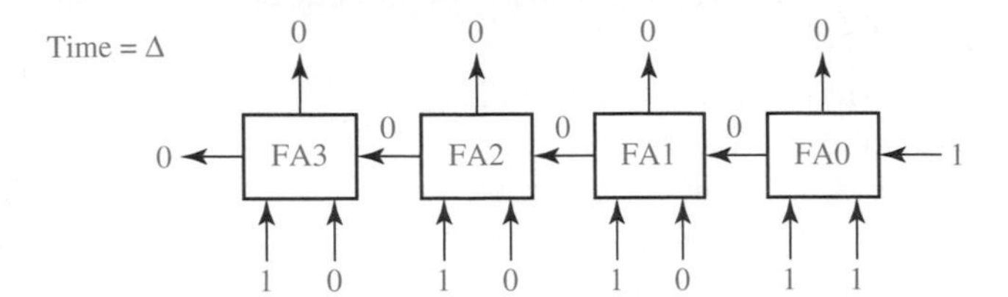


The sum and carry are computed by each FA and appear at the FA outputs 10 ns later:

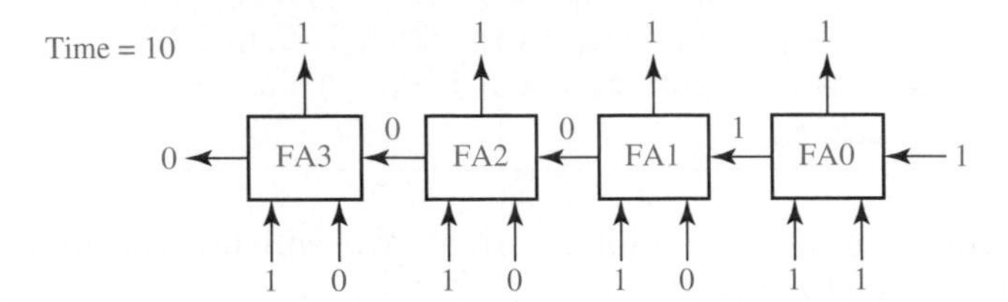


Because the inputs to FA1 have changed, the outputs change 10 ns later:

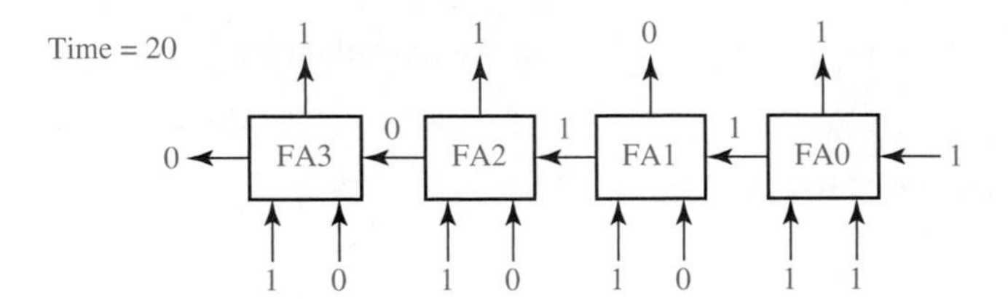


The final simulation results are:

1111 + 0001 + 1 = 0001 with a carry of 1 (at time = 40 ns) and

0101 + 1110 + 0 = 0011 with a carry of 1 (at time = 80 ns).

The simulation stops at 80 ns because no further changes occur after that time. For more details on how the simulator handles Δ delays, refer to Section 10.9.

In this section we have shown how to construct a VHDL module using an entity-architecture pair. The 4-bit adder module demonstrates the use of VHDL components to write structural VHDL code. Components used within the architecture are declared at the beginning of the architecture, using a component declaration of the form

```
component component-name
    port (list-of-interface-signals-and-their-types);
end component;
```

The port clause used in the component declaration has the same form as the port clause used in an entity declaration. The connections to each component used in a circuit are specified by using a component instantiation statement of the form

label: component-name **port map** (list-of-actual-signals);

The list of actual signals must correspond one-to-one to the list of interface signals specified in the component declaration.

## 10.4 Signals and Constants

Input and output signals for a module are declared in a port. Signals internal to a module are declared at the start of an architecture, before **begin**, and can be used only within that architecture. Port signals have an associated mode (usually in or out), but internal signals do not. A signal used within an architecture must be declared either in a port or in the declaration section of an architecture, but it cannot be declared in both places. A signal declaration has the form

**signal** list_of_signal_names: type_name [constraint] [:= initial_value];

The constraint can be an index range like (0 **to** 5) or (4 **downto** 1), or it can be a range of values such as **range** 0 to 7. Examples:

**signal** A, B, C: bit_vector(3 **downto** 0):= "1111";

A, B, and C are 4-bit vectors dimensioned 3 downto 0 and initialized to 1111.

**signal** E, F: integer **range** 0 **to** 15;

E and F are integers in the range 0 to 15, initialized by default to 0. The compiler or simulator will flag an error if we attempt to assign a value outside the specified range to E or F.

Constants declared at the start of an architecture can be used anywhere within that architecture. A constant declaration is similar to a signal declaration:

**constant** constant_name: type_name [constraint] [:= constant_value];

A constant named limit of type integer with a value of 17 can be defined as

**constant** limit : integer := 17;

A constant named delay1 of type time with the value of 5 ns can be defined as

**constant** delay1 : time := 5 ns;

This constant could then be used in an assignment statement

A <= B **after** delay1;

Once the value of a constant is defined in a declaration statement, unlike a signal, the value cannot be changed by using an assignment statement.

Signals and constants can have any one of the predefined VHDL types, or they can have a user-defined type. Some of the predefined types are

**Definition**

| | |
|---|---|
| bit | '0' or '1' |
| boolean | FALSE or TRUE |
| integer | an integer in the range $-(2^{31}-1)$ to $+(2^{31}-1)$ (some implementations support a wider range) |
| positive | an integer in the range 1 to $2^{31}-1$ (positive integers) |
| natural | an integer in the range 0 to $2^{31}-1$ (positive integers and zero) |
| real | floating-point number in the range −1.0E38 to + 1.0E38 |
| character | any legal VHDL character including upper- and lower case letters, digits, and special characters; each printable character must be enclosed in single quotes, e.g., 'd', '7', '+' |
| time | an integer with units fs, ps, ns, us, ms, sec, min, or hr |

Note that the integer range for VHDL is symmetrical even though the range for a 32-bit 2's complement integer is $-2^{31}$ to $+(2^{31}-1)$.

A common user-defined type is the enumeration type in which all of the values are enumerated. For example, the declarations

```
type state_type is (S0, S1, S2, S3, S4, S5);
signal state : state_type := S1;
```

define a signal called state which can have any one of the values S0, S1, S2, S3, S4, or S5 and which is initialized to S1. If no initialization is given, the default initialization is the left most element in the enumeration list, S0 in this example. If we declare the signal state as shown, the following assignment statement sets state to S3:

```
state <= S3;
```

VHDL is a strongly-typed language so signals of different types generally cannot be mixed in the same assignment statement, and no automatic type conversion is performed. Thus the statement A <= B **or** C is only valid if A, B, and C all have the same type or closely related types.

## 10.5 Arrays

In order to use an array in VHDL, we must first declare an array type, and then declare an array object. For example, the following declaration defines a one-dimensional array type named SHORT_WORD:

```
type SHORT_WORD is array (15 downto 0) of bit;
```

An array of this type has an integer index with a range from 15 downto 0, and each element of the array is of type bit.

Next, we will declare array objects of type SHORT_WORD:

```
signal DATA_WORD: SHORT_WORD;
signal ALT_WORD: SHORT_WORD := "0101010101010101";
constant ONE_WORD: SHORT_WORD := (others => '1');
```

DATA_WORD is a signal array of 16 bits, indexed 15 downto 0, which is initialized (by default) to all '0' bits. ALT_WORD is a signal array of 16 bits which is initialized to alternating 0's and 1's. ONE_WORD is a constant array of 16 bits; all bits are set to '1' by (**others** => '1'). Because none of the bits have been set individually,[1] in this case **others** applies to all of the bits.

We can reference individual elements of the array by specifying an index value. For example, ALT_WORD(0) accesses the far right bit of ALT_WORD. We can also specify a portion of the array by specifying an index range: ALT_WORD(5 **downto** 0) accesses the low order six bits of ALT_WORD, which have an initial value of 010101.

The array type and array object declarations illustrated above have the general forms:

```
type array_type_name is array index_range of element_type;
signal array_name: array_type_name [ := initial_values ];
```

In this declaration, **signal** may be replaced with **constant**.

Multidimensional array types may also be defined with two or more dimensions. The following example defines a two-dimensional array signal which is a matrix of integers with four rows and three columns:

```
type matrix4x3 is array (1 to 4, 1 to 3) of integer;
signal matrixA: matrix4x3 := ((1,2,3),(4,5,6),(7,8,9),(10,11,12));
```

The signal matrixA, will be initialized to

$$\begin{bmatrix} 1 & 2 & 3 \\ 4 & 5 & 6 \\ 7 & 8 & 9 \\ 10 & 11 & 12 \end{bmatrix}$$

The array element matrixA(3,2) references the element in the third row and second column, which has a value of 8. The statement B <= matrixA(2,3) assigns a value of 6 to B.

When an array type is declared, the dimensions of the array may be left undefined. This is referred to as an unconstrained array type. For example,

```
type intvec is array (natural range <>) of integer;
```

declares intvec as an array type which defines a one-dimensional array of integers with an unconstrained index range of natural numbers. The default type for array indices is integer, but another type may be specified. Because the index range is not specified in the unconstrained array type, the range must be specified when the array object is declared. For example,

```
signal intvec5: intvec(1 to 5) := (3,2,6,8,1);
```

defines a signal array named intvec5 with an index range of 1 to 5, which is initialized to 3, 2, 6, 8, 1. The following declaration defines matrix as a two-dimensional array with unconstrained row and column index ranges:

```
type matrix is array (natural range <> , natural range <>) of integer;
```

[1]See Reference [1, p. 86] for information on how to set individual bits.

Predefined unconstrained array types in VHDL include bit_vector and string, which are defined as follows:

```
type bit_vector is array (natural range <>) of bit;
type string is array (positive range <>) of character;
```

The characters in a string literal must be enclosed in double quotes. For example, "This is a string." is a string literal. The following example declares a constant string1 of type string:

```
constant string1: string(1 to 29) := "This string is 29 characters."
```

A bit_vector literal may be written either as a list of bits separated by commas or as a string. For example, ('1','0','1','1','0') and "10110" are equivalent forms. The following declares a constant A which is a bit_vector with a range 0 to 5.

```
constant A : bit_vector(0 to 5) := "101011";
```

A truth table can be implemented using a ROM (read-only memory) as illustrated in Figure 9-17. If we represent the ROM outputs by a bit_vector, F(0 **to** 3), we can represent the truth table that is stored in the ROM by an array of bit_vectors. The VHDL code for this ROM is given in Figure 10-13. The port declaration (line 4) defines the inputs and outputs for the ROM. The type declaration (line 7) defines an array with 8 rows where each row is 4 bits wide. Line 8 declares ROM1 to be an array of this type with binary data stored in each row. Line 9 declares an integer called index. This index will be used to select one of the 8 rows in the ROM1 array. In line 11, this index is formed by concatenating the three input bits to form a 3-bit vector, and this vector is converted to an integer. The data is read from the ROM1 array in line 13. For example, if A = '1', B = '0', and C = '1', index = 5, and "0001" is read from the ROM. Lines 1 and 2 allow us to use the vec2int function, which is defined in a library named BITLIB.

**FIGURE 10-13** VHDL Description of a ROM

```
1   library BITLIB;
2   use BITLIB.bit_pack.all;
3   entity ROM9_17 is
4       port (A, B, C: in bit; F: out bit_vector(0 to 3));
5   end entity;
6   architecture ROM of ROM9_17 is
7   type ROM8X4 is array (0 to 7) of bit_vector(0 to 3);
8   constant ROM1: ROM8X4 := ("1010", "1010", "0111", "0101", "1100", "0001", "1111", "0101");
9   signal index: Integer range 0 to 7;
10  begin
11      index <= vec2int(A&B&C);      -- A&B&C Is a 3-bit vector
12          -- vec2int is a function that converts this vector to an integer
13      F <= ROM1 (index);
14          -- this statement reads the output from the ROM
15  end ROM;
```

# 10.6 VHDL Operators

Predefined VHDL operators can be grouped into seven classes:

1. binary logical operators: **and or nand nor xor xnor**
2. relational operators: = /= < <= > >=
3. shift operators: **sll srl sla sra rol ror**
4. adding operators: + − & (concatenation)
5. unary sign operators: + −
6. multiplying operators: * / **mod rem**
7. miscellaneous operators: **not abs** **

When parentheses are not used, operators in class 7 have highest precedence and are applied first, followed by class 6, then class 5, etc. Class 1 operators have lowest precedence and are applied last. Operators in the same class have the same precedence and are applied from left to right in an expression. The precedence order can be changed by using parentheses. In the following expression, A, B, C, and D are bit_vectors:

```
not A or B and not C & D
```

In this expression, **not** is performed first, then **&** (concatenation), then **or**, and finally **and**. The equivalent expression using parentheses is

```
((not A) or B) and ((not C) &D)
```

The binary logical operators (class 1) as well as **not** can be applied to bits, booleans, bit_vectors, and boolean_vectors. The class 1 operators require two operands of the same type and size, and the result is of that type and size.

Relational operators (class 2) are used to compare two expressions and return a value of FALSE or TRUE. The two expressions must be of the same type and size. Equal (=) and not equal (/=) apply to any type, but the application of the other relational operators is more restricted. Note that "=" is always a relational operator, but "<=" also serves as an assignment operator. Example: If A = 5, B = 4, and C = 3 the expression

(A >= B) **and** (B <= C) evaluates to FALSE.

Figure 10-14 shows a comparator for two integers with a restricted range. C must be of type Boolean since the condition A <= B evaluates to TRUE or FALSE. If we implement the comparator in hardware, each integer would be represented by a 4-bit signal because the range is restricted to 0 to 15. C, D, and E would each be one bit (0 for FALSE or 1 for TRUE).

FIGURE 10-14 Comparator for Integers

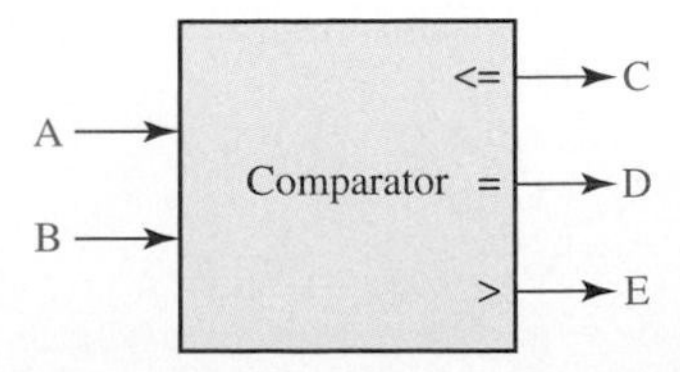


```
signal A,B: integer range 0 to 15;
signal C, D, E: Boolean;
--------------------------------
C <= A <= B;
D <= A = B;
E <= A > B;
```

The shift operators are used to shift or rotate a bit_vector. In the following examples, A is an 8- bit vector equal to "10010101":

| | |
|---|---|
| A **sll** 2 | is "01010100" (shift left logical, filled with '0') |
| A **srl** 3 | is "00010010" (shift right logical, filled with '0') |
| A **sla** 3 | is "10101111" (shift left arithmetic, filled with rightmost bit) |
| A **sra** 2 | is "11100101" (shift right arithmetic, filled with leftmost bit) |
| A **rol** 3 | is "10101100" (rotate left) |
| A **ror** 5 | is "10101100" (rotate right) |

We will not utilize these shift operators because some software used for synthesis uses different shift operators. Instead, we will do shifting using the concatenation operator. For example, if A in the above listing is dimensioned 7 downto 0, we can implement shift right arithmetic two places as follows:

```
A(7)&A(7)&A(7 downto 2) = '1'&'1'&"100101" = "11100101"
```

This makes two copies of the sign bit followed by the left 6 bits of A, which gives the same result as A **sra** 2.

The + and − operators can be applied to integer or real numeric operands. The & operator can be used to concatenate two vectors (or an element and a vector, or two elements) to form a longer vector. For example, "010" & '1' is "0101" and "ABC" & "DEF" is "ABCDEF."

The * and / operators perform multiplication and division on integer or floating-point operands. The **rem** and **mod** operators calculate the remainder and modulus for integer operands. (We will not use rem and mod; for further discussion of these operators see Reference [1].) The ** operator raises an integer or floating-point number to an integer power, and **abs** finds the absolute value of a numeric operand.

## 10.7 Packages and Libraries

Packages and libraries provide a convenient way of referencing frequently used functions and components. A package consists of a package declaration and an optional package body. The package declaration contains a set of declarations which may be shared by several design units. For example, it may contain type, signal, component, function, and procedure declarations. The package body usually contains component descriptions and the function and procedure bodies. The package and its associated compiled VHDL models may be placed in a library, so they can be accessed as required by different VHDL designs. A package declaration has the form:

```
package package-name is
    package declarations
end [package][package-name];
```

A package body has the form

```
package body package-name is
    package body declarations
end [package body][package name];
```

We have developed a package called bit_pack which is used in a number of examples in this book. This package contains commonly used components and functions which use signals of type bit and bit_vector. A complete listing of this package and associated component models is included on the CD-ROM that accompanies this text. Most of the components in this package have a default delay of 10 ns, but this delay can be changed by the use of generics. For an explanation of generics, refer to one of the VHDL references. We have compiled this package and the component models and placed the result in a library called BITLIB.

One of the components in the library is a two-input NOR gate named Nor2, which has default delay of 10 ns. The package declaration for bit_pack includes the component declaration

```
component Nor2
    port (A1, A2: in bit; Z: out bit);
end component;
```

The NOR gate is modeled using a concurrent statement. The entity-architecture pair for this component is

```
-- two-input NOR gate
entity Nor2 is
    port (A1, A2: in bit; Z: out bit);
end Nor2;

architecture concur of Nor2 is
begin
    Z <= not(A1 or A2) after 10 ns;
end concur;
```

To access components and functions within a package requires a **library** statement and a **use** statement. The statement

```
library BITLIB;
```

allows your design to access the BITLIB. The statement

```
use BITLIB.bit_pack.all;
```

allows your design to use the entire bit_pack package. A statement of the form

```
use BITLIB.bit_pack.Nor2;
```

may be used if you want to use a specific component (in this case Nor2) or function in the package.

When components from a library package are used, component declarations are not needed. Figure 10-15 shows a NOR-NOR circuit and the corresponding structural VHDL code. This code instantiates three copies of the Nor2 gate component from the package bit_pack and connects the gate inputs and outputs.

FIGURE 10-15
NOR-NOR Circuit and Structural VHDL Code Using Library Components

```
library BITLIB;
use BITLIB.bit_pack.all;
entity nor_nor is
   port (A,B,C,D: in bit; G: out bit);
end nor_nor;
architecture structural of nor_nor is
signal E,F,BN,CN: bit;  -- internal signals
begin
   BN <= not B;  CN <= not C;
   G1: Nor2 port map (A, BN, E);
   G2: Nor2 port map (CN, D, F);
   G3: Nor2 port map (E, F, G);
end structural;
```

# 10.8 IEEE Standard Logic

Use of two-valued logic (bits and bit vectors) is generally not adequate for simulation of digital systems. In addition to '0' and '1', values of 'Z' (high-impedance or no connection) and 'X' (unknown) are frequently used in digital system simulation. The IEEE Standard 1164 defines a std_logic type that actually has nine values ('U', 'X', '0', '1', 'Z', 'W', 'L', 'H', and '–'). We will only be concerned with the first five values in this text. 'U' stands for uninitialized. When a logic circuit is first turned on and before it is reset, the signals will be uninitialized. If these signals are represented by std_logic, they will have a value of 'U' until they are changed. Just as a group of bits is represented by a bit_vector, a group of std_logic signals is represented by a std_logic_vector.

Figure 10-16 shows how a tri-state buffer can be represented by a concurrent statement. When the buffer is enabled (B = '1'), the output is A, or else it is high impedance ('Z'). A and C could be std_logic_vectors instead of std_logic bits.

FIGURE 10-16
Tri-State Buffer

```
signal A,B,C: std_logic;
-------------------------------
C <= A when B = '1' else 'Z';
```

Figure 10-17 shows two tri-state buffers with their outputs connected together by a tri-state bus. If buffer 1 has an output of '1' and buffer 2 has a hi-Z output, the bus value is '1'. When both buffers are enabled, if buffer 1 drives '0' onto the bus and buffer 2 drives '1' onto the bus, the result is a bus conflict. In this case, the bus value is unknown, which we represent by an 'X'.

In the VHDL code, A, C, and F are std_logic_vectors and F represents the tri-state bus. The signal F is driven from two different sources. If the two concurrent statements

FIGURE 10-17
Tri-State Buffers Driving a Bus

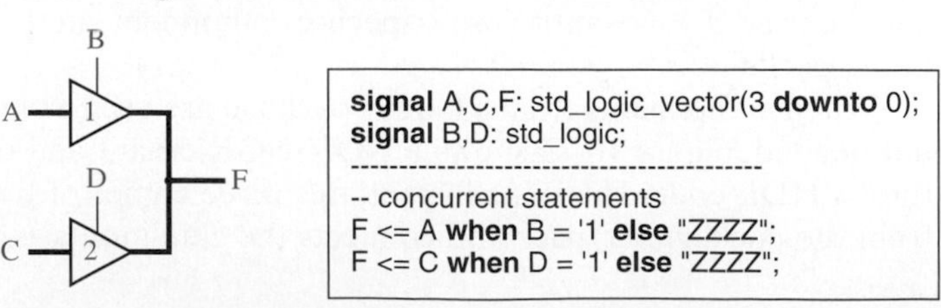


```
signal A,C,F: std_logic_vector(3 downto 0);
signal B,D: std_logic;
-------------------------------------------
-- concurrent statements
F <= A when B = '1' else "ZZZZ";
F <= C when D = '1' else "ZZZZ";
```

assign different values to F, VHDL automatically calls a *resolution function* to determine the resulting value. This is similar to the way the hardware works—if the two buffers have different output values, the hardware resolves the values and comes up with an appropriate value on the bus. VHDL uses the table of Figure 10-18 to resolve the bus value when two different std_logic signals, S1 and S2, drive the bus. (Only signal values 'U', 'X', '0', '1', and 'Z' are considered here.) This table is similar to Figure 9-10, which is used for four-valued logic simulation, except for the addition of a row and a column corresponding to 'U'. When an uninitialized signal is connected to any other signal, VHDL considers that the result is uninitialized.

**FIGURE 10-18** Resolution Function for Two Signals

| | S2 | | | | |
|---|---|---|---|---|---|
| S1 | U | X | 0 | 1 | Z |
| U | U | U | U | U | U |
| X | U | X | X | X | X |
| 0 | U | X | 0 | X | 0 |
| 1 | U | X | X | 1 | 1 |
| Z | U | X | 0 | 1 | Z |

If A, B, and F are bits (or bit_vectors) and we write the concurrent statements

```
F <= A;  F <= not B;
```

the compiler will flag an error because no resolution function exists for signals of type bit. If A, B, and F are std_logic bits or vectors, the compiler will generate a call to the resolution function and not report an error. If F is assigned conflicting values during simulation, then F will be set to 'X' (unknown).

In order to use signals of type std_logic and std_logic_vector in a VHDL module, the following declarations must be placed before the entity declaration:

```
library ieee;
use ieee.std_logic_1164.all;
```

The IEEE std_logic_1164 package defines std_logic and related types, logic operations on these types, and functions for working with these types.

The original IEEE standards for VHDL do not define arithmetic operations on bit_vectors or on std_logic vectors. Based on these standards, we cannot add, subtract, multiply, or divide bit_vectors or std_logic_vectors without first converting them to other types. For example, if A and B are bit_vectors, the expression A + B is not allowed. However, VHDL libraries and packages are available that define arithmetic and comparison operations on std_logic_vectors. The operators defined in these packages are referred to as *overloaded* operators. This means that the compiler will automatically use the proper definition of the operator depending on its context. For example, when evaluating the expression A + B, if A and B are integers, the compiler will use the integer arithmetic routine to do the addition. On the other hand, if A and B are of type std_logic_vector, the compiler will use the addition routine for standard logic vectors. In order to use overloaded operators, the appropriate library and use statements must be included in the VHDL code so that the compiler can locate the definitions of these operators.

In this text, we will use the std_logic_unsigned package, originally developed by Synopsis and now widely available. This package treats std_logic_vectors as unsigned numbers. The std_logic_unsigned package defines arithmetic operators (+, −, *) and comparison operators (<, <=, =, /=, >, >=) that operate on std_logic_vectors. For + , − , and comparison operators, if the two operands are of different length, the shorter operand is filled on the left end with zeros.

These operations can also be applied when the left operand is a std_logic_vector and the right operand is an integer. The arithmetic operations return a std_logic_vector, and the comparison operations return a Boolean. For example, if A is "10011", A + 7 returns a value of "11010", and A >= 5 returns TRUE. In these examples, + and >= are overloaded operators, and the compiler automatically calls the appropriate routine to add an integer to a std_logic_vector or to compare an integer with a std_logic_vector.

If A and B are 4-bit std_logic vectors, A + B gives their sum as a 4-bit vector, and any carry is lost. If the carry is needed, then A must be extended to five-bits before addition. This is accomplished by concatenating a '0' in front of A. Then '0' &A + B gives a 5-bit sum that can be split into a carry and a 4-bit sum.

Figure 10-19 shows a binary adder and its VHDL representation using the std_logic_unsigned package. Addout is a 5-bit sum that is split into Sum and Cout. For example, if A = "1011", B = "1001", and Cin = '1', Addout evaluates to "10101", which is then split into a sum "0101" with a carry out of '1'.

Figure 10-20 shows how to implement the bi-directional input-output pin and tri-state buffer of Figure 9-12 using IEEE std_logic. The I/O pin declared in the port

**FIGURE 10-19** VHDL Code for Binary Adder

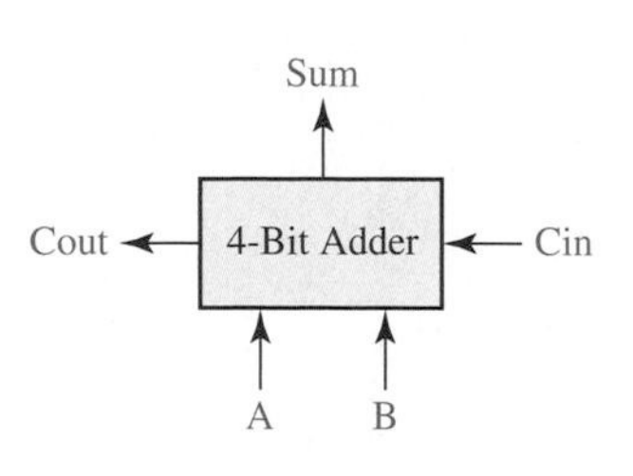


```
library IEEE;
use IEEE.std_logic_1164.all;
use IEEE.std_logic_unsigned.all;
-------------------------------------------
signal A,B,Sum: std_logic_vector(3 downto 0);
signal Addout: std_logic_vector(4 downto 0);
signal Cin,Cout: std_logic;
-----------------------------------
Addout <= '0'&A + B + Cin;
Sum <= Addout(3 downto 0);
Cout <= Addout(4);
```

**FIGURE 10-20** VHDL Code for Bi-Directional I/O Pin

```
entity IC_pin is
    port(IO_pin: inout std_logic);
end entity;
architecture bi_dir of IC_pin is
    component IC
        port(input: in std_logic; output: out std_logic);
    end component;
    signal input, output, en: std_logic;
begin          -- connections to bi-directional I/O pin
    IO_pin <= output when en = '1' else 'Z';
    input <= IO_pin;
    IC1: IC port map (input, output);
end bi_dir;
```

is of mode **inout**. The concurrent statements in the architecture connect the IC output to the pin via a tri-state buffer and also connect the pin to the IC input.

## 10.9 Compilation and Simulation of VHDL Code

After describing a digital system in VHDL, simulation of the VHDL code is important for two reasons. First, we need to verify the VHDL code correctly implements the intended design, and second, we need to verify that the design meets its specifications. Before the VHDL model of a digital system can be simulated, the VHDL code must first be compiled (see Figure 10-21). The VHDL compiler, also called an analyzer, first checks the VHDL source code to see that it conforms to the syntax and semantic rules of VHDL. If there is a syntax error such as a missing semicolon or a semantic error such as trying to add two signals of incompatible types, the compiler will output an error message. The compiler also checks to see that references to libraries are correct. If the VHDL code conforms to all of the rules, the compiler generates intermediate code which can be used by a simulator or by a synthesizer.

In preparation for simulation, the VHDL intermediate code must be converted to a form which can be used by the simulator. This step is referred to as *elaboration*. During elaboration, ports are created for each instance of a component, memory storage is allocated for the required signals, the interconnections among the port signals are specified, and a mechanism is established for executing the VHDL statements in the proper sequence. The resulting data structure represents the digital system being simulated. After an initialization phase, the simulator enters the execution phase. The simulator accepts simulation commands which control the simulation of the digital system and specify the desired simulator output.

Understanding the role of the delta (Δ) time delays is important when interpreting output from a VHDL simulator. Although the delta delays do not show up on waveform outputs from the simulator, they show up on listing outputs. The simulator uses delta delays to make sure that signals are processed in the proper sequence. Basically, the simulator works as follows: Whenever a component input changes, the output is scheduled to change after the specified delay or after Δ if no delay is specified. When all input changes have been processed, the simulated time is advanced to the next time at which an output change is specified. When time is advanced by a finite amount (1 ns for example), the Δ counter is reset, and simulation resumes. Real time does not advance again until all Δ delays associated with the current simulation time have been processed.

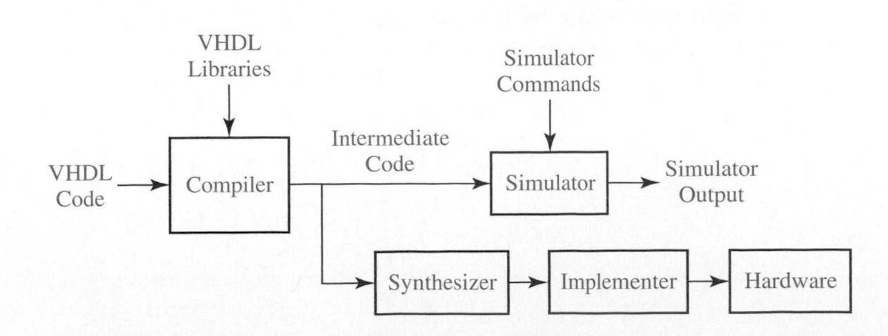


**FIGURE 10-21** Compilation, Simulation, and Synthesis of VHDL Code

The following example illustrates how the simulator works for the circuit of Figure 10-22. Suppose that A changes at time = 3 ns. Statement 1 executes, and B is scheduled to change at time 3 + Δ. Then time advances to 3 + Δ, and statement 2 executes. C is scheduled to change at time 3 + 2Δ. Time advances to 3 + 2Δ, and statement 3 executes. D is then scheduled to change at 8 ns. You may think the change should occur at (3 + 2Δ + 5) ns. However, when time advances a finite amount (as opposed to Δ, which is infinitesimal), the Δ counter is reset. For this reason, when events are scheduled a finite time in the future, the Δ's are ignored. Because no further changes are scheduled after 8 ns, the simulator goes into an idle mode and waits for another input change. The table gives the simulator output listing.

After the VHDL code for a digital system has been simulated to verify that it works correctly, the VHDL code can be synthesized to produce a list of required components and their interconnections. The synthesizer output can then be used to implement the digital system using specific hardware such as a CPLD or FPGA. The CAD software used for implementation generates the necessary information to program the CPLD or FPGA hardware. The synthesis and implementation of digital logic from VHDL code is discussed in more detail in Unit 17.

In this chapter, we have covered the basics of VHDL. We have shown how to use VHDL to model combinational logic and how to construct a VHDL module using an entity-architecture pair. Because VHDL is a hardware description language, it differs from an ordinary programming language in several ways. Most importantly, VHDL statements execute concurrently because they must model real hardware in which the components are all in operation at the same time.

**FIGURE 10-22** Simulation of VHDL Code

A B C D

```
1  B <= not A;
2  C <= not B;
3  D <= not C after 5 ns;
```

| ns | delta | A | B | C | D |
|---|---|---|---|---|---|
| 0 | +0 | 0 | 1 | 0 | 1 |
| 3 | +0 | 1 | 1 | 0 | 1 |
| 3 | +1 | 1 | 0 | 0 | 1 |
| 3 | +2 | 1 | 0 | 1 | 1 |
| 8 | +0 | 1 | 0 | 1 | 0 |

# Problems

**10.1** Write VHDL statements that represent the following circuit:
(a) Write a statement for each gate.
(b) Write one statement for the whole circuit.

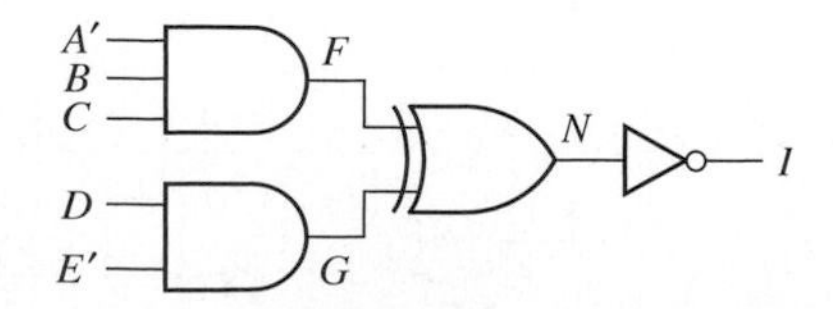

**10.2** Draw the circuit represented by the following VHDL statements:

```
F <= E and I;
I <= G or H;
G <= A and B;
H <= not C and D;
```

**10.3** (a) Implement the following VHDL conditional statement using two 2-to-1 MUXes:

```
F <= A when D = '1' else B when E = '1' else C;
```

(b) Implement the same statement using gates.

**10.4** Write the VHDL code for Figure 9-4 using a conditional signal assignment statement. Use bit_vectors for X, Y, and Z.

**10.5** Write a VHDL module that implements a full adder using an array of bit_vectors to represent the truth table.

**10.6** (a) Given that A = "00101101" and B = "10011", determine the value of F:

```
F <= not B & "0111" or A & '1' and '1'& A;
```

(b) Given A = "11000", B = "10011", and C = "0111", evaluate the following expression:

```
not A + C * 2 > B / 4 & "00"
```

**10.7** Write a VHDL module that finds the average value of four 16-bit unsigned numbers that are represented by std_logic_vectors. Division by four is best accomplished by shifting. Round off your answer to the nearest integer.

**10.8** Write VHDL code for the system shown in Figure 9-11. Use four concurrent statements to compute the signal on the tri-state bus.

**10.9** (a) Draw the circuit represented by the following VHDL statements:

```
T1 <= not A and not B and I0;
T2 <= not A and B and I1;
T3 <= A and not B and I2;
T4 <= A and B and I3;
F <= T1 or T2 or T3 or T4;
```

(b) Draw a MUX that implements F. Then write a selected signal assignment statement that describes the MUX.

**10.10** Assume that the following are concurrent VHDL statements:

```
(a) L <= P nand Q after 10 ns;
(b) M <= L nor N after 5 ns;
(c) R <= not M;
```

Initially at time $t = 0$ ns, P = 1, Q = 1, and N = 0. If Q becomes 0 at time $t = 4$ ns,

(1) At what time will statement (a) execute?

(2) At what time will L be updated?

(3) At what time will statement (c) execute?
(4) At what time will R be updated?

**10.11** (a) Write a single concurrent VHDL statement to represent the following circuit. Do not use parentheses in the statement.

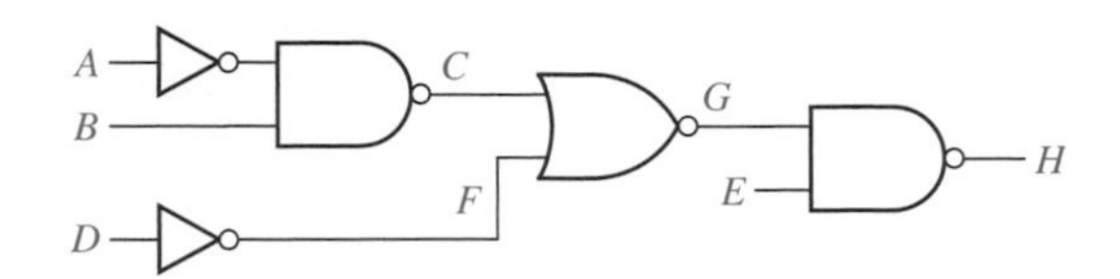


(b) Write individual statements to represent the circuit of part (a). Assume that all NAND gates have a delay of 10 ns, all NOR gates have a delay of 15 ns, and inverters have a delay of 5 ns.

**10.12** Draw a circuit that implements the following VHDL code.

```
V <= T and U;
U <= not R or S and P or not Q or S;
T <= not P or Q or R;
```

**10.13** Suppose L, M, and N are of type std_logic. If the following are concurrent statements, what are the values of L, M, and N? You can use the resolution function given in Figure 10-18.

```
L <= '1';  L <= '0';
M <= '1' when L = '0' else 'Z' when L = '1' else '0';
N <= M when L = '0' else not M;
N <= 'Z';
```

**10.14** (a) Given that D = "011001" and E = "110", determine the value of F.

```
F <= not E & "011" or "000100" and not D;
```

(b) Given A = "101" and B = "011", evaluate the following expression:

```
not (A & B) < (not B & A and not A & A)
```

**10.15** Write VHDL code to implement the following logic functions using a 16 words × 3 bits ROM.

$W = A'B'C + C'D + ACD'$
$X = A'C' + B'D$
$Y = BD' + B'C'D$

**10.16** The diagram shows an 8-bit-wide data bus that transfers data between a microprocessor and memory. Data on this bus is determined by the control signals mRead and mWrite. When mRead = '1', the data on the memory's internal bus 'membus' is output to the data bus. When mWrite = '1', the data on the processor's internal bus 'probus' is output to the data bus. When both control signals are '0', the data bus must be in a high-impedance state.

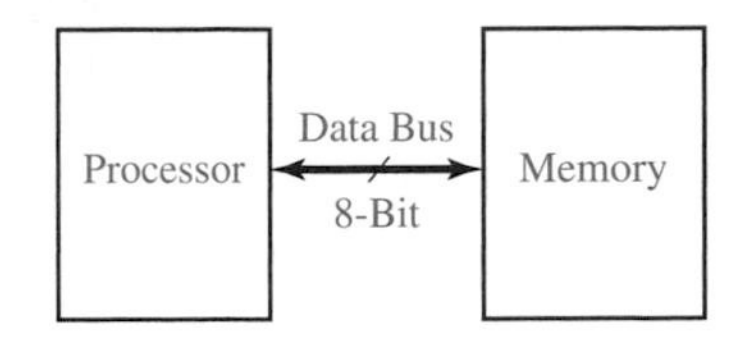


(a) Write VHDL statements to represent the data bus.
(b) Normally mRead = mWrite = '1' does not occur. But if it occurs, what value will the data bus take?

**10.17** (a) Write a selected signal assignment statement to represent the 4-to-1 MUX shown below. Assume that there is an inherent delay in the MUX that causes the change in output to occur 15 ns after a change in input.
(b) Repeat (a) using a conditional signal assignment statement.

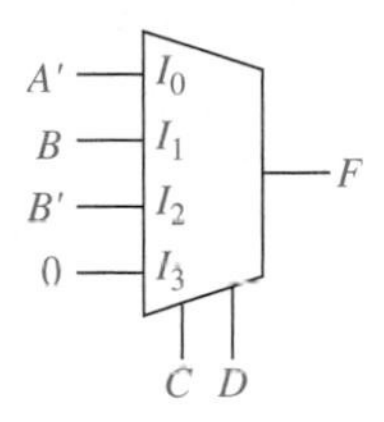


**10.18** (a) Write a complete VHDL module for a two-input NAND gate with 4-ns delay.
(b) Write a complete VHDL module for the following circuit that uses the NAND gate module of Part (a) as a component.

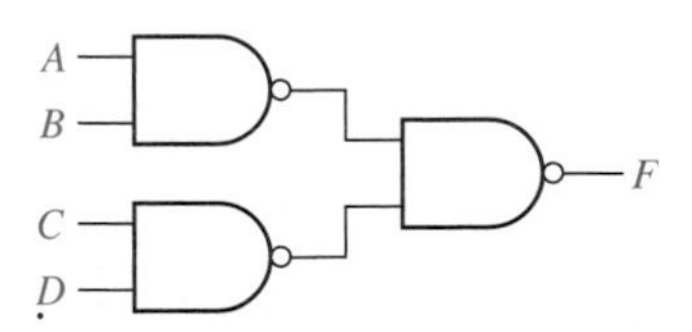


**10.19** In the following circuit, all gates, including the inverter, have an inertial delay of 10 ns.
(a) Write VHDL code that gives a dataflow description of the circuit. All delays should be inertial delays.
(b) Using the Direct VHDL simulator simulate the circuit. (Use a View Interval of 100 ns.) Initially set A = 1, B = 1 and C = 1, then run the simulator for 40 ns. Change B to 0, and run the simulator for 40 ns. Record the waveform.
(c) Change the VHDL code of Part (a) so that the inverter has a delay of 5 ns.
(d) Repeat Part (b).
(e) Change the VHDL code of Part (c) so that the output OR gate has a transport delay rather than an inertial delay.
(f) Repeat Part (b)
(g) Explain any differences between the waveforms for Parts (b), (d), and (f).

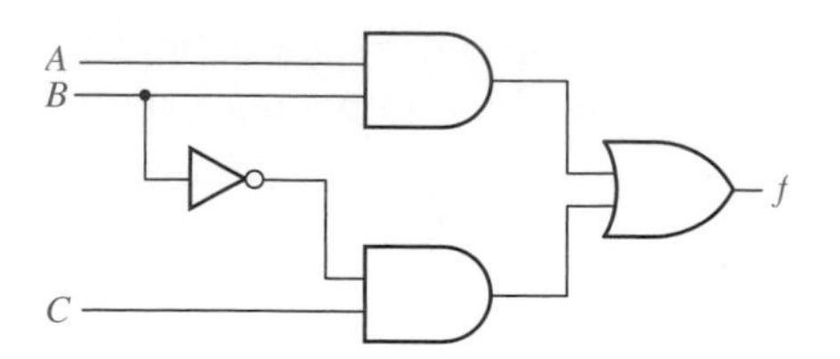


**10.20** In the following circuit, all gates, including the inverter, have an inertial delay of 10 ns except for gate 3, which has delay 40 ns.

(a) Write VHDL code that gives a dataflow description of the circuit. All delays should be inertial delays.

(b) Using the Direct VHDL simulator simulate the circuit. (Use a View Interval of 150 ns.) Initially set A = 1, B = 1, C = 1 and D = 0, then run the simulator for 60 ns. Change B to 0, and run the simulator for 60 ns. Record the waveform.

(c) Change the VHDL code of Part (a) so that the inverter has a delay of 5 ns.

(d) Repeat Part (b).

(e) Change the VHDL code of Part (c) so that gates 4 and 5 have a transport delay rather than an inertial delay.

(f) Repeat Part (b)

(g) Explain any differences between the waveforms for Parts (b), (d), and (f).

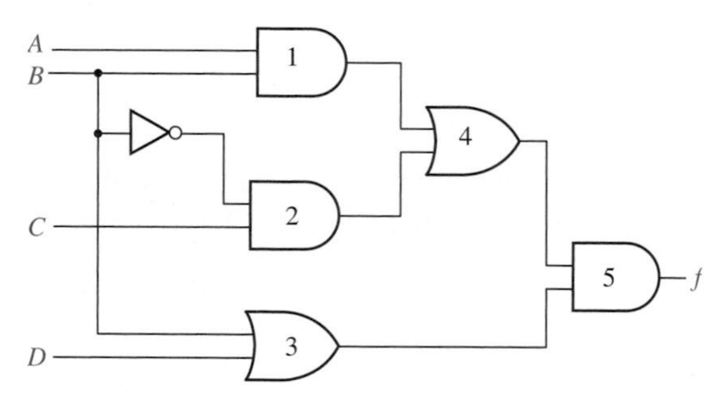


**10.21** Write VHDL code that gives a behavioral description of a circuit that converts the representation of decimal digits in BCD to the representation using the 2-4-2-1 weighted code, as follows:

| Digit | 2421 code |
|---|---|
| 0 | 0000 |
| 1 | 0001 |
| 2 | 0010 |
| 3 | 0011 |
| 4 | 0100 |
| 5 | 1011 |
| 6 | 1100 |
| 7 | 1101 |
| 8 | 1110 |
| 9 | 1111 |

For the six input combinations that do not represent valid BCD digits, the circuit output should be "XXXX". Make the inputs and outputs of type std_logic.
(a) Write the code using the **when else** assignment statement.
(b) Use the VHDL simulator to verify the code of Part (a) for the inputs $x = 0100$, 0101, 1001, and 1010.
(c) Write the code using the **with select when** assignment statement.
(d) Use the VHDL simulator to verify the code of Part (c) for the inputs $x = 0100$, 0101, 1001, and 1010.

**10.22** Write VHDL code that gives a behavioral description of a circuit that converts the representation of decimal digits in the weighted code with weights 8, 4, −2 and −1 to the representation using the excess-3 code.
(a) Write the code using the **when else** assignment statement.
(b) Use the VHDL simulator to verify the code of Part (a) for the inputs $x = 0011$, 0100, 1001, and 1010.
(c) Write the code using the **with select when** assignment statement.
(d) Use the VHDL simulator to verify the code of Part (c) for the inputs $x = 0100$, 0101, 1001, and 1010.

# Design Problems

**10.A** (a) Design a 4-to-1 MUX using only three 2-to-1 MUXes. Write an entity-architecture pair to implement a 2-to-1 MUX. Then write an entity-architecture pair to implement a 4-to-1 MUX using three instances of your 2-to-1 MUX.
[*Hint*: The equation for a 4-to-1 MUX can be rewritten as
$F = A' \, (I_0B' + I_1B) + A \, (I_2B' + I_3B)$].
Use the following port definitions:
For the 2-to-1 MUX:

```
port (i0, i1: in bit; sel: in bit; z: out bit);
```

For the 4-to-1 MUX:

```
port (i0, i1, i2, i3: in bit; a, b: in bit; f: out bit);
```

(b) Simulate your code and test it using the following inputs:
$I0 = I2 = 1, I1 = I3 = 0, AB = 00, 01, 11, 10$

**10.B** (a) Show how a BCD to Gray code converter can be designed using a 16 words × 4 bits ROM. Then write an entity-architecture pair to implement the converter using the ROM. For your code to function correctly, you will need to add the following two lines of code to the top of your program.

```
library BITLIB;
use BITLIB.bit_pack.all;
```

Use the port definition specified below for the ROM:

```
port (bcd: in bit_vector (3 downto 0);
   gray: out bit_vector (3 downto 0));
```

(b) Simulate your code and test it using the following inputs:
BCD = 0010, 0101, 1001

**10.C** (a) A half adder is a circuit that can add two bits at a time to produce a sum and a carry. Design a half adder using only two gates. Write an entity-architecture pair to implement the half adder. Now write an entity-architecture pair to implement a full adder using two instances of your half adder and an OR gate. Use the port definitions specified below:
For the half adder: **port** (a, b: **in bit;** s, c: **out bit**);
For the full adder: **port** (a, b, cin: **in bit;** sum, cout: **out bit**);
(b) Simulate your code and test it using the following inputs:
a b cin = 0 0 1, 0 1 1, 1 1 1, 1 1 0, 1 0 0

**10.D** (a) Using a 3-to-8 decoder and two four-input OR gates, design a circuit that has three inputs and a 2-bit output. The output of the circuit represents (in binary form) the number of 1's present in the input. For example, when the input is $ABC = 101$, the output will be Count = 10. Write an entity-architecture pair to implement a 3-to-8 decoder. Then write an entity-architecture pair for your circuit, using the decoder as a component. Use the port definitions specified below.
For the 3-to-8 decoder:

```
port (a, b, c: in bit;
    y0, y1, y2, y3, y4, y5, y6, y7: out bit);
```

For the main circuit: **port** (a, b, c: **in bit**; count: **out** bit_vector (1 **downto** 0));
(b) Simulate your code and test it using the following inputs:
a b c = 0 0 0, 0 1 0, 1 1 0, 1 1 1, 0 1 1

**10.E** (a) Show how a BCD to seven-segment LED code converter can be designed, using a 16 words × 7 bits ROM. Then write an entity-architecture pair to implement the converter using the ROM. Use the vec2int function in BITLIB for this problem. Use the port definition specified below for the ROM:

```
port (bcd: in bit_vector (3 downto 0);
    seven: out bit_vector (6 downto 0));
```

(b) Simulate your code and test it using the following inputs:
BCD = 0000, 0001, 1000, 1001

**10.F** (a) Using a 3-to-8 decoder, two three-input OR gates, and one two-input OR gate, design a circuit that has three inputs and a 1-bit output. The output of the circuit is 1 when the input 3-bit number is less than 3 or is greater than 4. Write an entity-architecture pair to implement a 3-to-8 decoder. Then write an entity-architecture pair for your circuit using the decoder as a component. Use the port definitions specified below.
For the 3-to-8 decoder:

```
port (a, b, c: in bit;
    y0, y1, y2, y3, y4, y5, y6, y7: out bit);
```

For the main circuit:

```
port (a, b, c: in bit; output : out bit);
```

(b) Simulate your code and test it using the following inputs:
a b c = 0 0 0, 1 0 0, 1 0 1, 0 0 1, 0 1 1

**10.G** (a) Write the VHDL code for a full subtracter, using logic equations. Assume that the full subtracter has a 5-ns delay.
(b) Write the VHDL code for a 4-bit subtracter using the module defined in (a) as a component.
(c) Simulate your code and test it using the following inputs:
1100 – 0101, 0110 – 1011

**10.H** (a) The diagram shows an 8-bit shifter that shifts its input one place to the left. Write a VHDL module for the shifter.

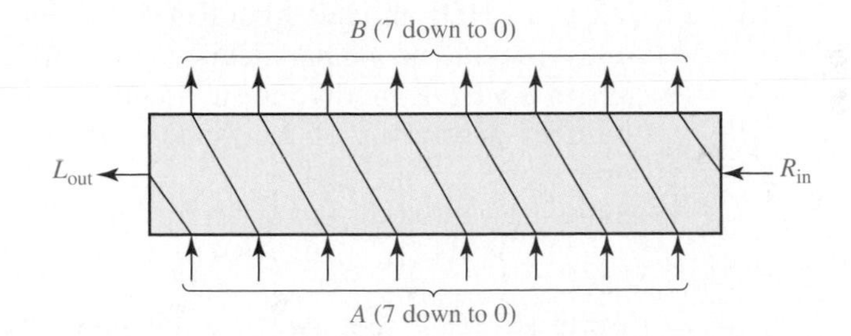


(b) Write a VHDL module that multiplies an 8-bit input (C) by $101_2$ to give a 11-bit product (D). This can be accomplished by shifting C two places to the left and adding the result to C. Use two of the modules written in (a) as components and an overloaded operator for addition.
(c) Simulate your code and test it using the following inputs:
10100101 11111111

**10.I** (a) Design a 4-to-2 priority encoder using gates [see Unit 9, Study Guide, Part 4(b)]. Write a VHDL module for your encoder. Use the port declaration
**Port** ( y : **in** std_logic_vector(0 **to** 3);
a1,b1,c1: **out** std_logic);
(b) Design an 8-to-3 priority encoder (Figure 9-16), using two instances of the 4-to-2 priority encoder you designed, two 2-to-1 multiplexers, and one OR gate. Write a VHDL module for the 8-to-3 encoder. Use the port declaration
**Port** ( y : **in** std_logic_vector(0 **to** 7);
a,b,c,d : **out** std_logic);
(*Hint*: In building the 8-to-3 encoder, use one 4-to-2 encoder for the four most significant bits, and the other for the four least significant bits. Outputs b and c of the 8-to-3 encoder should come from the multiplexers.)
(c) Simulate your code and test it using the following inputs:
00000000, 10000000, 11000000, ---, 11111111

**10.J** (a) Write a VHDL module for a 4-bit adder, with a carry-in and carry-out, using an overloaded addition operator and std_logic_vector inputs and outputs.

(b) Design an 8-bit subtracter with a borrow-out, using two of the 4-bit adders you designed in (a), along with any necessary gates or inverters. Write a VHDL module for the subtracter.

(c) Simulate your code and test it using the following inputs:
11011011 – 01110110, 01110110 – 11011011

**10.K** (a) Write a VHDL module for a tri-state buffer, with 6-bit data inputs and outputs and one control input.

(b) Design a 4-to-1 multiplexer with 6-bit data inputs and outputs and two control inputs. Use four tri-state buffers from part (a) and a 2-to-4 decoder.

(c) Simulate your code and test it for the following data inputs:
000111, 101010, 111000, 010101

**10.L** (a) Write a VHDL module for a ROM with four inputs and three outputs. The 3-bit output should be a binary number equal to the number of 1's in the ROM input.

(b) Write a VHDL module for a circuit that counts the number of 1's in a 12-bit number. Use three of the modules from (a) along with overloaded addition operators.

(c) Simulate your code and test it for the following data inputs:
111111111111, 010110101101, 100001011100

**10.M** (a) Write a VHDL module for a full subtracter using a ROM to implement the truth table.

(b) Write a VHDL module for a 3-bit subtracter using the module defined in part (a). Your module should have a borrow-in and a borrow-out.

(c) Simulate your code and test it for the following data:
110 – 010 with a borrow input of 1
011 – 101 with a borrow input of 0

**10.N** (a) Design a 4-to-2 priority encoder with an enable input, using gates. (See Unit 9, Study Guide Part 4(b)). When enable is 0, all outputs are 0. Write a VHDL module for the encoder. Use the following port declaration:

```
Port ( y : in std_logic_vector(0 to 3);
    enable : in std_logic; a1,b1,c1 : out std_logic);
```

(b) Design an 8-to-3 priority encoder (Figure 9-16) with an enable input, using two of the 4-to-2 priority encoders you designed in (a), three OR gates, an AND gate, and one inverter. Then write a VHDL module for this encoder. Use the port declaration:

```
Port ( y : in std_logic_vector(0 to 7);
    main_enable : in std_logic; a,b,c,d : out std_logic);
```

(*Hint*: In building the 8-to-3 encoder, use one 4-to-2 encoder for the four most significant bits, and another for the four least significant bits. Also, outputs b and c of the 8-to-3 encoder should come from OR gates. The enable input to the encoder for the least significant bits depends on the main_enable signal and the c1 output from the encoder for the most significant bits.)

(c) Simulate your code and test it using the following inputs:
00000000, 10000000, 11000000, ---, 11111111

# Latches and Flip-Flops

## Objectives

In this unit you will study one of the basic building blocks used in sequential circuits—the flip-flop. Some of the basic analysis techniques used for sequential circuits are introduced here. In particular, you will learn how to construct timing diagrams which show how each signal in the circuit varies as a function of time. Specific objectives are:

1. Explain in words the operation of S-R and gated D latches.
2. Explain in words the operation of D, D-CE, S-R, J-K, and T flip-flops.
3. Make a table and derive the characteristic (next-state) equation for such latches and flip-flops. State any necessary restrictions on the input signals.
4. Draw a timing diagram relating the input and output of such latches and flip-flops.
5. Show how latches and flip-flops can be constructed using gates. Analyze the operation of a flip-flop that is constructed of gates and latches.

# Study Guide

**1.** Review Section 8.3, *Gate Delays and Timing Diagrams*. Then study Section 11.1, *Introduction.*

(a) In the circuit shown, suppose that at some instant of time the inputs to both inverters are 0. Is this a stable condition of the circuit?

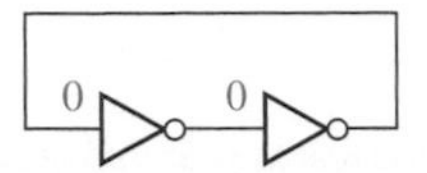


Assuming that the output of the left inverter changes before the output of the right inverter, what stable state will the circuit reach? (Indicate 0's and 1's on the inverters' inputs and outputs.)

(b) Work Problem 11.1.

**2.** Study Section 11.2, *Set-Reset Latch.*

(a) Build an S-R latch in *SimUaid*, using NOR gates as in Figure 11-3. Place switches on the inputs and probes on the outputs. Experiment with it. Describe in words the behavior of your S-R latch.

(b) For Figure 11-4(b), what values would $P$ and $Q$ assume if $S = R = 1$?

(c) What restriction is necessary on $S$ and $R$ so that the two outputs of the S-R latch are complements?

(d) State in words the meaning of the equation $Q^+ = S + R'Q$.

(e) Starting with $Q = 0$ and $\overline{S} = \overline{R} = 1$ in Figure 11-10(a), change $\overline{S}$ to 0 and trace signals through the latch until steady-state is reached. Then, change $\overline{S}$ to 1 and $\overline{R}$ to 0 and trace again.

(f) Work Problems 11.2 and 11.3.

**3.** Study Section 11.3, *Gated D Latch.*

(a) Build a gated $D$ latch in *SimUaid*. See Figure 11-11. (Construct the S-R latch as in Study Guide Section 2(a).) Place switches on the inputs and probes on the outputs. Experiment with it. Describe in words the behavior of your gated D latch.

(b) State in words the meaning of the equation $Q^+ = G'Q + GD$.

(c) Given a gated *D* latch with the following inputs, sketch the waveform for *Q*.

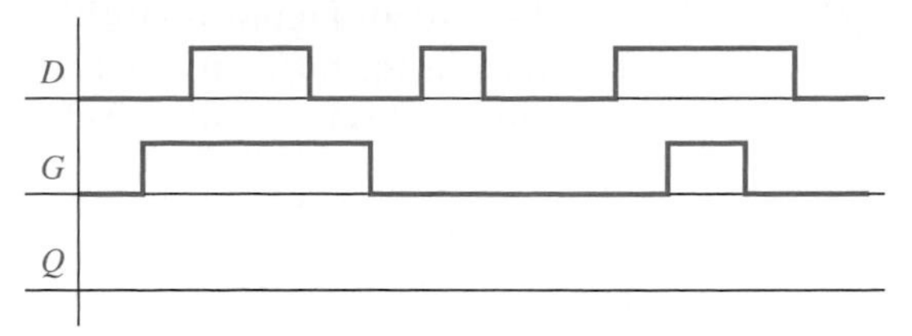


(d) Work Problem 11.4.

**4.** Study Section 11.4, *Edge-Triggered D Flip-Flop*.

(a) Experiment with a D flip-flop in *SimUaid*. Use the D flip-flop on the parts menu. Place switches on the inputs and probes on the outputs. Describe in words the behavior of your D flip-flop.

(b) Given a rising-edge-triggered D flip-flop with the following inputs, sketch the waveform for *Q*.

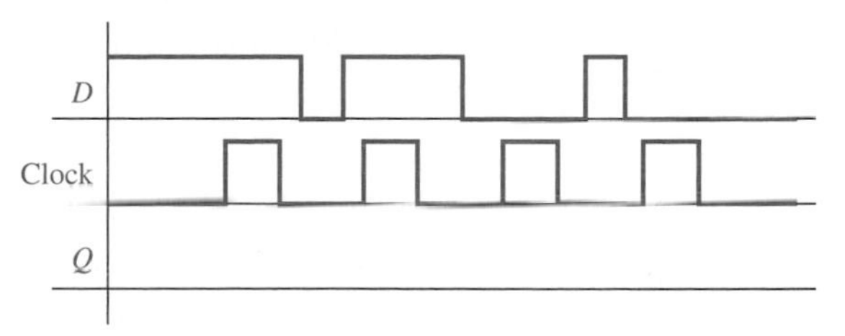


(c) Work Programmed Exercise 11.29.

(d) A D flip-flop with a falling-edge trigger is behaving erratically. It has a setup time of 2 ns and a hold time of 2 ns. The figure shows the inputs to the flip-flop over a typical clock cycle. Why might the flip-flop be behaving erratically?

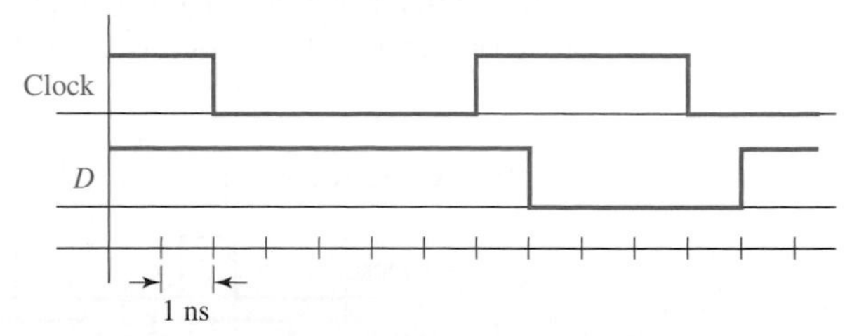


(e) Suppose that for the circuit of Figure 11-17, new semiconductor technology has allowed us to improve the delays and setup times. The propagation delay of the new inverter is 1.5 ns, and the propagation delay and setup times of the new flip-flop are 3.5 ns and 2 ns, respectively. What is the shortest clock period for the circuit of Figure 11-17(a) which will not violate the timing constraints?

(f) Work Problem 11.5.

**5.** Study Section 11.5, *S-R Flip-Flop*.

(a) Describe in words the behavior of an S-R flip-flop.

(b) Trace signals through the circuit of Figure 11-19(a) and verify the timing diagram of Figure 11-19(b).

(c) What is the difference between a master-slave flip-flop and an edge-triggered flip-flop? Assume that $Q$ changes on the rising clock edge in both cases.

(d) Work Problem 11.6.

**6.** Study Section 11.6, *J-K Flip-Flop.*

(a) Experiment with a *J-K* flip-flop in SimUaid. Use the *J-K* flip-flop in the parts menu. Place switches on the inputs and probes on the outputs. Describe in words the behavior of your *J-K* flip-flop.

(b) Derive the next-state equation for the J-K flip-flop.

(c) Examine Figures 11-19(a) and 11-21. Construct a J-K flip-flop, using a master-slave S-R flip-flop and two AND gates. (Do not draw the interior of the S-R flip-flop. Just use the symbol in Figure 11-18.)

(d) Work Problem 11.7.

**7.** Study Section 11.7, *T Flip-Flop.*

(a) Construct a T flip-flop in *SimUaid* from a D flip-flop as in Figure 11-24(b). Place switches on the inputs and probes on the outputs. Experiment with it. Describe in words the behavior of the T flip-flop.

(b) Complete the following timing diagram (assume that $Q = 0$ initially):

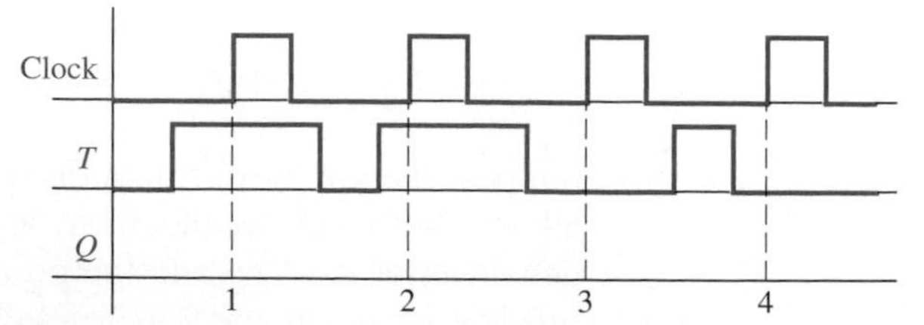


**8.** Study Section 11.8, *Flip-Flops with Additional Inputs.*

(a) To set the flip-flop of Figure 11-25 to $Q = 1$ without using the clock, the ClrN input should be set to ________ and the PreN input to ________. To reset this flip-flop to $Q = 0$ without using the clock, the ________ input should be set to ________ and the ________ input to ________.

(b) Complete the following timing diagram for a rising-edge-triggered D flip-flop with ClrN and PreN inputs. Assume $Q$ begins at 0.

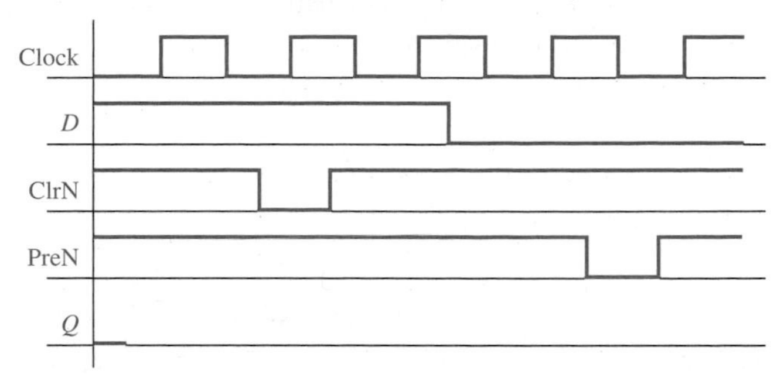


(c) In Figure 11-27(a), what would happen if En changed from 1 to 0 while CLK = 1?
What if En changed when CLK = 0?
In order to have $Q$ change synchronization with the clock, what restriction must be placed on the time at which En can change?

Why does this restriction not apply to Figures 11-27(b) and (c)?

(d) Make a table similar to Figure 11-25(b) that describes the operation of a D flip-flop with a falling-edge clock input, a clock enable input, and an asynchronous active-low clear input (ClrN), but no preset input.

(e) Work Problems 11.8 and 11.9.

**9.** Study Section 11.9, *Summary.*

(a) Given one of the flip-flops in this chapter or a similar flip-flop, you should be able to derive the characteristic equation which gives the next state of the flip-flop in terms of the present state and inputs. You should understand the meaning of each of the characteristic equations given in Section 11.9.

(b) An S-R flip-flop can be converted to a T flip-flop by adding gates at the $S$ and $R$ inputs. The $S$ and $R$ inputs must be chosen so that the flip-flop will change state whenever $T = 1$ and the clock is pulsed. In order to determine the $S$ and $R$ inputs, ask yourself the question, "Under what conditions must the flip-flop be set to 1, and under what conditions must it be reset?" The flip-flop must be set to 1 if $Q = 0$ and $T = 1$.

Therefore, $S =$ __________ . In a similar manner, determine the equation for $R$ and draw the circuit which converts an S-R flip-flop to a T flip-flop.

(c) Work Problem 11.10.

10. When you are satisfied that you can meet the objectives of this unit, take the readiness test.

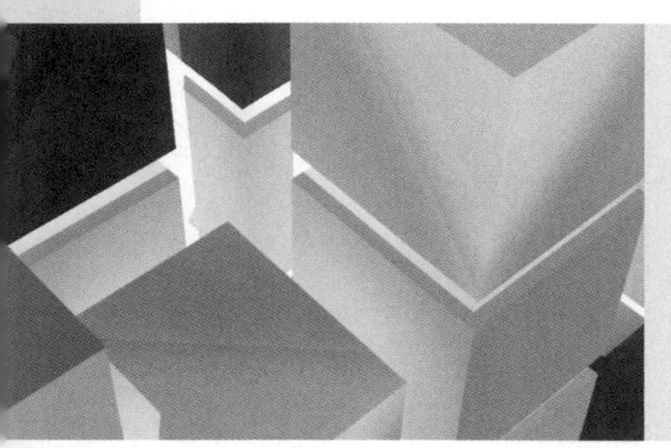

# Latches and Flip-Flops

## 11.1 Introduction

Sequential switching circuits have the property that the output depends not only on the present input but also on the past sequence of inputs. In effect, these circuits must be able to "remember" something about the past history of the inputs in order to produce the present output. Latches and flip-flops are commonly used memory devices in sequential circuits. Basically, latches and flip-flops are memory devices which can assume one of two stable output states and which have one or more inputs that can cause the output state to change. Several common types of latches and flip-flops are described in this unit.

In Units 12 through 16, we will discuss the analysis and design of synchronous digital systems. In such systems, it is common practice to synchronize the operation of all flip-flops by a common clock or pulse generator. Each of the flip-flops has a clock input, and the flip-flops can only change state in response to a clock pulse. The use of a clock to synchronize the operation of several flip-flops is illustrated in Units 12 and 13. A memory element that has no clock input is often called a latch, and we will follow this practice. We will then reserve the term flip-flop to describe a memory device that changes output state in response to a clock input and not in response to a data input.

The switching circuits that we have studied so far have not had feedback connections. By feedback we mean that the output of one of the gates is connected back into the input of another gate in the circuit so as to form a closed loop. In order to construct a switching circuit that has memory, such as a latch or flip-flop, we must introduce feedback into the circuit. For example, in the NOR-gate circuit of Figure 11-3(a), the output of the second NOR gate is fed back into the input of the first NOR gate.

FIGURE 11-1

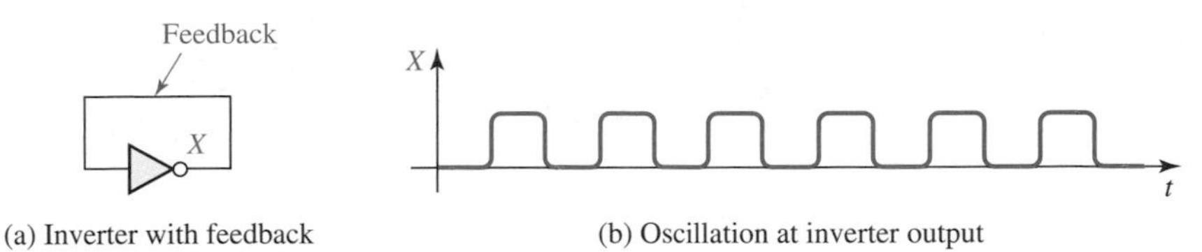


(a) Inverter with feedback (b) Oscillation at inverter output

In simple cases, we can analyze circuits with feedback by tracing signals through the circuit. For example, consider the circuit in Figure 11-1(a). If at some instant of time the inverter input is 0, this 0 will propagate through the inverter and cause the output to become 1 after the inverter delay. This 1 is fed back into the input, so after the propagation delay, the inverter output will become 0. When this 0 feeds back into the input, the output will again switch to 1, and so forth. The inverter output will continue to oscillate back and forth between 0 and 1, as shown in Figure 11-1(b), and it will never reach a stable condition. The rate at which the circuit oscillates is determined by the propagation delay in the inverter.

FIGURE 11-2

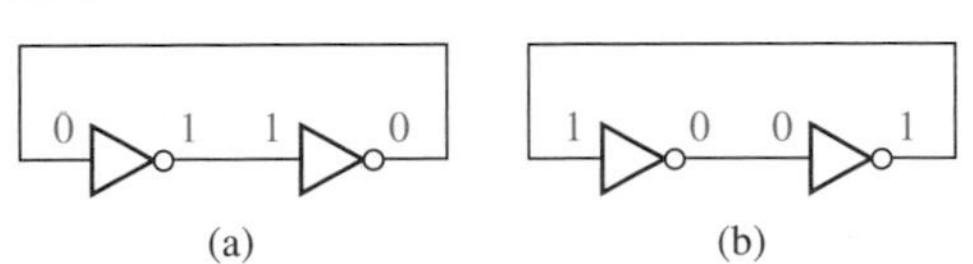


(a) (b)

Next, consider a feedback loop which has two inverters in it, as shown in Figure 11-2(a). In this case, the circuit has two stable conditions, often referred to as stable states. If the input to the first inverter is 0, its output will be 1. Then, the input to the second inverter will be 1, and its output will be 0. This 0 will feed back into the first inverter, but because this input is already 0, no changes will occur. The circuit is then in a stable state. As shown in Figure 11-2(b), a second stable state of the circuit occurs when the input to the first inverter is 1 and the input to the second inverter is 0.

## 11.2 Set-Reset Latch

We can construct a simple latch by introducing feedback into a NOR-gate circuit, as seen in Figure 11-3(a). As indicated, if the inputs are $S = R = 0$, the circuit can assume a stable state with $Q = 0$ and $P = 1$. Note that this is a stable condition of the circuit because $P = 1$ feeds into the second gate forcing the output to be $Q = 0$, and $Q = 0$ feeds into the first gate allowing its output to be 1. Now if we change $S$ to 1, $P$ will become 0. This is an unstable condition or state of the circuit because both the inputs and output of the second gate are 0; therefore $Q$ will change to 1, leading to the stable state shown in Figure 11-3(b).

FIGURE 11-3

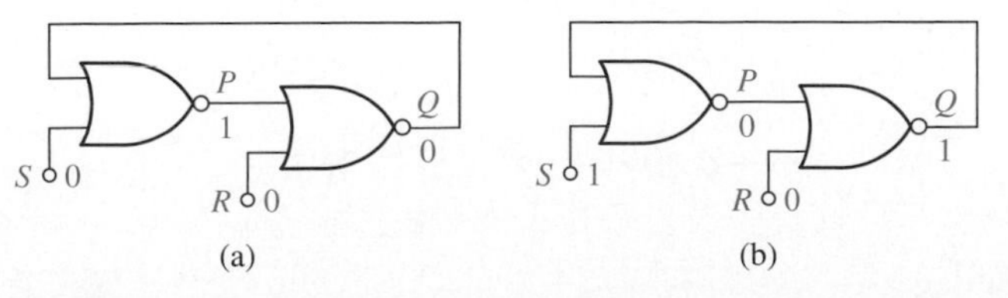


(a) (b)

FIGURE 11-4

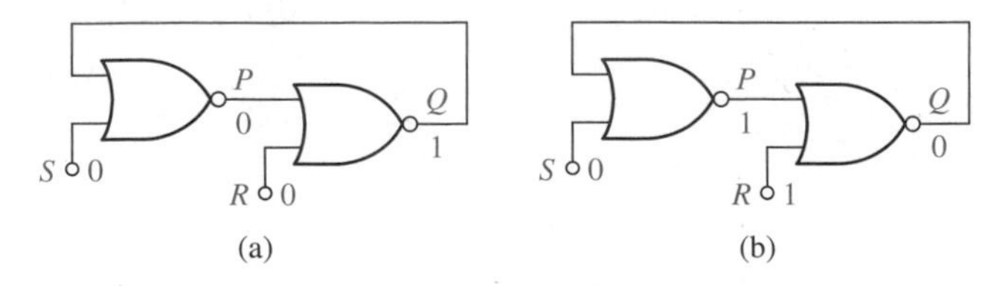


If $S$ is changed back to 0, the circuit will not change state because $Q = 1$ feeds back into the first gate, causing $P$ to remain 0, as shown in Figure 11-4(a). Note that the inputs are again $S = R = 0$, but the outputs are different than those with which we started. Thus, the circuit has two different stable states for a given set of inputs. If we now change $R$ to 1, $Q$ will become 0 and $P$ will then change back to 1, as seen in Figure 11-4(b). If we then change $R$ back to 0, the circuit remains in this state and we are back where we started.

This circuit is said to have memory because its output depends not only on the present inputs, but also on the past sequence of inputs. If we restrict the inputs so that $R = S = 1$ is not allowed, the stable states of the outputs $P$ and $Q$ are always complements, that is, $P = Q'$. To emphasize the symmetry between the operation of the two gates, the circuit is often drawn in *cross-coupled* form [see Figure 11-5(a)]. As shown in Figures 11-3(b) and 11-4(b), an input $S = 1$ *sets* the output to $Q = 1$, and an input $R = 1$ *resets* the output to $Q = 0$. When used with the restriction that $R$ and $S$ cannot be 1 simultaneously, the circuit is commonly referred to as a set-reset *(S-R)* latch and given the symbol shown in Figure 11-5(b). Note that although $Q$ comes out of the NOR gate with the $R$ input, the standard S-R latch symbol has $Q$ directly above the $S$ input.

If $S = R = 1$, the latch will not operate properly, as shown in Figure 11-6. The notation $1 \rightarrow 0$ means that the input is originally 1 and then changes to 0. Note that when $S$ and $R$ are both 1, $P$ and $Q$ are both 0. Therefore, $P$ is not equal to $Q'$, and this violates a basic rule of latch operation that requires the latch outputs to be complements. Furthermore, if $S$ and $R$ are simultaneously changed back to 0, $P$ and $Q$ may both change to 1. If $S = R = 0$ and $P = Q = 1$, then after the 1's propagate

FIGURE 11-5
S-R Latch

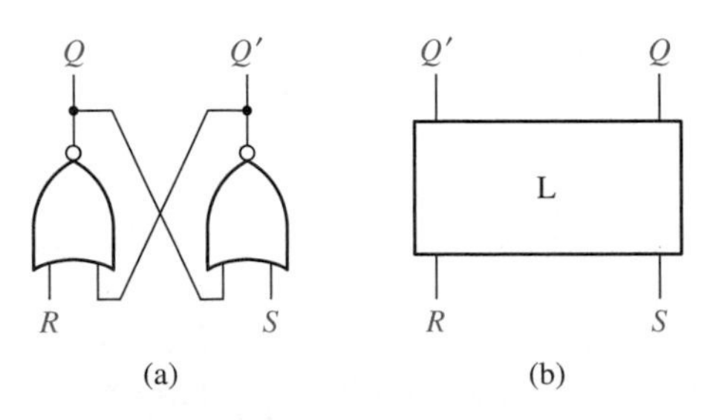


FIGURE 11-6
Improper S-R Latch Operation

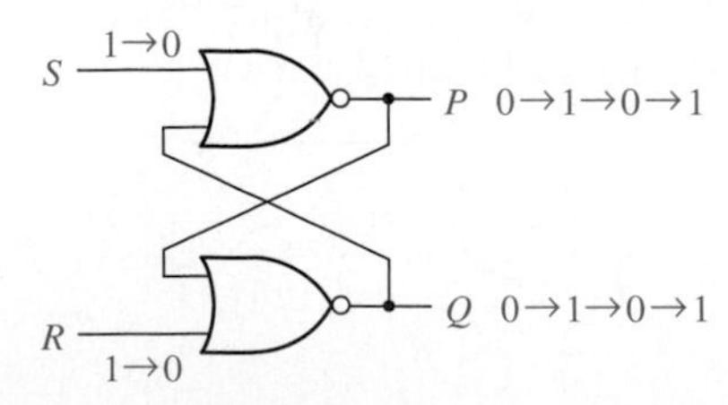

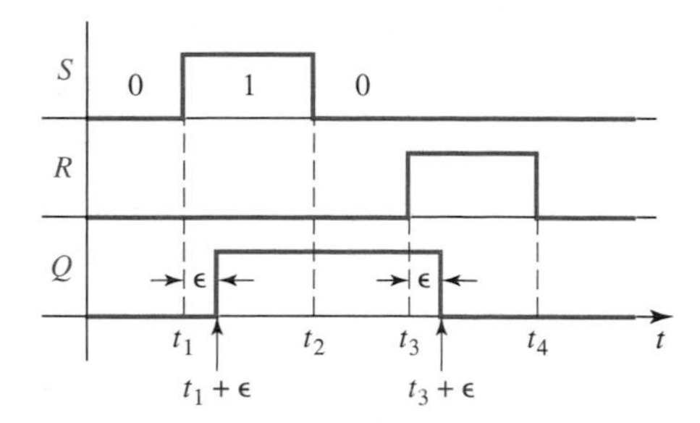


**FIGURE 11-7** Timing Diagram for S-R Latch

through the gates, $P$ and $Q$ will become 0 again, and the latch may continue to oscillate if the gate delays are equal.

Figure 11-7 shows a timing diagram for the S-R latch. Note that when $S$ changes to 1 at time $t_1$, $Q$ changes to 1 a short time ($\epsilon$) later. ($\epsilon$ represents the response time or delay time of the latch.) At time $t_2$, when $S$ changes back to 0, $Q$ does not change. At time $t_3$, $R$ changes to 1, and $Q$ changes back to 0 a short time ($\epsilon$) later. The duration of the $S$ (or $R$) input pulse must normally be at least as great as $\epsilon$ in order for a change in the state of $Q$ to occur. If $S = 1$ for a time less than $\epsilon$, the gate output will not change and the latch will not change state.

When discussing latches and flip-flops, we use the term *present state* to denote the state of the $Q$ output of the latch or flip-flop at the time any input signal changes, and the term *next state* to denote the state of the $Q$ output after the latch or flip-flop has reacted to the input change and stabilized. If we let $Q(t)$ represent the present state and $Q(t + \epsilon)$ represent the next state, an equation for $Q(t + \epsilon)$ can be obtained from the circuit by conceptually breaking the feedback loop at $Q$ and considering $Q(t)$ as an input and $Q(t + \epsilon)$ as the output. Then for the S-R latch of Figure 11-3

$$Q(t + \epsilon) = R(t)'[S(t) + Q(t)] = R(t)'S(t) + R(t)'Q(t) \tag{11-1}$$

and the equation for output $P$ is

$$P(t) = S(t)'Q(t)' \tag{11-2}$$

Normally we write the next-state equation without including time explicitly, using $Q$ to represent the present state of the latch and $Q^+$ to represent the next state:

$$Q^+ = R'S + R'Q \tag{11-3}$$

$$P = S'Q' \tag{11-4}$$

These equations are mapped in the next-state and output tables of Table 11-1. The stable states of the latch are circled. Note that for all stable states, $P = Q'$ except when $S = R = 1$. As discussed previously, this is one of the reasons why $S = R = 1$ is

**TABLE 11-1** S-R Latch Next State and Output

| Present State Q | Next State $Q^+$ SR 00 | SR 01 | SR 11 | SR 10 | Present Output $P$ SR 00 | SR 01 | SR 11 | SR 10 |
|---|---|---|---|---|---|---|---|---|
| 0 | ⓪ | ⓪ | ⓪ | 1 | 1 | 1 | 0 | 0 |
| 1 | ① | 0 | 0 | ① | 0 | 0 | 0 | 0 |

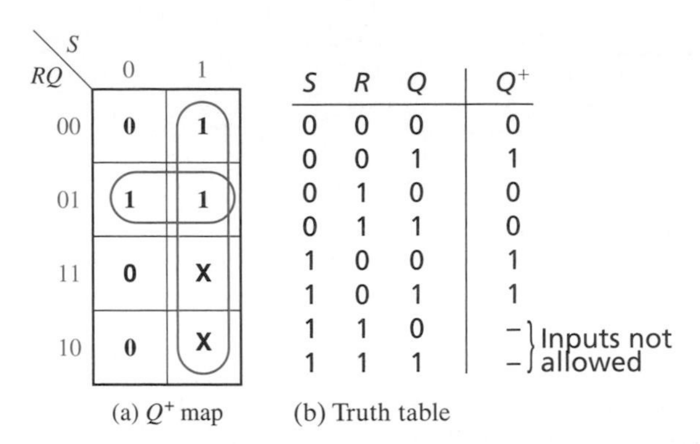


**FIGURE 11-8** Derivation of $Q^+$ for an S-R Latch

disallowed as an input combination to the S-R latch. Making $S = R = 1$ a don't-care combination allows simplifying the next-state equation, as shown in Figure 11-8(a). After plotting Equation (11-3) on the map and changing two entries to don't-cares, the next-state equation simplifies to

$$Q^+ = S + R'Q \qquad (SR = 0) \tag{11-5}$$

In words, this equation tells us that the next state of the latch will be 1 either if it is set to 1 with an $S$ input, or if the present state is 1 and the latch is not reset. The condition $SR = 0$ implies that $S$ and $R$ cannot both be 1 at the same time. An equation that expresses the next state of a latch in terms of its present state and inputs will be referred to as a *next-state equation*, or *characteristic equation*.

Another approach for deriving the characteristic equation for an S-R latch is based on constructing a truth table for the next state of $Q$. We previously discussed the latch operation by tracing signals through the gates, and the truth table in Figure 11-8(b) is based on this discussion. Plotting $Q^+$ on a Karnaugh map gives the same result as Figure 11-8(a).

The S-R latch is often used as a component in more complex latches and flip-flops and in asynchronous systems. Another useful application of the S-R latch is for debouncing switches. When a mechanical switch is opened or closed, the switch contacts tend to vibrate or bounce open and closed several times before settling down to their final position. This produces a noisy transition, and this noise can interfere with the proper operation of a logic circuit. The input to the switch in Figure 11-9 is connected to a logic 1 (+V). The pull-down resistors connected to contacts $a$ and $b$ assure that when the switch is between $a$ and $b$ the latch inputs $S$ and $R$ will always be at a logic 0, and the

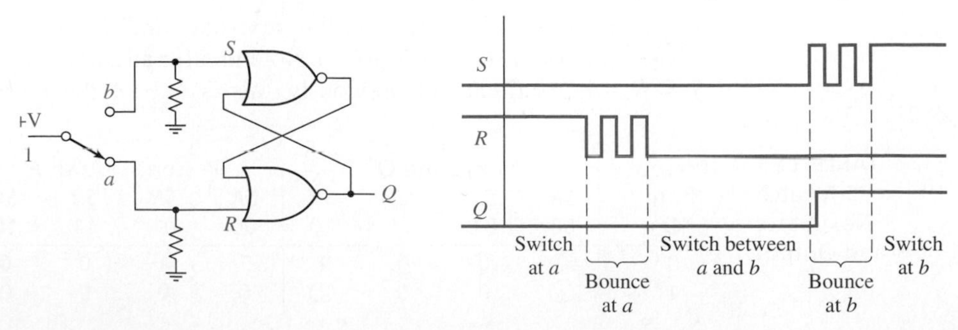


**FIGURE 11-9** Switch Debouncing with an S-R Latch

latch output will not change state. The timing diagram shows what happens when the switch is flipped from $a$ to $b$. As the switch leaves $a$, bounces occur at the $R$ input; when the switch reaches $b$, bounces occur at the $S$ input. After the switch reaches $b$, the first time $S$ becomes 1, after a short delay the latch switches to the $Q = 1$ state and remains there. Thus $Q$ is free of all bounces even though the switch contacts bounce. This debouncing scheme requires a *double throw* switch that switches between two contacts; it will not work with a *single throw* switch that switches between one contact and open.

An alternative form of the S-R latch uses NAND gates, as shown in Figure 11-10. We will refer to this circuit as an $\overline{\text{S}}$-$\overline{\text{R}}$ latch, and the table describes its operation. We have labeled the inputs to this latch $\overline{S}$ and $\overline{R}$ because $\overline{S} = 0$ will set $Q$ to 1 and $\overline{R} = 0$ will reset $Q$ to 0. If $\overline{S}$ and $\overline{R}$ are 0 at the same time, both the $Q$ and $Q'$ outputs are forced to 1. Therefore, for the proper operation of this latch, the condition $\overline{S} = \overline{R} = 0$ is not allowed.

**FIGURE 11-10** $\overline{\text{S}}$-$\overline{\text{R}}$ Latch

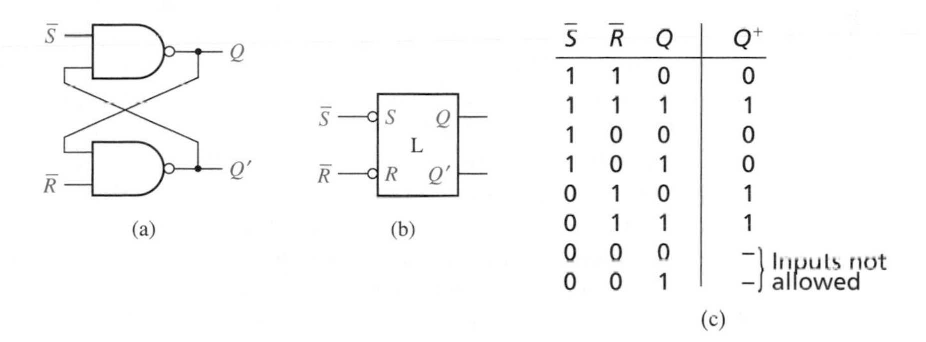


| $\overline{S}$ | $\overline{R}$ | $Q$ | $Q^+$ |
|---|---|---|---|
| 1 | 1 | 0 | 0 |
| 1 | 1 | 1 | 1 |
| 1 | 0 | 0 | 0 |
| 1 | 0 | 1 | 0 |
| 0 | 1 | 0 | 1 |
| 0 | 1 | 1 | 1 |
| 0 | 0 | 0 | – } Inputs not allowed |
| 0 | 0 | 1 | – |

## 11.3 Gated D Latch

A gated D latch (Figure 11-11) has two inputs—a data input ($D$) and a gate input ($G$). The D latch can be constructed from an S-R latch and gates (Figure 11-11(a)). When $G = 0$, $S = R = 0$, so $Q$ does not change. When $G = 1$ and $D = 1$, $S = 1$ and $R = 0$, so $Q$ is set to 1. When $G = 1$ and $D = 0$, $S = 0$ and $R = 1$, so $Q$ is reset to 0. In other words, when $G = 1$, the $Q$ output follows the $D$ input, and when $G = 0$, the $Q$ output holds the last value of $D$ (no state change). This type of latch is also referred to as a transparent latch because when $G = 1$, the $Q$ output is the same as the $D$ input. From the truth table (Figure 11-12), the characteristic equation for the latch is $Q^+ = G'Q + GD$.

**FIGURE 11-11** Gated D Latch

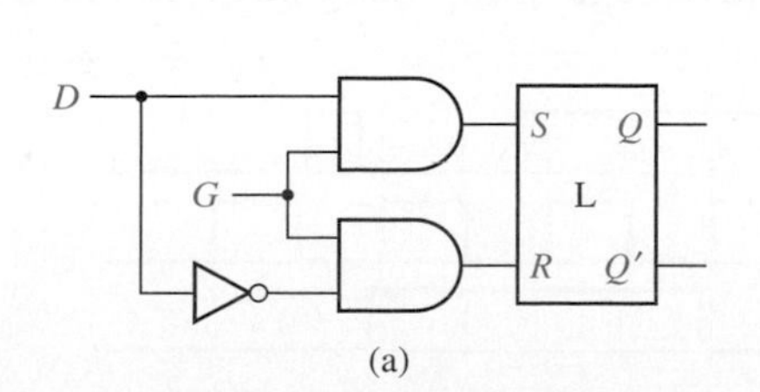


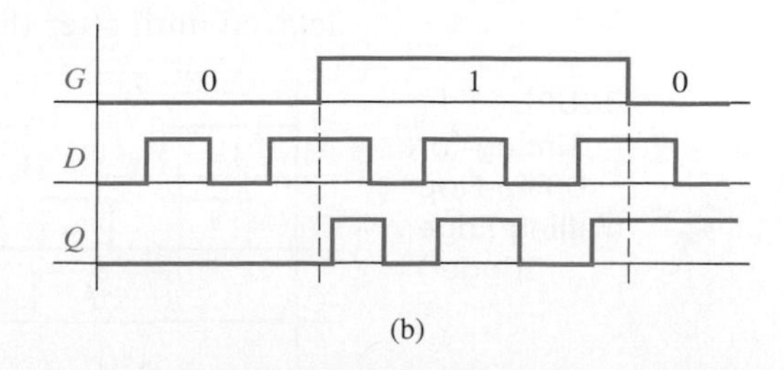

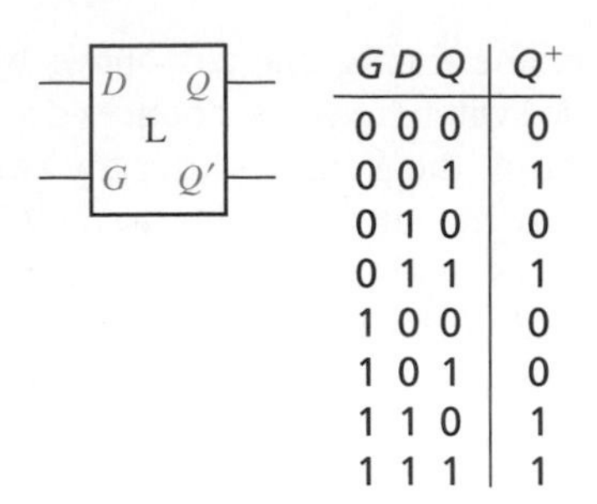


| G D Q | $Q^+$ |
|---|---|
| 0 0 0 | 0 |
| 0 0 1 | 1 |
| 0 1 0 | 0 |
| 0 1 1 | 1 |
| 1 0 0 | 0 |
| 1 0 1 | 0 |
| 1 1 0 | 1 |
| 1 1 1 | 1 |

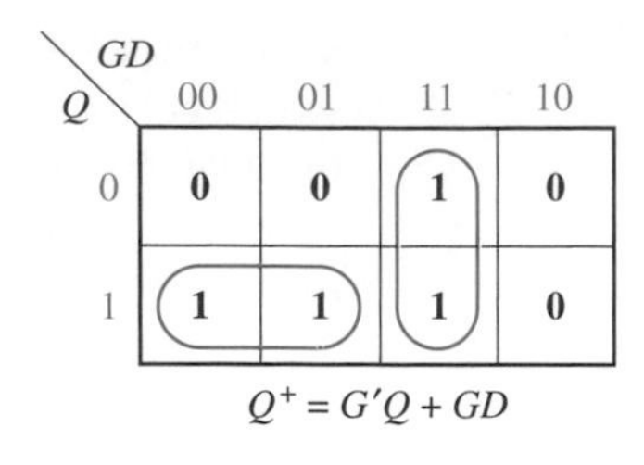


**FIGURE 11-12** Symbol and Truth Table for Gated Latch

## 11.4 Edge-Triggered D Flip-Flop

A D flip-flop (Figure 11-13) has two inputs, $D$ (data) and Ck (clock). The small arrowhead on the flip-flop symbol identifies the clock input. Unlike the $D$ latch, the flip-flop output changes only in response to the clock, not to a change in $D$. If the output can change in response to a 0 to 1 transition on the clock input, we say that the flip-flop is triggered on the *rising edge* (or positive edge) of the clock. If the output can change in response to a 1 to 0 transition on the clock input, we say that the flip-flop is triggered on the *falling edge* (or negative edge) of the clock. An inversion bubble on the clock input indicates a *falling-edge trigger* (Figure 11-13(b)), and no bubble indicates a *rising-edge trigger* [Figure 11-13(a)]. The term *active edge* refers to the clock edge (rising or falling) that triggers the flip-flop state change.

**FIGURE 11-13** D Flip-Flops

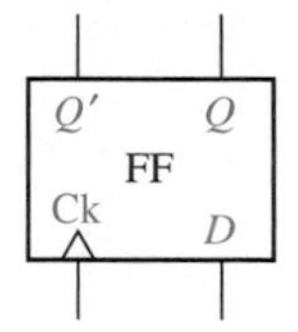


(a) Rising-edge trigger

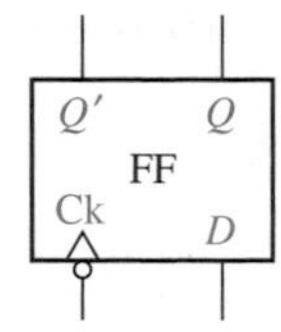


(b) Falling-edge trigger

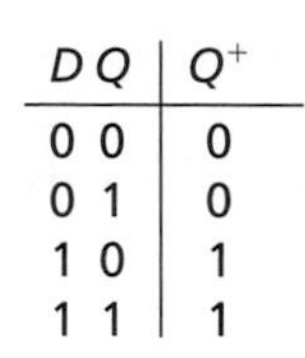

| D Q | $Q^+$ |
|---|---|
| 0 0 | 0 |
| 0 1 | 0 |
| 1 0 | 1 |
| 1 1 | 1 |

$Q^+ = D$

(c) Truth table

The state of a D flip-flop after the active clock edge ($Q^+$) is equal to the input ($D$) before the active edge. For example, if $D = 1$ before the clock pulse, $Q = 1$ after the active edge, regardless of the previous value of $Q$. Therefore, the characteristic equation is $Q^+ = D$. If $D$ changes at most once following each clock pulse, the output of the flip-flop is the same as the $D$ input, except that the output changes are delayed until after the active edge of the clock pulse, as illustrated in Figure 11-14.

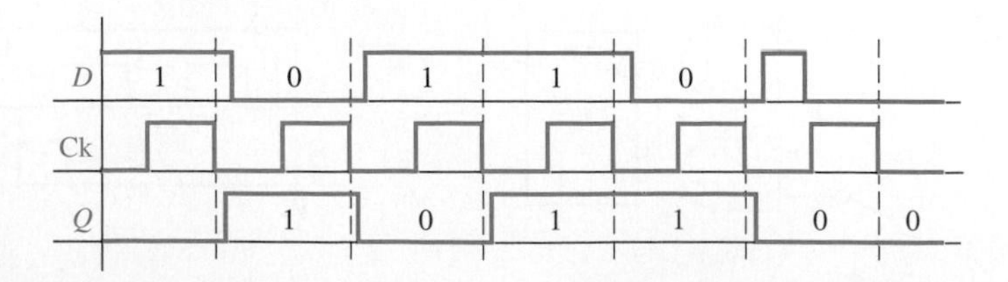


**FIGURE 11-14** Timing for D Flip-Flop (Falling-Edge Trigger)

**FIGURE 11-15**
D Flip-Flop (Rising-Edge Trigger)

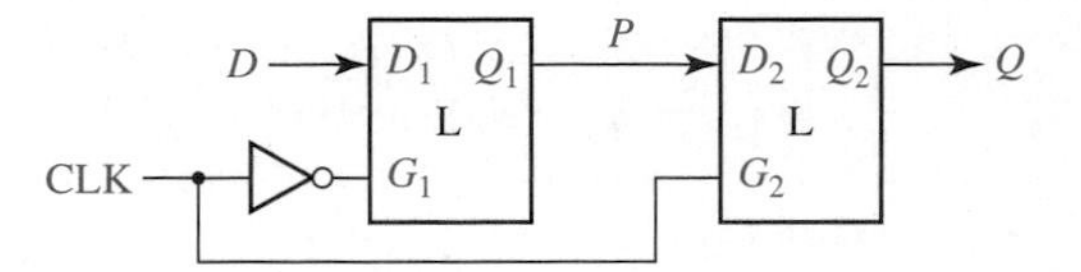


(a) Construction from two gated D latches

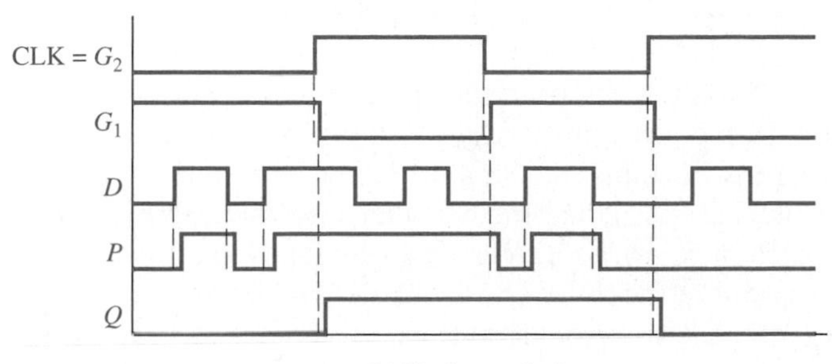


(b) Timing analysis

A rising-edge-triggered $D$ flip-flop can be constructed from two gated $D$ latches and an inverter, as shown in Figure 11-15(a). The timing diagram is shown in Figure 11-15(b). When CLK $= 0$, $G_1 = 1$, and the first latch is *transparent* so that the $P$ output follows the $D$ input. Because $G_2 = 0$, the second latch holds the current value of $Q$. When CLK changes to 1, $G_1$ changes to 0, and the current value of $D$ is stored in the first latch. Because $G_2 = 1$, the value of $P$ flows through the second latch to the $Q$ output. When CLK changes back to 0, the second latch takes on the value of $P$ and holds it and, then, the first latch starts following the $D$ input again. If the first latch starts following the $D$ input before the second latch takes on the value of $P$, the flip-flop will not function properly. Therefore, the circuit designers must pay careful attention to timing issues when designing edge-triggered flip-flops. With this circuit, output state changes occur only following the rising edge of the clock. The value of $D$ at the time of the rising edge of the clock determines the value of $Q$, and any extra changes in $D$ that occur between rising clock edges have no effect on $Q$.

Because a flip-flop changes state only on the active edge of the clock, the propagation delay of a flip-flop is the time between the active edge of the clock and the resulting change in the output. However, there are also timing issues associated with the $D$ input. To function properly, the $D$ input to an edge-triggered flip-flop must be held at a constant value for a period of time before and after the active edge of the clock. If $D$ changes at the same time as the active edge, the behavior is unpredictable. The amount of time that $D$ must be stable before the active edge is called the setup time ($t_{su}$), and the amount of time that $D$ must hold the same value after the active edge is the hold time ($t_h$). The times at which $D$ is allowed to change during the clock cycle are shaded in the timing diagram of Figure 11-16. The propagation delay ($t_p$) from the time the clock changes until the $Q$ output changes is also indicated. For Figure 11-15(a), the setup time allows a change in $D$ to propagate through the first latch before the rising edge of Clock. The hold time is required so that $D$ gets stored in the first latch before $D$ changes.

Using these timing parameters, we can determine the minimum clock period for a circuit which will not violate the timing constraints. Consider the circuit of

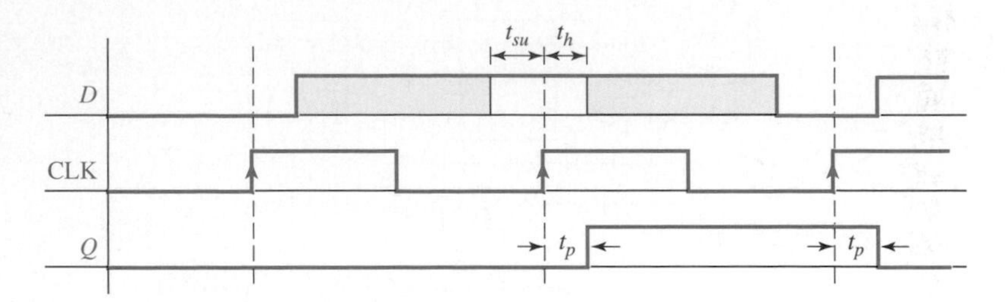


**FIGURE 11-16** Setup and Hold Times for an Edge-Triggered D Flip-Flop

Figure 11-17(a). Suppose the inverter has a propagation delay of 2 ns, and suppose the flip-flop has a propagation delay of 5 ns and a setup time of 3 ns. (The hold time does not affect this calculation.) Suppose, as in Figure 11-17(b), that the clock period is 9 ns, i.e., 9 ns is the time between successive active edges (rising edges for this figure). Then, 5 ns after a clock edge, the flip-flop output will change, and 2 ns after that, the output of the inverter will change. Therefore, the input to the flip-flop will change 7 ns after the rising edge, which is 2 ns before the next rising edge. But the setup time of the flip-flop requires that the input be stable 3 ns before the rising edge; therefore, the flip-flop may not take on the correct value.

Suppose instead that the clock period were 15 ns, as in Figure 11-17(c). Again, the input to the flip-flop will change 7 ns after the rising edge. However, because the clock is slower, this is 8 ns before the next rising edge. Therefore, the flip-flop will work properly. Note in Figure 11-17(c) that there is 5 ns of extra time between the time the *D* input is correct and the time when it must be correct for the setup time to be satisfied. Therefore, we can use a shorter clock period, and have less extra time, or no extra time. Figure 11-17(d) shows that 10 ns is the minimum clock period which will work for this circuit.

**FIGURE 11-17** Determination of Minimum Clock Period

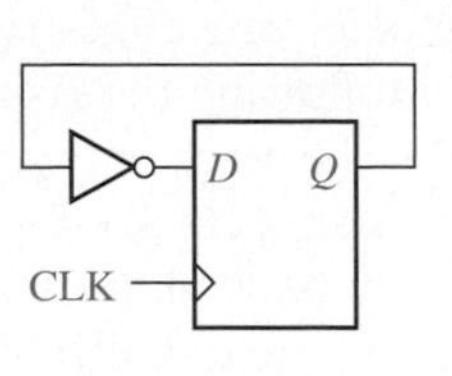


(a) Simple flip-flop circuit

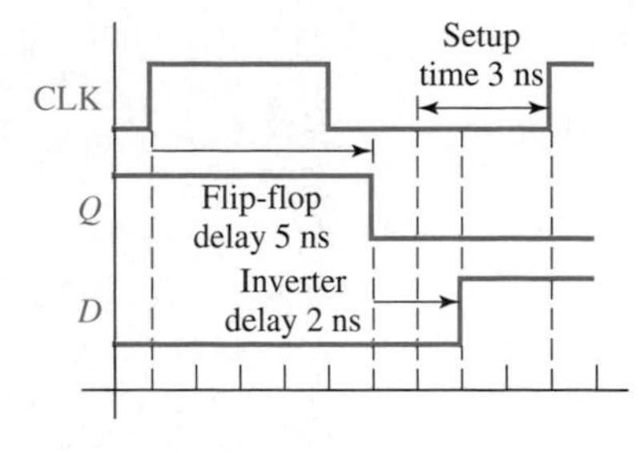


(b) Setup time not satisfied

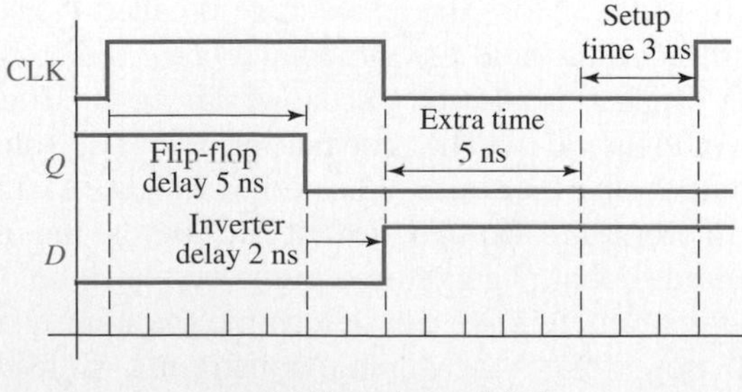


(c) Setup time satisfied

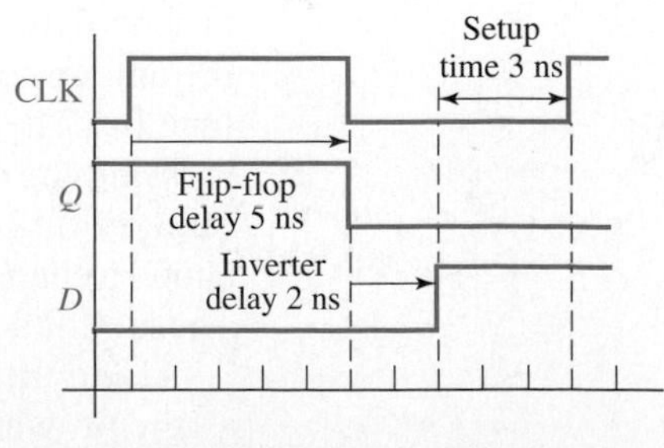


(d) Minimum clock period

## 11.5 S-R Flip-Flop

An S-R flip-flop (Figure 11-18) is similar to an S-R latch in that $S = 1$ sets the $Q$ output to 1, and $R = 1$ resets the $Q$ output to 0. The essential difference is that the flip-flop has a clock input, and the $Q$ output can change only after an active clock edge. The truth table and characteristic equation for the flip-flop are the same as for the latch, but the interpretation of $Q^+$ is different. For the latch, $Q^+$ is the value of $Q$ after the propagation delay through the latch, while for the flip-flop, $Q^+$ is the value that $Q$ assumes after the active clock edge.

Figure 11-19(a) shows an S-R flip-flop constructed from two S-R latches and gates. This flip-flop changes state after the rising edge of the clock. The circuit is often referred to as a master-slave flip-flop. When CLK = 0, the $S$ and $R$ inputs set the outputs of the master latch to the appropriate value while the slave latch holds the previous value of $Q$. When the clock changes from 0 to 1, the value of $P$ is held in the master latch and this value is transferred to the slave latch. The master latch holds the value of $P$ while CLK = 1, and, hence, $Q$ does not change. When the clock changes from 1 to 0, the $Q$ value is latched in the slave, and the master can process new inputs. Figure 11-19(b) shows the timing diagram. Initially, $S = 1$ and $Q$ changes to 1 at $t_1$. Then $R = 1$ and $Q$ changes to 0 at $t_3$.

**FIGURE 11-18**
S-R Flip-Flop

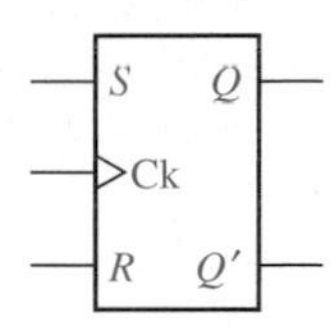


Operation summary:

| | |
|---|---|
| $S = R = 0$ | No state change |
| $S = 1, R = 0$ | Set $Q$ to 1 (after active Ck edge) |
| $S = 0, R = 1$ | Reset $Q$ to 0 (after active Ck edge) |
| $S = R = 1$ | Not allowed |

**FIGURE 11-19**
S-R Flip-Flop Implementation and Timing

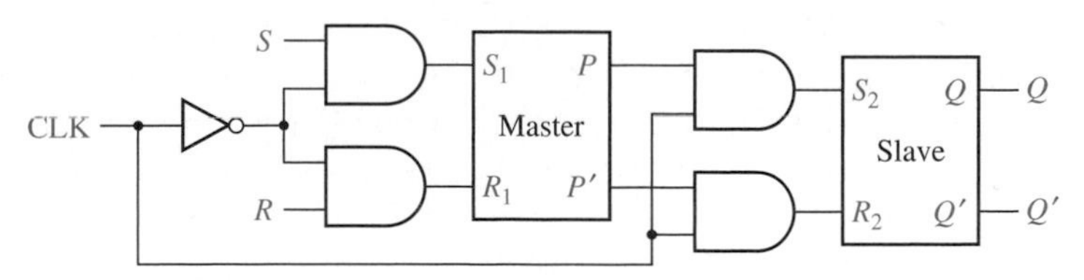


(a) Implementation with two latches

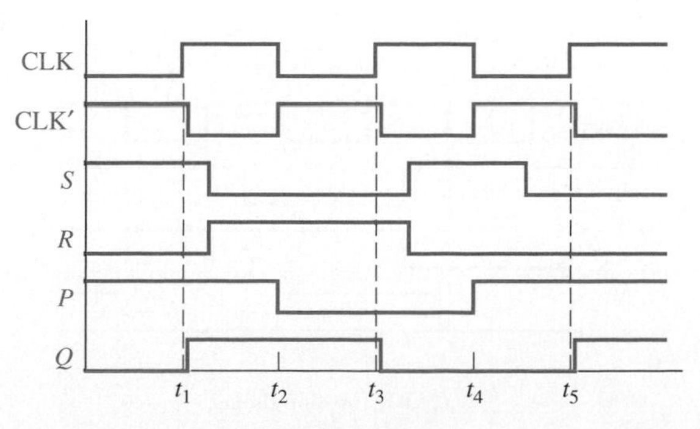


(b) Timing analysis

At first glance, this flip-flop appears to operate just like an edge-triggered flip-flop, but there is a subtle difference. For a rising-edge-triggered flip-flop the value of the inputs is sensed at the rising edge of the clock, and the inputs can change while the clock is low. For the master-slave flip-flop, if the inputs change while the clock is low, the flip-flop output may be incorrect. For example, in Figure 11-19(b) at $t_4$, $S = 1$ and $R = 0$, so $P$ changes to 1. Then $S$ changes to 0 at $t_5$, but $P$ does not change, so at $t_5$, $Q$ changes to 1 after the rising edge of CLK. However, at $t_5$, $S = R = 0$, so the state of $Q$ should not change. We can solve this problem if we only allow the $S$ and $R$ inputs to change while the clock is high.

## 11.6 J-K Flip-Flop

The J-K flip-flop (Figure 11-20) is an extended version of the S-R flip-flop. The J-K flip-flop has three inputs—$J$, $K$, and the clock (CK). The $J$ input corresponds to $S$, and $K$ corresponds to $R$. That is, if $J = 1$ and $K = 0$, the flip-flop output is set to $Q = 1$ after the active clock edge; and if $K = 1$ and $J = 0$, the flip-flop output is reset to $Q = 0$ after the active edge. Unlike the S-R flip-flop, a 1 input may be applied simultaneously to $J$ and $K$, in which case the flip-flop changes state after the active clock edge. When $J = K = 1$, the active edge will cause $Q$ to change from 0 to 1, or from 1 to 0. The next-state table and characteristic equation for the J-K flip-flop are given in Figure 11-20(b).

Figure 11-20(c) shows the timing for a J-K flip-flop. This flip-flop changes state a short time ($t_p$) after the rising edge of the clock pulse, provided that $J$ and $K$ have

FIGURE 11-20
J-K Flip-Flop
($Q$ Changes on the Rising Edge)

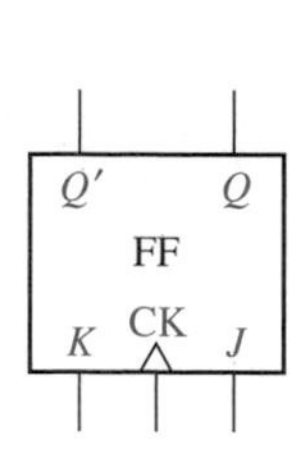


(a) J-K flip-flop

| J K Q | $Q^+$ |
|---|---|
| 0 0 0 | 0 |
| 0 0 1 | 1 |
| 0 1 0 | 0 |
| 0 1 1 | 0 |
| 1 0 0 | 1 |
| 1 0 1 | 1 |
| 1 1 0 | 1 |
| 1 1 1 | 0 |

$$Q^+ = JQ' + K'Q$$

(b) Truth table and characteristic equation

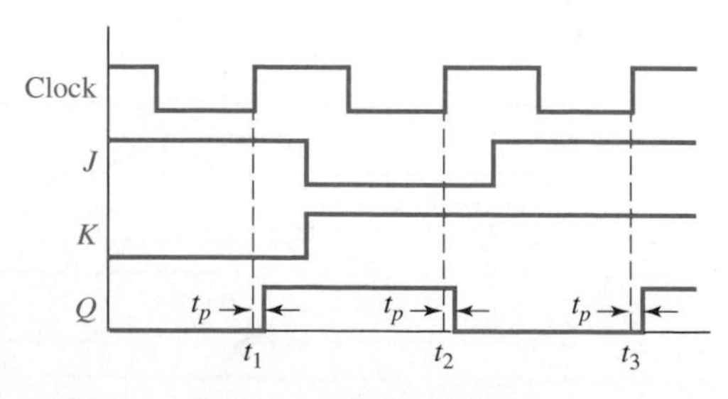


(c) J-K flip-flop timing

appropriate values. If $J = 1$ and $K = 0$ when Clock $= 0$, $Q$ will be set to 1 following the rising edge. If $K = 1$ and $J = 0$ when Clock $= 0$, $Q$ will be set to 0 after the rising edge. Similarly, if $J = K = 1$, $Q$ will change state after the rising edge. Referring to Figure 11-20(c), because $Q = 0$, $J = 1$, and $K = 0$ before the first rising clock edge, $Q$ changes to 1 at $t_1$. Because $Q = 1$, $J = 0$, and $K = 1$ before the second rising clock edge, $Q$ changes to 0 at $t_2$. Because $Q = 0$, $J = 1$, and $K = 1$ before the third rising clock edge, $Q$ changes to 1 at $t_3$.

One way to realize the J-K flip-flop is with two S-R latches connected in a master-slave arrangement, as shown in Figure 11-21. This is the same circuit as for the S-R master-slave flip-flop, except $S$ and $R$ have been replaced with $J$ and $K$, and the $Q$ and $Q'$ outputs are feeding back into the input gates. Because $S = J{\cdot}Q'{\cdot}\text{Clk}'$ and $R = K{\cdot}Q{\cdot}\text{Clk}'$, only one of $S$ and $R$ inputs to the first latch can be 1 at any given time. If $Q = 0$ and $J = 1$, then $S = 1$ and $R = 0$, regardless of the value of $K$. If $Q = 1$ and $K = 1$, then $S = 0$ and $R = 1$, regardless of the value of $J$.

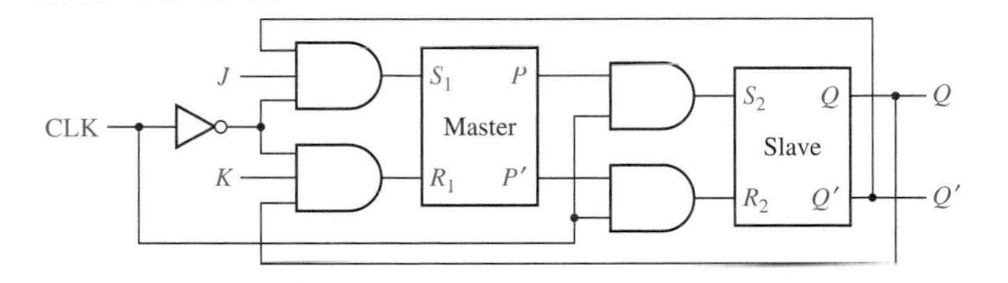


**FIGURE 11-21** Master-Slave J-K Flip-Flop (*Q* Changes on Rising Edge)

## 11.7 T Flip-Flop

The T flip-flop, also called the toggle flip-flop, is frequently used in building counters. Most CPLDs and FPGAs can be programmed to implement T flip-flops. The T flip-flop in Figure 11-22(a) has a $T$ input and a clock input. When $T = 1$ the flip-flop changes state after the active edge of the clock. When $T = 0$, no state change occurs. The next-state table and characteristic equation for the T flip-flop are given in Figure 11-22(b). The characteristic equation states that the next state of the flip-flop ($Q^+$) will be 1 iff the present state ($Q$) is 1 and $T = 0$ or the present state is 0 and $T = 1$.

Figure 11-23 shows a timing diagram for the T flip-flop. At times $t_2$ and $t_4$ the $T$ input is 1 and the flip-flop state ($Q$) changes a short time ($t_p$) after the falling edge of the clock pulse. At times $t_1$ and $t_3$ the $T$ input is 0, and the clock edge does not cause a change of state.

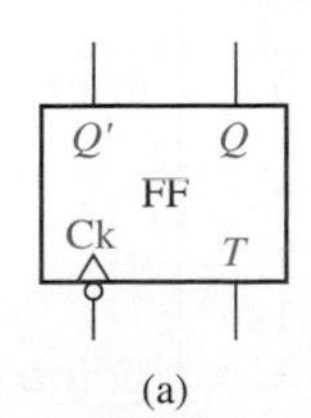


(a)

| T Q | $Q^+$ |
|---|---|
| 0 0 | 0 |
| 0 1 | 1 |
| 1 0 | 1 |
| 1 1 | 0 |

$$Q^+ = T'Q + TQ' = T \oplus Q$$

(b)

**FIGURE 11-22** T Flip-Flop

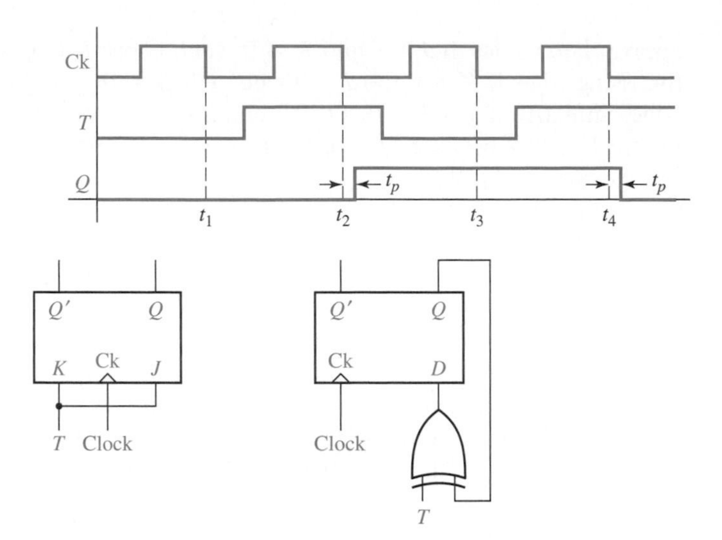


**FIGURE 11-23** Timing Diagram for T Flip-Flop (Falling-Edge Trigger)

**FIGURE 11-24** Implementation of T Flip-Flops

(a) Conversion of J-K to $T$ (b) Conversion of $D$ to $T$

One way to implement a T flip-flop is to connect the $J$ and $K$ inputs of a J-K flip-flop together, as shown in Figure 11-24(a). Substituting $T$ for $J$ and $K$ in the J-K characteristic equation gives

$$Q^+ = JQ' + K'Q = TQ' + T'Q$$

which is the characteristic equation for the T flip-flop. Another way to realize a T flip-flop is with a D flip-flop and an exclusive-OR gate [Figure 11-24(b)]. The $D$ input is $Q \oplus T$, so $Q^+ = Q \oplus T = TQ' + T'Q$, which is the characteristic equation for the T flip-flop.

## 11.8 Flip-Flops with Additional Inputs

Flip-flops often have additional inputs which can be used to set the flip-flops to an initial state independent of the clock. Figure 11-25 shows a D flip-flop with clear and preset inputs. The small circles (inversion symbols) on these inputs indicate that a logic 0 (rather than a 1) is required to clear or set the flip-flop. This type of input is often referred to as *active-low* because a low voltage or logic 0 will activate the clear

**FIGURE 11-25** D Flip-Flop with Clear and Preset

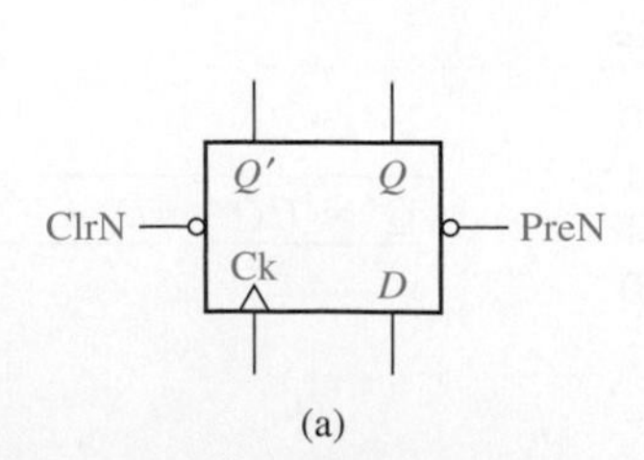


(a)

| Ck | $D$ | PreN | ClrN | $Q^+$ |
|---|---|---|---|---|
| X | X | 0 | 0 | (not allowed) |
| X | X | 0 | 1 | 1 |
| X | X | 1 | 0 | 0 |
| ↑ | 0 | 1 | 1 | 0 |
| ↑ | 1 | 1 | 1 | 1 |
| 0,1,↓ | X | 1 | 1 | $Q$ (no change) |

(b)

or preset function. We will use the notation ClrN or PreN to indicate active-low clear and preset inputs. Thus, a logic 0 applied to ClrN will reset the flip-flop to $Q = 0$, and a 0 applied to PreN will set the flip-flop to $Q = 1$. These inputs override the clock and $D$ inputs. That is, a 0 applied to the ClrN will reset the flip-flop regardless of the values of $D$ and the clock. Under normal operating conditions, a 0 should not be applied simultaneously to ClrN and PreN. When ClrN and PreN are both held at logic 1, the $D$ and clock inputs operate in the normal manner. ClrN and PreN are often referred to as asynchronous clear and preset inputs because their operation does not depend on the clock. The table in Figure 11-25(b) summarizes the flip-flop operation. In the table, ↑ indicates a rising clock edge, and X is a don't-care. The last row of the table indicates that if Ck is held at 0, held at 1, or has a falling edge, $Q$ does not change.

Figure 11-26 illustrates the operation of the clear and preset inputs. At $t_1$, ClrN = 0 holds the $Q$ output at 0, so the rising edge of the clock is ignored. At $t_2$ and $t_3$, normal state changes occur because ClrN and PreN are both 1. Then, $Q$ is set to 1 by PreN = 0, but $Q$ is cleared at $t_4$ by the rising edge of the clock because $D = 0$ at that time.

In synchronous digital systems, the flip-flops are usually driven by a common clock so that all state changes occur at the same time in response to the same clock edge. When designing such systems, we frequently encounter situations where we want some flip-flops to hold existing data even though the data input to the flip-flops may be changing. One way to do this is to gate the clock, as shown in Figure 11-27(a). When En = 0, the clock input to the flip-flop is 0, and $Q$ does not change. This method has two potential problems. First, gate delays may cause the clock to arrive at some flip-flops at different times than at other flip-flops, resulting in a loss of synchronization. Second, if En changes at the wrong time, the flip-flop may trigger due to the change in En instead of due to the change in the clock, again resulting in loss of synchronization. Rather than gating the clock, a better way is to use a flip-flop with a clock enable (CE). Such flip-flops are commonly used in CPLDs and FPGAs.

Figure 11-27(b) shows a D flip-flop with a clock enable, which we will call a D-CE flip-flop. When CE = 0, the clock is disabled and no state change occurs, so $Q^+ = Q$. When CE = 1, the flip-flop acts like a normal D flip-flop, so $Q^+ = D$. Therefore, the characteristic equation is $Q^+ = Q \bullet CE' + D \bullet CE$. The D-CE flip-flop is easily implemented using a D flip-flop and a multiplexer (Figure 11-27(c)). For this circuit, the MUX output is

$$Q^+ = D = Q \cdot CE' + D_{in} \cdot CE$$

Because there is no gate in the clock line, this cannot cause a synchronization problem.

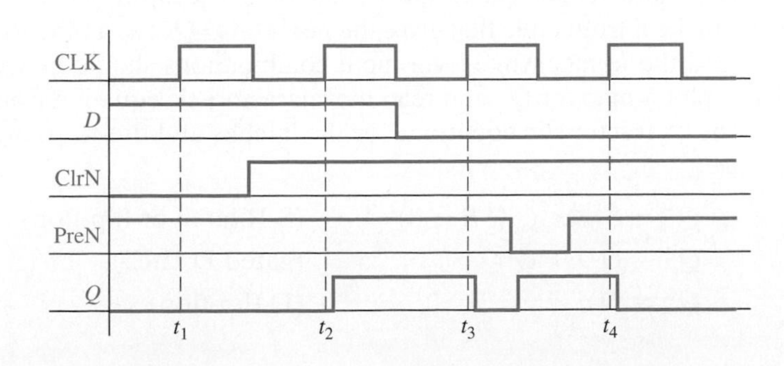


**FIGURE 11-26** Timing Diagram for D Flip-Flop with Asynchronous Clear and Preset

FIGURE 11-27
D Flip-Flop with Clock Enable

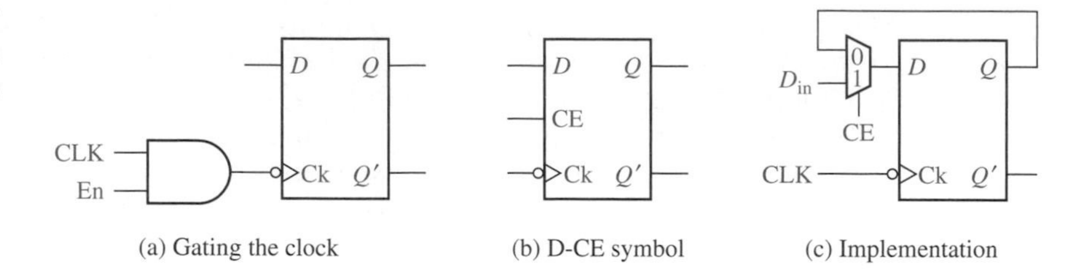


(a) Gating the clock (b) D-CE symbol (c) Implementation

# 11.9 Summary

In this unit, we have studied several types of latches and flip-flops. Flip-flops have a clock input, and the output changes only in response to a rising or falling edge of the clock. All of these devices have two output states: $Q = 0$ and $Q = 1$. For the S-R latch, $S = 1$ sets $Q$ to 1, and $R = 1$ resets $Q$ to 0. $S = R = 1$ is not allowed. The S-R flip-flop is similar except that $Q$ only changes after the active edge of the clock. The gated D latch transmits $D$ to the $Q$ output when $G = 1$. When $G$ is 0, the current value of $D$ is stored in the latch and $Q$ does not change. For the D flip-flop, $Q$ is set equal to $D$ after the active clock edge. The D-CE flip-flop works the same way, except the clock is only enabled when CE = 1. The J-K flip-flop is similar to the S-R flip-flop in that when $J = 1$ the active clock edge sets $Q$ to 1, and when $K = 1$, the active edge resets $Q$ to 0. When $J = K = 1$, the active clock edge causes $Q$ to change state. The T flip-flop changes state on the active clock edge when $T = 1$; otherwise, $Q$ does not change. Flip-flops can have asynchronous clear and preset inputs that cause $Q$ to be cleared to 0 or preset to 1 independently of the clock.

Flip-flops can be constructed using gate circuits with feedback. Analysis of such circuits can be accomplished by tracing signal changes through the gates. Analysis can also be done using flow tables and asynchronous sequential circuit theory, but that is beyond the scope of this text. Timing diagrams are helpful in understanding the time relationships between the input and output signals for a latch or flip-flops. In general, the inputs must be applied a specified time before the active clock edge (the setup time), and they must be held constant a specified time after the active edge (the hold time). The time after the active clock edge before $Q$ changes is the propagation delay.

The characteristic (next-state) equation for a flip-flop can be derived as follows: First, make a truth table that gives the next state ($Q^+$) as a function of the present state ($Q$) and the inputs. Any illegal input combinations should be treated as don't-cares. Then, plot a map for $Q^+$ and read the characteristic equation from the map.

The characteristic equations for the latches and flip-flops discussed in this chapter are:

$$Q^+ = S + R'Q \ (SR = 0) \qquad \text{(S-R latch or flip-flop)} \qquad (11\text{-}6)$$

$$Q^+ = GD + G'Q \qquad \text{(gated D latch)} \qquad (11\text{-}7)$$

$$Q^+ = D \qquad \text{(D flip-flop)} \qquad (11\text{-}8)$$

$$Q^+ = D{\cdot}CE + Q{\cdot}CE' \qquad \text{(D-CE flip-flop)} \qquad (11\text{-}9)$$

$$Q^+ = JQ' + K'Q \qquad \text{(J-K flip-flop)} \qquad (11\text{-}10)$$

$$Q^+ = T \oplus Q = TQ' + T'Q \qquad \text{(T flip-flop)} \qquad (11\text{-}11)$$

In each case, $Q$ represents an initial or present state of the flip-flop, and $Q^+$ represents the final or next state. These equations are valid only when the appropriate restrictions on the flip-flop inputs are observed. For the S-R flip-flop, $S = R = 1$ is forbidden. For the master-slave S-R flip-flop, $S$ and $R$ should not change during the half of the clock cycle preceding the active edge. Setup and hold time restrictions must also be satisfied.

The characteristic equations given above apply to both latches and flip-flops, but their interpretation is different for the two cases. For example, for the gated D latch, $Q^+$ represents the state of the flip-flop a short time after one of the inputs changes. However, for the D flip-flop, $Q^+$ represents the state of the flip-flop a short time after the active clock edge.

Conversion of one type of flip-flop to another is usually possible by adding external gates. Figure 11-24 shows how a J-K flip-flop and a D flip-flop can be converted to a T flip-flop.

## Problems

**11.1** Assume that the inverter in the given circuit has a propagation delay of 5 ns and the AND gate has a propagation delay of 10 ns. Draw a timing diagram for the circuit showing $X$, $Y$, and $Z$. Assume that $X$ is initially 0, $Y$ is initially 1, after 10 ns $X$ becomes 1 for 80 ns, and then $X$ is 0 again.

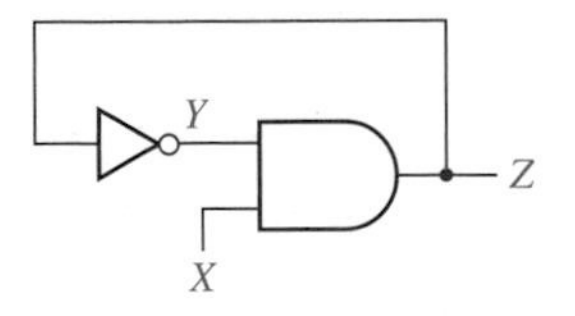


**11.2** A latch can be constructed from an OR gate, an AND gate, and an inverter connected as follows:

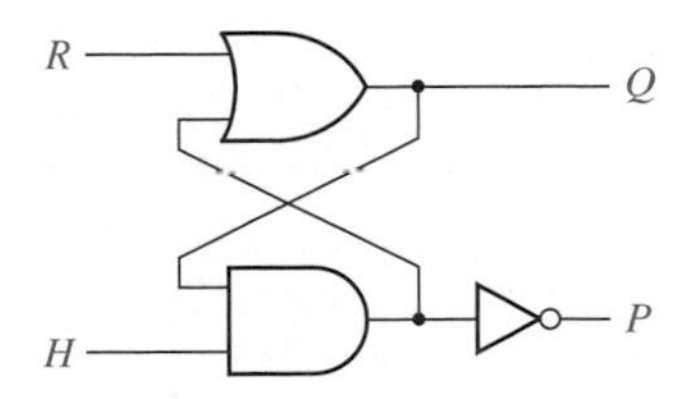


(a) What restriction must be placed on $R$ and $H$ so that $P$ will always equal $Q'$ (under steady-state conditions)?

(b) Construct a next-state table and derive the characteristic (next-state) equation for the latch.

(c) Complete the following timing diagram for the latch.

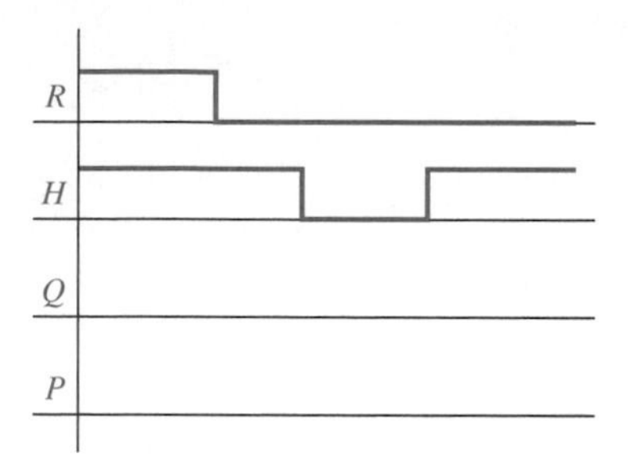


**11.3** This problem illustrates the improper operation that can occur if both inputs to an S-R latch are 1 and are then changed back to 0. For Figure 11-6, complete the following timing chart, assuming that each gate has a propagation delay of exactly 10 ns. Assume that initially $P = 1$ and $Q = 0$. Note that when $t = 100$ ns, $S$ and $R$ are both changed to 0. Then, 10 ns later, both $P$ and $Q$ will change to 1. Because these 1's are fed back to the gate inputs, what will happen after another 10 ns?

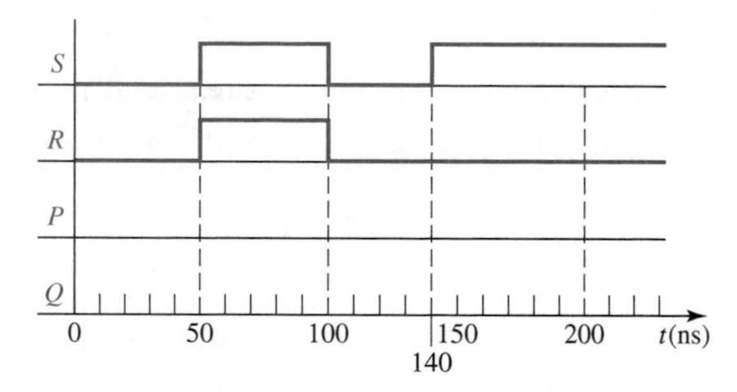


**11.4** Design a gated D latch using only NAND gates and one inverter.

**11.5** What change must be made to Figure 11-15(a) to implement a falling-edge-triggered D flip-flop? Complete the following timing diagram for the modified flip-flop.

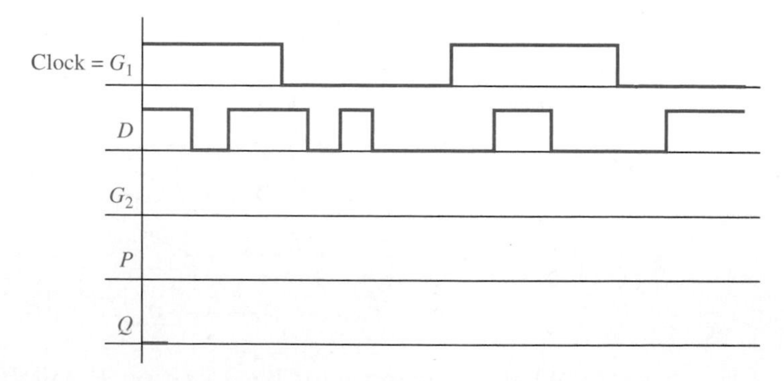


**11.6** A reset-dominant flip-flop behaves like an S-R flip-flop, except that the input $S = R = 1$ is allowed, and the flip-flop is reset when $S = R = 1$.

(a) Derive the characteristic equation for a reset-dominant flip-flop.

(b) Show how a reset-dominant flip-flop can be constructed by adding gate(s) to an S-R flip-flop.

**11.7** Complete the following timing diagram for the flip-flop of Figure 11-20(a).

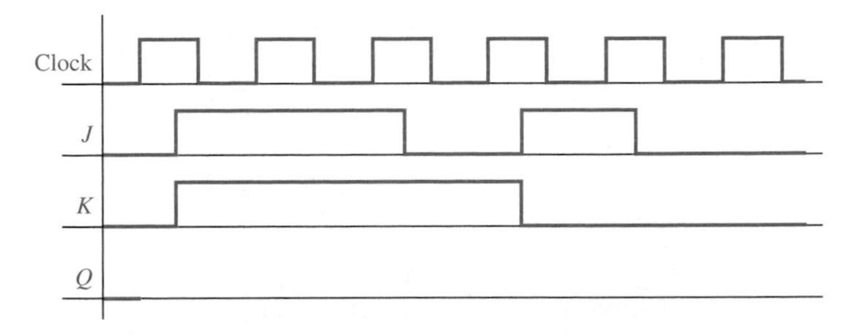


**11.8** Complete the following diagrams for the falling-edge-triggered D-CE flip-flop of Figure 11-27(c). Assume $Q$ begins at 1.

(a) First draw $Q$ based on your understanding of the behavior of a D flip-flop with clock enable.

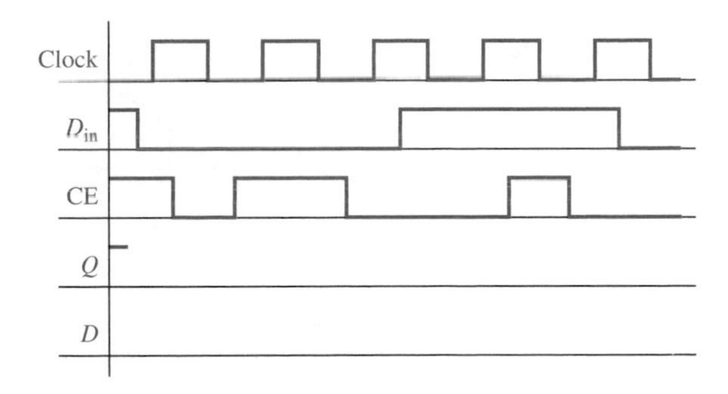


(b) Now draw in the internal signal $D$ from Figure 11-27(c), and confirm that this gives the same $Q$ as in (a).

**11.9** (a) Complete the following timing diagram for a J-K flip-flop with a falling-edge trigger and asynchronous ClrN and PreN inputs.

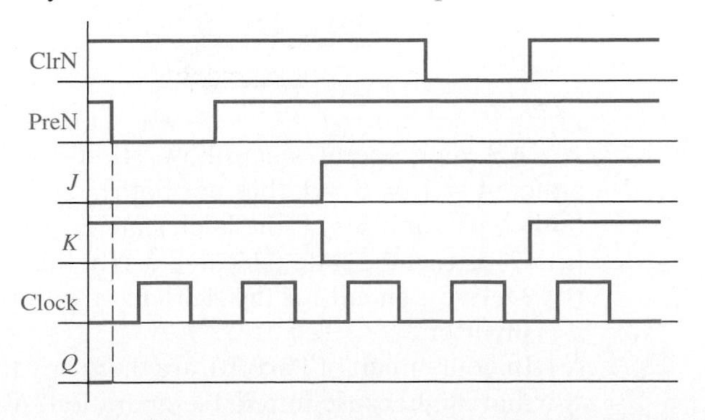

(b) Complete the timing diagram for the following circuit. Note that the Ck inputs on the two flip-flops are different.

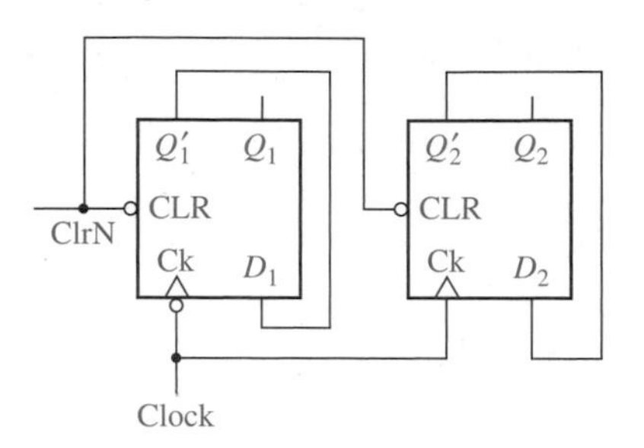


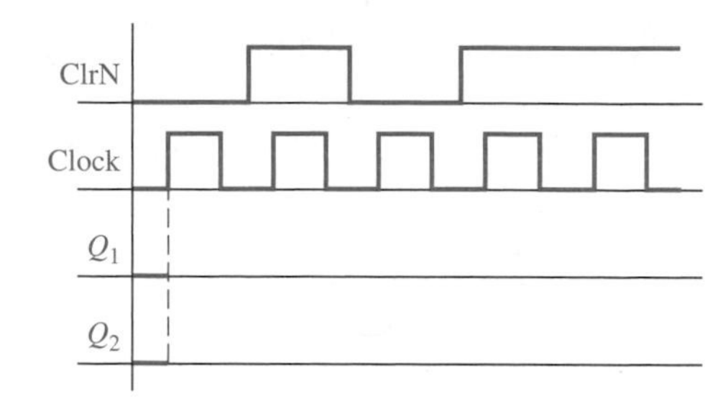


**11.10** Convert by adding external gates:

(a) a D flip-flop to a J-K flip-flop.

(b) a T flip-flop to a D flip-flop.

(c) a T flip-flop to a D flip-flop with clock enable.

**11.11** Complete the following timing diagram for an S-R latch. Assume $Q$ begins at 1.

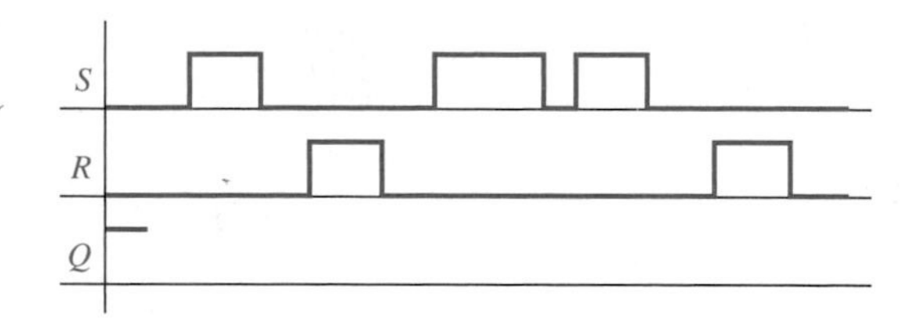


**11.12** Using a truth table similar to Figure 11-8(b), confirm that each of these circuits is an S-R latch. What happens when $S = R = 1$ for each circuit?

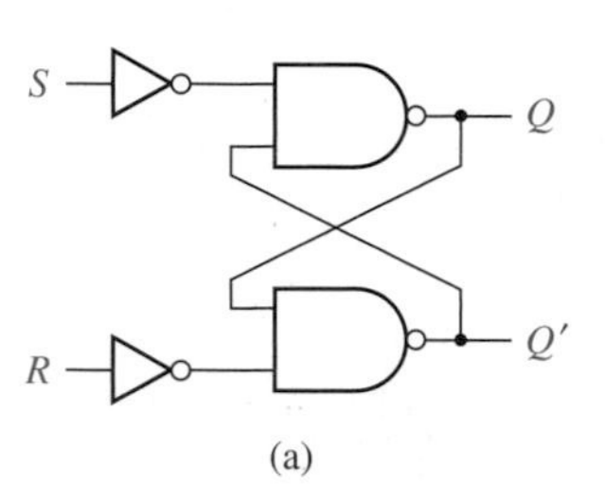


(a)

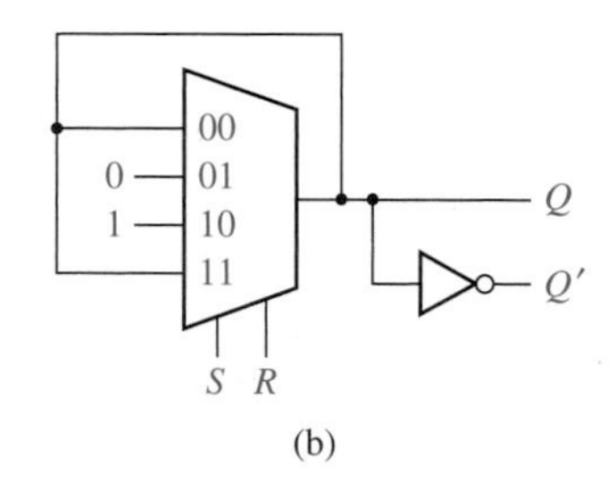


(b)

**11.13** An $AB$ latch operates as follows: If $A = 0$ and $B = 0$, the latch state is $Q = 0$; if either $A = 1$ or $B = 1$ (but not both), the latch output does not change; and when both $A = 1$ and $B = 1$, the latch state is $Q = 1$.

(a) Construct the state table and derive the characteristic equation for this $AB$ latch.

(b) Derive a circuit for the $AB$ latch that has four two-input NAND gates and two inverters.

(c) In your circuit of Part (b), are there any transitions between input combinations that might cause unreliable operation? Verify your answer.

(d) In your circuit of Part (b), is there a gate output that provides the signal $Q'$? Verify your answer.
(e) Derive a circuit for the $AB$ latch using four two-input NOR gates and two inverters.
(f) Answer Parts (c) and (d) for your circuit of Part (e).

**11.14** (a) Construct a state table for this circuit and identify the stable states of the circuit.
(b) Derive a Boolean algebra equation for the next value of the output $Q$ in terms of $Q$, $A$ and $B$.
(c) Analyze the behavior of the circuit. Is it a useful circuit? If not, explain why not; if yes, explain what it does.

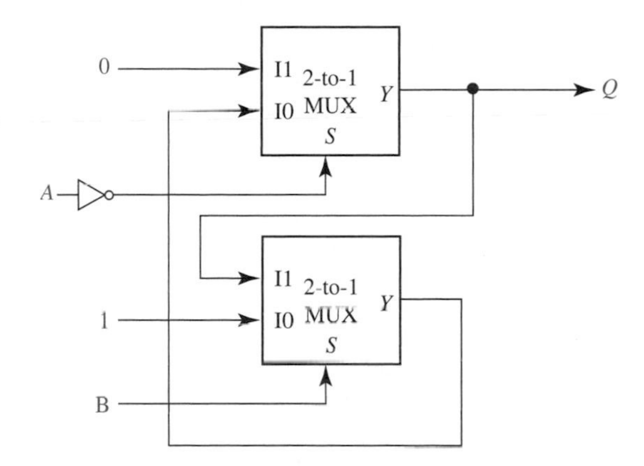


**11.15** The following circuit is intended to be a gated latch circuit where the signal $G$ is the gate.
(a) Derive the next-state equation for this circuit using $Q$ as the state variable and $P$ as an output.
(b) Construct the state table and output table for the circuit. Circle the stable states of the circuit.
(c) Are there any restrictions on the allowable input combinations on $M$ and $N$? Explain your answer.
(d) Is the output $P$ usable as the complement of $Q$? Verify your answer.
(e) Assume that Gate 1 has a propagation delay of 30 ns and Gates 2, 3, and 4 have propagation delays of 10 ns. Construct a timing diagram for the circuit for the following input change: $M = N = Q = 0$ with $G$ changing from 1 to 0.

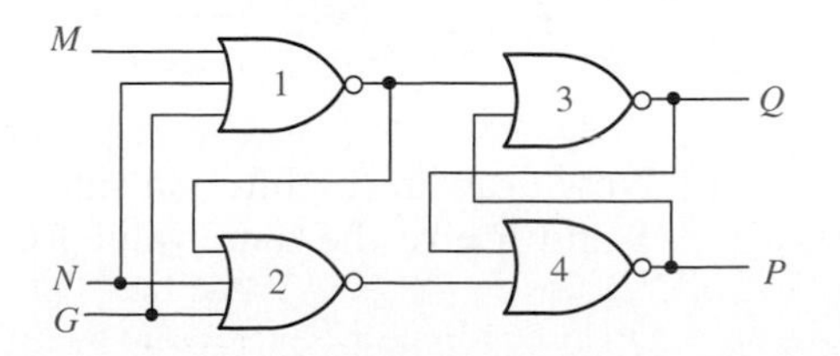

**11.16** Analyze the latch circuit shown.

(a) Derive the next-state equation for this circuit using $Q$ as the state variable and $P$ as an output.

(b) Construct the state table and output table for the circuit. Circle the stable states of the circuit.

(c) Are there any restrictions on the allowable input combinations on $A$ and $B$? Explain your answer.

(d) Is the output $P$ usable as the complement of $Q$? Verify your answer.

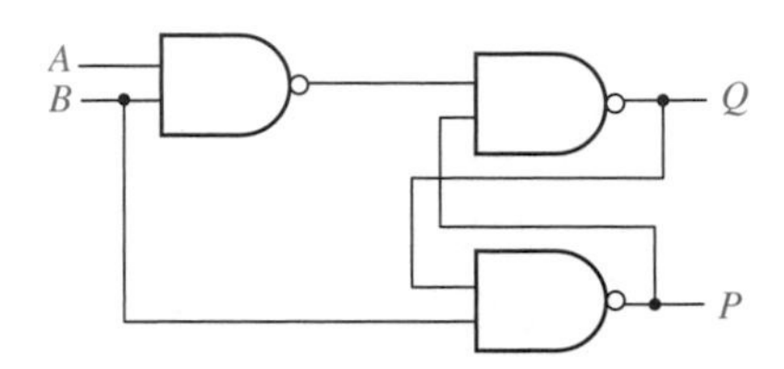


**11.17** Derive the characteristic equations for the following latches and flip-flops in product-of-sums form.

(a) S-R latch or flip-flop
(b) Gated D latch
(c) D flip-flop
(d) D-CE flip-flop
(e) J-K flip-flop
(f) T flip-flop

**11.18** Complete the following timing diagrams for a gated D latch. Assume $Q$ begins at 0.

(a) First draw $Q$ based on your understanding of the behavior of a gated D latch.

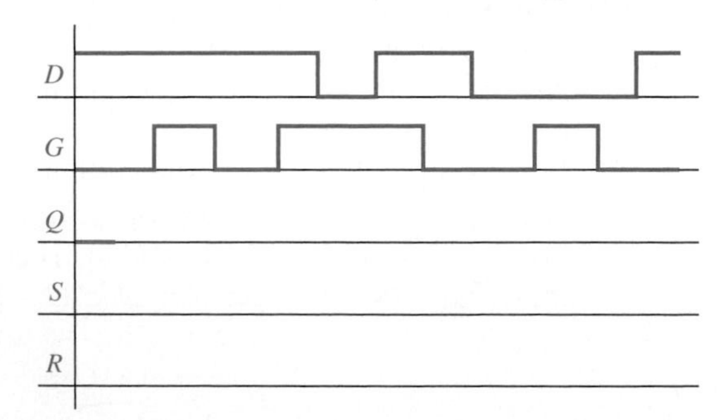


(b) Now draw in the internal signals $S$ and $R$ from Figure 11-11, and confirm that $S$ and $R$ give the same value for $Q$ as in (a).

**11.19** Complete the following diagrams for the rising-edge-triggered D flip-flop of Figure 11-15. Assume $Q$ begins at 1.

(a) First draw $Q$ based on your understanding of the behavior of a D flip-flop.

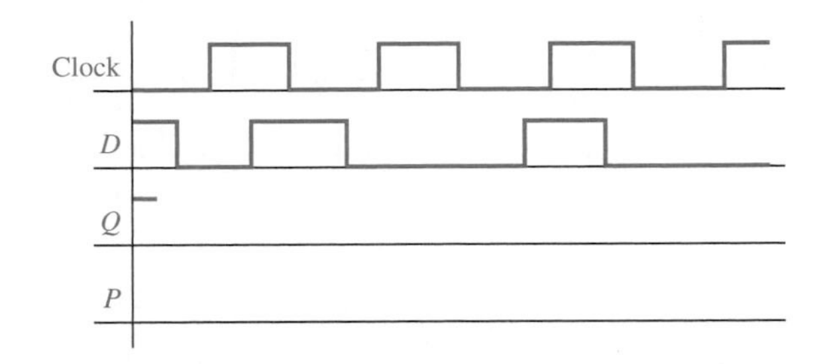


(b) Now draw in the internal signal $P$ from Figure 11-15, and confirm that $P$ gives the same $Q$ as in (a).

**11.20** A set-dominant flip-flop is similar to the reset-dominant flip-flop of Problem 11.6 except that the input combination $S = R = 1$ sets the flip-flop. Repeat Problem 11.6 for a set-dominant flip-flop.

**11.21** Fill in the timing diagram for a falling-edge-triggered S-R flip-flop. Assume $Q$ begins at 0.

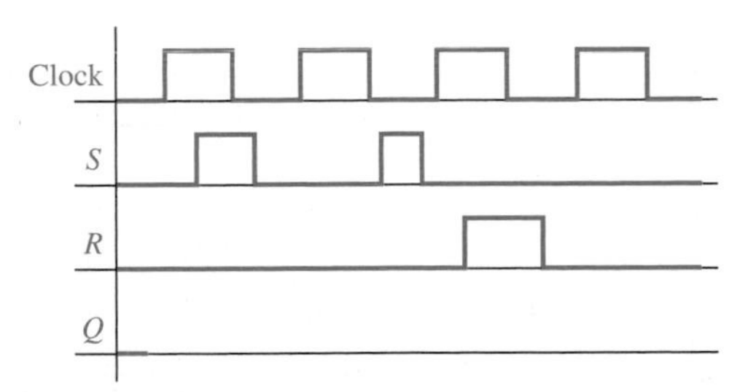


**11.22** Fill in the timing diagram for a falling-edge-triggered J-K flip-flop.

(a) Assume $Q$ begins at 0.

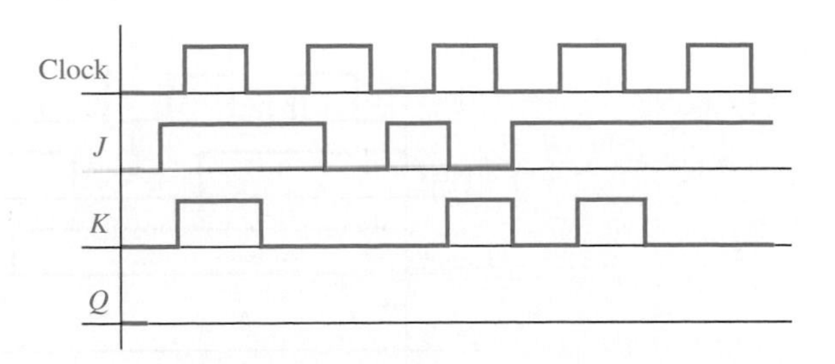


(b) Assume $Q$ begins at 1, but Clock, $J$, and $K$ are the same.

**11.23** (a) Find the input for a rising-edge-triggered D flip-flop that would produce the output $Q$ as shown. Fill in the timing diagram.

(b) Repeat for a rising-edge-triggered T flip-flop.

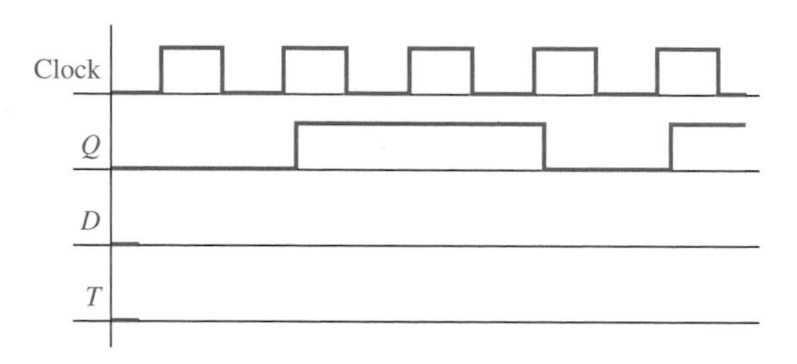


**11.24** Here is the diagram of a 3-bit ripple counter. Assume $Q_0 = Q_1 = Q_2 = 0$ at $t = 0$, and assume each flip-flop has a delay of 1 ns from the clock input to the $Q$ output. Fill in $Q_0$, $Q_1$, and $Q_2$ of the timing diagram. Flip-flop $Q_1$, will be triggered when $Q_0$ changes from 0 to 1.

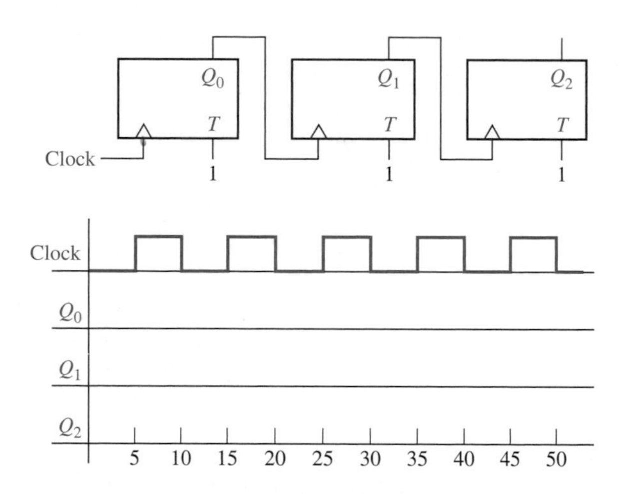


**11.25** Fill in the following timing diagram for a rising-edge-triggered T flip-flop with an asychronous active-low PreN input. Assume $Q$ begins at 1.

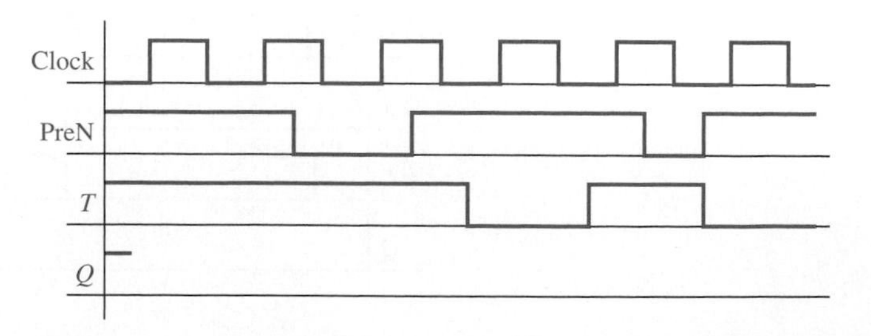


**11.26** The ClrN and PreN inputs introduced in Section 11.8 are called asynchronous because they operate independently of the clock (i.e., they are not synchronized with the clock). We can also make flip-flops with synchronous clears or preset inputs. A D-flip-flop with an active-low synchronous ClrN input may be constructed from a regular D flip-flop as follows.

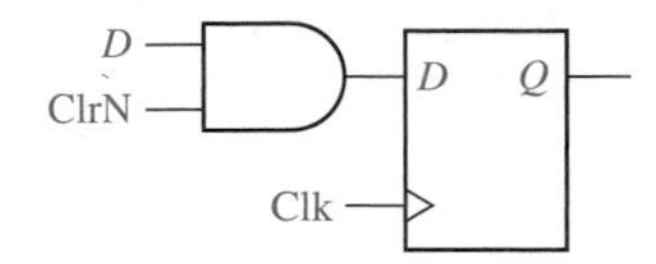


Fill in the timing diagram. For $Q_1$, assume a synchronous ClrN as above, and for $Q_2$, assume an asynchronous ClrN as in Section 11.8. Assume $Q_1 = Q_2 = 0$ at the beginning.

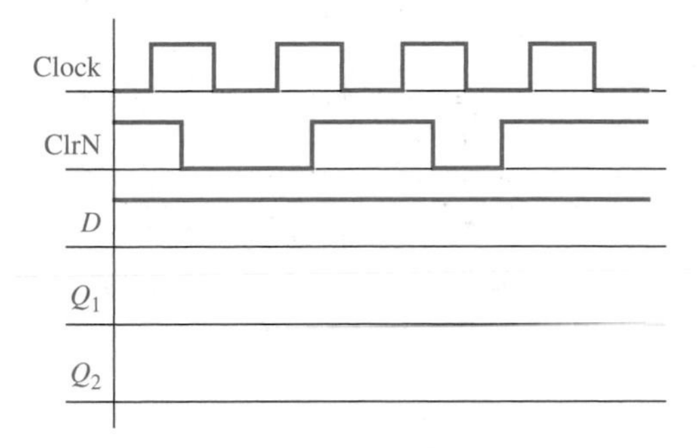


**11.27** (a) Construct a D flip-flop using an inverter and an S-R flip-flop.
(b) If the propagation delay and setup time of the S-R flip-flop in (a) are 2.5 ns and 1.5 ns, respectively, and if the inverter has a propagation delay of 1 ns, what are the propagation delay and setup time of the D flip-flop of Part (a)?

**11.28** Redesign the debouncing circuit of Figure 11-9 using the $\bar{S}$-$\bar{R}$ latch of Figure 11-10.

# Programmed Exercise 11.29

Cover the bottom part of each page with a sheet of paper and slide it down as you check your answers.

The internal logic diagram of a falling-edge-triggered D flip-flop follows. This flip-flop consists of two basic S-R latches with added gates. When the clock input ($CK$) is 1, the value of $D$ is stored in the first S-R latch ($P$). When the clock changes from 1 to 0, the value of $P$ is transferred to the output latch ($Q$). Thus, the operation is similar to that of the master-slave S-R flip-flop shown in Figure 11-19, except for the edges at which the data is stored.

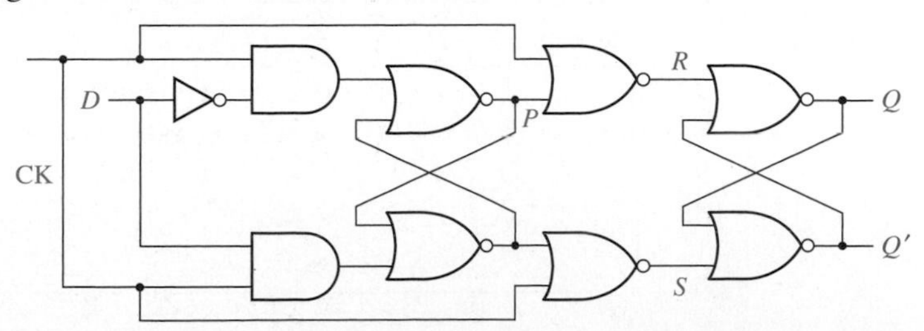

In this exercise you will be asked to analyze the operation of the D flip-flop shown above by filling in a table showing the values of CK, $D$, $P$, $S$, $R$, and $Q$ after each change of input. It will be helpful if you mark the changes in these values on the circuit diagram as you trace the signals. Initially, assume the following signal values:

| CK | $D$ | $P$ | $S$ | $R$ | $Q$ | |
|---|---|---|---|---|---|---|
| 0 | 0 | 0 | 0 | 1 | 0 | (stable) |

Verify by tracing signals through the circuit that this is a stable condition of the circuit; that is, no change will occur in $P$, $S$, $R$, or $Q$. Now assume that CK is changed to 1:

| | CK | $D$ | $P$ | $S$ | $R$ | $Q$ | |
|---|---|---|---|---|---|---|---|
| 1. | 0 | 0 | 0 | 0 | 1 | 0 | (stable) |
| 2. | 1 | 0 | 0 | 0 | 1 | 0 | ? |
| 3. | | | | | | | |

Trace the change in CK through the circuit to see if a change in $P$, $S$, or $R$ will occur. If a change does occur, mark row 2 of the preceding table "unstable" and enter the new values in row 3.

---

**Answer:**

| | CK | $D$ | $P$ | $S$ | $R$ | $Q$ | |
|---|---|---|---|---|---|---|---|
| 2. | 1 | 0 | 0 | 0 | 1 | 0 | (unstable) |
| 3. | 1 | 0 | 0 | 0 | 0 | 0 | (stable) |
| 4. | 1 | 1 | 0 | 0 | 0 | 0 | (unstable) |
| 5. | 1 | 1 | | | | | ? |

---

Verify that row 3 is stable; that is, by tracing signals show that no further change in $P$, $S$, $R$, or $Q$ will occur. Next $D$ is changed to 1 as shown in row 4. Verify that row 4 is unstable, fill in the new values in row 5, and indicate if row 5 is stable or unstable.

---

**Answer:**

| | CK | $D$ | $P$ | $S$ | $R$ | $Q$ | |
|---|---|---|---|---|---|---|---|
| 5. | 1 | 1 | 1 | 0 | 0 | 0 | (stable) |
| 6. | 0 | 1 | 1 | 0 | 0 | 0 | ? |
| 7. | 0 | 1 | | | | | ? |
| 8. | 0 | 1 | | | | | |

---

Then CK is changed to 0 (row 6). If row 6 is unstable, indicate the new value of $S$ in row 7. If row 7 is unstable, indicate the new value of $Q$ in row 8. Then determine whether row 8 is stable or not.

---

**Answer:**

| | CK | $D$ | $P$ | $S$ | $R$ | $Q$ | |
|---|---|---|---|---|---|---|---|
| 7. | 0 | 1 | 1 | 1 | 0 | 0 | (unstable) |
| 8. | 0 | 1 | 1 | 1 | 0 | 1 | (stable) |
| 9. | 0 | 0 | | | | | (stable) |
| 10. | 1 | 0 | | | | | |
| 11. | 1 | 0 | | | | | |

---

Next, $D$ is changed back to 0 (row 9). Fill in the values in row 9 and verify that it is stable. CK is changed to 1 in row 10. If row 10 is unstable, fill in row 11 and indicate whether it is stable or not.

---

**Answer:**

| | | | | | | | |
|---|---|---|---|---|---|---|---|
| 9. | 0 | 0 | 1 | 1 | 0 | 1 | (stable) |
| 10. | 1 | 0 | 1 | 1 | 0 | 1 | (unstable) |
| 11. | 1 | 0 | 0 | 0 | 0 | 1 | (stable) |
| 12. | 0 | 0 | | | | | |
| 13. | 0 | 0 | | | | | |
| 14. | 0 | 0 | | | | | |

---

CK is changed back to 0 in row 12. Complete the rest of the table.

---

**Answer:**

| | | | | | | | |
|---|---|---|---|---|---|---|---|
| 12. | 0 | 0 | 0 | 0 | 0 | 1 | (unstable) |
| 13. | 0 | 0 | 0 | 0 | 1 | 1 | (unstable) |
| 14. | 0 | 0 | 0 | 0 | 1 | 0 | (stable) |

---

Using the previous results, plot $P$ and $Q$ on the following timing diagram. Verify that your answer is consistent with the description of the flip-flop operation given in the first paragraph of this exercise.

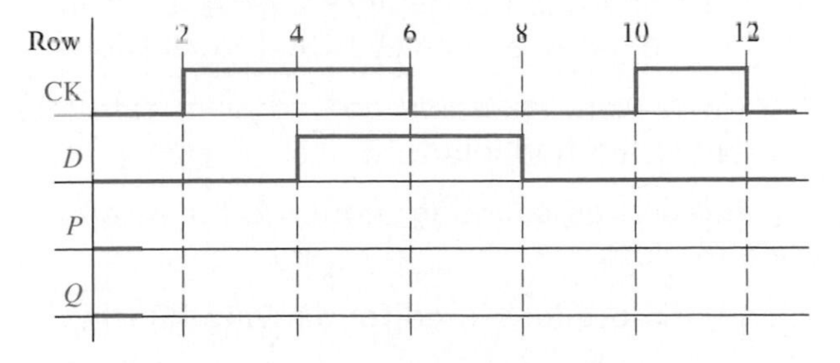


---

**Answer:**

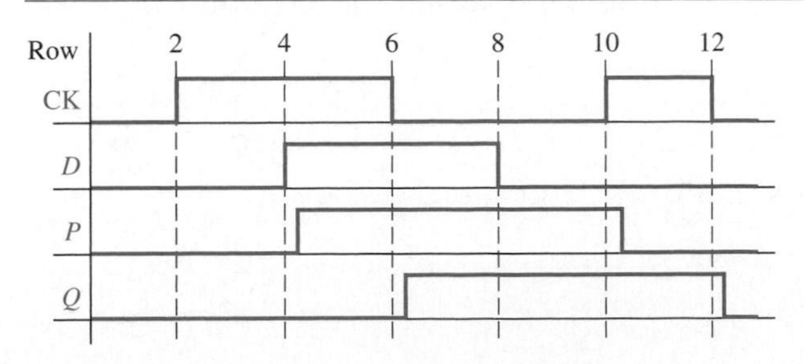


---

# Registers and Counters

## Objectives

1. Explain the operation of registers. Show how to transfer data between registers using a tri-state bus.
2. Explain the operation of shift registers, show how to build them using flip-flops, and analyze their operation. Construct a timing diagram for a shift register.
3. Explain the operation of binary counters, show how to build them using flip-flops and gates, and analyze their operation.
4. Given the present state and desired next state of a flip-flop, determine the required flip-flop inputs.
5. Given the desired counting sequence for a counter, derive the flip-flop input equations.
6. Explain the procedures used for deriving flip-flop input equations.
7. Construct a timing diagram for a counter by tracing signals through the circuit.

# Study Guide

1. **1.** Study Section 12.1, *Registers and Register Transfers.*
   (a) For the diagram of Figure 12-4, suppose registers $A$, $B$, $C$, and $D$ hold the 8-bit binary numbers representing 91, 70, 249, and 118, respectively. Suppose $G$ and $H$ are both initially 0. What are the contents of $G$ and $H$ (decimal equivalent) after the rising edge of the clock:

   (1) if $EF = 10$, $LdG = 0$, and $LdH = 1$ at the rising edge?
   (2) if $EF = 01$, $LdG = 0$, and $LdH = 1$ at the next rising edge?
   (3) if $EF = 11$, $LdG = 1$, and $LdH = 1$ at the next rising edge?
   (4) if $EF = 00$, $LdG = 1$, and $LdH = 0$ at the next rising edge?
   (5) if $EF = 10$, $LdG = 0$, and $LdH = 0$ at the next rising edge?

   (b) Work Problem 12.1.

2. **2.** Study Section 12.2, *Shift Registers.*

   (a) Compare the block diagrams for the shift registers of Figures 12-7 and 12-10. Which one changes state on the rising edge of the clock pulse? The falling edge?

   (b) Complete the following table and timing diagram (see next page) for the shift register of Figure 12-8.

| Clock Cycle Number | State of Shift Register When Clock = 1<br>$Q_7$ | $Q_6$ | $Q_5$ | $Q_4$ | $Q_3$ | $Q_2$ | $Q_1$ | $Q_0$ |
|---|---|---|---|---|---|---|---|---|
| 1 | 0 | 0 | 0 | 0 | 0 | 0 | 0 | 0 |
| 2 | | | | | | | | |
| 3 | | | | | | | | |
| 4 | | | | | | | | |
| 5 | | | | | | | | |
| 6 | | | | | | | | |
| 7 | | | | | | | | |
| 8 | | | | | | | | |
| 9 | | | | | | | | |
| 10 | | | | | | | | |
| 11 | | | | | | | | |
| 12 | | | | | | | | |
| 13 | | | | | | | | |
| 14 | | | | | | | | |
| 15 | | | | | | | | |
| 16 | | | | | | | | |

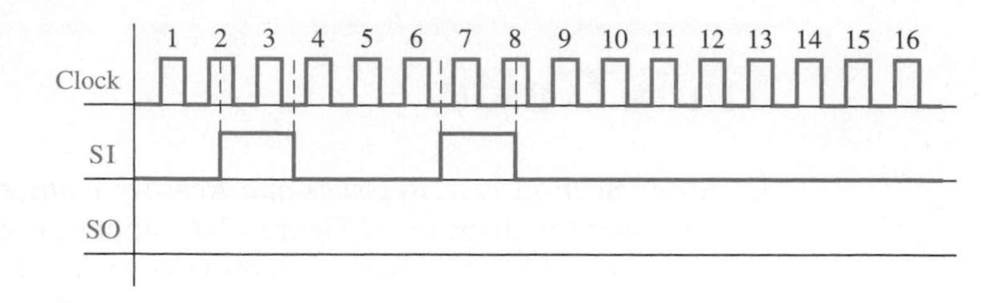


(c) Explain in words the function of the MUX on the $D$ input of flip-flop $Q_3$ in Figure 12-10(b). Explain in words the meaning of the first of Equations (12-1).

(d) Verify that Equations (12-1) are consistent with Table 12-1.
(e) Work Problem 12.2.

**3.** Study Section 12.3, *Design of Binary Counters*, and Section 12.4, *Counters for Other Sequences.*

(a) For Figure 12-13, if $CBA$ = 101, which of the $T$ inputs is 1?

(b) Complete the following timing diagram for the binary counter of Figure 12-13. The initial value of Clock is 1; this does not count as a rising edge.

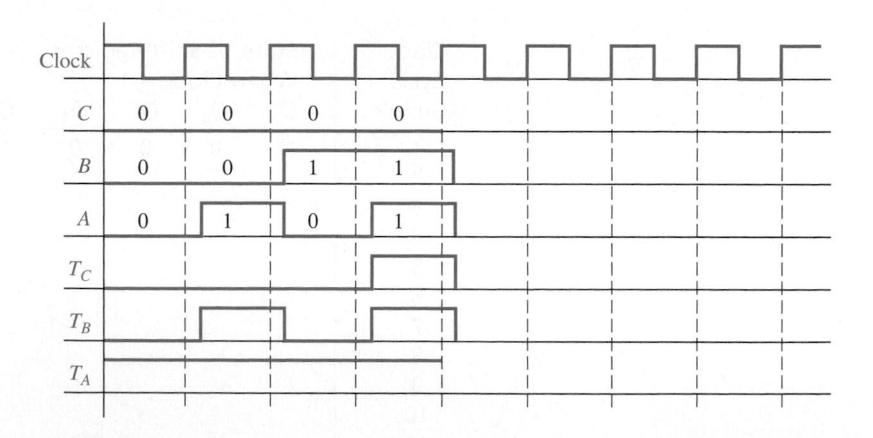


(c) Using the results of (b), draw a state graph for this binary counter (similar to Figure 12-21).

(d) Complete the following timing diagram for the binary counter of Figure 12-15.

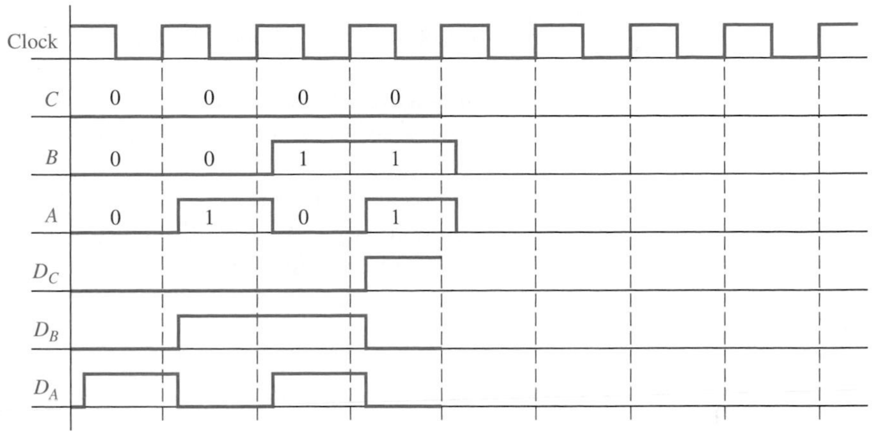

(e) Use Table 12-4 to verify that the values of $T_C$, $T_B$, and $T_A$ in Table 12-2 are correct.

(f) What happens if the circuit of Figure 12-23 is started in one of the don't-care states and, then, a clock pulse occurs? In particular, augment the state graph of Figure 12-25 to indicate the result for starting in states 101 and 110.

(g) What happens if the circuit of Figure 12-26 is started in one of the don't-care states and then a clock pulse occurs? In particular, augment the state graph of Figure 12-21 to indicate the result for starting in states 001, 101, and 110.

(h) Work Problems 12.3, 12.4, 12.5, 12.6, and 12.7.

**4.** Study Section 12.5, *Counter Design Using S-R and J-K Flip-Flops*.

(a) Referring to Table 12-5(c):
If $Q = Q^+ = 0$, explain in words why $R$ is a don't-care.

If $Q = Q^+ = 1$, explain in words why $S$ is a don't-care.

If $Q = 0$ and $Q^+ = 1$, what value should $S$ have and why?

If $Q = 1$ and $Q^+ = 0$, what value should $R$ have and why?

(b) For Figure 12-27, verify that the $R_B$ and $S_B$ maps are consistent with the $B^+$ map, and verify that the $R_c$ and $S_c$ maps are consistent with the $C^+$ map.

(c) In Figure 12-27, where do the gate inputs ($C$, $B$, $A$, etc.) come from?

(d) For Figure 12-27(c), which flip-flop inputs will be 1 if $CBA = 100$? What will be the state after the rising clock edge?

(e) Complete the following state graph by tracing signals in Figure 12-27(c). Compare your answer with Figure 12-21. What will happen if the counter is in state 110 and a clock pulse occurs?

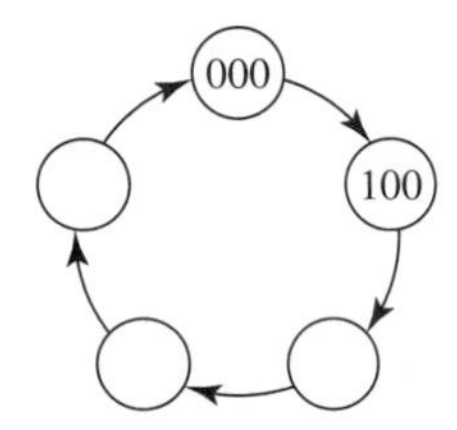


(f) Referring to Table 12-7(c).
If $Q = Q^+ = 0$, explain in words why $K$ is a don't-care.

If $Q = Q^+ = 1$, explain in words why $J$ is a don't-care.

If $Q = 0$ and $Q^+ = 1$, explain why both $JK = 10$ and $JK = 11$ will produce the required state change.

If $Q = 1$ and $Q^+ = 0$, give two sets of values for $J$ and $K$ which will produce the required state change, and explain why your answer is valid.

(g) Verify that the maps of Figure 12-28(b) can be derived from the maps of Figure 12-28(a).

(h) Compare the number of logic gates in Figures 12-27 and 12-28. The J-K realization requires fewer gates than the S-R realization because the J-K maps have more don't-cares than the S-R maps.

(i) Draw in the implied feedback connections on the circuit of Figure 12-28(c).

(j) By tracing signals through the circuit, verify that the state sequence for Figure 12-28(c) is correct.

(k) Find a minimum expression for $F_1$ and for $F_2$. (*Hint*: No variables are required.)

$F_1$:

| BC \ A | 0 | 1 |
|---|---|---|
| 00 | X | X |
| 01 | 1 | X |
| 11 | X | 1 |
| 10 | X | X |

$F_2$:

| BC \ A | 0 | 1 |
|---|---|---|
| 00 | X | X |
| 01 | 0 | X |
| 11 | X | X |
| 10 | X | X |

(l) Work Problems 12.8 and 12.9.

**5.** Study Section 12.6, *Derivation of Flip-Flop Input Equations—Summary*.

(a) Make sure that you know how to derive input equations for the different types of flip-flops. It is important that you understand the procedures for deriving the equations; merely memorizing the rules is not sufficient.

(b) Table 12-9 is provided mainly for reference. It is not intended that you memorize this table; instead you should understand the reasons for the entries in the table. If you understand the reasons why a given map entry is 0, 1, or X, you should be able to derive the flip-flop input maps without reference to a table.

**6.** Work the part of Problem 12.10 that you have been assigned. Bring your solution to this problem with you when you come to take the readiness test.

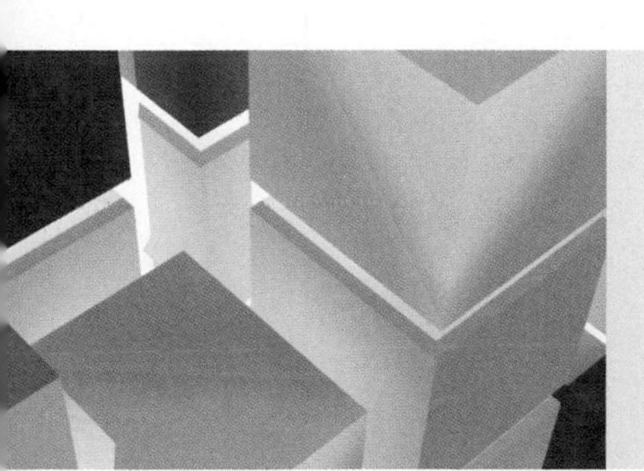

# Registers and Counters

A register consists of a group of flip-flops with a common clock input. Registers are commonly used to store and shift binary data. Counters are another simple type of sequential circuits. A counter is usually constructed from two or more flip-flops which change states in a prescribed sequence when input pulses are received. In this unit, you will learn procedures for deriving flip-flop input equations for counters. These procedures will be applied to more general types of sequential circuits in later units.

## 12.1 Registers and Register Transfers

Several D flip-flops may be grouped together with a common clock to form a register [Figure 12-1(a)]. Because each flip-flop can store one bit of information, this register can store four bits of information. This register has a load signal that is ANDed with the clock.

**FIGURE 12-1** 4-Bit D Flip-Flop Registers with Data, Load, Clear, and Clock Inputs

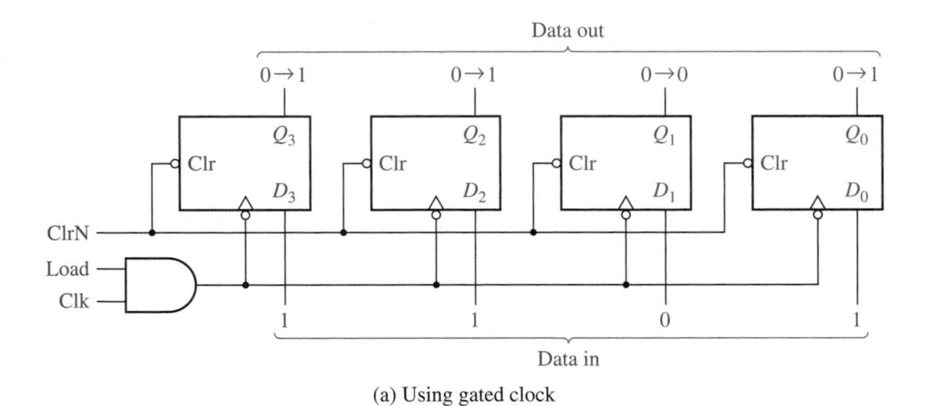


(a) Using gated clock

Data out
0→1 0→1 0→0 0→1
$Q_3$ $Q_2$ $Q_1$ $Q_0$
Clr CE $D_3$ Clr CE $D_2$ Clr CE $D_1$ Clr CE $D_0$
Load
ClrN
Clk
1 1 0 1
Data in

(b) With clock enable

4
$Q$
ClrN Clr
CE $D$
4
Load Clk

(c) Symbol

When Load = 0, the register is not clocked, and it holds its present value. When it is time to load data into the register, Load is set to 1 for one clock period. When Load = 1, the clock signal (Clk) is transmitted to the flip-flop clock inputs and the data applied to the $D$ inputs will be loaded into the flip-flops on the falling edge of the clock. For example, if the $Q$ outputs are 0000 ( $Q_3 = Q_2 = Q_1 = Q_0 = 0$) and the data inputs are 1101 ($D_3 = 1$, $D_2 = 1$, $D_1 = 0$ and $D_0 = 1$), after the falling edge $Q$ will change from 0000 to 1101 as indicated. (The notation $0 \rightarrow 1$ at the flip-flop outputs indicates a change from 0 to 1.)

The flip-flops in the register have asynchronous clear inputs that are connected to a common clear signal, ClrN. The bubble at the clear inputs indicates that a logic 0 is required to clear the flip-flops. ClrN is normally 1, and if it is changed momentarily to 0, the $Q$ outputs of all four flip-flops will become 0.

As discussed in Section 11.8, gating the clock with another signal can cause timing problems. If flip-flops with clock enable are available, the register can be designed as shown in Figure 12-1(b). The load signal is connected to all four CE inputs. When Load = 0, the clock is disabled and the register holds its data. When Load is 1, the clock is enabled, and the data applied to the $D$ inputs will be loaded into the flip-flops, following the falling edge of the clock. Figure 12-1(c) shows a symbol for the 4-bit register using bus notation for the $D$ inputs and $Q$ outputs. A group of wires that perform a common function is often referred to as a bus. A heavy line is used to represent a bus, and a slash with a number beside it indicates the number of bits in the bus.

Transferring data between registers is a common operation in digital systems. Figure 12-2 shows how data can be transferred from the output of one of two registers into a third register using tri-state buffers. If En = 1 and Load = 1, the output of register $A$ is enabled onto the tri-state bus and the data in register $A$ will be stored in $Q$ after the rising edge of the clock. If En = 0 and Load = 1, the output of register $B$ will be enabled onto the tri-state bus and stored in $Q$ after the rising edge of the clock.

**FIGURE 12-2** Data Transfer Between Registers

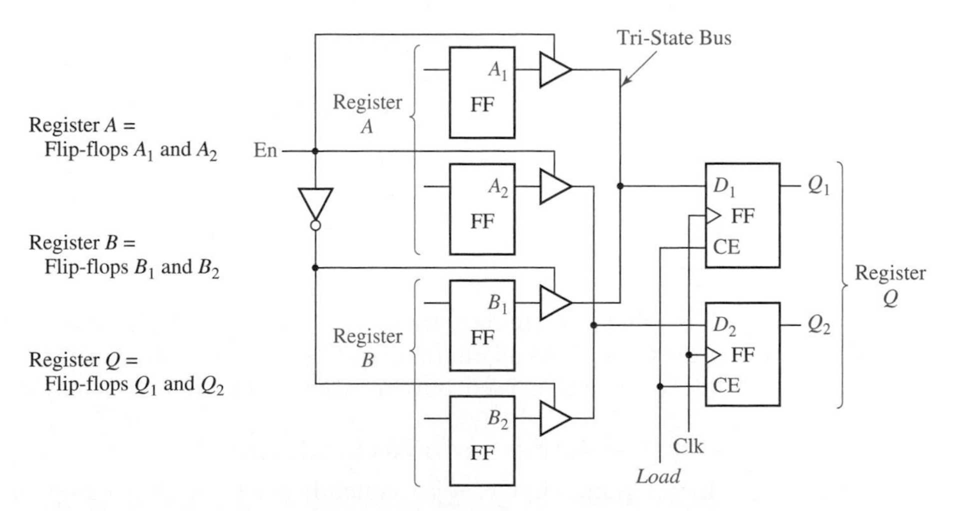


Figure 12-3(a) shows an integrated circuit register that contains eight D flip-flops with tri-state buffers at the flip-flop outputs. These buffers are enabled when En = 0. A symbol for this 8-bit register is shown in Figure 12-3(b).

Figure 12-4 shows how data can be transferred from one of four 8-bit registers into one of two other registers. Registers $A$, $B$, $C$, and $D$ are of the type shown in Figure 12-3.

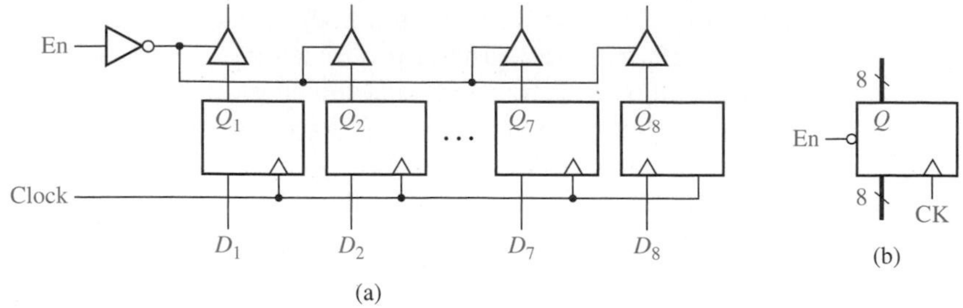

FIGURE 12-3
Logic Diagram for 8-Bit Register with Tri-State Output

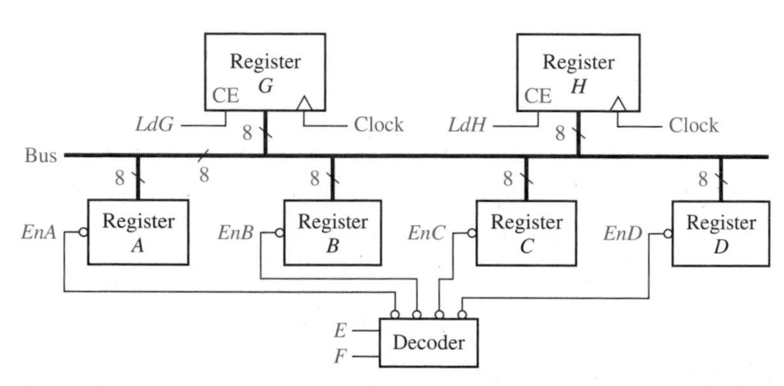


FIGURE 12-4
Data Transfer Using a Tri-State Bus

The outputs from these registers are all connected in parallel to a common tri-state bus. Registers $G$ and $H$ are similar to the register of Figure 12-1 except that they have eight flip-flops instead of four. The flip-flop inputs of registers $G$ and $H$ are also connected to the bus. When $EnA = 0$, the tri-state outputs of register $A$ are enabled onto the bus. If $LdG = 1$, these signals on the bus are loaded into register $G$ after the rising clock edge (or into register $H$ if $LdH = 1$). Similarly, the data in register $B$, $C$, or $D$ is transferred to $G$ (or $H$) when $EnB$, $EnC$, or $EnD$ is 0, respectively and $LdG = 1$ (or $LdH = 1$). If $LdG = LdH = 1$, both $G$ and $H$ will be loaded from the bus. The four enable signals may be generated by a decoder. The operation can be summarized as follows:

If $EF = 00$, $A$ is stored in $G$ (or $H$).
If $EF = 01$, $B$ is stored in $G$ (or $H$).
If $EF = 10$, $C$ is stored in $G$ (or $H$).
If $EF = 11$, $D$ is stored in $G$ (or $H$).

Note that 8 bits of data are transferred in parallel from register $A$, $B$, $C$, or $D$ to register $G$ or $H$. As an alternative to using a bus with tri-state logic, eight 4-to-1 multiplexers could be used, but this would lead to a more complex circuit.

## Parallel Adder with Accumulator

In computer circuits, it is frequently desirable to store one number in a register of flip-flops (called an accumulator) and add a second number to it, leaving the result stored in the accumulator. One way to build a parallel adder with an accumulator is to add a register to the adder of Figure 4-2, resulting in the circuit of Figure 12-5. Suppose that the number $X = x_n \ldots x_2x_1$ is stored in the accumulator. Then, the

FIGURE 12-5 *n*-Bit Parallel Adder with Accumulator

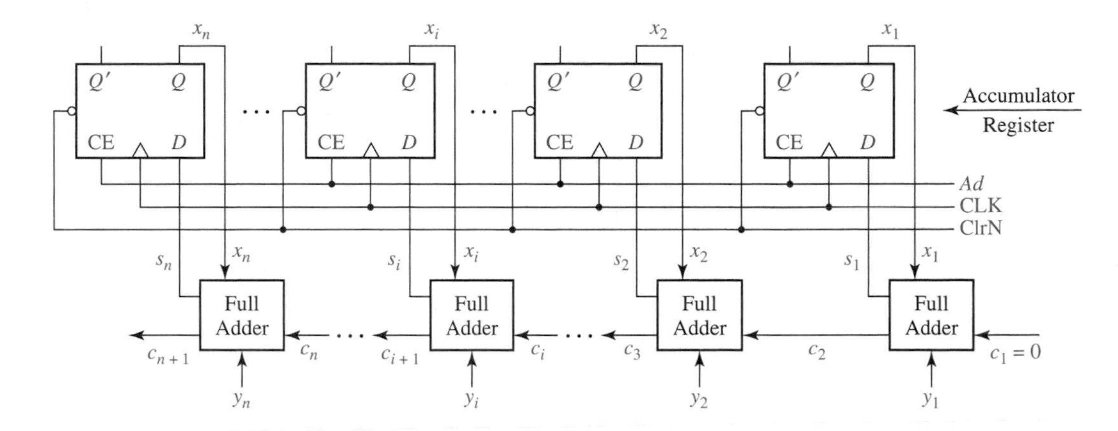


number $Y = y_n \ldots y_2y_1$ is applied to the full adder inputs, and after the carry has propagated through the adders, the sum of $X$ and $Y$ appears at the adder outputs. An add signal ($Ad$) is used to load the adder outputs into the accumulator flip-flops on the rising clock edge. If $s_i = 1$, the next state of flip-flop $x_i$ will be 1. If $s_i = 0$, the next state of flip-flop $x_i$ will be 0. Thus, $x_i^+ = s_i$, and if $Ad = 1$, the number $X$ in the accumulator is replaced with the sum of $X$ and $Y$, following the rising edge of the clock.

Observe that the adder with accumulator is an iterative structure that consists of a number of identical cells. Each cell contains a full adder and an associated accumulator flip-flop. Cell $i$, which has inputs $c_i$ and $y_i$ and outputs $c_{i+1}$ and $x_i$, is referred to as a typical cell.

Before addition can take place, the accumulator must be loaded with $X$. This can be accomplished in several ways. The easiest way is to first clear the accumulator using the asynchronous clear inputs on the flip-flops, and then put the $X$ data on the $Y$ inputs to the adder and add to the accumulator in the normal way. Alternatively, we could add multiplexers at the accumulator inputs so that we could select either the $Y$ input data or the adder output to load into the accumulator. This would eliminate the extra step of clearing the accumulator but would add to the hardware complexity. Figure 12-6

FIGURE 12-6 Adder Cell with Multiplexer

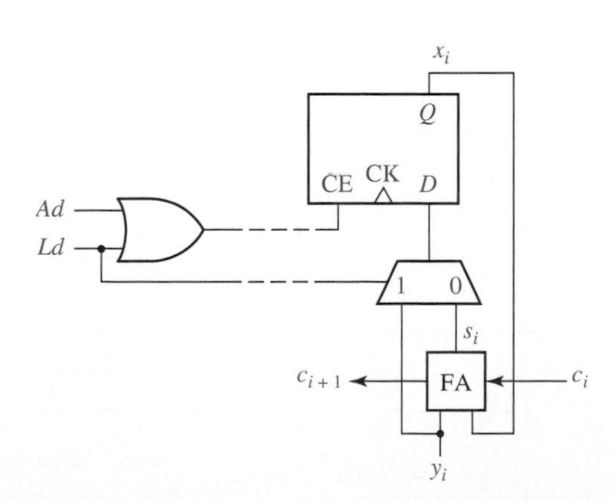

shows a typical cell of the adder where the accumulator flip-flop can either be loaded directly from $y_i$ or from the sum output ($s_i$). When $Ld = 1$ the multiplexer selects $y_i$, and $y_i$ is loaded into the accumulator flip-flop ($x_i$) on the rising clock edge. When $Ad = 1$ and $Ld = 0$, the adder output ($s_i$) is loaded into $x_i$. The $Ad$ and $Ld$ signals are ORed together to enable the clock when either addition or loading occurs. When $Ad = Ld = 0$, the clock is disabled and the accumulator outputs do not change.

## 12.2 Shift Registers

A shift register is a register in which binary data can be stored, and this data can be shifted to the left or right when a shift signal is applied. Bits shifted out one end of the register may be lost, or if the shift register is of cyclic type, bits shifted out one end are shifted back in the other end. Figure 12-7(a) illustrates a 4-bit right-shift register with serial input and output constructed from D flip-flops. When Shift = 1, the clock is enabled and shifting occurs on the rising clock edge. When Shift = 0, no shifting occurs and the data in the register is unchanged. The serial input (SI) is loaded into the first flip-flop ($Q_3$) by the rising edge of the clock. At the same time, the output of

**FIGURE 12-7** Right-Shift Register

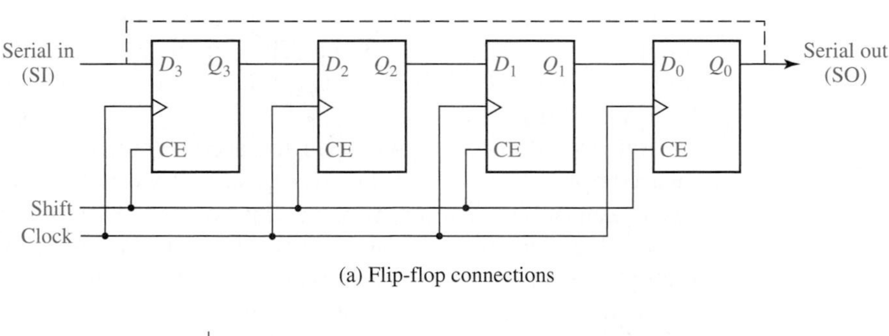


(a) Flip-flop connections

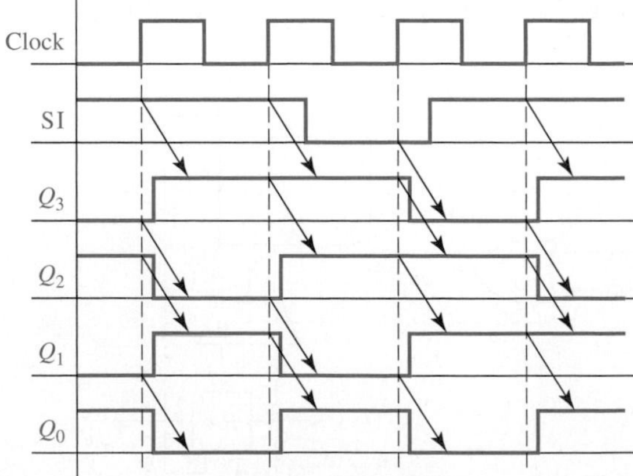


(b) Timing diagram

the first flip-flop is loaded into the second flip-flop, the output of the second flip-flop is loaded into the third flip-flop, and the output of the third flip-flop is loaded into the last flip-flop. Because of the propagation delay of the flip-flops, the output value loaded into each flip-flop is the value before the rising clock edge. Figure 12-7(b) illustrates the timing when the shift register initially contains 0101 and the serial input sequence is 1, 1, 0, 1. The sequence of shift register states is 0101, 1010, 1101, 0110, 1011.

If we connect the serial output to the serial input, as shown by the dashed line, the resulting cyclic shift register performs an end-around shift. If the initial contents of the register is 0111, after one clock cycle the contents is 1011. After a second pulse, the state is 1101, then 1110, and the fourth pulse returns the register to the initial 0111 state.

Shift registers with 4, 8, or more flip-flops are available in integrated circuit form. Figure 12-8 illustrates an 8-bit serial-in, serial-out shift register. *Serial in* means that data is shifted into the first flip-flop one bit at a time, and the flip-flops cannot be loaded in parallel. *Serial out* means that data can only be read out of the last flip-flop and the outputs from the other flip-flops are not connected to terminals of the integrated circuit. The inputs to the first flip-flop are $S$ = SI and $R$ = SI$'$. Thus, if SI = 1, a 1 is shifted into the register when it is clocked, and if SI = 0, a 0 is shifted in. Figure 12-9 shows a typical timing diagram.

Figure 12-10(a) shows a 4-bit parallel-in, parallel-out shift register. Parallel-in implies that all four bits can be loaded at the same time, and parallel-out

**FIGURE 12-8** 8-Bit Serial-in, Serial-out Shift Register

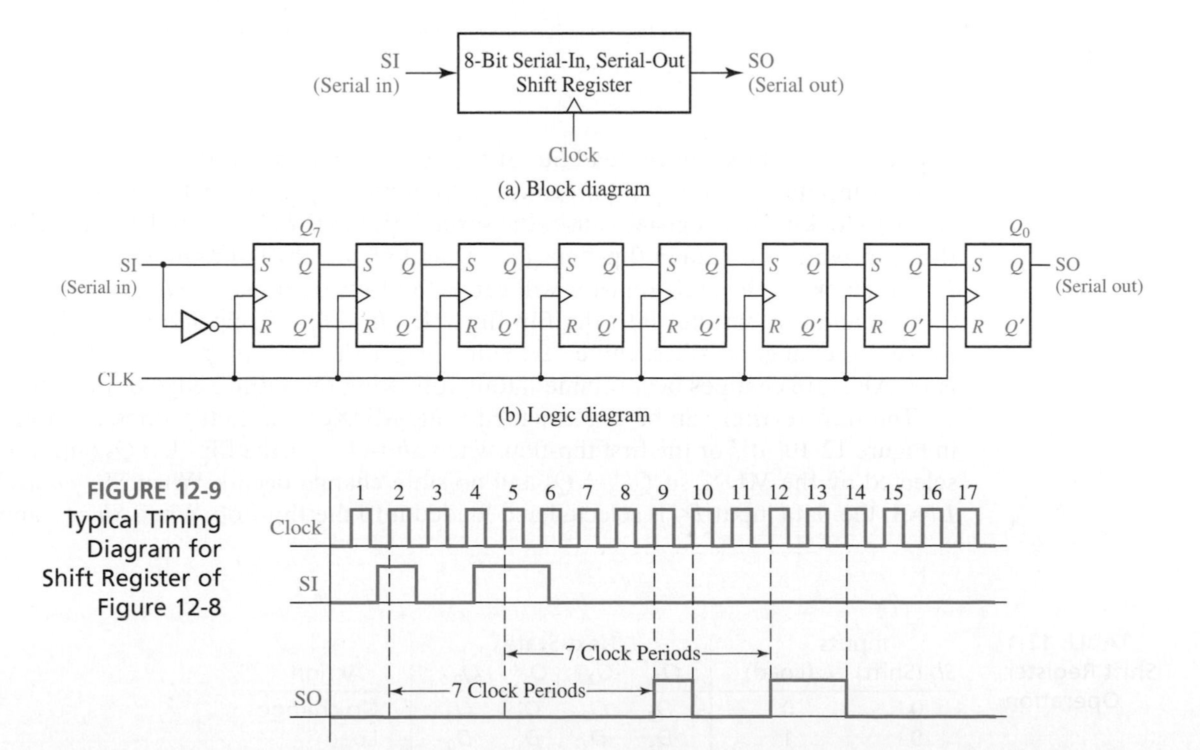


**FIGURE 12-9** Typical Timing Diagram for Shift Register of Figure 12-8

**FIGURE 12-10** Parallel-in, Parallel-Out Right Shift Register

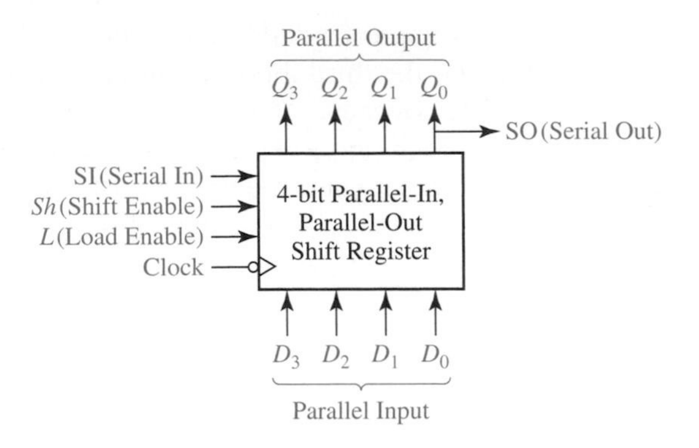


(a) Block diagram

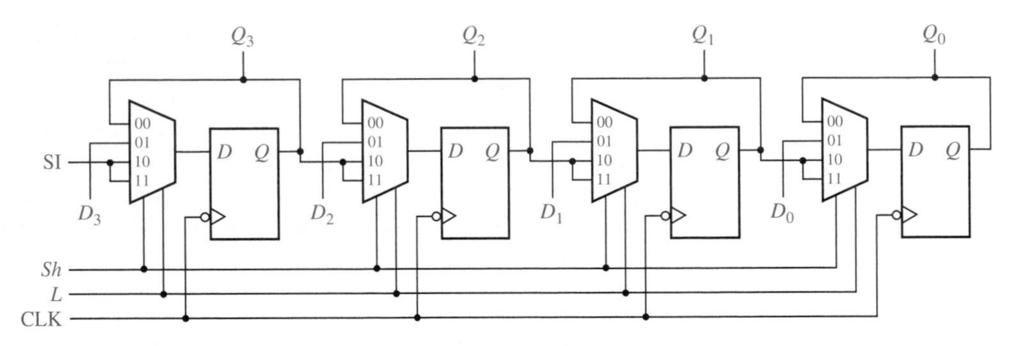


(b) Implementation using flip-flops and MUXes

implies that all bits can be read out at the same time. The shift register has two control inputs, shift enable ($Sh$) and load enable ($L$). If $Sh = 1$ (and $L = 1$ or $L = 0$), clocking the register causes the serial input (SI) to be shifted into the first flip-flop, while the data in flip-flops $Q_3$, $Q_2$, and $Q_1$ are shifted right. If $Sh = 0$ and $L = 1$, clocking the shift register will cause the four data inputs ($D_3$, $D_2$, $D_1$, $D_0$) to be loaded in parallel into the flip-flops. If $Sh = L = 0$, clocking the register causes no change of state. Table 12-1 summarizes the operation of this shift register. All state changes occur immediately following the falling edge of the clock.

The shift register can be implemented using MUXes and D flip-flops, as shown in Figure 12-10(b). For the first flip-flop, when $Sh = L = 0$, the flip-flop $Q_3$ output is selected by the MUX, so $Q_3^+ = Q_3$ and no state change occurs. When $Sh = 0$ and $L = 1$, the data input $D_3$ is selected and loaded into the flip-flop. When $Sh = 1$ and

**TABLE 12-1** Shift Register Operation

| Inputs | | Next State | | | | |
|---|---|---|---|---|---|---|
| $Sh$ (Shift) | $L$ (Load) | $Q_3^+$ | $Q_2^+$ | $Q_1^+$ | $Q_0^+$ | Action |
| 0 | 0 | $Q_3$ | $Q_2$ | $Q_1$ | $Q_0$ | No change |
| 0 | 1 | $D_3$ | $D_2$ | $D_1$ | $D_0$ | Load |
| 1 | X | SI | $Q_3$ | $Q_2$ | $Q_1$ | Right shift |

$L = 0$ or 1, SI is selected and loaded into the flip-flop. The second MUX selects $Q_2$, $D_2$, or $Q_3$, etc. The next-state equations for the flip-flops are

$$Q_3^+ = Sh'\cdot L'\cdot Q_3 + Sh'\cdot L\cdot D_3 + Sh\cdot \text{SI} \tag{12-1}$$
$$Q_2^+ = Sh'\cdot L'\cdot Q_2 + Sh'\cdot L\cdot D_2 + Sh\cdot Q_3$$
$$Q_1^+ = Sh'\cdot L'\cdot Q_1 + Sh'\cdot L\cdot D_1 + Sh\cdot Q_2$$
$$Q_0^+ = Sh'\cdot L'\cdot Q_0 + Sh'\cdot L\cdot D_0 + Sh\cdot Q_1$$

A typical application of this register is the conversion of parallel data to serial data. The output from the last flip-flop ($Q_0$) serves as a serial output as well as one of the parallel outputs. Figure 12-11 shows a typical timing diagram. The first clock pulse loads data into the shift register in parallel. During the next four clock pulses, this data is available at the serial output. Assuming that the register is initially clear ($Q_3Q_2Q_1Q_0 = 0000$), that the serial input is SI $= 0$ throughout, and that the data inputs $D_3D_2D_1D_0$ are 1011 during the load time ($t_0$), the resulting waveforms are as shown. Shifting occurs at the end of $t_1$, $t_2$, and $t_3$, and the serial output can be read during these clock times. During $t_4$, $Sh = L = 0$, so no state change occurs.

Figure 12-12(a) shows a 3-bit shift register with the $Q_1'$ output from the last flip-flop fed back into the $D$ input of the first flip-flop. If the initial state of the register is 000, the initial value of $D_3$ is 1, so after the first clock pulse, the register state is 100. Successive states are shown on the state graph of Figure 12-12(b). When the

**FIGURE 12-11** Timing Diagram for Shift Register

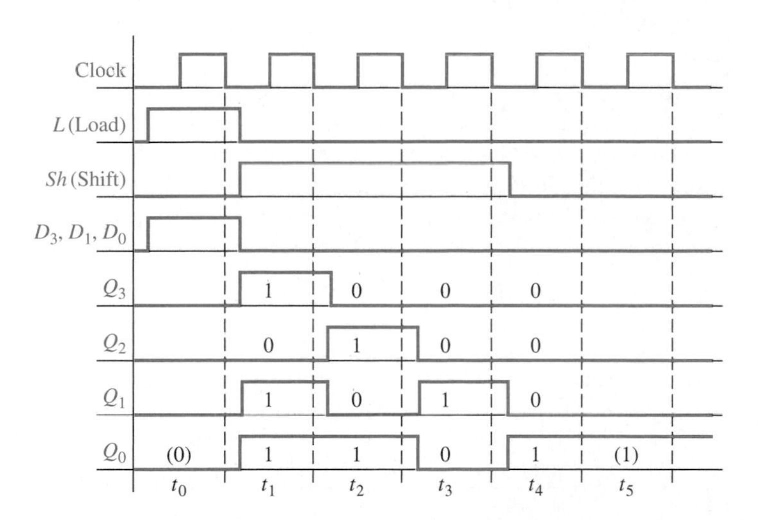


**FIGURE 12-12** Shift Register with Inverted Feedback

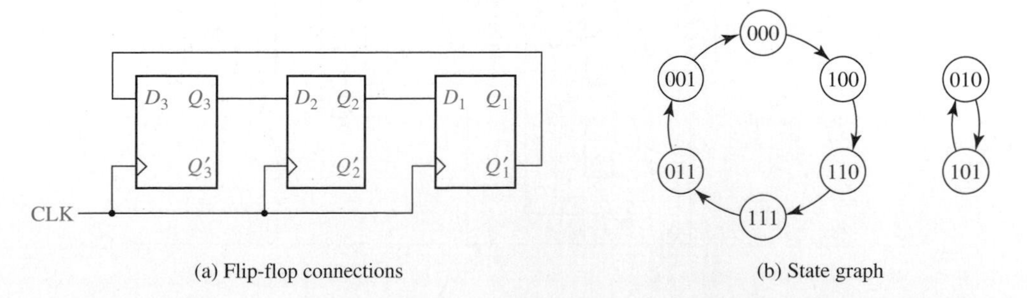


(a) Flip-flop connections

(b) State graph

register is in state 001, $D_3$ is 0, and the next register state is 000. Then, successive clock pulses take the register around the loop again. Note that states 010 and 101 are not in the main loop. If the register is in state 010, then a shift pulse takes it to 101 and vice versa; therefore, we have a secondary loop on the state graph. A circuit that cycles through a fixed sequence of states is called a *counter*, and a shift register with inverted feed back is often called a *Johnson counter*.

## 12.3 Design of Binary Counters

The counters discussed in this chapter are all synchronous counters. This means the operation of the flip-flops is synchronized by a common clock pulse so that when several flip-flops must change state, the state changes occur simultaneously. Ripple counters, in which the state change of one flip-flop triggers the next flip-flop in line, are not discussed in this text.

We will first construct a binary counter using three T flip-flops to count clock pulses (Figure 12-13). We will assume that all the flip-flops change state a short time following the rising edge of the input pulse. The state of the counter is determined by the states of the individual flip-flops; for example, if flip-flop $C$ is in state 0, $B$ in state 1, and $A$ in state 1, the state of the counter is 011. Initially, assume that all flip-flops are set to the 0 state. When a clock pulse is received, the counter will change to state 001; when a second pulse is received, the state will change to 010, etc. The sequence of flip-flop states is $CBA$ = 000, 001, 010, 011, 100, 101, 110, 111, 000, . . . Note that when the counter reaches state 111, the next pulse resets it to the 000 state, and then the sequence repeats.

First, we will design the counter by inspection of the counting sequence; then, we will use a systematic procedure which can be generalized to other types of counters. The problem is to determine the flip-flop inputs—$T_C$, $T_B$, and $T_A$. From the preceding counting sequence, observe that $A$ changes state every time a clock pulse is received. Because $A$ changes state on every rising clock edge, $T_A$ must equal 1. Next, observe that $B$ changes state only if $A = 1$. Therefore, $A$ is connected to $T_B$ as shown, so that if $A = 1$, $B$ will change state when a rising clock edge occurs. Similarly, $C$ changes state when a rising clock edge occurs only if $B$ and $A$ are both 1. Therefore, an AND gate is connected to $T_C$ so that $C$ will change state if $B = 1$ and $A = 1$ when a rising clock edge occurs.

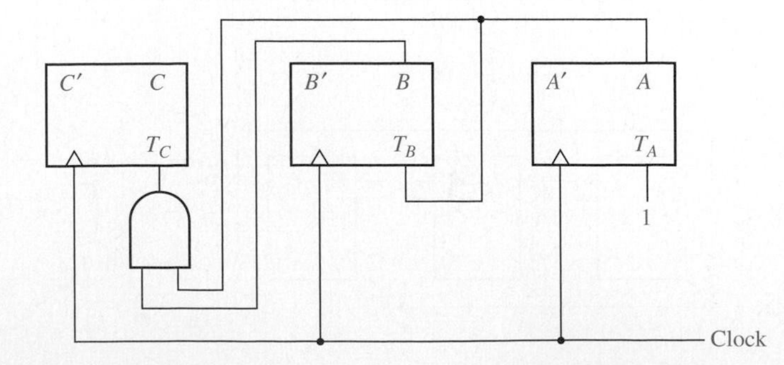


FIGURE 12-13 Synchronous Binary Counter

**TABLE 12-2** State Table for Binary Counter

| Present State | | | Next State | | | Flip-Flop Inputs | | |
|---|---|---|---|---|---|---|---|---|
| $C$ | $B$ | $A$ | $C^+$ | $B^+$ | $A^+$ | $T_C$ | $T_B$ | $T_A$ |
| 0 | 0 | 0 | 0 | 0 | 1 | 0 | 0 | 1 |
| 0 | 0 | 1 | 0 | 1 | 0 | 0 | 1 | 1 |
| 0 | 1 | 0 | 0 | 1 | 1 | 0 | 0 | 1 |
| 0 | 1 | 1 | 1 | 0 | 0 | 1 | 1 | 1 |
| 1 | 0 | 0 | 1 | 0 | 1 | 0 | 0 | 1 |
| 1 | 0 | 1 | 1 | 1 | 0 | 0 | 1 | 1 |
| 1 | 1 | 0 | 1 | 1 | 1 | 0 | 0 | 1 |
| 1 | 1 | 1 | 0 | 0 | 0 | 1 | 1 | 1 |

Now, we will verify that the circuit of Figure 12-13 counts properly by tracing signals through the circuit. Initially, $CBA = 000$, so only $T_A$ is 1 and the state will change to 001 when the first active clock edge arrives. Then, $T_B = T_A = 1$, and the state will change to 010 when the second active clock arrives. This process continues until finally when state 111 is reached, $T_C = T_B = T_A = 1$, and all flip-flops return to the 0 state.

Next, we will redesign the binary counter by using a state table (Table 12-2). This table shows the present state of flip-flops $C$, $B$, and $A$ (before a clock pulse is received) and the corresponding next state (after the clock pulse is received). For example, if the flip-flops are in state $CBA = 011$ and a clock pulse is received, the next state will be $C^+B^+A^+ = 100$. Although the clock is not explicit in the table, it is understood to be the input that causes the counter to go to the next state in sequence. A third column in the table is used to derive the inputs for $T_C$, $T_B$, and $T_A$. Whenever the entries in the $A$ and $A^+$ columns differ, flip-flop $A$ must change state and $T_A$ must be 1. Similarly, if $B$ and $B^+$ differ, $B$ must change state so $T_B$ must be 1. For example, if $CBA = 011$, $C^+B^+A^+ = 100$, all three flip-flops must change state, so $T_CT_BT_A = 111$.

$T_C$, $T_B$, and $T_A$ are now derived from the table as functions of $C$, $B$, and $A$. By inspection, $T_A = 1$. Figure 12-14 shows the Karnaugh maps for $T_C$ and $T_B$, from which $T_C = BA$ and $T_B = A$. These equations yield the same circuit derived previously for Figure 12-13.

Next, we will redesign the binary counter to use D flip-flops instead of T flip-flops. The easiest way to do this is to convert each D flip-flop to a T flip-flop by adding an XOR (exclusive-OR) gate, as shown in Figure 11-24(b). Figure 12-15

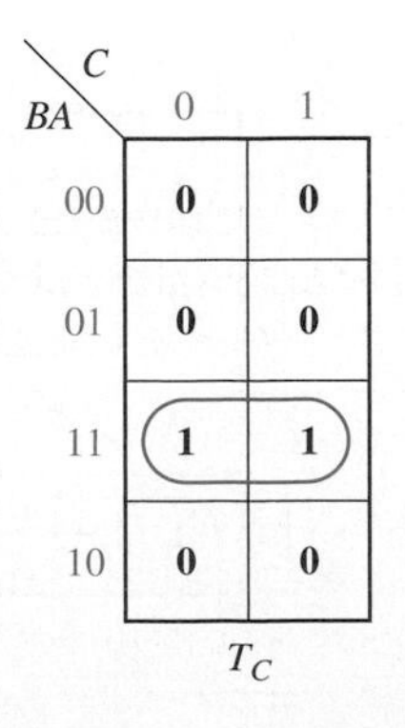

C
BA
0
1
00
01
11
10
0
0
0
0
1
1
0
0
$T_C$


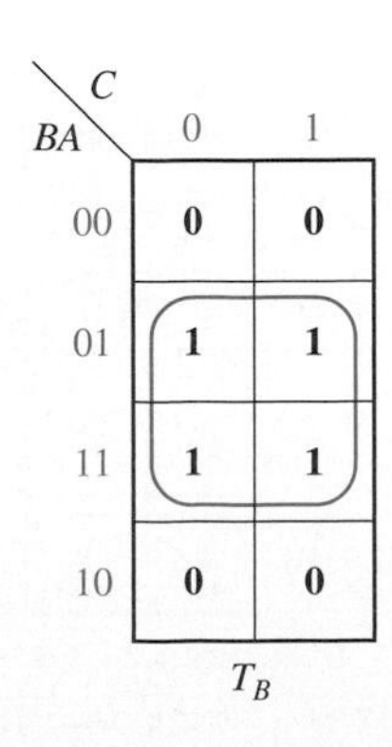

C
BA
0
1
00
01
11
10
0
0
1
1
1
1
0
0
$T_B$


**FIGURE 12-14** Karnaugh Maps for Binary Counter

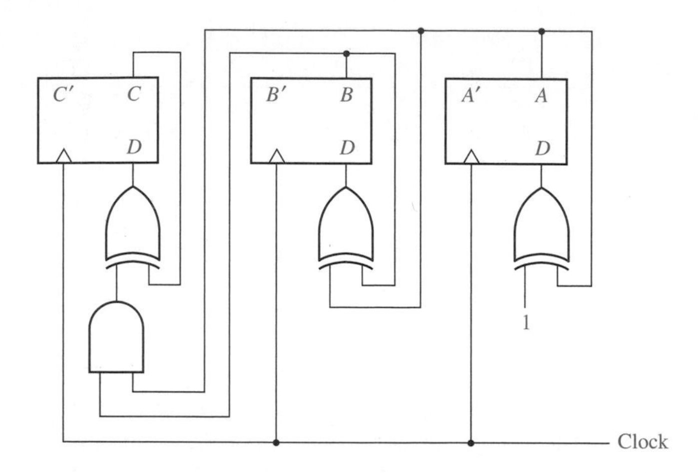


**FIGURE 12-15** Binary Counter with D Flip-Flops

shows the resulting counter circuit. The rightmost XOR gate can be replaced with an inverter because $A \oplus 1 = A'$.

We can also derive the D flip-flop inputs for the binary counter starting with its state table (Table 12-2). For a D flip-flop, $Q^+ = D$. By inspection of the table, $Q_A{}^+ = A'$, so $D_A = A'$. The maps for $Q_B{}^+$ and $Q_C{}^+$ are plotted in Figure 12-16. The $D$ input equations derived from the maps are

$$
\begin{aligned}
D_A &= A^+ = A' \\
D_B &= B^+ = BA' + B'A = B \oplus A \qquad (12\text{-}2) \\
D_C &= C^+ = C'BA + CB' + CA' = C'BA + C(BA)' = C \oplus BA
\end{aligned}
$$

which give the same logic circuit as was obtained by inspection.

Next, we will analyze an up-down binary counter. The state graph and table for an up-down counter are shown in Figure 12-17. When $U = 1$, the counter counts up in the sequence 000, 001, 010, 011, 100, 101, 110, 111, 000 . . . When $D = 1$, the counter counts down in the sequence 000, 111, 110, 101, 100, 011, 010, 001, 000 . . . When $U = D = 0$, the counter state does not change, and $U = D = 1$ is not allowed.

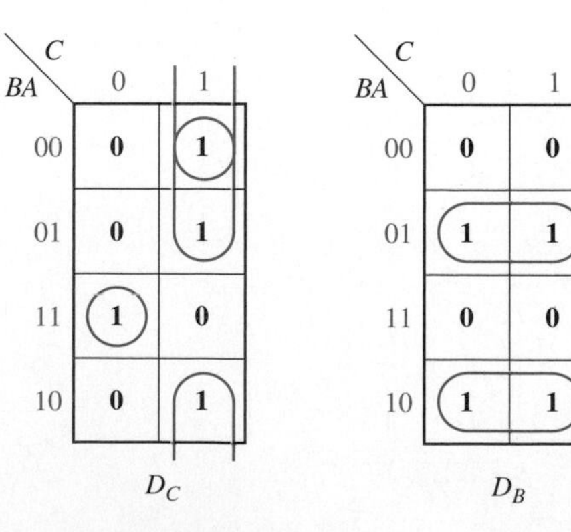


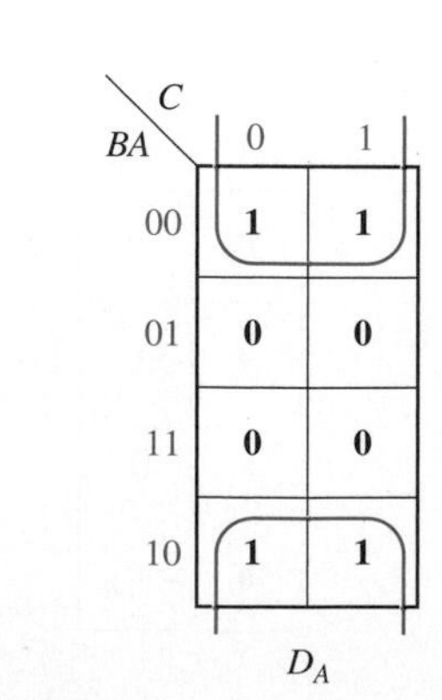


**FIGURE 12-16** Karnaugh Maps for D Flip-Flops

**FIGURE 12-17** State Graph and Table for Up-Down Counter

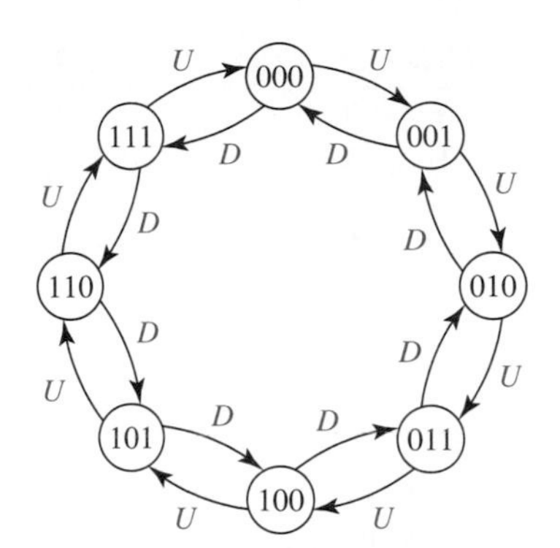


| CBA | $C^+B^+A^+$ U | D |
|---|---|---|
| 000 | 001 | 111 |
| 001 | 010 | 000 |
| 010 | 011 | 001 |
| 011 | 100 | 010 |
| 100 | 101 | 011 |
| 101 | 110 | 100 |
| 110 | 111 | 101 |
| 111 | 000 | 110 |

The up-down counter can be implemented using D flip-flops and gates, as shown in Figure 12-18. The corresponding logic equations are

$$D_A = A^+ = A \oplus (U + D)$$
$$D_B = B^+ = B \oplus (UA + DA')$$
$$D_C = C^+ = C \oplus (UBA + DB'A')$$

When $U = 1$ and $D = 0$, these equations reduce to equations for a binary up counter (Equations (12-2)).

When $U = 0$ and $D = 1$, these equations reduce to

$$D_A = A^+ = A \oplus 1 = A' \quad \text{(}A\text{ changes state every clock cycle)}$$
$$D_B = B^+ = B \oplus A' \quad \text{(}B\text{ changes state when }A = 0\text{)}$$
$$D_C = C^+ = C \oplus B'A' \quad \text{(}C\text{ changes state when }B = A = 0\text{)}$$

**FIGURE 12-18** Binary Up-Down Counter

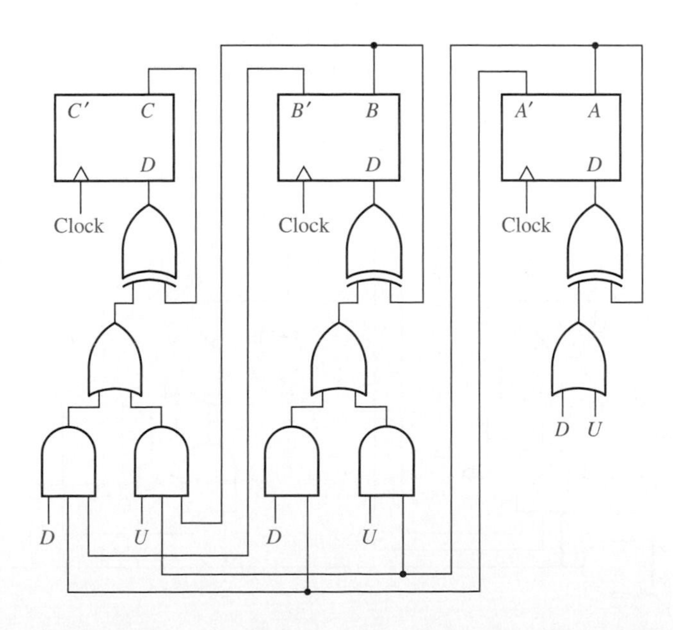

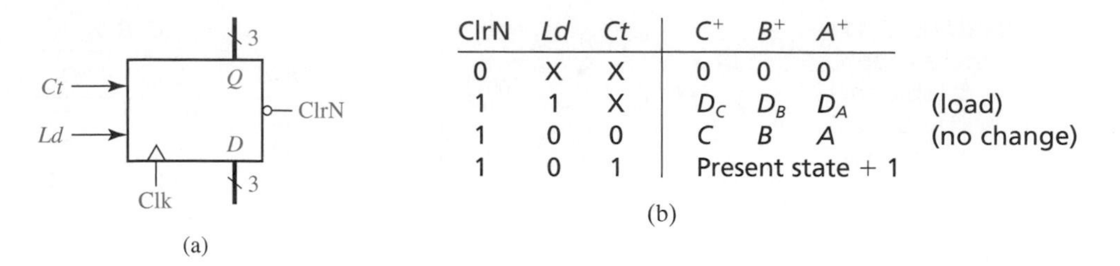


| ClrN | $Ld$ | $Ct$ | $C^+$ | $B^+$ | $A^+$ | |
|---|---|---|---|---|---|---|
| 0 | X | X | 0 | 0 | 0 | |
| 1 | 1 | X | $D_C$ | $D_B$ | $D_A$ | (load) |
| 1 | 0 | 0 | $C$ | $B$ | $A$ | (no change) |
| 1 | 0 | 1 | Present state + 1 | | | |

**FIGURE 12-19** Loadable Counter with Count Enable

By inspection of the table in Figure 12-17, we can verify that these are the correct equations for a down counter. For every row of the table, $A^+ = A'$, so $A$ changes state every clock cycle. For those rows where $A = 0, B^+ = B'$. For those rows where $B = 0$ and $A = 0, C^+ = C'$.

Next, we will design a loadable counter [Figure 12-19(a)]. This counter has two control signals $Ld$ (load) and $Ct$ (count). When $Ld = 1$ binary data is loaded into the counter on the rising clock edge, and when $Ct = 1$, the counter is incremented on the rising clock edge. When $Ld = Ct = 0$, the counter holds its present state. When $Ld = Ct = 1$, load overrides count, and data is loaded into the counter. The counter also has an asynchronous clear input that clears the counter when ClrN is 0. Figure 12-19(b) summarizes the counter operation. All state changes occur on the rising edge of the clock (except for the asynchronous clear).

Figure 12-20 shows how the loadable counter can be implemented using flip-flops, MUXes, and gates. When $Ld = 1$, each MUX selects a $D_i$ input, and because the output of each AND gate is 0, the output of each XOR gate is $D_i$, which gets stored in a flip-flop. When $Ld = 0$ and $Ct = 1$, each MUX selects one of the flip-flop outputs ($C$, $B$, or $A$). The circuit then becomes equivalent to Figure 12-15, and the counter is incremented on the rising clock edge.

The next-state equations for the counter of Figure 12-20 are

$$A^+ = D_A = (Ld' \cdot A + Ld \cdot D_{Ain}) \oplus Ld' \cdot Ct$$
$$B^+ = D_B = (Ld' \cdot B + Ld \cdot D_{Bin}) \oplus Ld' \cdot Ct \cdot A$$
$$C^+ = D_C = (Ld' \cdot C + Ld \cdot D_{Cin}) \oplus Ld' \cdot Ct \cdot B \cdot A$$

**FIGURE 12-20** Circuit for Figure 12-19

When $Ld = 0$ and $Ct = 1$, these equations reduce to $A^+ = A'$, $B^+ = B \oplus A$, and $C^+ = C \oplus BA$, which are the equations previously derived for a 3-bit counter.

## 12.4 Counters for Other Sequences

In some applications, the sequence of states of a counter is not in straight binary order. Figure 12-21 shows the state graph for such a counter. The arrows indicate the state sequence. If this counter is started in state 000, the first clock pulse will take it to state 100, the next pulse to 111, etc. The clock pulse is implicitly understood to be the input to the circuit and not shown on the graph. The corresponding state table for the counter is Table 12-3. Note that the next state is unspecified for the present states 001, 101, and 110.

We will design the counter specified by Table 12-3 using T flip-flops. We could derive $T_C$, $T_B$, and $T_A$ directly from this table, as in the preceding example. However, it is often more convenient to plot next-state maps showing $C^+$, $B^+$, and $A^+$ as functions of $C$, $B$, and $A$, and then derive $T_C$, $T_B$, and $T_A$ from these maps. The next-state maps in Figure 12-22(a) are easily plotted from inspection of Table 12-3. From the first row of the table, the $CBA = 000$ squares on the $C^+$, $B^+$, and $A^+$ maps are filled in with 1, 0, and 0, respectively. From the second row, the $CBA = 001$ squares on all three maps are filled in with don't-cares. From the third row, the $CBA = 010$ squares on the $C^+$, $B^+$, and $A^+$ maps are filled in with 0, 1, and 1, respectively. The next-state maps can be quickly completed by continuing in this manner.

Next, we will derive the maps for the $T$ inputs from the next-state maps. In the following discussion, the general symbol $Q$ represents the present state of the flip-flop ($C$, $B$, or $A$) under consideration, and $Q^+$ represents the next state ($C^+$, $B^+$, or $A^+$) of the same flip-flop. Given the present state of a T flip-flop ($Q$) and the desired next

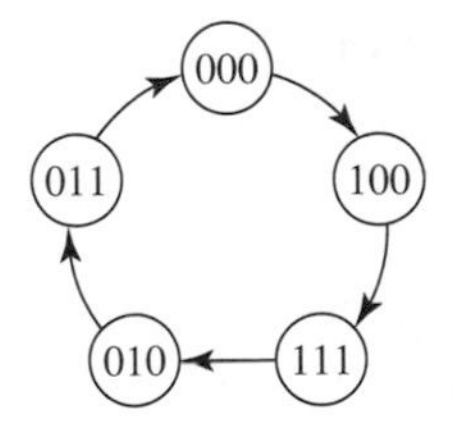


FIGURE 12-21 State Graph for Counter

TABLE 12-3 State Table for Figure 12.21

| C | B | A | $C^+$ | $B^+$ | $A^+$ |
|---|---|---|---|---|---|
| 0 | 0 | 0 | 1 | 0 | 0 |
| 0 | 0 | 1 | – | – | – |
| 0 | 1 | 0 | 0 | 1 | 1 |
| 0 | 1 | 1 | 0 | 0 | 0 |
| 1 | 0 | 0 | 1 | 1 | 1 |
| 1 | 0 | 1 | – | – | – |
| 1 | 1 | 0 | – | – | – |
| 1 | 1 | 1 | 0 | 1 | 0 |

FIGURE 12-22

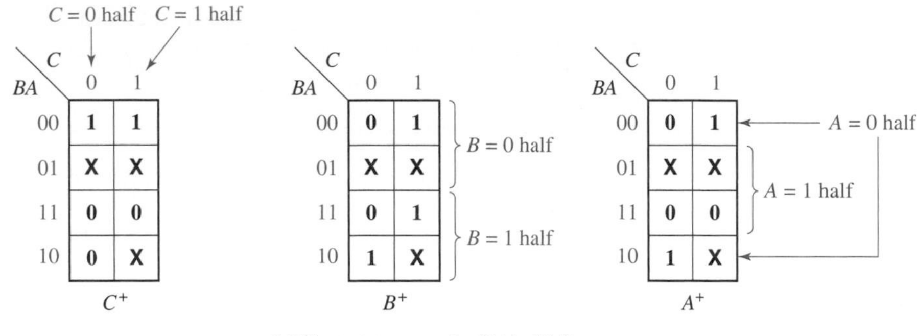


(a) Next-state maps for Table 12-3

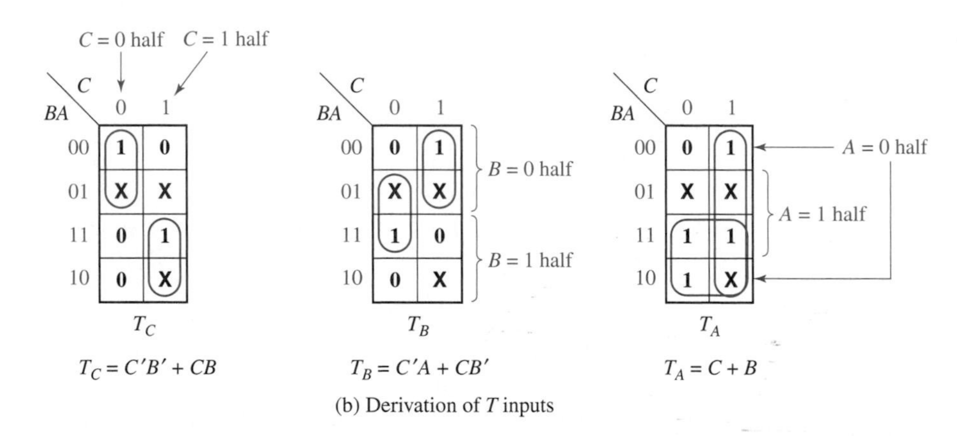


(b) Derivation of $T$ inputs

state ($Q^+$), the $T$ input must be 1 whenever a change of state is required. Thus, $T = 1$ whenever $Q^+ \neq Q$, as shown in Table 12-4.

In general, the next-state map for flip-flop $Q$ gives $Q^+$ as a function of $Q$ and several other variables. The value written in each square of the map gives the value of $Q^+$, while the value of $Q$ is determined from the row or column headings. Given the map for $Q^+$, we can then form the map for $T_Q$ by simply putting a 1 in each square of the $T_Q$ map for which $Q^+$ is different from $Q$. Thus, to form the $T_C$ map in Figure 12-22(b) from the $C^+$ map in Figure 12-22(a), we place a 1 in the $CBA = 000$ square of $T_C$ because $C = 0$ and $C^+ = 1$ for this square. We also place a 1 in the 111 square of $T_C$ because $C = 1$ and $C^+ = 0$ for this square.

If we don't care what the next state of a flip-flop is for some combination of variables, we don't care what the flip-flop input is for that combination of variables. Therefore, if the $Q^+$ map has a don't-care in some square, the $T_Q$ map will have a don't-care in the corresponding square. Thus, the $T_C$ map has don't-cares for $CBA$ = 001, 101, and 110 because $C^+$ has don't-cares in the corresponding squares.

TABLE 12-4
Input for T Flip-Flop

| $Q$ | $Q^+$ | $T$ |
|---|---|---|
| 0 | 0 | 0 |
| 0 | 1 | 1 |
| 1 | 0 | 1 |
| 1 | 1 | 0 |

$T = Q^+ \oplus Q$

Instead of transforming the $Q^+$ map into the $T_Q$ map one square at a time, we can divide the $Q^+$ map into two halves corresponding to $Q = 0$ and $Q = 1$, and transform each half of the map. From Table 12-4, whenever $Q = 0$, $T = Q^+$, and whenever $Q = 1$, $T = (Q^+)'$. Therefore, to transform the $Q^+$ map into a $T$ map, we copy the half for which $Q = 0$ and complement the half for which $Q = 1$, leaving the don't-cares unchanged.

We will apply this method to transform the $C^+$, $B^+$, and $A^+$ maps for our counter shown in Figure 12-22(a) into $T$ maps. For the first map, $C$ corresponds to $Q$ (and $C^+$ to $Q^+$), so to get the $T_C$ map from the $C^+$ map, we complement the second column (where $C = 1$) and leave the rest of the map unchanged. Similarly, to get $T_B$ from $B^+$, we complement the bottom half of the $B$ map, and to get $T_A$ from $A^+$, we complement the middle two rows. This yields the maps and equations of Figure 12-22(b) and the circuit shown in Figure 12-23. The clock input is connected to the clock (CK) input of each flip-flop so that the flip-flops can change state only in response to a clock pulse. The gate inputs connect directly to the corresponding flip-flop outputs as indicated by the dashed lines. To facilitate reading similar circuit diagrams, such connecting wires will be omitted in the remainder of the book.

The timing diagram of Figure 12-24, derived by tracing signals through the circuit, verifies that the counter functions according to the state diagram of Figure 12-21; for example, starting with $CBA = 000$, $T_C = 1$ and $T_B = T_A = 0$. Therefore, when the clock pulse comes along, only flip-flop $C$ changes state, and the new state is 100. Then, $T_C = 0$ and $T_B = T_A = 1$, so flip-flops $B$ and $A$ change state when the next clock pulse occurs, etc. Note that the flip-flops change state following the falling clock edge.

Although the original state table for the counter (Table 12-3) is not completely specified, the next states of states 001, 101, and 110 have been specified in the process of completing the circuit design. For example, if the flip-flops are initially set to $C = 0$, $B = 0$, and $A = 1$, tracing signals through the circuit shows that $T_C = T_B = 1$ and $T_A = 0$, so that the state will change to 111 when a clock pulse is applied. This behavior is indicated by the dashed line in Figure 12-25. Once state 111 is reached, successive clock pulses will cause the counter to continue in the original counting sequence as indicated on the state graph. When the power in a circuit is first turned

**FIGURE 12-23** Counter Using T Flip-Flops

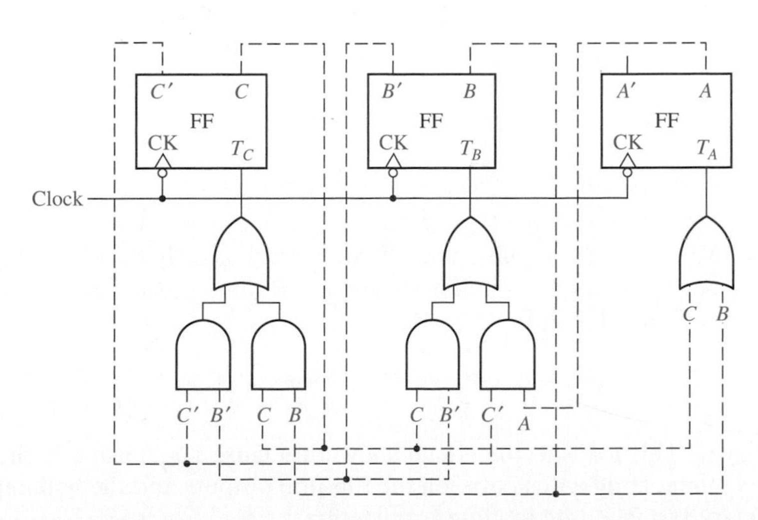

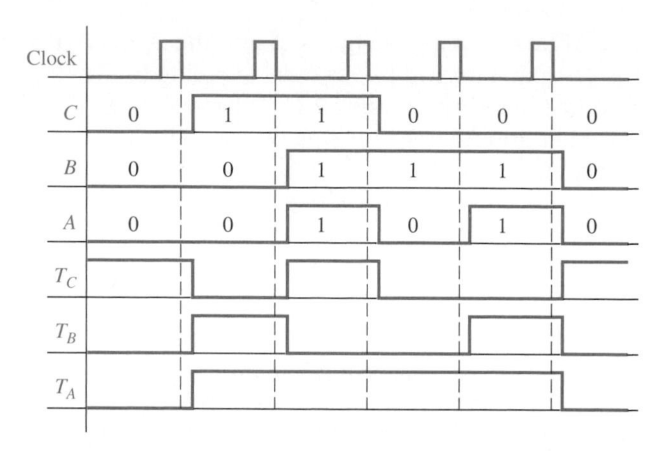


FIGURE 12-24 Timing Diagram for Figure 12-23

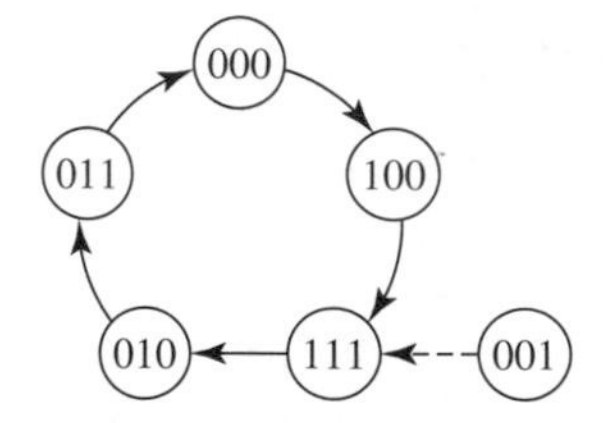


FIGURE 12-25 State Graph for Counter

on, the initial states of the flip-flops may be unpredictable. For this reason, all of the don't-care states in a counter should be checked to make sure that they eventually lead into the main counting sequence unless a power-up reset is provided.

In summary, the following procedure can be used to design a counter using T flip-flops:

1. Form a state table which gives the next flip-flop states for each combination of present flip-flop states.
2. Plot the next-state maps from the table.
3. Plot a $T$ input map for each flip-flop. When filling in the $T_Q$ map, $T_Q$ must be 1 whenever $Q^+ \neq Q$. This means that the $T_Q$ map can be formed from the $Q^+$ map by complementing the $Q = 1$ half of the map and leaving the $Q = 0$ half unchanged.
4. Find the $T$ input equations from the maps and realize the circuit.

## Counter Design Using D Flip-Flops

For a D flip-flop, $Q^+ = D$, so the $D$ input map is identical with the next-state map. Therefore, the equation for $D$ can be read directly from the $Q^+$ map. For the counter of Figure 12-21, the following equations can be read from the next-state maps shown in Figure 12-22(a):

$$D_C = C^+ = B' \qquad D_B = B^+ = C + BA'$$
$$D_A = A^+ = CA' + BA' = A'(C + B)$$

This leads to the circuit shown in Figure 12-26 using D flip-flops. Note that the connecting wires between the flip-flop outputs and the gate inputs have been omitted to facilitate reading the diagram.

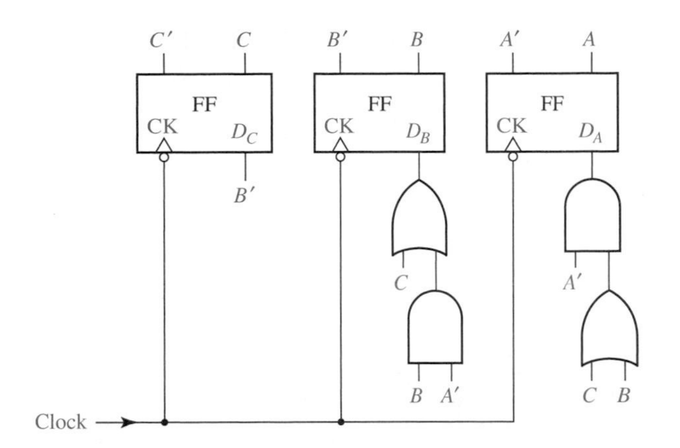


FIGURE 12-26
Counter of Figure 12-21 Using D Flip-Flops

## 12.5 Counter Design Using S-R and J-K Flip-Flops

The procedures used to design a counter with S-R flip-flops are similar to the procedures discussed in Sections 12.3 and 12.4. However, instead of deriving an input equation for each D or T flip-flop, the $S$ and $R$ input equations must be derived. We will now develop methods for deriving these $S$ and $R$ flip-flop input equations.

Table 12-5(a) describes the behavior of the S-R flip-flop. Given $S$, $R$, and $Q$, we can determine $Q^+$ from this table. However, the problem we must solve is to determine $S$ and $R$ given the present state $Q$ and the desired next state $Q^+$. If the present state of the flip-flop is $Q = 0$ and the desired next state is $Q^+ = 1$, a 1 must be applied to the $S$ input to set the flip-flop to 1. If the present state is 1, and the desired next state is 0, a 1 must be applied to the $R$ input to reset the flip-flop to 0. Restrictions on the flip-flop inputs require that $S = 0$ if $R = 1$, and $R = 0$ if $S = 1$. Thus, when forming Table 12-5(b), the rows corresponding to $QQ^+ = 01$ and 10 are filled in with $SR = 10$ and 01, respectively. If the present state and next state are both 0, $S$ must be 0 to prevent setting the flip-flop to 1. However, $R$ may be either 0 or 1 because when $Q = 0$, $R = 1$ has no effect on the flip-flop state. Similarly, if the present state and next state are both 1, $R$ must be 0 to prevent resetting the flip-flop, but $S$ may be either 0 or 1. The required $S$ and $R$ inputs are summarized in Table 12-5(b). Table 12-5(c) is the same as 12-5(b), except the alternative choices for $R$ and $S$ have been indicated by don't-cares.

TABLE 12-5
S-R Flip-Flop Inputs

(a)

| $S$ | $R$ | $Q$ | $Q^+$ |
|---|---|---|---|
| 0 | 0 | 0 | 0 |
| 0 | 0 | 1 | 1 |
| 0 | 1 | 0 | 0 |
| 0 | 1 | 1 | 0 |
| 1 | 0 | 0 | 1 |
| 1 | 0 | 1 | 1 |
| 1 | 1 | 0 | – } inputs not allowed |
| 1 | 1 | 1 | – |

(b)

| $Q$ | $Q^+$ | $S$ | $R$ |
|---|---|---|---|
| 0 | 0 | {0 | 0 |
| | | {0 | 1 |
| 0 | 1 | 1 | 0 |
| 1 | 0 | 0 | 1 |
| 1 | 1 | {0 | 0 |
| | | {1 | 0 |

(c)

| $Q$ | $Q^+$ | $S$ | $R$ |
|---|---|---|---|
| 0 | 0 | 0 | X |
| 0 | 1 | 1 | 0 |
| 1 | 0 | 0 | 1 |
| 1 | 1 | X | 0 |

**TABLE 12-6**

| $C$ $B$ $A$ | $C^+$ $B^+$ $A^+$ | $S_C$ | $R_C$ | $S_B$ | $R_B$ | $S_A$ | $R_A$ |
|---|---|---|---|---|---|---|---|
| 0 0 0 | 1 0 0 | 1 | 0 | 0 | X | 0 | X |
| 0 0 1 | – – – | X | X | X | X | X | X |
| 0 1 0 | 0 1 1 | 0 | X | X | 0 | 1 | 0 |
| 0 1 1 | 0 0 0 | 0 | X | 0 | 1 | 0 | 1 |
| 1 0 0 | 1 1 1 | X | 0 | 1 | 0 | 1 | 0 |
| 1 0 1 | – – – | X | X | X | X | X | X |
| 1 1 0 | – – – | X | X | X | X | X | X |
| 1 1 1 | 0 1 0 | 0 | 1 | X | 0 | 0 | 1 |

Next, we will redesign the counter of Figure 12-21 using S-R flip-flops. Table 12-3 is repeated in Table 12-6 with columns added for the $S$ and $R$ flip-flop inputs. These columns can be filled in using Table 12-5(c). For $CBA = 000$, $C = 0$ and $C^+ = 1$, so $S_C = 1$, $R_C = 0$. For $CBA = 010$ and $011$, $C = 0$ and $C^+ = 0$, so $S_C = 0$ and $R_C = \text{X}$. For $CBA = 100$, $C = 1$ and $C^+ = 1$, so $S_C = \text{X}$ and $R_C = 0$. For row 111, $C = 1$ and $C^+ = 0$, so $S_C = 0$ and $R_C = 1$. For $CBA = 001$, 101, and 110, $C^+ = \text{X}$, so $S_C = R_C = \text{X}$. Similarly, the values of $S_B$ and $R_B$ are derived from the values of $B$ and $B^+$, and $S_A$ and $R_A$ are derived from $A$ and $A^+$. The resulting flip-flop input functions are mapped in Figure 12-27(b).

It is generally faster and easier to derive the S-R flip-flop input maps directly from the next-state maps than to derive them from the state table as was done in Table 12-6. For each flip-flop, we will derive the $S$ and $R$ input maps from the next-state ($Q^+$) map using Table 12-5(c) to determine the values for $S$ and $R$. Just as we did for the T flip-flop, we will use the next-state maps for $C^+$, $B^+$, and $A^+$ in Figure 12-22(a) as a starting point for deriving the S-R flip-flop input equations. For convenience, these maps are repeated in Figure 12-27(a). We will consider one-half of each next-state map at a time when deriving the input maps. We will start with flip-flop $C$ ($Q = C$ and $Q^+ = C^+$) and consider the $C = 0$ column of the map. From Table 12-5(c), if $C = 0$ and $C^+ = 1$, then $S = 1$ and $R = 0$. Therefore, for every square in the $C = 0$ column where $C^+ = 1$, we plot $S_C = 1$ and $R_C = 0$ (or blank) in the corresponding squares of the input maps. Similarly, for every square in the $C = 0$ column where $C^+ = 0$, we plot $S_C = 0$ and $R_C = \text{X}$ on the input maps. For the $C = 1$ column, if $C^+ = 0$, we plot $S_C = 0$ and $R_C = 1$; if $C^+ = 1$, we plot $S_C = \text{X}$ and $R_C = 0$. Don't-cares on the $C^+$ map remain don't-cares on the $S_C$ and $R_C$ maps, because if we do not care what the next state is, we do not care what the input is. In a similar manner, we can derive the $S_B$ and $R_B$ maps from the $B^+$ map by working with the $B = 0$ (top) half of the map and the $B = 1$ (bottom) half of the map. As before, 1's are placed on the $S$ or $R$ map when the flip-flop must be set or reset. $S$ is a don't-care if $Q = 1$ and no state change is required, and $R = \text{X}$ if $Q = 0$ and no state change is required. Finally, $S_A$ and $R_A$ are derived from the $A^+$ map. Figure 12-27(c) shows the resulting circuit.

The procedure used to design a counter with J-K flip-flops is very similar to that used for S-R flip-flops. The J-K flip-flop is similar to the S-R flip-flop except that $J$ and $K$ can be 1 simultaneously, in which case the flip-flop changes state. Table 12-7(a) gives the next state ($Q^+$) as a function of $J$, $K$, and $Q$. Using this table, we can derive the required input conditions for $J$ and $K$ when $Q$ and $Q^+$ are given. Thus if a change from $Q = 0$ to $Q^+ = 1$ is required, either the flip-flop can be set to 1 by using $J = 1$

FIGURE 12-27 Counter of Figure 12-21 Using S-R Flip-Flops

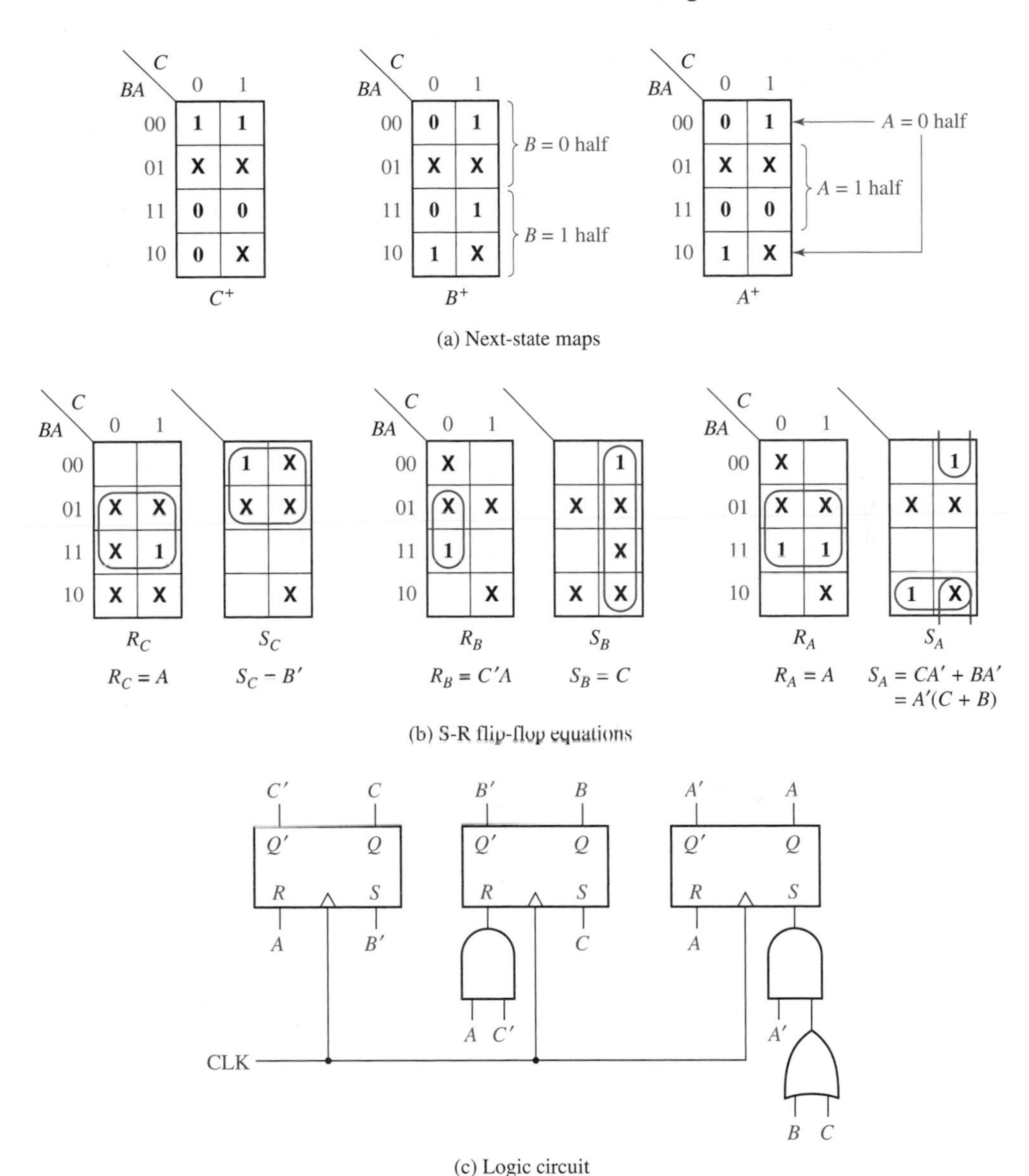


(and $K = 0$) or the state can be changed by using $J = K = 1$. In other words, $J$ must be 1, but $K$ is a don't-care. Similarly, a state change from 1 to 0 can be accomplished by resetting the flip-flop with $K = 1$ (and $J = 0$) or by changing the flip-flop state with $J = K = 1$. When no state change is required, the inputs are the same as the corresponding inputs for the S-R flip-flops. The J-K flip-flop input requirements are summarized in Tables 12-7(b) and 12-7(c).

We will now redesign the counter of Figure 12-21 using J-K flip-flops. Table 12-3 is repeated in Table 12-8 with columns added for the $J$ and $K$ flip-flop inputs. We will fill in these columns using Table 12-7(c). For $CBA = 000$, $C = 0$ and $C^+ = 1$,

**TABLE 12-7** J-K Flip-Flop Inputs

(a)

| $J$ | $K$ | $Q$ | $Q^+$ |
|---|---|---|---|
| 0 | 0 | 0 | 0 |
| 0 | 0 | 1 | 1 |
| 0 | 1 | 0 | 0 |
| 0 | 1 | 1 | 0 |
| 1 | 0 | 0 | 1 |
| 1 | 0 | 1 | 1 |
| 1 | 1 | 0 | 1 |
| 1 | 1 | 1 | 0 |

(b)

| $Q$ | $Q^+$ | $J$ | $K$ |
|---|---|---|---|
| 0 | 0 | {0 | 0 |
| | | 0 | 1} |
| 0 | 1 | {1 | 0 |
| | | 1 | 1} |
| 1 | 0 | {0 | 1 |
| | | 1 | 1} |
| 1 | 1 | {0 | 0 |
| | | 1 | 0} |

(c)

| $Q$ | $Q^+$ | $J$ | $K$ |
|---|---|---|---|
| 0 | 0 | 0 | X |
| 0 | 1 | 1 | X |
| 1 | 0 | X | 1 |
| 1 | 1 | X | 0 |

**TABLE 12-8**

| $C$ | $B$ | $A$ | $C^+$ | $B^+$ | $A^+$ | $J_C$ | $K_C$ | $J_B$ | $K_B$ | $J_A$ | $K_A$ |
|---|---|---|---|---|---|---|---|---|---|---|---|
| 0 | 0 | 0 | 1 | 0 | 0 | 1 | X | 0 | X | 0 | X |
| 0 | 0 | 1 | – | – | – | X | X | X | X | X | X |
| 0 | 1 | 0 | 0 | 1 | 1 | 0 | X | X | 0 | 1 | X |
| 0 | 1 | 1 | 0 | 0 | 0 | 0 | X | X | 1 | X | 1 |
| 1 | 0 | 0 | 1 | 1 | 1 | X | 0 | 1 | X | 1 | X |
| 1 | 0 | 1 | – | – | – | X | X | X | X | X | X |
| 1 | 1 | 0 | – | – | – | X | X | X | X | X | X |
| 1 | 1 | 1 | 0 | 1 | 0 | X | 1 | X | 0 | X | 1 |

so $J_C = 1$ and $K_C = \text{X}$. For $CBA = 010$ and $011$, $C = 0$ and $C^+ = 0$, so $J_C = 0$ and $K_C = \text{X}$. The remaining table entries are filled in similarly. The resulting J-K flip-flop input functions are plotted in Figure 12-28(b) on the next page. After deriving the flip-flop input equations from the J-K maps, we can draw the logic circuit of Figure 12-28(c).

## 12.6 Derivation of Flip-Flop Input Equations—Summary

The input equation for the flip-flops in a sequential circuit may be derived from the next-state equations by using truth tables or by using Karnaugh maps. For circuits with three to five variables, it is convenient to first plot maps for the next-state equations, and then transform these maps into maps for the flip-flop inputs.

Given the present state of a flip-flop ($Q$) and the desired next state ($Q^+$), Table 12-9 gives the required inputs for various types of flip-flops. For the D flip-flop, the input is the same as the next state. For the T flip-flop, the input is 1 whenever a state change is required. For the S-R flip-flop, $S$ is 1 whenever the flip-flop must be set to 1 and $R$ is 1 when it must be reset to 0. We do not care what $S$ is if the flip-flop state is 1 and must remain 1; we do not care what $R$ is if the flip-flop state is 0 and must remain 0. For a J-K flip-flop, the $J$ and $K$ inputs are the same

FIGURE 12-28 Counter of Figure 12-21 Using J-K Flip-Flops

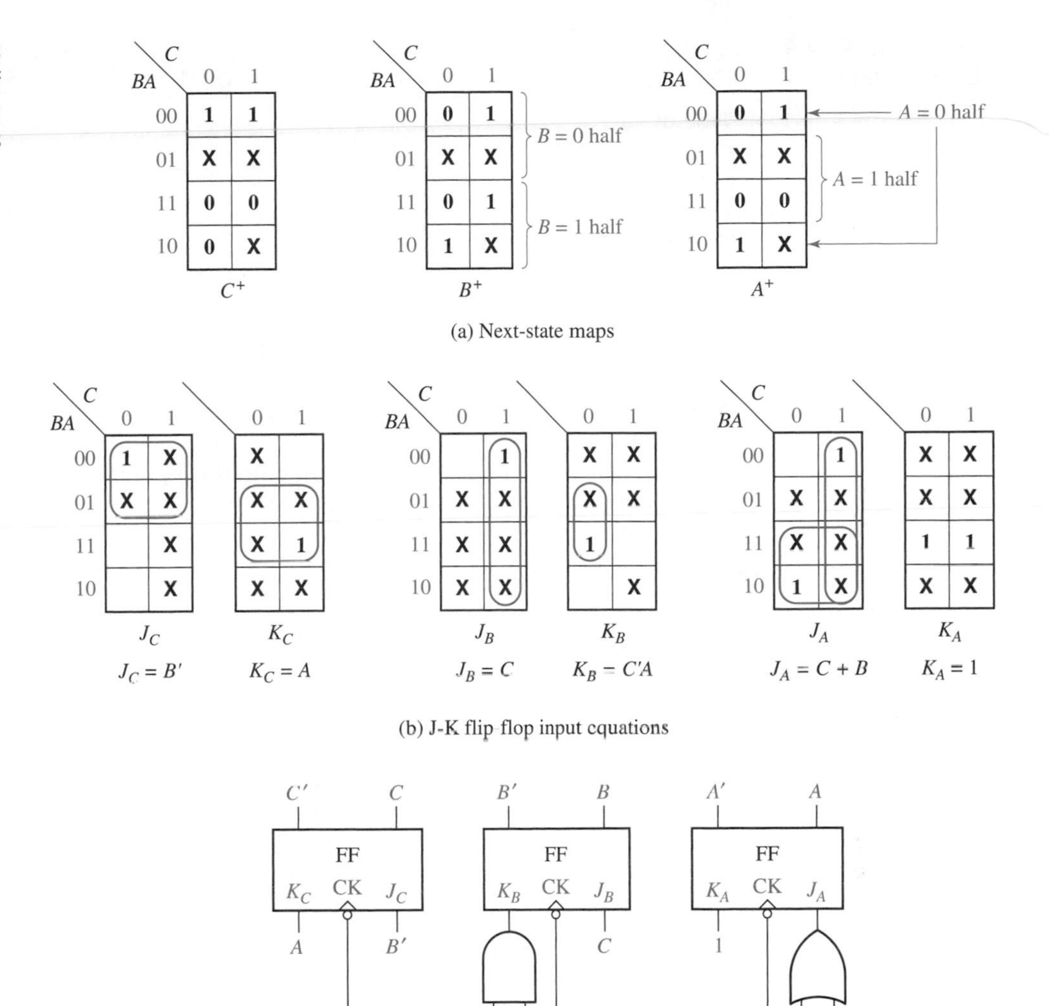


as $S$ and $R$, respectively, except that when one input is 1 the other input is X. This difference arises because $S = R = 1$ is not allowed, but $J = K = 1$ causes a change of state.

Table 12-9 summarizes the rules for transforming next-state maps into flip-flop input maps. Before applying these rules, we must copy any don't-cares from the next-state maps onto the input maps. Then, we must work with the $Q = 0$ and $Q = 1$ halves of each next-state map separately. The rules given in Table 12-9 are easily derived by comparing the values of $Q^+$ with the corresponding input values. For example, in the $Q = 0$ column of the table, we see that $J$ is the same as $Q^+$, so the $Q = 0$ half of the $J$ map is the same as the $Q^+$ map. In the $Q = 1$ column, $J = X$ (independent of $Q^+$), so we fill in the $Q = 1$ half of the $J$ map with X's.

**TABLE 12-9** Determination of Flip-Flop Input Equations from Next-State Equations Using Karnaugh Maps

| | | $Q = 0$ | | $Q = 1$ | | Rules for Forming Input Map From Next-State Map* | |
|---|---|---|---|---|---|---|---|
| Type of Flip-Flop | Input | $Q^+ = 0$ | $Q^+ = 1$ | $Q^+ = 0$ | $Q^+ = 1$ | $Q = 0$ Half of Map | $Q = 1$ Half of Map |
| Delay | $D$ | 0 | 1 | 0 | 1 | no change | no change |
| Toggle | $T$ | 0 | 1 | 1 | 0 | no change | complement |
| Set-Reset | $S$ | 0 | 1 | 0 | X | no change | replace 1's with X's** |
| | $R$ | X | 0 | 1 | 0 | replace 0's with X's** | complement |
| J-K | $J$ | 0 | 1 | X | X | no change | fill in with X's |
| | $K$ | X | X | 1 | 0 | fill in with X's | complement |

$Q^+$ means the next state of Q
X is a don't-care
*Always copy X's from the next-state map onto the input maps first.
**Fill in the remaining squares with 0's.

***Example (illustrating the use of Table 12-9)***

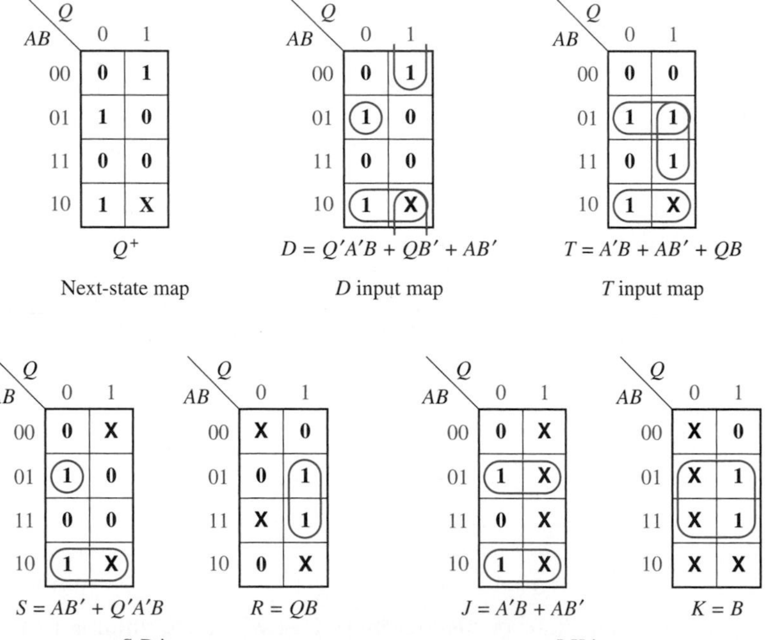


For the S-R flip-flop, note that when $Q = 0$, $R = X$ if $Q^+ = 0$; and when $Q = 1$, $R = 1$ if $Q^+ = 0$. Therefore, to form the $R$ map from the $Q^+$ map, replace 0's with X's on the $Q = 0$ half of the map and replace 0's with 1's on the $Q = 1$ half (and fill in 0's for the remaining entries). Similarly, to form the $S$ map from the $Q^+$ map, copy the 1's on the $Q = 0$ half of the map, and replace the 1's with X's on the $Q = 1$ half.

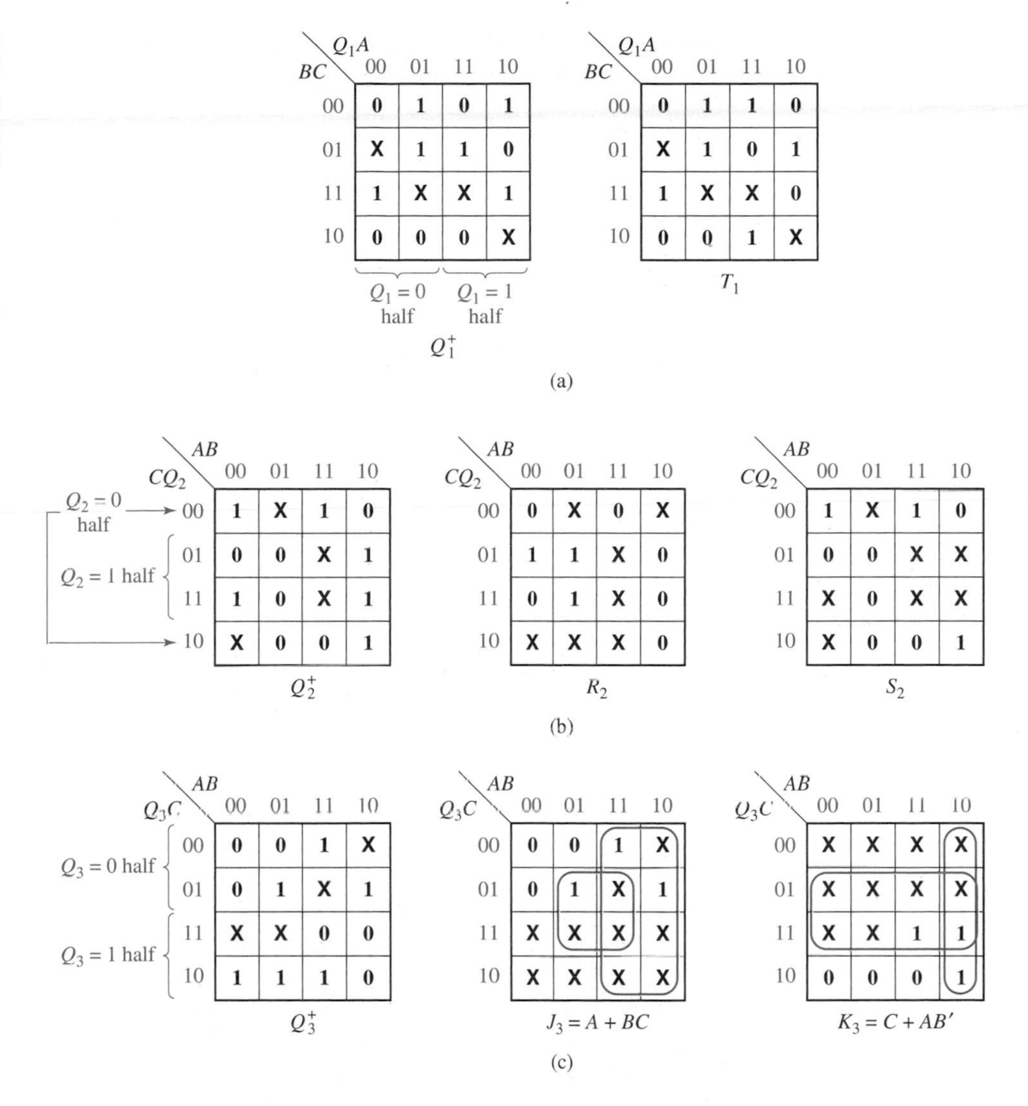


**FIGURE 12-29** Derivation of Flip-Flop Input Equations Using 4-Variable Maps

Examples of deriving 4-variable input maps are given in Figure 12-29. In each case, $Q_i$ represents the flip-flop for which input equations are being derived $A$, $B$, and $C$ represent other variables on which the next state depends. As shown in Figure 12-29(a), a 1 is placed on the $T_1$ map whenever $Q_1$ must change state. In Figure 12-29(b), 1's are placed on the $Q_2 = 0$ half of the $S_2$ map whenever $Q_2$ must be set to 1, and 1's are placed on the $Q_2 = 1$ half of the $R_2$ map whenever $Q_2$ must be reset. Figure 12-29(c) illustrates derivation of $J_3$ and $K_3$ by using separate $J$ and $K$ maps. As will be seen in Unit 14, the methods used to derive flip-flop input equations for counters are easily extended to general sequential circuits.

The procedures for deriving flip-flop input equations discussed in this unit can be extended to other types of flip-flops. If we want to derive input equations for a different type of flip-flop, the first step is to construct a table which gives the next state ($Q^+$) as a function of the present state ($Q$) and the flip-flop inputs. From this table, we can construct another table which gives the required flip-flop input combinations for each of the four possible pairs of values of $Q$ and $Q^+$. Then, using this table, we can plot a Karnaugh map for each input function and derive minimum expressions from the maps.

# Problems

**12.1** Consider a 6-bit adder with an accumulator, as in Figure 12-5. Suppose the $X$ register contains a number from a previous calculation. We do not want this number. Instead, we want $X$ to equal $3 \times Y$. ($X = x_5x_4x_3x_2x_1x_0$ and $Y = y_5y_4y_3y_2y_1y_0$.) On the timing diagram, give values for Ad and ClrN so that we will have $X = 3 \times Y$ held in the accumulator.

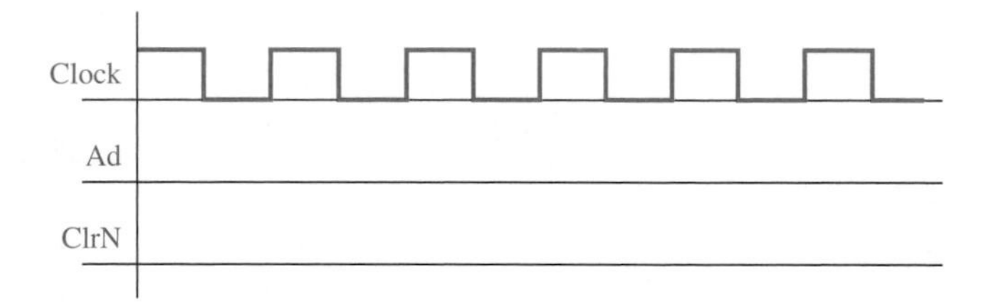


**12.2** The shift register of Figure 12-10 can be made to shift to the left by adding external connections between the $Q$ outputs and $D$ inputs. Draw a block diagram like the one in Figure 12-10(a) and indicate the appropriate connections. Which input line would serve as a serial input in this case? With the connections you have made, what should *Sh* and *Ld* be for a left shift? For a right shift?

**12.3** Show how to modify the internal circuitry of the shift register of Figure 12-10 so that it will also shift to the left without external connections as in Problem 12.2. Replace *Sh* and *L* with *A* and *B* and let the register operate according to the following table:

| Inputs | | Next State | | | | |
|---|---|---|---|---|---|---|
| $A$ | $B$ | $Q_3^+$ | $Q_2^+$ | $Q_1^+$ | $Q_0^+$ | Action |
| 0 | 0 | $Q_3$ | $Q_2$ | $Q_1$ | $Q_0$ | no change |
| 0 | 1 | SI | $Q_3$ | $Q_2$ | $Q_1$ | right shift |
| 1 | 0 | $Q_2$ | $Q_1$ | $Q_0$ | SI | left shift |
| 1 | 1 | $D_3$ | $D_2$ | $D_1$ | $D_0$ | load |

**12.4** (a) Design a 4-bit synchronous binary counter using T flip-flops.
(*Hint*: Add one flip-flop, with necessary gates, to the left side of Figure 12-13. Verify that the gates for the other three flip-flops do not change.)
(b) Repeat (a) using D flip-flops. See Figure 12-15.

**12.5** Repeat Problem 12.4(a) using D flip-flops, but implement each $D$ input as a sum of products, without using XOR gates. (*Hint*: Use Equations (12-2). As in Problem 12.4, you will need one more equation.)

**12.6** Design a circuit using D flip-flops that will generate the sequence 0, 0, 1, 0, 1, 1 and repeat. Do this by designing a counter for any sequence of states such that the first flip-flop takes on this sequence. There are many correct answers, but do not duplicate states, because each state can have only one next state.

**12.7** Design a 3-bit counter which counts in the sequence:
001, 011, 010, 110, 111, 101, 100, (repeat) 001, . . .
(a) Use D flip-flops
(b) Use T flip-flops
In each case, what will happen if the counter is started in state 000?

**12.8** Design a 3-bit counter which counts in the sequence:
001, 011, 010, 110, 111, 101, 100, (repeat) 001, . . .
(a) Use J-K flip-flops
(b) Use S-R flip-flops
In each case, what will happen if the counter is started in state 000?

**12.9** An M-N flip-flop works as follows:
If $MN = 00$, the next state of the flip-flop is 0.
If $MN = 01$, the next state of the flip-flop is the same as the present state.
If $MN = 10$, the next state of the flip-flop is the complement of the present state.
If $MN = 11$, the next state of the flip-flop is 1.
(a) Complete the following table (use don't-cares when possible):

| Present State $Q$ | Next State $Q^+$ | $M$ $N$ |
|---|---|---|
| 0 | 0 | |
| 0 | 1 | |
| 1 | 0 | |
| 1 | 1 | |

(b) Using this table and Karnaugh maps, derive and minimize the input equations for a counter composed of three M-N flip-flops which counts in the following sequence: $CBA$ = 000, 001, 011, 111, 101, 100, (repeat) 000, . . .

**12.10** Design a counter which counts in the sequence that has been assigned to you. Use D flip-flops and NAND gates. Simulate your design using *SimUaid*.
(a) 000, 001, 011, 101, 111, 010, (repeat) 000, . . .
(b) 000, 011, 101, 111, 010, 110, (repeat) 000, . . .
(c) 000, 110, 111, 100, 101, 001, (repeat) 000, . . .
(d) 000, 100, 001, 110, 101, 111, (repeat) 000, . . .
(e) 000, 010, 111, 101, 011, 110, (repeat) 000, . . .
(f) 000, 100, 001, 111, 110, 101, (repeat) 000, . . .

(g) 000, 010, 111, 101, 001, 110, (repeat) 000, . . .
(h) 000, 101, 010, 011, 001, 110, (repeat) 000, . . .
(i) 000, 100, 010, 001, 110, 111, (repeat) 000, . . .
(j) 000, 001, 111, 010, 110, 011, (repeat) 000, . . .
(k) 000, 100, 010, 001, 101, 111, (repeat) 000, . . .
(l) 000, 011, 111, 110, 001, 100, (repeat) 000, . . .
(m) 000, 100, 111, 110, 010, 011, (repeat) 000, . . .
(n) 000, 011, 111, 110, 010, 100, (repeat) 000, . . .

**12.11** Redesign the right-shift register circuit of Figure 12-10 using four D flip-flops with clock enable, four 2-to-1 MUXes, and a single OR gate.

**12.12** Design a left-shift register similar to that of Figure 12-10. Your register should shift left if $Sh = 1$, load if $Sh = 0$ and $Ld = 1$, and hold its state if $Sh = Ld = 0$.
(a) Draw the circuit using four D flip-flops and four 4-to-1 MUXes.
(b) Give the next-state equations for the flip-flops.

**12.13** A 74178 shift register is described by the given table. All state changes occur on the 1-0 transition of the clock. The shift register is connected as shown. Complete the timing diagram.

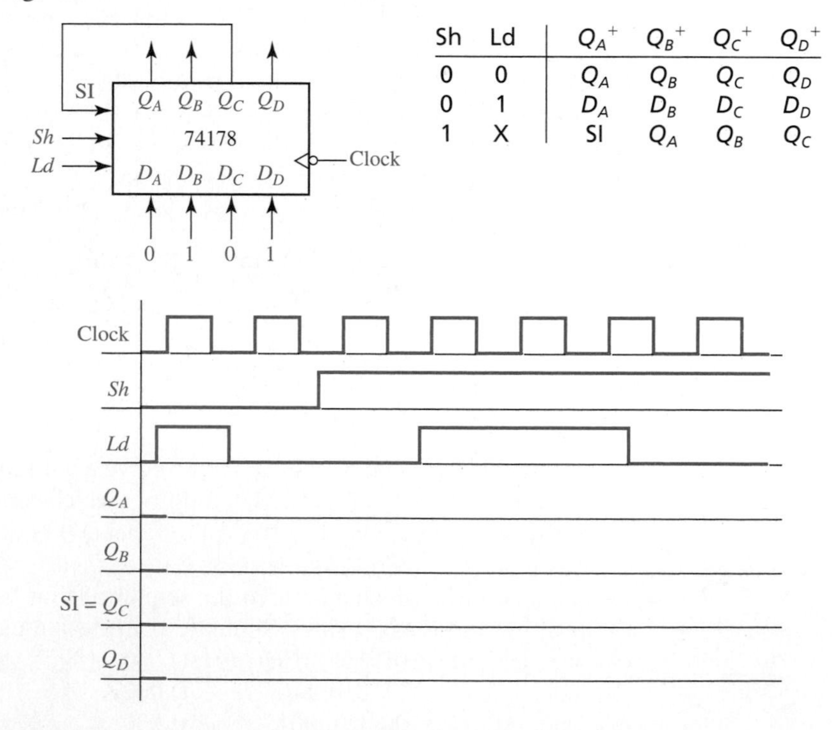


| Sh | Ld | $Q_A^+$ | $Q_B^+$ | $Q_C^+$ | $Q_D^+$ |
|---|---|---|---|---|---|
| 0 | 0 | $Q_A$ | $Q_B$ | $Q_C$ | $Q_D$ |
| 0 | 1 | $D_A$ | $D_B$ | $D_C$ | $D_D$ |
| 1 | X | SI | $Q_A$ | $Q_B$ | $Q_C$ |

**12.14** Design a 5-bit synchronous binary counter. (*Hint*: See Problem 12.4.)
(a) Use T flip-flops.
(b) Use D flip-flops.

**12.15** Construct a 4-bit Johnson counter using J-K flip-flops. (See Figure 12-12 for a Johnson counter.) What sequence of states does the counter go through if it is started in state 0000? State 0110?

**12.16** Design a 3-bit binary up-down counter which functions the same as the up-down counter of Figures 12-17 and 12-18. Use a 3-bit register of D flip-flops, a 3-bit adder, and one OR gate. (If you are clever enough, you can do it without the OR gate.) (*Hint*: To subtract one, add 111.)

**12.17** Design a decade counter which counts in the sequence:
0000, 0001, 0010, 0011, 0100, 0101, 0110, 0111, 1000, 1001, 0000, . . .
(a) Use D flip-flops.
(b) Use J-K flip-flops.
(c) Use S-R flip-flops.
(d) Use T flip-flops.
(e) Draw a complete state diagram for the counter of (b) showing what happens when the counter is started in each of the unused states.

**12.18** Repeat Problem 12.17 for the downward decade sequence:
0000, 1001, 1000, 0111, 0110, 0101, 0100, 0011, 0010, 0001, 0000, . . .

**12.19** Design a 3-bit counter which counts in the sequence:
001, 100, 101, 111, 110, 010, 011, 001, . . .
(a) Use D flip-flops.
(b) Use J-K flip-flops.
(c) Use T flip-flops.
(d) Use S-R flip-flops.
(e) What will happen if the counter of (a) is started in state 000?

**12.20** Design a decade counter using the following 2-4-2-1 weighted code for decimal digits. Use NAND gates and the indicated flip-flop types.
(a) Use D flip-flops.
(b) Use J-K flip-flops.
(c) Use T flip-flops.
(d) Use S-R flip-flops.

| Digit | *ABCD* |
|---|---|
| 0 | 0000 |
| 1 | 0001 |
| 2 | 0010 |
| 3 | 0011 |
| 4 | 0100 |
| 5 | 1011 |
| 6 | 1100 |
| 7 | 1101 |
| 8 | 1110 |
| 9 | 1111 |

**12.21** Repeat Problem 12.20 using NOR gates instead of NAND gates.

**12.22** Design a decade counter using the excess-3 code for decimal digits. Use NAND gates and the indicated flip-flop types.
(a) Use D flip-flops.
(b) Use J-K flip-flops.
(c) Use T flip-flops.
(d) Use S-R flip-flops.

**12.23** Repeat Problem 12.22 using NOR gates instead of NAND gates.

**12.24** The following binary counter increments on each rising clock edge unless the external clear (ClrN) control input is low.
(a) Implement a modulo 12 counter using this binary counter assuming the Clr control input is a synchronous control input.
(b) Repeat Part (a) assuming Clr is an asynchronous control input.

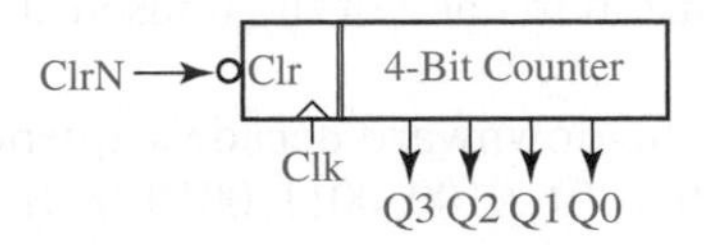


**12.25** The following binary counter operates according to the function table given. Using this binary counter, implement a decimal counter that uses the 2-4-2-1 weighted code for representing decimal digits. Minimize the gate logic required by using the parallel load inputs only to change the counting sequence from straight binary to 2-4-2-1 code.

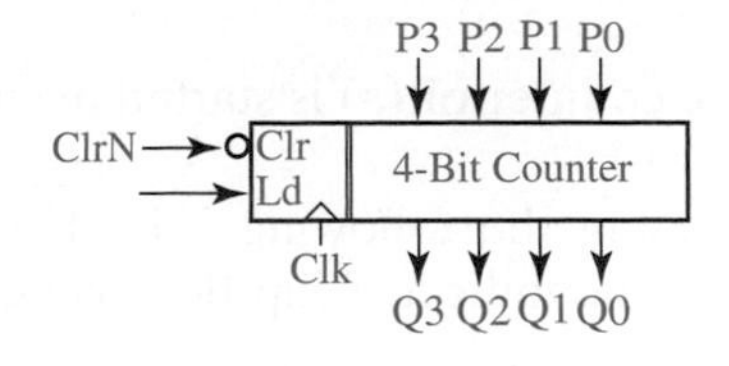


| ClN | Ld | Function |
|---|---|---|
| 0 | — | Clear |
| 1 | 1 | Parallel Load |
| 1 | 0 | Increment |

**12.26** The general form of a shift register counter is shown. The inputs to the logic are the shift register outputs, and the output from the logic is the serial input to the shift register. If the gate logic contains only exclusive-OR gates, then this is a linear shift register counter. For each value of $N$, there exists an exclusive-OR circuit so that the counter cycles through $2^N - 1$ counts.
(a) For $N = 3$, construct the state diagram for the counter if $S_{in} = Q_2 \oplus Q_1$. (The shift register stages are numbered $Q_0$, $Q_1$, $Q_2$ from left to right.)
(b) For $N = 4$, find an exclusive-OR circuit so that the counter cycles through 15 counts.

(c) Make a simple modification the logic of Part (b) so that the counter cycles through 16 counts. (The counter is no longer linear.)

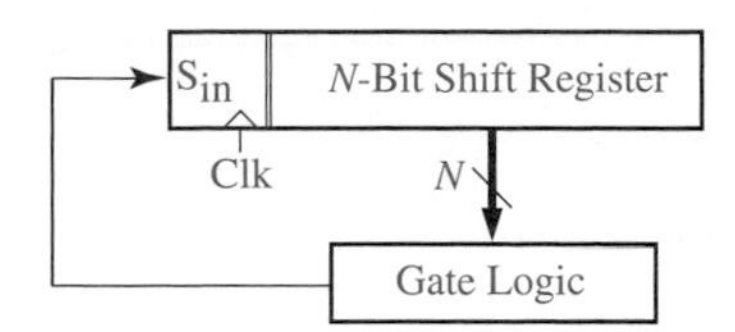


**12.27** Binary up counters can be designed using J-K flip-flops by noting that the least significant stage, $Q_0$, always toggles and stage $Q_i$ always toggles when stages 0, . . . $(i - 1)$ are 1. This approach can be modified to design counters with a shorter cycle and obtaining nearly minimum equations. Note that an optimum solution may not have the same equations for $J$ and $K$.

(a) Modify the binary up counter design to obtain a BCD decade up counter using J-K flip-flops.

(b) Modify the binary up counter design to obtain an excess-3 decade up counter using J-K flip-flops.

(c) Modify the design for Part (b) so that the counters can be cascaded to obtain excess-3 counters that can count to 99, 999, etc.

**12.28** A three-stage binary up-down counter has control input $U$; when $U = 0$, the counter counts down and when $U = 1$, the counter counts up. Design this counter with a minimum number of NAND gates, using

(a) reset-dominant S-R flip-flops.

(b) D-CE flip-flops.

**12.29** A two-stage counter has two input control lines, $M$ and $N$. The count sequences are as follows:

| *MN* | Sequence |
|---|---|
| 00 | 0, 1, 2, 3, 0, . . . . |
| 01 | 0, 1, 0, 1, 0, 1, . . . |
| 10 | 2, 0, 2, 0, 2, 0, . . . |
| 11 | 1, 2, 1, 2, 1, 2, . . . |

(a) Design the counter assuming the outputs come directly from J-K flip-flops.

(b) Design the counter assuming a two-stage binary counter is used with the J-K flip-flop outputs decoded.

**12.30** A pulse-generating circuit generates eight repetitive pulses as shown in the figure. Implement the pulse-generating circuit using the counter circuits listed and a minimum of gate logic. Use J-K flip-flops for the counters that trigger on the falling edge of a clock that has a frequency eight times the frequency of one of the pulses. The pulses must be free of glitches; explain any restrictions on the propagation delays of gates and flip-flops so that the pulses will be glitch free.

(a) Ring counter (A ring counter is a shift register with end-to-end feedback.)

(b) Johnson counter
(c) Binary counter

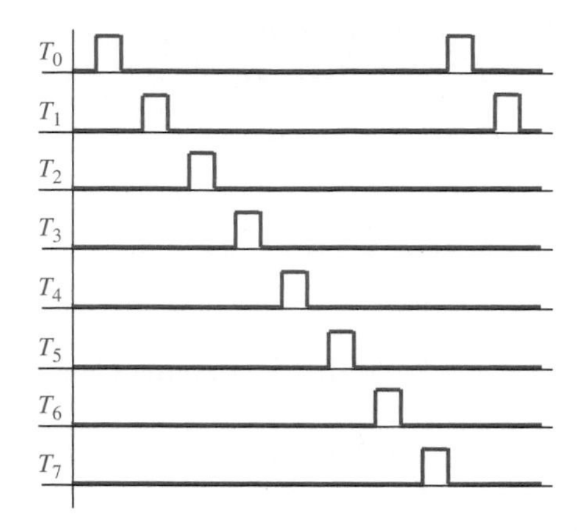


**12.31** A U-V flip-flop behaves as follows:
If $UV = 00$, the flip-flop does not change state.
If $UV = 10$, the flip-flop is set to $Q = 0$.
If $UV = 11$, the flip-flop changes state.
The input combination $UV = 01$ is not allowed.
(a) Give the characteristic (next-state) equation for this flip-flop.
(b) Complete the following table, using don't-cares where possible.

| $Q$ | $Q^+$ | $U$ | $V$ |
|---|---|---|---|
| 0 | 0 | | |
| 0 | 1 | | |
| 1 | 0 | | |
| 1 | 1 | | |

(c) Realize the following next-state equation for $Q$ using a U-V flip-flop: $Q^+ = A + BQ$. Find equations for $U$ and $V$.

**12.32** A M-F flip-flop behaves as follows:
If $MF = 01$, the flip-flop changes state.
If $MF = 11$, the flip-flop is set to $Q = 0$.
If $MF = 00$, the flip-flop is set to $Q = 1$.
The input combination $MF = 10$ is not allowed.
(a) Give the characteristic (next-state) equation for this flip-flop.
(b) Complete the table, using don't-cares where possible.

| $Q$ | $Q^+$ | $M$ | $F$ |
|---|---|---|---|
| 0 | 0 | | |
| 0 | 1 | | |
| 1 | 0 | | |
| 1 | 1 | | |

(c) Realize the following next-state equation for $Q$ using a $MF$ flip-flop: $Q^+ = CQ + DQ'$. Find equations for $M$ and $F$.

**12.33** An L-M flip-flop works as follows:
If $LM = 00$, the next state of the flip-flop is 1.
If $LM = 01$, the next state of the flip-flop is the same as the present state.
If $LM = 10$, the next state of the flip-flop is the complement of the present state.
If $LM = 11$, the next state of the flip-flop is 0.
(a) Complete the following table (use don't-cares when possible):

| Present State $Q$ | Next State $Q^+$ | $L$ | $M$ |
|---|---|---|---|
| 0 | 0 | | |
| 0 | 1 | | |
| 1 | 0 | | |
| 1 | 1 | | |

(b) Using this table and Karnaugh maps, derive and minimize the input equations for a counter composed of three L-M flip-flops which counts in the following sequence: $ABC = 000, 100, 101, 111, 011, 001, 000, \ldots$

**12.34** A sequential circuit contains a register of four flip-flops. Initially a binary number $N$ ($0000 \le N \le 1100$) is stored in the flip-flops. After a single clock pulse is applied to the circuit, the register should contain $N + 0011$. In other words, the function of the sequential circuit is to add 3 to the contents of a 4-bit register. Design the circuit using J-K flip-flops.

**12.35** When an adder is part of a larger digital system, an arrangement like the given figure often works well. For the control signals and the input data in the following table, give the value of the addend, the accumulator, and the bus at the end of each clock cycle (i.e., immediately before the active clock edge). Express the register and bus values in decimal.

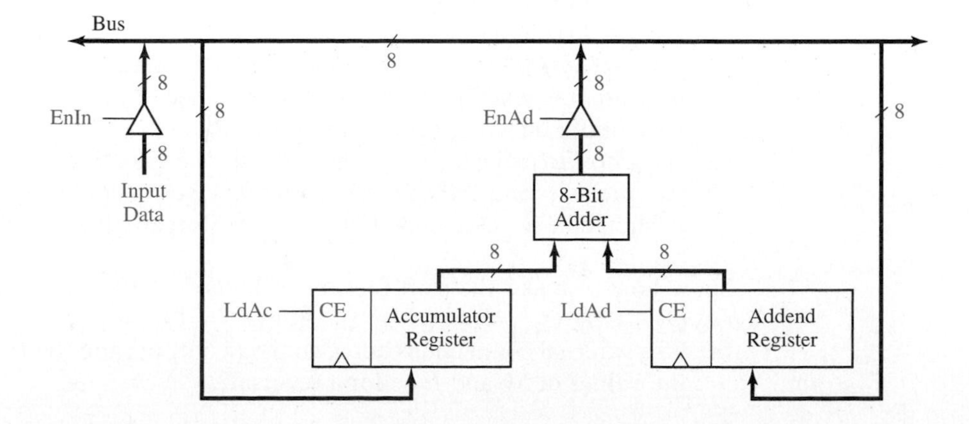

| Clock Cycle | Input Data | EnIn | EnAd | LdAc | LdAd | Accumulator Register | Addend Register | Bus |
|---|---|---|---|---|---|---|---|---|
| 0 | 18 | 1 | 0 | 1 | 0 | 0 | 0 | 18 |
| 1 | 13 | 1 | 0 | 0 | 1 | | | |
| 2 | 15 | 0 | 1 | 1 | 0 | | | |
| 3 | 93 | 1 | 0 | 0 | 1 | | | |
| 4 | 47 | 0 | 1 | 1 | 0 | | | |
| 5 | 22 | 1 | 0 | 0 | 1 | | | |
| 6 | 0 | 0 | 1 | 0 | 0 | | | |

**12.36** A digital system can perform any 4-variable bitwise logic function, but it may take several clock cycles. (A bitwise logic function performs the same logic function on each bit.) Recall that the NAND operation is functionally complete, i.e., we can do any logic function by a series of NAND operations. On the following 8-bit registers, En is a tri-state buffer enable as in Figure 12-3, and CE is a clock enable as in Figure 12-1.

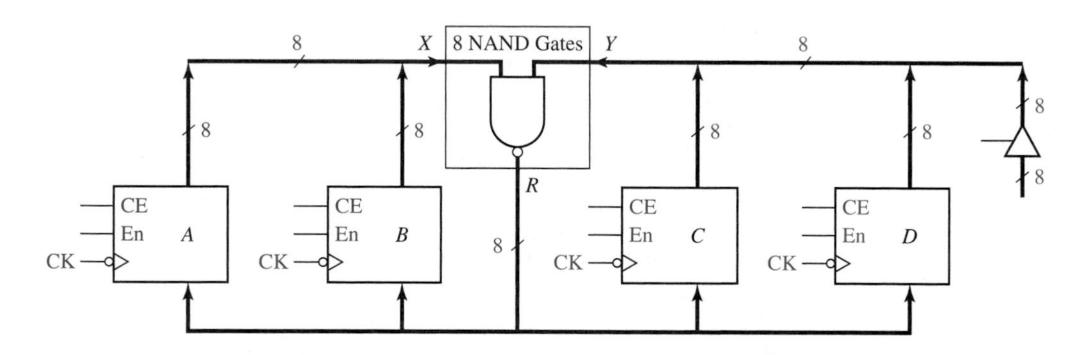


(a) Show how to connect a 2-to-4 decoder (with inverting outputs) so that the next rising edge of the clock will load the result into register $A$, $B$, $C$, or $D$ for control inputs $G_0G_1$ = 00, 01, 10, or 11, respectively.

(b) Show how to connect three control signals, $E_0$, $E_1$, and $E_2$, to the registers so that $E_0 = 0$ places the $A$ register contents on the $X$ bus, $E_0 = 1$ places $B$ onto the $X$ bus, $E_1E_2 = 00$ places $C$ onto the $Y$ bus, $E_1E_2 = 01$ places $D$ onto the $Y$ bus, $E_1E_2 = 10$ places 00000000 on the $Y$ bus, and $E_1E_2 = 11$ places 11111111 on the $Y$ bus. You may use a few additional gates. (*Hint*: Connect $E_2$ to all 8 data inputs on the tri-state buffer on the right side of the circuit.)

(c) Show how to make the bits in the $C$ register be the OR of the corresponding bits in the $A$ register and in the $D$ register, in four clock cycles. Tell what $G_0$, $G_1$, $E_0$, $E_1$, and $E_2$ should be for each cycle. [*Hint*: Use DeMorgan's law and $X' = (X \text{ NAND } 1)$.]

**12.37** Show how to make the shift register of Figure 12-10 reverse the order of its bits, i.e., $Q_3^+ = Q_0$, $Q_2^+ = Q_1$, $Q_1^+ = Q_2$, and $Q_0^+ = Q_3$.

(a) Use external connections between the $Q$ outputs and the $D$ inputs. What should the values of $Sh$ and $L$ be for a reversal?

(b) Change the internal circuitry to allow bit reversal, so that the $D$ inputs may be used for other purposes. Replace $Sh$ and $L$ with $A$ and $B$, and let the register operate according to the following table:

| Inputs | | Next State | | | | |
|---|---|---|---|---|---|---|
| $A$ | $B$ | $Q_3^+$ | $Q_2^+$ | $Q_1^+$ | $Q_0^+$ | Action |
| 0 | 0 | $Q_3$ | $Q_2$ | $Q_1$ | $Q_0$ | No change |
| 0 | 1 | SI | $Q_3$ | $Q_2$ | $Q_1$ | Right shift |
| 1 | 0 | $D_3$ | $D_2$ | $D_1$ | $D_0$ | Load |
| 1 | 1 | $Q_0$ | $Q_1$ | $Q_2$ | $Q_3$ | Reverse bits |

# Analysis of Clocked Sequential Circuits

## Objectives

1. Analyze a sequential circuit by signal tracing.
2. Given a sequential circuit, write the next-state equations for the flip-flops and derive the state graph or state table. Using the state graph, determine the state sequence and output sequence for a given input sequence.
3. Explain the difference between a Mealy machine and a Moore machine.
4. Given a state table, construct the corresponding state graph, and conversely.
5. Given a sequential circuit or a state table and an input sequence, draw a timing chart for the circuit. Determine the output sequence from the timing chart, neglecting any false outputs.
6. Draw a general model for a clocked Mealy or Moore sequential circuit. Explain the operation of the circuit in terms of these models. Explain why a clock is needed to ensure proper operation of the circuit.

# Study Guide

1. Study Section 13.1, *A Sequential Parity Checker.*
   (a) Explain how parity can be used for error detection.

   (b) Verify that the parity checker (Figure 13-4) will produce the output waveform given in Figure 13-2 when the input waveform is as shown.

2. Study Section 13.2, *Analysis by Signal Tracing and Timing Charts.*
   (a) What is the difference between a Mealy machine and a Moore machine?

   (b) For normal operation of clocked sequential circuits of the types discussed in this section, when should the inputs be changed?

   When do the flip-flops change state?

   At what times can the output change for a Moore circuit?

   At what times can the output change for a Mealy circuit?

   (c) At what time (with respect to the clock) should the output of a Mealy circuit be read?

   (d) Why can false outputs appear in a Mealy circuit and not in a Moore circuit?

   What can be done to eliminate the false outputs?

   If the output of a Mealy circuit is used as an input to another Mealy circuit synchronized by the same clock, will false outputs cause any problem? Explain.

   (e) Examine the timing diagram of Figure 13-8. The value of $Z$ will always be correct just before the falling (active) clock edge that causes the state change. Note there are two types of false outputs. A false 0 output occurs if $Z$ is 1 just before two successive falling clock edges, and $Z$ goes to 0 between the clock edges. A false 1 output occurs if $Z$ is 0 just before two successive falling clock edges and $Z$ goes to 1 between the edges. When the output is 0 (or 1) just before an active clock edge and 1 (or 0) just before the next, the output may be temporarily incorrect after the state changes following the first active edge but before the input has changed to its next value. In this case, we will not say that a false output has occurred because the sequence of outputs is still correct.

**3.** Study Section 13.3, *State Tables and Graphs*.

(a) In Equations (13-1) through (13-5), at what time (with respect to the clock) is the right-hand side evaluated?

What does $Q^+$ mean?

(b) Derive the timing chart of Figure 13-6 using Table 13-2(a).

(c) What is the difference between the state graphs for Mealy and Moore machines?

(d) For a state table, Table 13-3(b) for example, what do the terms "*present* state," "*next* state," and "*present* output" mean with respect to the active clock edge?

(e) Why does a Moore state table have only one output column?

(f) For ease in making state tables from Karnaugh maps and vice versa, state transition tables with three or four states are often written with states in the order 00, 01, 11, 10. However, this is not necessary. (In fact, for sequential circuits with five or more states, it is impossible.) For example, the following table is equivalent to Table 13-2, because it represents the circuit of Figure 13-5.

| | $A^+B^+$ | | |
|---|---|---|---|
| $AB$ | $X = 0$ | $X = 1$ | $Z$ |
| 00 | 10 | 01 | 0 |
| 01 | 00 | 11 | 1 |
| 10 | 11 | 01 | 1 |
| 11 | 01 | 11 | 0 |

**4.** The following timing chart was derived from the circuit of Figure 13-7.

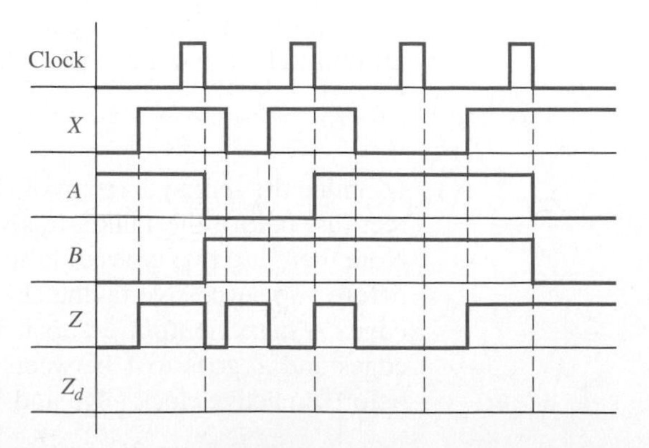

(a) Noting that extra input changes which occur between clock pulses cannot affect the state of the circuit, what is the effective input sequence seen by the flip-flops in the circuit?

(b) Using Table 13-3, verify the waveforms given for $A$, $B$, and $Z$.

(c) Indicate any false outputs. What is the correct output sequence from the circuit?

(d) Using the effective input sequence from (a), determine the output sequence from the state graph (Figure 13-11). This output sequence should be the same as your answer to (c).

(e) The output $Z$ is fed into a clocked D flip-flop, using the same clock (CK) as the circuit. Sketch the waveform for $Z_d$. Does $Z_d$ have any false outputs?

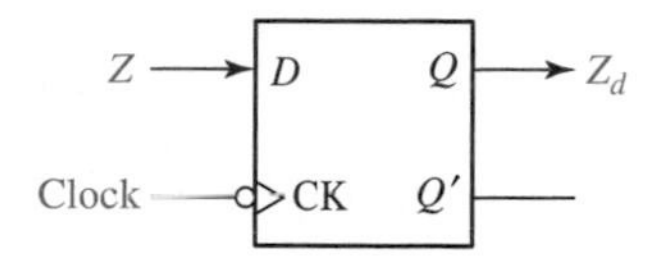


(f) Starting with Figure 13-11, construct the corresponding state table. Verify that your answer is the same as Table 13-3(b). Note that the output label on a given arrow of the graph is associated with the state from which the arrow originates.

(g) Assume that the flip-flops in Figure 13-7 are changed to flip-flops which trigger on the rising edge of the clock; that is, the inversion circles are removed from the clock inputs. Also, the clock waveform in Figure 13-8 is replaced with

The input waveform is left unchanged. What changes, if any, would occur in the remainder of the timing diagram? Explain.

**5.** Consider the following state tables:

Mealy

| P.S. | N.S. $X = 0$ | N.S. $1$ | $Z$ $X = 0$ | $Z$ $1$ |
|---|---|---|---|---|
| $S_0$ | $S_1$ | $S_0$ | 0 | 0 |
| $S_1$ | $S_0$ | $S_2$ | 1 | 0 |
| $S_2$ | $S_0$ | $S_0$ | 1 | 0 |

Moore

| P.S. | N.S. $X = 0$ | N.S. $1$ | $Z$ |
|---|---|---|---|
| $S_0$ | $S_1$ | $S_0$ | 0 |
| $S_1$ | $S_3$ | $S_2$ | 0 |
| $S_2$ | $S_3$ | $S_0$ | 0 |
| $S_3$ | $S_3$ | $S_0$ | 1 |

(a) Draw the corresponding state graphs.

(b) Show that the same output sequence is obtained from both state graphs when the input sequence is 010 (ignore the initial output for the Moore circuit).

(c) Using the state tables, complete the following timing diagrams for the two circuits. Note that the Mealy circuit has a false output, but the Moore does not. Also note that the output from the Moore circuit is delayed with respect to the Mealy.

Mealy

$X$ 0 1 0

Clock

State $S_0$ $S_1$ $S_2$ $S_0$

$Z$ 0 1

Moore

$X$ 0 1 0

Clock

State $S_0$

$Z$

(d) Work Programmed Exercise 13.1.
(e) Work Problems 13.2 and 13.3.

**6.** Study Section 13.4, *General Models for Sequential Circuits.*

(a) A Mealy sequential circuit has the form shown. The combinational circuit realizes the following equations:

$$Q_1^+ = X_1'Q_1 + X_1Q_1'Q_2' \qquad Z_1 = X_1Q_1$$
$$Q_2^+ = X_1Q_2' + X_2'Q_1 \qquad Z_2 = X_1'Q_1 + X_2Q_2'$$

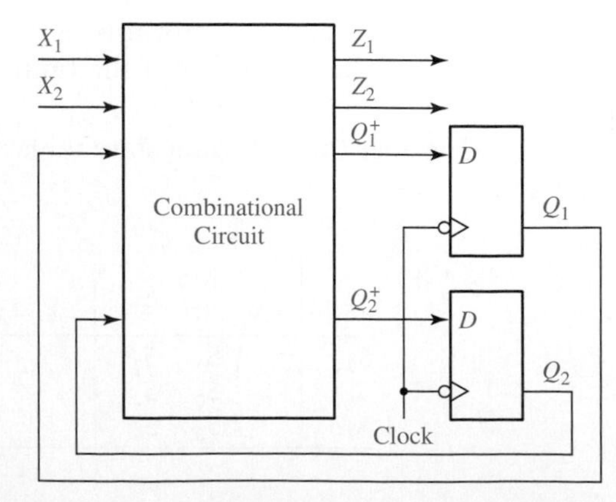

Initially, $X_1 = X_2 = 1$ and $Q_1 = Q_2 = 0$ as shown.

(1) Before the falling clock edge, show the values of the four combinational circuit outputs on the preceding diagram and on the following timing chart.
(2) Show the signal values on the circuit and timing chart immediately after the falling edge.
(3) Show any further changes in signal values which will occur after the new values of $Q_1$ and $Q_2$ have propagated through the circuit.
(4) Next change $X_1$ to 0 and repeat steps (1), (2), and (3). Show the values for each step on the circuit and on the timing chart.
(5) Next change the inputs to $X_1 = 1$ and $X_2 = 0$ and repeat steps (1), (2), and (3).
(6) Change $X_2$ to 1 and repeat.

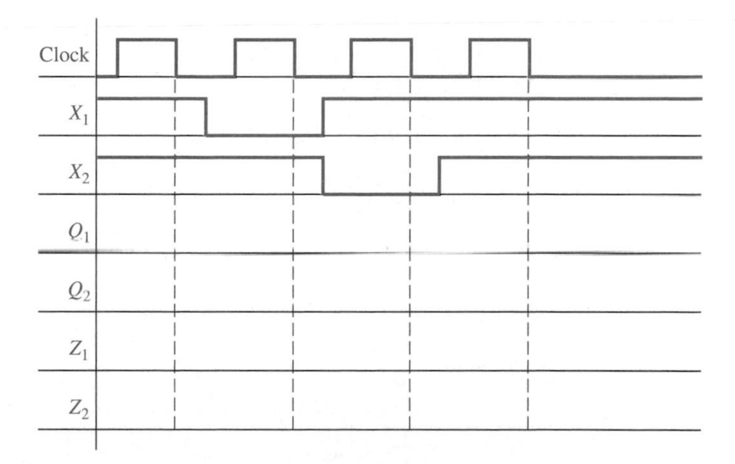


(b) Draw a block diagram for a general model of a Mealy circuit, using J-K flip-flops as memory elements. If the circuit has $n$ output variables and $k$ flip-flops, how many outputs will the combinational subcircuit have?
(c) If the circuit of Figure 13-5 were not synchronized using a clock, but instead, the flip-flops were updated continuously, and if the XOR gate had a longer delay than the OR gate, what problem could appear?

(d) The minimum clock period for a Moore circuit is determined the same way as for a Mealy circuit. Should $t_c$ be determined by the combinational subcircuit for the flip-flop inputs or for the outputs? (See Figure 13-19.)

(e) We can think of the binary counter of Figure 12-15 as a Moore circuit if we say the outputs $Z_1$, $Z_2$, and $Z_3$ are $Z_1 = A$, $Z_2 = B$, and $Z_3 = C$. (The combinational subcircuit for outputs has no gates, but that is okay.) If the XOR gates have a propagation delay of 4 ns and the AND gate has a propagation delay of 2 ns, what is the longest total propagation delay through the combinational subcircuit for flip-flops (i.e., the XOR gates and the AND gate) to the $D$ inputs of the flip-flops? If the flip-flops have $t_{su} = 3$ ns and $t_p = 3$ ns, what is the minimum clock period for the binary counter?

(f) In Equations (13-6) and (13-7), what do the symbols $\delta$ and $\lambda$ mean?

Equation (13-7) is for a Mealy circuit. What is the corresponding equation for a Moore circuit?

(g) For Table 13-5,

$\delta(S_3, 1) =$ __________ $\lambda(S_3, 1) =$ __________

$\delta(S_1, 2) =$ __________ $\lambda(S_1, 2) =$ __________

**7.** Work Problems 13.4 through 13.6.

**8.** When you are satisfied that you can meet the objectives, take the readiness test.

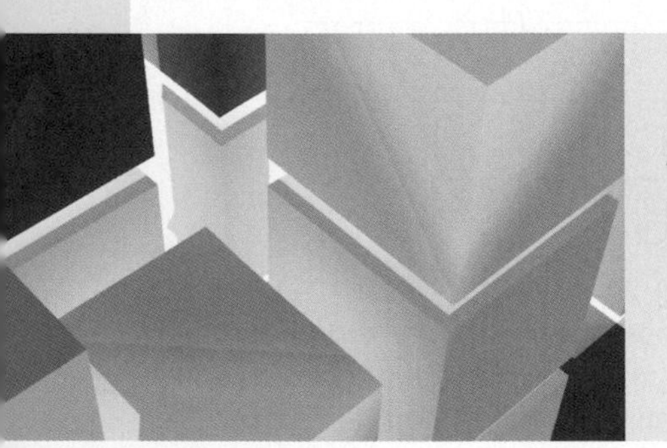

# Analysis of Clocked Sequential Circuits

The sequential circuits which we discussed in Chapter 12 perform simple functions such as shifting or counting. The counters we designed go through a fixed sequence of states and have no inputs other than a clock pulse that causes the state to change. We will now consider sequential circuits that have additional inputs. In general, the sequence of outputs and the sequence of flip-flop states for such circuits will depend on the input sequence which is applied to the circuit. Given a sequential circuit and an input sequence, we can analyze the circuit to determine the flip-flop state sequence and the output sequence by tracing the 0 and 1 signals through the circuit. Although signal tracing may be adequate for small circuits, for larger circuits it is better to construct a state graph or state table which represents the behavior of the circuit. Then, we can determine the output and state sequences from the graph or table. Such graphs and tables are also useful for the design of sequential circuits.

In this chapter we will also study the timing relationships between the inputs, the clock, and the outputs for sequential circuits by constructing timing diagrams. These timing relationships are very important when a sequential circuit is used as part of a larger digital system. After analyzing several specific sequential circuits, we will discuss a general model for a sequential circuit which consists of a combinational circuit together with flip-flops that serve as memory.

## 13.1 A Sequential Parity Checker

When binary data is transmitted or stored, an extra bit (called a parity bit) is frequently added for purposes of error detection. For example, if data is being transmitted in groups of 7 bits, an eighth bit can be added to each group of 7 bits to make the total number of 1's in each block of 8 bits an odd number. When the total number of 1 bits in the block (including the parity bit) is odd, we say that the parity is odd. Alternately, the parity bit could be chosen such that the total number of 1's in the block is even, in which case we would have even parity. Some examples of 8-bit words with odd parity are

```
7 Data Bits   ┌─Parity Bits
0000000|1
0000001|0
0110110|1
1010101|1
0111000|0
8-Bit Word
```

If any single bit in the 8-bit word is changed from 0 to 1 or from 1 to 0, the parity is no longer odd. Thus, if any single bit error occurs in transmission of a word with odd parity, the presence of this error can be detected because the number of 1 bits in the word has been changed from odd to even.

As a simple example of a sequential circuit which has one input in addition to the clock, we will design a parity checker for serial data. (Serial implies that the data enters the circuit sequentially, one bit at a time.) This circuit has the form shown in Figure 13-1. When a sequence of 0's and 1's is applied to the $X$ input, the output of

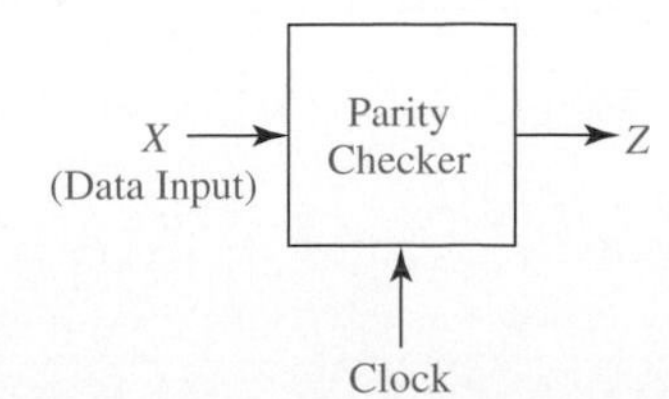


FIGURE 13-1 Block Diagram for Parity Checker

the circuit should be $Z = 1$ if the total number of 1 inputs received is odd; that is, the output should be 1 if the input parity is odd. Thus, if data which originally had odd parity is transmitted to the circuit, a final output of $Z = 0$ indicates that an error in transmission has occurred.

The value of $X$ is read at the time of the active clock edge. The $X$ input must be synchronized with the clock so that it assumes its next value before the next active clock edge. The clock input is necessary in order to distinguish consecutive 0's or consecutive 1's on the $X$ input. Typical input and output waveforms are shown in Figure 13-2.

We will start the design by constructing a state graph (Figure 13-3). The sequential circuit must "remember" whether the total number of 1 inputs received is even or odd; therefore, only two states are required. We will designate these states as $S_0$ and $S_1$, corresponding respectively to an even number of 1's received and an odd number of 1's received. We will start the circuit in state $S_0$ because initially zero 1's have been received, and zero is an even number. As indicated in Figure 13-3, if the circuit is in state $S_0$ (even number of 1's received) and $X = 0$ is received, the circuit must stay in $S_0$ because the number of 1's received is still even. However, if $X = 1$ is received, the circuit goes to state $S_1$ because the number of 1's received is then odd. Similarly, if the circuit is in state $S_1$ (odd number of 1's received) a 0 input causes no state change, but a 1 causes a change to $S_0$ because the number of 1's received is then even. The output $Z$ should be 1 whenever the circuit is in state $S_1$ (odd number of 1's received). The output is listed below the state on the state graph.

Table 13-1(a) gives the same information as the state graph in tabular form. For example, the table shows that if the present state is $S_0$, the output is $Z = 0$, and if the input is $X = 1$, the next state will be $S_1$.

Because only two states are required, a single flip-flop ($Q$) will suffice. We will let $Q = 0$ correspond to state $S_0$ and $Q = 1$ correspond to $S_1$. We can then set up a table which shows the next state of flip-flop $Q$ as a function of the present state and $X$. If we use a T flip-flop, $T$ must be 1 whenever $Q$ and $Q^+$ differ. From Table 13-1(b), the $T$ input must be 1 whenever $X = 1$. Figure 13-4 shows the resulting circuit.

Figure 13-2 shows the output waveform for the circuit. When $X = 1$, the flip-flop changes state after the falling edge of the clock. Note that the final value of $Z$ is 0 because an even number of 1's was received. If the number of 1's received had been odd,

FIGURE 13-2 Waveforms for Parity Checker

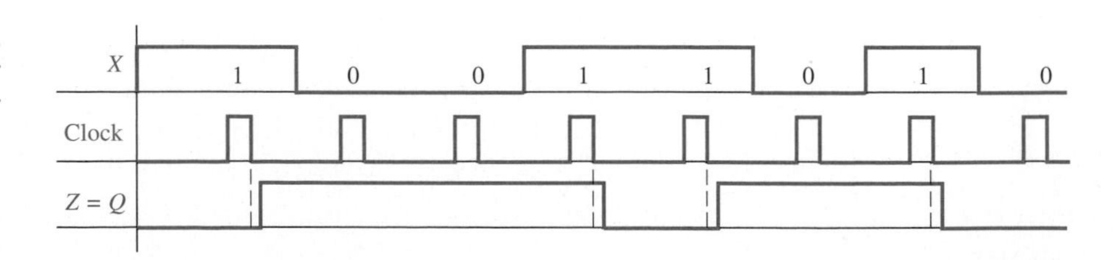


FIGURE 13-3 State Graph for Parity Checker

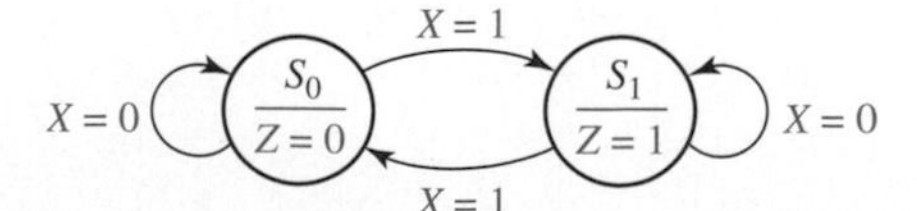

TABLE 13-1
State Table for Parity Checker

(a)

| Present State | Next State $X = 0$ | $X = 1$ | Present Output |
|---|---|---|---|
| $S_0$ | $S_0$ | $S_1$ | 0 |
| $S_1$ | $S_1$ | $S_0$ | 1 |

(b)

| $Q$ | $Q^+$ $X = 0$ | $X = 1$ | $T$ $X = 0$ | $X = 1$ | $Z$ |
|---|---|---|---|---|---|
| 0 | 0 | 1 | 0 | 1 | 0 |
| 1 | 1 | 0 | 0 | 1 | 1 |

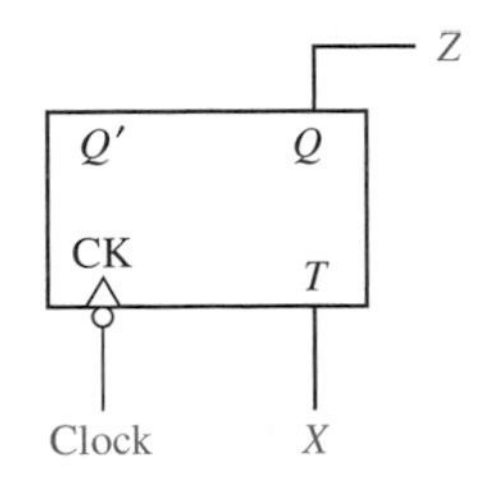


FIGURE 13-4
Parity Checker

the final value of $Z$ would be 1. In this case, it would be necessary to reset the flip-flop to the proper initial state ($Q = 0$) before checking the parity of another input sequence.

## 13.2 Analysis by Signal Tracing and Timing Charts

In this section we will analyze clocked sequential circuits to find the output sequence resulting from a given input sequence by tracing 0 and 1 signals through the circuit. The basic procedure is as follows:

1. Assume an initial state of the flip-flops (all flip-flops reset to 0 unless otherwise specified).
2. For the first input in the given sequence, determine the circuit output(s) and flip-flop inputs.
3. Determine the new set of flip-flop states after the next active clock edge.
4. Determine the output(s) that corresponds to the new states.
5. Repeat 2, 3, and 4 for each input in the given sequence.

As we carry out the analysis, we will construct a timing chart which shows the relationship between the input signal, the clock, the flip-flop states, and the circuit output. We have already seen how to construct timing charts for flip-flops (Unit 11) and counters (Unit 12).

In this unit we will use edge-triggered flip-flops that change state shortly after the active edge (rising or falling edge) of the clock. We will assume that the flip-flop inputs are stable a sufficient time before and after the active clock edge so that setup and hold time requirements are met. When the state of the sequential circuit changes, the change will always occur in response to the active clock edge. The circuit output may change at the time the flip-flops change state or at the time the input changes depending on the type of circuit.

Two types of clocked sequential circuits will be considered—those in which the output depends only on the present state of the flip-flops and those in which the output depends on both the present state of the flip-flops and on the value of the circuit inputs. If the output of a sequential circuit is a function of the present state only (as in Figures 13-4 and 13-5), the circuit is often referred to as a *Moore machine*. The state graph for a Moore machine has the output associated with the state (as in Figures 13-3 and 13-9). If the output is a function of both the present state and the input (as in Figure 13-7), the circuit is referred to as a *Mealy machine*. The state graph for a Mealy machine has the output associated with the arrow going between states (as in Figure 13-11).

As an example of a Moore circuit, we will analyze Figure 13-5 using an input sequence $X = 01101$. In this circuit, the initial state is $A = B = 0$, and all state changes occur after the rising edge of the clock, as shown in Figure 13-6. The $X$ input is synchronized with the clock so that it assumes its next value after each rising edge. Because $Z$ is a function only of the present state (in this case, $Z = A \oplus B$) the output will only change when the state changes. Initially, $X = 0$, so $D_A = 1$ and $D_B = 0$, and the state will change to $A = 1$ and $B = 0$ after the first rising clock edge. Then $X$ is changed to 1, so $D_A = 0$, $D_B = 1$ and the state changes to $AB = 01$ after the second rising clock edge. After the state change, $X$ remains 1, so $D_A = D_B = 1$, and the next rising edge causes the state to change to 11. When $X$ changes to 0, $D_A = 0$ and $D_B = 1$, and the state changes to $AB = 01$ on the fourth rising edge. Then, with $X = 1$, $D_A = D_B = 1$, so the

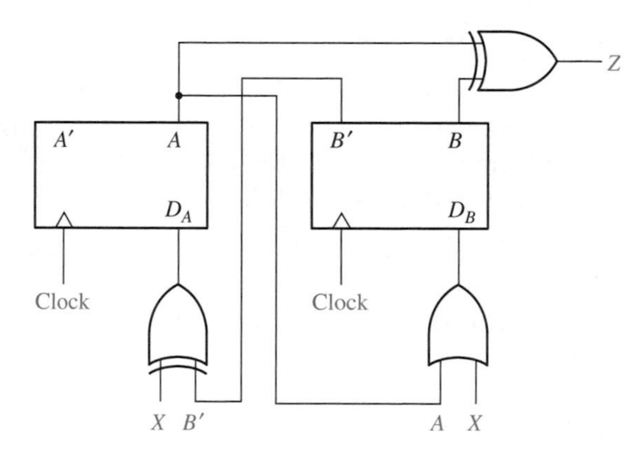


FIGURE 13-5 Moore Sequential Circuit to be Analyzed

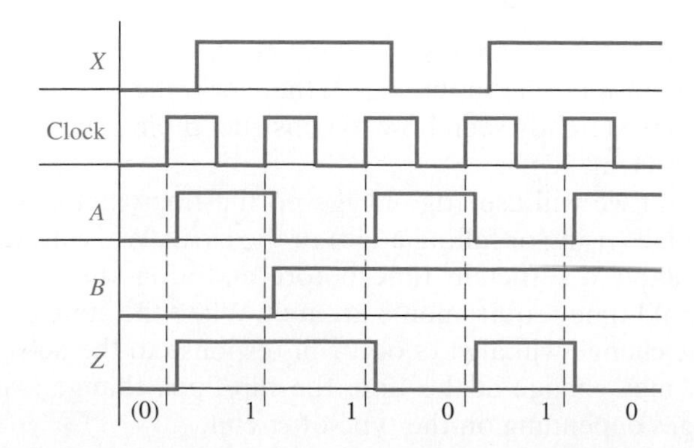


FIGURE 13-6 Timing Chart for Figure 13-5

fifth rising clock edge causes the state to change to $AB = 11$. The input, state, and output sequences are plotted on the timing chart of Figure 13-6 and are also listed below.

$$
\begin{array}{rcccccc}
X = & 0 & 1 & 1 & 0 & 1 & \\
A = & 0 & 1 & 0 & 1 & 0 & 1 \\
B = & 0 & 0 & 1 & 1 & 1 & 1 \\
Z = & (0) & 1 & 1 & 0 & 1 & 0
\end{array}
$$

When the circuit is reset to its initial state ($A = B = 0$), the initial output is $Z = 0$. Because this initial 0 is not in response to any $X$ input, it should be ignored. The resulting output sequence is $Z = 11010$. Note that for the *Moore* circuit, the output which results from application of a given input does not appear until after the active clock edge; therefore, the output sequence is displaced in time with respect to the input sequence.

As an example of a Mealy circuit, we will analyze Figure 13-7 and construct a timing chart using the input sequence $X = 10101$. The input is synchronized with the clock so that input changes occur after the falling edge, as shown in Figure 13-8. In this example, the output depends on both the input ($X$) and the flip-flop states ($A$ and $B$), so $Z$ may change either when the input changes or when the flip-flops change state. Initially, assume that the flip-flop states are $A = 0, B = 0$. If $X = 1$, the output is $Z = 1$ and $J_B = K_A = 1$. After the falling edge of the first clock pulse, $B$ changes to 1 so $Z$ changes to 0. If the input is changed to $X = 0$, $Z$ will change back to 1. All flip-flop inputs are then 0, so no state change occurs with the second falling edge. When $X$ is changed to 1, $Z$ becomes 0 and $J_A = K_A = J_B = 1$. $A$ changes to 1 on the third falling clock edge, at which time $Z$ changes to 1. Next, $X$ is changed to 0 so $Z$ becomes 0, and no state change occurs with the fourth clock pulse. Then, $X$ is changed to 1, and $Z$ becomes 1. Because $J_A = K_A = J_B = K_B = 1$, the fifth clock pulse returns the circuit to the initial state. The input, state, and output sequences are plotted on the timing chart of Figure 13-8 and are also listed below

$$
\begin{array}{rcccccc}
X = & 1 & 0 & 1 & 0 & 1 & \\
A = & 0 & 0 & 0 & 1 & 1 & 0 \\
B = & 0 & 1 & 1 & 1 & 1 & 0 \\
Z = & 1(0) & 1 & 0(1) & 0 & 1 &
\end{array}
$$

(False outputs are indicated in parentheses.)

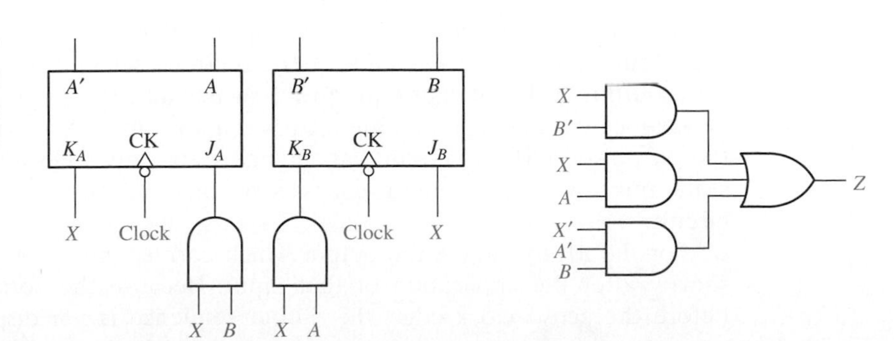


**FIGURE 13-7** Mealy Sequential Circuit to be Analyzed

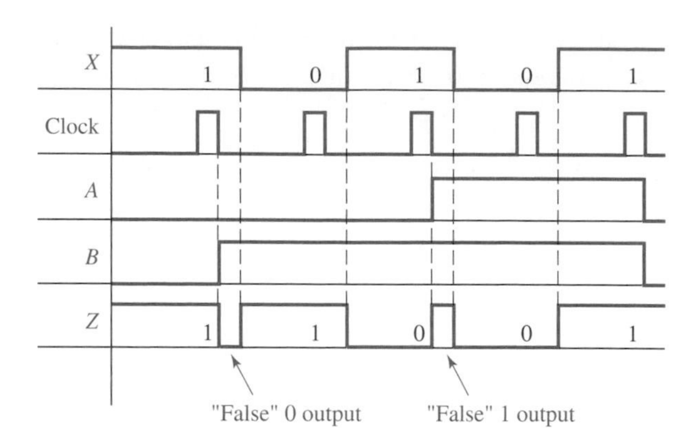


**FIGURE 13-8** Timing Chart for Circuit of Figure 13-7

A careful interpretation of the output waveform ($Z$) of the Mealy circuit is necessary. After the circuit has changed state and before the input is changed, the output may temporarily assume an incorrect value, which we call a *false* output. As indicated on the timing chart, this false value arises when the circuit has assumed a new state but the old input associated with the previous state is still present.

For a clocked sequential circuit, the value of the input immediately preceding the active clock edge determines the next state of the flip-flops. Extra input changes which might occur between active clock edges do not affect the state of the flip-flops. In a similar manner, the output from a Mealy circuit is only of interest immediately preceding the active clock edge, and extra output changes (false outputs) which might occur between active clock edges should be ignored.

Two types of false outputs can occur, as indicated in Figure 13-8. In one case the output $Z$ momentarily goes to 0 and returns to 1 before the active clock edge. In the other case the output $Z$ momentarily goes to 1 and returns to 0 before the active edge. These false outputs are often referred to as *glitches* and *spikes*. In both cases, two changes of output occur when no change is expected. Ignoring the false outputs by reading the output just before the falling clock edge, the output sequence for the circuit is $Z = 11001$. If circuit delays are negligible, the false outputs could be eliminated if the input $X$ was allowed to change only at the same time as the falling edge of the clock. If the output of the circuit is fed into a second sequential circuit which uses the same clock, the false outputs will not cause any problem because the inputs to the second circuit can cause a change of state only when a falling clock edge occurs. Because the output of a Moore circuit can change state only when the flip-flops change state and not when the input changes, no false outputs can appear in a Moore circuit.

For the *Mealy* circuit, the output which corresponds to a given input appears shortly after the application of that input. Because the correct output appears before the active clock edge, the output sequence is *not* displaced in time with respect to the input sequence as was the case for the Moore circuit.

## 13.3 State Tables and Graphs

In the previous section we analyzed clocked sequential circuits by signal tracing and the construction of timing charts. Although this is satisfactory for small circuits and short input sequences, the construction of state tables and graphs provides a more systematic approach which is useful for the analysis of larger circuits and which leads to a general synthesis procedure for sequential circuits.

The state table specifies the next state and output of a sequential circuit in terms of its present state and input. The following method can be used to construct the state table:

1. Determine the flip-flop input equations and the output equations from the circuit.
2. Derive the next-state equation for each flip-flop from its input equations, using one of the following relations:

| | | |
|---|---|---|
| D flip-flop | $Q^+ = D$ | (13-1) |
| D-CE flip-flop | $Q^+ = D \cdot CE + Q \cdot CE'$ | (13-2) |
| T flip-flop | $Q^+ = T \oplus Q$ | (13-3) |
| S-R flip-flop | $Q^+ = S + R'Q$ | (13-4) |
| J-K flip-flop | $Q^+ = JQ' + K'Q$ | (13-5) |

3. Plot a next-state map for each flip-flop.
4. Combine these maps to form the state table. Such a state table, which gives the next state of the flip-flops as a function of their present state and the circuit inputs, is frequently referred to as a transition table.

As an example of this procedure, we will derive the state table for the circuit of Figure 13-5:

1. The flip-flop input equations and output equation are

$$D_A = X \oplus B' \quad D_B = X + A \qquad Z = A \oplus B$$

2. The next-state equations for the flip-flops are

$$A^+ = X \oplus B' \quad B^+ = X + A$$

3. The corresponding maps are

| AB \ X | 0 | 1 |
|---|---|---|
| 00 | 1 | 0 |
| 01 | 0 | 1 |
| 11 | 0 | 1 |
| 10 | 1 | 0 |

$A^+$

| AB \ X | 0 | 1 |
|---|---|---|
| 00 | 0 | 1 |
| 01 | 0 | 1 |
| 11 | 1 | 1 |
| 10 | 1 | 1 |

$B^+$

**TABLE 13-2** Moore State Tables for Figure 13-5

(a)

| $AB$ | $A^+B^+$ $X = 0$ | $X = 1$ | $Z$ |
|---|---|---|---|
| 00 | 10 | 01 | 0 |
| 01 | 00 | 11 | 1 |
| 11 | 01 | 11 | 0 |
| 10 | 11 | 01 | 1 |

(b)

| Present State | Next State $X = 0$ | $X = 1$ | Present Output ($Z$) |
|---|---|---|---|
| $S_0$ | $S_3$ | $S_1$ | 0 |
| $S_1$ | $S_0$ | $S_2$ | 1 |
| $S_2$ | $S_1$ | $S_2$ | 0 |
| $S_3$ | $S_2$ | $S_1$ | 1 |

4. Combining these maps yields the transition table in Table 13-2(a), which gives the next state of both flip-flops ($A^+B^+$) as a function of the present state and input. The output function $Z$ is then added to the table. In this example, the output depends only on the present state of the flip-flops and not on the input, so only a single output column is required.

Using Table 13-2(a), we can construct the timing chart of Figure 13-6 or any other timing chart for some given input sequence and specified initial state. Initially $AB = 00$ and $X = 0$, so $Z = 0$ and $A^+B^+ = 10$. This means that *after* the rising clock edge, the flip-flop state will be $AB = 10$. Then, with $AB = 10$, the output is $Z = 1$. The next input is $X = 1$, so $A^+B^+ = 01$ and the state will change after the next rising clock edge. Continuing in this manner, we can complete the timing chart.

If we are not interested in the individual flip-flop states, we can replace each combination of flip-flop states with a single symbol which represents the state of the circuit. Replacing 00 with $S_0$, 01 with $S_1$, 11 with $S_2$, and 10 with $S_3$ in Table 13-2(a) yields Table 13-2(b). The $Z$ column is labeled Present Output because it is the output associated with the Present State. The state graph of Figure 13-9 represents Table 13-2(b). Each node of the graph represents a state of the circuit, and the corresponding output is placed in the circle below the state symbol. The arc joining two nodes is labeled with the value of $X$ which will cause a state change between these nodes. Thus, if the circuit is in state $S_0$ and $X = 1$, a clock edge will cause a transition to state $S_1$.

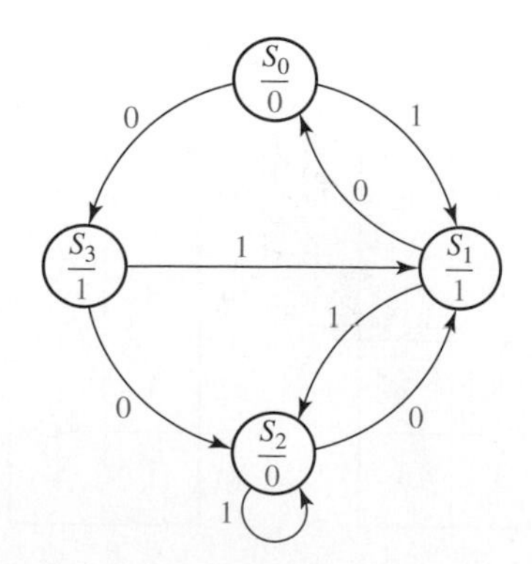


**FIGURE 13-9** Moore State Graph for Figure 13-5

Next, we will construct the state table and graph for the Mealy machine of Figure 13-7. The next-state and output equations are

$$A^+ = J_A A' + K'_A A = XBA' + X'A$$
$$B^+ = J_B B' + K'_B B = XB' + (AX)'B = XB' + X'B + A'B$$
$$Z = X'A'B + XB' + XA$$

The next-state and output maps (Figure 13-10) combine to form the transition table in Table 13-3(a). Given values for $A$, $B$, and $X$, the current value of the output is determined from the $Z$ column of this table, and the states of the flip-flops after the active clock edge are determined from the $A^+B^+$ columns.

We can construct the timing chart of Figure 13-8 using Table 13-3(a). Initially with $A = B = 0$ and $X = 1$, the table shows that $Z = 1$ and $A^+B^+ = 01$. Therefore, after the falling clock edge, the state of flip-flop $B$ will change to 1, as indicated in Figure 13-8. Now, from the 01 row of the table, if $X$ is still 1, the output will be 0 until the input is changed to $X = 0$. Then, the output is $Z = 1$, and the next falling clock edge produces no state change. Finish stepping through the state table in this manner and verify that $A$, $B$, and $Z$ are as given in Figure 13-8.

If we let $AB = 00$ correspond to circuit state $S_0$, 01 to $S_1$, 11 to $S_2$, and 10 to $S_3$, we can construct the state table in Table 13-3(b) and the state graph of Figure 13-11. In Table 13-3(b), the Present Output column gives the output associated with the present state and present input. Thus, if the present state is $S_0$ and the input changes from 0 to 1, the output will immediately change from 0 to 1. However, the state will not change to the next state ($S_1$) until after the clock pulse. For Figure 13-11, the

FIGURE 13-10

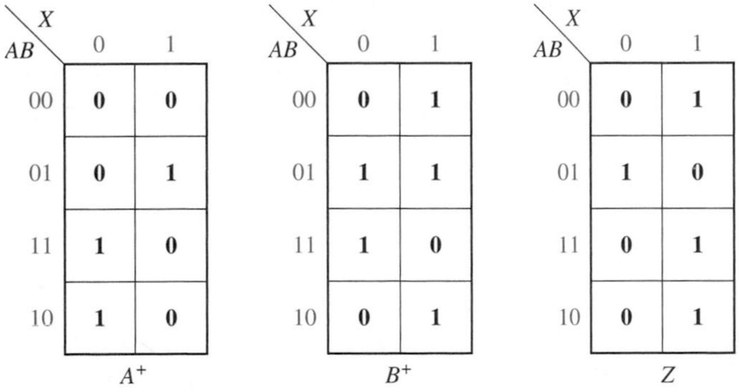

$A^+$:

| AB \ X | 0 | 1 |
|---|---|---|
| 00 | 0 | 0 |
| 01 | 0 | 1 |
| 11 | 1 | 0 |
| 10 | 1 | 0 |

$B^+$:

| AB \ X | 0 | 1 |
|---|---|---|
| 00 | 0 | 1 |
| 01 | 1 | 1 |
| 11 | 1 | 0 |
| 10 | 0 | 1 |

$Z$:

| AB \ X | 0 | 1 |
|---|---|---|
| 00 | 0 | 1 |
| 01 | 1 | 0 |
| 11 | 0 | 1 |
| 10 | 0 | 1 |

TABLE 13-3 Mealy State Tables for Figure 13-7

(a)

| AB | $A^+B^+$ $X = 0$ | 1 | $Z$ $X = 0$ | 1 |
|---|---|---|---|---|
| 00 | 00 | 01 | 0 | 1 |
| 01 | 01 | 11 | 1 | 0 |
| 11 | 11 | 00 | 0 | 1 |
| 10 | 10 | 01 | 0 | 1 |

(b)

| Present State | Next State $X = 0$ | 1 | Present Output $X = 0$ | 1 |
|---|---|---|---|---|
| $S_0$ | $S_0$ | $S_1$ | 0 | 1 |
| $S_1$ | $S_1$ | $S_2$ | 1 | 0 |
| $S_2$ | $S_2$ | $S_0$ | 0 | 1 |
| $S_3$ | $S_3$ | $S_1$ | 0 | 1 |

FIGURE 13-11 Mealy State Graph for Figure 13-7

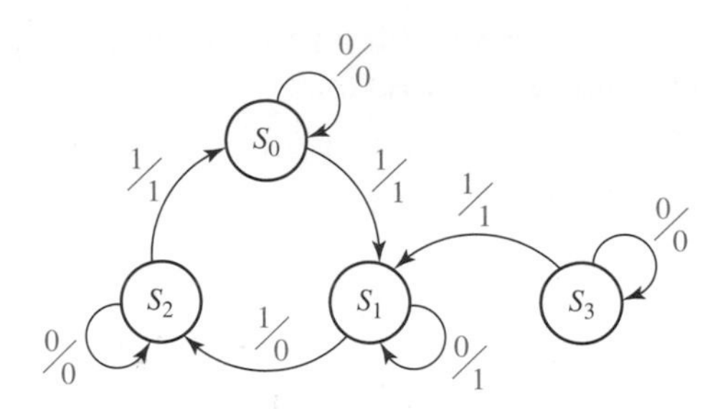


labels on the arrows between states are of the form $X/Z$, where the symbol before the slash is the input and the symbol after the slash is the corresponding output. Thus, in state $S_0$ an input of 0 gives an output of 0, and an input of 1 gives an output of 1. For any given input sequence, we can easily trace out the state and output sequences on the state graph. For the input sequence $X = 10101$, verify that the corresponding output sequence is 11001. This agrees with Figure 13-8 if the false outputs are ignored. Note that the false outputs do not show on the state graph because the inputs are read at the active clock edge, and no provision is made for extra input changes between active edges.

Next, we will analyze the operation of a serial adder [Figure 13-12(a)] that adds two $n$-bit binary numbers $X = x_{n-1} \ldots x_1 x_0$ and $Y = y_{n-1} \ldots y_1 y_0$. The operation of the serial adder is similar to the parallel adder of Figure 4-2 except that the binary numbers are fed in serially, one pair of bits at a time, and the sum is read out serially, one bit at a time. First, $x_0$ and $y_0$ are fed in; a sum digit $s_0$ is generated, and the carry $c_1$ is stored. At the next clock time, $x_1$ and $y_1$ are fed in and added to $c_1$ to give the next sum digit $s_1$ and the new carry $c_2$, which is stored. This process continues until all bits have been added. A full adder is used to add the $x_i$, $y_i$, and $c_i$ bits to form $c_{i+1}$ and $s_i$. A D flip-flop is used to store the carry $(c_{i+1})$ on the rising edge of the clock. The $x_i$ and $y_i$ inputs must be synchronized with the clock.

Figure 13-13 shows a timing diagram for the serial adder. In this example we add $10011 + 00110$ to give a sum of 11001 and a final carry of 0. Initially the carry flip-flop must be cleared so that $c_0 = 0$. We start by adding the least-significant (rightmost) bits in each word. Adding $1 + 0 + 0$ gives $s_0 = 1$ and $c_1 = 0$, which is stored in the flip-flop

FIGURE 13-12 Serial Adder

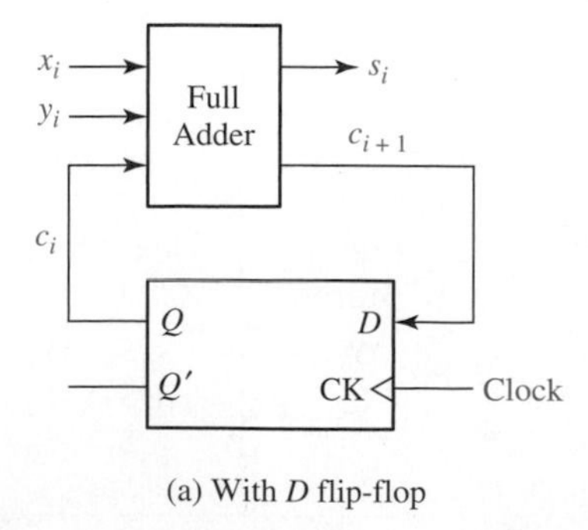


(a) With $D$ flip-flop

| $x_i$ | $y_i$ | $c_i$ | $c_{i+1}$ | $s_i$ |
|---|---|---|---|---|
| 0 | 0 | 0 | 0 | 0 |
| 0 | 0 | 1 | 0 | 1 |
| 0 | 1 | 0 | 0 | 1 |
| 0 | 1 | 1 | 1 | 0 |
| 1 | 0 | 0 | 0 | 1 |
| 1 | 0 | 1 | 1 | 0 |
| 1 | 1 | 0 | 1 | 0 |
| 1 | 1 | 1 | 1 | 1 |

(b) Truth table

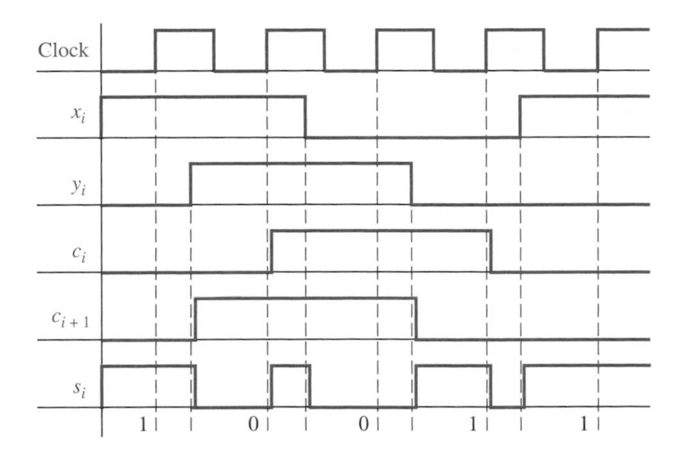


**FIGURE 13-13** Timing Diagram for Serial Adder

at the rising clock edge. Because $y_1$ is 1, adding $1 + 1 + 0$ gives $s_1 = 0$ and $c_2 = 1$, which is stored in the flip-flop on the rising clock edge. This process continues until the addition is completed. Reading the sum output just before the rising edge of the clock gives the correct result.

The truth table for the full adder (Table 4-4) is repeated in Figure 13-12(b) in modified form. Using this table, we can construct a state graph (Figure 13-14) for the serial adder. The serial adder is a Mealy machine with inputs $x_i$ and $y_i$ and output $s_i$. The two states represent a carry ($c_i$) of 0 and 1, respectively. From the table, $c_i$ is the present state of the sequential circuit, and $c_{i+1}$ is the next state. If we start in $S_0$ (no carry), and $x_i\,y_i = 11$, the output is $s_i = 0$ and the next state is $S_1$. This is indicated by the arrow going from state $S_0$ to $S_1$.

Table 13-4 shows a state table for a Mealy sequential circuit with two inputs and two outputs. Figure 13-15 shows the corresponding state graph. The notation 00, 01/00 on the arc from $S_3$ to $S_2$ means if $X_1 = X_2 = 0$ or $X_1 = 0$ and $X_2 = 1$, then $Z_1 = 0$ and $Z_2 = 0$.

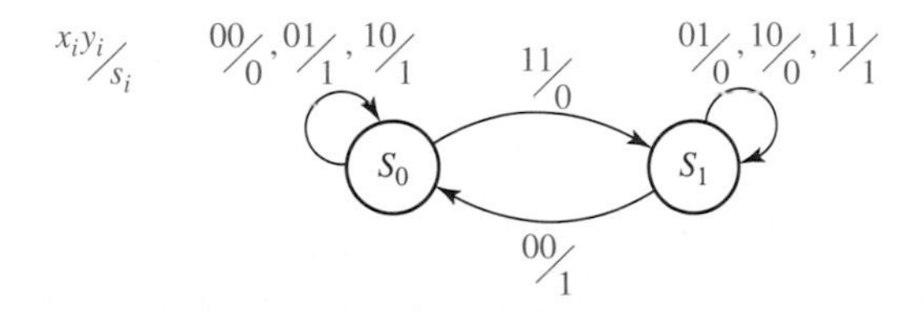


**FIGURE 13-14** State Graph for Serial Adder

**TABLE 13-4** A State Table with Multiple Inputs and Outputs

| Present State | Next State $X_1X_2 = 00$ | 01 | 10 | 11 | Present Output $(Z_1Z_2)$ $X_1X_2 = 00$ | 01 | 10 | 11 |
|---|---|---|---|---|---|---|---|---|
| $S_0$ | $S_3$ | $S_2$ | $S_1$ | $S_0$ | 00 | 10 | 11 | 01 |
| $S_1$ | $S_0$ | $S_1$ | $S_2$ | $S_3$ | 10 | 10 | 11 | 11 |
| $S_2$ | $S_3$ | $S_0$ | $S_1$ | $S_1$ | 00 | 10 | 11 | 01 |
| $S_3$ | $S_2$ | $S_2$ | $S_1$ | $S_0$ | 00 | 00 | 01 | 01 |

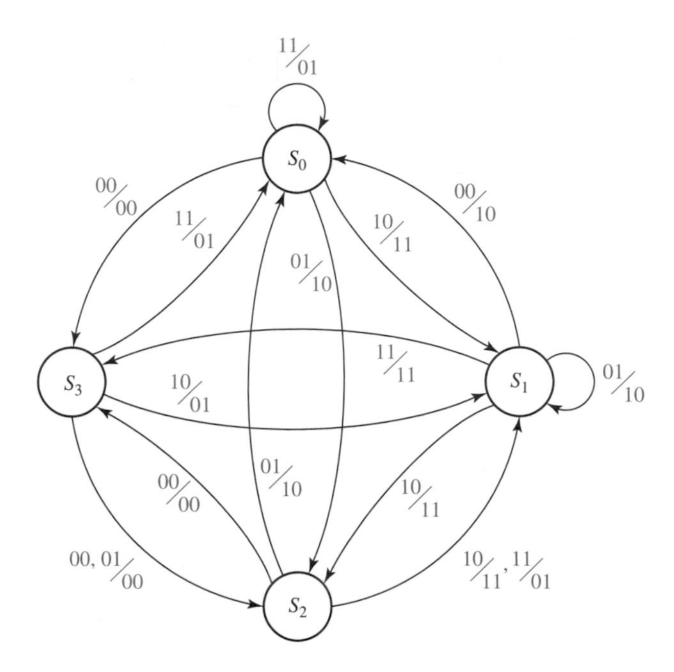


FIGURE 13-15
State Graph for Table 13-4

## Construction and Interpretation of Timing Charts

Several important points concerning the construction and interpretation of timing charts are summarized as follows:

1. When constructing timing charts, note that a state change can only occur after the rising (or falling) edge of the clock, depending on the type of flip-flop used.
2. The input will normally be stable immediately before and after the active clock edge.
3. For a Moore circuit, the output can change only when the state changes, but for a Mealy circuit, the output can change when the input changes as well as when the state changes. A false output may occur between the time the state changes and the time the input is changed to its new value. (In other words, if the state has changed to its next value, but the old input is still present, the output may be temporarily incorrect.)
4. False outputs are difficult to determine from the state graph, so use either signal tracing through the circuit or use the state table when constructing timing charts for Mealy circuits.
5. When using a Mealy state table for constructing timing charts, the procedure is as follows:
   (a) For the first input, read the present output and plot it.
   (b) Read the next state and plot it (following the active edge of the clock pulse).

(c) Go to the row in the table which corresponds to the next state and read the output under the old input column and plot it. (This may be a false output.)

(d) Change to the next input and repeat steps (a), (b), and (c).
(Note: If you are just trying to read the correct output sequence from the table, step (c) is naturally omitted.)

**6.** For Mealy circuits, the best time to read the output is just before the active edge of the clock, because the output should always be correct at that time.

The example in Figure 13-16 shows a state graph, a state table, a circuit that implements the table, and a timing chart. When the state is $S_0$ and the input is $X = 0$, the output from the state graph, state table, circuit, and timing chart is $Z = 1$ (labeled $A$ on the figure). Note that this output occurs *before* the rising edge of the clock. In a Mealy circuit, the output is a function of the present state and input; therefore, the output should be read just before the clock edge that causes the state to change.

FIGURE 13-16

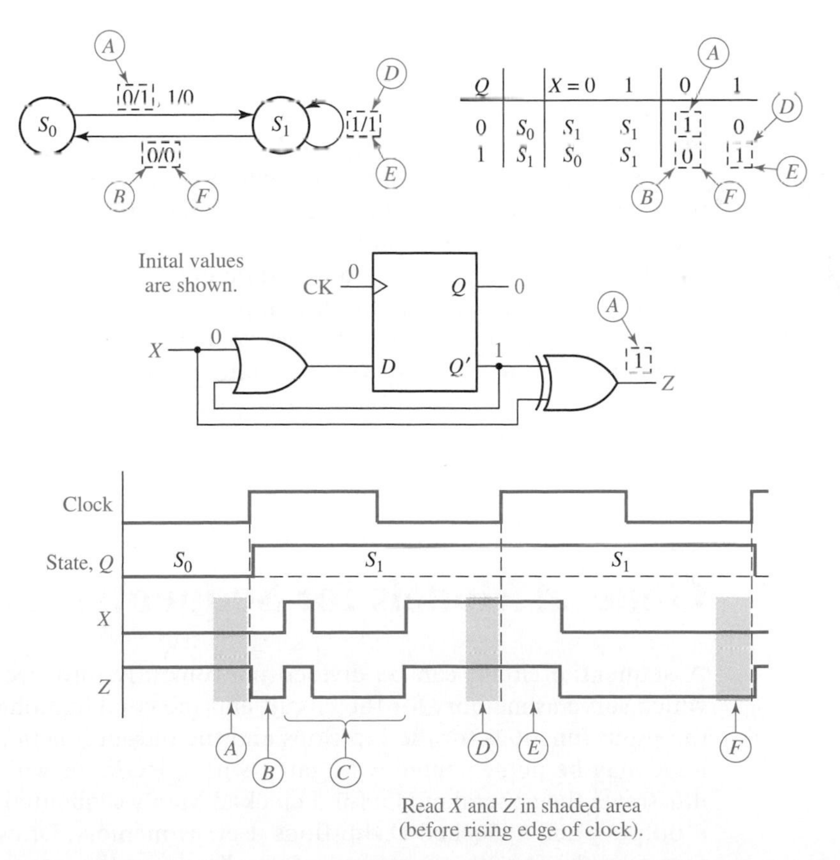

As you continue to study this example, each time the input $X$ changes, trace the changes on the state graph, the state table, the circuit, and the timing chart. Because the input $X$ was 0 before the first rising edge of the clock, the state changes to $S_1$ after the first rising edge of the clock. Because of the state change, the output also changes (B on the timing chart), but because the input has not yet changed to its new value, the output value may not be correct. We refer to this as a *false output* or *glitch*. If the input changes several times before it assumes its correct value, the output may also change several times (C). The input must assume its correct value before the rising edge of the clock, and the output should be read at this time (D). After the rising clock edge, the state stays the same and the output stays the same for this particular example. In general, the state may change after a rising edge of the clock, and the state change may result in an output change. Again, the output value may be wrong because the input still has the old value (E). When the input is changed to its new value, the output changes to its new value (F), and this value should be read before the next rising clock edge.

If we look at the input and output just before each rising edge of the clock, we find the following sequences:

$$X = 0\ 1\ 0$$

$$Z = 1\ 1\ 0$$

You should be able to verify the sequence for $Z$ using the state graph, using the state table, and using the circuit diagram.

The synthesis procedure for sequential circuits, discussed in detail in Units 14 through 16, is just the opposite of the procedure used for analysis. Starting with the specifications for the sequential circuit to be synthesized, a state graph is constructed. This graph is then translated to a state table, and the flip-flop output values are assigned for each state. The flip-flop input equations are then derived, and finally, the logic diagram for the circuit is drawn. For example, to synthesize the circuit in Figure 13-7, we would start with the state graph of Figure 13-11. Then, we would derive Table 13-3(b), Table 13-3(a), the next-state and output equations, and, finally, the circuit of Figure 13-7.

## 13.4 General Models for Sequential Circuits

A sequential circuit can be divided conveniently into two parts—the flip-flops which serve as memory for the circuit and the combinational logic which realizes the input functions for the flip-flops and the output functions. The combinational logic may be implemented with gates, with a ROM, or with a PLA. Figure 13-17 illustrates the general model for a clocked Mealy sequential circuit with $m$ inputs, $n$ outputs, and $k$ clocked D flip-flops used as memory. Drawing the model in this form emphasizes the presence of feedback in the sequential circuit because the flip-flop outputs are fed back as inputs to the combinational subcircuit.

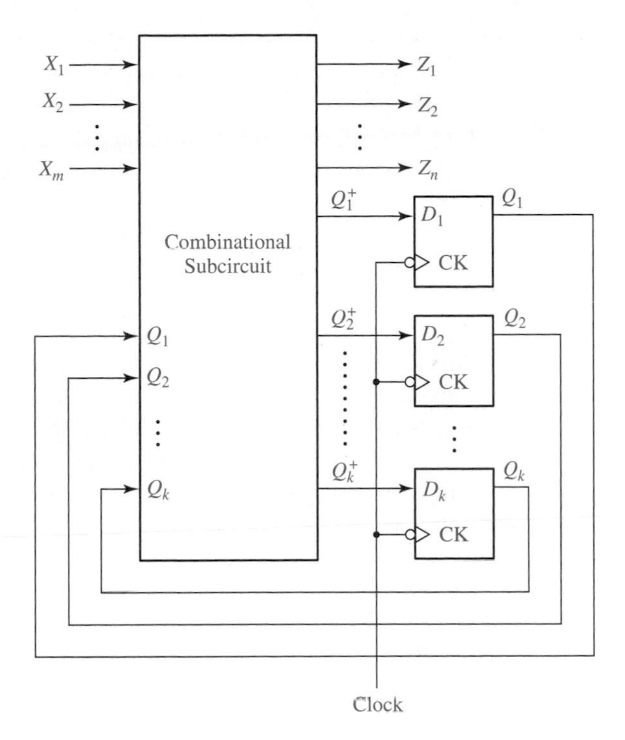


**FIGURE 13-17** General Model for Mealy Circuit Using Clocked D Flip-Flops

The combinational subcircuit realizes the $n$ output functions and the $k$ next-state functions, which serve as inputs to the D flip-flops:

$$\left.\begin{aligned} Z_1 &= f_1(X_1, X_2, \dots, X_m, Q_1, Q_2, \dots, Q_k) \\ Z_2 &= f_2(X_1, X_2, \dots, X_m, Q_1, Q_2, \dots, Q_k) \\ &\vdots \\ Z_n &= f_n(X_1, X_2, \dots, X_m, Q_1, Q_2, \dots, Q_k) \end{aligned}\right\} n \text{ output functions}$$

$$\left.\begin{aligned} Q_1^+ &= D_1 = g_1(X_1, X_2, \dots, X_m, Q_1, Q_2, \dots, Q_k) \\ Q_2^+ &= D_2 = g_2(X_1, X_2, \dots, X_m, Q_1, Q_2, \dots, Q_k) \\ &\vdots \\ Q_k^+ &= D_k = g_k(X_1, X_2, \dots, X_m, Q_1, Q_2, \dots, Q_k) \end{aligned}\right\} k \text{ next-state functions}$$

When a set of inputs is applied to the circuit, the combinational subcircuit generates the outputs $(Z_1, Z_2, \dots, Z_n)$ and the flip-flop inputs $(D_1, D_2, \dots, D_k)$. Then, a clock pulse is applied and the flip-flops change to the proper next state. This process is repeated for each set of inputs. Note that at a given point in time, the outputs of the flip-flops represent the present state of the circuit $(Q_1, Q_2, \dots, Q_k)$. These $Q_i$'s feed back into the combinational circuit, which generates the flip-flop inputs using the $Q_i$'s and the $X$ inputs. When D flip-flops are used, $D_i = Q_i^+$; therefore, the combinational circuit outputs are labeled $Q_1^+$, $Q_2^+$, etc. Although the model in Figure 13-17 uses D flip-flops, a similar model may be used for other types of clocked flip-flops, in which case the combinational circuit must generate the appropriate flip-flop inputs instead of the next-state functions.

The clock synchronizes the operation of the flip-flops and prevents timing problems. The gates (or other logic) in the combinational subcircuit have finite propagation delays, so when the inputs to the circuit are changed, a finite time is required before the flip-flop inputs reach their final values. Because the gate delays are not all the same, the flip-flop input signals may contain transients, and they may change at different times. If the next active clock edge does not occur until all flip-flop input signals have reached their final steady-state values, the unequal gate delays will not cause any timing problems. All flip-flops which must change state do so at the same time in response to the active edge of the clock. When the flip-flops change state, the new flip-flop outputs are fed back into the combinational subcircuit. However, no further change in the flip-flop states can occur until the next clock pulse.

We can determine the fastest clock speed (the minimum clock period) from the general model of the Mealy circuit in Figure 13-17. The computation of the minimum clock period is similar to that of Figure 11-17, except that we must also consider the effect of the $X$ inputs. Figure 13-18 shows the sequence of events during one clock period. Following the active edge of the clock the flip-flops change state, and the flip-flop output is stable after the propagation delay ($t_p$). The new values of $Q$ then propagate through the combinational circuit so that the $D$ values are stable after the combinational circuit delay ($t_c$). Then, the flip-flop setup time ($t_{su}$) must elapse before the next active clock edge. Thus, the propagation delay in the flip-flops, the propagation delay in the combinational subcircuit, and the setup time for the flip-flops determine how fast the sequential circuit can operate, and the minimum clock period is

$$t_{\text{clk}}\ (min) = t_p + t_c + t_{su}$$

The preceding discussion assumes that the $X$ inputs are stable no later than $t_c + t_{su}$ before the next active clock edge. If this is not the case, then we must calculate the minimum clock period by

$$t_{\text{clk}}\ (min) = t_x + t_c + t_{su}$$

where $t_x$ is the time after the active clock edge at which the $X$ inputs are stable.

The general model for the clocked Moore circuit (Figure 13-19) is similar to the clocked Mealy circuit. The output subcircuit is drawn separately for the Moore circuit because the output is only a function of the present state of the flip-flops and not a function of the circuit inputs. Operation of the Moore circuit is similar to that of the Mealy except when a set of inputs is applied to the Moore circuit, the resulting outputs do not appear until after the clock causes the flip-flops to change state.

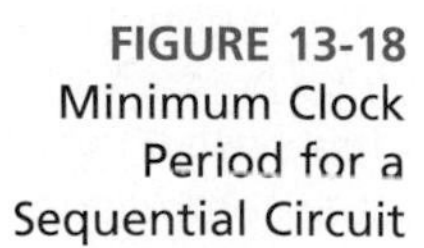

**FIGURE 13-18** Minimum Clock Period for a Sequential Circuit

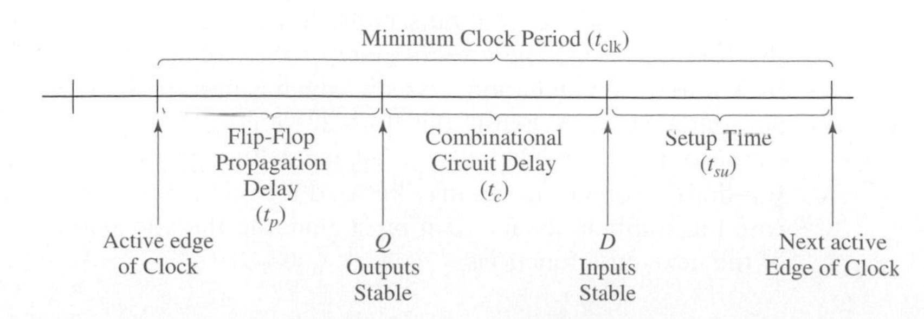

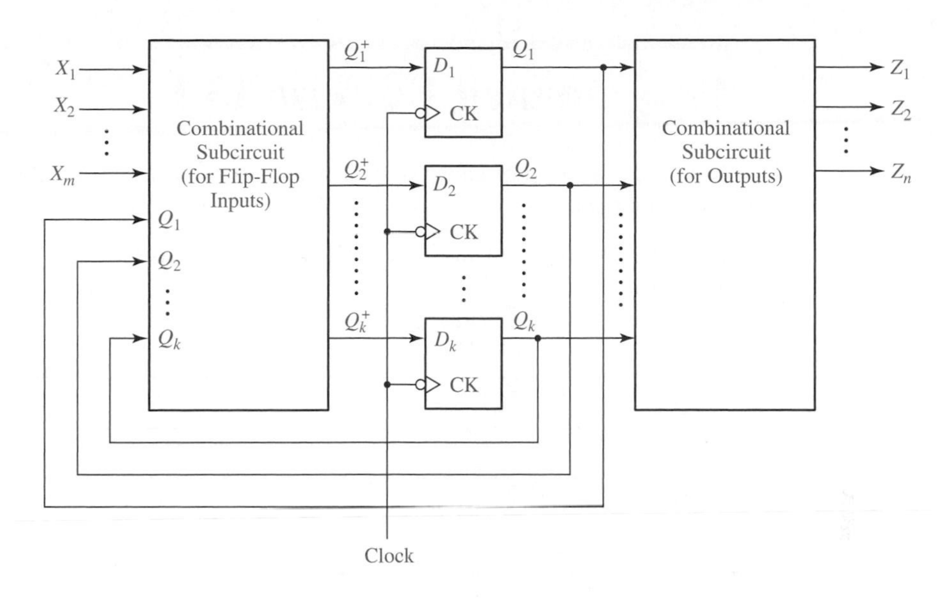


**FIGURE 13-19** General Model for Moore Circuit Using Clocked D Flip-Flops

**TABLE 13-5** State Table with Multiple Inputs and Outputs

| Present State | Next State $X = 0$ | 1 | 2 | 3 | Present Output ($Z$) $X = 0$ | 1 | 2 | 3 |
|---|---|---|---|---|---|---|---|---|
| $S_0$ | $S_3$ | $S_2$ | $S_1$ | $S_0$ | 0 | 2 | 3 | 1 |
| $S_1$ | $S_0$ | $S_1$ | $S_2$ | $S_3$ | 2 | 2 | 3 | 3 |
| $S_2$ | $S_3$ | $S_0$ | $S_1$ | $S_1$ | 0 | 2 | 3 | 1 |
| $S_3$ | $S_2$ | $S_2$ | $S_1$ | $S_0$ | 0 | 0 | 1 | 1 |

To facilitate the study of sequential circuits with multiple inputs and outputs, the assignment of symbols to represent each combination of input values and each combination of output values is convenient. For example, we can replace Table 13-4 with Table 13-5 if we let $X = 0$ represent the input combination $X_1X_2 = 00$, $X = 1$ represent $X_1X_2 = 01$, etc., and similarly let $Z = 0$ represent the output combination $Z_1Z_2 = 00$, $Z = 1$ represent $Z_1Z_2 = 01$, etc. In this way we can specify the behavior of any sequential circuit in terms of a single input variable $X$ and a single output variable $Z$.

Table 13-5 specifies two functions, the next-state function and the output function. The next-state function, designated $\delta$, gives the next state of the circuit (i.e., the state after the clock pulse) in terms of the present state ($S$) and the present input ($X$):

$$S^+ = \delta\,(S, X) \tag{13-6}$$

The output function, designated $\lambda$, gives the output of the circuit ($Z$) in terms of the present state ($S$) and input ($X$):

$$Z = \lambda\,(S, X) \tag{13-7}$$

Values of $S^+$ and $Z$ can be determined from the state table. From Table 13-5, we have

$$\delta\,(S_0, 1) = S_2 \qquad \delta\,(S_2, 3) = S_1$$
$$\lambda\,(S_0, 1) = 2 \qquad \lambda\,(S_2, 3) = 1$$

We will use the $\lambda$ and $\delta$ notation when we discuss equivalent sequential circuits in Unit 15.

# Programmed Exercise 13.1

Cover the bottom of each page with a sheet of paper and slide it down as you check your answers.

**13.1(a)** In this exercise you will analyze the following sequential circuit using a state table and a timing chart.
Derive the next-state and output equations.

$A^+ =$ ____________________________

$B^+ =$ ____________________________

$Z =$ ____________________________

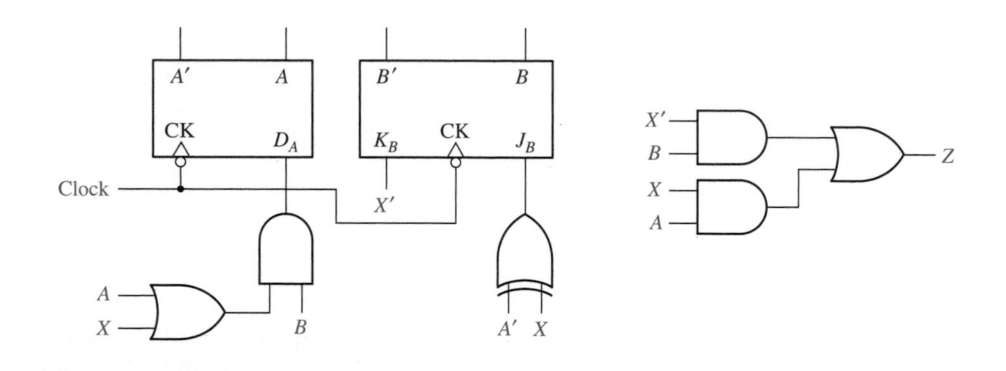


**Answer** $Z = XA + X'B,\ B^+ = (A' \oplus X)B' + XB = A'B'X' + AB'X + XB$
$A^+ = B(A + X)$

**13.1(b)** Plot these equations on maps and complete the transition table.

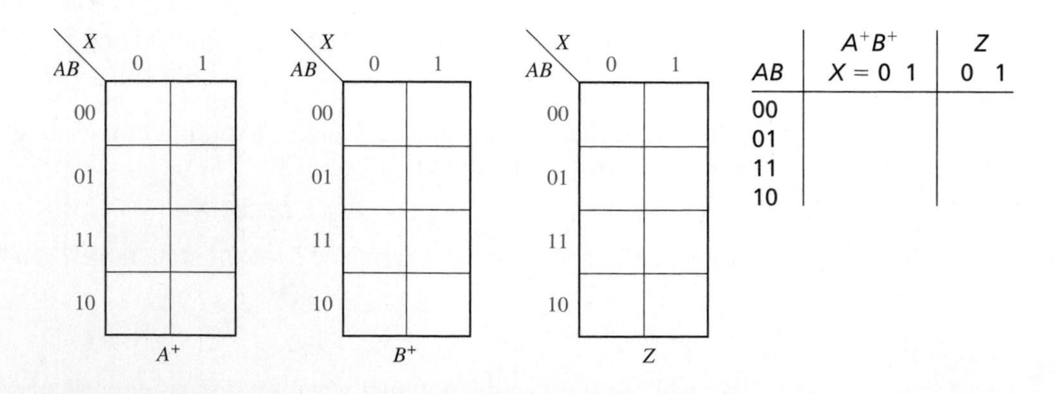


| AB | $A^+B^+$ X = 0 1 | Z 0 1 |
|---|---|---|
| 00 | | |
| 01 | | |
| 11 | | |
| 10 | | |

**Answer to 13.1(b)**

| | $AB$ | $A^+B^+$ $X=0$ | $A^+B^+$ $1$ | $Z$ $0$ | $Z$ $1$ |
|---|---|---|---|---|---|
| $S_0$ | 00 | 01 | 00 | 0 | 0 |
| $S_1$ | 01 | 00 | 11 | 1 | 0 |
| $S_2$ | 11 | 10 | 11 | 1 | 1 |
| $S_3$ | 10 | 00 | 01 | 0 | 1 |

**13.1(c)** Convert your transition table to a state table using the given state numbering.

| | Next State $X=0$ | Next State $1$ | Output $0$ | Output $1$ |
|---|---|---|---|---|
| $S_0$ | | | | |
| $S_1$ | | | | |
| $S_2$ | | | | |
| $S_3$ | | | | |

**Answer to 13.1(c)**

| | $X=0$ | $1$ | $0$ | $1$ |
|---|---|---|---|---|
| $S_0$ | $S_1$ | $S_0$ | 0 | 0 |
| $S_1$ | $S_0$ | $S_2$ | 1 | 0 |
| $S_2$ | $S_3$ | $S_2$ | 1 | 1 |
| $S_3$ | $S_0$ | $S_1$ | 0 | 1 |

**13.1(d)** Complete the corresponding state graph.

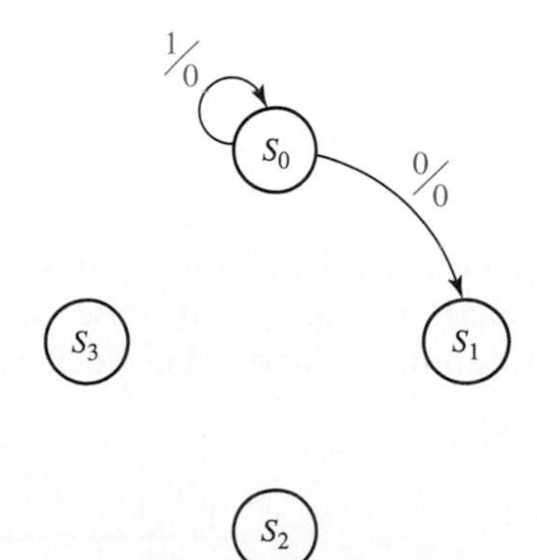

**Answer to 13.1(d)**

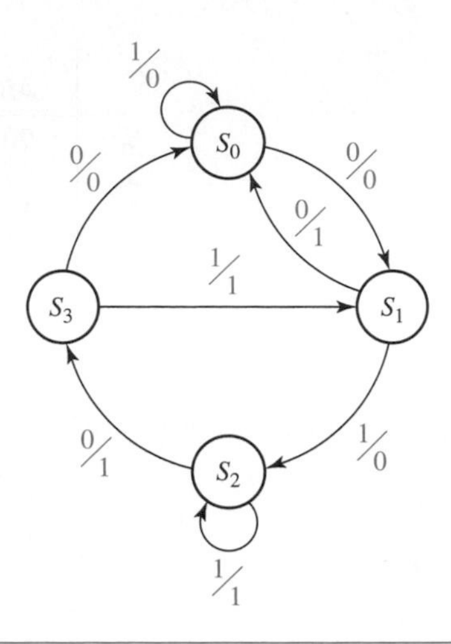


**13.1(e)** Using this graph, determine the state sequence and output sequence if the initial state is $S_0$ and the input sequence is $X = 0, 1, 0, 1$.

(1) The initial output with $X = 0$ in state $S_0$ is $Z =$ __________ and the next state is __________.

(2) The output in this state when the next input ($X = 1$) is applied is $Z =$ __________ and the next state is__________.

(3) When the third input ($X = 0$) is applied, the output is $Z =$ __________ and the next state is __________.

(4) When the last input is applied, Z = __________ and the final state is __________.

In summary, the state sequence is $S_0$,__________, __________, __________, __________. The output sequence is $Z =$ __________.

**Answer to 13.1(e)** $S_0, S_1, S_2, S_3, S_1$ $\quad Z = 0011$

**13.1(f)** This sequence for $Z$ is the correct output sequence. Next, we will determine the timing chart including any false outputs for $Z$. Assuming that $X$ changes midway between falling and rising clock edges, draw the waveform for $X$ ($X = 0, 1, 0, 1$).

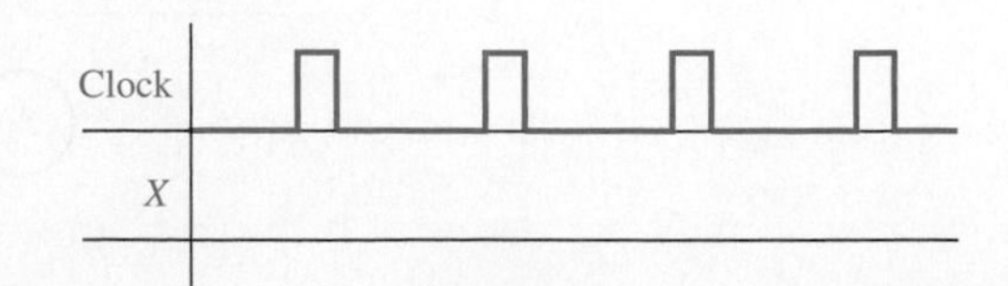

**Answer**

Clock

$X$

$A$

$B$

| $AB$ | $A^+B^+$ $X = 0$ | 1 |
|---|---|---|
| 00 | 01 | 00 |
| 01 | 00 | 11 |
| 11 | 10 | 11 |
| 10 | 00 | 01 |

**13.1(g)** Referring to the transition table, sketch the waveforms for $A$ and $B$ assuming that initially $A = B = 0$. The state sequence is

$AB = 00$, ________, ________, ________, ________ .

**Answer** (Note that $A$ and $B$ change immediately after the falling clock edge.)

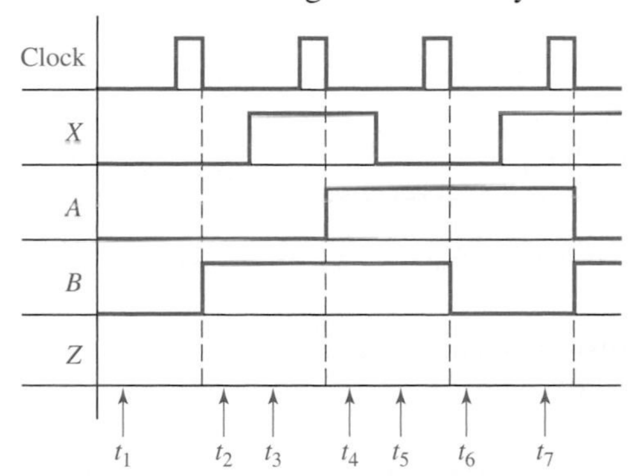


| $AB$ | $Z$ $X = 0$ | 1 |
|---|---|---|
| 00 | 0 | 0 |
| 01 | 1 | 0 |
| 11 | 1 | 1 |
| 10 | 0 | 1 |

Check your state sequence against the answer to 13.1(e), noting that $S_0 = 00$, $S_1 = 01$, $S_2 = 11$ and $S_3 = 10$.

**13.1(h)** Using the output table, sketch the waveform for $Z$. At time $t_1$, $X = A = B = 0$, so $Z =$ ________. At time $t_2$, $X =$ ________ and $AB =$ ________, so $Z =$ ________. At time $t_3$, $X =$ ________ and $AB =$ ________, so $Z =$ ________. Complete the waveform for $Z$, showing the output at $t_4$, $t_5$, etc.

**Answer** (Note that $Z$ can change immediately following the change in $X$ or immediately following the falling clock edge.)

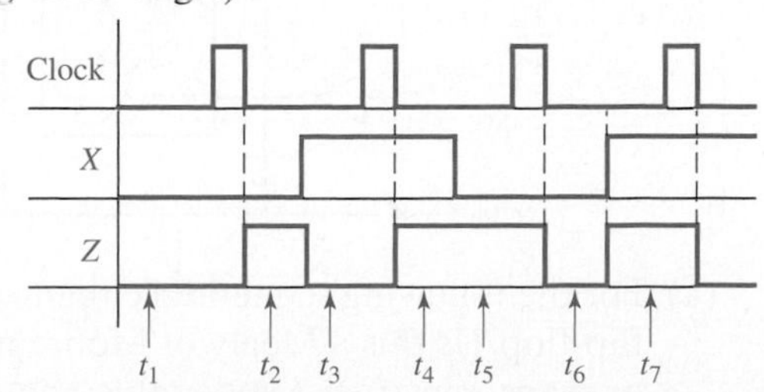

**13.1(i)** (1) Because this is a Mealy circuit, the correct times to read the output are during intervals $t_1$, ___________, ___________, and ___________ .
(2) The correct output sequence is therefore $Z$ = ___________ .
(3) False outputs may occur during intervals ___________, ___________, and ___________ .
(4) In two of these intervals, false outputs actually occur. These intervals are ___________ and ___________.

---

**Answer** (1) $t_1$, $t_3$, $t_5$, and $t_7$
(2) Check your $Z$ sequence against the answer to 13.1(e).
(3) $t_2$, $t_4$, and $t_6$.
(4) $t_2$ and $t_6$ (output during $t_4$ is not false because it is the same as $t_5$).

---

**13.1(j)** Finally, we will verify part of the timing chart by signal tracing on the original circuit (see 13.1(a)).

(1) Initially, $A = B = 0$ and $X = 0$, so $D_A$ = ___________, $J_B$ = ___________, $K_B$ = ___________, and $Z$ = ___________ .
(2) After the clock pulse $A$ = ___________, $B$ = ___________, and $Z$ = ___________ .
(3) After $X$ is changed to 1, $D_A$ = ___________, $J_B$ = ___________, $K_B$ = ___________, and $Z$ = ___________ .
(4) After the clock pulse, $A$ = ___________, $B$ = ___________, and $Z$ = ___________ .

Check your answers against the timing chart. Answer to (1) corresponds to $t_1$, (2) to $t_2$, (3) to $t_3$, and (4) to $t_4$.

# Problems

**13.2** Construct a state graph for the shift register shown. ($X$ is the input, and $Z$ is the output.) Is this a Mealy or Moore machine?

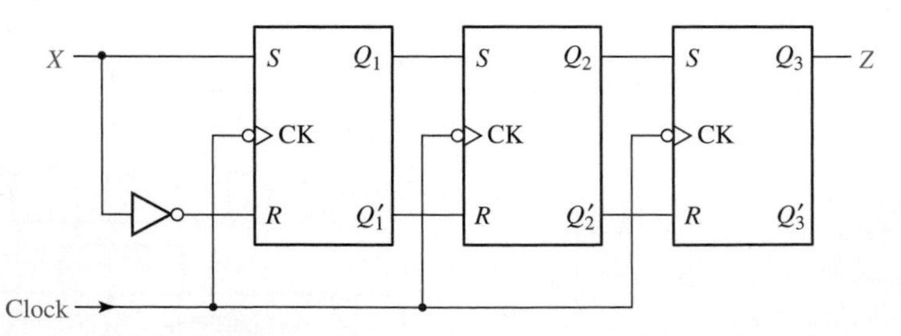


**13.3** (a) For the following sequential circuit, find the next-state equation or map for each flip-flop. (Is this a Mealy or Moore machine?) Using these next-state equations or maps, construct a state table and graph for the circuit.
(b) What is the output sequence when the input sequence is $X = 01100$?

(c) Draw a timing diagram for the input sequence in (b). Show the clock, $X$, $A$, $B$, and $Z$. Assume that the input changes between falling and rising clock edges.

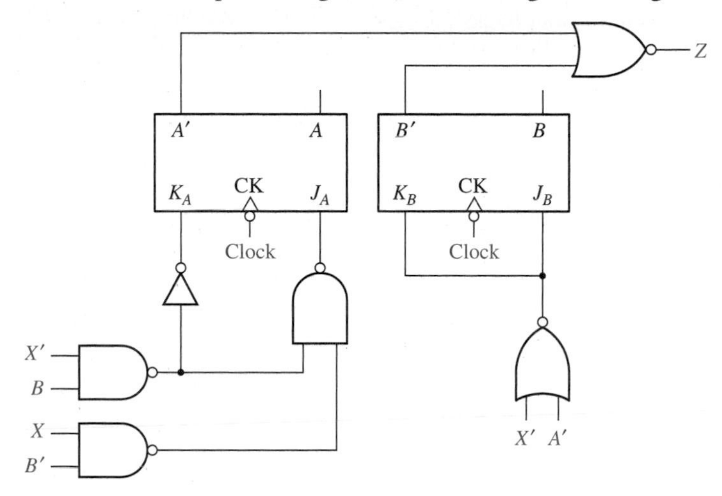


**13.4** A sequential circuit has the form shown in Figure 13-17, with

$$D_1 = Q_2Q_3' \qquad D_3 = Q_2' + X$$
$$D_2 = Q_3 \qquad Z = XQ_2' + X'Q_2$$

(a) Construct a state table and state graph for the circuit. (Is this a Mealy or Moore machine?)

(b) Draw a timing diagram for the circuit showing the clock, $X$, $Q_1$, $Q_2$, $Q_3$, and $Z$. Use the input sequence $X = 01011$. Change $X$ between clock edges so that we can see false outputs, and indicate any false outputs on the diagram.

(c) Compare the output sequence obtained from the timing diagram with that from the state graph.

(d) At what time with respect to the clock should the input be changed in order to eliminate the false output(s)?

**13.5** Below is a state transition table with the outputs missing. The output should be $Z = X'B' + XB$.

(a) Is this a Mealy machine or Moore machine?

(b) Fill in the outputs on the state transition table.

(c) Give the state graph.

(d) For an input sequence of $X = 10101$, give a timing diagram for the clock, $X$, $A$, $B$, $C$, and $Z$. State changes occur on the rising clock edge. What is the correct output sequence for $Z$? Change $X$ between rising and falling clock edges so that we can see false outputs, and indicate any false outputs on the diagram.

| | $A^+B^+C^+$ | | |
|---|---|---|---|
| $ABC$ | $X = 0$ | $X = 1$ | |
| 000 | 011 | 010 | |
| 001 | 000 | 100 | |
| 010 | 100 | 100 | |
| 011 | 010 | 000 | |
| 100 | 100 | 001 | |

**13.6** A sequential circuit of the form shown in Figure 13-17 is constructed using a ROM and two rising-edge-triggered D flip-flops. The contents of the ROM are given in the table. Assume the propagation delay of the ROM is 8 ns, the setup time for the flip-flops is 2 ns, and the propagation delay of the flip-flops is 4 ns.

| $Q_1$ | $Q_2$ | $X$ | $D_1$ | $D_2$ | $Z$ |
|---|---|---|---|---|---|
| 0 | 0 | 0 | 1 | 0 | 0 |
| 0 | 0 | 1 | 1 | 0 | 0 |
| 0 | 1 | 0 | 0 | 0 | 0 |
| 0 | 1 | 1 | 1 | 1 | 0 |
| 1 | 0 | 0 | 1 | 1 | 0 |
| 1 | 0 | 1 | 0 | 1 | 1 |
| 1 | 1 | 0 | 0 | 1 | 1 |
| 1 | 1 | 1 | 1 | 1 | 1 |

(a) What is the minimum clock period for this circuit?

(b) Draw a timing diagram for this circuit, using the given delays and the minimum clock period of Part (a). Give the clock, $X$, $D_1$, $D_2$, $Q_1$, $Q_2$, and $Z$. Assume $Q_1Q_2 = 00$ to start with and assume $X$ takes on its new value 4 ns after each rising edge. Use the input sequence $X = 0, 1, 1, 0$. Specify the correct output sequence for $Z$.

(c) Construct a state table and a state graph for the circuit.

**13.7** (a) Construct a state table and graph for the circuit shown.

(b) Construct a timing chart for the circuit for an input sequence $X = 10111$. (Assume that initially $Q_1 = Q_2 = 0$ and that $X$ changes midway between the rising and falling clock edges.)

(c) List the output values produced by the input sequence.

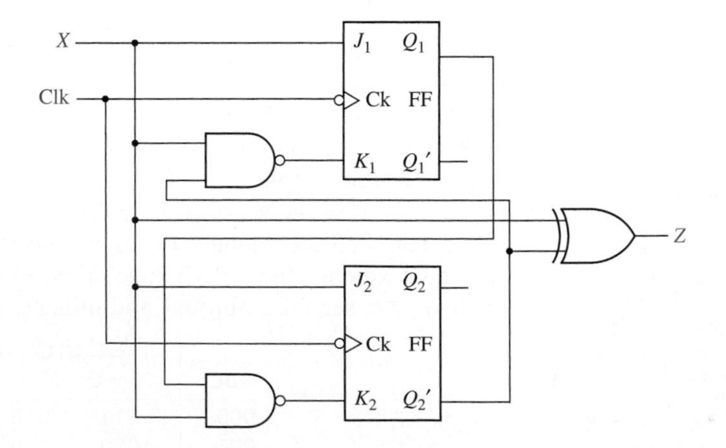

**13.8** (a) Construct a state table and graph for the circuit shown.
(b) Construct a timing chart for the input sequence $X = 10101$. (Assume that initially $Q_1 = Q_2 = 0$ and that $X$ changes midway between the rising and falling clock edges.) Indicate the times $Z$ has the correct value.
(c) List the output values produced by the input sequence.

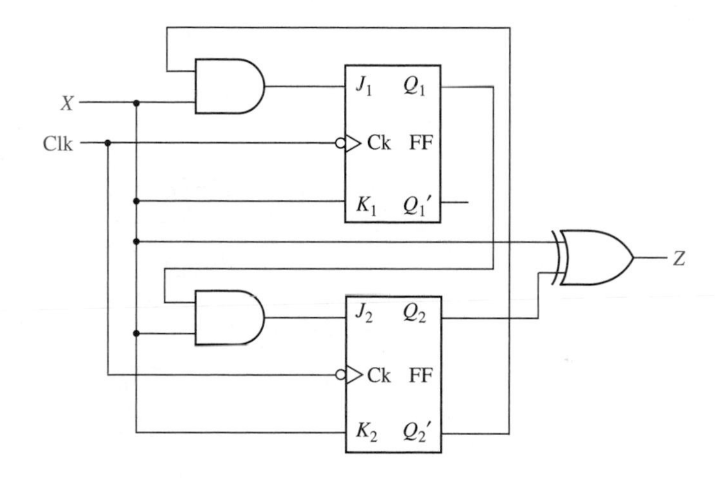


**13.9** (a) Construct a state table and graph for the circuit shown.
(b) Construct a timing chart for the input sequence $X_1X_2 = 11, 11, 01, 10, 10, 00$. (Assume that initially $Q_1 = Q_2 = 0$ and that $X_1$ and $X_2$ change midway between the rising and falling clock edges.)
(c) List the output values produced by the input sequence.

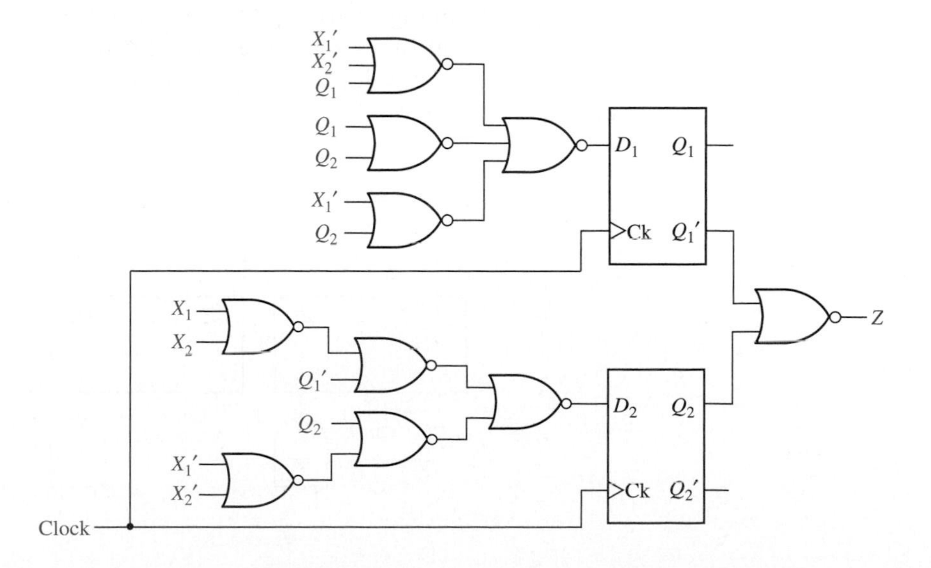

**13.10** (a) Construct a state table and graph for the circuit shown.
(b) Construct a timing chart for the input sequence $X_1X_2 = 01, 10, 01, 11, 11, 01$. (Assume that initially $Q_1 = Q_2 = 0$ and that $X_1$ and $X_2$ change midway between the rising and falling clock edges.)
(c) List the output values produced by the input sequence.

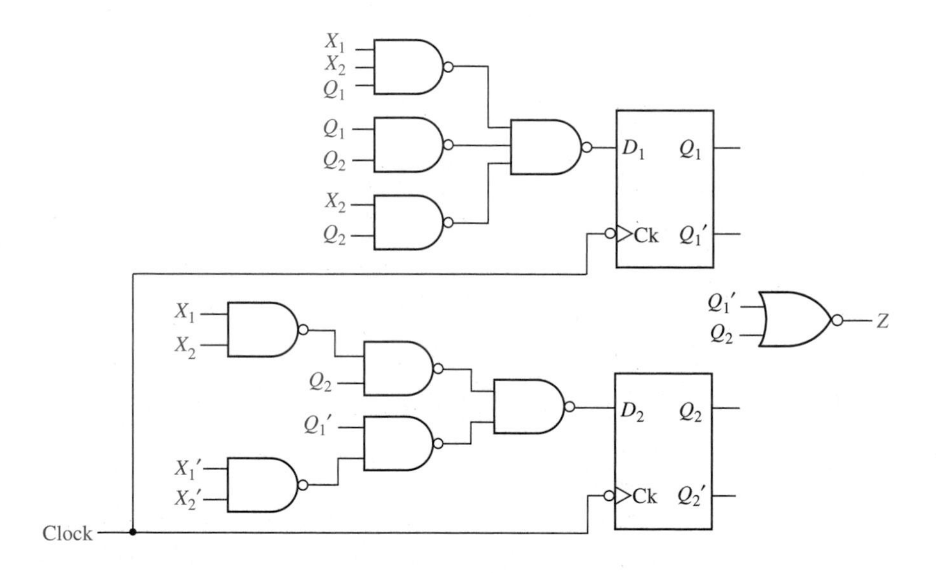


**13.11** (a) Construct a state table and graph for the given circuit.
(b) Construct a timing chart for the circuit for an input sequence $X = 10011$. Indicate at what times $Z$ has the correct value and specify the correct output sequence. (Assume that $X$ changes midway between falling and rising clock edges.) Initially, $Q_1 = Q_2 = 0$.

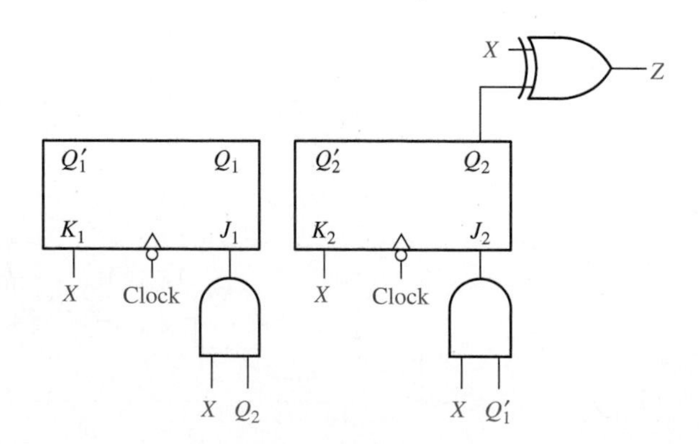

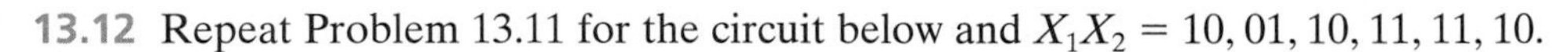

**13.12** Repeat Problem 13.11 for the circuit below and $X_1X_2 = 10, 01, 10, 11, 11, 10$.

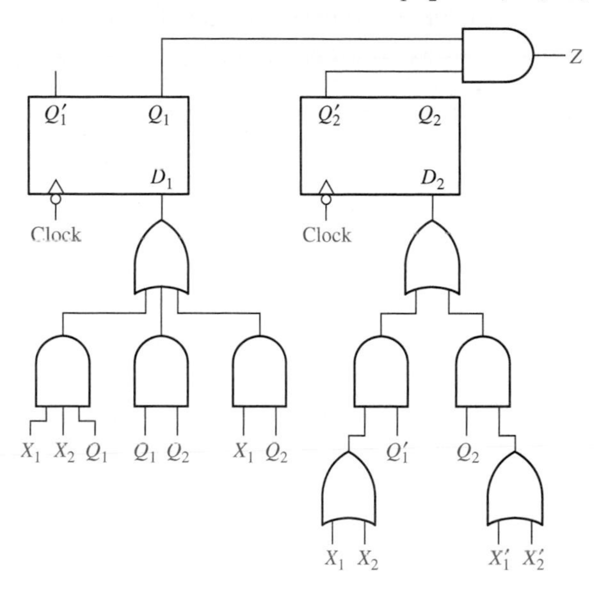


**13.13** A sequential circuit has one input $X$, one output $Z$, and three flip-flops $Q_1$, $Q_2$, and $Q_3$. The transition and output tables for the circuit follow:

| Present State | Next State $X = 0$ | Next State $X = 1$ | Output ($Z$) $X = 0$ | Output ($Z$) $X = 1$ |
|---|---|---|---|---|
| 000 | 100 | 101 | 1 | 0 |
| 001 | 100 | 101 | 0 | 1 |
| 010 | 000 | 000 | 1 | 0 |
| 011 | 000 | 000 | 0 | 1 |
| 100 | 111 | 110 | 1 | 0 |
| 101 | 110 | 110 | 0 | 1 |
| 110 | 011 | 010 | 1 | 0 |
| 111 | 011 | 011 | 0 | 1 |

(a) Construct a timing chart for the input sequence $X = 0101$ and initial state $Q_1Q_2Q_3 = 000$. Identify any false outputs. (Assume that the flip-flops are rising-edge triggered and that the input changes midway between the rising and falling edges of the clock.)

(b) List the output values produced by the input sequence.

**13.14** Repeat Problem 13.13 for the input sequence $X = 1001$ and initial state $Q_1Q_2Q_3 = 000$.

**13.15** A sequential circuit has the form shown in Figure 13-17 with

$$D_1 = Q_2Q_3' + XQ_1' \qquad D_3 = Q_2' + X$$
$$D_2 = Q_3 + X'Q_2 \qquad Z = XQ_2' + X'Q_2$$

(a) Construct a state table and state graph for the circuit.
(b) Draw a timing diagram for the circuit showing the clock, $X$, $Q_1$, $Q_2$, $Q_3$, and $Z$. Use the input sequence $X = 01011$ and assume that $X$ changes midway between falling and rising clock edges. Indicate any false outputs on the diagram.
(c) Compare the output sequence obtained from the timing diagram with that from the state graph.
(d) At what time with respect to the clock should the input be changed in order to eliminate the false output(s)?

**13.16** Repeat Problem 13.15 for the given equations and the input sequence $X = 01100$.

$$D_1 = Q_3'X' \qquad D_3 = Q_2'X + Q_1Q_2$$
$$D_2 = Q_3'Q_1 + XQ_2' \qquad Z = XQ_3 + X'Q_3'$$

**13.17** Consider the circuit shown.
(a) Construct a state table and graph for the following circuit. Is the circuit a Mealy or Moore circuit? Does the circuit have any unused states? Assume 00 is the initial state.
(b) Draw a timing diagram for the input sequence $X = 01100$.
(c) What is the output sequence for the input sequence?

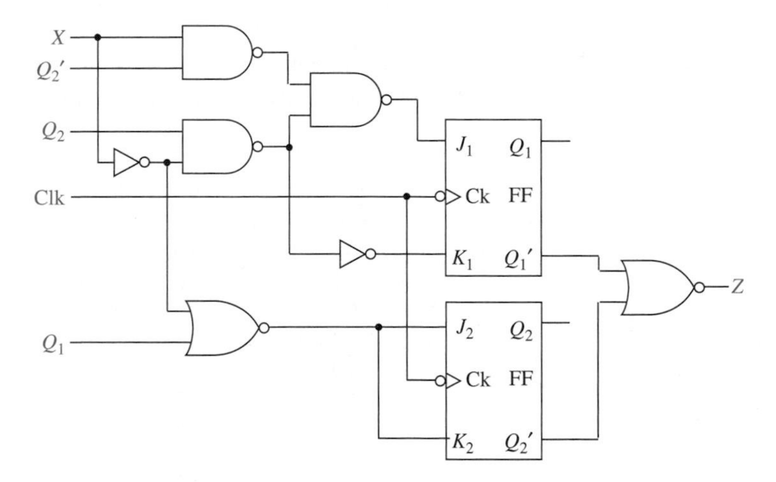


**13.18** A Mealy sequential circuit has one input, one output, and two flip-flops. A timing diagram for the circuit follows. Construct a state table and state graph for the circuit.

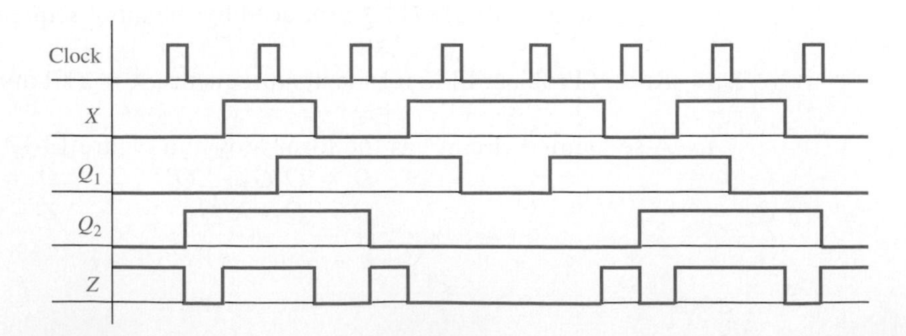

**13.19** Repeat Problem 13.18 for the following timing diagram.

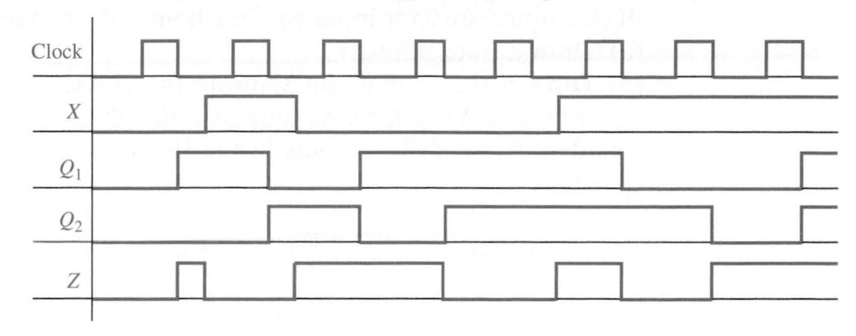


**13.20** Given the following timing chart for a sequential circuit, construct as much of the state table as possible. Is this a Mealy or Moore circuit?

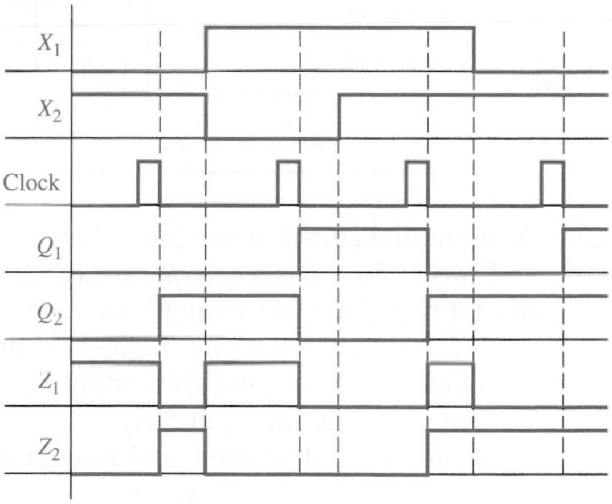

**13.21** Given the following timing chart for a sequential circuit, construct as much of the state table as possible.

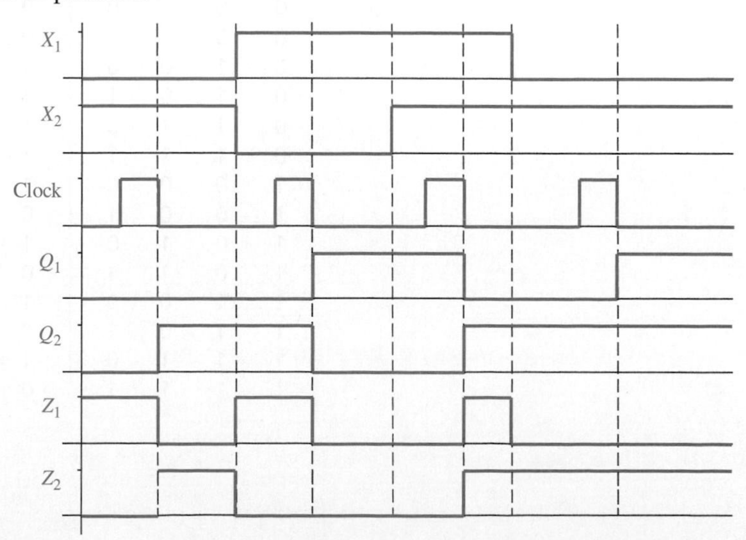

**13.22** For the following sequential circuit, the table gives the contents of the PLA. (All PLA outputs are 0 for input combinations not listed in the table.)

(a) Draw a state graph.

(b) Draw a timing diagram showing the clock, $X$, $Q_1$, $Q_2$, and $Z$ for the input sequence $X = 10011$. Assume that initially $Q_1 = Q_2 = 0$.

(c) Identify any false outputs in the timing diagram. What is the correct output sequence for $Z$?

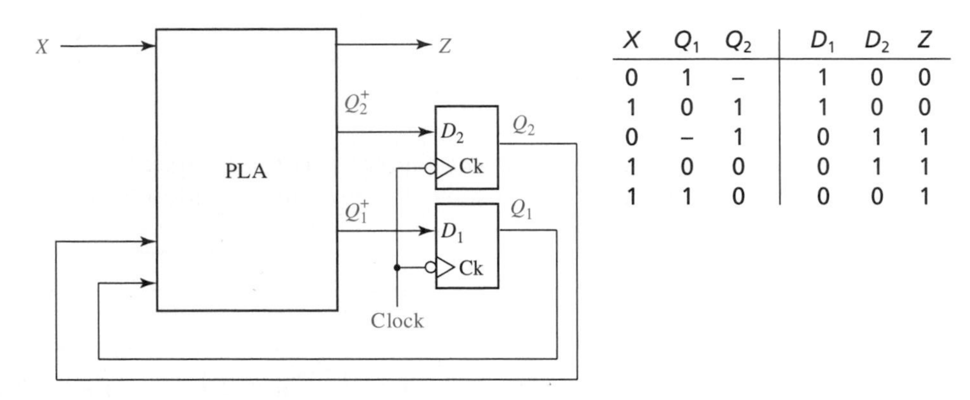


| $X$ | $Q_1$ | $Q_2$ | $D_1$ | $D_2$ | $Z$ |
|---|---|---|---|---|---|
| 0 | 1 | – | 1 | 0 | 0 |
| 1 | 0 | 1 | 1 | 0 | 0 |
| 0 | – | 1 | 0 | 1 | 1 |
| 1 | 0 | 0 | 0 | 1 | 1 |
| 1 | 1 | 0 | 0 | 0 | 1 |

**13.23** A sequential circuit of the form shown in Figure 13-17 is constructed using a ROM and two D flip-flops. The contents of the ROM are given in the table.

(a) Draw a timing diagram for the circuit for the input sequence $X_1X_2 = 10, 01, 11, 10$. Assume that input changes occur midway between rising and falling clock edges. Indicate any false outputs on the diagram, and specify the correct output sequence for $Z_1$ and $Z_2$.

(b) Construct a state table and state graph for the circuit.

| $Q_1$ | $Q_2$ | $X_1$ | $X_2$ | $D_1$ | $D_2$ | $Z_1$ | $Z_2$ |
|---|---|---|---|---|---|---|---|
| 0 | 0 | 0 | 0 | 0 | 0 | 1 | 0 |
| 0 | 0 | 0 | 1 | 0 | 0 | 1 | 0 |
| 0 | 0 | 1 | 0 | 0 | 1 | 1 | 0 |
| 0 | 0 | 1 | 1 | 0 | 1 | 1 | 0 |
| 0 | 1 | 0 | 0 | 1 | 1 | 0 | 0 |
| 0 | 1 | 0 | 1 | 1 | 1 | 1 | 0 |
| 0 | 1 | 1 | 0 | 1 | 0 | 0 | 1 |
| 0 | 1 | 1 | 1 | 1 | 0 | 1 | 1 |
| 1 | 0 | 0 | 0 | 1 | 1 | 0 | 0 |
| 1 | 0 | 0 | 1 | 0 | 0 | 0 | 0 |
| 1 | 0 | 1 | 0 | 1 | 1 | 0 | 1 |
| 1 | 0 | 1 | 1 | 0 | 0 | 0 | 1 |
| 1 | 1 | 0 | 0 | 1 | 0 | 0 | 0 |
| 1 | 1 | 0 | 1 | 0 | 1 | 0 | 0 |
| 1 | 1 | 1 | 0 | 1 | 0 | 0 | 0 |
| 1 | 1 | 1 | 1 | 0 | 1 | 0 | 0 |

**13.24** For the following state graph, give the state table. Then, give the timing diagram for the input sequence $X = 101001$. Assume $X$ changes midway between the falling and rising edges of the clock, and that the flip-flops are falling-edge triggered. What is the correct output sequence?

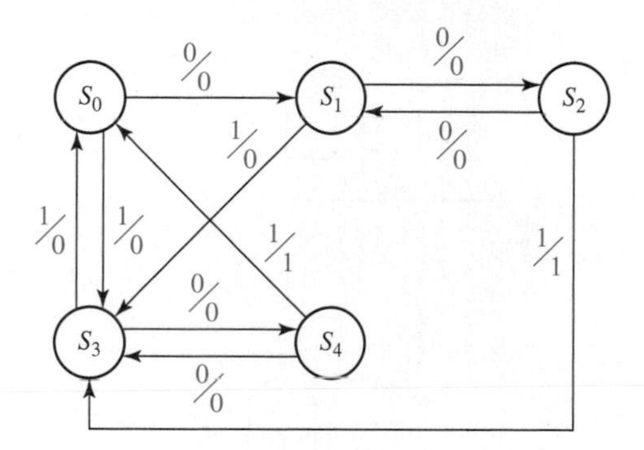


**13.25** For the circuit of Problem 13.3, assume the delays of the NAND gates and NOR gates are 3 ns, and assume the delay of the inverter is 2 ns. Assume the propagation delays and setup times for the J-K flip-flops are 4 ns and 2 ns, respectively.

(a) Fill in the given timing diagram. The clock period is 15 ns, and 1-ns increments are marked on the clock signal. Does the circuit operate properly with these timing parameters?

(b) What is the minimum clock period for this circuit, if $X$ is changed early enough? How late may $X$ change with this clock period without causing improper operation of the circuit?

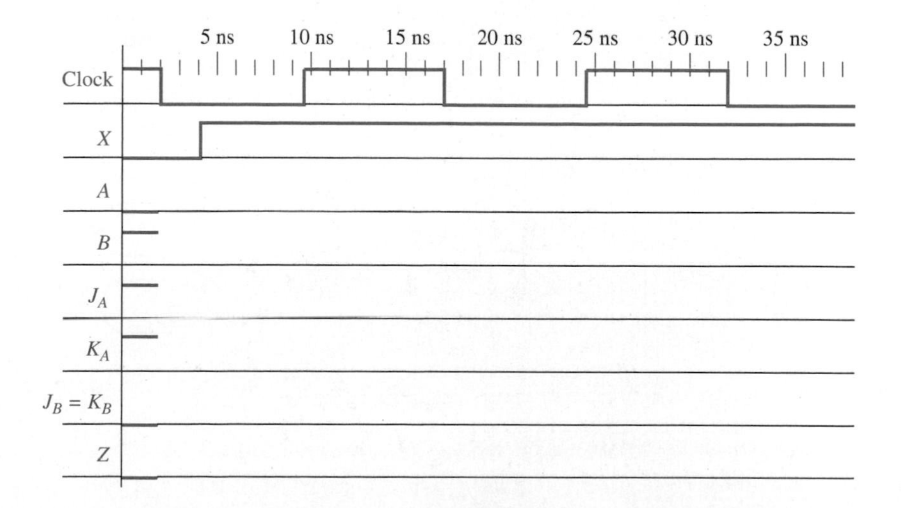

**13.26** Draw a timing diagram for the following circuit starting with an initial state $ABC = 000$ and using an input sequence $X = 01010$. Assume that the input changes occur midway between the falling and rising clock edges. Give the output sequence, and indicate false outputs, if any. Verify that your answer is correct by making a state table for the circuit.

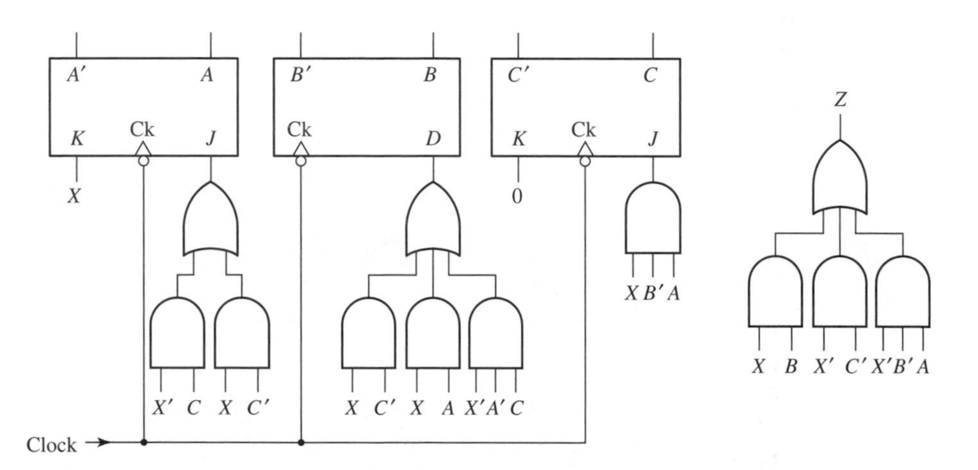


**13.27** (a) For the following sequential circuit, write the next-state equations for flip-flops $A$ and $B$.

(b) Using these equations, find the state table and draw the state graph.

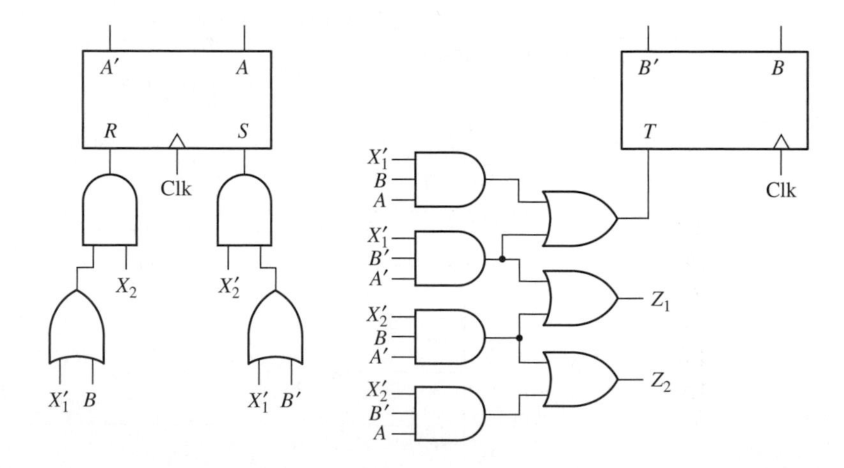

# Derivation of State Graphs and Tables

## Objectives

1. Given a problem statement for the design of a Mealy or Moore sequential circuit, find the corresponding state graph and table.
2. Explain the significance of each state in your graph or table in terms of the input sequences required to reach that state.
3. Check your state graph using appropriate input sequences.

# Study Guide

1. Study Section 14.1, *Design of a Sequence Detector.*
   (a) Verify that the state graph in Figure 14-4 will produce the correct output sequence for $Z$ when the input sequence for $X$ is as given in Equation (14-1).
   (b) Using the equations from the Karnaugh maps on p. 396, construct the next-state table for the circuit and verify that it is the same as given in Table 14-2, except that the new table will have four states because the don't-cares were assigned in the process of designing the circuit.
   (c) Complete the design of the Moore sequential circuit whose transition table is given by Table 14-4. Use clocked J-K flip-flops for $A$ and $B$.
   (d) Verify that the state graph of Figure 14-6 gives the correct output sequence when the input sequence (14-1) is applied. (Ignore the initial output for the Moore graph.)

2. Study Section 14.2, *More Complex Design Problems.*

3. Study Section 14.3, *Guidelines for Construction of State Graphs.* Study the examples carefully and observe how some of the guidelines were applied.

4. Work through Programmed Exercises 14.1, 14.2, and 14.3.

5. A very important part of deriving state tables or state graphs is knowing how to tell when your answer is right (or wrong!). One way to do this is to make up a suitable list of test sequences, apply them to the state graph, and check the resulting output sequences.

6. To gain proficiency in construction of state tables or graphs requires a fair amount of practice. Work Problems 14.4, 14.5, 14.6, 14.7, and 14.8. The problems on the readiness tests will be about the same order of difficulty as these problems, so make sure that you can work them in a reasonable time.

   *Note*: Do not look at the answers to these problems in the back of the book until you have tried the problems and checked your answers using the following test sequences:

   14.4 $X = \;\;\;\; 0\;1\;1\;1\;0\;1\;0\;1$
   $Z = (0)\;0\;0\;0\;0\;1\;1\;1\;1$
   (Your solution should have five self-loops. A self-loop is an arrow which starts at one state and goes back to the same state.)

   14.5 $X = 1\;0\;1\;0\;1\;0\;0\;1\;0\;0\;0\;1\;0\;0$
   $Z_1 = 0\;0\;0\;1\;0\;1\;0\;0\;0\;0\;0\;0\;0\;0$
   $Z_2 = 0\;0\;0\;0\;0\;0\;1\;0\;0\;1\;0\;0\;0\;1$
   (Your solution should have four self-loops.)

   14.6 $X_1 = \;\;\;\; 1\;0\;0\;1\;0\;1\;1\;0\;0\;0$
   $X_2 = \;\;\;\; 1\;0\;0\;0\;0\;1\;0\;0\;1\;0$
   $Z = (0)\;0\;1\;1\;1\;0\;0\;0\;1\;1\;0$
   (Your solution should have at least four self-loops.)

14.7 (a) $X = 0\ 0\ 1\ 1\ 0\ 1\ 0\ 1\ 0\ 1\ 1$
$Z = 1\ 1\ 0\ 0\ 0\ 1\ 1\ 0\ 0\ 0\ 1$
(Your solution should have three self-loops.)
(b) $X = 1\ 1\ 1\ 0\ 0\ 1\ 0\ 1\ 0\ 1$
$Z = 0\ 0\ 0\ 0\ 1\ 0\ 0\ 0\ 0\ 1$

14.8 (a) $X_1 = 0\ 0\ 1\ 1\ 1\ 1\ 0\ 0\ 1\ 0\ 0$
$X_2 = 0\ 1\ 0\ 1\ 1\ 0\ 1\ 0\ 0\ 1\ 1$
$Z_1 = 0\ 1\ 1\ 1\ 0\ 0\ 0\ 0\ 1\ 0\ 0$
$Z_2 = 0\ 0\ 0\ 0\ 0\ 1\ 1\ 1\ 0\ 1\ 0$
(b) You should get the same sequences as in (a) after an initial output of $Z_1Z_2 = 00$.

**7.** If you have the *LogicAid* program available, use it to check your state tables. This has several advantages over looking at the answers in the back of the book. First, *LogicAid* will determine whether or not your solution is correct even if your states are numbered differently from those in the solution, or even if the number of states is different. Second, if your solution is wrong, *LogicAid* will find a short input sequence for which your state table fails, and you can use this sequence to help locate the error in your solution. If you are having trouble learning to derive state graphs, *LogicAid* has a state graph tutor mode which can be used to check partial state graphs. By using the partial graph checker, you can check your graph after adding each state, and then correct any errors before proceeding to the next state.

**8.** Read Section 14.4, *Serial Data Code Conversion.*

(a) Complete the following timing diagram, showing waveforms for the NRZ, NRZI, RZ, and Manchester coding schemes:

| Bit sequence | 0 | 1 | 0 | 0 | 1 | 1 | 1 | 0 |
|---|---|---|---|---|---|---|---|---|
| NRZ | | | | | | | | |
| NRZI | | | | | | | | |
| RZ | | | | | | | | |
| Manchester | | | | | | | | |
| Clock | | | | | | | | |

(b) The timing chart of Figure 14-20(b) shows several glitches. By referring to the state table, explain why the second glitch is present.

(c) Consider Figure 14-20. If an error in data transmission occurs, the input sequence $X = 01$ or $10$ could occur. Add an Error state to the state diagram. The circuit should go to this error state if such an error occurs.

(d) Work Problem 14.9.

**9.** Read Section 14.5, *Alphanumeric State Graph Notation.*

(a) Sometimes all outputs are 0 for a given state or arc. We denote this by placing a 0 in the place of the output. For example, a Moore state with all outputs being 0 might be labeled $S_3/0$, and a Mealy arc with all outputs being 0 might be labeled $X'Y/0$.

(b) Try to write the row of a Mealy state table that describes state $S_3$ in the following partial state graph. You cannot, because there are two contradictory directions when $S = N = 1$. Also, the row is not completely specified. Redraw state $S_3$ so that $S$ takes priority over $N$, and so that the circuit stays in state $S_3$ with no output if no directions are specified by the partial state graph. Then, give the state table row. Show that it has no contradictions and that it is completely specified.

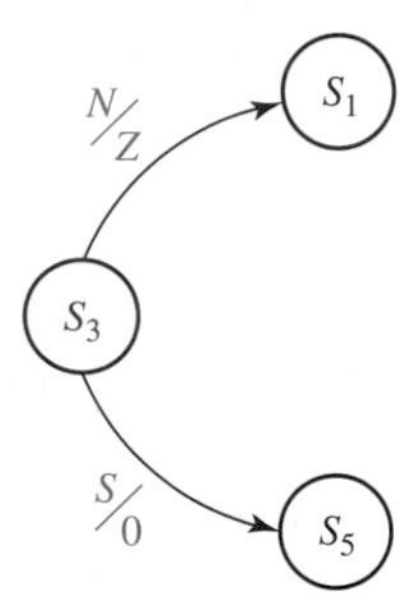


(c) Work Problems 14.10 and 14.11.

**10.** When you are satisfied that you can meet all of the objectives, take the readiness test.

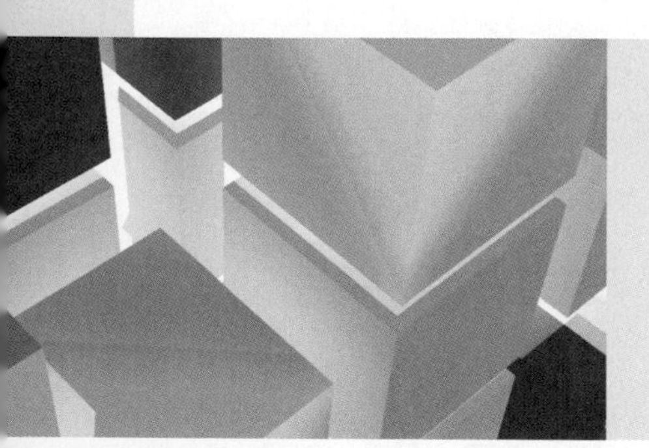

# Derivation of State Tables

In Unit 13 we analyzed sequential circuits using timing charts and state graphs. Now, we will consider the design of sequential circuits starting from a problem statement which specifies the desired relationship between the input and output sequences. The first step in the design is to construct a state table or graph which specifies the desired behavior of the circuit. Flip-flop input equations and output equations can then be derived from this table. Construction of the state table or graph, one of the most important and challenging parts of sequential circuit design, is discussed in detail in this unit.

# 14.1 Design of a Sequence Detector

To illustrate the design of a clocked Mealy sequential circuit, we will design a sequence detector. The circuit has the form shown in Figure 14-1.

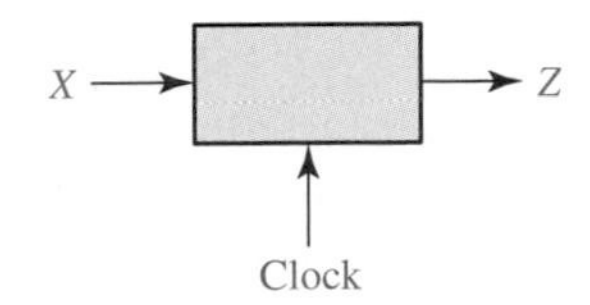


FIGURE 14-1 Sequence Detector to be Designed

The circuit will examine a string of 0's and 1's applied to the $X$ input and generate an output $Z = 1$ only when a prescribed input sequence occurs. It will be assumed that the input $X$ can only change between clock pulses. Specifically, we will design the circuit so that any input sequence ending in 101 will produce an output $Z = 1$ coincident with the last 1. The circuit does not reset when a 1 output occurs. A typical input sequence and the corresponding output sequence are

$$\begin{array}{llllllllllllllllll} X = & 0 & 0 & 1 & 1 & 0 & 1 & 1 & 0 & 0 & 1 & 0 & 1 & 0 & 1 & 0 & 0 \\ Z = & 0 & 0 & 0 & 0 & 0 & 1 & 0 & 0 & 0 & 0 & 0 & 1 & 0 & 1 & 0 & 0 \\ (\text{time:} & 0 & 1 & 2 & 3 & 4 & 5 & 6 & 7 & 8 & 9 & 10 & 11 & 12 & 13 & 14 & 15) \end{array} \quad (14\text{-}1)$$

Initially, we do not know how many flip-flops will be required, so we will designate the circuit states as $S_0$, $S_1$, etc., and later assign flip-flop states to correspond to the circuit states. We will construct a state graph to show the sequence of states and outputs which occur in response to different inputs. Initially, we will start the circuit in a reset state designated $S_0$. If a 0 input is received, the circuit can stay in $S_0$ because the input sequence we are looking for does not start with 0. However, if a 1 is received, the circuit must go to a new state ($S_1$) to "remember" that the first input in the desired sequence has been received (Figure 14-2). The labels on the graph are of the form $X/Z$, where the symbol before the slash is the input and the symbol after the slash is the corresponding output.

When in state $S_1$, if we receive a 0, the circuit must change to a new state ($S_2$) to remember that the first two inputs of the desired sequence (10) have been received. If a 1 is received in state $S_2$, the desired input sequence (101) is complete and the output should be 1. The question arises whether the circuit should then go to a new state or back to $S_0$ or $S_1$. Because the circuit is not supposed to reset when an output occurs, we cannot go back to $S_0$. However, because the last 1 in a sequence can also be the first 1 in a new sequence, we can return to $S_1$, as indicated in Figure 14-3.

FIGURE 14-2

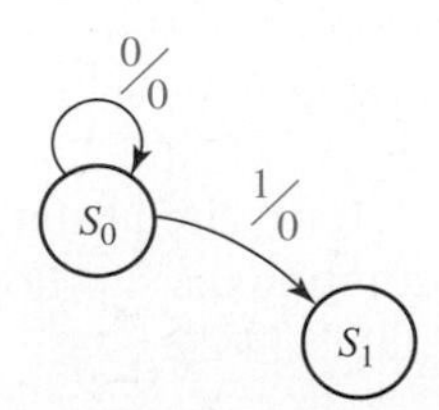

FIGURE 14-3

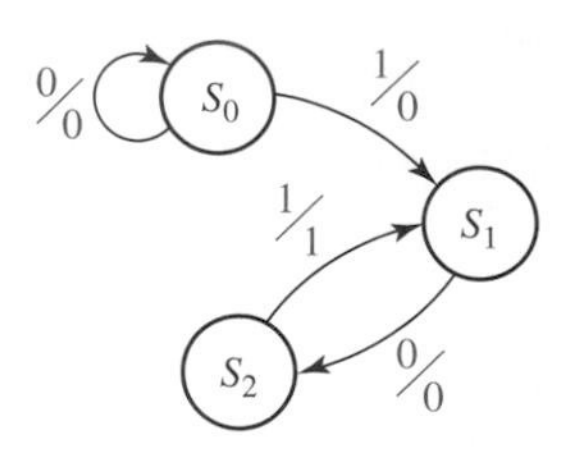


The graph of Figure 14-3 is still incomplete. If a 1 input occurs when in state $S_1$, we can stay in $S_1$ because the sequence is simply restarted. If a 0 input occurs in state $S_2$, we have received two 0's in a row and must reset the circuit to state $S_0$ because 00 is not part of the desired input sequence, and going to one of the other states could lead to an incorrect output. The final state graph is given in Figure 14-4. Note that for a single input variable each state must have two exit lines (one for each value of the input variable) but may have any number of entry lines, depending on the circuit specifications.

FIGURE 14-4 Mealy State Graph for Sequence Detector

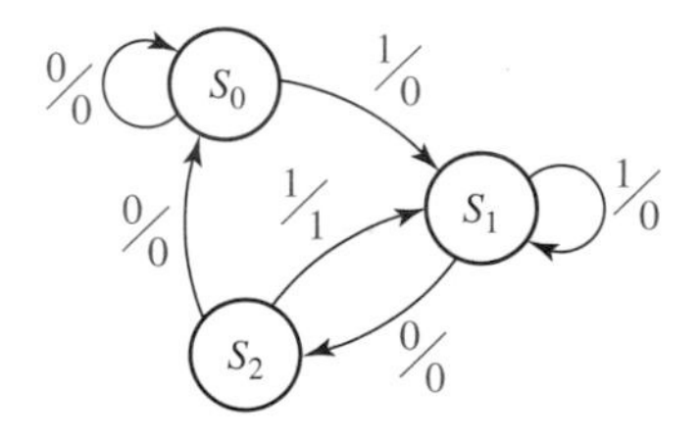


State $S_0$ is the starting state, state $S_1$ indicates that a sequence ending in 1 has been received, and state $S_2$ indicates that a sequence ending in 10 has been received. An alternative way to start the solution would be to first define states in this manner and then construct the state graph. Converting the state graph to a state table yields Table 14-1. For example, the arc from $S_2$ to $S_1$ is labeled 1/1. This means that when the present state is $S_2$ and $X = 1$, the present output is 1. This 1 output is present as soon as $X$ becomes 1, that is, *before* the state change occurs. Therefore, the 1 is placed in the $S_2$ row of the table.

TABLE 14-1

| Present State | Next State $X = 0$ | Next State $X = 1$ | Present Output $X = 0$ | Present Output $X = 1$ |
|---|---|---|---|---|
| $S_0$ | $S_0$ | $S_1$ | 0 | 0 |
| $S_1$ | $S_2$ | $S_1$ | 0 | 0 |
| $S_2$ | $S_0$ | $S_1$ | 0 | 1 |

At this point, we are ready to design a circuit which has the behavior described by the state table. Because one flip-flop can have only two states, two flip-flops are needed to represent the three states. Designate the two flip-flops as $A$ and $B$. Let flip-flop states $A = 0$ and $B = 0$ correspond to circuit state $S_0$; $A = 0$ and $B = 1$ correspond to $S_1$; and $A = 1$ and $B = 0$ correspond to circuit state $S_2$. Each circuit state is then represented by a unique combination of flip-flop states. Substituting the flip-flop states for $S_0$, $S_1$ and $S_2$ in the state table yields the transition table (Table 14-2).

TABLE 14-2

| AB | $A^+B^+$ $X = 0$ | $X = 1$ | Z $X = 0$ | $X = 1$ |
|---|---|---|---|---|
| 00 | 00 | 01 | 0 | 0 |
| 01 | 10 | 01 | 0 | 0 |
| 10 | 00 | 01 | 0 | 1 |

From this table, we can plot the next-state maps for the flip-flops and the map for the output function $Z$:

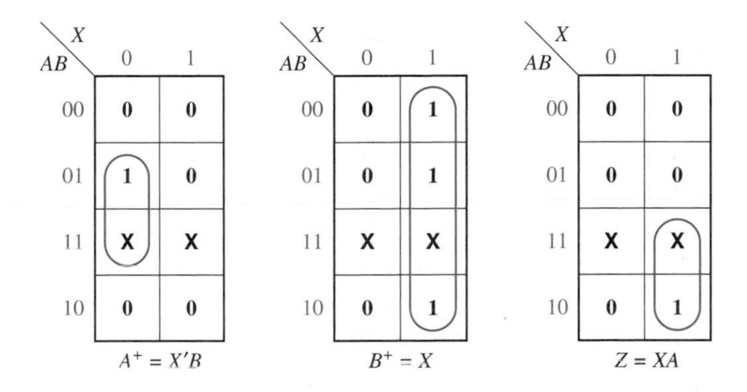


The flip-flop inputs are then derived from the next-state maps using the same method that was used for counters (Section 12.4). If D flip-flops are used, $D_A = A^+ = X'B$ and $D_B = B^+ = X$, which leads to the circuit shown in Figure 14-5. Initially, we will reset both flip-flops to the 0 state. By tracing signals through the circuit, you can verify that an output $Z = 1$ will occur when an input sequence ending in 101 occurs. To avoid reading false outputs, always read the value of $Z$ after the input has changed and before the active clock edge.

FIGURE 14-5

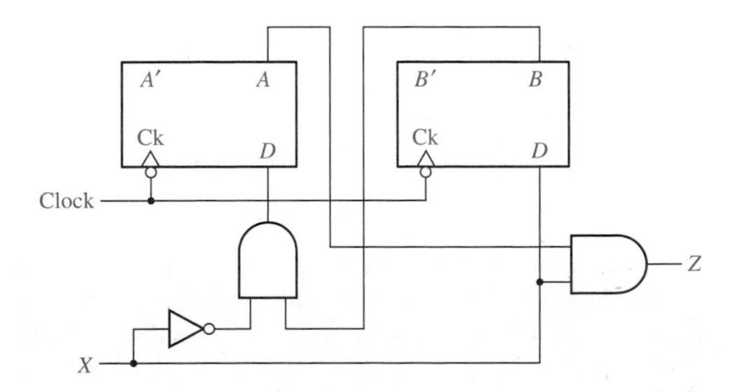


The procedure for finding the state graph for a Moore machine is similar to that used for a Mealy machine, except that the output is written with the state instead of with the transition between states. We will rework the previous example as a Moore machine to illustrate this procedure. The circuit should produce an output of 1 only if an input sequence ending in 101 has occurred. The design is similar to that for the

Mealy machine up until the input sequence 10 has occurred, except that 0 output is associated with states $S_0$, $S_1$, and $S_2$:

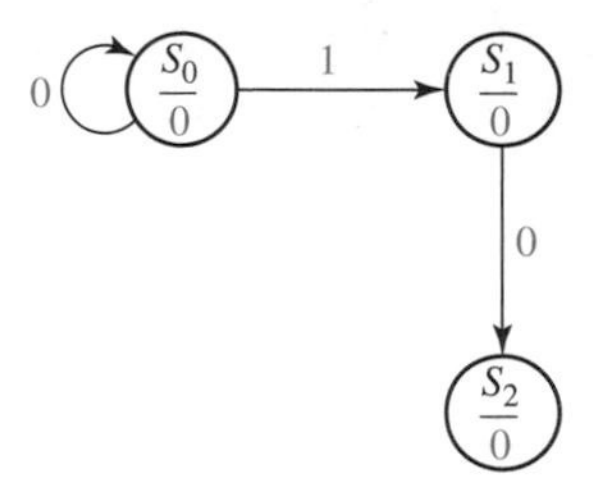


Now, when a 1 input occurs to complete the 101 sequence, the output must become 1; therefore, we cannot go back to state $S_1$ and must create a new state $S_3$ with a 1 output:

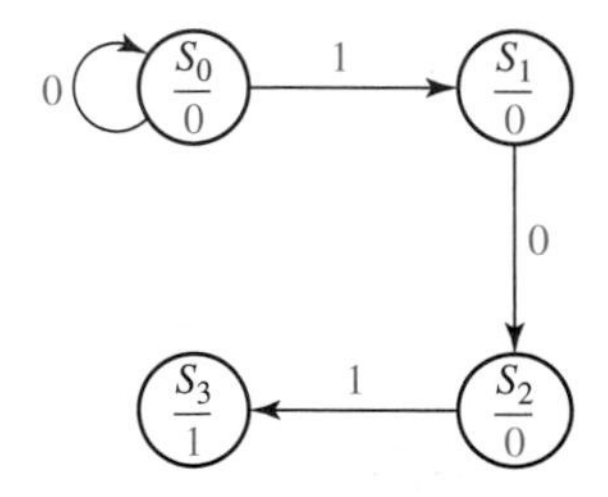


We now complete the graph, as shown in Figure 14-6. Note the sequence 100 resets the circuit to $S_0$. A sequence 1010 takes the circuit back to $S_2$ because another 1 input should cause $Z$ to become 1 again.

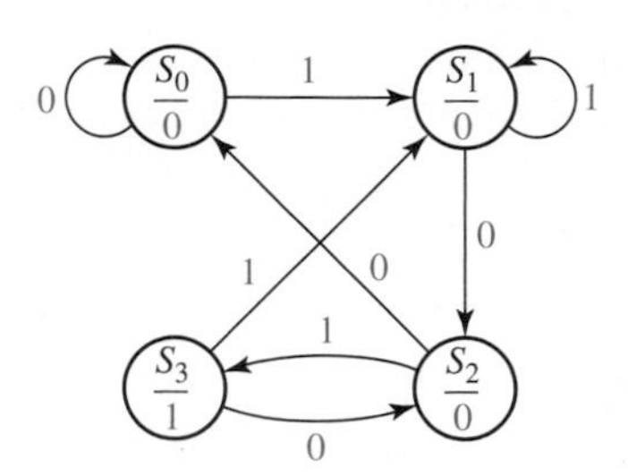


FIGURE 14-6 Moore State Graph for Sequence Detector

The state table corresponding to the circuit is given by Table 14-3. Note that there is a single column for the output because the output is determined by the present state and does not depend on $X$. Note that in this example the Moore machine requires one more state than the Mealy machine which detects the same input sequence.

TABLE 14-3

| Present State | Next State $X = 0$ | $X = 1$ | Present Output($Z$) |
|---|---|---|---|
| $S_0$ | $S_0$ | $S_1$ | 0 |
| $S_1$ | $S_2$ | $S_1$ | 0 |
| $S_2$ | $S_0$ | $S_3$ | 0 |
| $S_3$ | $S_2$ | $S_1$ | 1 |

Because there are four states, two flip-flops are required to realize the circuit. Using the state assignment $AB = 00$ for $S_0$, $AB = 01$ for $S_1$, $AB = 11$ for $S_2$, and $AB = 10$ for $S_3$, the following transition table for the flip-flops results (Table 14-4):

TABLE 14-4

| $AB$ | $A^+B^+$ $X = 0$ | $X = 1$ | $Z$ |
|---|---|---|---|
| 00 | 00 | 01 | 0 |
| 01 | 11 | 01 | 0 |
| 11 | 00 | 10 | 0 |
| 10 | 11 | 01 | 1 |

The output function is $Z = AB'$. Note that $Z$ depends only on the flip-flop states and is independent of $X$, while for the corresponding Mealy machine, $Z$ was a function of $X$. The derivation of the flip-flop input equations is straightforward and will not be given here.

## 14.2 More Complex Design Problems

In this section we will derive a state graph for a sequential circuit of somewhat greater complexity than the previous examples. The circuit to be designed again has the form shown in Figure 14-1. The output $Z$ should be 1 if the input sequence ends in either 010 or 1001, and $Z$ should be 0 otherwise. Before attempting to draw the state graph, we will work out some typical input-output sequences to make sure that we have a clear understanding of the problem statement. We will determine the desired output sequence for the following input sequence:

$$
\begin{array}{rcccccccccccccccccc}
X = & 0 & 0 & 1 & 0 & 1 & 0 & 0 & 1 & 0 & 0 & 0 & 1 & 0 & 0 & 1 & 1 & 0 \\
 & & & & \uparrow & & \uparrow & & & \uparrow & \uparrow & & & & \uparrow & & \uparrow & \\
 & & & & a & & b & & & c & d & & & & e & & f & \\
Z = & 0 & 0 & 0 & 1 & 0 & 1 & 0 & 1 & 1 & 0 & 0 & 0 & 1 & 0 & 1 & 0 & 0
\end{array}
$$

At point $a$, the input sequence ends in 010, one of the sequences for which we are looking, so the output is $Z = 1$. At point $b$, the input again ends in 010, so $Z = 1$. Note that overlapping sequences are allowed because the problem statement does not say anything about resetting the circuit when a 1 output occurs. At point $c$, the input sequence ends in 1001, so $Z$ is again 1. Why do we have a 1 output at points $d$, $e$, and $f$? This is just one of many input sequences. A state machine that gives the correct output for this sequence will not necessarily give the correct output for all other sequences.

We will start construction of the state graph by working with the two sequences which lead to a 1 output. Then, we will later add arrows and states as required to make sure that the output is correct for other cases. We start off with a reset state $S_0$ which corresponds to having received no inputs. Whenever an input is received that corresponds to part of one of the sequences for which we are looking, the circuit should go to a new state to "remember" having received this input. Figure 14-7 shows a partial state graph which gives a 1 output for the sequence 010. In this graph $S_1$ corresponds

FIGURE 14-7

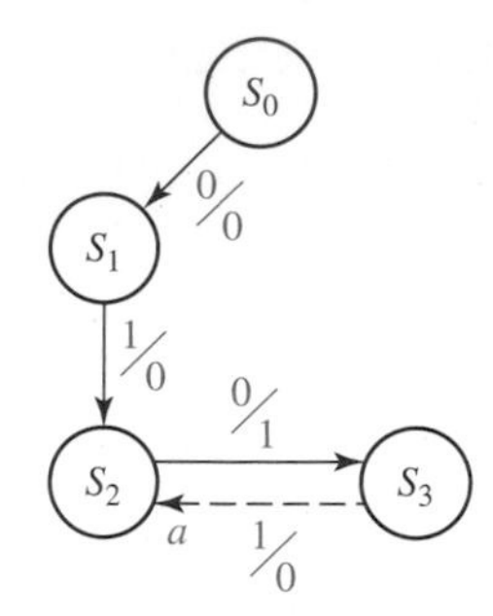


| State | Sequence Received |
|---|---|
| $S_0$ | Reset |
| $S_1$ | 0 |
| $S_2$ | 01 |
| $S_3$ | 010 |

to having received a sequence ending in 0, $S_2$ to a sequence ending in 01, and $S_3$ to a sequence ending in 010. Now, if a 1 input is received in state $S_3$, we again have a sequence ending in 01, which is part of the input sequence for which we are looking. Therefore, we can go back to state $S_2$ (arrow *a*) because $S_2$ corresponds to having received a sequence ending in 01. Then, if we get another 0 in state $S_2$, we go to $S_3$ with a 1 output. This is correct because the sequence again ends in 010.

Next, we will construct the part of the graph corresponding to the sequence 1001. Again, we start in the reset state $S_0$, and when we receive a 1 input, we go to $S_4$ (Figure 14-8, arrow *b*) to remember that we have received the first 1 in the sequence 1001. The next input in the sequence is 0, and when this 0 is received, we should ask the question: Should we create a new state to correspond to a sequence ending in 10, or can we go to one of the previous states on the state graph? Because $S_3$ corresponds to a sequence ending in 10, we can go to $S_3$ (arrow *c*). The fact that we did not have an initial 0 this time does not matter because 10 starts off the sequence for which we are looking. If we get a 0 input when in $S_3$, the input sequence received will end in 100 regardless of the path we took to get to $S_3$. Because there is so far no state corresponding to the sequence 100, we create a new state $S_5$ to indicate having received a sequence ending in 100.

If we get a 1 input when in state $S_5$, this completes the sequence 1001 and gives a 1 output as indicated by arrow *e*. Again, we ask the question: Can we go back to one of the previous states or do we have to create a new state? Because the end of the sequence 1001 is 01, and $S_2$ corresponds to a sequence ending in 01, we can go back to $S_2$ (Figure 14-9). If we get another 001, we have again completed the sequence 1001 and get another 1 output.

FIGURE 14-8

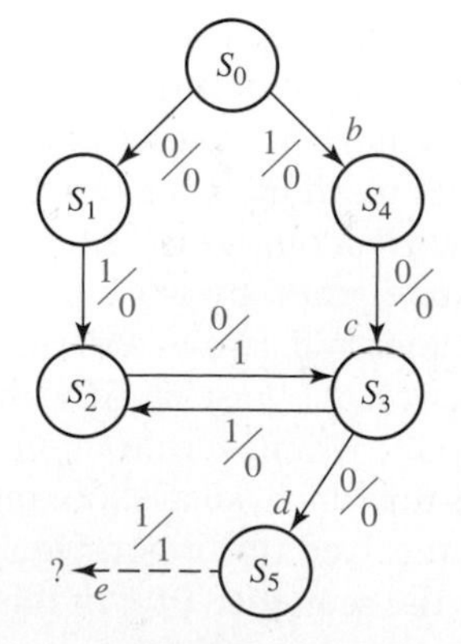


| State | Sequence Ends in |
|---|---|
| $S_0$ | Reset |
| $S_1$ | 0 (but not 10) |
| $S_2$ | 01 |
| $S_3$ | 10 |
| $S_4$ | 1 (but not 01) |
| $S_5$ | 100 |

FIGURE 14-9

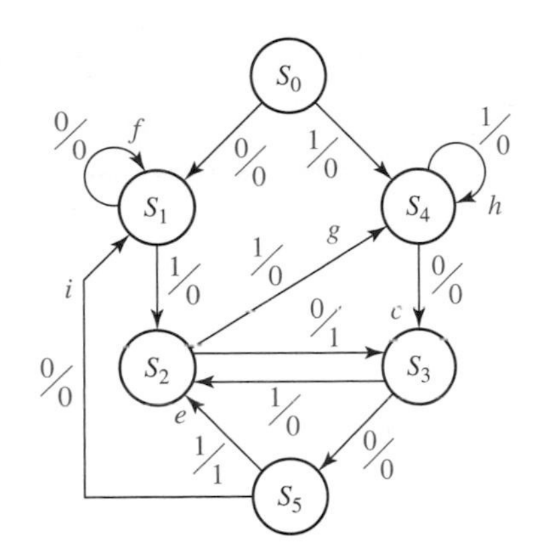

| State | Sequence Ends in |
|---|---|
| $S_0$ | Reset |
| $S_1$ | 0 (but not 10) |
| $S_2$ | 01 |
| $S_3$ | 10 |
| $S_4$ | 1 (but not 01) |
| $S_5$ | 100 |

We have now taken care of putting out a 1 when either the sequence 010 or 1001 is completed. Next, we will go back and complete the state graph to take care of the other input sequences, for which we have not already accounted. In state $S_1$, we have accounted for a 1 input but not a 0 input. If we are in $S_1$ and get a 0 input, to which state should we go? If a 0 input occurs in $S_1$, we have a sequence ending in 00. Because 00 is not part of either of the input sequences for which we are looking, we can ignore the extra 0 and stay in $S_1$ (arrow $f$). No matter how many extra 0's occur, we still have a sequence ending in 0, and we stay in $S_1$ until a 1 input occurs. In $S_2$, we have taken care of the 0 input case but not the 1 input case. If a 1 is received, the input sequence ends in 11. Because 11 is not part of either the sequence 010 or 1001, we do not need a state which corresponds to a sequence ending in 11. We cannot stay in $S_2$ because $S_2$ corresponds to a sequence ending in 01. Therefore, we go to $S_4$, which corresponds to having received a sequence ending in 1 (arrow $g$). $S_3$ already has arrows corresponding to 0 and 1 inputs, so we examine $S_4$ next. If a 1 is received in $S_4$, the input sequence ends in 11. We can stay in $S_4$ and ignore the extra 1 (arrow $h$) because 11 is not part of either sequence for which we are looking. In $S_5$, if we get a 0 input, the sequence ends in 000. Because 000 is not contained in either 010 or 1001, we can go back to $S_1$, because $S_1$ corresponds to having received a sequence ending in one (or more) 0's. This completes the state graph because every state has arrows leaving it which correspond to both 0 and 1 inputs. We should now go back and check the state graph against the original input sequences to make sure that a 1 output is always obtained for a sequence ending in 010 or 1001 and that a 1 output does not occur for any other sequence.

Next, we will derive the state graph for a Moore sequential circuit with one input $X$ and one output $Z$. The output $Z$ is to be 1 if the total number of 1's received is odd and at least two consecutive 0's have been received. A typical input and output sequence is

$$
\begin{array}{lcccccccc}
X = & 1 & 0 & 1 & 1 & 0 & 0 & 1 & 1 \\
 & \uparrow & & & \uparrow & & \uparrow & \uparrow & \uparrow \\
 & a & & & b & & c & d & e \\
Z = (0) & 0 & 0 & 0 & 0 & 0 & 1 & 0 & 1
\end{array}
$$

We have shifted the $Z$ sequence to the right to emphasize that for a Moore circuit an input change does not affect $Z$ immediately, but $Z$ can change only after the next active clock edge. The initial 0 in parentheses is the output associated with the reset state. At points $a$ and $b$ in the preceding sequence, an odd number of 1's has been received, but two consecutive 0's have not been received, so the output remains 0. At points $c$ and $e$, an odd number of 1's and two consecutive 0's have been received, so $Z = 1$. At point $d$, $Z = 0$ because the number of 1's is even.

We start construction of the Moore state graph (Figure 14-10) with the reset state $S_0$, and we associate a 0 output with this state. First, we will consider keeping track of whether the number of 1's is even or odd. If we get a 1 input in $S_0$, we will go to state $S_1$ to indicate an odd number of 1's received. The output for $S_1$ is 0 because two consecutive 0's have not been received. When a second 1 is received, should we go to a new state or go back to $S_0$? For this problem, it is unnecessary to distinguish between an even number of 1's and no 1's received, so we can go back to $S_0$. A third 1 then takes us to $S_1$ (odd number of 1's), a fourth 1 to $S_0$ (even 1's), and so forth.

FIGURE 14-10

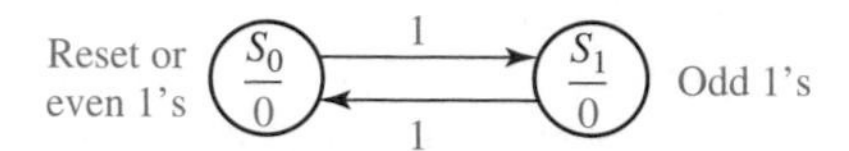


If a 0 is received in $S_0$, this starts a sequence of two consecutive 0's, so we go to $S_2$ (0 output) in Figure 14-11. Another 0 then takes us to $S_3$ to indicate two consecutive 0's received. The output is still 0 in $S_3$ because the number of 1's received is even. Now if we get a 1 input, we have received an odd number of 1's and go to $S_4$. (Why can we not go to $S_1$?) In $S_4$ we have received two consecutive 0's and an odd number of 1's, so the output is 1.

FIGURE 14-11

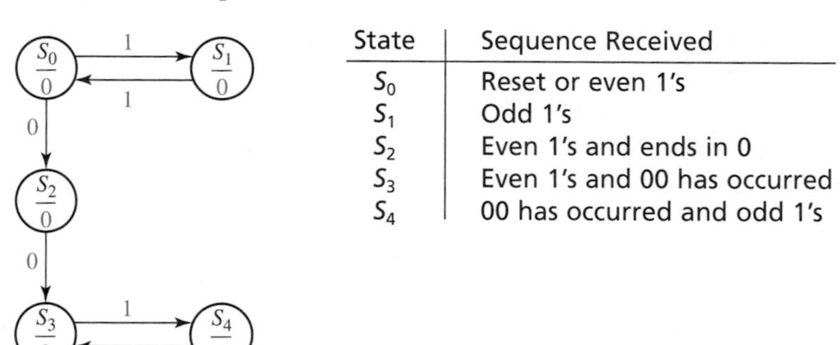


| State | Sequence Received |
|---|---|
| $S_0$ | Reset or even 1's |
| $S_1$ | Odd 1's |
| $S_2$ | Even 1's and ends in 0 |
| $S_3$ | Even 1's and 00 has occurred |
| $S_4$ | 00 has occurred and odd 1's |

If we receive a 1 in $S_4$, we have an even number of 1's and two consecutive 0's, so we can return to $S_3$ (arrow $a$). The output in $S_3$ is 0, and when we get another 1 input, the number of 1's is odd, so we again go to $S_4$ with a 1 output. Now, suppose that we are in $S_1$ (odd number of 1's received), and we get a 0. We cannot go to $S_2$ (Why?), so we go to a new state $S_5$ (Figure 14-12, arrow $b$) which corresponds to an odd number of 1's followed by a 0. Another 0 results in two consecutive 0's, and we can go to $S_4$ (arrow $c$) which gives us a 1 output.

Now, we must go back and complete the state graph by making sure that there are two arrows leaving each state. In $S_2$, a 1 input means that we have received an odd number of 1's. Because we have not received two consecutive 0's, we must return to $S_1$ (arrow $d$) and start counting 0's over again. Similarly, if we receive a 1 in $S_5$, we return to $S_0$ (Why?). Now, what should happen if we receive a 0 in $S_3$? Referring to the original problem statement, we see that once two consecutive 0's have been received, additional

FIGURE 14-12

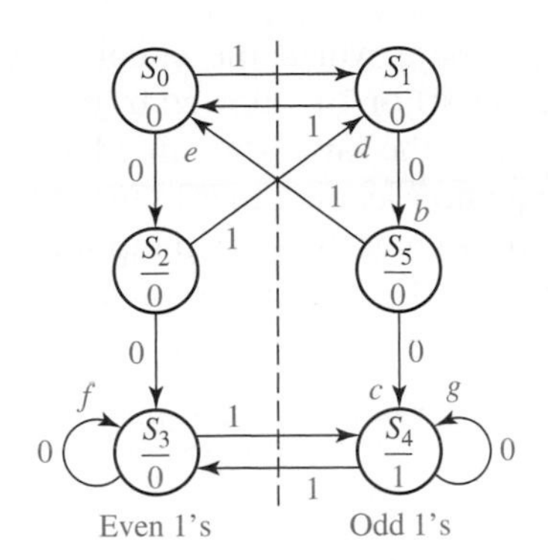


| State | Input Sequences |
|---|---|
| $S_0$ | Reset or even 1's |
| $S_1$ | Odd 1's |
| $S_2$ | Even 1's and ends in 0 |
| $S_3$ | Even 1's and 00 has occurred |
| $S_4$ | Odd 1's and 00 has occurred |
| $S_5$ | Odd 1's and ends in 0 |

0's can be ignored. Therefore, we can stay in $S_3$ (arrow $f$). Similarly, extra 0 inputs can be ignored in $S_4$ (arrow $g$). This completes the Moore state diagram, and we should go back and verify that the correct output sequence is obtained for various input sequences.

# 14.3 Guidelines for Construction of State Graphs

Although there is no one specific procedure which can be used to derive state graphs or tables for every problem, the following guidelines should prove helpful:

1. First, construct some sample input and output sequences to make sure that you understand the problem statement.
2. Determine under what conditions, if any, the circuit should reset to its initial state.
3. If only one or two sequences lead to a nonzero output, a good way to start is to construct a partial state graph for those sequences.
4. Another way to get started is to determine what sequences or groups of sequences must be remembered by the circuit and set up states accordingly.
5. Each time you add an arrow to the state graph, determine whether it can go to one of the previously defined states or whether a new state must be added.
6. Check your graph to make sure there is one and only one path leaving each state for each combination of values of the input variables.
7. When your graph is complete, test it by applying the input sequences formulated in part 1 and making sure the output sequences are correct.

Several examples of deriving state graphs or tables follow.

***Example 1*** A sequential circuit has one input ($X$) and one output ($Z$). The circuit examines groups of four consecutive inputs and produces an output $Z = 1$ if the input sequence 0101 or 1001 occurs. The circuit resets after every four inputs. Find the Mealy state graph.

***Solution*** A typical sequence of inputs and outputs is

$X = 0101 \mid 0010 \mid 1001 \mid 0100$
$Z = 0001 \mid 0000 \mid 0001 \mid 0000$

The vertical bars indicate the points at which the circuit resets to the initial state. Note that an input sequence of either 01 or 10 followed by 01 will produce an output of $Z = 1$. Therefore, the circuit can go to the same state if either 01 or 10 is received. The partial state graph for the two sequences leading to a 1 output is shown in Figure 14-13.

Note that the circuit resets to $S_0$ when the fourth input is received. Next, we add arrows and labels to the graph to take care of sequences which do not give a 1 output, as shown in Figure 14-14.

**FIGURE 14-13** Partial State Graph for Example 1

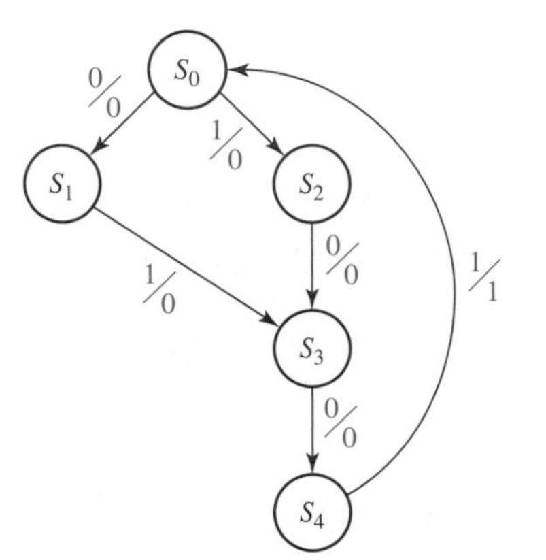


| State | Sequence Received |
|---|---|
| $S_0$ | Reset |
| $S_1$ | 0 |
| $S_2$ | 1 |
| $S_3$ | 01 or 10 |
| $S_4$ | 010 or 100 |

**FIGURE 14-14** Complete State Graph for Example 1

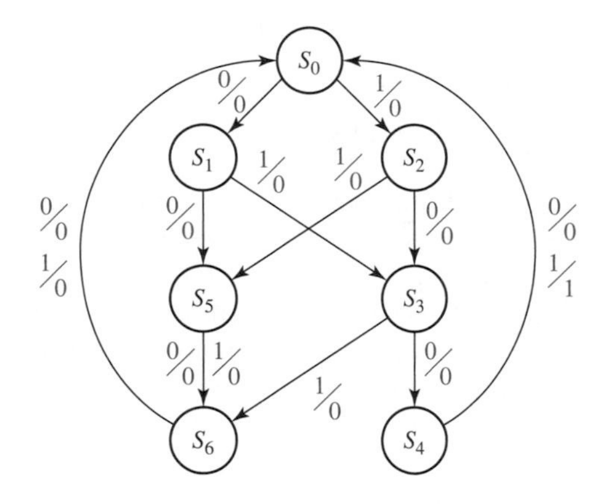


| State | Sequence Received |
|---|---|
| $S_0$ | Reset |
| $S_1$ | 0 |
| $S_2$ | 1 |
| $S_3$ | 01 or 10 |
| $S_4$ | 010 or 100 |
| $S_5$ | Two inputs received, no 1 output is possible |
| $S_6$ | Three inputs received, no 1 output is possible |

The addition of states $S_5$ and $S_6$ was necessary so that the circuit would not reset to $S_0$ before four inputs were received. Note that once a 00 or 11 input sequence has been received (state $S_5$), no output of 1 is possible until the circuit is reset.

### *Example 2*

A sequential circuit has one input ($X$) and two outputs ($Z_1$ and $Z_2$). An output $Z_1 = 1$ occurs every time the input sequence 100 is completed, provided that the sequence 010 has never occurred. An output $Z_2 = 1$ occurs every time the input sequence 010 is completed. Note that once a $Z_2 = 1$ output has occurred, $Z_1 = 1$ can never occur but not vice versa. Find a Mealy state graph and table.

***Solution*** A typical sequence of inputs and outputs is:

$$X = 1\ 0\ 0\ 1\ 1\ 0\ 0\ 1\ 0 \mid 1\ 0\ 1\ 0\ 0\ 1\ 0\ 1\ 1\ 0\ 1\ 0\ 0$$
$$Z_1 = 0\ 0\ 1\ 0\ 0\ 0\ 1\ 0\ 0 \mid 0\ 0\ 0\ 0\ 0\ 0\ 0\ 0\ 0\ 0\ 0\ 0\ 0$$
$$Z_2 = 0\ 0\ 0\ 0\ 0\ 0\ 0\ 0\ 1 \mid 0\ 1\ 0\ 1\ 0\ 0\ 1\ 0\ 0\ 0\ 0\ 1\ 0$$

Note that the sequence 100 occurs twice before 010 occurs, and $Z_1 = 1$ each time. However, once 010 occurs and $Z_2 = 1$, $Z_1 = 0$ even when 100 occurs again. $Z_2 = 1$ all five times that 010 occurs. Because we were not told to reset the circuit, 01010 means that 010 occurred twice.

We can begin to solve this problem by constructing the part of the state graph which will give the correct outputs for the sequences 100 and 010. Figure 14-15(a) shows this portion of the state graph.

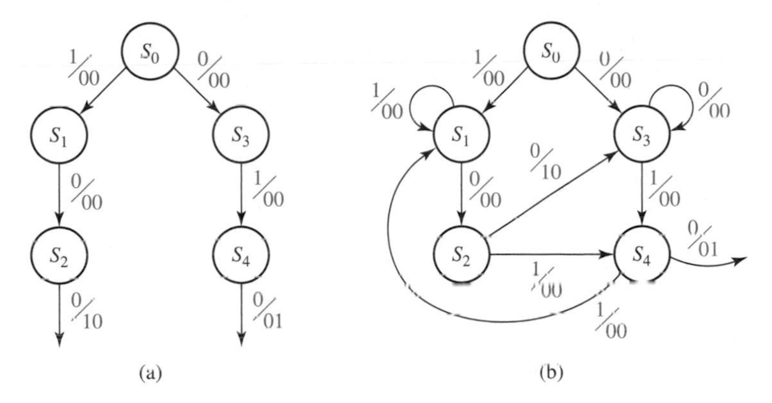


**FIGURE 14-15** Partial Graphs for Example 2

An important question to ask at this point is, what does this circuit need to remember to give the correct outputs? The circuit will need to remember how much progress has been made on the sequence 010, so it will know when to output $Z_2 = 1$. The circuit will also need to remember how much progress has been made on the sequence 100 and whether 010 has ever occurred, so it will know when to output $Z_1 = 1$.

Keeping track of what is remembered by each state will help us make the correct state graph. Table 14-5 will help us to do this. State $S_0$ is the initial state of the circuit, so there is no progress on either sequence, and 010 has never occurred. State $S_1$ is the state we go to when a 1 is received from $S_0$, so in state $S_1$, we have made progress on the sequence 100 by getting a 1. In state $S_2$, we have made progress on the sequence 100 by getting 10. Similarly, states $S_3$ and $S_4$ represent progress of 0 and 01 toward 010. In $S_1$,

**TABLE 14-5** State Descriptions for Example 2

| State | Description | | |
|---|---|---|---|
| $S_0$ | No progress on 100 | No progress on 010 | |
| $S_1$ | Progress of 1 on 100 | No progress on 010 | |
| $S_2$ | Progress of 10 on 100 | Progress of 0 on 010 | 010 has never occurred |
| $S_3$ | No progress on 100 | Progress of 0 on 010 | |
| $S_4$ | Progress of 1 on 100 | Progress of 01 on 010 | |
| $S_5$ | | Progress of 0 on 010 | |
| $S_6$ | | Progress of 01 on 010 | 010 has occurred |
| $S_7$ | | No progress on 010 | |

there is no progress toward the sequence 010, and in $S_3$, there is no progress toward the sequence 100. However, in $S_2$, we have received 10, so if the next two inputs are 1 and 0, the sequence 010 will be completed. Therefore, in $S_2$, we have not only made progress of 10 toward 100, but we have also made progress of 0 toward 010. Similarly, in $S_4$, we have made progress of 1 toward 100, as well as progress of 01 toward 010.

Using this information, we can fill in more of the state graph to get Figure 14-15(b). If the circuit is in state $S_1$ and a 1 is received, then the last two inputs are 11. The previous 1 is of no use toward the sequence 100. However, the circuit will need to remember the new 1, and there is a progress of 1 toward the sequence 100. There is no progress on the sequence 010, and 010 has never occurred, but this is the same situation as state $S_1$. Therefore, the circuit should return to state $S_1$. Similarly, if a 0 is received in state $S_3$, the last two inputs are 00. There is a progress only of 0 toward the sequence 010, there is no progress toward 100, and 010 has never occurred, so the circuit should return to state $S_3$. In state $S_2$, if a 0 is received, the sequence 100 is complete and the circuit should output $Z_1 = 1$. Then, there is no progress on another sequence of 100, and 010 has still not occurred. However, the last input is 0, so there is progress of 0 toward the sequence 010. We can see from Table 14-5 that this is the same situation as $S_3$, so the circuit should go to state $S_3$. If, in state $S_2$, a 1 is received, we have made progress of 01 toward 010 and progress of 1 toward 100, and 010 has still not occurred. We can see from Table 14-5 that the circuit should go to state $S_4$.

If a 0 is received in state $S_4$, the sequence 010 is complete, and we should output $Z_2 = 1$. At this point we must go to a new state ( $S_5$) to remember that 010 has been received so that $Z_1 = 1$ can never occur again. When $S_5$ is reached, we stop looking for 100 and only look for 010. Figure 14-16(a) shows a partial state graph that outputs $Z_2 = 1$ when the input sequence ends in 010. In $S_5$ we have progress of 0 toward 010 and additional 0's can be ignored by looping back to $S_5$. In $S_6$ we have progress of 01 toward 010. If a 0 is received, the sequence is completed, $Z_2 = 1$ and we can go back to $S_5$ because this 0 starts the 010 sequence again.

FIGURE 14-16 State Graphs for Example 2

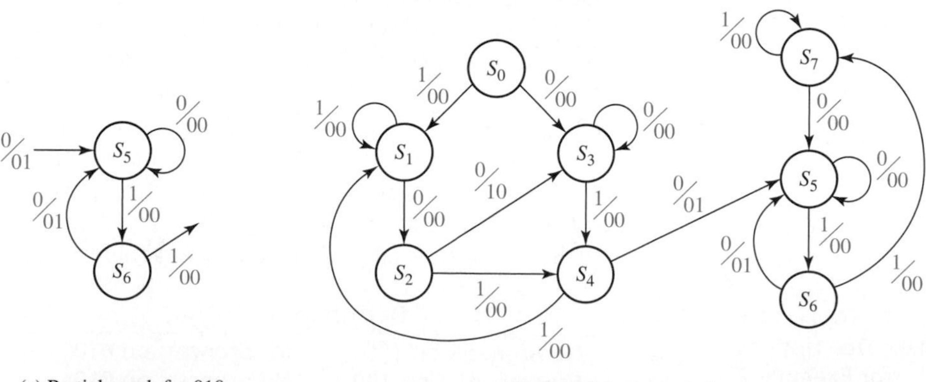


(a) Partial graph for 010

(b) Complete state graph

If we receive a 1 in state $S_6$, the 010 sequence is broken and we must add a new state ($S_7$) to start looking for 010 again. In state $S_7$ we ignore additional 1's, and when a 0 is received, we go back to $S_5$ because this 0 starts the 010 sequence over again. Figure 14-16(b) shows the complete state graph, and the corresponding table is Table 14-6.

TABLE 14-6

| Present State | Next State $X = 0$ | $X = 1$ | Output $(Z_1Z_2)$ $X = 0$ | $X = 1$ |
|---|---|---|---|---|
| $S_0$ | $S_3$ | $S_1$ | 00 | 00 |
| $S_1$ | $S_2$ | $S_1$ | 00 | 00 |
| $S_2$ | $S_3$ | $S_4$ | 10 | 00 |
| $S_3$ | $S_3$ | $S_4$ | 00 | 00 |
| $S_4$ | $S_5$ | $S_1$ | 01 | 00 |
| $S_5$ | $S_5$ | $S_6$ | 00 | 00 |
| $S_6$ | $S_5$ | $S_7$ | 01 | 00 |
| $S_7$ | $S_5$ | $S_7$ | 00 | 00 |

### Example 3

A sequential circuit has two inputs $(X_1, X_2)$ and one output $(Z)$. The output remains a constant value unless one of the following input sequences occurs:

(a) The input sequence $X_1\ X_2 = 01, 11$ causes the output to become 0.
(b) The input sequence $X_1\ X_2 = 10, 11$ causes the output to become 1.
(c) The input sequence $X_1\ X_2 = 10, 01$ causes the output to change value.

(The notation $X_1X_2 = 01, 11$ means $X_1 = 0, X_2 = 1$ followed by $X_1 = 1, X_2 = 1$.) Derive a Moore state graph for the circuit.

**Solution** The only sequences of input pairs which affect the output are of length two. Therefore, the previous and present inputs will determine the output, and the circuit must remember only the previous input pair. At first, it appears that three states are required, corresponding to the last input received being $X_1X_2 = 01, 10$ and (00 or 11). Note that it is unnecessary to use a separate state for 00 and 11 because neither input starts a sequence which leads to an output change. However, for each of these states the output could be either 0 or 1, so we will initially define six states as follows:

| Previous Input $(X_1X_2)$ | Output $(Z)$ | State Designation |
|---|---|---|
| 00 or 11 | 0 | $S_0$ |
| 00 or 11 | 1 | $S_1$ |
| 01 | 0 | $S_2$ |
| 01 | 1 | $S_3$ |
| 10 | 0 | $S_4$ |
| 10 | 1 | $S_5$ |

Using this state designation, we can then set up a state table (Table 14-7). The six-row table given here can be reduced to five rows, using the methods given in Unit 15.

TABLE 14-7

| Present State | $Z$ | Next State $X_1X_2 = 00$ | 01 | 11 | 10 |
|---|---|---|---|---|---|
| $S_0$ | 0 | $S_0$ | $S_2$ | $S_0$ | $S_4$ |
| $S_1$ | 1 | $S_1$ | $S_3$ | $S_1$ | $S_5$ |
| $S_2$ | 0 | $S_0$ | $S_2$ | $S_0$ | $S_4$ |
| $S_3$ | 1 | $S_1$ | $S_3$ | $S_0$ | $S_5$ |
| $S_4$ | 0 | $S_0$ | $S_3$ | $S_1$ | $S_4$ |
| $S_5$ | 1 | $S_1$ | $S_2$ | $S_1$ | $S_5$ |

The $S_4$ row of this table was derived as follows. If 00 is received, the input sequence has been 10, 00, so the output does not change, and we go to $S_0$ to remember that the last input received was 00. If 01 is received, the input sequence has been 10, 01, so the output must change to 1, and we go to $S_3$ to remember that the last input received was 01. If 11 is received, the input sequence has been 10, 11, so the output should become 1, and we go to $S_1$. If 10 is received, the input sequence has been 10, 10, so the output does not change, and we remain in $S_4$. Verify for yourself that the other rows in the table are correct. The state graph is shown in Figure 14-17.

FIGURE 14-17 State Graph for Example 3

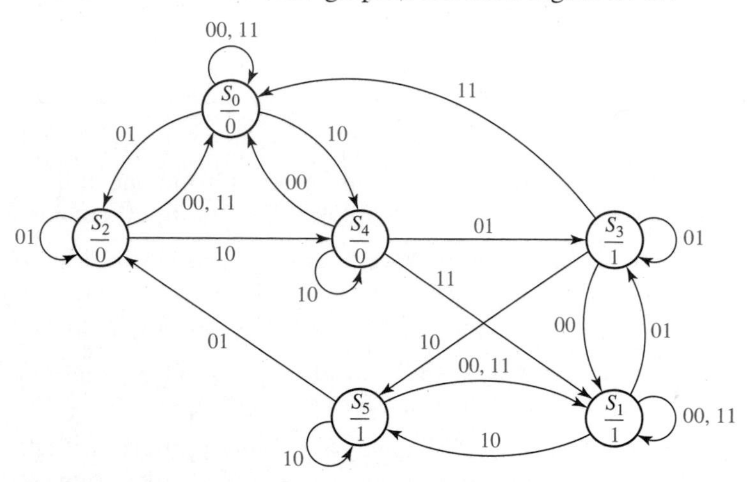


## 14.4 Serial Data Code Conversion

As a final example of state graph construction, we will design a converter for serial data. Binary data is frequently transmitted between computers as a serial stream of bits. As shown in Figure 14-18(a), a clock signal is often transmitted along with the data,

FIGURE 14-18 Serial Data Transmission

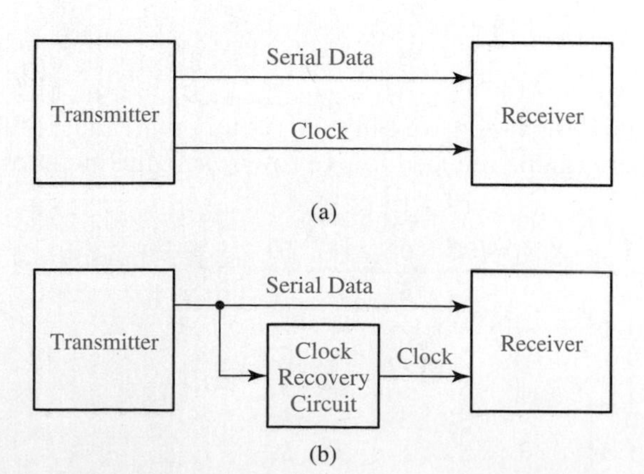

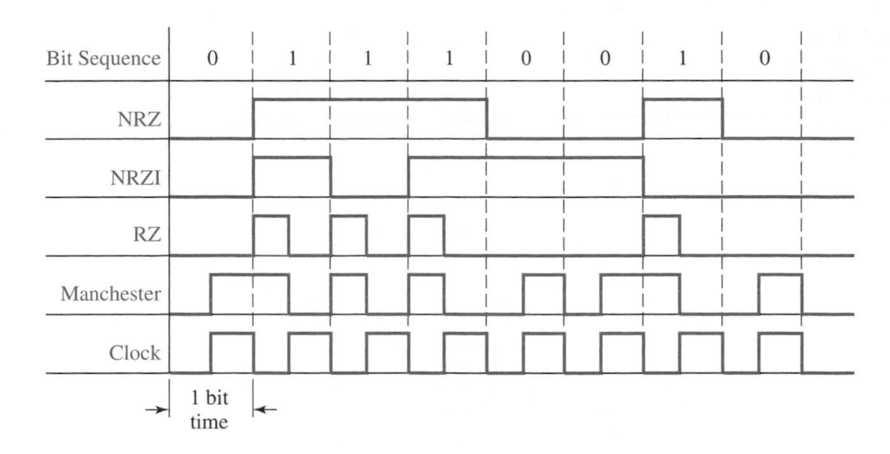


**FIGURE 14-19** Coding Schemes for Serial Data Transmission

so the receiver can read the data at the proper time. Alternatively [Figure 14-18(b)], only the serial data is transmitted, and a clock recovery circuit (called a digital phase-locked loop) is used to regenerate the clock signal at the receiver.

Figure 14-19 shows four different coding schemes for serial data together with the clock used to synchronize the data transmission. The example shows the transmission of the bit sequence 0, 1, 1, 1, 0, 0, 1, 0. With the NRZ (non-return-to-zero) code, each bit is transmitted for one bit time without any change. With the NRZI (non-return-to-zero-inverted) code, the data is encoded by the presence or absence of transitions in the output signal. For each 0 in the original sequence, the bit transmitted is the same as the previous bit transmitted. For each 1 in the original sequence, the bit transmitted is the complement of the previous bit transmitted. Thus, the preceding sequence is encoded as 0, 1, 0, 1, 1, 1, 0, 0. In other words, a 0 is encoded by no change in the transmitted value, and a 1 is encoded by inverting the previous transmitted value. For the RZ (return-to-zero) code, a 0 is transmitted as a 0 for one full bit time, but a 1 is transmitted as a 1 for the first half of the bit time and, then, the signal returns to 0 for the second half. For the Manchester code, a 0 is transmitted as 0 for the first half of the bit time and 1 for the second half, but a 1 is transmitted as 1 for the first half and 0 for the second half. Thus, the encoded bit always changes in the middle of the bit time. When the original bit sequence has a long string of 1's and 0's, the Manchester code has more transitions. This makes it easier to recover the clock signal.

We will design a sequential circuit which converts an NRZ-coded bit stream to a Manchester-coded bit stream [Figure 14-20(a)]. In order to do this, we will use a clock, Clock2, that is twice the frequency of the basic clock [Figure 14-20(b)]. In this way, all output changes will occur on the same edge of Clock2, and we can use the standard synchronous design techniques which we have been using in this unit. First, we will design a Mealy circuit to do the code conversion. Note that if the NRZ bit is 0, it will be 0 for two Clock2 periods. Similarly, if the NRZ bit is 1, it will be 1 for two Clock2 periods. Thus, starting in the reset state [$S_0$ in Figure 14-20(c)], the only two possible input sequences are 00 and 11. For the sequence 00, when the first 0 is received, the output is 0. At the end of the first Clock2 period, the circuit goes to $S_1$. The input is still 0, so the output becomes 1 and remains 1 for one Clock2 period, and then the circuit resets to $S_0$. For the sequence 11, when the first 1 is received, the

**FIGURE 14-20**
Mealy Circuit for NRZ to Manchester Conversion

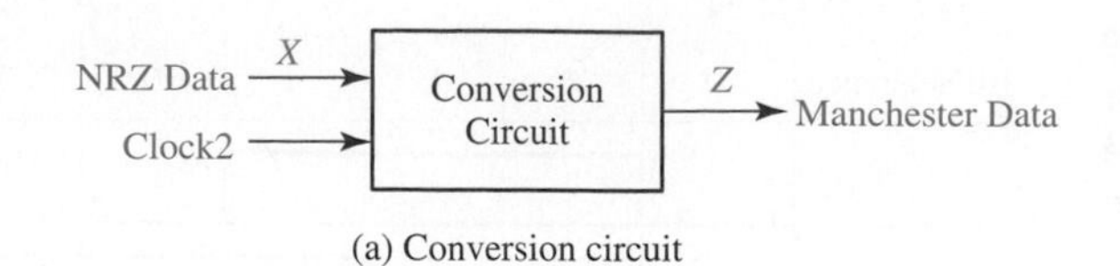


(a) Conversion circuit

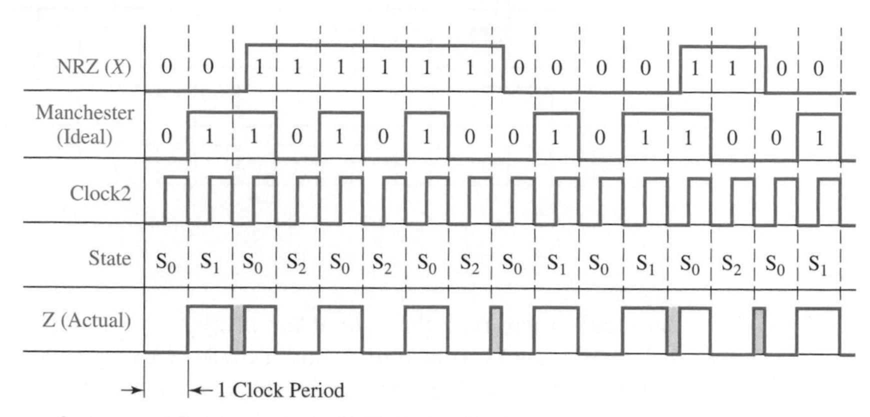


(b) Timing chart

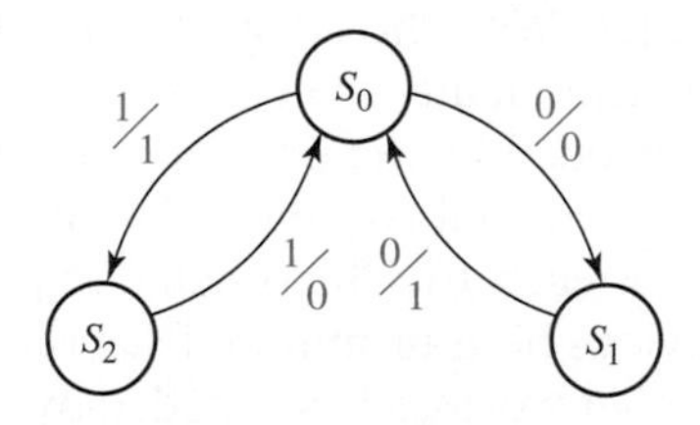


(c) State graph

| Present State | Next State $X=0$ | Next State $X=1$ | Output ($Z$) $X=0$ | Output ($Z$) $X=1$ |
|---|---|---|---|---|
| $S_0$ | $S_1$ | $S_2$ | 0 | 1 |
| $S_1$ | $S_0$ | – | 1 | – |
| $S_2$ | – | $S_0$ | – | 0 |

(d) State table

output is 1 for one Clock2 period and, then, the circuit goes to $S_2$. Then, the output is 0 for one Clock2 period, and the circuit resets to $S_0$.

When we convert the Mealy graph to a state table [Figure 14-20(d)], the next state of $S_1$ with an input of 1 is not specified and is represented by a dash. Similarly, the next state of $S_2$ with a 0 input is not specified. The dashes are like don't-cares, in that we do not care what the next state will be because the corresponding input sequence never occurs. A careful timing analysis for the Mealy circuit shows some possible glitches (false outputs) in the output waveform [Figure 14-20(b)]. The input waveform may not be exactly synchronized with the clock, and we have exaggerated this condition in the figure by shifting the input waveform to the right so that the input changes do not line up with the clock edges. For this situation, we will use the state table to analyze the occurrence of glitches in the $Z$ output. The first glitch shown in the timing chart occurs when the circuit is in state $S_1$, with an input $X = 0$. The state table shows that the output is $Z = 1$, and when the clock goes low, the state changes to $S_0$. At this time, the input is still $X = 0$, so $Z$ becomes 0. Then $X$ changes to 1, $Z$ becomes 1 again, so a glitch has occurred in the output during the time interval between the clock change and the input change. The next glitch occurs in $S_2$ with $X = 1$ and $Z = 0$. When the clock goes low, the output momentarily becomes 1 until $X$ is changed to 0.

To overcome the possible glitch problem with the Mealy circuit, we will redesign the circuit in Moore form (Figure 14-21). Because the output of a Moore circuit cannot change until after the active edge of the clock, the output will be delayed by one clock period. Starting in $S_0$, the input sequence 00 takes us to state $S_1$ with a 0 output and, then, to $S_2$ with a 1 output. Starting in $S_0$, 11 takes us to $S_3$ with a 1 output, and the second 1 can take us back to $S_0$ which has a 0 output. To complete the graph, we add the two arrows starting in $S_2$. Note that a 1 input cannot occur in $S_1$, and a 0 output cannot occur in $S_3$, so the corresponding state table has two don't-cares.

**FIGURE 14-21** Moore Circuit for NRZ to Manchester Conversion

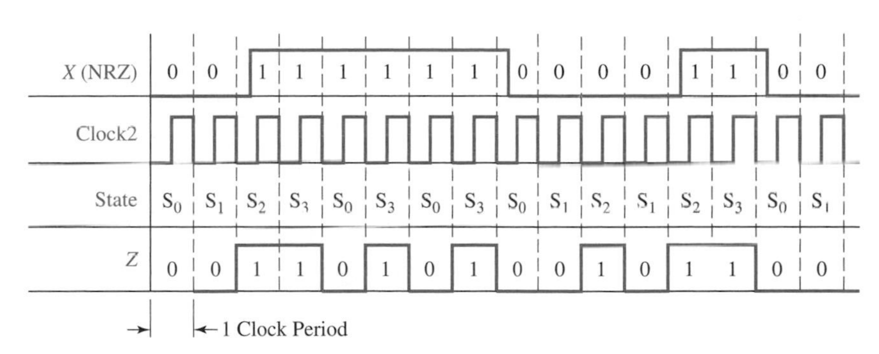


(a) Timing chart

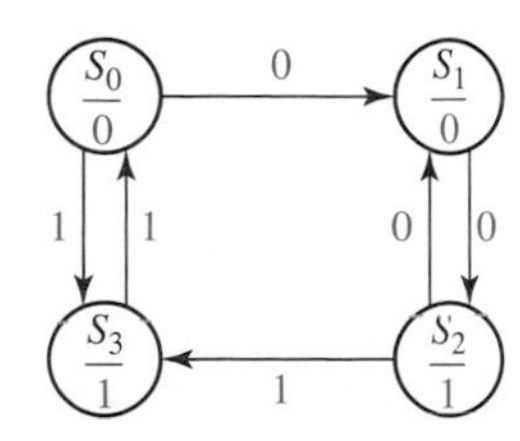


(b) State graph

| Present State | Next State $X = 0$ | $X = 1$ | Present Output ($Z$) |
|---|---|---|---|
| $S_0$ | $S_1$ | $S_3$ | 0 |
| $S_1$ | $S_2$ | – | 0 |
| $S_2$ | $S_1$ | $S_3$ | 1 |
| $S_3$ | – | $S_0$ | 1 |

(c) State table

## 14.5 Alphanumeric State Graph Notation

When a state sequential circuit has several inputs, it is often convenient to label the state graph arcs with alphanumeric input variable names instead of 0's and 1's. This makes it easier to understand the state graph and often leads to a simpler state graph. Consider the following example: A sequential circuit has two inputs ($F$ = forward, $R$ = reverse) and three outputs ($Z_1$, $Z_2$, and $Z_3$). If the input sequence is all $F$'s, the output sequence is $Z_1Z_2Z_3Z_1Z_2Z_3$ . . . ; if the input sequence is all $R$'s, the output sequence is $Z_3Z_2Z_1Z_3Z_2Z_1$ . . . . Figure 14-22(a) shows a preliminary Moore state graph that gives the specified output sequences. An arc label $F$ means that the corresponding state transition occurs when $F = 1$. The notation $Z_1$ within a state means that the output $Z_1$ is 1, and the other outputs ($Z_2$ and $Z_3$) are 0. As long as $F$ is 1, the graph cycles through the states $S_0$, $S_1$, $S_2$, $S_0$, . . . which gives the output sequence $Z_1Z_2Z_3Z_1$ . . . . When $R = 1$ the state and output sequences occur in reverse order.

FIGURE 14-22 State Graphs with Variable Names on Arc Labels

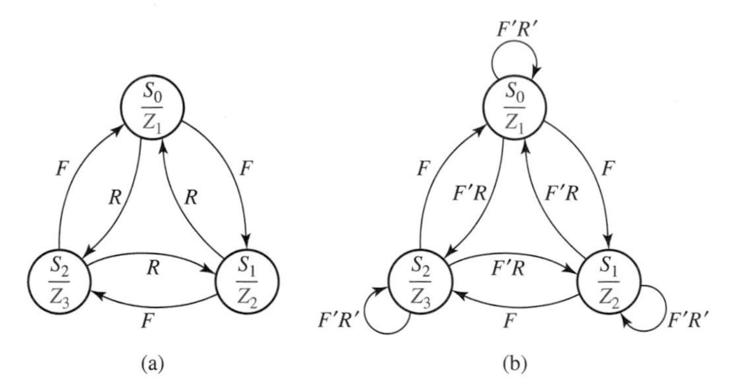


At this point the state graph is not completely specified. What happens if both inputs are 0? What happens if both are 1 at the same time? For example, in state $S_0$ if $F = R = 1$, does the circuit go to state $S_1$ or to $S_2$? Because the circuit can only be in one state at a time, we must assign a priority. We will assume that input $F$ takes priority over input $R$. We can then modify the state graph to implement this priority. By replacing $R$ with $F'R$, this means that the corresponding state transition only occurs if $R = 1$ and $F = 0$. When F = R = 0, we will assume that the output should not change. This can be accomplished by adding a self-loop to each state with an arc label $F'R'$. The resulting state graph [Figure 14-22(b)] is completely specified for all combinations of values of $F$ and $R$, and if both inputs are 1, $F$ takes precedence over $R$. If we convert the graph to a table, the result is Table 14-8.

When we construct a state graph using input variable names on the arcs, we should be careful to make sure that the graph is properly specified. To do this, we

TABLE 14-8 State Table for Figure 14-22

| PS | NS | | | | Output |
|---|---|---|---|---|---|
| | $FR = 00$ | 01 | 10 | 11 | $Z_1Z_2Z_3$ |
| $S_0$ | $S_0$ | $S_2$ | $S_1$ | $S_1$ | 1 0 0 |
| $S_1$ | $S_1$ | $S_0$ | $S_2$ | $S_2$ | 0 1 0 |
| $S_2$ | $S_2$ | $S_1$ | $S_0$ | $S_0$ | 0 0 1 |

can check the labels on all the arcs emanating from each state. For state $S_0$, if we OR together all of the arc labels, we simplify the result to get

$$F + F'R + F'R' = F + F' = 1$$

This result indicates that for any combination of values of the input variables, one of the labels must be 1.

If we AND together every possible pair of arc labels emanating from $S_0$ we get

$$F \cdot F'R = 0, \qquad F \cdot F'R' = 0, \qquad F'R \cdot F'R' = 0$$

This result indicates that for any combination of input values, only one arc label can have a value of 1.

In general, a completely specified state graph has the following properties: (1) When we OR together all input labels on arcs emanating from a state, the result reduces to 1. (2) When we AND together any pair of input labels on arcs emanating from a state, the result is 0. Property (1) ensures that for every input combination, at least one next state is defined. Property (2) ensures that for every input combination, no more than one next state is defined. If both properties are true, then exactly one next state is defined, and the graph is properly specified. If we know that certain input combinations cannot occur, then an incompletely specified graph may be acceptable.

We will use the following notation on Mealy state graphs for sequential circuits: $X_iX_j/Z_pZ_q$ means if inputs $X_i$ and $X_j$ are 1 (we don't care what the other input values are), the outputs $Z_p$ and $Z_q$ are 1 (and the other outputs are 0). That is, for a circuit with four inputs ($X_1$, $X_2$, $X_3$, and $X_4$) and four outputs ($Z_1$, $Z_2$, $Z_3$, and $Z_4$), $X_1X_4'/Z_2Z_3$ is equivalent to 1--0/0110. This type of notation is very useful for large sequential circuits where there are many inputs and outputs.

We will use a dash to indicate that all inputs are don't-cares. For example, an arc label $-/Z_1$ means that for any combination of input values, the indicated state transition will occur and the output $Z_1$ will be 1.

# Programmed Exercise 14.1

Cover the lower part of each page with a sheet of paper and slide it down as you check your answers. *Write* your answer in the space provided before looking at the correct answers.

***Problem*:** A clocked Mealy sequential circuit with one input ($X$) and one output ($Z$) is to be designed. The output is to be 0, unless the input is 0 following a sequence *of exactly* two 0 inputs followed by a 1 input.

To make sure you understand the problem statement, specify the output sequence for each of the following input sequences:

(a) $X = 0010$

$Z =$ ______________

(b) $X = \ldots 1\,0\,0\,1\,0$ (... means any input sequence not ending in 00)

$Z = \ldots$ ______________

(c) $X = \ldots 00010$ $\quad Z = \ldots$ ______________

(d) $X = 00100100010$

$Z =$ ______________

(e) Does the circuit reset after a 1 output occurs?

---

**Answers** (a) $Z = 0001$ (b) $Z = \ldots 00001$ (c) $Z = \ldots 00000$
(d) $Z = 00010010000$ (e) No
Note that no 1 output occurs in answer (c) because there are three input 0's in a row.

---

Add arrows to the following graph so that the sequence $X = 0010$ gives the correct output (do not add another state).

---

**Answer**

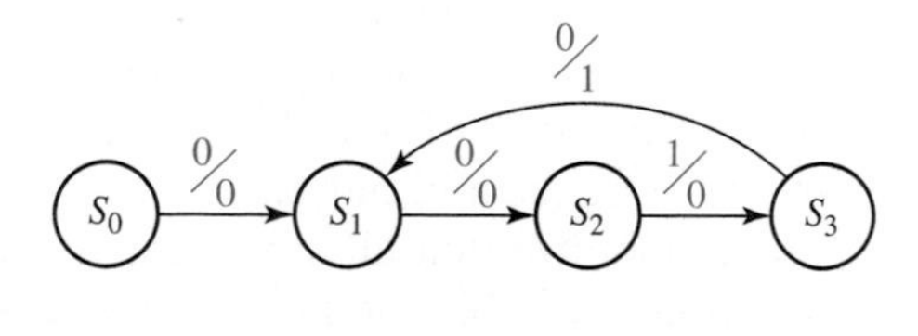


| State | Sequence Received |
|---|---|
| $S_0$ | (Reset) |
| $S_1$ | 0 or 0010 |
| $S_2$ | |
| $S_3$ | |
| $S_4$ | |

Note that the arrow from $S_3$ returns to $S_1$ so that an additional input of 010 will produce another 1 output.

---

Add a state to the preceding graph which corresponds to "three or more consecutive 0's received." Also complete the preceding table to indicate the sequence received which corresponds to each state.

---

**Answer**

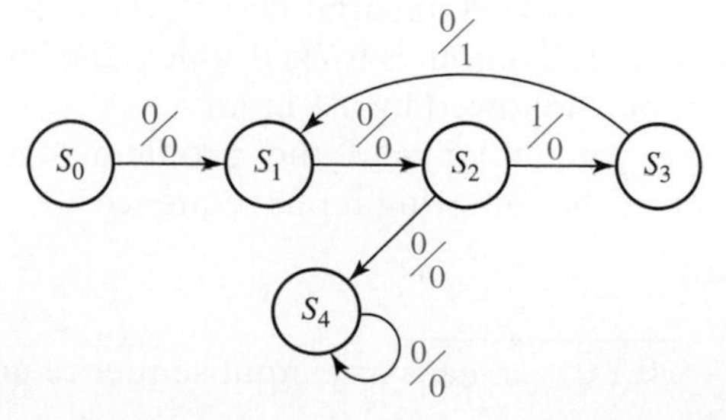


| State | Sequence Received |
|---|---|
| $S_0$ | (Reset) |
| $S_1$ | 0 or 0010 |
| $S_2$ | 00 |
| $S_3$ | 001 |
| $S_4$ | 3 (or more) consecutive 0's |

The preceding state graph is not complete because there is only one arrow leaving most states. Complete the graph by adding the necessary arrows. Return to one of the previously used states when possible.

Answer

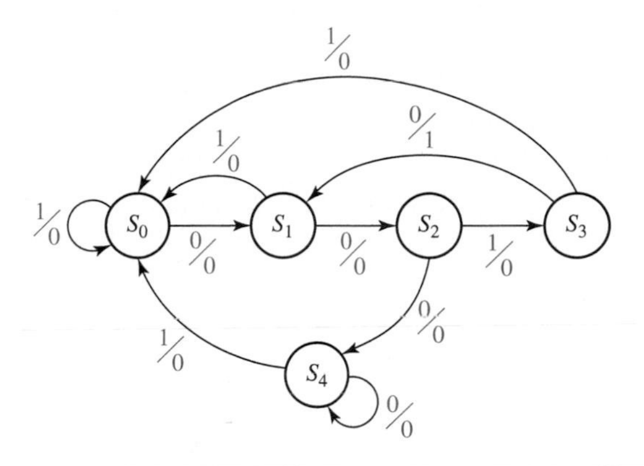


Verify that this state graph gives the proper output sequences for the input sequences listed at the start of this exercise. Write down the Mealy state table which corresponds to the preceding graph.

Answer

| Present State | Next State 0 | Next State 1 | Output 0 | Output 1 |
|---|---|---|---|---|
| $S_0$ | $S_1$ | $S_0$ | 0 | 0 |
| $S_1$ | $S_2$ | $S_0$ | 0 | 0 |
| $S_2$ | $S_4$ | $S_3$ | 0 | 0 |
| $S_3$ | $S_1$ | $S_0$ | 1 | 0 |
| $S_4$ | $S_4$ | $S_0$ | 0 | 0 |

# Programmed Exercise 14.2

***Problem*:** A clocked Moore sequential circuit should have an output of $Z = 1$ if the total number of 0's received is an even number greater than zero, provided that two consecutive 1's have never been received.

To make sure that you understand the problem statement, specify the output sequence for the following input sequence:

$X =$ 0 0 0 0 1 0 1 0 1 1 0 0 0 0
$Z =$ (0)____________________________

↖—— this 0 is the initial output before any inputs have been received

---

**Answer** Z = (0)01011001100000
Note that once two consecutive 1's have been received, the output can never become 1 again.

---

To start the state graph, consider only 0 inputs and construct a Moore state graph which gives an output of 1 if the total number of 0's received is an even number greater than zero.

---

**Answer**

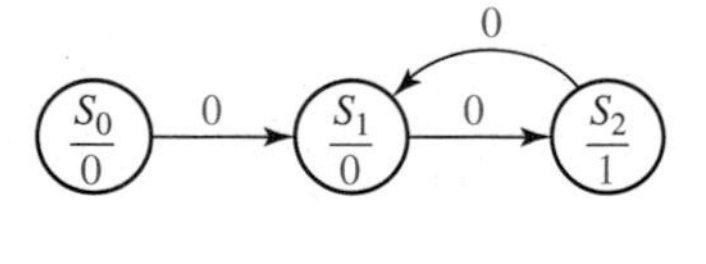


| State | Sequence received |
|---|---|
| $S_0$ | (Reset) |
| $S_1$ | Odd number of 0's |
| $S_2$ | |
| $S_3$ | |
| $S_4$ | |

---

Now add states to the above graph so that starting in $S_0$, if two consecutive 1's are received followed by any other sequence, the output will remain 0. Also, complete the preceding table to indicate the sequence received that corresponds to each state.

---

**Answer**

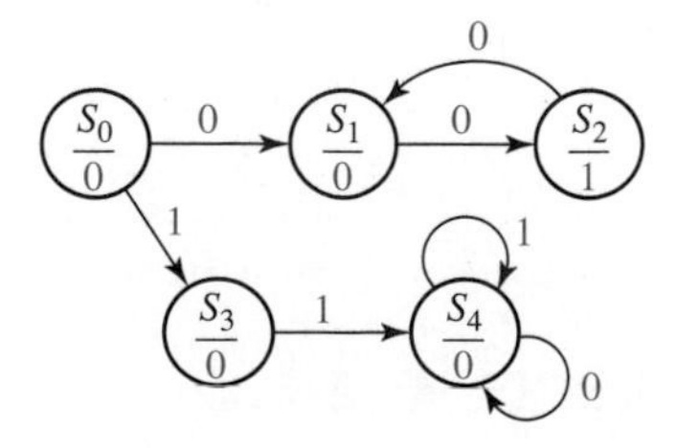


| State | Sequence Received |
|---|---|
| $S_0$ | (Reset) |
| $S_1$ | Odd number of 0's |
| $S_2$ | Even number of 0's |
| $S_3$ | 1 |
| $S_4$ | 11 (followed by any sequence) |
| $S_5$ | |
| $S_6$ | |

---

Now complete the graph so that each state has both a 0 and 1 arrow leading away from it. Add as few extra states to the graph as possible. Also, complete the preceding table.

**Answer**

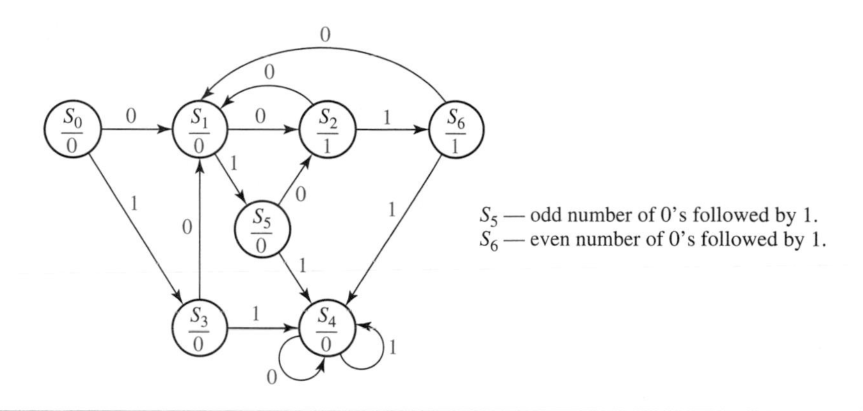


Verify that this state graph gives the proper output sequence for each input sequence at the start of this exercise. Write down the Moore state table which corresponds to the preceding graph. (Note that a Moore table has only one output column.)

**Answer**

| Present State | Next State 0 | Next State 1 | Output |
|---|---|---|---|
| $S_0$ | $S_1$ | $S_3$ | 0 |
| $S_1$ | $S_2$ | $S_5$ | 0 |
| $S_2$ | $S_1$ | $S_6$ | 1 |
| $S_3$ | $S_1$ | $S_4$ | 0 |
| $S_4$ | $S_4$ | $S_4$ | 0 |
| $S_5$ | $S_2$ | $S_4$ | 0 |
| $S_6$ | $S_1$ | $S_4$ | 1 |

# Programmed Exercise 14.3

Derive the state graph and table for a Moore sequential circuit which has an output of 1 iff (1) an even number of 0's have occurred as inputs *and* (2) an odd number of (non overlapping) *pairs* of 1's have occurred. For purposes of this problem, a pair of 1's consists of two consecutive 1's. If three consecutive 1's occur followed by a 0, the third 1 is ignored. If four consecutive 1's occur, this counts as two pairs, etc.

(a) The first step is to analyze the problem and make sure that you understand it. Note that both condition (1) and condition (2) must be satisfied in order to have a 1 output. Consider condition (1) by itself. Would condition (1) be satisfied if zero 0's occurred? ___________

If one 0 occurred? ___________ Two 0's? ___________ Three 0's? ___________.

(*Hint*: Is zero an even or odd number? ___________ )

(b) How many states would it take to determine if condition (1) by itself is satisfied, and what would be the meaning of each state?

_______________________________________________

(c) Now consider condition (2) by itself. For each of the following patterns, determine whether condition (2) is satisfied:

010___________ 0110___________ 01110___________
011110___________ 01010___________ 011010___________
0110110___________

Now check your answers to (a), (b), and (c).

---

**Answers to (a)** yes, no, yes, no, even

**Answers to (b)** two states: even number of 0's, odd number of 0's

**Answers to (c)** From left to right: no, yes, yes, no, no, yes, no

---

(d) Consider condition (2) by itself and consider an input sequence of consecutive 1's. Draw a Moore state diagram (with only 1 inputs) which will give a 1 output when condition (2) is satisfied. State the meaning of each of the four states in your diagram (for example, odd pairs of 1's).

**Answer to (d)**

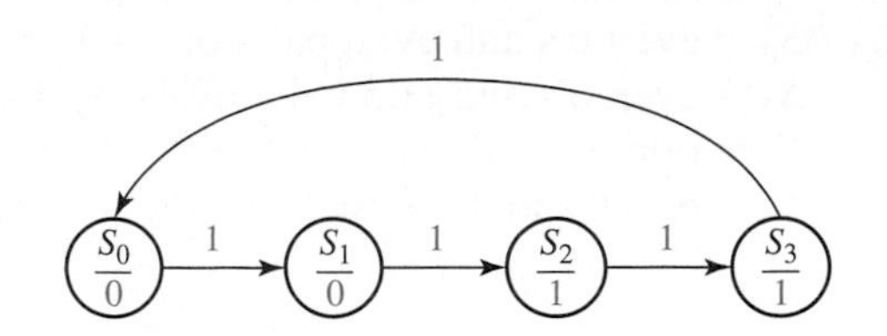


$S_0$ = even pairs of 1's, $S_1$ = even pairs of 1's + one 1,
$S_2$ = odd pairs of 1's, $S_3$ = odd pairs of 1's + one 1

(e) For the original problem, determine the sequence for $Z$ for the following example:

$X =$ 1 1 0 0 1 1 1 1 1 0 0 0 1 1 1 1 0

$Z =$ 0 ____________________

Now turn to the next page and check your answer.

**Answer to (e)**

$X =$ 1 1 0 0 1 1 1 1 1 0 0 0 1 1 1 1 0
$Z =$ 0 0 1 0 1 1 0 0 1 1 0 1 0 0 0 0 0 1

(f) Considering that we must keep track of both even or odd 0's, and even or odd pairs of 1's, how many states should the final graph have? __________

(g) Construct the final Moore state graph. Draw the graph in a symmetric manner with even 0's on the top side and odd 0's on the bottom side. List the meanings of the states such as
$S_0$ = even 0's & even pairs of 1's.

(h) Check your answer using the test sequence from part (e). Then, check your answers below.

**Answer to (f)** Eight states

**Answer to (g)** $S_0$ = even 0's and even pairs of 1's, $S_1$ = even 0's and even pairs of 1's + one 1, $S_2$ = even 0's and odd pairs of 1's, $S_3$ = even 0's and odd pairs of 1's + one 1, $S_4$ = odd 0's and even pairs of 1's, $S_5$ = odd 0's and even pairs of 1's + one 1, $S_6$ = odd 0's and odd pairs of 1's, $S_7$ = odd 0's and odd pairs of 1's + one 1

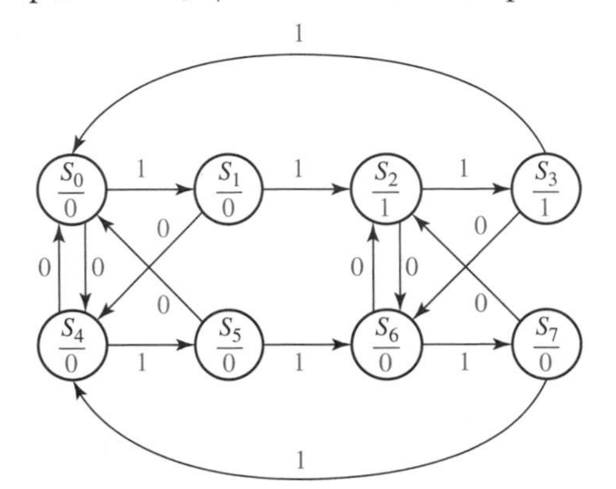


# Problems

**14.4** A sequential circuit has one input and one output. The output becomes 1 and remains 1 thereafter when at least two 0's *and* at least two 1's have occurred as inputs, regardless of the order of occurrence. Draw a state graph (Moore type) for the circuit (nine states are sufficient). Your final state graph should be neatly drawn with no crossed lines.

**14.5** A sequential circuit has one input ($X$) and two outputs ($Z_1$ and $Z_2$). An output $Z_1 = 1$ occurs every time the input sequence 010 is completed, provided that the sequence 100 has never occurred. An output $Z_2 = 1$ occurs every time the input 100 is completed. Note that once a $Z_2 = 1$ output has occurred, $Z_1 = 1$ can never occur but *not* vice versa. Find a Mealy state graph and state table (minimum number of states is eight).

**14.6** A sequential circuit has two inputs ($X_1$ and $X_2$) and one output ($Z$). The output begins as 0 and remains a constant value unless one of the following input sequences occurs:
(a) The input sequence $X_1X_2 = 01, 00$ causes the output to become 0.
(b) The input sequence $X_1X_2 = 11, 00$ causes the output to become 1.
(c) The input sequence $X_1X_2 = 10, 00$ causes the output to change value.
Derive a Moore state table.

**14.7** A sequential circuit has one input ($X$) and one output ($Z$).
Draw a Mealy state graph for each of the following cases:
(a) The output is $Z = 1$ iff the total number of 1's received is divisible by 3. (*Note*: 0, 3, 6, 9, . . . are divisible by 3.)

(b) The output is $Z = 1$ iff the total number of 1's received is divisible by 3 *and* the total number of 0's received is an even number greater than zero (nine states are sufficient).

**14.8** A sequential circuit has two inputs and two outputs. The inputs ($X_1$ and $X_2$) represent a 2-bit binary number, $N$. If the present value of $N$ is greater than the previous value, then $Z_1$ is 1. If the present value of $N$ is less than the previous value, then $Z_2$ is 1. Otherwise, $Z_1$ and $Z_2$ are 0. When the first pair of inputs is received, there is no previous value of $N$, so we cannot determine whether the present $N$ is greater than or less than the previous value; therefore, the "otherwise" category applies.

(a) Find a Mealy state table or graph for the circuit (minimum number of states, including starting state, is five).

(b) Find a Moore state table for the circuit (minimum number of states is 11).

**14.9** (a) Derive the state graph and table for a Mealy sequential circuit which converts a serial stream of bits from NRZ code to NRZI code. Assume that the clock period is the same as the bit time as in Figure 14-19.

(b) Repeat (a) for a Moore sequential circuit.

(c) Draw a timing diagram for your answer to (a), using the NRZ waveform in Figure 14-19 as the input waveform to your circuit. If the input changes occur slightly after the clock edge, indicate places in the output waveform where glitches (false outputs) can occur.

(d) Draw the timing diagram for your answer to (b), using the same input waveform as in (c).

**14.10** For the following state graph, construct the state table, and demonstrate that it is completely specified.

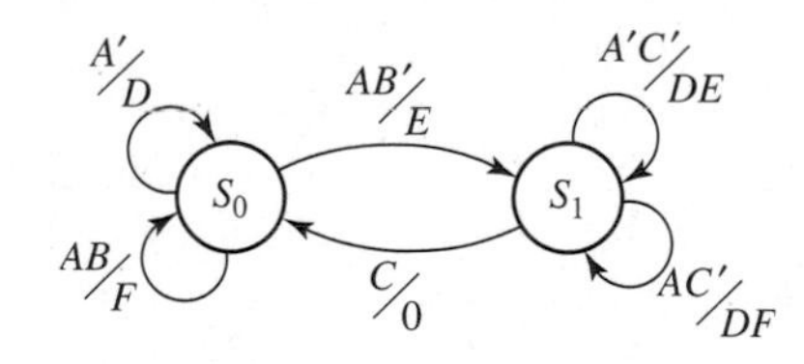


**14.11** Design a sequential circuit which will output $Z = 1$ for exactly four clock cycles each time a person pushes a button (which sets $X = 1$). The clock for a digital circuit is usually much faster than a person's finger! The person probably will not have released the button by the time four clock cycles have passed, so $X$ may still be 1 when the four $Z = 1$ outputs have been generated. Therefore, after $Z$ is 1 for four clock cycles, $Z$ should go to 0, until $X$ returns to 0 and then becomes 1 again. Design a Mealy state graph for this circuit, using the alphanumeric state graph notation given in Section 14.5.

**14.12** (a) A Moore sequential circuit has one input ($x$) and one output ($z$). $z = 1$ if and only if the most recent input <u>was</u> 1 and it was preceded by exactly two 0's. Derive a state table for the circuit.

(b) Repeat for a Mealy circuit, i.e., $z = 1$ if and only if the most recent input <u>is</u> 1 and it was preceded by exactly two 0's. Derive a state table for the circuit.

**14.13** (a) A Mealy sequential circuit has one input ($x$) and one output ($z$). $z$ can be 1 when the fourth, eighth, twelfth, etc. inputs are present, and $z = 1$ if and only if the most recent input combined with the preceding three inputs was not a valid BCD encoding for a decimal digit; otherwise, $z = 0$. Assume the BCD digits are received most significant bit first. Derive a state table for the circuit. (Eight states are sufficient.)

(b) Repeat for a Moore circuit, i.e., $z = 1$ if and only if, after the fourth, eighth, twelfth, etc. inputs have been received, the previous four inputs were not a valid BCD digit. (Nine states are sufficient.)

(c) Is it possible for a Moore circuit to generate the correct output while the fourth input bit is present rather than after it has been received? Explain your answer.

**14.14** (a) A Mealy sequential circuit has one input ($x$) and one output ($z$). $z = 1$ if and only if the most recent input, combined with the preceding three inputs, was not a valid BCD encoding of a decimal digit; otherwise, $z = 0$. Assume the BCD digits are received most significant bit first. Derive a state table for the circuit. (Seven states are sufficient.)

(b) Repeat for a Moore circuit, i.e., $z = 1$ if and only if the previous four inputs were not a valid BCD digit. (Thirteen states are sufficient.)

(c) Is it possible for a Moore circuit to generate the correct output while the fourth input bit is present rather than after it has been received? Explain your answer.

**14.15** (a) A Mealy sequential circuit has one input ($x$) and one output ($z$). $z$ can be 1 when the fourth, eighth, twelfth, etc. inputs are present, and $z = 1$ if and only if the most recent input combined with the preceding three inputs was not a valid BCD encoding of a decimal digit; otherwise, $z = 0$. Assume the BCD digits are received least significant bit first. Derive a state table for the circuit. (Six states are sufficient.)

(b) Repeat for a Moore circuit, i.e., $z = 1$ if and only if, after the fourth, eighth, twelfth, etc. inputs have been received, the previous four inputs were not a valid BCD digit.

(c) Is it possible for a Moore circuit to generate the correct output while the fourth input bit is present rather than after it has been received? Explain your answer.

**14.16** (a) A Mealy sequential circuit has one input ($x$) and one output ($z$). $z = 1$ if and only if the most recent input, combined with the preceding three inputs, was not a valid BCD encoding of a decimal digit; otherwise, $z = 0$. Assume the BCD digits are received least significant bit first. Derive a state table for the circuit. (Three states are sufficient.)

(b) Repeat for a Moore circuit, i.e., $z = 1$ if and only if the previous four inputs were not a valid BCD digit. (Four states are sufficient.)

(c) Is it possible for a Moore circuit to generate the correct output while the fourth input bit is present rather than after it has been received? Explain your answer.

**14.17** (a) A Mealy sequential circuit has one input ($x$) and one output ($z$). $z$ can be 1 when the fourth, eighth, twelfth, etc. inputs are present, and $z = 1$ if and only if the most recent input, combined with the preceding three inputs, was not a valid excess-3 encoding of a decimal digit; otherwise, $z = 0$. Assume the excess-3 digits are received most significant bit first. Derive a state table for the circuit. (Ten states are sufficient.)

(b) Repeat for a Moore circuit, i.e., $z = 1$ if and only if, after the fourth, eighth, twelfth, etc. inputs have been received, the previous four inputs were not a valid excess-3 digit. (Eleven states are sufficient.)

(c) Is it possible to for a Moore circuit to generate the correct output while the fourth input bit is present rather than after it has been received? Explain your answer.

**14.18** (a) A Mealy sequential circuit has one input ($x$) and one output ($z$). $z = 1$ if and only if the most recent input, combined with the preceding three inputs, was not a valid excess-3 encoding of a decimal integer; otherwise, $z = 0$. Assume the excess-3 digits are received most significant bit first. Derive a state table for the circuit. (Eight states are sufficient.)

(b) Repeat for a Moore circuit, i.e., $z = 1$ if and only if the previous four inputs were not a valid excess-3 digit. (Fourteen states are sufficient.)

(c) Is it possible for a Moore circuit to generate the correct output while the fourth input bit is present rather than after it has been received? Explain your answer.

**14.19** (a) A Mealy sequential circuit has one input ($x$) and one output ($z$). $z$ can be 1 when the fourth, eighth, twelfth, etc. inputs are present, and $z = 1$ if and only if the most recent input, combined with the preceding three inputs, was not a valid excess-3 encoding of a decimal digit; otherwise, $z = 0$. Assume the excess-3 digits are received least significant bit first. Derive a state table for the circuit. (Nine states are sufficient.)

(b) Repeat for a Moore circuit, i.e., $z = 1$ if and only if, after the fourth, eighth, twelfth, etc. inputs have been received, the previous four inputs were not a valid excess-3 digit. (Ten states are sufficient.)

(c) Is it possible to for a Moore circuit to generate the correct output while the fourth input bit is present rather than after it has been received? Explain your answer.

**14.20** (a) A Mealy sequential circuit has one input ($x$) and one output ($z$). $z = 1$ if and only if the most recent input combined with the preceding three inputs was not a valid excess-3 encoding of a decimal digit; otherwise, $z = 0$. Assume the excess-3 digits are received least significant bit first. Derive a state table for the circuit. (Six states are sufficient.)

(b) Repeat for a Moore circuit, i.e., $z = 1$ if and only if the previous four inputs were not a valid excess-3 digit. (Eight states are sufficient.)

(c) Is it possible for a Moore circuit to generate the correct output while the fourth input bit is present rather than after it has been received? Explain your answer.

**14.21** A sequential circuit has one input and one output. The output becomes 1 and remains 1 thereafter when at least one 1 and three 0's have occurred as inputs, regardless of the order of occurrence. Draw a state graph (Moore type) for the circuit (eight states are sufficient). Your final state graph should be neatly drawn with no crossed lines.

**14.22** A sequential circuit has one input ($X$) and two outputs ($Z_1$ and $Z_2$). An output $Z_1 = 1$ occurs every time the input sequence 100 is completed provided that the sequence 011 has never occurred. An output $Z_2 = 1$ occurs every time the input 011 is completed. Note that once a $Z_2 = 1$ output has occurred, $Z_1 = 1$ can never occur but *not* vice versa. Find a Mealy state graph and state table (minimum number of states is eight).

**14.23** A sequential circuit has two inputs ($X_1$ and $X_2$) and one output ($Z$). The output begins as 0 and remains a constant value unless one of the following input sequences occurs:
(a) The input sequence $X_1X_2 = 11, 10$ causes the output to become 0.
(b) The input sequence $X_1X_2 = 00, 10$ causes the output to become 1.
(c) The input sequence $X_1X_2 = 01, 10$ causes the output to toggle.
Derive a Moore state table and state graph.

**14.24** A sequential circuit has one input ($X$) and one output ($Z$).
Draw a Mealy state graph for each of the following cases:
(a) The output is $Z = 1$ iff the total number of 1's received is divisible by 4. (*Note*: 0, 4, 8, 12, . . . are divisible by 4.)
(b) The output is $Z = 1$ iff the total number of 1's received is divisible by 4 and the total number of 0's received is an odd number (eight states are sufficient).

**14.25** A sequential circuit has two inputs and two outputs. The inputs ($X_1$ and $X_2$) represent a 2-bit binary number, $N$. If the present value of $N$ plus the previous value of $N$ is greater than 2, then the $Z_1$ is 1. If the present value of $N$ times the previous value of $N$ is greater than 2, then $Z_2$ is 1. Otherwise, $Z_1$ and $Z_2$ are 0. When the first pair of inputs is received, use 0 as the previous value of $N$.
(a) Find a Mealy state table or graph for the circuit (minimum number of states is four).
(b) Find a Moore state table for the circuit (minimum number of states is 10, but any correct answer with 16 or fewer states is acceptable).

**14.26** A Moore sequential circuit has one input and one output. When the input sequence 011 occurs, the output becomes 1 and remains 1 until the sequence 011 occurs again in which case the output returns to 0. The output then remains 0 until 011 occurs a third time, etc. For example, the input sequence

$$X = \ 0\ 1\ 0\ 1\ 1\ 0\ 1\ 0\ 1\ 1\ 0\ 1\ 0\ 0\ 1\ 1\ 1$$

has the output

$$Z = (0)\ 0\ 0\ 0\ 0\ 1\ 1\ 1\ 1\ 1\ 0\ 0\ 0\ 0\ 0\ 0\ 1\ 1$$

Derive the state graph (six states minimum).

**14.27** Work Problem 14.26 if the input sequence 101 causes the output to change value. For example, the input sequence

$$X = \ 0\ 1\ 0\ 1\ 0\ 1\ 0\ 0\ 1\ 0\ 1\ 0\ 1\ 1\ 0\ 1\ 0$$

has the output

$$Z = (0)\ 0\ 0\ 0\ 1\ 1\ 0\ 0\ 0\ 0\ 0\ 1\ 1\ 0\ 0\ 0\ 1\ 1$$

(six states minimum)

**14.28** A Mealy sequential circuit has two inputs and one output. If the total number of 0's received is $\geq 4$ and at least three pairs of inputs have occurred, then the output should be 1 coincident with the last input pair in the sequence. Whenever a 1 output occurs, the circuit resets. Derive a state graph and state table. Specify the meaning of each state. For example, $S_0$ means reset, $S_1$ means one pair of inputs received but no 0's received, etc.

Example:

| | | |
|---|---|---|
| Input sequence | $X_1 =$ | 1 1 1 0 0 0 1 1 1 0 0 0 1 1 0 0 0 1 0 |
| | $X_2 =$ | 1 0 0 0 0 0 1 1 1 1 1 1 0 1 0 0 0 1 0 |
| Output sequence: | $Z =$ | 0 0 0 1 0 0 1 0 0 0 0 0 1 0 0 1 0 0 1 |

**14.29** A Moore sequential circuit has one input and one output. The output should be 1 if the total number of 1's received is odd and the total number of 0's received is an even number greater than 0. Derive the state graph and table (six states).

**14.30** A Mealy sequential circuit has one input ($X$) and two outputs ($Z_1$ and $Z_2$). The circuit produces an output of $Z_1 = 1$ whenever the sequence 011 is completed, and an output of $Z_2 = 1$ whenever the sequence 0111 is completed. Derive the state graph and table.

**14.31** A Moore sequential circuit has two inputs ($X_1$ and $X_2$) and one output ($Z$). $Z$ begins at 0. It becomes 1 when $X_1 = 1$ and $X_2 = 1$ either concurrently, or one after the other (in either order). $Z$ returns to zero when $X_1 = X_2 = 0$. The following input and output sequence should help you understand the problem:

$$X_1 = \quad 0\ 1\ 0\ 0\ 1\ 0\ 0\ 0\ 1\ 1\ 0\ 1\ 1\ 0$$
$$X_2 = \quad 0\ 0\ 1\ 1\ 0\ 0\ 1\ 1\ 0\ 0\ 0\ 1\ 0\ 0$$
$$Z = (0)\ 0\ 0\ 1\ 1\ 1\ 0\ 0\ 0\ 1\ 1\ 0\ 1\ 1\ 0$$

Give the Moore state graph and table.

**14.32** A Mealy sequential circuit has one input ($X$) and one output ($Z$). The circuit should transmit its input, except that it should prevent the sequence 00110 from occurring. So $Z$ should be the same as $X$, except that if the input sequence 00110 occurs, $Z$ should be 1 rather than 0 when the last 0 is received, so that the sequence $X = 00110$ is replaced with $Z = 00111$. Derive the state graph and table.

**14.33** A Moore sequential circuit has one input and one output. The output is 1 if and only if both of the following conditions are met:

(a) The input sequence contains exactly two groups of 1's, and
(b) Each of these groups contains exactly two 1's.

Each group of 1's must be separated by at least one 0. A single 1 is considered a group of 1's containing one 1. For example, the sequence

$$X = 0\ 1\ 1\ 0\ 0\ 0\ 1\ 1\ 0\ 1\ 1\ 1\ 0$$

satisfies both conditions after the first two pairs of 1's. However, when more 1's appear, condition (a) is no longer satisfied. Therefore, the output sequence should be

$$Z = (0)\ 0\ 0\ 0\ 0\ 0\ 0\ 0\ 1\ 1\ 0\ 0\ 0\ 0$$

On the other hand, the sequence

$$X = 1\ 0\ 1\ 1\ 0\ 1\ 1\ 0$$

never satisfies condition (b), because the first group of 1's contains only one 1. Besides, after the second pair of 1's, (a) is no longer satisfied because the input sequence contains three groups of 1's. Therefore, the output should always be 0.

$$Z = (0)\ 0\ 0\ 0\ 0\ 0\ 0\ 0\ 0$$

Derive a state graph and table.

**14.34** A sequential circuit has an input ($X$) and an output ($Z$). The output is the same as the input was two clock periods previously. For example,

$$X = 0\ 1\ 0\ 1\ 1\ 0\ 1\ 0\ 1\ 1\ 0\ 1\ 0\ 0\ 0\ 1$$
$$Z = 0\ 0\ 0\ 1\ 0\ 1\ 1\ 0\ 1\ 0\ 1\ 1\ 0\ 1\ 0\ 0$$

The first two values of $Z$ are 0. Find a Mealy state graph and table for the circuit.

**14.35** A sequential circuit has an input ($X$) and an output ($Z$). The output is the same as the input was three clock periods previously. For example,

$$X = 0\ 1\ 0\ 1\ 1\ 0\ 1\ 0\ 1\ 1\ 0\ 1\ 0\ 0\ 0\ 1$$
$$Z = 0\ 0\ 0\ 0\ 1\ 0\ 1\ 1\ 0\ 1\ 0\ 1\ 1\ 0\ 1\ 0$$

The first three values of $Z$ are 0. Find a Mealy state graph and table for the circuit.

**14.36** (a) Construct a Moore state table for the circuit of Problem 14.34. The initial outputs are 0.

(b) How many states are required in a Moore state table for the circuit of Problem 14.35? Explain

**14.37** A sequential circuit has an input ($X$) and two outputs ($S$ and $V$). $X$ represents a 4-bit binary number $N$ which is input least significant bit first. $S$ represents a 4-bit binary number equal to $N + 2$, which is output least significant bit first. At the time the fourth input occurs, $V = 1$ if $N + 2$ is too large to be represented by four bits; otherwise, $V = 0$. The circuit always resets after the fourth bit of $X$ is received. Find a Mealy state graph and table for the circuit.

Example: $X = 0111$ (binary 14 with the least significant bit first)
$S = 0000$ (because $14 + 2 = 16$, and 16 requires 5 bits)
$V = 0001$

**14.38** A sequential circuit has an input ($X$) and two outputs ($D$ and $B$). $X$ represents a 4-bit binary number $N$ which is input least significant bit first. $D$ represents a 4-bit binary number equal to $N - 2$, which is output least significant bit first. At the time the fourth input occurs, $B = 1$ if $N - 2$ is less than 0; otherwise $B = 0$. The circuit always resets after the fourth bit of $X$ is received. Find a Mealy state graph and table for the circuit.

Example:

$X = 0001 \quad 1000 \quad 1100$
$D = 0110 \quad 1111 \quad 1000$
$B = 0000 \quad 0001 \quad 0000$

**14.39** A sequential circuit has an input ($X$) and outputs ($Y$ and $Z$). $YZ$ represents a 2-bit binary number equal to the number of 1's that have been received as inputs. The circuit resets when the total number of 1's received is 3, or when the total number of 0's received is 3. Find a Moore state graph and table for the circuit.

**14.40** A sequential circuit has an input $X$ and outputs ($Y$ and $Z$). $YZ$ represents a 2-bit binary number equal to the number of pairs of adjacent 1's that have been received as inputs. For example, the input sequence 0110 contains one pair, the sequence 01110 two pairs, and the sequence 0110111 contains three pairs of adjacent 1's. The circuit *resets* when the total number of pairs of 1's received reaches four. Find a Moore state graph and table for the circuit.

Examples:

Input sequence: $X = 0\,1\,0\,1\,1\,0\,1\,1\,1\,0\,0\,1\,0\,1\,0\,1\,0\,1\,1\,1\,0\,1\,1\,0\,1\,1\,0\,0\,1\,0$
Output sequences: $Y = 0\,0\,0\,0\,0\,0\,0\,1\,1\,1\,1\,1\,1\,1\,1\,1\,1\,1\,0\,0\,0\,0\,0\,0\,0\,1\,1\,1\,1\,1$
$Z = 0\,0\,0\,0\,1\,1\,1\,0\,1\,1\,1\,1\,1\,1\,1\,1\,1\,1\,1\,0\,0\,0\,0\,1\,1\,1\,0\,0\,0\,0\,0$

Input sequence: $X = 1\,1\,1\,1\,1\,1\,1\,1$
Output sequences: $Y = 0\,0\,1\,1\,0\,0\,0\,1$
$Z = 0\,1\,0\,1\,0\,0\,1\,0$

(*Hint*: Be sure that the circuit resets as shown in the examples.)

**14.41** A sequential circuit with one input and one output is used to stretch the first two bits of a 4-bit sequence as follows:

| Input | Output |
|---|---|
| 00XX | 0000 |
| 01XX | 0011 |
| 10XX | 1100 |
| 11XX | 1111 |

After every 4 bits, the circuit resets. Find a Mealy state graph and table for the circuit. The third and fourth bits of the input sequence can be either 1 or 0, so make sure that the circuit will work for all possible combinations.

**14.42** A sequential circuit is to be used to control the operation of a vending machine which dispenses a \$0.25 product. The circuit has three inputs ($N$, $D$, and Q) and two outputs ($R$ and $C$). The coin detector mechanism in the vending machine is synchronized with the same clock as the sequential circuit you are to design. The coin detector outputs a single 1 to the $N$, $D$, or $Q$ input for every nickel, dime, or quarter, respectively, that the customer inserts. Only one input will be 1 at a time. When the customer has inserted at least $0.25 in any combination of nickels, dimes, and quarters, the vending machine must give change and dispense the product. The coin return mechanism gives change by returning nickels to the customer. For every 1 output on $C$, the coin return mechanism will return one nickel to the customer. The product is dispensed when the circuit outputs a single 1 on output R. The circuit should reset after dispensing the product.

Example: The customer inserts a nickel, a dime, and a quarter. The circuit inputs and outputs could look like this:

Inputs: $N = 0\ 0\ 0\ 1\ 0\ 0\ 0\ 0\ 0\ 0\ 0\ 0\ 0\ 0\ 0\ 0\ 0$
$D = 0\ 0\ 0\ 0\ 0\ 0\ 0\ 1\ 0\ 0\ 0\ 0\ 0\ 0\ 0\ 0\ 0$
$Q = 0\ 0\ 0\ 0\ 0\ 0\ 0\ 0\ 0\ 0\ 0\ 1\ 0\ 0\ 0\ 0\ 0$

Outputs: $R = 0\ 0\ 0\ 0\ 0\ 0\ 0\ 0\ 0\ 0\ 0\ 0\ 0\ 0\ 1\ 0\ 0$
$C = 0\ 0\ 0\ 0\ 0\ 0\ 0\ 0\ 0\ 0\ 0\ 0\ 1\ 1\ 1\ 0\ 0\ 0$

Note that any number of 0's can occur between 1 inputs.
Derive a Moore state table for the sequential circuit, and for each state indicate how much money the customer has inserted or how much change is due.

**14.43** (a) Derive the state graph and table for a Mealy sequential circuit that converts a serial stream of bits from Manchester code to NRZ code. Assume that a double frequency clock (Clock2) is available.
(b) Repeat (a) for a Moore sequential circuit.
(c) Draw a timing diagram similar to Figure 14-20(b) for your answer to (a), using the Manchester waveform in Figure 14-20(b) as the input waveform to your circuit. If the input changes occur slightly after the clock edge, indicate places in the output waveform where glitches (false outputs) can occur. If possible, assign the don't-cares in the output part of your state table to eliminate some of the glitches.
(d) Draw the timing diagram for your answer to (b), using the same input waveform as in (c).

**14.44** Design a sequential circuit to control a phone answering machine. The circuit should have three inputs ($R$, $A$, and $S$) and one output ($Z$). $R = 1$ for one clock cycle at the end of each phone ring. $A = 1$ when the phone is answered. $S$ selects whether the machine should answer the phone after two rings ($S = 0$) or four rings ($S = 1$). To cause the tape recorder to answer the phone, the circuit should set the output $Z = 1$ after the end of the second ($S = 0$) or fourth ($S = 1$) ring, and hold $Z = 1$ until the recorder circuit answers the phone (i.e., when $A$ goes to 1). If a person answers the phone at any point, $A$ will become 1, and the circuit should reset. Assume that $S$ is not changed while the phone is counting rings. Give a Moore state graph for this circuit, using the alphanumeric state graph notation given in Section 14.5.

**14.45** For the following state graph, derive the state table.

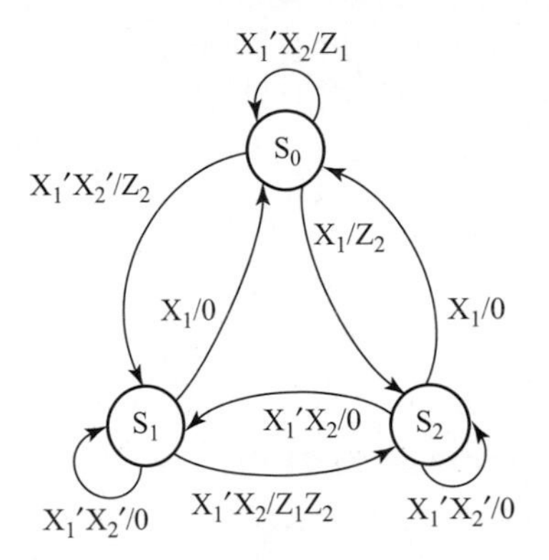


**14.46** There are two errors in the state graph shown. One state is not completely specified for one combination of $X_1$ and $X_2$. In another state, there is a contradiction for one combination of $X_1$ and $X_2$. Correct the state graph by making two minor changes. Demonstrate that the modified state graph is completely specified.

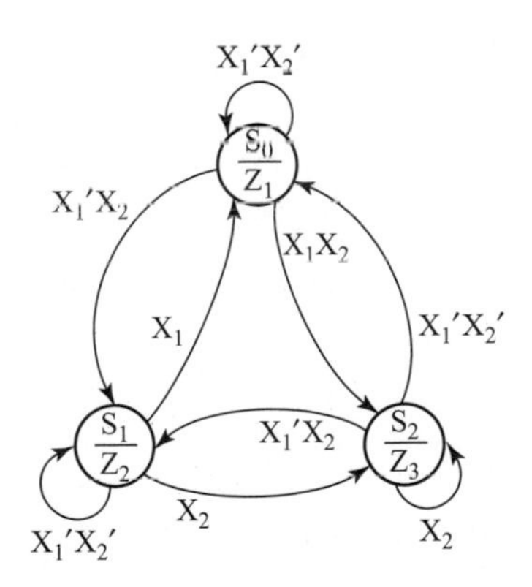

# Reduction of State Tables State Assignment

## Objectives

1. Define equivalent states, state several ways of testing for state equivalence, and determine if two states are equivalent.
2. Define equivalent sequential circuits and determine if two circuits are equivalent.
3. Reduce a state table to a minimum number of rows.
4. Specify a suitable set of state assignments for a state table, eliminating those assignments which are equivalent with respect to the cost of realizing the circuit.
5. State three guidelines which are useful in making state assignments, and apply these to making a good state assignment for a given state table.
6. Given a state table and assignment, form the transition table and derive flip-flop input equations.
7. Make a one-hot state assignment for a state graph and write the next-state and output equations by inspection.

# Study Guide

1. Study Section 15.1, *Elimination of Redundant States*.
2. Study Section 15.2, *Equivalent States*.
   (a) State in words the meaning of $\lambda_1(p, \underline{X}) = \lambda_2(q, \underline{X})$.

   (b) Assuming that $N_1$ and $N_2$ are identical circuits with the following state graph, use Definition 15.1 to show that $p$ is *not* equivalent to $q$. [Calculate $\lambda(p, \underline{X})$ and $\lambda(q, \underline{X})$ for $\underline{X} = 0, \underline{X} = 1, \underline{X} = 00, \underline{X} = 01$, etc.]

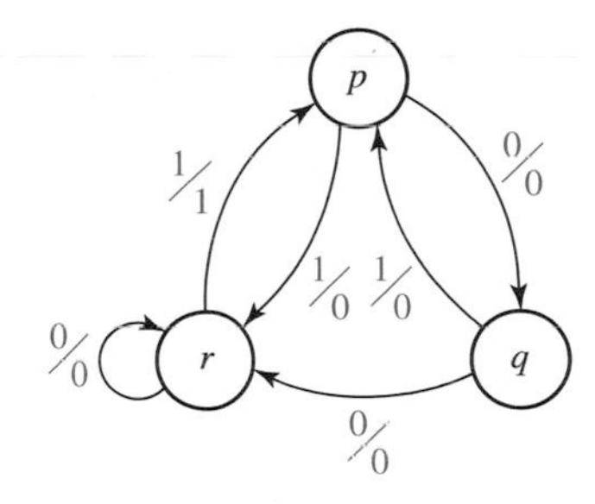


   (c) Suppose you were given two sequential circuits ($N_1$ and $N_2$) in black boxes with only input and output terminals available. Each box has a reset button. The button on $N_1$ resets it to state $p$ and the button on $N_2$ resets it to state $q$. Could you experimentally determine if $p = q$ using Definition 15.1? Explain.

   (d) Apply Theorem 15.1 to show that in Table 15-6, $S_2 \not\equiv S_3$.

   (e) Note the difference between the *definition* of state equivalence (Definition 15.1) and the state equivalence *theorem* (Theorem 15.1). The definition requires an examination of output sequences but *not* next states, while the theorem requires looking at *both* the output and next state for each single input. Make sure that you know both the definition and the theorem. Write out the definition of equivalent states:

   Write out the state equivalence theorem:

   When you check your answers, note that the theorem requires *equal* outputs and *equivalent* next states. This distinction between equal and equivalent is very important. For example, in the following state table, no two states have equal next states, but we can still reduce the table to two states, because some next states are equivalent. Note that the state equivalence theorem tells us that $S_3 \equiv S_0$ if $S_3 \equiv S_0$. When this happens, we may say $S_3 \equiv S_0$. What other pair of states are equivalent?

| Present State | Next State | | Z |
|---|---|---|---|
| $S_0$ | $S_1$ | $S_0$ | 0 |
| $S_1$ | $S_0$ | $S_2$ | 1 |
| $S_2$ | $S_3$ | $S_2$ | 1 |
| $S_3$ | $S_1$ | $S_3$ | 0 |

**3.** Study Section 15.3, *Determination of State Equivalence Using an Implication Table.*

(a) Fill in the following implication chart to correspond to the given table (first pass only).

| | Next State $X = 0$ | 1 | Present Output $X = 0$ | 1 |
|---|---|---|---|---|
| a | a | b | 0 | 0 |
| b | d | a | 0 | 1 |
| c | a | b | 0 | 1 |
| d | g | f | 0 | 0 |
| f | d | g | 0 | 1 |
| g | d | f | 0 | 1 |

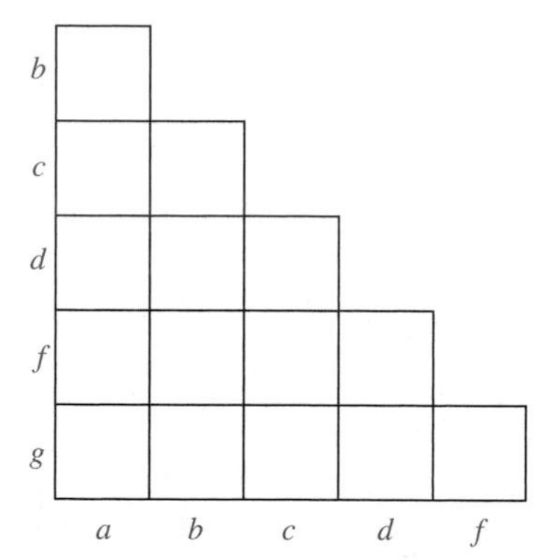


Your answer should have eight squares with X's, two squares with one implied pair, and four squares with two implied pairs. There should be a check in square *f-g* because the only nontrivial implication of *f-g* is *f-g* itself.

(b) Now go through your chart and eliminate all nonequivalent pairs (several passes may be required). What is the only equivalent state pair? According to the state equivalence theorem, why is $b \not\equiv d$? Why is $a \not\equiv b$?

(c) Find all of the equivalent states in the following table using an implication table:

| | Next State $X = 0$ | 1 | Present Output $X = 0$ | 1 |
|---|---|---|---|---|
| a | b | c | 0 | 1 |
| b | d | b | 0 | 0 |
| c | e | a | 0 | 1 |
| d | d | e | 0 | 0 |
| e | e | e | 0 | 0 |

(You should have found four pairs of equivalent states. If you found only two pairs, reread Section 15.3).
Reduce the table to two rows.

4. Study Section 15.4, *Equivalent Sequential Circuits*. Define equivalent sequential circuits. (Make sure you know the difference between equivalent *states* and equivalent *circuits*.)

5. Work Problems 15.1, 15.2, and 15.3 using the methods of Sections 15.3 and 15.4. When forming the implication charts for state equivalence, follow the convention used in the text. That is, label the bottom of the chart starting with the first state and ending with the next-to-last state. Then, label the left side of the chart starting with the second state at the top and ending with the last state at the bottom.

6. Study Section 15.5, *Incompletely Specified State Tables*.
   (a) State two reasons why a state table might be incompletely specified.

   (b) For Table 15-5(a), fill in the don't-care outputs in the $X = 0$ column as 1 and 0 (instead of 0 and 1). Show that with this choice of outputs, the minimum number of states is three.

7. Read Section 15.6, *Derivation of Flip-Flop Input Equations*.
   (a) Derive $J_C$ and $K_C$ from the $C^+$ map of Figure 15-9(a).

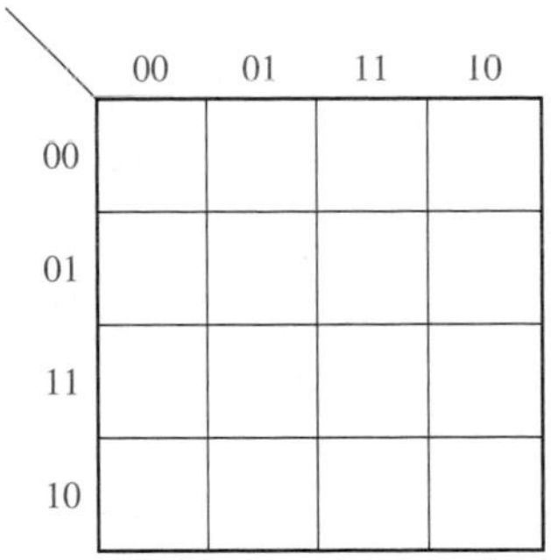

| | 00 | 01 | 11 | 10 |
|---|---|---|---|---|
| 00 | | | | |
| 01 | | | | |
| 11 | | | | |
| 10 | | | | |

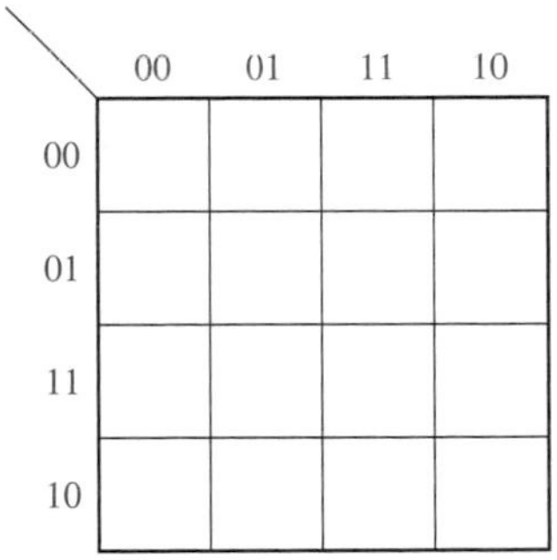

| | 00 | 01 | 11 | 10 |
|---|---|---|---|---|
| 00 | | | | |
| 01 | | | | |
| 11 | | | | |
| 10 | | | | |

   (b) Plot the map for the output function ($Z$) from the transition table of Table 15-6(b) and derive the minimum equation for $Z$.

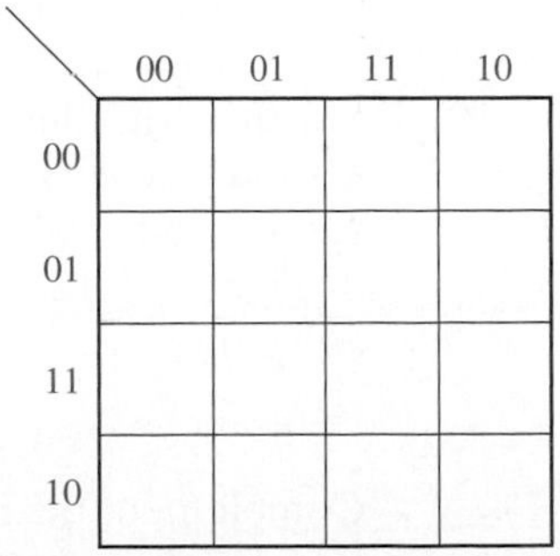

| | 00 | 01 | 11 | 10 |
|---|---|---|---|---|
| 00 | | | | |
| 01 | | | | |
| 11 | | | | |
| 10 | | | | |

(c) Derive the $J$-$K$ input equations for flip-flop $A$ from the next-state map of Figure 15-10. Your answers should be

$J_A = X_2B, \qquad K_A = X_2'B'$

(d) Work Problem 15.4.

**8.** Study Section 15.7, *Equivalent State Assignments.*

(a) Fill in the missing assignments (numbered 8 through 18) in Table 15-8. First, list the remaining assignments with 01 in the first row and then the assignments with 10 in the first row.

(b) Why is it unnecessary to try all possible state assignments to be assured of finding a minimum cost circuit?

(c) For symmetrical flip-flops, why is it always possible to assign all 0's to the starting state and still obtain a minimum circuit?

(d) Complete the following transition table for Table 15-9 using assignment $A$. Then, complete the next-state maps and derive $D_1$ and $D_2$.

| $Q_1Q_2$ | $X = 0$ | 1 |
|---|---|---|
| 00 | 00 | 10 |
| 01 | | |
| 10 | | |

$Q_1^+ = D_1$

| $Q_1Q_2$ \ $X$ | 0 | 1 |
|---|---|---|
| 00 | | |
| 01 | | |
| 11 | X | X |
| 10 | | |

$Q_2^+ = D_2$

| $Q_1Q_2$ \ $X$ | 0 | 1 |
|---|---|---|
| 00 | | |
| 01 | | |
| 11 | X | X |
| 10 | | |

Starting with the equations for assignment $A$, replace all of the 1's with 2's and all 2's with 1's. Verify that the resulting equations are the same as those for assignment $B$.

Starting with the $J$ and $K$ equations for assignment $A$, replace each $Q$ with $Q'$ and vice versa. Then, replace the equations for $J$ with the corresponding $K$ equations and vice versa. (This corresponds to the transformation given in Figure 15-12.) Verify that the resulting equations are the same as for assignment $C$.

Complement the right-hand side of the $D$ equations for assignment $A$ and, then, replace each $Q$ with $Q'$ and vice versa. (This corresponds to the transformation given in Figure 15-13.) Verify that the resulting equations are the same as for assignment *C*.

(e) Show that each of the assignments in Table 15-8 is equivalent to one of the assignments in Table 15-10.

(f) Why are the following two state assignments equivalent in cost?

| | | |
|---|---|---|
| *A* | 000 | 011 |
| *B* | 001 | 111 |
| *C* | 011 | 101 |
| *D* | 101 | 110 |
| *E* | 100 | 010 |
| *F* | 010 | 001 |
| *G* | 110 | 000 |

(g) Show that each of the following assignments can be generated from Table 15-10 by permuting and/or complementing columns:

| | | |
|---|---|---|
| 10 | 11 | 01 |
| 01 | 01 | 11 |
| 00 | 00 | 00 |
| 11 | 10 | 10 |

(h) Why is the trial-and-error method of state assignment of limited usefulness?

(i) Read Problem 15.5, and then answer the following questions regarding state assignments before you work the problem:
(1) Why should a column *not* be assigned all 0's or all 1's?

(2) Why should two columns *not* be given the same assignment?

(3) Does interchanging two columns affect the cost of realizing the circuit?

(4) Does interchanging two rows affect the cost?

(5) Why is an assignment which has two identical rows invalid?

(6) Consider the following two assignments (the number at the top of each column is the decimal equivalent of the binary number in the column):

| (1) | (3) | (5) | | (3) | (1) | (5) |
|---|---|---|---|---|---|---|
| 0 | 0 | 0 | | 0 | 0 | 0 |
| 0 | 0 | 1 | | 0 | 0 | 1 |
| 0 | 1 | 0 | | 1 | 0 | 0 |
| 1 | 1 | 1 | | 1 | 1 | 1 |

If we try the column assignment (1) (3) (5), why is it unnecessary to try (3) (1) (5)?

Why is it desirable to assign the column values in increasing numerical order?

**9.** Study Section 15.8, *Guidelines for State Assignment.*

(a) Why do the guidelines for making state assignments help in making an economical assignment?

(b) What should be done if all the adjacencies specified by the guidelines cannot be satisfied?

(c) The state assignment guidelines for Figure 15-14(a) indicate that the following sets of states should be given adjacent assignments:

(1) $(S_0, S_1, S_3, S_5)$ $(S_3, S_5)$ $(S_4, S_6)$ $(S_0, S_2, S_4, S_6)$
(2) $(S_1, S_2)$ $(S_2, S_3)$ $(S_1, S_4)$ $(S_2, S_5)2X$ $(S_1, S_6)2X$

Because the adjacencies from guideline 1 are generally most important, we will start by placing one of the largest groups from guideline 1 in four adjacent squares:

| | |
|---|---|
| $S_0$ | |
| $S_1$ | |
| $S_3$ | |
| $S_5$ | |

Note that $(S_3, S_5)$ is also satisfied by this grouping. Place $S_2$, $S_4$, and $S_6$ in the remaining squares to satisfy as many of the remaining guidelines as possible. Keeping in mind that groups labeled $2X$ should be given preference over groups which are not repeated. Compare your answer with Figure 15-14(b).

(d) Complete the transition table for the state table of Figure 15-16(a), using the assignment of Figure 15-16(b).

| | $Q_1$ | $Q_2$ | $Q_3$ | $Q_1^+Q_2^+Q_3^+$ $X = 0$ | 1 | 0 | 1 |
|---|---|---|---|---|---|---|---|
| a | 0 | 0 | 0 | 000 | 100 | 0 | 0 |
| b | 1 | 1 | 1 | 011 | 110 | 0 | 1 |
| c | 1 | 0 | 0 | 100 | 000 | 0 | 0 |

(e) Complete the next-state and output maps, and verify that the cost of realizing the corresponding equations with an AND-OR gate circuit is 13 gates and 35 gate inputs.

(f) Find $J_1$ and $K_1$ from the $Q_1^+$ map.

$J_1$ = ____________________

$K_1$ = ____________________

| $Q_2Q_3$ \ $XQ_1$ | 00 | 01 | 11 | 10 |
|---|---|---|---|---|
| 00 | 0 | 1 | 0 | 1 |
| 01 | X | | | X |
| 11 | | 0 | 1 | |
| 10 | | | | |

$Q_1^+$

| $Q_2Q_3$ \ $XQ_1$ | 00 | 01 | 11 | 10 |
|---|---|---|---|---|
| 00 | 0 | 0 | 0 | 0 |
| 01 | X | | | X |
| 11 | | 1 | 1 | |
| 10 | | | | |

$Q_2^+$

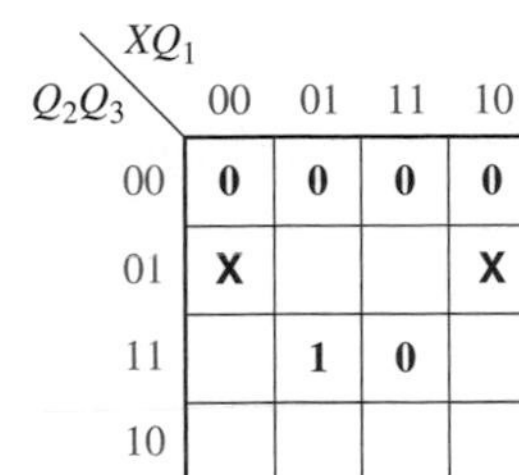

| $Q_2Q_3$ \ $XQ_1$ | 00 | 01 | 11 | 10 |
|---|---|---|---|---|
| 00 | 0 | 0 | 0 | 0 |
| 01 | X | | | X |
| 11 | | 1 | 0 | |
| 10 | | | | |

$Q_3^+$

| $Q_2Q_3$ \ $XQ_1$ | 00 | 01 | 11 | 10 |
|---|---|---|---|---|
| 00 | 0 | 0 | 0 | 0 |
| 01 | X | | | X |
| 11 | | 0 | 1 | |
| 10 | | | | |

$Z$

$D_1 =$ ______________

$D_2 =$ ______________

$D_3 =$ ______________

$Z =$ ______________

**10.** Work Problems 15.6, 15.7, and 15.8.

**11.** Study Section 15.9, *Using a One-Hot State Assignment.*

(a) A one-hot state assignment does not usually give a solution that uses less hardware because of the extra flip-flops required. In what situation is it often advantageous to use it anyway?

(b) It is easy to derive flip-flop input equations directly from a state graph for a one-hot state assignment by inspecting the arcs leading into a given state. Give the next-state equation for $Q_5$. Only the parts of the state graph which are needed to find $Q_5^+$ are given.

$S_2$ —— $X/0$, $X'Y/P$ ——→ $S_5$ (self-loop $X'/0$)

(c) For the state graph in Figure 15-19 and the one-hot state assignment shown, determine the next-state equations for $Q_2$ and $Q_3$.

$Q_2^+ =$ ______________ and $Q_3^+ =$ ______________

(d) For the state graph in Figure 15-19 and the one-hot state assignment shown, determine the output equations for $Ad$ and Done.

$Ad =$ ______________ and Done $=$ ______________

(e) Work Problem 15.9.

**12.** When you are satisfied that you can meet all of the objectives, take the readiness test.

# Reduction of State Tables State Assignment

Given a description of the desired input-output behavior of a sequential circuit, the first step in designing the circuit is to derive a state table using methods similar to the ones discussed in the previous unit. Before we realize this state table using flip-flops and logic gates, reduction of the state table to a minimum number of states is desirable. In general, reducing the number of states in a table will reduce the amount of logic required, and the number of flip-flops may also be reduced. For example, if a table with nine states is reduced to eight states, the number of flip-flops required is reduced from four to three, with a possible corresponding reduction in the amount of input logic for the flip-flops. If the table is further reduced to six states, three flip-flops are still required, but the presence of more don't-cares in the flip-flop input equations will probably further reduce the required logic.

Given the reduced state table, the next step in synthesizing the circuit is to assign binary flip-flop states to correspond to the circuit states. The way in which this assignment is made will determine the amount of logic required for the circuit. The problem of finding a good state assignment which leads to an economical circuit is a difficult one, but some guidelines for achieving this are discussed in Sections 15.7–15.8.

The next step in designing the sequential circuit is to derive the flip-flop input equations. We have already done this for counters in Unit 12, and we will show how to apply these techniques to more general sequential circuits.

## 15.1 Elimination of Redundant States

In Unit 14, we were careful to avoid introducing unnecessary states when setting up a state graph or table. We will now approach the problem of deriving the state graph somewhat differently. Initially, when first setting up the state table, we will not be overly concerned with inclusion of extra states, but when the table is complete, we will eliminate any redundant states. In previous units, we have used the notation $S_0, S_1, S_2, \ldots$ to represent states in a sequential circuit. In this unit, we will frequently use $A$, $B$, $C$, ... (or $a$, $b$, $c$, ...) to represent these states.

We will rework Example 1 in Section 14.3. Initially, we will set up enough states to remember the first three bits of every possible input sequence. Then, when the fourth bit comes in, we can determine the correct output and reset the circuit to the

**TABLE 15-1**
State Table for Sequence Detector

| Input Sequence | Present State | Next State X = 0 | Next State X = 1 | Present Output X = 0 | Present Output X = 1 |
|---|---|---|---|---|---|
| reset | *A* | *B* | *C* | 0 | 0 |
| 0 | *B* | *D* | *E* | 0 | 0 |
| 1 | *C* | *F* | *G* | 0 | 0 |
| 00 | *D* | *H* | *I* | 0 | 0 |
| 01 | *E* | *J* | *K* | 0 | 0 |
| 10 | *F* | *L* | *M* | 0 | 0 |
| 11 | *G* | *N* | *P* | 0 | 0 |
| 000 | *H* | *A* | *A* | 0 | 0 |
| 001 | *I* | *A* | *A* | 0 | 0 |
| 010 | *J* | *A* | *A* | 0 | 1 |
| 011 | *K* | *A* | *A* | 0 | 0 |
| 100 | *L* | *A* | *A* | 0 | 1 |
| 101 | *M* | *A* | *A* | 0 | 0 |
| 110 | *N* | *A* | *A* | 0 | 0 |
| 111 | *P* | *A* | *A* | 0 | 0 |

initial state. As indicated in Table 15-1, we will designate state $A$ as the reset state. If we receive a 0, we go to state $B$; if we receive a 1, we go to state $C$. Similarly, starting in state $B$, a 0 takes us to state $D$ to indicate that the sequence 00 has been received, and a 1 takes us to state $E$ to indicate that 01 has been received. The remaining states are defined in a similar manner. When the fourth input bit is received, we return to the reset state. The output is 0 unless we are in state $J$ or $L$ and receive a 1, which corresponds to having received 0101 or 1001.

Next, we will attempt to eliminate redundant states from the table. The input sequence information was only used in setting up the table and will now be disregarded. Looking at the table, we see that there is no way of telling states $H$ and $I$ apart. That is, if we start in state $H$, the next state is $A$ and the output is 0; similarly, if we start in state $I$, the next state is $A$ and the output is 0. Hence, there is no way of telling states $H$ and $I$ apart, and we can replace $I$ with $H$ where it appears in the next-state portion of the table. Having done this, there is no way to reach state $I$, so row $I$ can be removed from the table. We say that $H$ is *equivalent* to $I$ ($H \equiv I$). Similarly, rows $K$, $M$, $N$, and $P$ have the same next state and output as $H$, so $K$, $M$, $N$, and $P$ can be replaced by $H$, and these rows can be deleted. Also, the next states and outputs are the same for rows $J$ and $L$, so $J \equiv L$. Thus, $L$ can be replaced with $J$ and eliminated from the table. The result is shown in Table 15-2.

Having made these changes in the table, rows $D$ and $G$ are identical and so are rows $E$ and $F$. Therefore, $D \equiv G$, and $E \equiv F$, so states $F$ and $G$ can be eliminated. Figure 15-1 shows a state diagram for the final reduced table. Note that this is identical to the state graph of Figure 14-14, except for the designations for the states. The procedure used to find equivalent states in this example is known as *row matching*. In general, row matching is *not* sufficient to find all equivalent states, except in the special case where the circuit resets to the starting state after receiving a fixed number of inputs.

**TABLE 15-2** State Table for Sequence Detector

| Present State | Next State $X = 0$ | Next State $X = 1$ | Present Output $X = 0$ | Present Output $X = 1$ |
|---|---|---|---|---|
| $A$ | $B$ | $C$ | 0 | 0 |
| $B$ | $D$ | $E$ | 0 | 0 |
| $C$ | ~~$F$~~ $E$ | ~~$G$~~ $D$ | 0 | 0 |
| $D$ | $H$ | ~~$I$~~ $H$ | 0 | 0 |
| $E$ | $J$ | ~~$K$~~ $H$ | 0 | 0 |
| ~~$F$~~ | ~~$L$ $J$~~ | ~~$M$ $H$~~ | ~~0~~ | ~~0~~ |
| ~~$G$~~ | ~~$N$ $H$~~ | ~~$P$ $H$~~ | ~~0~~ | ~~0~~ |
| $H$ | $A$ | $A$ | 0 | 0 |
| ~~$I$~~ | ~~$A$~~ | ~~$A$~~ | ~~0~~ | ~~0~~ |
| $J$ | $A$ | $A$ | 0 | 1 |
| ~~$K$~~ | ~~$A$~~ | ~~$A$~~ | ~~0~~ | ~~0~~ |
| ~~$L$~~ | ~~$A$~~ | ~~$A$~~ | ~~0~~ | ~~1~~ |
| ~~$M$~~ | ~~$A$~~ | ~~$A$~~ | ~~0~~ | ~~0~~ |
| ~~$N$~~ | ~~$A$~~ | ~~$A$~~ | ~~0~~ | ~~0~~ |
| ~~$P$~~ | ~~$A$~~ | ~~$A$~~ | ~~0~~ | ~~0~~ |

**FIGURE 15-1** Reduced State Table and Graph for Sequence Detector

| Present State | Next State $X = 0$ | Next State $X = 1$ | Output $X = 0$ | Output $X = 1$ |
|---|---|---|---|---|
| $A$ | $B$ | $C$ | 0 | 0 |
| $B$ | $D$ | $E$ | 0 | 0 |
| $C$ | $E$ | $D$ | 0 | 0 |
| $D$ | $H$ | $H$ | 0 | 0 |
| $E$ | $J$ | $H$ | 0 | 0 |
| $H$ | $A$ | $A$ | 0 | 0 |
| $J$ | $A$ | $A$ | 0 | 1 |

(a)

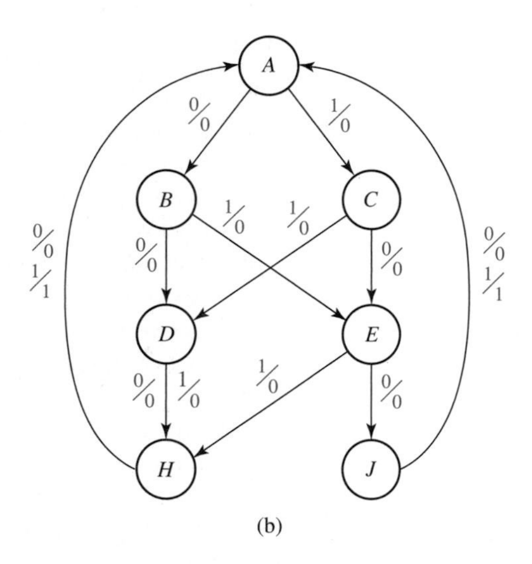


(b)

## 15.2 Equivalent States

As we have seen in the previous example, state tables can be reduced by eliminating equivalent states. A state table with fewer rows often requires fewer flip-flops and logic gates to realize; therefore, the determination of equivalent states is important in order to obtain economical realizations of sequential circuits.

FIGURE 15-2

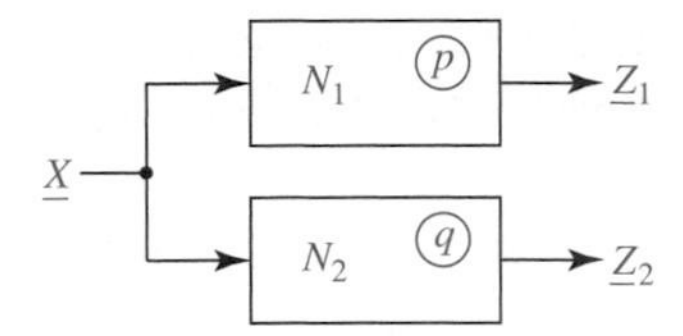


Let us now consider the general problem of state equivalence. Basically, two states are equivalent if there is no way of telling them apart through observation of the circuit inputs and outputs. Consider two sequential circuits (these may be different circuits or two copies of the same circuit), one which is started in state $p$ and one which is started in state $q$ (Figure 15-2): Let $\underline{X}$ represent a sequence of inputs $X_1, X_2, \ldots, X_n$. Feed the same input sequence $\underline{X}$ into both circuits and observe the output sequences $\underline{Z}_1$ and $\underline{Z}_2$. If these output sequences are the same, so far so good. Then, reset the circuits to the states $p$ and $q$ and try a different input sequence for $\underline{X}$ and again compare output sequences. If, for every possible input sequence $\underline{X}$, these output sequences are the same, then there is no way of telling states $p$ and $q$ apart by observing the terminal behavior of the circuits, and we say $p$ is equivalent to $q$ ($p \equiv q$). On the other hand, if, for some input sequence $\underline{X}$, the output sequences $\underline{Z}_1$ and $\underline{Z}_2$ are different, then we can distinguish between states $p$ and $q$, and they are not equivalent. Because the output sequence is a function of the initial state and the input sequence, we will write

$$\underline{Z}_1 = \lambda_1 (p, \underline{X}) \qquad \underline{Z}_2 = \lambda_2 (q, \underline{X})$$

We can then state formally the definition of state equivalence as follows:

**Definition 15.1** Let $N_1$ and $N_2$ be sequential circuits (not necessarily different). Let $\underline{X}$ represent a sequence of inputs of arbitrary length. Then state $p$ in $N_1$ is equivalent to state $q$ in $N_2$ iff $\lambda_1(p, \underline{X}) = \lambda_2(q, \underline{X})$ for every possible input sequence $\underline{X}$.

To apply Definition 15.1 directly, we should first test the circuits with $\underline{X} = 0$ and $\underline{X} = 1$. Then, we should test with all input sequences of length 2: $\underline{X} = 00, 01, 10,$ and 11. Next, we should test with all input sequences of length 3: $\underline{X} = 000, 001, 010, 011, 100,$ 101, 110, and 111. We should then continue this process with all input sequences of length 4, length 5, and so forth. Definition 15.1 is not practical to apply directly in practice because it requires testing the circuit with an infinite number of input sequences in order to prove that two states are equivalent. A more practical way of testing for state equivalence uses the following theorem:

**Theorem15.1**[1] Two states $p$ and $q$ of a sequential circuit are equivalent iff for every single input $X$, the outputs are the same and the next states are equivalent, that is,

$$\lambda(p, X) = \lambda(q, X) \qquad \text{and} \qquad \delta(p, X) \equiv \delta(q, X)$$

where $\lambda(p, X)$ is the output given the present state $p$ and input $X$, and $\delta(p, X)$ is the next state given the present state $p$ and input $X$. Note that the next states do not have to be equal, just equivalent. For example, in Table 15-1, $D \equiv G$, but the next states ($H$ and $N$ for $X = 0$, and $I$ and $P$ for $X = 1$) are not equal.

The row matching procedure previously discussed is a special case of Theorem 15.1 where the next states are actually the same instead of just being equivalent. We will

[1]See Appendix D for proof.

use this theorem to show that Table 13-4 has no equivalent states. By inspection of the output part of the table, the only possible pair of equivalent states is $S_0$ and $S_2$. From the table,

$$S_0 \equiv S_2 \quad \text{iff} \quad (S_3 \equiv S_3, S_2 \equiv S_0, S_1 \equiv S_1, \text{and } S_0 \equiv S_1)$$

But $S_0 \not\equiv S_1$ (because the outputs differ), so the last condition is not satisfied and $S_0 \not\equiv S_2$.

## 15.3 Determination of State Equivalence Using an Implication Table

In this section we will discuss a procedure for finding all of the equivalent states in a state table. If the equivalent states found by this procedure are eliminated, then the table can be reduced to a minimum number of states. We will use an implication table (sometimes referred to as a pair chart) to check each pair of states for possible equivalence. The nonequivalent pairs are systematically eliminated until only the equivalent pairs remain.

We will use the example of Table 15-3 to illustrate the implication table method. The first step is to construct a chart of the form shown in Figure 15-3. This chart has a square for every possible pair of states. A square in column *i* and row *j* corresponds to state pair *i-j*. Thus, the squares in the first column correspond to state pairs *a-b*, *a-c*, etc. Note that the squares above the diagonal are not included in the chart because if $i \equiv j$, $j \equiv i$, and only one of the state pairs *i-j* and *j-i* is needed. Also, squares corresponding to pairs *a-a*, *b-b*, etc., are omitted. To fill in the first column of the chart, we compare row *a* of Table 15-3 with each of the other rows. Because the output for row *a* is different than the output for row *c*, we place an X in the *a-c* square of the chart to indicate that $a \not\equiv c$. Similarly, we place X's in squares *a-e*, *a-f*, and *a-h* to indicate that $a \not\equiv e$, $a \not\equiv f$, and $a \not\equiv h$ because of output differences. States *a* and *b* have the same outputs, and thus, by Theorem 15.1,

$$a \equiv b \quad \text{iff} \quad d \equiv f \quad \text{and} \quad c \equiv h$$

To indicate this, we place the *implied pairs*, *d-f* and *c-h*, in the *a-b* square. Similarly, because *a* and *d* have the same outputs, we place *a-d* and *c-e* in the *a-d* square to indicate that

$$a \equiv d \quad \text{iff} \quad a \equiv d \quad \text{and} \quad c \equiv e$$

**TABLE 15-3**

| Present State | Next State X = 0 | 1 | Present Output |
|---|---|---|---|
| a | d | c | 0 |
| b | f | h | 0 |
| c | e | d | 1 |
| d | a | e | 0 |
| e | c | a | 1 |
| f | f | b | 1 |
| g | b | h | 0 |
| h | c | g | 1 |

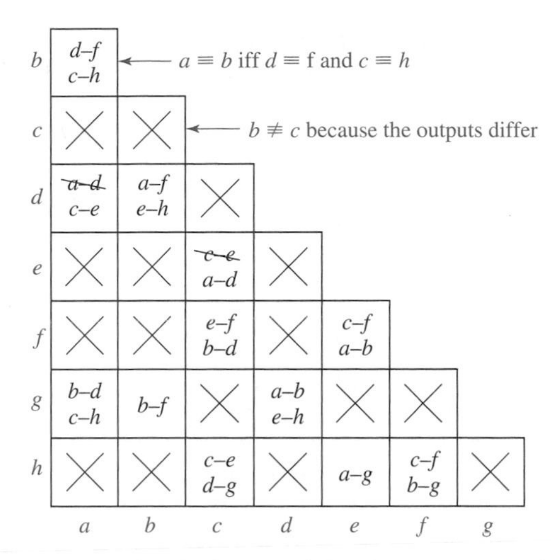


**FIGURE 15-3** Implication Chart for Table 15-3

The entries *b-d* and *c-h* in the *a-g* square indicate that

$$a \equiv g \qquad \text{iff} \qquad b \equiv d \qquad \text{and} \qquad c \equiv h$$

Next, row *b* of the state table is compared with each of the remaining rows of the table, and column *b* of the implication chart is filled in. Similarly, the remaining columns in the chart are filled in to complete Figure 15-3. Self-implied pairs are redundant, so *a-d* can be eliminated from square *a-d*, and *c-e* from square *c-e*.

At this point, each square in the implication table has either been filled in with an X to indicate that the corresponding state pair is not equivalent (because the outputs are different) or filled in with implied pairs. We now check each implied pair. If one of the implied pairs in square *i-j* is not equivalent, then by Theorem 15.1, $i \not\equiv j$. The *a-b* square of Figure 15-3 contains two implied pairs (*d-f* and c-*h*). Because $d \not\equiv f$ (the *d-f* square has an X in it), $a \not\equiv b$ and we place an X in the *a-b* square, as shown in Figure 15-4. Continuing to check the first column, we note that the *a-d* square contains the implied pair *c-e*. Because square *c-e* does not contain an X, we cannot determine at this point whether or not $a \equiv d$. Similarly, because neither square *b-d*

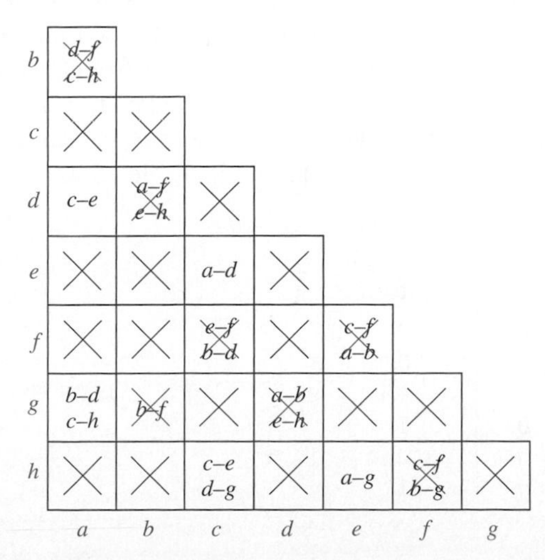


**FIGURE 15-4** Implication Chart After First Pass

**FIGURE 15-5** Implication Chart After Second Pass

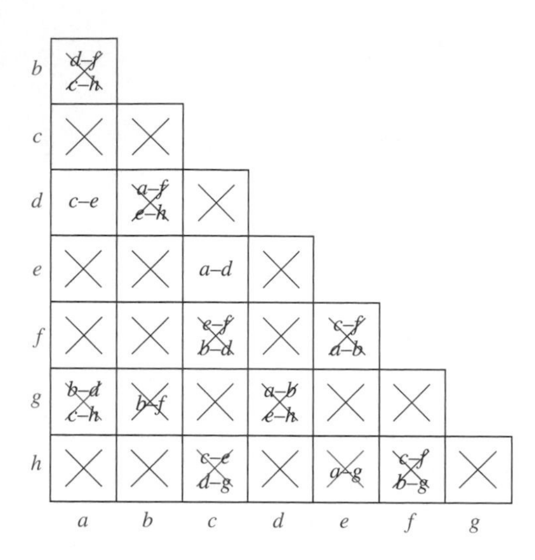


nor $c$-$h$ contains an X, we cannot determine immediately whether $a \equiv g$ or not. Going on to the second column, we place X's in squares $b$-$d$ and $b$-$g$ because we have already shown $a \not\equiv f$ and $b \not\equiv f$. In a similar manner, we check each of the remaining columns and X out squares $c$-$f$, $d$-$g$, $e$-$f$, and $f$-$h$. Figure 15-4 shows the resulting chart.

In going from Figure 15-3 to Figure 15-4, we found several additional nonequivalent state pairs. Therefore, we must go through the chart again to see if the added X's make any other pairs nonequivalent. Rechecking column $a$, we find that we can place an X in square $a$-$g$ because square $b$-$d$ has an X. Checking the remaining columns, we X out squares $c$-$h$ and $e$-$h$ because $d$-$g$ and $a$-$g$ have X's. This completes the second pass through the implication table, as shown in Figure 15-5. Because we added some X's on the second pass, a third pass is required.

No new X's are added on the third pass through the table, so all squares which correspond to nonequivalent state pairs have been Xed out. The coordinates of the remaining squares must then correspond to equivalent state pairs. Because square $a$-$d$ (in column $a$, row $d$) does not contain an X, we conclude that $a \equiv d$. Similarly, square $c$-$e$ does not contain an X, so $c \equiv e$. All other squares contain X's, so there are no other equivalent state pairs. Note that we determined equivalent states from the column-row coordinates of the squares without X's, *not* by reading the implied pairs contained within the squares.

If we replace $d$ with $a$ and $e$ with $c$ in Table 15-3, we can eliminate rows $d$ and $e$, and the table reduces to six rows, as shown in Table 15-4.

**TABLE 15-4**

| Present State | Next State X = 0 | 1 | Output |
|---|---|---|---|
| a | a | c | 0 |
| b | f | h | 0 |
| c | c | a | 1 |
| f | f | b | 1 |
| g | b | h | 0 |
| h | c | g | 1 |

The implication table method of determining state equivalence can be summarized as follows:

1. Construct a chart which contains a square for each pair of states.
2. Compare each pair of rows in the state table. If the outputs associated with states $i$ and $j$ are different, place an X in square $i$-$j$ to indicate that $i \not\equiv j$. If the outputs are the same, place the implied pairs in square $i$-$j$. (If the next states of $i$ and $j$ are $m$ and $n$ for some input $x$, then $m$-$n$ is an implied pair.) If the outputs and next states are the same (or if $i$-$j$ only implies itself), place a check (√) in square $i$-$j$ to indicate that $i \equiv j$.
3. Go through the table square-by-square. If square $i$-$j$ contains the implied pair $m$-$n$, and square $m$-$n$ contains an X, then $i \not\equiv j$, and an X should be placed in square $i$-$j$.
4. If any X's were added in step 3, repeat step 3 until no more X's are added.
5. For each square $i$-$j$ which does not contain an X, $i \equiv j$.

If desired, row matching can be used to partially reduce the state table before constructing the implication table. Although we have illustrated this procedure for a Moore table, the same procedure applies to a Mealy table.

## 15.4 Equivalent Sequential Circuits

In the last section, we found the equivalent states within a single state table so that we could reduce the number of rows in the table. Reducing the number of rows usually leads to a sequential circuit with fewer gates and flip-flops. In this section, we will consider equivalence between sequential circuits. Essentially, two sequential circuits are equivalent if they are capable of doing the same "work." Equivalence between sequential circuits is defined as follows:

**Definition 15.2** Sequential circuit $N_1$ is equivalent to sequential circuit $N_2$ if for each state $p$ in $N_1$, there is a state $q$ in $N_2$ such that $p \equiv q$, and conversely, for each state $s$ in $N_2$, there is a state $t$ in $N_1$ such that $s \equiv t$.

Thus, if $N_1 \equiv N_2$, for every starting state $p$ in $N_1$, we can find a corresponding starting state $q$ such that $\lambda_1(p, \underline{X}) \equiv \lambda_2(q, \underline{X})$ for all input sequences $\underline{X}$ (i.e., the output sequences are the same for the same input sequence). Then, in a given application, $N_1$ could be replaced with its equivalent circuit $N_2$.

If $N_1$ and $N_2$ have only a few states, one way to show that $N_1 \equiv N_2$ is to match up pairs of equivalent states by inspection and, then, show that Theorem 15.1 is satisfied for each pair of equivalent states. If both $N_1$ and $N_2$ have a minimum number of states and $N_1 \equiv N_2$, then $N_1$ and $N_2$ must have the same number of states. Otherwise, one circuit would have a state left over which was not equivalent to any state in the other circuit, and Definition 15.2 would not be satisfied.

Figure 15-6 shows two reduced state tables and their corresponding state graphs. By inspecting the state graphs, it appears that if the circuits are equivalent, we must have $A$ equivalent to either $S_2$ or $S_3$ because these are the only states in $N_2$ with self-loops.

**FIGURE 15-6** Tables and Graphs for Equivalent Circuits

| | $N_1$ | | | |
|---|---|---|---|---|
| | $X = 0$ | $1$ | $X = 0$ | $1$ |
| $A$ | $B$ | $A$ | 0 | 0 |
| $B$ | $C$ | $D$ | 0 | 1 |
| $C$ | $A$ | $C$ | 0 | 1 |
| $D$ | $C$ | $B$ | 0 | 0 |

| | $N_2$ | | | |
|---|---|---|---|---|
| | $X = 0$ | $1$ | $X = 0$ | $1$ |
| $S_0$ | $S_3$ | $S_1$ | 0 | 1 |
| $S_1$ | $S_3$ | $S_0$ | 0 | 0 |
| $S_2$ | $S_0$ | $S_2$ | 0 | 0 |
| $S_3$ | $S_2$ | $S_3$ | 0 | 1 |

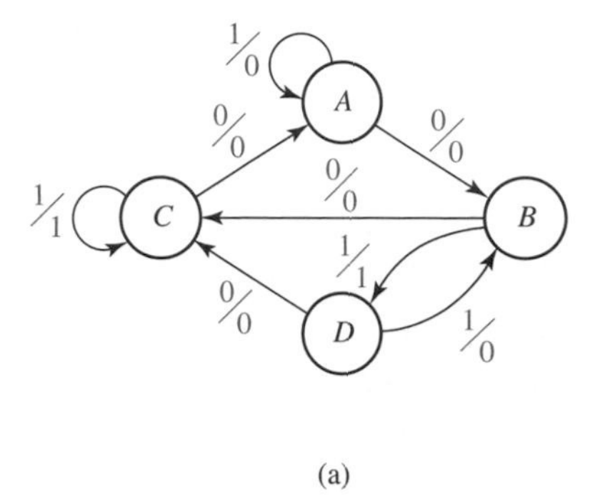

(a)

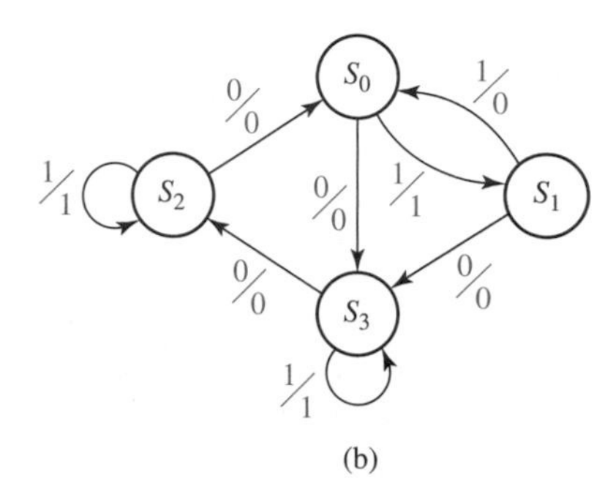

(b)

Because the outputs of $A$ and $S_2$ correspond, the only possibility is $A \equiv S_2$. If we assume that $A \equiv S_2$, this implies that we must have $B \equiv S_0$ which in turn implies that we must have $D \equiv S_1$ and $C \equiv S_3$. Using the state tables, we can verify that these assumptions are correct because for every pair of assumed equivalent states, the next states are equivalent and the outputs are equal when $X \equiv 0$ and also when $X \equiv 1$. This verifies that $N_1 \equiv N_2$.

The implication table can easily be adapted for determining the equivalence of sequential circuits. Because the states of one circuit must be checked for equivalence against states of the other circuit, an implication chart is constructed with rows corresponding to states of one circuit and columns corresponding to states of the other. For example, for the circuits of Figure 15-6 we can set up the implication table of Figure 15-7(a). The first column of Figure 15-7(a) is filled in by comparing row $A$ of the state table in Figure 15-6(a) with each of the rows in Figure 15-6(b). Because states $A$ and $S_0$ have different outputs, an X is placed in the $A$-$S_0$ square. Because states $A$ and $S_1$ have the same outputs, the implied next-state pairs ($B$-$S_3$ and $A$-$S_0$) are placed in the $A$-$S_1$ square, etc. The remainder of the table is filled in similarly.

In the next step [Figure 15-7(b)], squares corresponding to additional nonequivalent state pairs are crossed out. Thus, square $A$-$S_1$ is crossed out because $A \not\equiv S_0$. Similarly, square $B$-$S_3$ is crossed out because $C \not\equiv S_2$, square $C$-$S_0$ because $A \not\equiv S_3$, and

**FIGURE 15-7** Implication Tables for Determining Circuit Equivalence

| | $A$ | $B$ | $C$ | $D$ |
|---|---|---|---|---|
| $S_0$ | X | $C$–$S_3$ $D$–$S_1$ | $A$–$S_3$ $C$–$S_1$ | X |
| $S_1$ | $B$–$S_3$ $A$–$S_0$ | X | X | $C$–$S_3$ $B$–$S_0$ |
| $S_2$ | $B$–$S_0$ $A$–$S_2$ | X | X | $C$–$S_0$ $B$–$S_2$ |
| $S_3$ | X | $C$–$S_2$ $D$–$S_3$ | $A$–$S_2$ $C$–$S_3$ | X |

(a)

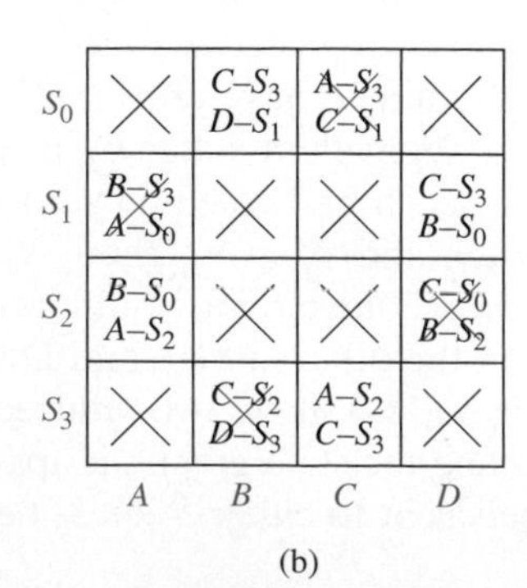

| | $A$ | $B$ | $C$ | $D$ |
|---|---|---|---|---|
| $S_0$ | X | $C$–$S_3$ $D$–$S_1$ | ~~$A$–$S_3$ $C$–$S_1$~~ | X |
| $S_1$ | ~~$B$–$S_3$ $A$–$S_0$~~ | X | X | $C$–$S_3$ $B$–$S_0$ |
| $S_2$ | $B$–$S_0$ $A$–$S_2$ | X | X | ~~$C$–$S_0$ $B$–$S_2$~~ |
| $S_3$ | X | ~~$C$–$S_2$ $D$–$S_3$~~ | $A$–$S_2$ $C$–$S_3$ | X |

(b)

square $D$-$S_2$ because $B \not\equiv S_2$. Another pass through the table reveals no additional nonequivalent pairs; therefore, the remaining equivalent state pairs are

$$A \equiv S_2 \qquad B \equiv S_0 \qquad C \equiv S_3 \qquad D \equiv S_1$$

Because each state in $N_1$ has an equivalent state in $N_2$ and conversely, $N_1 \equiv N_2$.

# 15.5 Incompletely Specified State Tables

When a sequential circuit is used as part of a larger digital system, it frequently happens that certain sequences will never occur as inputs to the sequential circuit. In other cases, the output of the sequential circuit is only observed at certain times rather than at every clock time. Such restrictions lead to unspecified next states or outputs in the state table. When such don't-cares are present, we say that the state table is incompletely specified. Just as don't-cares in a truth table can be used to simplify the resulting combinational circuit, don't-cares in a state table can be used to simplify the sequential circuit.

The following example illustrates how don't-cares arise in a state table. Assume that circuit $A$ (Figure 15-8) can only generate two possible output sequences, $X = 100$ and $X = 110$. Thus, the sequential circuit subsystem ($B$) has only two possible input sequences. When the third input in the sequence is received, the output of $B$ is to be $Z = 0$ if 100 was received and $Z = 1$ if 110 was received. Assume that circuit $C$ ignores the value of $Z$ at other times so that we do not care what $Z$ is during the first two inputs in the $X$ sequence. The possible input-output sequences for circuit $B$ are listed in the following table, where $t_0$, $t_1$, and $t_2$ represent three successive clock times:

| | $t_0$ | $t_1$ | $t_2$ | | $t_0$ | $t_1$ | $t_2$ |
|---|---|---|---|---|---|---|---|
| $X =$ | 1 | 0 | 0 | $Z =$ | – | – | 0 |
| | 1 | 1 | 0 | | – | – | 1 |

(– is a don't-care output)

State Table 15-5(a) will produce the required outputs. Note that the next-state entry for $S_0$ with $X = 0$ is a don't-care because 0 can never occur as the first input in the sequence. Similarly, the next-state entries for $S_2$ and $S_3$ with $X = 1$ are don't-cares because $X = 1$ cannot occur as the third input in the sequence. If we fill in the don't-cares in the state table, as indicated in Table 15-5(b), we can use row matching to reduce the table to two states, as shown in Table 15-5(c).

FIGURE 15-8

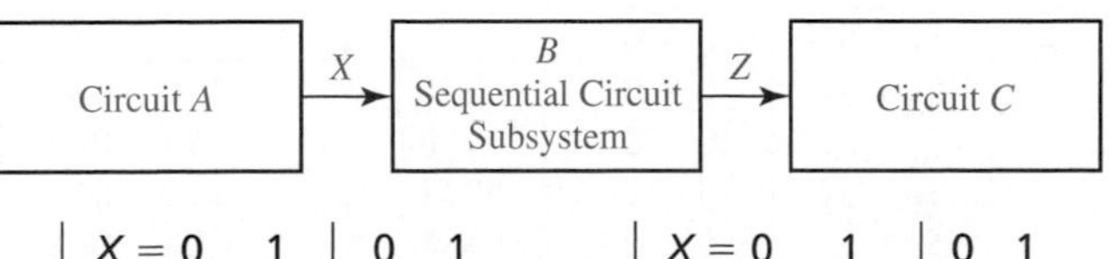


TABLE 15-5 Incompletely Specified State Table

| | $X = 0$ | 1 | 0 | 1 |
|---|---|---|---|---|
| $S_0$ | - | $S_1$ | - | - |
| $S_1$ | $S_2$ | $S_3$ | - | - |
| $S_2$ | $S_0$ | - | 0 | - |
| $S_3$ | $S_0$ | - | 1 | - |

(a)

| | $X = 0$ | 1 | 0 | 1 |
|---|---|---|---|---|
| $S_0$ | $(S_0)$ | $S_1$ | (0) | - |
| $S_1$ | ~~$S_2$~~ $S_0$ | $S_3$ | (1) | - |
| $S_2$ | $S_0$ | $(S_1)$ | 0 | - |
| $S_3$ | $S_0$ | $(S_3)$ | 1 | - |

(b)

$S_0 \equiv S_2$, $S_1 \equiv S_3$

| | $X = 0$ | 1 | 0 | 1 |
|---|---|---|---|---|
| $S_0$ | $S_0$ | $S_1$ | 0 | - |
| $S_1$ | $S_0$ | $S_1$ | 1 | - |

(c)

As illustrated in Table 15-5, one method of reducing incompletely specified state tables is to fill in the don't-cares in an appropriate manner and, then, reduce the table, using one of the methods which apply to completely specified state tables. This procedure may be applied to small tables or to tables with only a few don't-cares, but, in general, it does not lead to a minimum-row reduced table. Determining the best way to fill in the don't-cares may require considerable trial and error, and even if the best way of filling in the don't-cares is found, the resulting table cannot always be reduced to a minimum-row table. General procedures are known which will reduce an incompletely specified state table to a minimum number of rows,[2] but the discussion of such procedures is beyond the scope of this text.

## 15.6 Derivation of Flip-Flop Input Equations

After the number of states in a state table has been reduced, the following procedure can be used to derive the flip-flop input equations:

1. Assign flip-flop state values to correspond to the states in the reduced table.
2. Construct a transition table which gives the next states of the flip-flops as a function of the present states and inputs.
3. Derive the next-state maps from the transition table.
4. Find flip-flop input maps from the next-state maps using the techniques developed in Unit 12 and find the flip-flop input equations from the maps.

As an example, we will design a sequential circuit to realize Table 15-6(a). Because there are seven states, we will need three flip-flops. We will designate the flip-flop outputs as $A$, $B$, and $C$.

We could make a straight binary state assignment for which $S_0$ is represented by flip-flop states $ABC = 000$, $S_1$ by $ABC = 001$, $S_2$ by $ABC = 010$, etc. However, because the correspondence between flip-flop states and the state names is arbitrary, we could use many different state assignments. Using a different assignment may

**TABLE 15-6**

(a) State table

| | $X = 0$ | 1 | 0 | 1 |
|---|---|---|---|---|
| $S_0$ | $S_1$ | $S_2$ | 0 | 0 |
| $S_1$ | $S_3$ | $S_2$ | 0 | 0 |
| $S_2$ | $S_1$ | $S_4$ | 0 | 0 |
| $S_3$ | $S_5$ | $S_2$ | 0 | 0 |
| $S_4$ | $S_1$ | $S_6$ | 0 | 0 |
| $S_5$ | $S_5$ | $S_2$ | 1 | 0 |
| $S_6$ | $S_1$ | $S_6$ | 0 | 1 |

(b) Transition table

| | $A^+B^+C^+$ | | $Z$ | |
|---|---|---|---|---|
| $ABC$ | $X = 0$ | 1 | 0 | 1 |
| 000 | 110 | 001 | 0 | 0 |
| 110 | 111 | 001 | 0 | 0 |
| 001 | 110 | 011 | 0 | 0 |
| 111 | 101 | 001 | 0 | 0 |
| 011 | 110 | 010 | 0 | 0 |
| 101 | 101 | 001 | 1 | 0 |
| 010 | 110 | 010 | 0 | 1 |

[2]See, for example, Edward I. McClushey, *Logic Design Principles* (Prentice-Hall, 1986), Chap. 9.

lead to simpler or more complex flip-flop input equations. As an example, we will use the following assignment for the states of flip-flops $A$, $B$, and $C$:

$$S_0 = 000, S_1 = 110, S_2 = 001, S_3 = 111, S_4 = 011, S_5 = 101, S_6 = 010 \quad (15\text{-}1)$$

This state assignment is derived in Section 15.8, and the reasons why it leads to an economical solution are given in that section. Starting with Table 15-6(a), we substitute 000 for $S_0$, 110 for $S_1$, 001 for $S_2$, etc. Table 15-6(b) shows the resulting transition table. This table gives the next states of flip-flops $A$, $B$, and $C$ in terms of the present states and the input $X$. We can fill in the next-state maps, Figure 15-9(a), directly from this table. For $XABC = 0000$ the next-state entry is 110, so we fill in $A^+ = 1$, $B^+ = 1$, and $C^+ = 0$; for $XABC = 1000$ the next-state entry is 001, so we fill in $A^+ = 0$, $B^+ = 0$, and $C^+ = 1$; etc. Because the state assignment $ABC = 100$ is not used, the map squares corresponding to $XABC = 0100$ and 1100 are filled with don't-cares.

Once the next-state maps have been plotted from the transition table, the flip-flop input equations can be derived using the techniques developed in Unit 12. As shown in Figure 15-9(a), the D flip-flop input equations can be derived directly from the next-state maps because $D_A = A^+$, $D_B = B^+$, and $D_C = C^+$. If J-K flip-flops are used, the $J$ and $K$ input equations can be derived from the next-state maps as illustrated in Figure 15-9(b). As was shown in Section 12.5, the $A = 0$ half of the $J_A$ map is the same as the $A^+$ map and the $A = 1$ half is all don't-cares. The $A = 1$ half of the $K_A$ map is the complement of the $A = 1$ half of the $A^+$ map, and the $A = 0$ half is all don't-cares. We can

**FIGURE 15-9** Next-State Maps for Table 15-6

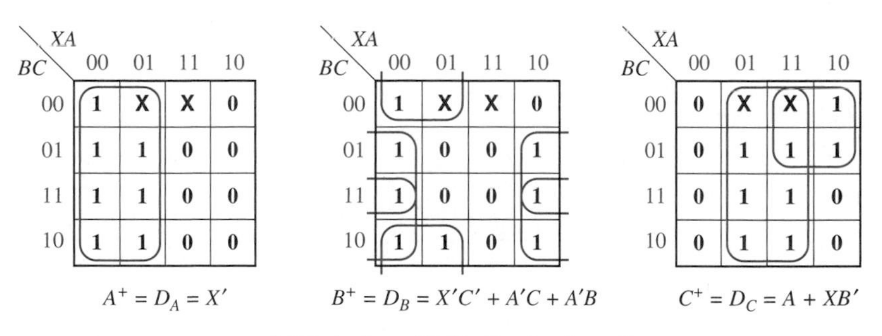


(a) Derivation of D flip-flop input equations

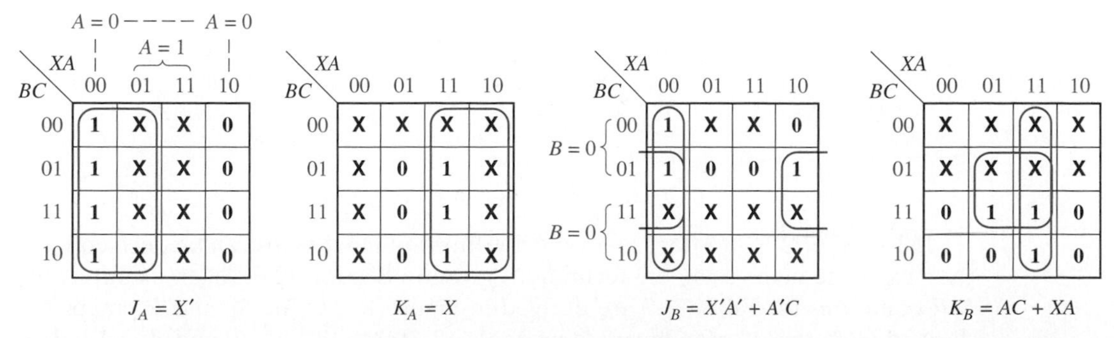


(b) Derivation of J-K flip-flop input equations

**TABLE 15-7**

(a) State table

| P.S. | Next State $X_1X_2 =$ 00 | 01 | 11 | 10 | Outputs $(Z_1Z_2)$ $X_1X_2 =$ 00 | 01 | 11 | 10 |
|---|---|---|---|---|---|---|---|---|
| $S_0$ | $S_0$ | $S_0$ | $S_1$ | $S_1$ | 00 | 00 | 01 | 01 |
| $S_1$ | $S_1$ | $S_3$ | $S_2$ | $S_1$ | 00 | 10 | 10 | 00 |
| $S_2$ | $S_3$ | $S_3$ | $S_2$ | $S_2$ | 11 | 11 | 00 | 00 |
| $S_3$ | $S_0$ | $S_3$ | $S_2$ | $S_0$ | 00 | 00 | 00 | 00 |

(b) Transition table

| $AB$ | $A^+B^+$ $X_1X_2 =$ 00 | 01 | 11 | 10 | Outputs $(Z_1Z_2)$ $X_1X_2 =$ 00 | 01 | 11 | 10 |
|---|---|---|---|---|---|---|---|---|
| 00 | 00 | 00 | 01 | 01 | 00 | 00 | 01 | 01 |
| 01 | 01 | 10 | 11 | 01 | 00 | 10 | 10 | 00 |
| 11 | 10 | 10 | 11 | 11 | 11 | 11 | 00 | 00 |
| 10 | 00 | 10 | 11 | 00 | 00 | 00 | 00 | 00 |

plot the $J_B$ and $K_B$ in a similar manner by looking at the $B = 0$ and $B = 1$ halves of the $B^+$ map. Derivation of the $J_C$ and $K_C$ maps from the $C^+$ map is left as an exercise.

Table 15-7(a) represents a sequential circuit with two inputs ($X_1$ and $X_2$) and two outputs ($Z_1$ and $Z_2$). Note that the column headings are listed in Karnaugh map order because this will facilitate derivation of the flip-flop input equations. Because the table has four states, two flip-flops ($A$ and $B$) are required to realize the table. We will use the state assignment $AB = 00$ for $S_0$, $AB = 01$ for $S_1$, $AB = 11$ for $S_2$, and $AB = 10$ for $S_3$. By substituting the corresponding values of $AB$ for the state names, we obtain the transition table, Table 15-7(b). We can then fill in the next-state and output maps (Figure 15-10) from the transition table. For example, when $X_1X_2AB = 0011$, $A^+B^+ = 10$, and $Z_1Z_2 = 11$; therefore, we fill in the 0011 squares of the $A^+$, $B^+$, $Z_1$, and $Z_2$ maps with 1, 0, 1, and 1, respectively. We can read the D flip-flop input equations directly from the next-state maps.

**FIGURE 15-10** Next-State Maps for Table 15-7

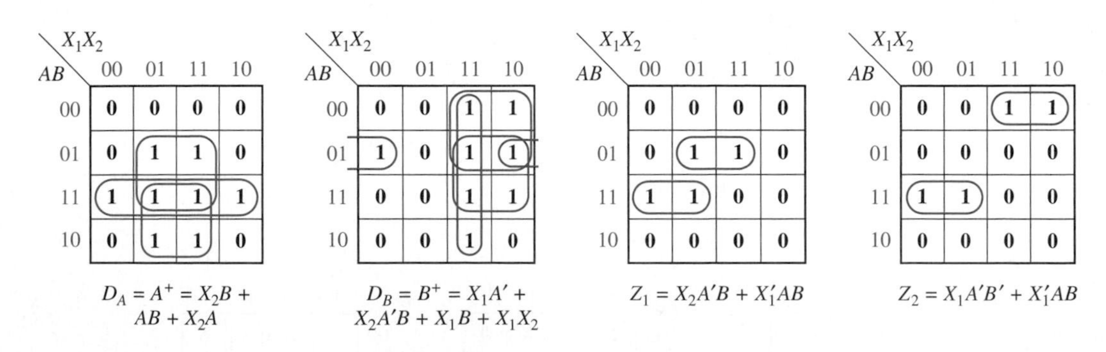

$D_A = A^+ = X_2B + AB + X_2A$
$D_B = B^+ = X_1A' + X_2A'B + X_1B + X_1X_2$
$Z_1 = X_2A'B + X_1'AB$
$Z_2 = X_1A'B' + X_1'AB$


If J-K, T, or S-R flip-flops are used, the flip-flop input maps can be derived from the next-state maps using the techniques given in Section 12.6. As an example, the $S$-$R$ equations for Table 15-7 are derived in Figure 15-11. The $S_A$ and $R_A$ maps are derived from the $A^+$ map by applying Table 12-5(c) to the $A = 0$ and $A = 1$ halves of the map. $S_B$ and $R_B$ are derived in a similar manner.

FIGURE 15-11 Derivation of S-R Equations for Table 15-7

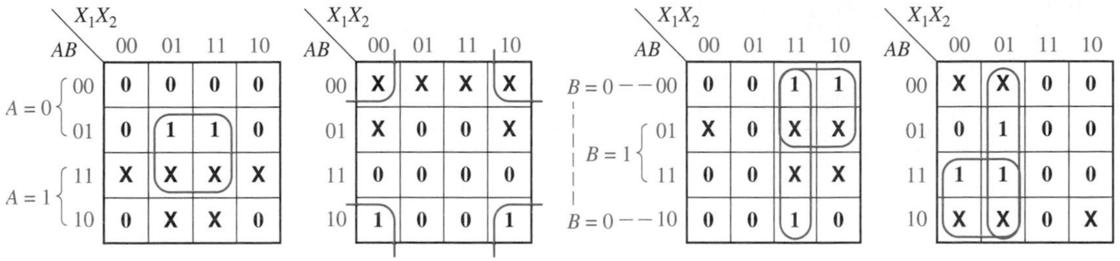

| $AB \backslash X_1X_2$ | 00 | 01 | 11 | 10 |
|---|---|---|---|---|
| $A=0$ 00 | 0 | 0 | 0 | 0 |
| $A=0$ 01 | 0 | 1 | 1 | 0 |
| $A=1$ 11 | X | X | X | X |
| $A=1$ 10 | 0 | X | X | 0 |

$S_A = X_2B$

| $AB \backslash X_1X_2$ | 00 | 01 | 11 | 10 |
|---|---|---|---|---|
| 00 | X | X | X | X |
| 01 | X | 0 | 0 | X |
| 11 | 0 | 0 | 0 | 0 |
| 10 | 1 | 0 | 0 | 1 |

$R_A = X_2'B'$

| $AB \backslash X_1X_2$ | 00 | 01 | 11 | 10 |
|---|---|---|---|---|
| $B=0$ 00 | 0 | 0 | 1 | 1 |
| $B=1$ 01 | X | 0 | X | X |
| $B=1$ 11 | 0 | 0 | X | X |
| $B=0$ 10 | 0 | 0 | 1 | 0 |

$S_B = X_1X_2 + X_1A'$

| $AB \backslash X_1X_2$ | 00 | 01 | 11 | 10 |
|---|---|---|---|---|
| 00 | X | X | 0 | 0 |
| 01 | 0 | 1 | 0 | 0 |
| 11 | 1 | 1 | 0 | 0 |
| 10 | X | X | 0 | X |

$R_B = X_1'X_2 + X_1'A$

# 15.7 Equivalent State Assignments

After the number of states in a state table has been reduced, the next step in realizing the table is to assign flip-flop states to correspond to the states in the table. The cost of the logic required to realize a sequential circuit is strongly dependent on the way this state assignment is made. Several methods for choosing state assignments to obtain economical realizations are discussed in this chapter. The trial-and error method described next is useful for only a small number of states. The guideline method discussed in Section 15.8 produces good solutions for some problems, but it is not entirely satisfactory in other cases.

If the number of states is small, it may be feasible to try all possible state assignments, evaluate the cost of the realization for each assignment, and choose the assignment with the lowest cost. Consider a state table with three states ($S_0$, $S_1$, and $S_2$) as in Table 14-1. Two flip-flops ($A$ and $B$) are required to realize this table. The four possible assignments for state $S_0$ are $AB = 00$, $AB = 01$, $AB = 10$, and $AB = 11$. Choosing one of these assignments leaves three possible assignments for state $S_1$ because each state must have a unique assignment. Then, after state $S_1$ is assigned, we have two possible assignments for state $S_2$. Thus, there are $4 \times 3 \times 2 = 24$ possible state assignments for the three states, as shown in Table 15-8. As an example, for assignment 7, the entry 01 in the $S_0$ row means that flip-flops $A$ and $B$ are assigned values 0 and 1, respectively.

Trying all 24 of these assignments is not really necessary. If we interchange two columns in one of the given assignments, the cost of realization will be unchanged because interchanging columns is equivalent to relabeling the flip-flop variables. For example, consider assignment 1 in Table 15-8. The first column of this assignment shows that flip-flop $A$ is assigned the values 0, 0, and 1 for states $S_0$, $S_1$, and $S_2$, respectively. Similarly, the second column shows that $B$ is assigned the values 0, 1, and 0. If we interchange the two columns, we get assignment 3, for which $A$ has the

TABLE 15-8 State Assignments for 3-Row Tables

| | 1 | 2 | 3 | 4 | 5 | 6 | 7 | | 19 | 20 | 21 | 22 | 23 | 24 |
|---|---|---|---|---|---|---|---|---|---|---|---|---|---|---|
| $S_0$ | 00 | 00 | 00 | 00 | 00 | 00 | 01 | ... | 11 | 11 | 11 | 11 | 11 | 11 |
| $S_1$ | 01 | 01 | 10 | 10 | 11 | 11 | 00 | | 00 | 00 | 01 | 01 | 10 | 10 |
| $S_2$ | 10 | 11 | 01 | 11 | 01 | 10 | 10 | | 01 | 10 | 00 | 10 | 00 | 01 |

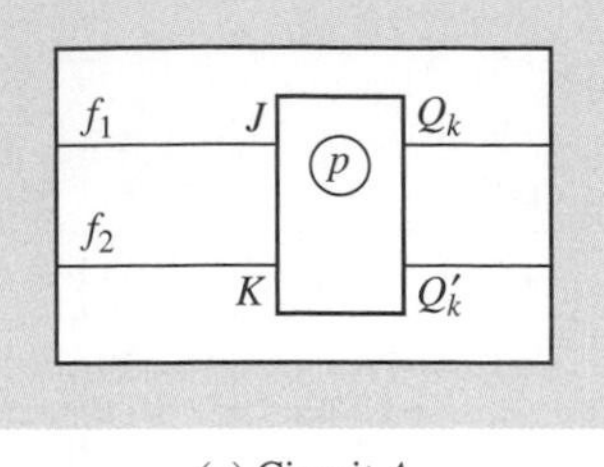


(a) Circuit $A$

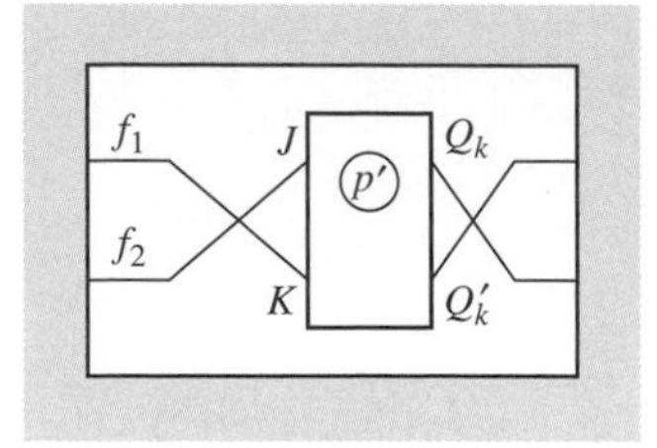


(b) Circuit $B$ (identical to $A$ except leads to flip-flop $Q_k$ are crossed)

**FIGURE 15-12** Equivalent Circuits Obtained by Complementing $Q_k$

values 0, 1, and 0 and $B$ has the values 0, 0, and 1. We could have achieved the same result by using assignment 1 and labeling the flip-flop variables $BA$ instead of $AB$. If we interchange the columns of assignment 2, we get assignment 4, so assignments 2 and 4 have the same cost. Similarly, assignments 5 and 6 have the same cost. Interchanging rows, however, will usually change the cost of realization. Thus, assignments 4 and 6 will have a different cost for many state tables.

If symmetrical flip-flops such as T, J-K, or S-R are used, complementing one or more columns of the state assignment will have no effect on the cost of realization. Consider a J-K flip-flop imbedded in a circuit, Figure 15-12(a). Leave the circuit unchanged and interchange the $J$ and $K$ input connections and the $Q_k$ and $Q'_k$ output connections, Figure 15-12(b). If circuit $A$ is started with $Q_k = p$ and circuit $B$ with $Q_k = p'$, the behavior of the two circuits will be identical, except the value of $Q_k$ will always be complemented in the second circuit because whenever $J$ is 1 in the first circuit, $K$ will be 1 in the second and conversely. The state table for the second circuit is therefore the same as for the first, except the value of $Q_k$ is complemented for the second circuit. This implies that complementing one or more columns in the state assignment will not affect the cost of the realization when J-K flip-flops are used. Similar reasoning applies to T and S-R flip-flops. Thus, in Table 15-8, assignments 2 and 7 have the same cost, and so do assignments 6 and 19.

If unsymmetrical flip-flops are used such as a D flip-flop, it is still true that permuting (i.e., rearranging the order of) columns in the state assignment will not affect the cost; however, complementing a column may require adding an inverter to the circuit, as shown in Figure 15-13. If different types of gates are available, the circuit can generally be redesigned to eliminate the inverter and use the same number of gates as the original. If circuit $A$ in Figure 15-13 is started with $Q_k = p$ and circuit $B$ with $Q_k = p'$, the

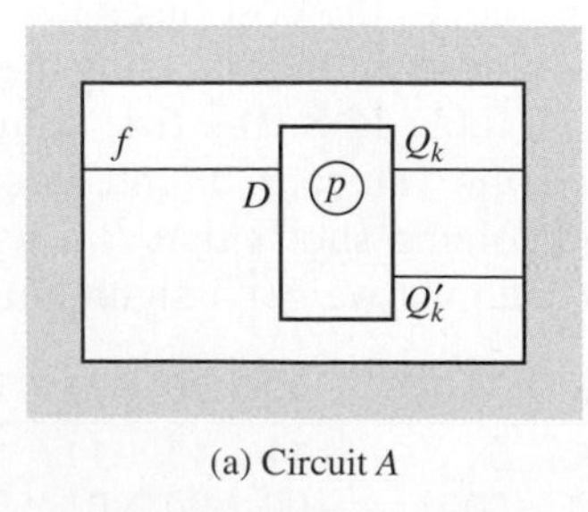


(a) Circuit $A$

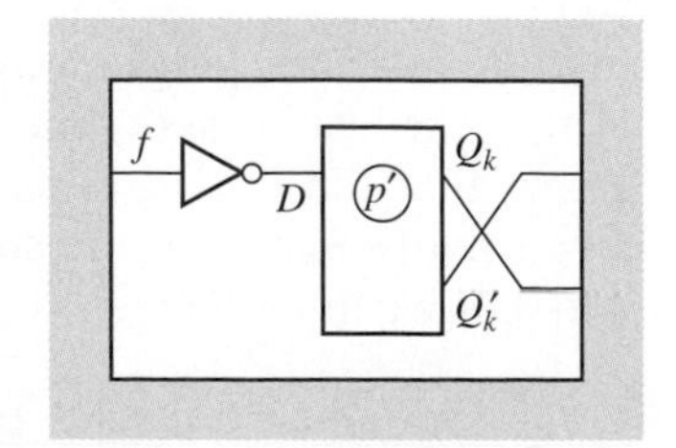


(b) Circuit $B$ (identical to $A$ except for connections to flip-flop $Q_k$)

**FIGURE 15-13** Equivalent Circuits Obtained by Complementing $Q_k$

**TABLE 15-9**

| Assignments $A_3$ | $B_3$ | $C_3$ | Present State | Next State $X = 0$ | 1 | Output 0 | 1 |
|---|---|---|---|---|---|---|---|
| 00 | 00 | 11 | $S_1$ | $S_1$ | $S_3$ | 0 | 0 |
| 01 | 10 | 10 | $S_2$ | $S_2$ | $S_1$ | 0 | 1 |
| 10 | 01 | 01 | $S_3$ | $S_2$ | $S_3$ | 1 | 0 |

behavior of the two circuits will be identical, except the value of $Q_k$ will always be complemented in circuit $B$ because $f$ is the same in both circuits and $D = f'$ for circuit $B$.

Table 15-9 illustrates the effect of interchanging or complementing state assignment columns on the equations for realizing a specific state table.

The J-K and D flip-flop input equations for the three assignments can be derived, using Karnaugh maps as explained in Unit 12 and Section 15.6. The resulting J and K input equations are:

| Assignment $A$ | Assignment $B$ | Assignment $C$ |
|---|---|---|
| $J_1 = XQ_2'$ | $J_2 = XQ_1'$ | $K_1 = XQ_2$ |
| $K_1 = X'$ | $K_2 = X'$ | $J_1 = X'$ |
| $J_2 = X'Q_1$ | $J_1 = X'Q_2$ | $K_2 = X'Q_1'$ |
| $K_2 = X$ | $K_1 = X$ | $J_2 = X$ |
| $Z = X'Q_1 + XQ_2$ | $Z = X'Q_2 + XQ_1$ | $Z = X'Q_1' + XQ_2'$ |
| $D_1 = XQ_2'$ | $D_2 = XQ_1'$ | $D_1 = X' + Q_2'$ |
| $D_2 = X'(Q_1 + Q_2)$ | $D_1 = X'(Q_2 + Q_1)$ | $D_2 = X + Q_1Q_2$ |

Note that assignment $B$ in Table 15-9 was obtained by interchanging the columns of $A$. The corresponding equations for assignment $B$ are the same as for $A$, except that subscripts 1 and 2 are interchanged. Assignment $C$ was obtained by complementing the columns of $A$. The $Z$ equation for $C$ is the same as for $A$, except that $Q_1$ and $Q_2$ are complemented. The $K$ and $J$ equations for $C$ are the same, respectively, as the $J$ and $K$ equations for $A$ with the $Q$'s complemented. The $D$ equations for $C$ can be obtained by complementing those for $A$ and, then, complementing the $Q$'s. Thus, the cost of realizing Table 15-9 using J-K flip-flops and any kind of logic gates will be exactly the same for all three assignments. If both AND and OR (or NAND and NOR) gates are available, the cost of realizing the three sets of $D$ equations will be the same. If only NOR gates are available, for example, then realizing $D_1$ and $D_2$ for assignment $C$ would require two additional inverters compared with $A$ and $B$.

By complementing one or more columns, any state assignment can be converted to one in which the first state is assigned all 0's. If we eliminate assignments which can be obtained by permuting or complementing columns of another state assignment, Table 15-8 reduces to three assignments (Table 15-10). Thus, when realizing a

**TABLE 15-10** Nonequivalent Assignments for Three and Four States

| | 3-State Assignments | | | 4-State Assignments | | |
|---|---|---|---|---|---|---|
| States | 1 | 2 | 3 | 1 | 2 | 3 |
| *a* | 00 | 00 | 00 | 00 | 00 | 00 |
| *b* | 01 | 01 | 11 | 01 | 01 | 11 |
| *c* | 10 | 11 | 01 | 10 | 11 | 01 |
| *d* | – | – | – | 11 | 10 | 10 |

**TABLE 15-11**
Number of Distinct (Nonequivalent) State Assignments

| Number of States | Minimum Number of State Variables | Number of Distinct Assignments |
|---|---|---|
| 2 | 1 | 1 |
| 3 | 2 | 3 |
| 4 | 2 | 3 |
| 5 | 3 | 140 |
| 6 | 3 | 420 |
| 7 | 3 | 840 |
| 8 | 3 | 840 |
| 9 | 4 | 10,810,800 |
| ⋮ | ⋮ | ⋮ |
| 16 | 4 | $\approx 5.5 \times 10^{10}$ |

three-state sequential circuit with symmetrical flip-flops, it is only necessary to try three different state assignments to be assured of a minimum cost realization. Similarly, only three different assignments must be tried for four states.

We will say that two state assignments are *equivalent* if one can be derived from the other by permuting and complementing columns. Two state assignments which are not equivalent are said to be *distinct*. Thus, a four-row table has three distinct state assignments, and any other assignment is equivalent to one of these three. Unfortunately, the number of distinct assignments increases very rapidly with the number of states, as shown in Table 15-11. Hand solution is feasible for two, three, or four states; computer solution is feasible for five through eight states; but for more than nine states it is not practical to try all assignments even if a high-speed computer is used.

## 15.8 Guidelines for State Assignment

Because trying all nonequivalent state assignments is not practical in most cases, other methods of state assignment are needed. The next method to be discussed involves trying to choose an assignment which will place the 1's on the flip-flop input maps in adjacent squares so that the corresponding terms can be combined. This method does not apply to all problems, and even when applicable, it does not guarantee a minimum solution.

Assignments for two states are said to be adjacent if they differ in only one variable. Thus, 010 and 011 are adjacent, but 010 and 001 are not. The following *guidelines* are useful in making assignments which will place 1's together (or 0's together) on the next-state maps:

1. States which have the same next state for a given input should be given adjacent assignments.
2. States which are the next states of the same state should be given adjacent assignments.

A third guideline is used for simplification of the output function:

**3.** States which have the same output for a given input should be given adjacent assignments.

The application of Guideline 3 will place 1's together on the output maps.

When using the state assignment guidelines, the first step is to write down all of the sets of states which should be given adjacent assignments according to the guidelines. Then, using a Karnaugh map, try to satisfy as many of these adjacencies as possible. A fair amount of trial and error may be required to fill in the map so that the maximum number of desired state adjacencies is obtained. When filling in the map, keep in mind the following:

(a) Assign the starting state (reset state) to the "0" square on the map. (For an exception to this rule, see the one-hot assignment in Section 15.9.) Nothing is to be gained by trying to put the starting state in different squares on the map because the same number of adjacencies can be found no matter where you put the starting state. Usually, assigning "0" to the starting state simplifies the initialization of the circuit using the clear inputs on the flip-flops.

(b) Adjacency conditions from Guideline 1 and adjacency conditions from Guideline 2 that are required two or more times should be satisfied first.

(c) When guidelines require that three or four states be adjacent, these states should be placed within a group of four adjacent squares on the assignment map.

(d) If the output table is to be considered, then Guideline 3 should also be applied. The priority given to adjacency conditions from Guideline 3 should generally be less than that given to Guidelines 1 and 2 if a single output function is being derived. If there are two or more output functions, a higher priority for Guideline 3 may be appropriate.

The following example should clarify the application of Guidelines 1 and 2. The state table from Table 15-6 is repeated in Figure 15-14(a) so that we can illustrate derivation of the state assignment. According to Guideline 1, $S_0$, $S_2$, $S_4$, and $S_6$ should be given adjacent assignments because they all have $S_1$ as a next state (with input 0). Similarly, $S_0$, $S_1$, $S_3$, and $S_5$ should have adjacent assignments because they have $S_2$ as a next state (with input 1); also, $S_3$ and $S_5$ should have adjacent assignments and so should $S_4$ and $S_6$. The application of Guideline 2 indicates that $S_1$ and

FIGURE 15-14

| ABC | | X = 0 | 1 | 0 | 1 |
|---|---|---|---|---|---|
| 000 | $S_0$ | $S_1$ | $S_2$ | 0 | 0 |
| 110 | $S_1$ | $S_3$ | $S_2$ | 0 | 0 |
| 001 | $S_2$ | $S_1$ | $S_4$ | 0 | 0 |
| 111 | $S_3$ | $S_5$ | $S_2$ | 0 | 0 |
| 011 | $S_4$ | $S_1$ | $S_6$ | 0 | 0 |
| 101 | $S_5$ | $S_5$ | $S_2$ | 1 | 0 |
| 010 | $S_6$ | $S_1$ | $S_6$ | 0 | 1 |

(a) State table

| BC \ A | 0 | 1 |
|---|---|---|
| 00 | $S_0$ | |
| 01 | $S_2$ | $S_5$ |
| 11 | $S_4$ | $S_3$ |
| 10 | $S_6$ | $S_1$ |

| BC \ A | 0 | 1 |
|---|---|---|
| 00 | $S_0$ | |
| 01 | $S_1$ | $S_6$ |
| 11 | $S_3$ | $S_4$ |
| 10 | $S_5$ | $S_2$ |

(b) Assignment maps

$S_2$ should be given adjacent assignments because they are both next states of $S_0$. Similarly, $S_2$ and $S_3$ should have adjacent assignments because they are both next states of $S_1$. Further application of Guideline 2 indicates that $S_1$ and $S_4$, $S_2$ and $S_5$ (two times), and $S_1$ and $S_6$ (two times) should be given adjacent assignments. In summary, the sets of adjacent states specified by Guidelines 1 and 2 are

1. $(S_0, S_1, S_3, S_5)$ $(S_3, S_5)$ $(S_4, S_6)$ $(S_0, S_2, S_4, S_6)$
2. $(S_1, S_2)$ $(S_2, S_3)$ $(S_1, S_4)$ $(S_2, S_5)2x$ $(S_1, S_6)2x$

We will attempt to fulfill as many of these adjacency conditions as possible. A Karnaugh map will be used to make the assignments so that states with adjacent assignments will appear in adjacent squares on the map. If the guidelines require that three or four states be adjacent, these states should be placed within a group of four adjacent squares on the assignment map. Two possible ways of filling in the assignment maps are shown in Figure 15-14(b). These maps were filled in by trial and error, attempting to fulfill as many of the preceding adjacency conditions as possible. The conditions from Guideline 1 are given preference to conditions from Guideline 2. The conditions which are required two times (such as $S_2$ adjacent to $S_5$, and $S_1$ adjacent to $S_6$) are given preference over conditions which are required only once (such as $S_1$ adjacent to $S_2$, and $S_2$ adjacent to $S_3$).

The left assignment map in Figure 15-14(b) implies an assignment for the states of flip-flops $A$, $B$, and $C$ which is listed to the left of the state table in Figure 15-14(a). This assignment is the same as the one given in Equations (15-1). We derived the D flip-flop input equations and $J$ and $K$ input equations for this assignment in Section 15.6. The cost of realizing the D flip-flop input equations given in Figure 15-9(a) is six gates and 13 inputs. If a straight binary assignment ($S_0 = 000, S_1 = 001, S_2 = 010$, etc.) were used instead, the cost of realizing the flip-flop input equations would be 10 gates and 39 inputs. Although application of the guidelines gives good results in this example, this is not always the case.

Next, we will explain why the guidelines help to simplify the flip-flop equations when the assignment of Figure 15-14(a) is used. Figure 15-15 shows a next-state map which was constructed using this assignment. Note that if $X = 0$ and $ABC = 000$, the next state is $S_1$; if $X = 1$ and $ABC = 000$, the next state is $S_2$. Because Guideline 1 was used in making the state assignment, $S_1$ appears in four adjacent squares on the next-state map, $S_5$ appears in two adjacent squares, etc.

The next-state maps for the individual flip-flops, Figure 15-15(b), can be derived in the usual manner from a transition table, or they can be derived directly from Figure 15-15(a). Using the latter approach, wherever $S_1$ appears in Figure 15-15(a), it is replaced with 110 so that 1, 1, and 0 are plotted on the corresponding squares of the $A^+$, $B^+$, and $C^+$ maps, respectively. The other squares on the next-state maps are filled in similarly.

Because four $S_1$'s are adjacent in Figure 15-15(a), the corresponding squares on the $A^+$, $B^+$, and $C^+$ maps have four adjacent 1's or four adjacent 0's as indicated by the blue shading. This illustrates why Guideline 1 helps to simplify the flip-flop equations. Each time Guideline 2 is applied, two out of the three next-state maps will have an additional pair of adjacent 1's or adjacent 0's. This occurs because two of the three state variables are the same for adjacent assignments.

**FIGURE 15-15** Next-State Maps for Figure 15-14

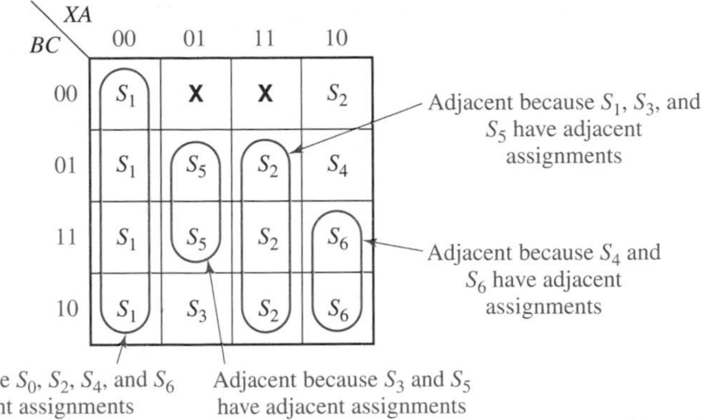


(a) Next-state maps for Figure 15-14

These pairs are adjacent because $S_2$ and $S_5$ have adjacent assignments

| BC \ XA | 00 | 01 | 11 | 10 |
|---|---|---|---|---|
| 00 | 1 | X | X | 0 |
| 01 | 1 | 1 | 0 | 0 |
| 11 | 1 | 1 | 0 | 0 |
| 10 | 1 | 1 | 0 | 0 |

$A^+ = D_A = X'$

| BC \ XA | 00 | 01 | 11 | 10 |
|---|---|---|---|---|
| 00 | 1 | X | X | 1 |
| 01 | 1 | 0 | 0 | 1 |
| 11 | 1 | 0 | 0 | 1 |
| 10 | 1 | 1 | 0 | 1 |

$B^+ = D_B = X'C' + A'C + A'B$

| BC \ XA | 00 | 01 | 11 | 10 |
|---|---|---|---|---|
| 00 | 0 | X | X | 1 |
| 01 | 0 | 1 | 1 | 1 |
| 11 | 0 | 1 | 1 | 0 |
| 10 | 0 | 1 | 1 | 0 |

$C^+ = D_C = A + XB'$

(b) Next-state maps for Figure 15-14 (cont.)

Next, we will apply the state assignment guidelines to Figure 15-16(a). First, we list the sets of adjacent states specified by each Guideline:

1. $(b, d)$ $(c, f)$ $(b, e)$
2. $(a, c)2x$ $(d, f)$ $(b, d)$ $(b, f)$ $(c, e)$
3. $(a, c)$ $(b, d)$ $(e, f)$

Next, we try to arrange the states on a map so as to satisfy as many of these pairs as possible, but giving preference to the duplicated pairs $(b, d)$ and $(a, c)$. Two such arrangements and the corresponding assignments are given in Figures 15-16(b) and (c). For Figure 15-16(c), all adjacencies are satisfied except $(b, f)$, $(c, e)$, and $(e, f)$. We will derive D flip-flop input equations for this assignment. First, we construct the transition table (Table 15-12) from the state table [Figure 15-16(a)] by replacing *a* with 100, *b* with 111, *c* with 000, etc. Then, we plot the next-state and output maps

**FIGURE 15-16** State Table and Assignments

| | $X = 0$ | 1 | $X = 0$ | 1 |
|---|---|---|---|---|
| a | a | c | 0 | 0 |
| b | d | f | 0 | 1 |
| c | c | a | 0 | 0 |
| d | d | b | 0 | 1 |
| e | b | f | 1 | 0 |
| f | c | e | 1 | 0 |

(a)

| $Q_2Q_3$ \ $Q_1$ | 0 | 1 |
|---|---|---|
| 00 | *a* | *c* |
| 01 | | *e* |
| 11 | *d* | *b* |
| 10 | | *f* |

$a = 000$, $b = 111$, $c = 100$, $d = 011$, $e = 101$, $f = 110$

(b)

| $Q_2Q_3$ \ $Q_1$ | 0 | 1 |
|---|---|---|
| 00 | *c* | *a* |
| 01 | | *e* |
| 11 | *d* | *b* |
| 10 | *f* | |

$a = 100$, $b = 111$, $c = 000$, $d = 011$, $e = 101$, $f = 010$

(c)

**TABLE 15-12** Transition Table for Figure 15-16(a)

| | $Q_1^+Q_2^+Q_3^+$ | | | |
|---|---|---|---|---|
| $Q_1Q_2Q_3$ | $X = 0$ | 1 | $X = 0$ | 1 |
| 1 0 0 | 100 | 000 | 0 | 0 |
| 1 1 1 | 011 | 010 | 0 | 1 |
| 0 0 0 | 000 | 100 | 0 | 0 |
| 0 1 1 | 011 | 111 | 0 | 1 |
| 1 0 1 | 111 | 010 | 1 | 0 |
| 0 1 0 | 000 | 101 | 1 | 0 |

(Figure 15-17) from the transition table. The D flip-flop input equations can be read directly from these maps:

$$D_1 = Q_1^+ = X'Q_1Q_2' + XQ_1'$$
$$D_2 = Q_2^+ = Q_3$$
$$D_3 = Q_3^+ = XQ_1'Q_2 + X'Q_3$$

and the output equation is

$$Z = XQ_2Q_3 + X'Q_2'Q_3 + X\,Q_2Q_3'$$

The cost of realizing these equations is 10 gates and 26 gate inputs.

**FIGURE 15-17** Next-State and Output Maps for Table15-12

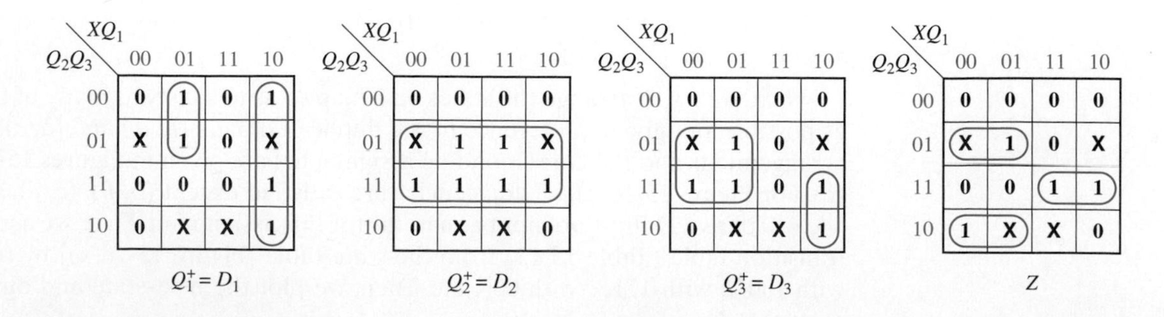

The assignment of Figure 15-16(b) satisfies all of the guidelines except $(d, f)$ and $(e, f)$. Using this assignment, the cost of realizing the state table with D flip-flops is 13 gates and 35 gate inputs. We would expect that this assignment would produce better results than Figure 15-16(c) because it satisfies one more of the adjacencies given by the guidelines, but just the opposite is true. As illustrated by this example, the assignment which satisfies the most guidelines is not necessarily the best assignment. In general, it is a good idea to try several assignments which satisfy most of the guidelines and choose the one which gives the lowest cost solution.

The guidelines work best for D flip-flops and J-K flip-flops. They do not work as well for T and S-R flip-flops. In general, the best assignment for one type of flip-flop is not the best for another type.

## 15.9 Using a One-Hot State Assignment

When designing with CPLDs and FPGAs, we should keep in mind that each logic cell contains one or more flip-flops. These flip-flops are there whether we use them or not. This means that it may not be important to minimize the number of flip-flops used in the design. Instead, we should try to reduce the total number of logic cells used and try to reduce the interconnections between cells. When several cells must be cascaded to realize a function as in Figure 9-36(b), the propagation delay is longer, and the logic runs slower. In order to design faster logic, we should try to reduce the number of cells required to realize each equation. Using a *one-hot* state assignment may help to accomplish this.

The one-hot assignment uses one flip-flop for each state, so a state machine with $N$ states requires $N$ flip-flops. Exactly one of the flip-flops is set to one in each state. For example, a system with four states ($S_0$, $S_1$, $S_2$, and $S_3$) could use four flip-flops ($Q_0$, $Q_1$, $Q_2$, and $Q_3$) with the following state assignment:

$$S_0\colon Q_0\, Q_1\, Q_2\, Q_3 = 1000, \quad S_1\colon 0100, \quad S_2\colon 0010, \quad S_3\colon 0001 \tag{15-2}$$

The other 12 combinations are not used.

We can write next-state and output equations by inspecting the state graph. Consider the partial state graph given in Figure 15-18. Because four arcs lead into $S_3$, there are four conditions under which the next state is $S_3$. These conditions are as

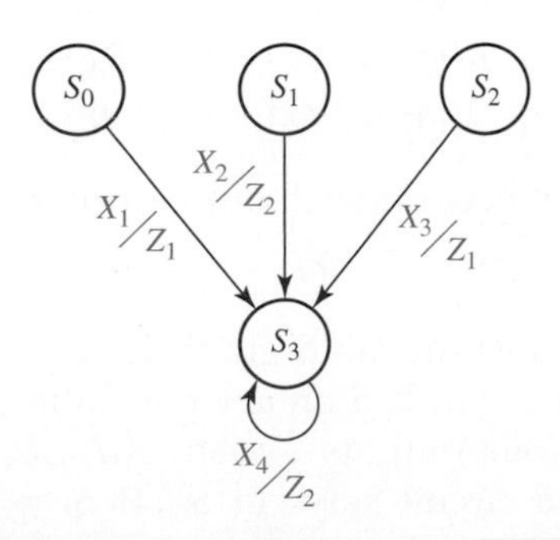


FIGURE 15-18
Partial State Graph

follows: present state (PS) $= S_0$ and $X_1 = 1$, PS $= S_1$ and $X_2 = 1$, PS $= S_2$ and $X_3 = 1$, PS $= S_3$ and $X_4 = 1$. The next state of flip-flop $Q_3$ is 1 under these four conditions (and 0 otherwise). Therefore, the next-state equation for $Q_3$ can be written as:

$$Q_3^+ = X_1 (Q_0 Q_1' Q_2' Q_3') + X_2 (Q_0' Q_1 Q_2' Q_3') + X_3 (Q_0' Q_1' Q_2 Q_3') + X_4 (Q_0' Q_1' Q_2' Q_3)$$

However, because $Q_0 = 1$ implies $Q_1 = Q_2 = Q_3 = 0$, the $Q_1' Q_2' Q_3'$ term is redundant and can be eliminated. Similarly, all of the primed state variables can be eliminated from the other terms, so the next-state equation reduces to

$$Q_3^+ = X_1Q_0 + X_2Q_1 + X_3Q_2 + X_4Q_3$$

In general, when a one-hot state assignment is used, each term in the next-state equation for each flip-flop contains exactly one state variable, and the reduced equation can be written by inspecting the state graph.

Similarly, each term in each reduced output equation contains exactly one state variable. Because $Z_1 = 1$ when PS $= S_0$ and $X_1 = 1$, and also when PS $= S_2$ and $X_3 = 1$, we can write $Z_1 = X_1Q_0 + X_3Q_2$. By inspecting the state graph, we can also write $Z_2 = X_2Q_1 + X_4Q_3$.

When a one-hot assignment is used, resetting the system requires that one flip-flop be set to 1 instead of resetting all flip-flops to 0. If the flip-flops used do not have a preset input, then we can modify the one-hot assignment by replacing $Q_0$ with $Q_0'$ throughout. For the Assignment (15-2), the modification is

$$S_0: Q_0\, Q_1\, Q_2\, Q_3 = 0000, \quad S_1: 1100, \quad S_2: 1010, \quad S_3: 1001 \tag{15-3}$$

and the modified equations are:

$$Q_3^+ = X_1Q_0' + X_2Q_1 + X_3Q_2 + X_4Q_3$$
$$Z_1 = X_1Q_0' + X_3Q_2, \quad Z_2 = X_2Q_1 + X_4Q_3$$

For the Moore machine of Figure 14-22(b), we will make the following one-hot assignment for flip-flops $Q_0\, Q_1\, Q_2$: $S_0 = 100$, $S_1 = 010$, and $S_2 = 001$. When $Q_0 = 1$, the state is $S_0$; when $Q_1 = 1$, the state is $S_1$; and when $Q_2 = 1$, the state is $S_2$. By inspection, because three arcs lead into each state, the next-state equations are

$$Q_0^+ = F'R'Q_0 + FQ_2 + F'RQ_1$$
$$Q_1^+ = F'R'Q_1 + FQ_0 + F'RQ_2$$
$$Q_2^+ = F'R'Q_2 + FQ_1 + F'RQ_0$$

The output equations are trivial because each output occurs in only one state:

$$Z_1 = Q_0, \quad Z_2 = Q_1, \quad Z_3 = Q_2$$

As another example, consider the state graph in Figure 15-19, which represents a sequential circuit that controls a binary multiplier. The circuit has three inputs (*St*, *M*, and *K*), and four outputs (Load, *Ad*, *Sh*, and Done). Starting in state $S_0$, if $St = 0$, then the circuit stays in $S_0$. If $St = 1$, the circuit outputs

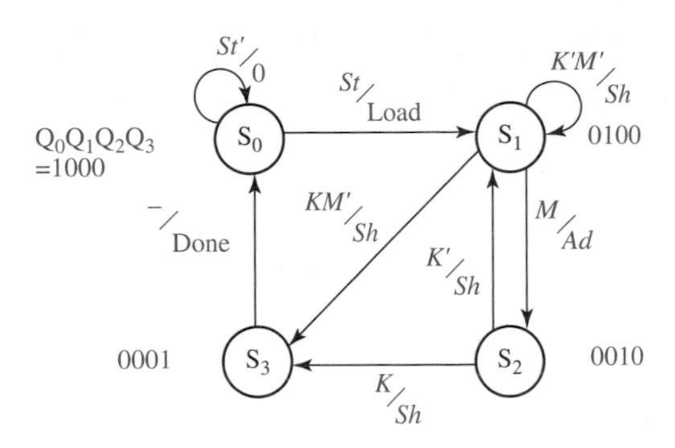


FIGURE 15-19 Multiplier Control State Graph

Load = 1 (the other outputs are 0), and the next state is $S_1$. In $S_1$, if $M = 1$, then $Ad = 1$ and the next state is $S_2$. If $M = 0$ and $K = 0$, the output is $Sh = 1$ and the next state is $S_1$. If $M = 0$ and $K = 1$, the output is $Sh = 1$ and the next state is $S_3$. In $S_2$, the output is $Sh = 1$ for both $K = 0$ and $K = 1$. If $K = 0$, the next state is $S_1$, and if $K = 1$, the next state is $S_3$. In $S_3$, the output is Done = 1 and the next state is $S_0$.

Because there are four states, a one-hot state assignment requires four flip-flops. The one-hot assignments for each state are shown on the state graph. Only $Q_0$ is 1 in $S_0$ and the other $Q$'s are 0. Similarly, only $Q_1$ is 1 in $S_1$ and the other $Q$'s are 0, etc. To determine the next-state equation for $Q_0$, note that two arrows lead into state $S_0$. The loop from state $S_0$ back to itself indicates $Q_0^+ = 1$ if $Q_0 = 1$ and $St = 0$. The arrow from $S_3$ to $S_0$ indicates $Q_0^+ = 1$ if $Q_3 = 1$. Therefore,

$$Q_0^+ = Q_0St' + Q_3$$

Because three arrows lead into $S_1$, $Q_1^+$ has three terms:

$$Q_1^+ = Q_0St + Q_1K'M' + Q_2K'$$

We can also determine the output functions by inspection of the state graph. In $S_0$, Load = 1 when $St = 1$ and Load = 0 for all other states and inputs; therefore, Load = $Q_0St$. The output $Sh$ appears in four places on the graph. $Sh = 1$ in $S_1$ if $K'M' = 1$ or if $KM' = 1$; also, $Sh = 1$ in $S_2$ if $K' = 1$ or $K = 1$. Therefore,

$$Sh = Q_1(K'M' + KM') + Q_2(K' + K) = Q_1M' + Q_2$$

When designing with CPLDs or FPGAs, you should try both an assignment with a minimum number of state variables and a one-hot assignment to see which one leads to a design with the smallest number of logic cells. Alternatively, if the speed of operation is important, the design which leads to the fastest logic should be chosen. When a one-hot assignment is used, more next-state equations are required, but for some state graphs both the next-state and output equations may contain fewer variables. An equation with fewer variables may require fewer logic cells to realize. The more cells which are cascaded, the longer the propagation delay, and the slower the operation.

# Problems

**15.1** (a) Reduce the following state table to a minimum number of states.

| Present State | Next State $X = 0$ | Next State $X = 1$ | Present Output $X = 0$ | Present Output $X = 1$ |
|---|---|---|---|---|
| $A$ | $A$ | $E$ | 1 | 0 |
| $B$ | $C$ | $F$ | 0 | 0 |
| $C$ | $B$ | $H$ | 0 | 0 |
| $D$ | $E$ | $F$ | 0 | 0 |
| $E$ | $D$ | $A$ | 0 | 0 |
| $F$ | $B$ | $F$ | 1 | 0 |
| $G$ | $D$ | $H$ | 0 | 0 |
| $H$ | $H$ | $G$ | 1 | 0 |

(b) You are given two identical sequential circuits which realize the preceding state table. One circuit is initially in state $B$ and the other circuit is initially in state $G$. Specify an input sequence of length three which could be used to distinguish between the two circuits and give the corresponding output sequence from each circuit.

**15.2** Reduce the following state table to a minimum number of states.

| Present State | Next State $X = 0$ | Next State $X = 1$ | Present Output ($Z$) |
|---|---|---|---|
| $a$ | $e$ | $e$ | 1 |
| $b$ | $c$ | $e$ | 1 |
| $c$ | $i$ | $h$ | 0 |
| $d$ | $h$ | $a$ | 1 |
| $e$ | $i$ | $f$ | 0 |
| $f$ | $e$ | $g$ | 0 |
| $g$ | $h$ | $b$ | 1 |
| $h$ | $c$ | $d$ | 0 |
| $i$ | $f$ | $b$ | 1 |

**15.3** Digital engineer B. I. Nary has just completed the design of a sequential circuit which has the following state table:

| Present State | Next State $X = 0$ | Next State $X = 1$ | Output 0 | Output 1 |
|---|---|---|---|---|
| $S_0$ | $S_5$ | $S_1$ | 0 | 0 |
| $S_1$ | $S_5$ | $S_6$ | 0 | 0 |
| $S_2$ | $S_2$ | $S_6$ | 0 | 0 |
| $S_3$ | $S_0$ | $S_1$ | 1 | 0 |
| $S_4$ | $S_4$ | $S_3$ | 0 | 0 |
| $S_5$ | $S_0$ | $S_1$ | 0 | 0 |
| $S_6$ | $S_5$ | $S_1$ | 1 | 0 |

His assistant, F. L. Ipflop, who has just completed this course, claims that his design can be used to replace Mr. Nary's circuit. Mr. Ipflop's design has the following state table:

| | Next State $X = 0$ | 1 | Output 0 | 1 |
|---|---|---|---|---|
| a | a | b | 0 | 0 |
| b | a | c | 0 | 0 |
| c | a | b | 1 | 0 |

(a) Is Mr. Ipflop correct? (Prove your answer.)
(b) If Mr. Nary's circuit is always started in state $S_0$, is Mr. Ipflop correct? (Prove your answer by showing equivalent states, etc.)

**15.4** Realize the following state table using a minimum number of AND and OR gates together with
(a) a D flip-flop
(b) an S-R flip-flop

| | $X_1X_2X_3$ 000 | 001 | 010 | 011 | 100 | 101 | 110 | 111 | Z |
|---|---|---|---|---|---|---|---|---|---|
| A | A | A | B | B | B | B | A | A | 0 |
| B | A | B | B | A | A | B | B | A | 1 |

**15.5** It is sometimes possible to save logic by using more than the minimum number of flip-flops. For both (a) and (b), fill in each state assignment by columns and, then, check for duplicate rows instead of filling in the assignments by rows and checking for permuted columns. If the columns are assigned in ascending numerical order and the first row is all 0's, then equivalent assignments will not be generated. Do not list degenerate assignments for which two columns are identical or complements of each other, or assignments where one column is all 0's or all 1's.
(a) Consider a state table with three states to be realized using three J-K flip-flops. To be sure of getting the minimum amount of logic, how many different state assignments must be tried? Enumerate these assignments.
(b) For four states and three flip-flops, 29 assignments must be tried. Enumerate 10 of these, always assigning 000 to the first state.

**15.6** A sequential circuit with one input and one output has the following state table:

| Present State | Next State $X = 0$ | $X = 1$ | Present Output |
|---|---|---|---|
| $S_1$ | $S_5$ | $S_4$ | 0 |
| $S_2$ | $S_1$ | $S_6$ | 1 |
| $S_3$ | $S_7$ | $S_8$ | 1 |
| $S_4$ | $S_7$ | $S_1$ | 0 |
| $S_5$ | $S_2$ | $S_3$ | 1 |
| $S_6$ | $S_4$ | $S_2$ | 0 |
| $S_7$ | $S_6$ | $S_8$ | 0 |
| $S_8$ | $S_5$ | $S_3$ | 1 |

(a) For this part of the problem, do not consider the flip-flop input equations (this means that you can ignore the next-state part of the table). Make a state assignment which might minimize the output equation, and derive the minimum output equation for your assignment.

(b) Forget about your solution to (a). Apply Guidelines 1 and 2 to make a state assignment, assigning 000 to $S_1$. Derive input equations for D flip-flops using this assignment.

**15.7** The following table is to be realized using D flip-flops.

(a) Find a good state assignment using the three guidelines (do not reduce the table first.) Try to satisfy as many of the adjacency conditions as possible.

(b) Using this assignment, derive the D flip-flop input equations and the output equations.

| | $X = 0$ | 1 | $Z$ $X = 0$ | 1 |
|---|---|---|---|---|
| A | F | D | 0 | 0 |
| B | D | B | 0 | 0 |
| C | A | C | 0 | 1 |
| D | F | D | 0 | 0 |
| E | A | C | 0 | 1 |
| F | F | B | 0 | 0 |

**15.8** (a) For the following state table, use the three guidelines to determine which of the three possible nonequivalent state assignments should give the best solution.

| | $X_1X_2 = 00$ | 01 | 11 | 10 | $Z_1Z_2$ $X_1X_2 = 00$ | 01 | 11 | 10 |
|---|---|---|---|---|---|---|---|---|
| A | A | C | B | D | 00 | 00 | 00 | 00 |
| B | B | B | D | D | 00 | 00 | 10 | 10 |
| C | C | A | C | A | 01 | 01 | 01 | 01 |
| D | B | B | C | A | 01 | 01 | 10 | 10 |

(b) Using your answer to (a), derive the T flip-flop input equations and the output equations.

**15.9** Implement the given state graph using D flip-flops and gates. Use a one-hot assignment and write down the logic equations by inspecting the state graph.

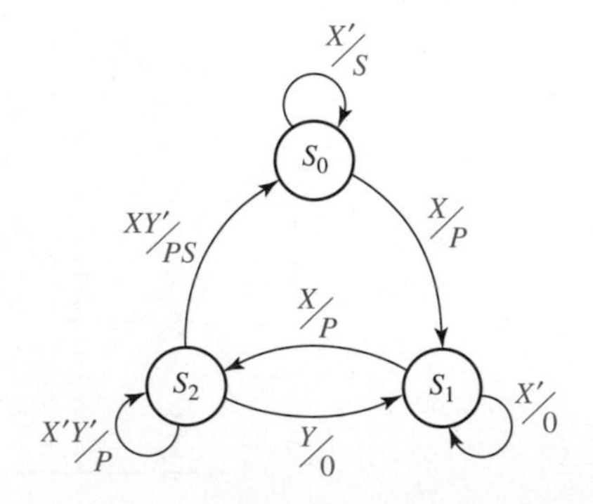

**15.10** (a) Reduce the following state table to a minimum number of states.

| Present State | Next State X = 0 | 1 | Present Output X = 0 | 1 |
|---|---|---|---|---|
| a | h | c | 1 | 0 |
| b | c | d | 0 | 1 |
| c | h | b | 0 | 0 |
| d | f | h | 0 | 0 |
| e | c | f | 0 | 1 |
| f | f | g | 0 | 0 |
| g | g | c | 1 | 0 |
| h | a | c | 1 | 0 |

(b) You are given two identical sequential circuits which realize this state table. One circuit is initially in state *d*, and the other circuit is initially in state *c*. Specify an input sequence of length two which could be used to distinguish between the two circuits, and give the corresponding output sequence from each circuit.

**15.11** For the following state table:

(a) Reduce the table to a minimum number of states.

(b) Using the basic definition of state equivalence, show that state *a* is not equivalent to state *b*.

| Present State | Next State X = 0 | X = 1 | Present Output X = 0 | X = 1 |
|---|---|---|---|---|
| a | e | g | 0 | 1 |
| b | d | f | 0 | 1 |
| c | e | c | 1 | 0 |
| d | b | f | 0 | 1 |
| e | g | f | 0 | 1 |
| f | b | d | 1 | 0 |
| g | e | c | 1 | 0 |

**15.12** A Moore sequential circuit has a single input ($X$) and a single output ($Z$). $Z$ is 1 if the most recent four inputs contained exactly two consecutive 1's or exactly two consecutive 0's, i.e., the input sequences 0011, 1001, 1100, 0110, 0100, 1011, and 1101. (The initial state $S_0$ acts as if the preceding inputs were all 0's.) The following state table was constructed using a sufficient number of states to remember the last four inputs and the output for each state assigned according to the sequence remembered by that state.

(a) Reduce the table to a minimum number of states. (*Hint:* First use the simple example of state equivalence used in Section 15.1 to eliminate as many states as possible.)

(b) For each state in the reduced table, give the input pattern remembered by that state.

(c) Convert the reduced table from Part (a) into a Mealy state table that produces the same outputs.

(d) Reduce the Mealy state table to a minimum number of states.

| Input Pattern | Present State | Next State $X = 0$ | $X = 1$ | Output $Z$ |
|---|---|---|---|---|
| 0000 | $S_0$ | $S_0$ | $S_1$ | 0 |
| 0001 | $S_1$ | $S_2$ | $S_3$ | 0 |
| 0010 | $S_2$ | $S_4$ | $S_5$ | 0 |
| 0011 | $S_3$ | $S_6$ | $S_7$ | 1 |
| 0100 | $S_4$ | $S_8$ | $S_9$ | 1 |
| 0101 | $S_5$ | $S_{10}$ | $S_{11}$ | 0 |
| 0110 | $S_6$ | $S_{12}$ | $S_{13}$ | 1 |
| 0111 | $S_7$ | $S_{14}$ | $S_{15}$ | 0 |
| 1000 | $S_8$ | $S_0$ | $S_1$ | 0 |
| 1001 | $S_9$ | $S_2$ | $S_3$ | 1 |
| 1010 | $S_{10}$ | $S_4$ | $S_5$ | 0 |
| 1011 | $S_{11}$ | $S_6$ | $S_7$ | 1 |
| 1100 | $S_{12}$ | $S_8$ | $S_9$ | 1 |
| 1101 | $S_{13}$ | $S_{10}$ | $S_{11}$ | 1 |
| 1110 | $S_{14}$ | $S_{12}$ | $S_{13}$ | 0 |
| 1111 | $S_{15}$ | $S_{14}$ | $S_{15}$ | 0 |

**15.13** A sequential circuit has a single input ($X$) and a single output ($Z$). The circuit examines each disjoint block of four inputs and determines whether the block is a valid BCD representation of a decimal digit; if not, $Z = 1$. State $S_0$ is the initial state, and the circuit enters state $S_0$ after the fourth input. The BCD digits are received most significant bit first. The following state table was constructed as a Mealy table using a sufficient number of states to remember the last three inputs with the output produced when the fourth input bit of a block is received.

(a) Does the resulting table specify a Mealy or a Moore circuit?
(b) Reduce the state table to a minimum number of states. (*Hint:* Use the simple example of state equivalence used in Section 15.1 to eliminate as many states as possible.)
(c) For each state in the reduced table, give the input pattern remembered by that state.

| Input Pattern | Present State | Next State $X = 0$ | $X = 1$ | Present Output $Z$ $X = 0$ | $X = 1$ |
|---|---|---|---|---|---|
| – | $S_1$ | $S_2$ | $S_3$ | 0 | 0 |
| 0 | $S_2$ | $S_4$ | $S_5$ | 0 | 0 |
| 1 | $S_3$ | $S_6$ | $S_7$ | 0 | 0 |
| 00 | $S_4$ | $S_8$ | $S_9$ | 0 | 0 |
| 01 | $S_5$ | $S_{10}$ | $S_{11}$ | 0 | 0 |
| 10 | $S_6$ | $S_{12}$ | $S_{13}$ | 0 | 0 |
| 11 | $S_7$ | $S_{14}$ | $S_{15}$ | 0 | 0 |
| 000 | $S_8$ | $S_1$ | $S_1$ | 0 | 0 |
| 001 | $S_9$ | $S_1$ | $S_1$ | 0 | 0 |
| 010 | $S_{10}$ | $S_1$ | $S_1$ | 0 | 0 |
| 011 | $S_{11}$ | $S_1$ | $S_1$ | 0 | 0 |
| 100 | $S_{12}$ | $S_1$ | $S_1$ | 0 | 0 |
| 101 | $S_{13}$ | $S_1$ | $S_1$ | 1 | 1 |
| 110 | $S_{14}$ | $S_1$ | $S_1$ | 1 | 1 |
| 111 | $S_{15}$ | $S_1$ | $S_1$ | 1 | 1 |

**15.14** A sequential circuit has a single input ($X$) and a single output ($Z$). The circuit examines each disjoint block of four inputs and determines whether the block is a valid BCD representation of a decimal digit; if not, $Z = 1$. State $S_0$ is the initial state, and the circuit enters state $S_0$ after the fourth input. The BCD digits are received least significant bit first. A Mealy state table can be constructed using a sufficient number of states to remember the last three inputs with the output produced when the fourth input bit of a block is received.

(a) Using the method indicated by Problem 15.13, construct a state table for this circuit.

(b) Reduce the state table to a minimum number of states. (*Hint:* Use the simple example of state equivalence used in Section 15.1 to eliminate as many states as possible.)

(c) For each state in the reduced table, give the input pattern(s) remembered by that state.

**15.15** Reduce each of the following state tables to a minimum number of states:

(a)

| | XY = 00 | 01 | 11 | 10 | Z |
|---|---|---|---|---|---|
| a | a | c | e | d | 0 |
| b | d | e | e | a | 0 |
| c | e | a | f | b | 1 |
| d | b | c | c | b | 0 |
| e | c | d | f | a | 1 |
| f | f | b | a | d | 1 |

(b)

| | X = 0 | 1 | 0 | 1 |
|---|---|---|---|---|
| a | b | c | 1 | 0 |
| b | e | d | 1 | 0 |
| c | g | d | 1 | 1 |
| d | e | b | 1 | 0 |
| e | f | g | 1 | 0 |
| f | h | b | 1 | 1 |
| g | h | i | 0 | 1 |
| h | g | i | 0 | 1 |
| i | a | a | 0 | 1 |

**15.16** Reduce each of the following tables to a minimum number of states:

(a)

| | XY = 00 | 01 | 11 | 10 | Z |
|---|---|---|---|---|---|
| a | b | i | c | g | 0 |
| b | b | c | f | g | 0 |
| c | h | d | d | f | 1 |
| d | h | c | e | g | 1 |
| e | b | c | i | g | 0 |
| f | f | i | i | k | 0 |
| g | j | k | g | h | 0 |
| h | e | f | c | g | 0 |
| i | i | i | i | d | 0 |
| j | b | f | c | g | 0 |
| k | a | c | e | g | 1 |

| (b) | XY = 00 | 01 | 11 | 10 | XY = 00 | 01 | 11 | 10 |
|---|---|---|---|---|---|---|---|---|
| a | a | a | g | k | 1 | 0 | 0 | 0 |
| b | c | f | g | d | 0 | 0 | 0 | 0 |
| c | g | c | a | i | 1 | 0 | 0 | 0 |
| d | a | d | g | i | 1 | 0 | 0 | 0 |
| e | f | h | g | a | 0 | 0 | 0 | 0 |
| f | g | c | d | k | 1 | 0 | 0 | 0 |
| g | c | j | g | e | 0 | 1 | 0 | 0 |
| h | g | h | d | k | 1 | 0 | 0 | 0 |
| i | h | h | g | d | 0 | 0 | 0 | 0 |
| j | j | j | g | k | 1 | 0 | 0 | 0 |
| k | c | c | g | d | 0 | 0 | 0 | 0 |

**15.17** Circuits $N$ and $M$ have the state tables that follow.

(a) Without first reducing the tables, determine whether circuits $N$ and $M$ are equivalent.

(b) Reduce each table to a minimum number of states, and then show that $N$ is equivalent to $M$ by inspecting the reduced tables.

*M*

| | X = 0 | 1 | |
|---|---|---|---|
| $S_0$ | $S_3$ | $S_1$ | 0 |
| $S_1$ | $S_0$ | $S_1$ | 0 |
| $S_2$ | $S_0$ | $S_2$ | 1 |
| $S_3$ | $S_0$ | $S_3$ | 1 |

*N*

| | X = 0 | 1 | |
|---|---|---|---|
| A | E | A | 1 |
| B | F | B | 1 |
| C | E | D | 0 |
| D | E | C | 0 |
| E | B | D | 0 |
| F | B | C | 0 |

**15.18** Below is an incompletely specified state table.

| Present State | Next State X = 0 | 1 | Present Output (Z) |
|---|---|---|---|
| $S_0$ | $S_1$ | $S_0$ | 0 |
| $S_1$ | $S_0$ | $S_2$ | 0 |
| $S_2$ | $S_3$ | $S_4$ | 1 |
| $S_3$ | $S_0$ | $S_3$ | 0 |
| $S_4$ | $S_0$ | – | 0 |

(a) Reduce the state table to four states in two different ways by filling in the don't-care in the state table in different ways.

(b) Show that your two state tables in Part (a) are not equivalent, using an implication table similar to Figure 15-7.

(c) Show that your two state tables in (a) are not equivalent by giving a short input sequence which gives different outputs for the two state tables.

**15.19** Repeat 15.18 for this state table (four states).

| Present State | Next State $X = 0$ | $1$ | Present Output ($Z$) $X = 0$ | $1$ |
|---|---|---|---|---|
| $S_0$ | $S_1$ | $S_5$ | 0 | 0 |
| $S_1$ | $S_3$ | $S_2$ | 1 | 1 |
| $S_2$ | $S_2$ | $S_4$ | 0 | 1 |
| $S_3$ | $S_4$ | $S_2$ | 1 | 1 |
| $S_4$ | $S_4$ | $S_2$ | – | 1 |
| $S_5$ | $S_5$ | $S_2$ | 0 | 1 |

**15.20** The following are possible state assignments for a six-state sequential circuit.

| | (i) | (ii) | (iii) | (iv) | (v) |
|---|---|---|---|---|---|
| $S_0$ | 000 | 010 | 100 | 110 | 001 |
| $S_1$ | 001 | 111 | 101 | 010 | 111 |
| $S_2$ | 010 | 001 | 000 | 111 | 101 |
| $S_3$ | 011 | 110 | 001 | 011 | 011 |
| $S_4$ | 100 | 000 | 111 | 000 | 000 |
| $S_5$ | 101 | 100 | 110 | 100 | 010 |

(a) Which two state assignments are equivalent?

(b) For each assignment (except (i)), give an equivalent assignment for which state $S_0$ is assigned to 000.

(c) Give a state assignment which is not equivalent to any of the assignments.

**15.21** (a) For an eight-state sequential circuit using three flip-flops, give three state assignments that assign 000 to $S_0$ and are equivalent to a straight binary assignment.

(b) Give three state assignments that assign 111 to $S_0$ and are not equivalent to a straight binary assignment or to each other.

**15.22** A sequential circuit with one input and one output has the following state table:

| Present State | Next State $X = 0$ | $X = 1$ | Present Output |
|---|---|---|---|
| $A$ | $D$ | $G$ | 1 |
| $B$ | $E$ | $H$ | 0 |
| $C$ | $B$ | $F$ | 1 |
| $D$ | $F$ | $G$ | 0 |
| $E$ | $C$ | $A$ | 1 |
| $F$ | $H$ | $C$ | 0 |
| $G$ | $E$ | $A$ | 1 |
| $H$ | $D$ | $B$ | 0 |

(a) For this part of the problem, do not consider the flip-flop input equations (this means that you can ignore the next-state part of the table). Make a state assignment which will minimize the output equation, and derive the minimum output equation for your assignment.

(b) Forget about your solution to (a). Apply Guidelines 1 and 2 to make a state assignment, assigning 000 to $A$. Derive input equations for D flip-flops using this assignment.

**15.23** (a) For the following state table, use the three guidelines to determine which of the three possible nonequivalent state assignments should give the best solution.

| | $X_1X_2 = 00$ | 01 | 11 | 10 | $Z_1Z_2$: $X_1X_2 = 00$ | 01 | 11 | 10 |
|---|---|---|---|---|---|---|---|---|
| A | A | A | C | C | 01 | 01 | 01 | 01 |
| B | B | D | B | D | 11 | 11 | 11 | 11 |
| C | A | A | B | D | 11 | 11 | 00 | 00 |
| D | D | B | A | C | 01 | 01 | 01 | 01 |

(b) Using your answer to (a), derive J-K flip-flop input equations and the output equations.

**15.24** Consider the following Moore sequential circuit.
(a) Derive the equations for a one-hot state assignment.
(b) Use Guidelines 1 and 2 to make a "good" state assignment using three state variables. Derive the next-state equations assuming D flip-flops are used.

| Present State | Next State $X = 0$ | $X = 1$ | Present Output $Z$ |
|---|---|---|---|
| A | B | A | 0 |
| B | C | A | 0 |
| C | E | D | 0 |
| D | B | A | 1 |
| E | E | A | 0 |

**15.25** (a) Reduce the following state table to a minimum number of states using implication charts.
(b) Use the guideline method to determine a suitable state assignment for the reduced table.
(c) Realize the table using D flip-flops.
(d) Realize the table using J-K flip-flops.

| | $X = 0$ | 1 | $Z$ |
|---|---|---|---|
| A | A | B | 1 |
| B | C | E | 0 |
| C | F | G | 1 |
| D | C | A | 0 |
| E | I | G | 1 |
| F | H | I | 1 |
| G | C | F | 0 |
| H | F | B | 1 |
| I | C | E | 0 |

**15.26** Repeat Problem 15.25 for the following table:

| | $X = 0$ | $1$ | $Z$ |
|---|---|---|---|
| $A$ | $I$ | $C$ | 1 |
| $B$ | $B$ | $I$ | 1 |
| $C$ | $C$ | $G$ | 1 |
| $D$ | $I$ | $C$ | 0 |
| $E$ | $D$ | $E$ | 0 |
| $F$ | $I$ | $C$ | 0 |
| $G$ | $E$ | $F$ | 0 |
| $H$ | $H$ | $A$ | 1 |
| $I$ | $A$ | $C$ | 1 |

**15.27** Make a suitable state assignment and realize the state graph of Figure 14-9 using:
(a) D flip-flops (b) S-R flip-flops

**15.28** Make a suitable state assignment and realize the state graph of Figure 14-12 using:
(a) J-K flip-flops (b) T flip-flops

**15.29** Make a suitable state assignment and realize the state table of Problem 14.22 using D flip-flops and NAND gates.

**15.30** Make a suitable state assignment and realize the state table of Problem 14.5 using J-K flip-flops and NAND gates.

**15.31** Reduce the state table of Problem 14.6 to a minimum number of rows. Then, make a suitable state assignment and realize the state table using D flip-flops.

**15.32** Reduce the state table of Problem 14.23 to a minimum number of rows. Then, make a suitable state assignment and realize the state table using D flip-flops.

**15.33** A logic designer who had not taken this course designed a sequential circuit with an input $W$ using three T flip-flops, $A$, $B$, and $C$. The input equations for these flip-flops are

$$T_A = W'A'B + W'BC' + A'BC' + AB'C + WB'C + WAC$$
$$T_B = W'A'C + W'A'B + A'BC + AB'C' + WB'C' + WAC'$$
$$T_C = W'AC + W'B'C' + WBC + WA'C'$$

and the output equation is $Z = W'BC'$. Find an equivalent sequential circuit which uses fewer states. Realize it, trying to minimize the amount of logic required.

**15.34** Modify the given state graph so that it is completely specified. Assume that if $X = Y = 1$, $X$ takes precedence. Then implement the state graph using D flip-flops and gates. Use a one-hot assignment and write down the logic equations by inspecting the state graph.

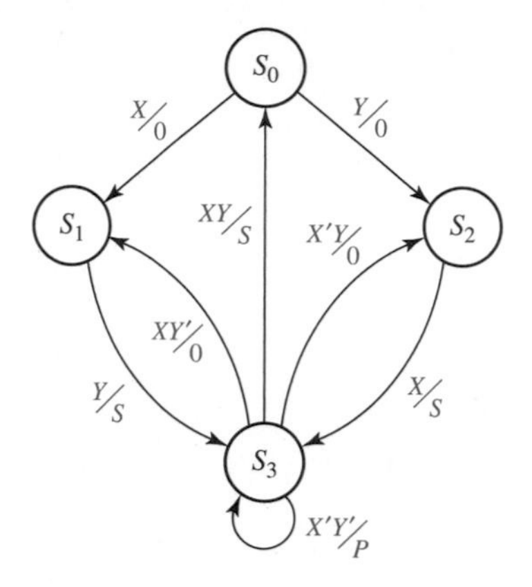


**15.35** Implement the following state graph using D flip-flops and gates. Use a one-hot assignment and write down the logic equations by inspecting the state graph.

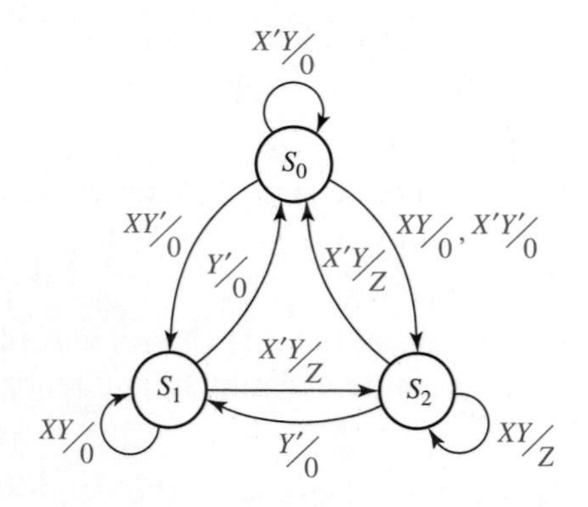


**15.36** A state graph for a single-input sequential circuit is given. Implement the circuit using a three-bit parallel loading counter that has the given operation table. Label the counter outputs $Q_2,Q_1,Q_0$, where $Q_0$ is the least significant bit and the parallel inputs $P_2,P_1,P_0$. (*Hint*: Because the Ld signal overrides the Cnt signal, the counting sequence can easily be changed by doing a parallel load at the appropriate times.)

| Clr | Ld | Cnt | Function |
|---|---|---|---|
| 0 | — | — | Clear |
| 1 | 1 | — | Parallel Load |
| 1 | 0 | 1 | Increment |
| 1 | 0 | 0 | Hold (No change) |

**15.37** Consider the following Mealy sequential circuit.

| Present State | Next State $X = 0$ | Next State $X = 1$ | Present Output $X = 0$ | Present Output $X = 1$ |
|---|---|---|---|---|
| $A$ | $B$ | $A$ | 0 | 0 |
| $B$ | $C$ | $A$ | 0 | 0 |
| $C$ | $D$ | $A$ | 0 | 1 |
| $D$ | $D$ | $A$ | 0 | 0 |

(a) Use a one-hot state assignment, and implement the circuit using D flip-flops.
(b) Use the state assignment $A = 00$, $B = 01$, $C = 11$, and $D = 10$, and implement the circuit using D flip-flops.
(c) Implement the circuit using a 4-bit parallel loading counter instead of flip-flops for memory. Assume the synchronous counter controls are as follows:

| $S_1$ | $S_0$ | Function |
|---|---|---|
| 0 | 0 | Hold |
| 0 | 1 | Increment |
| 1 | 0 | Parallel load |
| 1 | 1 | Clear |

$Q_3Q_2Q_1Q_0$ are the outputs; $P_3P_2P_1P_0$ are the parallel inputs; and $Q_0$ is the least significant bit of the counter. (With the proper state assignment, this can be done without using the parallel load function of the counter.)
(d) Implement the circuit using a 4-bit parallel loading shift register instead of flip-flops for memory. Assume the synchronous shift register controls are as follows:

| $S_1$ | $S_0$ | Function |
|---|---|---|
| 0 | 0 | Hold |
| 0 | 1 | Shift right |
| 1 | 0 | Parallel load |
| 1 | 1 | Clear |

$S_{in}$ is the input for the shift; $Q_3Q_2Q_1Q_0$ are the outputs; $P_3P_2P_1P_0$ are the parallel inputs. When shifting, $S_{in} \rightarrow Q_3$, $Q_3 \rightarrow Q_2$, $Q_2 \rightarrow Q_1$, $Q_1 \rightarrow Q_0$. (With the proper state assignment, this can be done without using the parallel load function of the shift register.)

**15.38** A sequential circuit contains two D flip-flops; the excitation equations for the flip-flops are $D_1 = XQ_1 + XQ_2$ and $D_2 = XQ_1 + XQ_2'$
(a) Convert the circuit into an equivalent one where each D flip-flop is replaced by a T flip-flop. Do this by converting the next-state equations into the form for a T flip-flop. [*Hint:* $MQ + NQ' = (M'Q + NQ')'Q + (M'Q + NQ')Q'$.]
(b) Repeat Part (a) by constructing an excitation table for the T flip-flops, i.e., a truth table for $T_1$ and $T_2$ as a function of $X$, $Q_1$, and $Q_2$.
(c) Convert the circuit into an equivalent one where each D flip-flop is replaced by a J-K flip-flop. Do this by converting the next-state equations into the form for a J-K flip-flop.

(d) Repeat Part (c) by constructing an excitation table for the J-K flip-flops, i.e., a truth table for the J and K flip-flop inputs as a function of $X$, $Q_1$, and $Q_2$.

**15.39** A sequential circuit contains two J-K flip-flops; the excitation equations for the flip-flops are $J_1 = Q_2, K_1 = Q_1, J_2 = X + Q_1'$, and $K_2 = 1$.

(a) Convert the circuit into an equivalent one where each J-K flip-flop is replaced by a T flip-flop. Do this by converting the next-state equations into the form for a T flip-flop. [*Hint:* $MQ + NQ' = (M'Q + NQ')'Q + (M'Q + NQ')Q'$.]

(b) Repeat Part (a) by constructing an excitation table for the T flip-flops, i.e., a truth table for $T_1$ and $T_2$ as a function of $X$, $Q_1$, and $Q_2$.

(c) Convert the circuit into an equivalent one where each D flip-flop is replaced by an S-R flip-flop. Do this by converting the next-state equations into the form for a S-R flip-flop.

(d) Repeat Part (c) by constructing an excitation table for the S-R flip-flops, i.e., a truth table for the S and R flip-flop inputs as a function of $X$, $Q_1$, and $Q_2$.

UNIT 16

# Sequential Circuit Design

## Objectives

1. Design a sequential circuit using gates and flip-flops.
2. Test your circuit by simulating it and by implementing it in lab.
3. Design a unilateral iterative circuit. Explain the relationship between iterative and sequential circuits, and convert from one to the other.
4. Show how to implement a sequential circuit using a ROM or PLA and flip-flops.
5. Explain the operation of CPLDs and FPGAs and show how they can be used to implement sequential logic.

# Study Guide

1. **1.** Study Sections 16.1, *Summary of Design Procedure for Sequential Circuits*, and 16.2, *Design Example—Code Converter*.
   (a) Why are the states in the next-state part of Table 16-2 listed in a different order from the states in Table 15-1?

   (b) Consider the design of a sequential circuit to convert an 8-4-2-1 code to a 6-3-1-1 code (see Table 1-2). If the least significant bit of an 8-4-2-1 coded digit is fed into the circuit at $t_0$, can the least significant bit of the 6-3-1-1 coded digit be determined immediately? Explain. Why can the technique described in this section not be used to design the 8-4-2-1 to 6-3-1-1 code converter?

**2.** Study Section 16.3, *Design of Iterative Circuits*

(a) Draw a state graph for the comparator of Table 16-4. Compare several pairs of binary numbers using the scheme represented by Table 16-4 and make sure you understand why this method works. Draw a circuit similar to Figure 16-6 with five cells. Show the values of all the cell inputs and outputs if $X = 10101$ and $Y = 10011$.

(b) If the state table for a typical cell of an iterative circuit has $n$ states, what is the minimum number of signals required between each pair of adjacent cells?

(c) Work Problem 16.17.

**3.** Study Section 16.4, *Design of Sequential Circuits Using ROMs and PLAs.*

(a) Review Section 9.5, *Read-Only Memories*, and Section 9.6, *Programmable Logic Devices*.

(b) What size ROM would be required to realize a state table with 13 states, two input variables, and three output variables?

(c) In going from Table 16-6(b) to 16-6(c), note that for $X = 0$, $Q_1Q_2Q_3 = 000$, $Z = 1$, and $Q_1^+Q_2^+Q_3^+ = D_1D_2D_3 = 001$; therefore, 1001 is entered in the first row of the truth table. Verify that the other truth table entries are correct.

(d) Continue the analysis of the PLA realization of the code converter which was started in the paragraph following Table 16-7. In particular, if $Q_1Q_2Q_3 = 100$ and $X = 1$, what will be the PLA outputs? What will the state be after the clock?

(e) Work Problems 16.15 and 16.16.

**4.** Study Section 16.5, *Sequential Circuit Design Using CPLDs*, and Section 16.6, *Sequential Circuit Design Using FPGAs*.

(a) How many macrocells of a CoolRunner-II are needed to implement a Moore machine with six states, one input, and two outputs? With two inputs?

(b) How many LUT's of a Virtex/Spartan II are needed to implement each of these Moore machines? How many CLB's?

(c) Rewrite the equations for $Q_2^+$, $Q_1^+$, and $Q_0^+$ of Equations 12-1 to fit into one LUT each, as we did for $Q_3^+$ in Equation (16-2), using CE $= Ld + Sh$.

**5.** Study Section 16.7, *Simulation and Testing of Sequential Circuits*.

(a) Observe the simulator output of Figure 16-23(b), and note the times at which the $Z$ output changes. Assuming that each gate and flip-flop in Figure 16-22 has a 10-ns delay, explain the $Z$ waveform.

(b) Suppose that you are testing the circuit of Figure 16-4, and that when you set $X = 0$ and $Q_1Q_2Q_3 = 011$ and pulse the clock, the circuit goes to state 100 instead of 000. What would you do to determine the cause of the malfunction?

**6.** Read Section 16.8, *Overview of Computer-Aided Design*, for general information.

**7.** Work out your assigned design problem by hand. Then, use *LogicAid* to check your state table using the state table checker, and then verify that your logic equations are correct. Try at least two different state assignments and choose the one which requires the smallest number of logic gates.

**8.** Answer the following questions before you simulate your circuit or test it in lab. At which of the following times will the output of your circuit be correct? (If you are not absolutely sure that your answer is correct, review Section 13.2, paying particular attention to the timing charts for Mealy circuits.)

(a) Just before the rising clock edge

(b) Just after the rising clock edge (after the state has changed but before the input is changed to the next value)

(c) After the input has been changed to the next value, but before the next rising edge occurs

**9.** (a) Explain how it is possible to get false outputs from your circuit even though the circuit is correctly designed and working properly.

(b) If the output of your circuit was fed into another sequential circuit using the same clock, would the false outputs cause any problems? Explain.

**10.** When you get your circuit working properly, determine the output sequences for the given test sequences. Demonstrate the operation of your circuit to a proctor and have him or her check your output sequences. After successful completion of the project, turn in your design and the test results. (No readiness test is required.)

# Sequential Circuit Design

We have already studied the various steps in sequential circuit design—derivation of state tables (Unit 14), state table reduction (Unit 15), state assignment (Unit 15), and derivation of flip-flop input equations (Units 12 and 15). This unit contains a summary of the design procedure, a comprehensive design example, and procedures for testing your circuit in lab.

## 16.1 Summary of Design Procedure for Sequential Circuits

1. Given the problem statement, determine the required relationship between the input and output sequences and derive a state table. For many problems, it is easiest to first construct a state graph.
2. Reduce the table to a minimum number of states. First, eliminate duplicate rows by row matching and, then, form an implication table and follow the procedure in Section 15.3.

3. If the reduced table has $m$ states ($2^{n-1} < m \le 2^n$), $n$ flip-flops are required. Assign a unique combination of flip-flop states to correspond to each state in the reduced table. The guidelines given in Section 15.8 may prove helpful in finding an assignment which leads to an economical circuit.
4. Form the transition table by substituting the assigned flip-flop states for each state in the reduced state table. The resulting transition table specifies the next states of the flip-flops, and the output in terms of the present states of the flip-flops and the input.
5. Plot next-state maps and input maps for each flip-flop and derive the flip-flop input equations. (Depending on the type of gates to be used, either determine the sum-of-products form from the 1's on the map or the product-of-sums form from the 0's on the map.) Derive the output functions.
6. Realize the flip-flop input equations and the output equations using the available logic gates.
7. Check your design by signal tracing, computer simulation, or laboratory testing.

## 16.2 Design Example–Code Converter

We will design a sequential circuit to convert BCD to excess-3 code. This circuit adds three to a binary-coded-decimal digit in the range 0 to 9. The input and output will be serial with the least significant bit first. A list of allowed input and output sequences is shown in Table 16-1.

Table 16-1 lists the desired inputs and outputs at times $t_0$, $t_1$, $t_2$, and $t_3$. After receiving four inputs, the circuit should reset to the initial state, ready to receive another group of four inputs. It is not clear at this point whether a sequential circuit can actually be realized to produce the output sequences as specified in Table 16-1 without delaying the output.

TABLE 16-1

| $X$ Input (BCD) | | | | $Z$ Output (excess-3) | | | |
|---|---|---|---|---|---|---|---|
| $t_3$ | $t_2$ | $t_1$ | $t_0$ | $t_3$ | $t_2$ | $t_1$ | $t_0$ |
| 0 | 0 | 0 | 0 | 0 | 0 | 1 | 1 |
| 0 | 0 | 0 | 1 | 0 | 1 | 0 | 0 |
| 0 | 0 | 1 | 0 | 0 | 1 | 0 | 1 |
| 0 | 0 | 1 | 1 | 0 | 1 | 1 | 0 |
| 0 | 1 | 0 | 0 | 0 | 1 | 1 | 1 |
| 0 | 1 | 0 | 1 | 1 | 0 | 0 | 0 |
| 0 | 1 | 1 | 0 | 1 | 0 | 0 | 1 |
| 0 | 1 | 1 | 1 | 1 | 0 | 1 | 0 |
| 1 | 0 | 0 | 0 | 1 | 0 | 1 | 1 |
| 1 | 0 | 0 | 1 | 1 | 1 | 0 | 0 |

For example, if at $t_0$ some sequences required an output $Z = 0$ for $X = 0$ and other sequences required $Z = 1$ for $X = 0$, it would be impossible to design the circuit without delaying the output. For Table 16-1 we see that at $t_0$ if the input is 0 the output is always 1, and if the input is 1 the output is always 0; therefore, there is no conflict at $t_0$. At time $t_1$ the circuit will have available only the inputs received at $t_1$ and $t_0$. There will be no conflict at $t_1$ if the output at $t_1$ can be determined only from the inputs received at $t_1$ and $t_0$. If 00 has been received at $t_1$ and $t_0$, the output should be 1 at $t_1$ in all three cases where 00 occurs in the table. If 01 has been received, the output should be 0 at $t_1$ in all three cases where 01 occurs. For sequences 10 and 11 the outputs at $t_1$ should be 0 and 1, respectively. Therefore, there is no output conflict at $t_1$. In a similar manner we can check to see that there is no conflict at $t_2$, and at $t_3$ all four inputs are available, so there is no problem.

We will now proceed to set up the state table (Table 16-2), using the same procedure as in Section 15.1. The arrangement of next states in the table is different from that in Table 15-1 because in this example the input sequences are received with least significant bit first, while for Table 15-1 the first input bit received is listed first in the sequence. Dashes (don't-cares) appear in this table because only 10 of the 16 possible 4-bit sequences can occur as inputs to the code converter. The output part of the table is filled in, using the reasoning discussed in the preceding paragraph. For example, if the circuit is in state $B$ at $t_1$ and a 1 is received, this means that the sequence 10 has been received and the output should be 0.

Next, we will reduce the table using row matching. When matching rows which contain dashes (don't-cares), a dash will match with any state or with any output value. By matching rows in this manner, we have $H \equiv I \equiv J \equiv K \equiv L$ and $M \equiv N \equiv P$. After eliminating $I$, $J$, $K$, $L$, $N$, and $P$, we find $E \equiv F \equiv G$ and the table reduces to seven rows (Table 16-3).

**TABLE 16-2** State Table for Code Converter

| Time | Input Sequence Received (Least Significant Bit First) | Present State | Next State $X = 0$ | Next State $X = 1$ | Present Output ($Z$) $X = 0$ | Present Output ($Z$) $X = 1$ |
|---|---|---|---|---|---|---|
| $t_0$ | reset | $A$ | $B$ | $C$ | 1 | 0 |
| $t_1$ | 0 | $B$ | $D$ | $F$ | 1 | 0 |
| | 1 | $C$ | $E$ | $G$ | 0 | 1 |
| $t_2$ | 00 | $D$ | $H$ | $L$ | 0 | 1 |
| | 01 | $E$ | $I$ | $M$ | 1 | 0 |
| | 10 | $F$ | $J$ | $N$ | 1 | 0 |
| | 11 | $G$ | $K$ | $P$ | 1 | 0 |
| $t_3$ | 000 | $H$ | $A$ | $A$ | 0 | 1 |
| | 001 | $I$ | $A$ | $A$ | 0 | 1 |
| | 010 | $J$ | $A$ | – | 0 | – |
| | 011 | $K$ | $A$ | – | 0 | – |
| | 100 | $L$ | $A$ | – | 0 | – |
| | 101 | $M$ | $A$ | – | 1 | – |
| | 110 | $N$ | $A$ | – | 1 | – |
| | 111 | $P$ | $A$ | – | 1 | – |

**TABLE 16-3** Reduced State Table for Code Converter

| Time | Present State | Next State $X = 0$ | Next State $X = 1$ | Present Output ($Z$) $X = 0$ | Present Output ($Z$) $X = 1$ |
|---|---|---|---|---|---|
| $t_0$ | $A$ | $B$ | $C$ | 1 | 0 |
| $t_1$ | $B$ | $D$ | $E$ | 1 | 0 |
| | $C$ | $E$ | $E$ | 0 | 1 |
| $t_2$ | $D$ | $H$ | $H$ | 0 | 1 |
| | $E$ | $H$ | $M$ | 1 | 0 |
| $t_3$ | $H$ | $A$ | $A$ | 0 | 1 |
| | $M$ | $A$ | – | 1 | – |

An alternate approach to deriving Table 16-2 is to start with a state graph. The state graph (Figure 16-1) has the form of a tree. Each path starting at the reset state represents one of the ten possible input sequences. After the paths for the input sequences have been constructed, the outputs can be filled in by working backwards along each path. For example, starting at $t_3$, the path 0 0 0 0 has outputs 0 0 1 1 and the path 1 0 0 0 has outputs 1 0 1 1. Verify that Table 16-2 corresponds to this state graph.

Three flip-flops are required to realize the reduced table because there are seven states. Each of the states must be assigned a unique combination of flip-flop states. Some assignments will lead to economical circuits with only a few gates, while other assignments will require many more gates. Using the guidelines given in Section 15.8, states $B$ and $C$, $D$ and $E$, and $H$ and $M$ should be given adjacent assignments in order to simplify the next-state functions. To simplify the output function, states ($A$, $B$, $E$, and $M$) and ($C$, $D$, and $H$) should be given adjacent assignments. A good assignment for this example is given on the map and table in Figure 16-2. After the state assignment has been made, the transition table is filled in according to the assignment, and the next-state maps are plotted as shown in Figure 16-3. The $D$ input equations are then read off the $Q^+$ maps as indicated. Figure 16-4 shows the resulting sequential circuit.

**FIGURE 16-1** State Graph for Code Converter

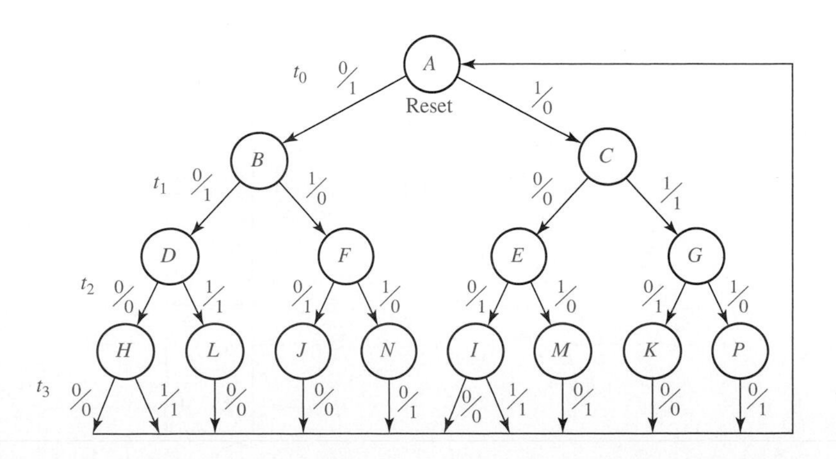

**FIGURE 16-2** Assignment Map and Transition Table for Flip-Flops

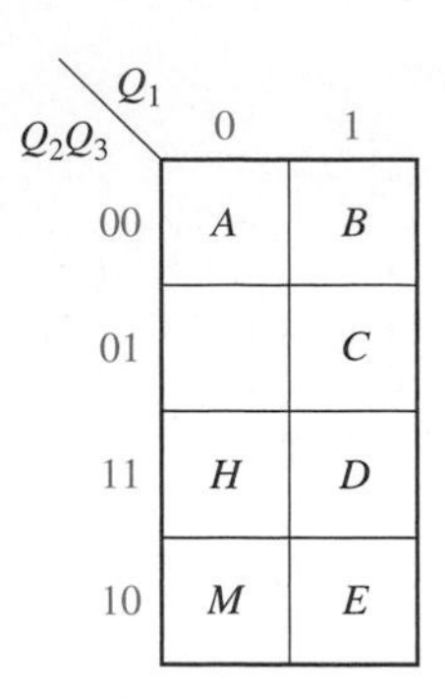

$Q_1$
$Q_2Q_3$
0
1
00
01
11
10
A
B
C
H
D
M
E


(a) Assignment map

| | $Q_1Q_2Q_3$ | $Q_1^+Q_2^+Q_3^+$ $X = 0$ | $X = 1$ | $Z$ $X = 0$ | $X = 1$ |
|---|---|---|---|---|---|
| A | 000 | 100 | 101 | 1 | 0 |
| B | 100 | 111 | 110 | 1 | 0 |
| C | 101 | 110 | 110 | 0 | 1 |
| D | 111 | 011 | 011 | 0 | 1 |
| E | 110 | 011 | 010 | 1 | 0 |
| H | 011 | 000 | 000 | 0 | 1 |
| M | 010 | 000 | xxx | 1 | x |
| – | 001 | xxx | xxx | x | x |

(b) Transition table

**FIGURE 16-3** Karnaugh Maps for Code Converter Design

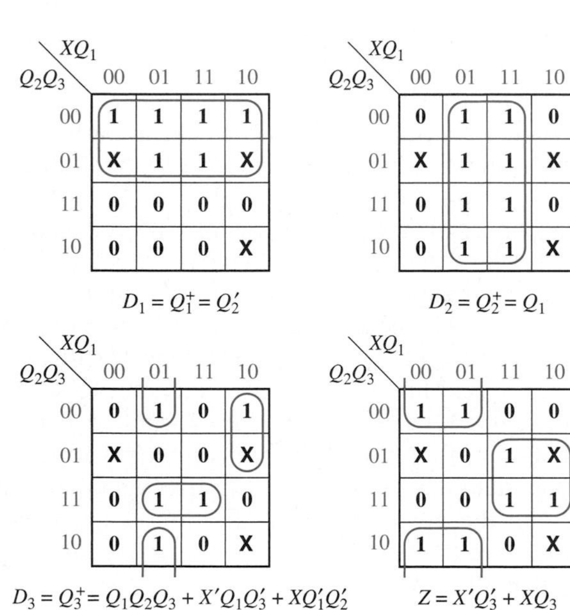

$XQ_1$
$Q_2Q_3$
00 01 11 10
$D_1 = Q_1^+ = Q_2'$
$D_2 = Q_2^+ = Q_1$
$D_3 = Q_3^+ = Q_1Q_2Q_3 + X'Q_1Q_3' + XQ_1'Q_2'$
$Z = X'Q_3' + XQ_3$


**FIGURE 16-4** Code Converter Circuit

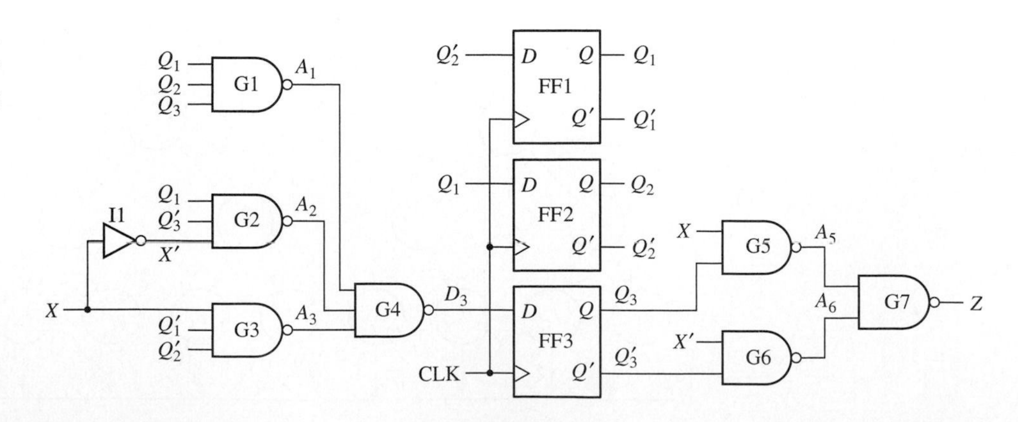

# 16.3 Design of Iterative Circuits

Many of the design procedures used for sequential circuits can be applied to the design of iterative circuits. An iterative circuit consists of a number of identical cells interconnected in a regular manner. Some operations, such as binary addition, naturally lend themselves to realization with an iterative circuit because the same operation is performed on each pair of input bits. The regular structure of an iterative circuit makes it easier to fabricate in integrated circuit form than circuits with less regular structures.

The simplest form of an iterative circuit consists of a linear array of combinational cells with signals between cells traveling in only one direction (Figure 16-5). Each cell is a combinational circuit with one or more primary inputs ($x_i$) and possibly one or more primary outputs ($z_i$). In addition, each cell has one or more secondary inputs ($a_i$) and one or more secondary outputs ($a_{i+1}$). The $a_i$ signals carry information about the "state" of one cell to the next cell.

The primary inputs to the cells ($x_1, x_2, \ldots, x_n$) are applied in parallel; that is, they are all applied at the same time. The $a_i$ signals then propagate down the line of cells. Because the circuit is combinational, the time required for the circuit to reach a steady-state condition is determined only by the delay times of the gates in the cells. As soon as steady state is reached, the outputs may be read. Thus, the iterative circuit can function as a parallel-input, parallel-output device, in contrast with the sequential circuit in which the input and output are serial. One can think of the iterative circuit as receiving its inputs as a sequence in space in contrast with the sequential circuit which receives its inputs as a sequence in time. The parallel adder of Figure 4-3 is an example of an iterative circuit that has four identical cells. The serial adder of Figure 13-12 uses the same full adder cell as the parallel adder, but it receives its inputs serially and stores the carry in a flip-flop instead of propagating it from cell to cell.

## Design of a Comparator

As an example, we will design a circuit which compares two $n$-bit binary numbers and determines if they are equal or which one is larger if they are not equal. Direct design as a $2n$-input combinational circuit is not practical for $n$ larger than 4 or 5, so we will try the iterative approach. Designate the two binary numbers to be compared as

$$X = x_1x_2 \ldots x_n \qquad \text{and} \qquad Y = y_1y_2 \ldots y_n$$

We have numbered the bits from left to right, starting with $x_1$ as the most significant bit because we plan to do the comparison from left to right.

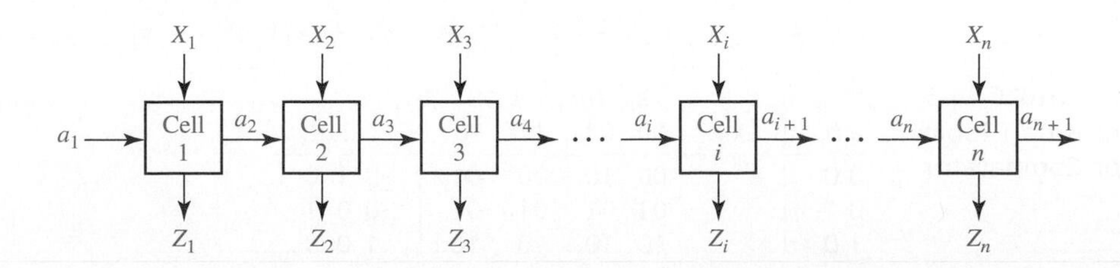


FIGURE 16-5 Unilateral Iterative Circuit

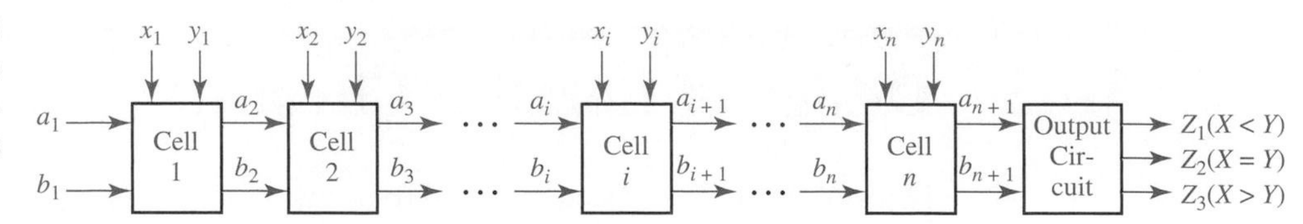


**FIGURE 16-6** Form of Iterative Circuit for Comparing Binary Numbers

Figure 16-6 shows the form of the iterative circuit, although the number of leads between each pair of cells is not yet known. Comparison proceeds from left to right. The first cell compares $x_1$ and $y_1$ and passes on the result of the comparison to the next cell, the second cell compares $x_2$ and $y_2$, etc. Finally, $x_n$ and $y_n$ are compared by the last cell, and the output circuit produces signals to indicate if $X = Y$, $X > Y$, or $X < Y$.

We will now design a typical cell for the comparator. To the left of cell $i$, three conditions are possible: $X = Y$ so far $(x_1 x_2 \ldots x_{i-1} = y_1 y_2 \ldots y_{i-1})$, $X > Y$ so far, and $X < Y$ so far. We designate these three input conditions as states $S_0$, $S_1$, and $S_2$, respectively. Table 16-4 shows the output state at the right of the cell $(S_{i+1})$ in terms of the $x_i y_i$ inputs and the input state at the left of the cell $(S_i)$. If the numbers are equal to the left of cell $i$ and $x_i = y_i$, the numbers are still equal including cell $i$, so $S_{i+1} = S_0$. However, if $S_i = S_0$ and $x_i y_i = 10$, then $x_1 x_2 \ldots x_i > y_1 y_2 \ldots y_i$ and $S_{i+1} = S_1$. If $X > Y$ to the left of cell $i$, then regardless of the values of $x_i$ and $y_i$, $x_1 x_2 \ldots x_i > y_1 y_2 \ldots y_i$ and $S_{i+1} = S_1$. Similarly, if $X < Y$ to the left of cell $i$, then $X < Y$ including the inputs to cell $i$, and $S_{i+1} = S_2$.

**TABLE 16-4** State Table for Comparator

| | $S_i$ | $S_{i+1}$: $x_i y_i = 00$ | 01 | 11 | 10 | $Z_1 Z_2 Z_3$ |
|---|---|---|---|---|---|---|
| $X = Y$ | $S_0$ | $S_0$ | $S_2$ | $S_0$ | $S_1$ | 0 1 0 |
| $X > Y$ | $S_1$ | $S_1$ | $S_1$ | $S_1$ | $S_1$ | 0 0 1 |
| $X < Y$ | $S_2$ | $S_2$ | $S_2$ | $S_2$ | $S_2$ | 1 0 0 |

The logic for a typical cell is easily derived from the state table. Because there are three states, two intercell signals are required. Using the guidelines from Section 15.8 leads to the state assignment $a_i b_i = 00$ for $S_0$, 01 for $S_1$, and 10 for $S_2$. Substituting this assignment into the state table yields Table 16-5. Figure 16-7 shows the Karnaugh maps, next-state equations, and the realization of a typical cell using NAND gates. Inverters must be included in the cell because only $a_i$ and $b_i$ and not their complements are transmitted between cells.

The $a_1 b_1$ inputs to the left end cell must be 00 because we must assume that the numbers are equal (all 0) to the left of the most significant bit. The equations for the first cell can then be simplified if desired:

$$a_2 = a_1 + x_1' y_1 b_1' = x_1' y_1$$
$$b_2 = b_1 + x_1 y_1' a_1' = x_1 y_1'$$

**TABLE 16-5** Transition Table for Comparator

| $a_i b_i$ | $a_{i+1} b_{i+1}$: $x_i y_i = 00$ | 01 | 11 | 10 | $Z_1 Z_2 Z_3$ |
|---|---|---|---|---|---|
| 0 0 | 00 | 10 | 00 | 01 | 0 1 0 |
| 0 1 | 01 | 01 | 01 | 01 | 0 0 1 |
| 1 0 | 10 | 10 | 10 | 10 | 1 0 0 |

**FIGURE 16-7** Typical Cell for Comparator

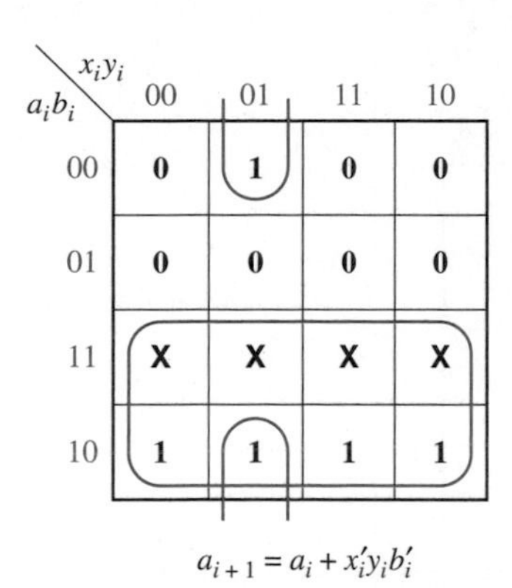


| $a_ib_i$ \ $x_iy_i$ | 00 | 01 | 11 | 10 |
|---|---|---|---|---|
| 00 | 0 | 0 | 0 | 1 |
| 01 | 1 | 1 | 1 | 1 |
| 11 | X | X | X | X |
| 10 | 0 | 0 | 0 | 0 |

$a_{i+1} = a_i + x_i'y_ib_i'$ $\qquad$ $b_{i+1} = b_i + x_iy_i'a_i'$

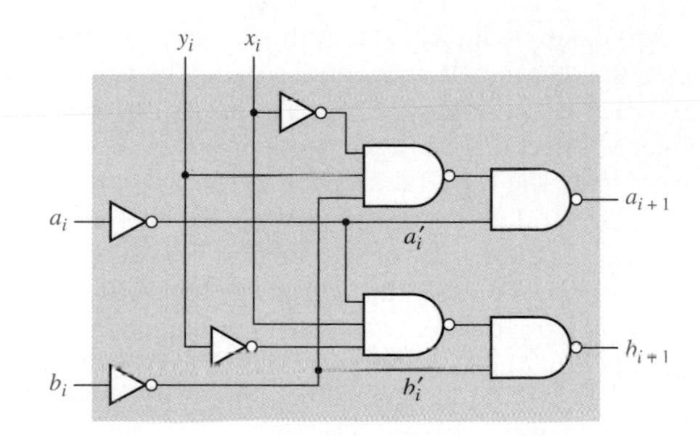


**FIGURE 16-8** Output Circuit for Comparator

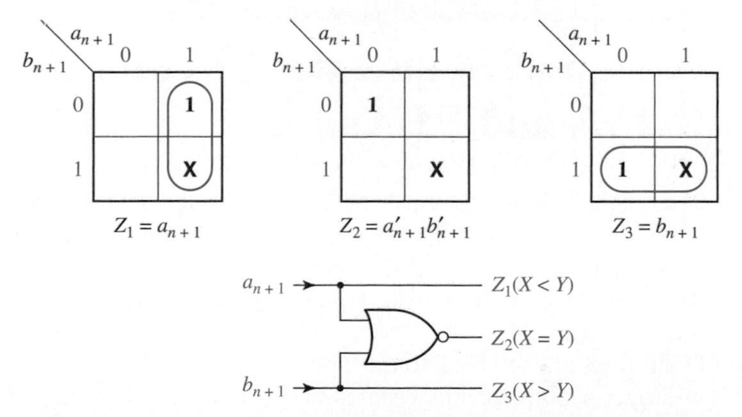


For the output circuit, let $Z_1 = 1$ if $X < Y$, $Z_2 = 1$ if $X = Y$, $Z_3 = 1$ if $X > Y$. Figure 16-8 shows the output maps, equations, and circuit.

Conversion to a sequential circuit is straightforward. If $x_i$ and $y_i$ inputs are received serially instead of in parallel, Table 16-4 is interpreted as a state table for a sequential circuit, and the next-state equations are the same as in Figure 16-7. If D flip-flops are used, the typical cell of Figure 16-7 can be used as the combinational part of the sequential circuit, and Figure 16-9 shows the resulting circuit. After all of the inputs have been read in, the output is determined from the state of the two flip-flops.

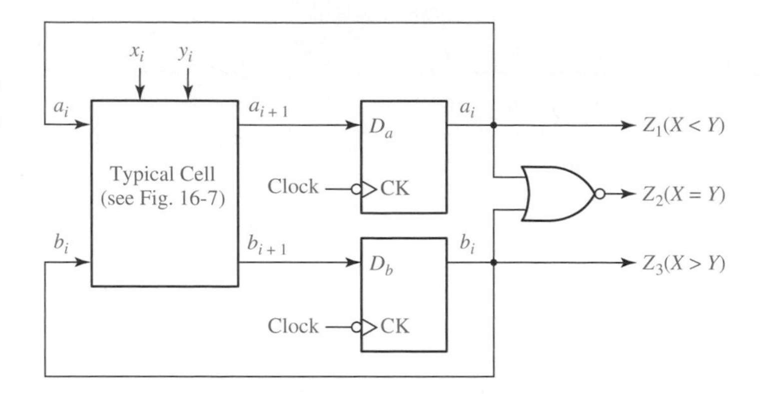


**FIGURE 16-9** Sequential Comparator for Binary Numbers

This example indicates that the design of a unilateral iterative circuit is very similar to the design of a sequential circuit. The principal difference is that for the iterative circuit the inputs are received in parallel as a sequence in space, while for the sequential circuit the inputs are received serially as a sequence in time. For the iterative circuit, the state table specifies the output state of a typical cell in terms of its input state and primary inputs, while for the corresponding sequential circuit, the same table specifies the next state (in time) in terms of the present state and inputs. If D flip-flops are used, the typical cell for the iterative circuit can serve as the combinational logic for the corresponding sequential circuit. If other flip-flop types are used, the input equations can be derived in the usual manner.

## 16.4 Design of Sequential Circuits Using ROMs and PLAs

A sequential circuit can easily be designed using a ROM (read-only memory) and flip-flops. Referring to the general model of a Mealy sequential circuit given in Figure 13-17, the combinational part of the sequential circuit can be realized using a ROM. The ROM can be used to realize the output functions ($Z_1, Z_2, \ldots, Z_n$) and the next-state functions ($Q_1^+, Q_2^+, \ldots, Q_k^+$). The state of the circuit can then be stored in a register of D flip-flops and fed back to the input of the ROM. Thus, a Mealy sequential circuit with $m$ inputs, $n$ outputs, and $k$ state variables can be realized using $k$ D flip-flops and a ROM with $m + k$ inputs ($2^{m+k}$ words) and $n + k$ outputs. The Moore sequential circuit of Figure 13-19 can be realized in a similar manner. The next-state and output combinational subcircuits of the Moore circuit can be realized using two ROMs. Alternatively, a single ROM can be used to realize both the next-state and output functions.

Use of D flip-flops is preferable to J-K flip-flops because use of two-input flip-flops would require increasing the number of outputs from the ROM. The fact that the D flip-flop input equations would generally require more gates than the J-K equations is of no consequence because the size of the ROM depends only on the number of inputs and outputs and not on the complexity of the equations being realized. For this reason, the state assignment which is used is also of little importance, and, generally, a state assignment in straight binary order is as good as any.

In Section 16.2, we realized a code converter using gates and D flip-flops. We will now realize this converter using a ROM and D flip-flops. The state table for the converter is reproduced in Table 16-6(a). Because there are seven states, three D flip-flops are required. Thus, a ROM with four inputs ($2^4$ words) and four outputs is required, as shown in Figure 16-10. Using a straight binary state assignment, we can construct the transition table, seen in Table 16-6(b), which gives the next state of the flip-flops as a function of the present state and input. Because we are using D flip-flops, $D_1 = Q_1^+, D_2 = Q_2^+$, and $D_3 = Q_3^+$. The truth table for the ROM, shown in Table 16-6(c), is easily constructed from the transition table. This table gives the ROM outputs ($Z$, $D_1$, $D_2$, and $D_3$) as functions of the ROM inputs ($X$, $Q_1$, $Q_2$, and $Q_3$).

Sequential circuits can also be realized using PLAs (programmable logic arrays) and flip-flops in a manner similar to using ROMs and flip-flops. However, in the case of PLAs, the state assignment may be important because the use of a

**TABLE 16-6**

(a) State table

| Present State | Next State X = 0 | 1 | Present Output (Z) X = 0 | 1 |
|---|---|---|---|---|
| A | B | C | 1 | 0 |
| B | D | E | 1 | 0 |
| C | E | E | 0 | 1 |
| D | H | H | 0 | 1 |
| E | H | M | 1 | 0 |
| H | A | A | 0 | 1 |
| M | A | – | 1 | – |

(b) Transition table

| | $Q_1Q_2Q_3$ | $Q_1^+Q_2^+Q_3^+$ X = 0 | X = 1 | Z X = 0 | X = 1 |
|---|---|---|---|---|---|
| A | 0 0 0 | 001 | 010 | 1 | 0 |
| B | 0 0 1 | 011 | 100 | 1 | 0 |
| C | 0 1 0 | 100 | 100 | 0 | 1 |
| D | 0 1 1 | 101 | 101 | 0 | 1 |
| E | 1 0 0 | 101 | 110 | 1 | 0 |
| H | 1 0 1 | 000 | 000 | 0 | 1 |
| M | 1 1 0 | 000 | – | 1 | – |

(c) Truth table

| X | $Q_1$ | $Q_2$ | $Q_3$ | Z | $D_1$ | $D_2$ | $D_3$ |
|---|---|---|---|---|---|---|---|
| 0 | 0 | 0 | 0 | 1 | 0 | 0 | 1 |
| 0 | 0 | 0 | 1 | 1 | 0 | 1 | 1 |
| 0 | 0 | 1 | 0 | 0 | 1 | 0 | 0 |
| 0 | 0 | 1 | 1 | 0 | 1 | 0 | 1 |
| 0 | 1 | 0 | 0 | 1 | 1 | 0 | 1 |
| 0 | 1 | 0 | 1 | 0 | 0 | 0 | 0 |
| 0 | 1 | 1 | 0 | 1 | 0 | 0 | 0 |
| 0 | 1 | 1 | 1 | x | x | x | x |
| 1 | 0 | 0 | 0 | 0 | 0 | 1 | 0 |
| 1 | 0 | 0 | 1 | 0 | 1 | 0 | 0 |
| 1 | 0 | 1 | 0 | 1 | 1 | 0 | 0 |
| 1 | 0 | 1 | 1 | 1 | 1 | 0 | 1 |
| 1 | 1 | 0 | 0 | 0 | 1 | 1 | 0 |
| 1 | 1 | 0 | 1 | 1 | 0 | 0 | 0 |
| 1 | 1 | 1 | 0 | x | x | x | x |
| 1 | 1 | 1 | 1 | x | x | x | x |

**FIGURE 16-10** Realization of Table 16.6(a) Using a ROM

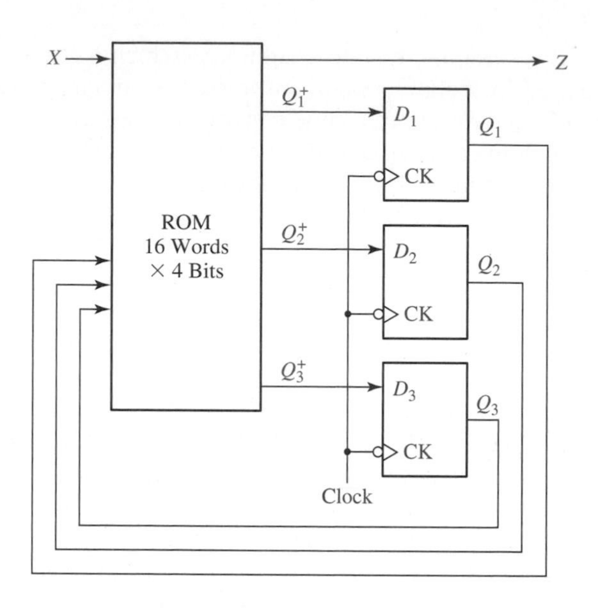


good state assignment can reduce the required number of product terms and, hence, reduce the required size of the PLA.

As an example, we will consider realizing the state table of Table 16-6(a) using a PLA and three D flip-flops. The circuit configuration is the same as Figure 16-10, except that the ROM is replaced with a PLA of appropriate size. Using a straight binary assignment leads to the truth table given in Table 16-6(c). This table could be stored in a PLA with four inputs, 13 product terms, and four outputs, but this would offer little reduction in size compared with the 16-word ROM solution discussed earlier.

If the state assignment of Figure 16-2 is used, the resulting output equation and D flip-flop input equations, derived from the maps in Figure 16-3, are

$$\begin{aligned} D_1 &= Q_1^+ = Q_2' \\ D_2 &= Q_2^+ = Q_1 \\ D_3 &= Q_3^+ = Q_1Q_2Q_3 + X'Q_1Q_3' + XQ_1'Q_2' \\ Z &= X'Q_3' + XQ_3 \end{aligned} \tag{16-1}$$

The PLA table which corresponds to these equations is in Table 16-7. Realization of this table requires a PLA with four inputs, seven product terms, and four outputs.

**TABLE 16-7**

| $X$ | $Q_1$ | $Q_2$ | $Q_3$ | $Z$ | $D_1$ | $D_2$ | $D_3$ |
|---|---|---|---|---|---|---|---|
| – | – | 0 | – | 0 | 1 | 0 | 0 |
| – | 1 | – | – | 0 | 0 | 1 | 0 |
| – | 1 | 1 | 1 | 0 | 0 | 0 | 1 |
| 0 | 1 | – | 0 | 0 | 0 | 0 | 1 |
| 1 | 0 | 0 | – | 0 | 0 | 0 | 1 |
| 0 | – | – | 0 | 1 | 0 | 0 | 0 |
| 1 | – | – | 1 | 1 | 0 | 0 | 0 |

Next, we will verify the operation of the circuit of Figure 16-4 using a PLA which corresponds to Table 16-7. Initially, assume that $X = 0$ and $Q_1Q_2Q_3 = 000$. This selects rows --0- and 0--0 in the table, so $Z = 1$ and $D_1D_2D_3 = 100$. After the active clock edge, $Q_1Q_2Q_3 = 100$. If the next input is $X = 1$, then rows --0- and -1-- are selected, so $Z = 0$ and $D_1D_2D_3 = 110$. After the active clock edge, $Q_1Q_2Q_3 = 110$. Continuing in this manner, we can verify the transition table of Figure 16-2.

PALs also provide a convenient way of realizing sequential circuits. PALs are available which contain D flip-flops that have their inputs driven from programmable array logic. Figure 16-11 shows a segment of a sequential PAL. The D flip-flop is driven from an OR gate which is fed by two AND gates. The flip-flop output is fed back to the programmable AND array through a buffer. Thus, the AND gate inputs can be connected to $A$, $A'$, $B$, $B'$, $Q$, or $Q'$. The X's on the diagram show the connections required to realize the next-state equation

$$Q^+ = D = A'BQ' + AB'Q$$

The flip-flop output is connected to an inverting tri-state buffer, which is enabled when En = 1.

**FIGURE 16-11** Segment of a Sequential PAL

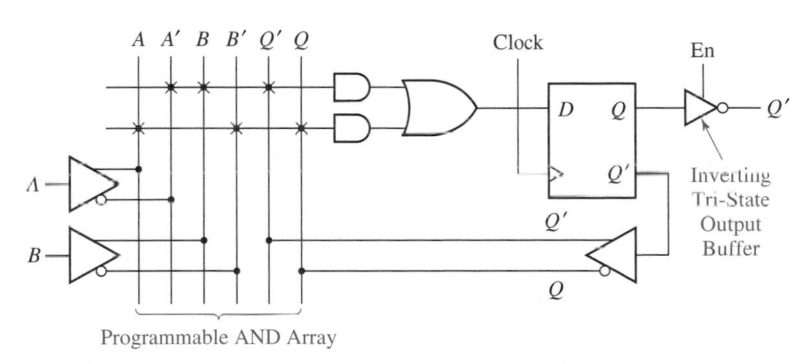


# 16.5 Sequential Circuit Design Using CPLDs

As discussed in Section 9.7, a typical CPLD contains a number of macrocells that are grouped into function blocks. Connections between the function blocks are made through an interconnection array. Each macrocell contains a flip-flop and an OR gate, which has its inputs connected to an AND gate array. Some CPLDs are based on PALs, in which case each OR gate has a fixed set of AND gates associated with it. Other CPLDs are based on PLAs, in which case any AND gate output within a function block can be connected to any OR gate input in that block.

Figure 16-12 shows the structure of a Xilinx CoolRunner II CPLD, which uses a PLA in each function block. This CPLD family is available in sizes from two to 32 function blocks (32 to 512 macrocells). Each function block has 16 inputs from the AIM (advanced interconnection matrix) and up to 40 outputs to the AIM. Each function block PLA contains the equivalent of 56 AND gates.

FIGURE 16-12 CoolRunner-II Architecture[1] (Figure based on figures and text owned by Xilinx, Inc., Courtesy of Xilinx, Inc. © Xilinx, Inc. 1999–2003. All rights reserved.)

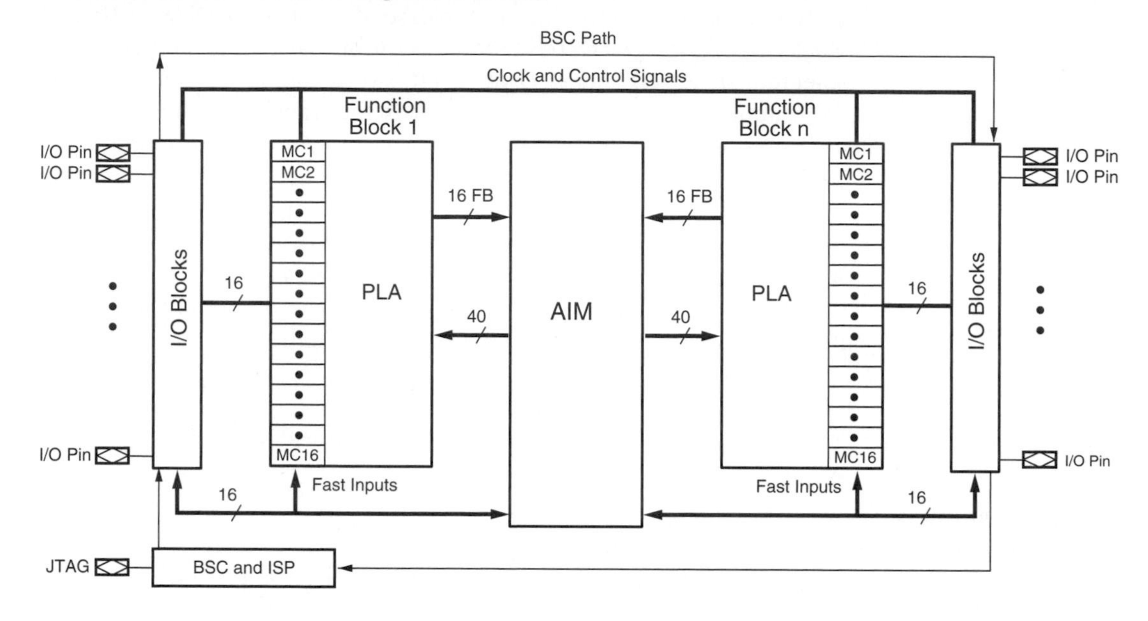


The basic CoolRunner II architecture is similar to that shown in Figure 9-29. Figure 16-13 represents a CoolRunner-II macrocell and the associated AND array. Box (1) represents the AND array which is driven by signals from the AIM. Each of the 56 product terms (P-terms) generated by the AND array (2) can have up to 40 variables. Box (3) represents the OR array which selects the AND gates for each macrocell. The OR gate (4) in a specific macrocell can have any subset of the P-terms as inputs. The MUXes on the diagram do not have control inputs shown because each MUX is programmed to select one of its inputs. For example, MUX (5) can be programmed to select a product term, the complement of a product term, a logic 1, or a logic 0 for the MUX output. If logic 1 is selected, the XOR gate complements the OR gate output; if logic 0 is selected, the XOR gate passes the OR gate output without change. By complementing or not complementing the OR gate output, a function can be implemented as either a product of sums or as a sum of products.

The XOR gate output can be routed directly to an I/O block or to the macrocell flip-flop input. The flip-flop can be programmed as a D-CE flip-flop or as a T flip-flop. The flip-flop can be programmed as an ordinary flip-flop (F/F), a latch, or a dual-edge triggered flip-flop, which can change state on either clock edge. The CK input and the asynchronous $S$ and $R$ inputs can each be programmed to come from several different sources. MUX (6) can invert the clock input or not, so that the flip-flop can trigger on either clock edge. MUX (7) selects either the flip-flop output or the XOR gate output and passes it to an I/O block.

Figure 16-14 shows how a Mealy sequential machine with two inputs, two outputs, and two flip-flops can be implemented by a CPLD. Four macrocells are required, two to

[1]Additional data on Xilinx CPLDs and FPGAs is available from www.Xilinx.com.

FIGURE 16-13 CoolRunner-II Macrocell (Figure based on figures and text owned by Xilinx, Inc., Courtesy of Xilinx, Inc. )

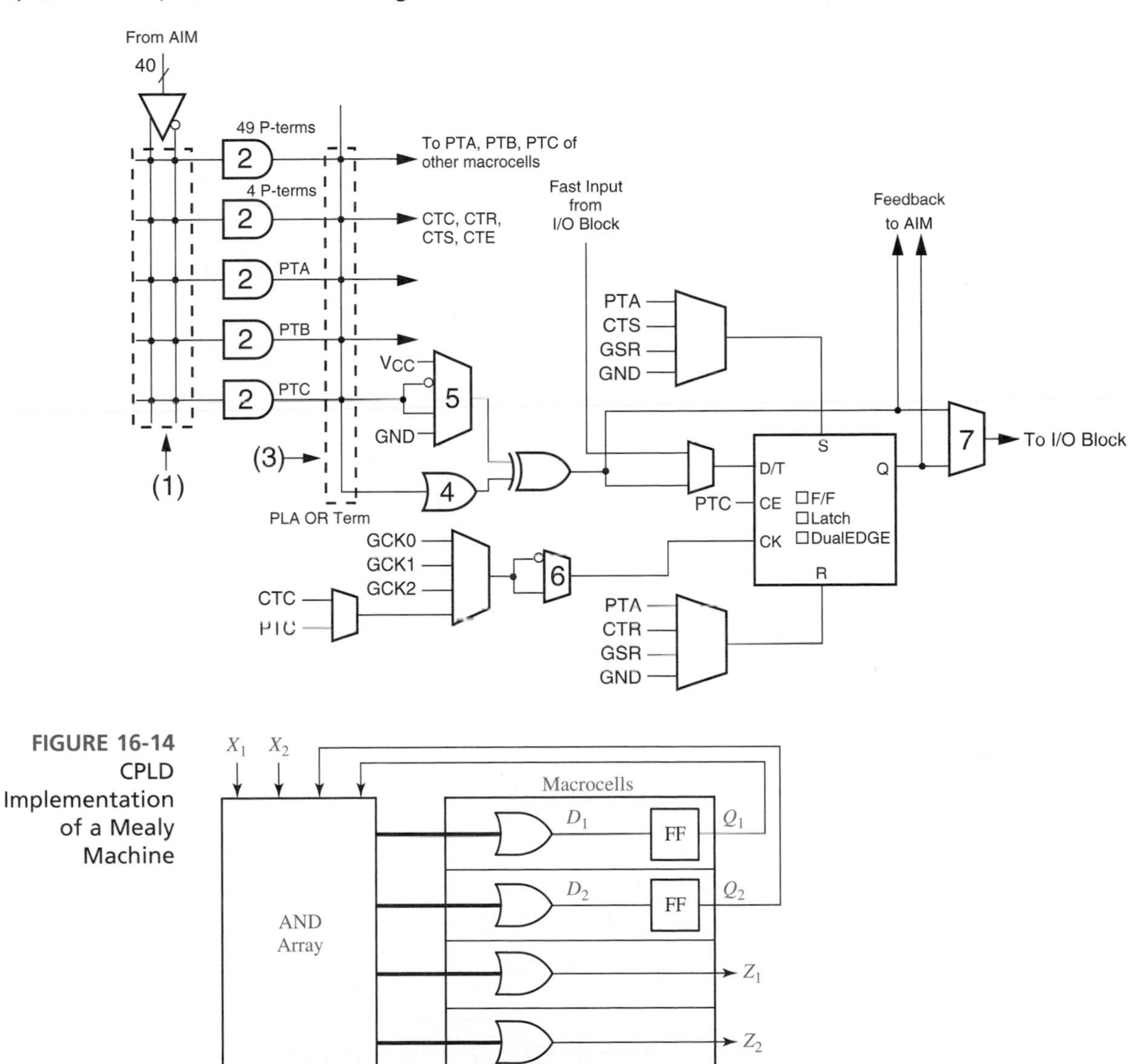


FIGURE 16-14 CPLD Implementation of a Mealy Machine

generate the D inputs to the flip-flops and two to generate the $Z$ outputs. The flip-flop outputs are fed back to the AND array inputs via the interconnection matrix (not shown). The number of product terms required depends on the complexity of the equations for the $D$'s and the $Z$'s.

Figure 16-15 shows how the 4-bit loadable right-shift register of Figure 12-15 can be implemented using four macrocells of a CPLD. The four OR-gate outputs implement the $D$ inputs specified by Equations (12-1). A total of 12 product terms are required. The $Q$ outputs are fed back to the AND array via the interconnection matrix (not shown).

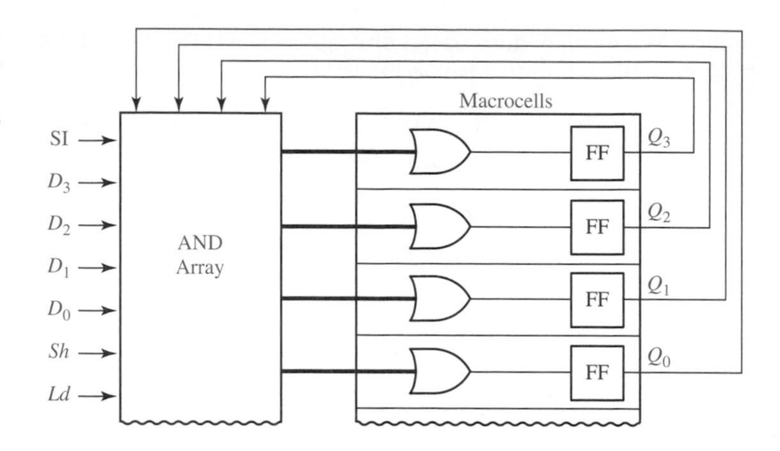


**FIGURE 16-15** CPLD Implementation of a Shift Register

Figure 16-16 shows how three bits of the parallel adder with accumulator of Figure 12-5 can be implemented using a CPLD. Each bit of the adder requires two macrocells. One of the macrocells implements the sum function and an accumulator flip-flop. The other macrocell implements the carry, which is fed back into the AND array. The *Ad* signal can be connected to the CE input of each flip-flop via an AND gate (not shown). Each bit of the adder requires eight product terms (four for the sum, three for the carry, and one for CE). If the flip-flops are programmed as T flip-flops, then the logic for the sum can be simplified. For each accumulator flip-flop

$$X_i^+ = X_i \oplus Y_i \oplus C_i$$

Then, the $T$ input is

$$T_i = X_i^+ \oplus X_i = Y_i \oplus C_i$$

which requires only two product terms.

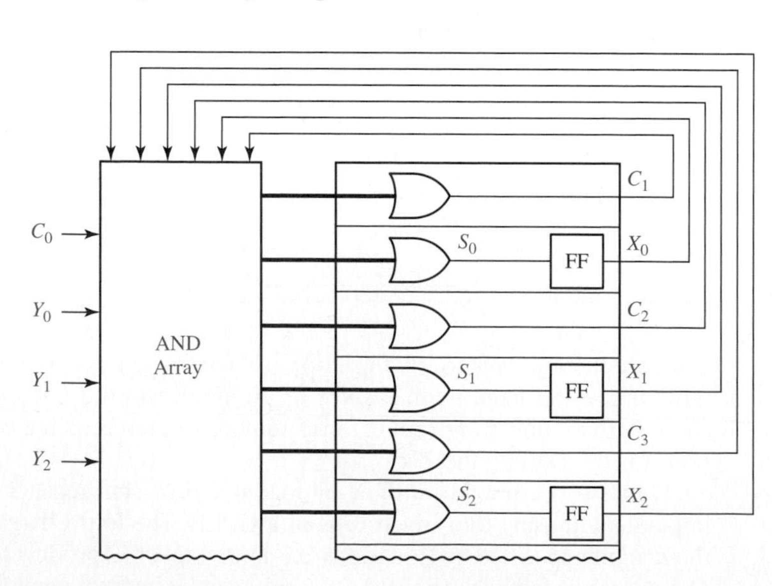


**FIGURE 16-16** CPLD Implementation of a Parallel Adder with Accumulator

The add signal can be ANDed with the $T_i$ input so that the flip-flop state can change only when $Ad = 1$:

$$T_i = Ad\,(Y_i \oplus C_i) = Ad\,Y_i\,C_i' + Ad\,Y_i'\,C_i$$

# 16.6 Sequential Circuit Design Using FPGAs

As discussed in Section 9.8, an FPGA usually consists of an array of configurable logic blocks (CLBs) surrounded by a ring of I/O blocks. The FPGA may also contain other components such as memory blocks, clock generators, tri-state buffers, etc. A typical CLB contains two or more function generators, often referred to as look-up tables or LUTs, programmable multiplexers, and D-CE flip-flops (see Figure 9-33). The I/O blocks usually contain additional flip-flops for storing inputs or outputs and tri-state buffers for driving the I/O pins.

Figure 16-17 shows a simplified block diagram for a Xilinx Virtex or Spartan II CLB. This CLB is divided into two nearly identical *slices*. Each slice contains two 4-variable function generators (LUTs), two D-CE flip-flops, and additional logic for carry and control. This additional logic includes MUXes for selecting the flip-flop inputs and for multiplexing the LUT outputs to form functions of five or more variables.

Figure 16-18 shows how a Mealy sequential machine with two inputs, two outputs, and two flip-flops can be implemented by a FPGA. Four LUTs (FGs or function generators) are required, two to generate the $D$ inputs to the flip-flops and two to generate the $Z$ outputs. The flip-flop outputs are fed back to the CLB inputs via interconnections external to the CLB. The entire circuit fits into one Virtex CLB. This implementation works because each $D$ and $Z$ is a function of only four variables

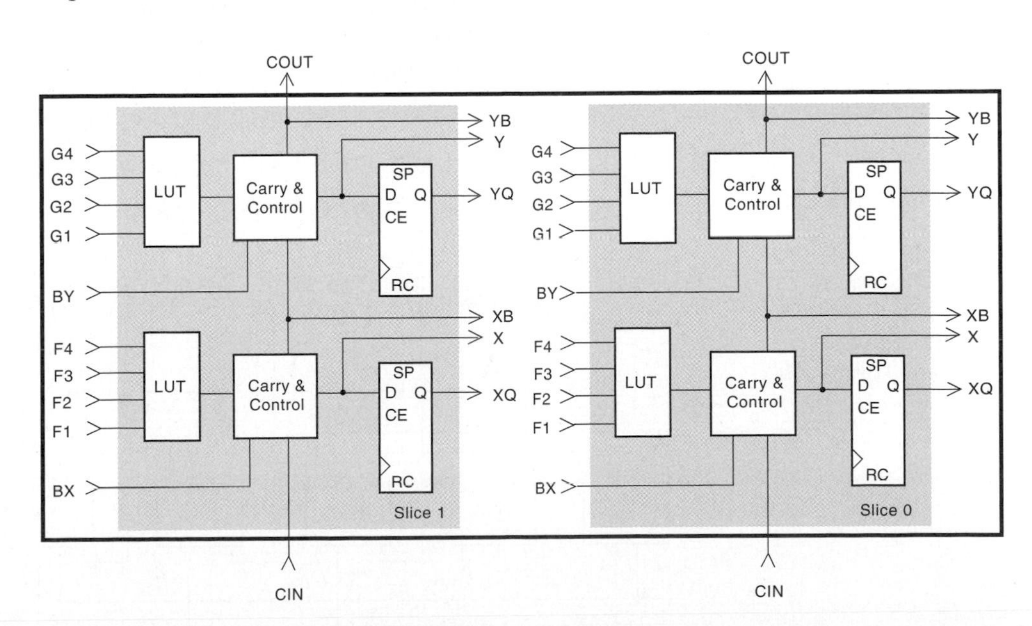


**FIGURE 16-17** Xilinx Virtex/Spartan II CLB (Figure based on figures and text owned by Xilinx, Inc., Courtesy of Xilinx, Inc. © Xilinx, Inc. 1999–2003. All rights reserved.)

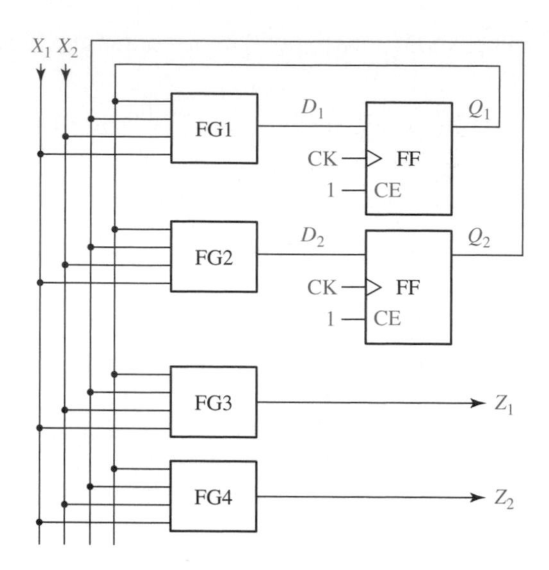


**FIGURE 16-18** FPGA Implementation of a Mealy Machine

($X_1, X_2, Q_1$, and $Q_2$). If more flip-flops or inputs are needed, the $D$ or $Z$ functions may have to be decomposed to use additional function generators as in Figure 9-36.

Figure 16-19 shows how the 4-bit loadable right-shift register of Figure 12-15 can be implemented using an FPGA. Four LUTs are used to generate the $D$ inputs to the flip-flops, and a fifth LUT generates the CE input. If we had implemented Equations (12-1) directly without using the CE input, we would need to implement four 5-variable functions. This would require eight LUTs because each 5-variable function requires two 4-variable function generators (see Figure 9-36(a)). However, if we set CE $= Ld + Sh$, then CE $= 0$ when $Ld = Sh = 0$ and the flip-flops hold their current values. Therefore, we do not need the first term in each of Equations (12-1), and the flip-flop $D$ input equations fit into 4-variable function generators. We can rewrite Equation (12-1(a)) in terms of CE as follows:

$$Q_3^+ = \text{CE}'Q_3 + \text{CE}\ D_{3\text{f}} = Ld'Sh'Q_3 + (Ld + Sh)(Sh'D_3 + Sh\ \text{SI}) \quad (16\text{-}2)$$

where $D_{3\text{f}}$ is the $D$ input to flip-flop 3. The $D$ input to the $Q_3$ flip-flop is therefore

$$D_{3\text{f}} = Sh'D_3 + Sh\ \text{SI}$$

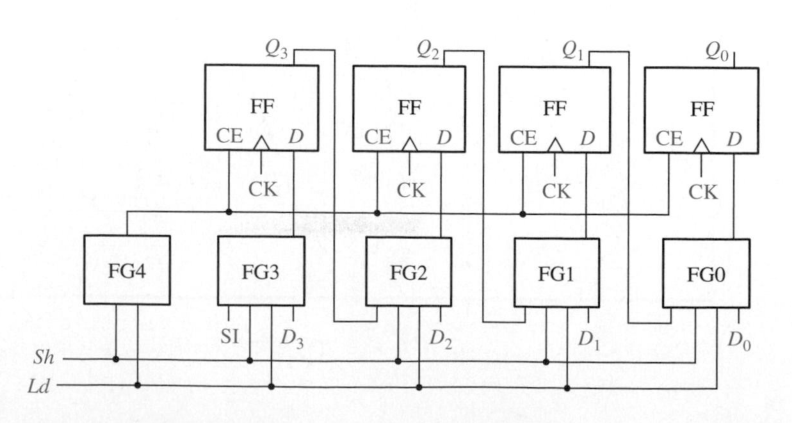


**FIGURE 16-19** FPGA Implementation of a Shift Register

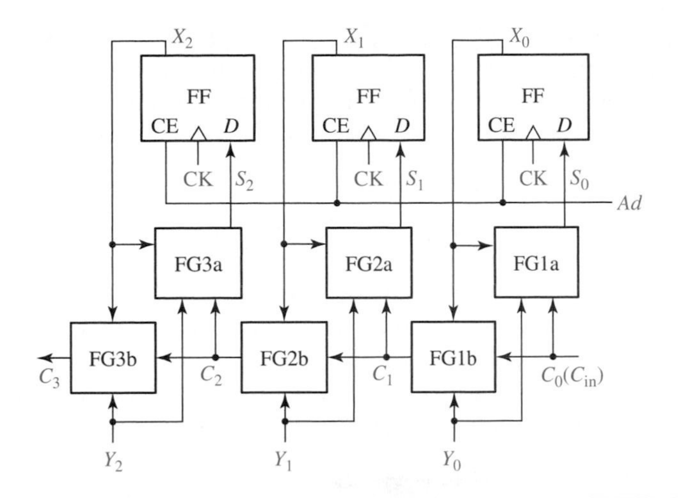


**FIGURE 16-20** FPGA Implementation of a Parallel Adder with Accumulator

which is a 3-variable function. We can determine the other three flip-flop $D$ inputs in a similar way.

Figure 16-20 shows how three bits of the parallel adder with accumulator of Figure 12-5 can be implemented using an FPGA. Each bit of the adder can be implemented with two 3-variable function generators, one for the sum and one for the carry. The $Ad$ signal can be connected to the CE input of each flip-flop so that the sum is loaded by the rising clock edge when $Ad = 1$. The arrangement for generating the carries, shown in Figure 16-20, is rather slow because the carry signal must propagate through a function generator and its external interconnections for each bit. Because adders are frequently used in FPGAs, most FPGAs have built-in fast carry logic in addition to the function generators. If the fast carry logic is used, the bottom row of function generators in Figure 16-20 is not needed, and a parallel adder with an accumulator can be implemented using only one function generator for each bit.

## 16.7 Simulation and Testing of Sequential Circuits

Simulation of a digital system can take place at several levels of detail. At the functional level, system operation is described in terms of a sequence of transfers between registers, adders, memories, and other functional units. Simulation at this level may be used to verify the high-level system design. At the logic level, the system is described in terms of logic elements such as gates and flip-flops and their interconnections. Logic level simulation may be used to verify the correctness of the logic design and to analyze the timing. At the circuit level, each gate is described in terms of its circuit components such as transistors, resistances, and capacitances. Circuit level simulation gives detailed information about voltage levels and switching speeds. In this text, we will consider simulation at the logic level as well as system level simulation using VHDL.

Simulation of sequential circuits is similar to the simulation of combinational circuits described in Section 8.5. However, for sequential circuits, the propagation delays associated with the individual logic elements must be taken into account, and the presence of feedback may cause complications. The simulator output usually includes timing diagrams which show the times at which different signals in the circuit change. The delays in the gates and flip-flops may be modeled in several ways. The simplest method is to assume that each element has one unit of delay. The use of this unit delay model is generally sufficient to verify that the design is logically correct. If a more detailed timing analysis is required, each logic element may be assigned a nominal delay value. The nominal or typical delays for a device are usually provided by the device manufacturer on the specification sheets.

In practice, no two gates of a given type will have *exactly* the same delay, and the value of the delay may change depending on temperature and voltage levels. For these reasons, manufacturers often specify a minimum and maximum delay value for each type of logic element. Some simulators can take the minimum and maximum delay values into account. Instead of showing the exact time at which a signal changes, the simulator output indicates a time interval in which the signal may change. Figure 16-21 shows the output from an inverter which has a nominal delay of 10 ns, a minimum delay of 5 ns, and a maximum delay of 15 ns. The shaded region indicates that the inverter output may change at any time during the interval. Min-max delay simulators can be used to verify that a digital system will operate correctly as long as the delay in each element is within its specified range.

Testing of sequential circuits is generally more difficult than testing combinational circuits. If the flip-flop outputs can be observed, then the state table can be verified directly on a row-by-row basis. The state table can be checked out with a simulator or in lab as follows:

1. Using the direct set and clear inputs, set the flip-flop states to correspond to one of the present states in the table.
2. For a Moore machine, check to see that the output is correct. For a Mealy machine, check to see that the output is correct for each input combination.
3. For each input combination, clock the circuit and check to see that the next state of the flip-flops is correct. (Reset the circuit to the proper state before each input combination is applied.)
4. Repeat steps 1, 2, and 3 for each of the present states in the table.

In many cases when a sequential circuit is implemented as part of an integrated circuit, only the inputs and outputs are available at the IC pins, and observing the state of the internal flip-flops is impossible. In this case, testing must be done by applying input sequences to the circuit and observing the output sequences. Determining a small set of input sequences which will completely test the circuit is generally a difficult problem

**FIGURE 16-21** Simulator Output for an Inverter

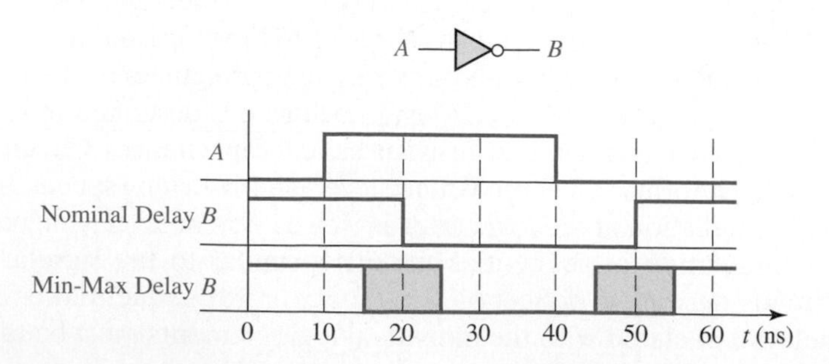

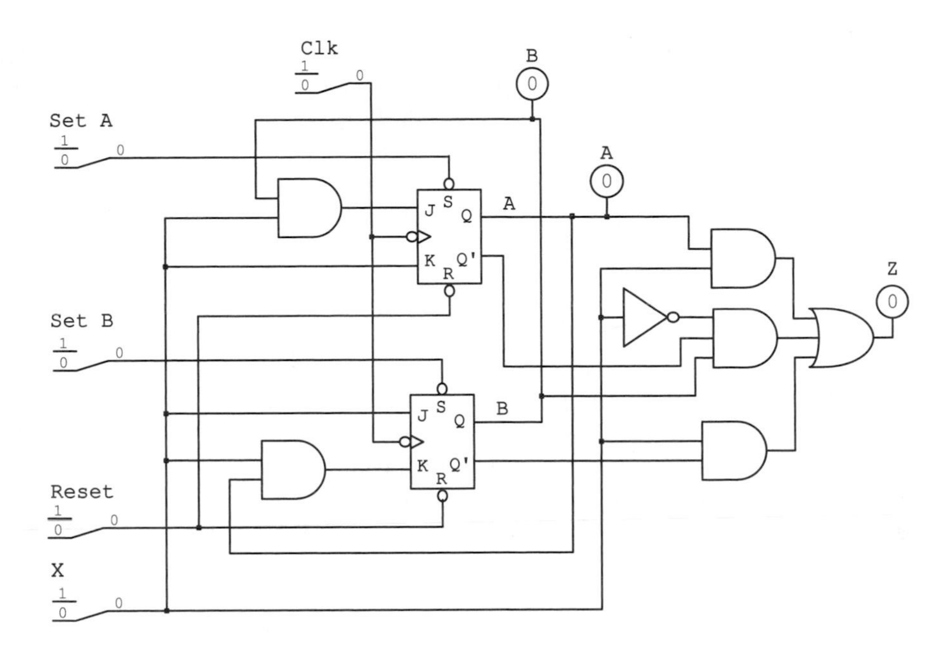


FIGURE 16-22 Simulation Screen for Figure 13.7

that is beyond the scope of this text. The set of test sequences must traverse all arcs on the state graph, but this is generally not a sufficient test.

Figure 16-22 shows a simulator screen for testing the Mealy sequential circuit of Figure 13-7. To step through the circuit one input at a time, switches are used for the Clock and $X$ inputs. Another switch is used to reset both flip-flops, and two switches are used to set flip-flops $A$ and $B$. Probes are used to observe the $Z$ output and the state of the flip-flops. After $X$ has been set to the desired value, the clock cycle is simulated by flipping the Clock switch to 1 and back to 0. For a Mealy machine, the output should be read just *before* the active edge of the clock.

If an incorrect $Z$ output is found in the process of verifying the state table, the output circuit can be checked using the techniques discussed in Section 8.5. If one of the next states is wrong, this may be due to an incorrect flip-flop input. After determining which flip-flop goes to the wrong state, the circuit should be reset to the proper present state, and the flip-flop inputs should be checked before applying another clock pulse.

***Example***

Assume that you have built the circuit of Figure 16-4 to implement the state table of Figure 16-2. Suppose that when you set the flip-flop states to 100, set $X = 1$, and pulse the clock, the circuit goes to state 111 instead of 110. This indicates that flip-flop $Q_3$ went to the wrong state. You should then reset the flip-flops to state 100 and observe the inputs to flip-flop $Q_3$. Because the flip-flop is supposed to remain in state 0, $D_3$ should be 0. If $D_3 = 1$, this indicates that either $D_3$ was derived wrong or that the $D_3$ circuit has a problem. Check the $D_3$ map and equation to make sure that $D_3 = 0$ when $X = 1$ and $Q_1Q_2Q_3 = 100$. If the map and equation are correct, then the $D_3$ circuit should be checked using the procedure in Section 8.5.

After you have verified that the circuit works according to your state table, you must then check the circuit to verify that it works according to the problem statement. To do this, you must apply appropriate input sequences and observe the resulting output sequence. When testing a Mealy circuit, you must be careful to read the outputs at the proper time to avoid reading false outputs (see Section 13.2). The output should be read just before the active edge of the clock. If the output is read immediately following the active clock edge, a false output may be read. See Figure 13-8 for an example.

Instead of manually stepping through the input sequence, simulated input waveforms may be defined for $X$ and Clock. Figure 16-23 shows the simulator input waveform for the example of Figure 16-22, using the test sequence $X = 10101$. When the simulator is run, the timing chart for $A$, $B$, and $Z$ will be generated as shown. Note that the simulator output is very similar to the timing chart of Figure 13-8. The simulator output in Figure 16-23(a) assumes the unit delay model, that is, each gate or flip-flop has one unit of delay. Figure 16-23(b) shows the same simulation using a nominal delay of 10 ns for each gate and flip-flop.

So far in our discussion of sequential circuits, we have assumed that the inputs are properly synchronized with the clock. This means that one input in the sequence occurs for each clock cycle, and all input changes satisfy setup and hold time specifications. Synchronization is no problem in the laboratory if we use a manual clock because we can easily change the inputs between active clock edges. However, if we operate our circuits at a high clock rate, then synchronization becomes a problem. We must either generate our input sequences in synchronization with the clock, or we must use a special circuit to synchronize the inputs with the clock. The former can be accomplished

FIGURE 16-23

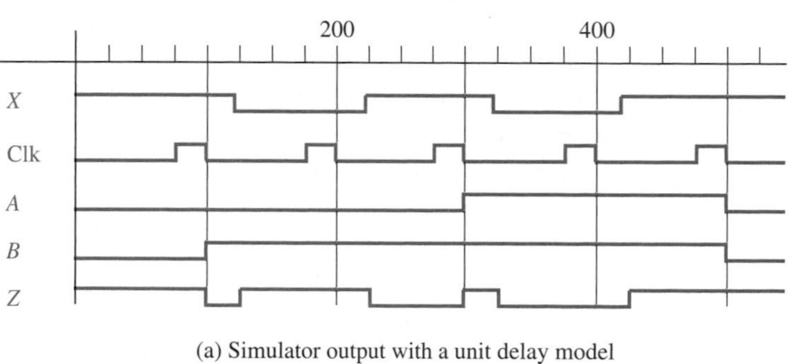


(a) Simulator output with a unit delay model

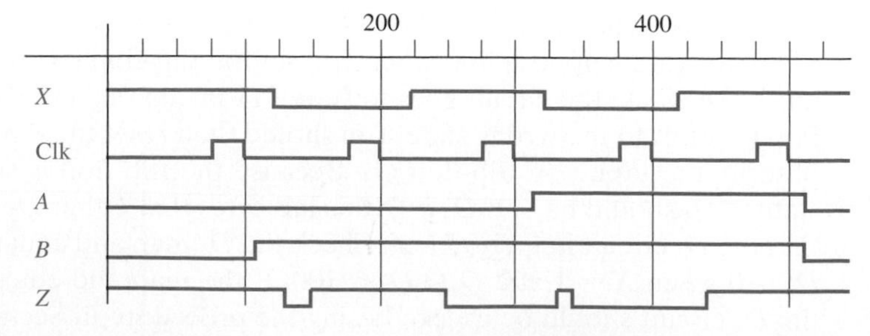


(b) Simulator output with a nominal delay of 10 ns

**FIGURE 16-24** Using a Shift Register to Generate Synchronized Inputs

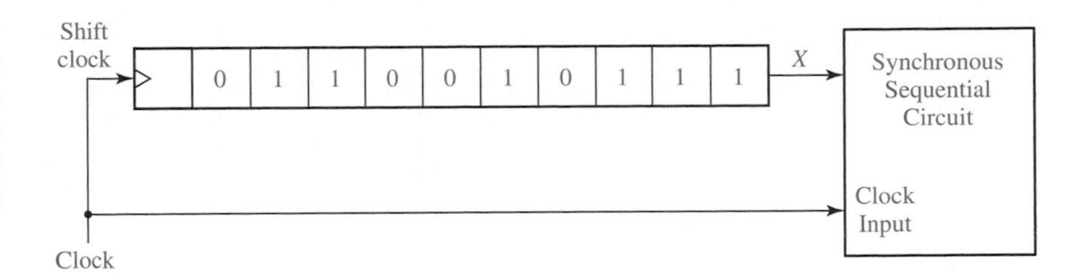


by loading the inputs into a shift register, and then using the circuit clock to shift them into the circuit one at a time, as shown in Figure 16-24.

If the input changes are not synchronized with the clock, edge-triggered D flip-flops can be used to synchronize them, as shown in Figure 16-25(a). In this figure, although $X_1$ and $X_2$ change at arbitrary times with respect to the clock, $X_{1S}$ and $X_{2S}$ change after the rising clock edge, and the inputs to the sequential circuit should be properly synchronized, as shown in Figure 16-25(b). However, this design has an inherent problem and may occasionally fail to operate properly. If a $D$ input changes very close to the rising clock edge so that setup and hold times are not satisfied (see Figure 11-16), one of the flip-flops may malfunction.

Figure 16-26 shows a more reliable synchronizer[2] that uses two D flip-flops to synchronize a single asynchronous input, $X$. If $X$ changes from 0 to 1 in the critical region where the setup or hold time is not satisfied, several outcomes could occur: the flip-flop $Q_1$ output might change to 1; it might remain 0; it might start to change to 1 and then change back or it might oscillate between 0 and 1 for a short time and then settle down to 0 or 1. This region of uncertainty is indicated by the shading on the $Q_1$ waveform. We will assume that the clock period is chosen so that $Q_1$ will be settled in either the 0 or 1 state by $t_2$. If $Q_1 = 1$, $Q_2$ will change to 1 shortly after $t_2$. If $Q_1 = 0$, $Q_1$ will change to 1 shortly after $t_2$, and $Q_2$ will change to 1 shortly after $t_3$. Because $X$ is an asynchronous input, normally it will not matter whether $X_{1S}$ is delayed by one or two clock periods. The important thing is that $X_{1S}$ is a clean signal that is synchronized with the clock.

**FIGURE 16-25**

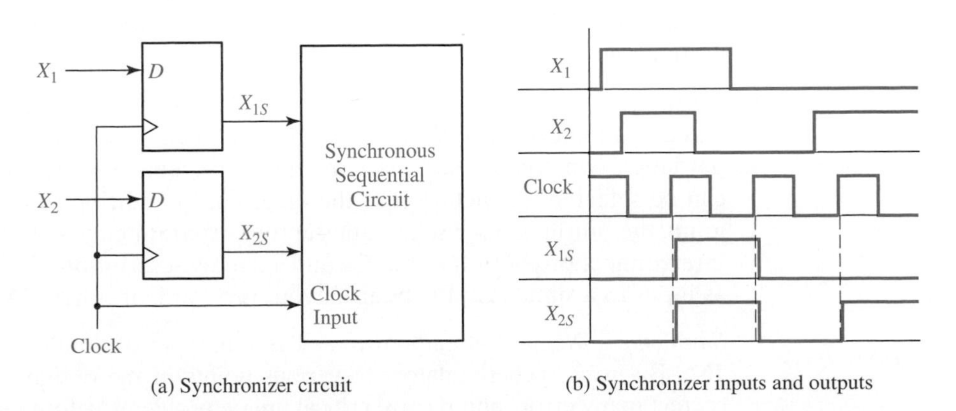


(a) Synchronizer circuit (b) Synchronizer inputs and outputs

[2]For more detailed discussion of synchronizer design, see John F. Wakerly, *Digital Design Principles and Practices*, 4th ed, (Prentice-Hall, 2006).

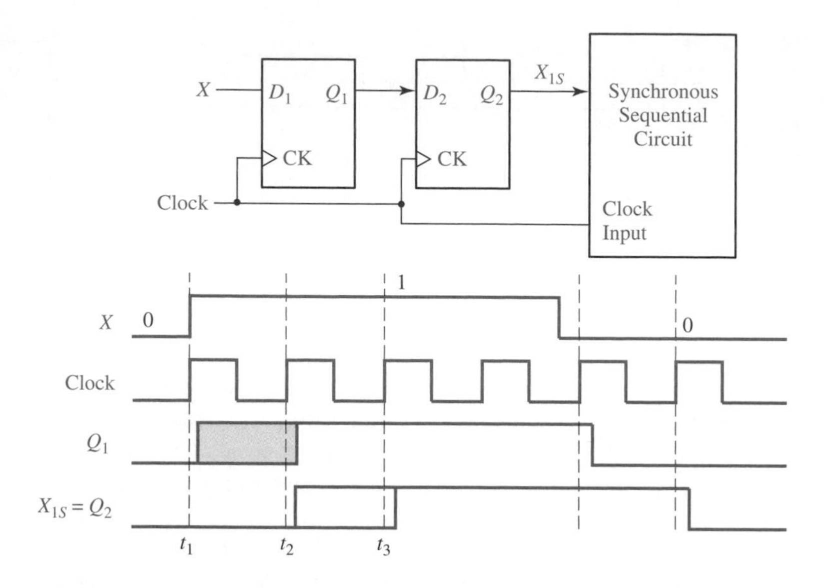


**FIGURE 16-26** Synchronizer with Two D Flip-flops

# 16.8 Overview of Computer-Aided Design

A wide variety of computer-aided design (CAD) software tools are available to assist in the design of digital systems. Many of these CAD programs will run on a personal computer, but others require a more powerful workstation for execution. Several functions performed by these CAD tools are discussed below.

*Generation and minimization of logic equations.* Programs of this type accept truth tables, state tables, or state graphs as input and generate minimized logic equations. *LogicAid* is an example of this type of program.

*Generation of bit patterns for programming PLDs.* These programs generate a file which can be downloaded to a PLD programmer to program PALs and other programmable logic devices.

*Schematic capture.* This type of program allows the designer to interactively enter and edit a logic diagram for a digital design. The program provides libraries of standard logic components such as gates, flip-flops, registers, adders, counters, etc., which can be selected for inclusion in the diagram. In addition to a plot of the logic diagram, the output from a schematic capture program may include a parts list, a list of interconnections between the ICs, and a circuit description file. This file may be used as input to a simulator, PC board layout program, or other CAD programs.

*Simulation.* We have already discussed several types of simulators in Sections 10.3 and 16.5. By using such simulators at various points in the design process, designers can correct many errors and resolve critical timing problems before any hardware is actually built. Use of a simulator is essential when an IC is being designed, because the correction of design errors after the IC has been fabricated is very time-consuming and costly.

*SimUaid* performs the schematic capture and simulation functions for small digital systems. It also automatically generates a structural VHDL description from the schematic.

*Synthesis tools.* Synthesis software accepts as input a description of a desired digital system written in VHDL, Verilog, or another hardware description language. The HDL code is analyzed and translated into a circuit description that specifies the needed logic components and the connections between these components. The synthesizer output is then fed into software that implements the circuit for a specific target device such as an FPGA, CPLD, or ASIC (application-specific integrated circuit). More details of synthesis and implementation of VHDL code are given in Section 17.5.

*IC design and layout.* A digital integrated circuit is typically composed of interconnected transistors which are fabricated on a chip of silicon. Such ICs are usually made of several layers of conducting material separated by layers of insulating material with appropriate connections between layers. The patterns for paths on each layer are transferred into the layers during the fabrication process using masks which are similar to photographic negatives. CAD tools for IC design facilitate the process of specifying the geometries of the transistors, placing the transistors on the chip, and routing the interconnections between them. Libraries of standard modules are available for inclusion in the chip designs. Automatic checking of the designs is provided to verify consistency with design rules. The output from the IC design program includes the mask patterns necessary for fabricating the IC.

*Test generation.* As digital systems become more complex, testing the finished product becomes increasingly difficult. It is not practical to test the system using all possible combinations or sequences of inputs. Automatic test generation programs are available which attempt to generate a relatively small set of input patterns that will adequately test the system in a reasonable length of time.

*PC board layout.* Most digital systems are built by mounting the integrated circuit components on a printed circuit board. The wiring on such PC boards is made up of thin metallic strips which interconnect the ICs. In order to make all of the required connections, these boards typically have two, three, or more layers of interconnect wiring. PC board layout programs perform two main functions—they determine the placement of the ICs on the board, and they route the connections between the ICs. The output of the layout program includes a set of plots which show the wiring on each layer of the PC board.

Many CAD systems integrate several of these CAD tools into a single package so that you can, for example, input a logic diagram, simulate its operation, and then lay out a PC board or IC. The design of large, complex integrated circuits and digital systems would not be feasible without the use of appropriate CAD tools.

One method of designing a small digital system with an FPGA uses the following steps:

1. Draw a block diagram of the digital system. Define the required control signals and construct a state graph that describes the required sequence of operations.
2. Work out a detailed logic design of the system using gates, flip-flops, registers, counters, adders, etc.

3. Construct a logic diagram of the system, using a schematic capture program.
4. Simulate and debug the logic diagram and make any necessary corrections to the design.
5. Run an implementation program that fits the design into the target FPGA. This program carries out the following steps:
   (a) Partition the logic diagram into pieces that will fit into CLBs of the target FPGA.
   (b) Place the CLBs within the logic cell array of the FPGA and route the connections between the logic cells.
   (c) Generate the bit pattern necessary to program the FPGA.
6. Run a timing simulation of the completed design to verify that it meets specifications. Make any necessary corrections and repeat the process as necessary.
7. Download the bit pattern into the internal configuration memory cells in the FPGA and test the operation of the FPGA.

When a hardware description language is used, steps 2 and 3 are replaced with writing HDL code. The HDL code is then simulated and debugged in step 4.

# Design Problems

The following problems require the design of a Mealy sequential circuit of the form shown in Figure 16-27. For purposes of testing, the input $X$ will come from a toggle switch, and the clock pulse will be supplied manually from a push button or switch.

FIGURE 16-27

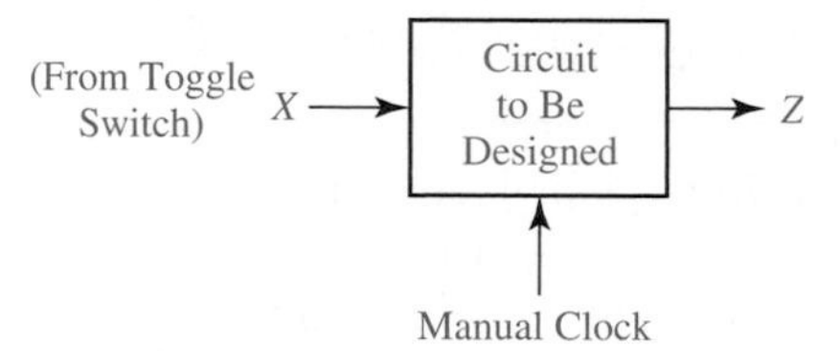


**16.1** Design a Mealy sequential circuit (Figure 16-27) which investigates an input sequence $X$ and will produce an output of $Z = 1$ for any input sequence ending in 0010 or 100.

*Example:*

$X = 1\ 1\ 0\ 0\ 1\ 0\ 0\ 1\ 0\ 1\ 0\ 0\ 1\ 0\ 1$
$Z = 0\ 0\ 0\ 1\ 0\ 1\ 1\ 0\ 1\ 0\ 0\ 1\ 0\ 1\ 0$

Notice that the circuit does not reset to the start state when an output of $Z = 1$ occurs. However, your circuit should have a start state and should be provided with a method for manually resetting the flip-flops to the start state. A minimum solution requires six states. Design your circuit using NAND gates, NOR gates, and three D flip-flops. Any solution which is minimal for your state assignment and uses 10 or fewer gates and inverters is acceptable. (Assign 000 to the start state.)

*Test Procedure:* First, check out your state table by starting in each state and making sure that the present output and next state are correct for each input. Then, starting in the proper initial state, determine the output sequence for each of the following input sequences:

(1) 0 0 1 1 0 1 0 0 1 0 1 0 1 0 0 0 1 0 0 1 0 0 1 0
(2) 1 1 0 0 1 1 0 0 1 0 1 0 1 0 0 1 0 1 0 1 0 0 1 0

**16.2** Design a Mealy sequential circuit (Figure 16-27) which investigates an input sequence $X$ and will produce an output of $Z = 1$ for any input sequence ending in 1101 or 011.

*Example:*

$X = 0\ 0\ 1\ 1\ 0\ 1\ 1\ 0\ 1\ 0\ 1\ 1\ 0\ 1\ 0$
$Z = 0\ 0\ 0\ 1\ 0\ 1\ 1\ 0\ 1\ 0\ 0\ 1\ 0\ 1\ 0$

Notice that the circuit does not reset to the start state when an output of $Z = 1$ occurs. However, your circuit should have a start state and should be provided with a method for manually resetting the flip-flops to the start state. A minimum solution requires six states. Design your circuit using NAND gates, NOR gates, and three D flip-flops. Any solution which is minimal for your state assignment and uses nine or fewer gates and inverters is acceptable. (Assign 000 to the start state.)

*Test Procedure:* First, check out your state table by starting in each state and making sure that the present output and next state are correct for each input. Then, starting in the proper initial state, determine the output sequence for each of the following input sequences:

(1) 1 1 0 0 1 0 1 1 0 1 0 1 0 1 1 1 0 1 1 0 1 1 0 1
(2) 0 0 1 1 0 0 1 1 0 1 0 1 0 1 1 0 1 0 1 0 1 1 0 1

**16.3** Design a sequential circuit (Figure 16-27) to convert excess-3 code to BCD code. The input and output should be serial with the least significant bit first. The input $X$ represents an excess-3 coded decimal digit, and the output $Z$ represents the corresponding BCD code. Design your circuit using three D flip-flops, NAND gates, and NOR gates. Any solution which is minimal for your state assignment and uses eight or fewer gates and inverters is acceptable. (Assign 000 to the reset state.)

*Test Procedure:* First, check out your state table by starting in each state and making sure that the present output and next state are correct for each input. Then, starting in the reset state, determine the output sequence for each of the ten possible input sequences and make a table.

**16.4** Design a sequential circuit (Figure 16-27) which adds six to a binary number in the range 0000 through 1001. The input and output should be serial with the least significant bit first. Find a state table with a minimum number of states. Design the circuit using NAND gates, NOR gates, and three D flip-flops. Any solution which is minimal for your state assignment and uses 10 or fewer gates and inverters is acceptable. (Assign 000 to the reset state.)

*Test Procedure:* First, check out your state table by starting in each state and making sure that the present output and next state are correct for each input. Then, starting in the reset state, determine the output sequence for each of the ten possible input sequences and make a table.

**16.5** Design a Mealy sequential circuit (Figure 16-27) which investigates an input sequence $X$ and will produce an output of $Z = 1$ for any input sequence ending in 0110 or 101.

*Example:*

$X = 0\ 1\ 0\ 1\ 1\ 0\ 1$
$Z = 0\ 0\ 0\ 1\ 0\ 1\ 1$

Notice that the circuit does not reset to the start state when an output of $Z = 1$ occurs. However, your circuit should have a start state and should be provided with a method for manually resetting the flip-flops to the start state. A minimum solution requires six states. Design your circuit using NAND gates, NOR gates, and three D flip-flops. Any solution which is minimal for your state assignment and uses eight or fewer gates and inverters is acceptable. (Assign 000 to the start state.)

*Test Procedure:* First, check out your state table by starting in each state and making sure that the present output and next state are correct for each input. Then, starting in the proper initial state, determine the output sequence for each of the following input sequences:

(1) 0 0 1 1 0 1 1 1 1 0 0 1 0 1 0 0
(2) 1 0 1 0 0 0 1 1 1 1 0 1 1 0 0 0

**16.6** Design a Mealy sequential circuit which investigates an input sequence $X$ and which will produce an output of $Z = 1$ for any input sequence ending in 0101 provided that the sequence 110 has never occurred.

*Example:*

$X = 0\ 1\ 0\ 1\ 0\ 1\ 1\ 0\ 1\ 0\ 1$
$Z = 0\ 0\ 0\ 1\ 0\ 1\ 0\ 0\ 0\ 0\ 0$

Notice that the circuit does not reset to the start state when an output of $Z = 1$ occurs. However, your circuit should have a start state and should be provided with a method for manually resetting the flip-flops to the start state. A minimum solution requires six states. Design your circuit using NAND gates, NOR gates, and three D flip-flops. Any solution which is minimal for your state assignment and uses eight or fewer gates and inverters is acceptable. (Assign 000 to the start state.)

*Test Procedure:* First, check out your state table by starting in each state and making sure that the present output and next state are correct for each input. Then, starting in the proper initial state, determine the output sequence for the following input sequences:

(1) $X = 0\ 1\ 0\ 1\ 0\ 0\ 0\ 1\ 0\ 1\ 1\ 0$
(2) $X = 1\ 0\ 1\ 0\ 1\ 0\ 1\ 1\ 0\ 1\ 0\ 1$

**16.7** Design a Mealy sequential circuit which investigates an input sequence $X$ and which will produce an output of $Z = 1$ if the total number of 1's received is even (consider zero 1's to be an even number of 1's) and the sequence 00 has occurred at least once. *Note*: The *total* number of 1's received includes those received before and after 00.

*Example:*

$X = 1\ 0\ 1\ 0\ 1\ 0\ 0\ 1\ 1\ 0\ 1$
$Z = 0\ 0\ 0\ 0\ 0\ 0\ 0\ 1\ 0\ 0\ 1$

Notice that the circuit does not reset to the start state when an output of $Z = 1$ occurs. However, your circuit should have a start state and should be provided with a method of manually resetting the flip-flops to the start state. A minimum solution requires six states. Design your circuit using NAND gates, NOR gates, and three D flip-flops. Any solution which is minimal for your state assignment and uses 12 or fewer gates and inverters is acceptable; the best known solution uses seven. (Assign 000 to the start state.)

*Test Procedure:* First, check out your state table by starting in each state and making sure that the present output and next state arc correct for each input. Then, starting in the proper initial state, determine the output sequence for each of the following input sequences:

(1) $X = 0\ 1\ 1\ 0\ 0\ 1\ 0\ 1\ 0\ 0$
(2) $X = 1\ 0\ 1\ 1\ 1\ 1\ 0\ 0\ 1\ 1\ 1\ 0$

**16.8** Design a Mealy sequential circuit (Figure 16-27) which investigates an input sequence $X$ and will produce an output of $Z = 1$ for any input sequence ending in 0011 or 110.

*Example:*

$X = 1\ 0\ 1\ 0\ 0\ 1\ 1\ 0\ 0\ 1\ 1$
$Z = 0\ 0\ 0\ 0\ 0\ 0\ 1\ 1\ 0\ 0\ 1$

Notice that the circuit does not reset to the start state when an output of $Z = 1$ occurs. However, your circuit should have a start state and should be provided with a method for manually resetting the flip-flops to the start state. Design your circuit using NAND gates, NOR gates, and three D flip-flops. Any solution which is minimal for your state assignment and uses 10 or fewer gates and inverters is acceptable; the best known solution uses six. (Assign 000 to the start state.)

*Test Procedure:* First, check out your state table by starting in each state and making sure that the present output and next state are correct for each input. Then, starting in the reset state, determine the output sequence for each of the following input sequences:

(1) $X = 0\ 0\ 0\ 1\ 0\ 0\ 0\ 1\ 1\ 0\ 1\ 0$
(2) $X = 1\ 1\ 1\ 0\ 0\ 1\ 0\ 0\ 0\ 1\ 1\ 0$

**16.9** Design a Mealy sequential circuit which investigates an input sequence $X$ and produces an output $Z$ which is determined by two rules. The initial output from the circuit is $Z = 0$. Thereafter, the output $Z$ will equal the *preceding* value of $X$ (rule 1) until the input sequence 001 occurs. Starting with the next input after 001, the output $Z$ will equal the *complement* of the *present* value of $X$ (rule 2) until the sequence

100 occurs. Starting with the next input after 100, the circuit output is again determined by rule 1, etc. Note that overlapping 001 and 100 sequences may occur.

*Example:*

Rule: 1 1 1 1 2 2 2 2 2 1 1 2
$X = 1\ 0\ 0\ 1\ 1\ 0\ 1\ 0\ 0\ 0\ 1\ 1$
$Z = 0\ 1\ 0\ 0\ 0\ 1\ 0\ 1\ 1\ 0\ 0\ 0$

Design your circuit using NAND gates, NOR gates, and three D flip-flops. Your circuit should be provided with a method for manually resetting the flip-flops to the start state. A minimum solution requires six states. Any solution which is minimal for your state assignment and uses 12 or fewer gates and inverters is acceptable. (Assign 000 to the start state.)

*Test Procedure:* First, check out your state table by starting in each state and making sure that the present output and next state are correct for each input. Then, starting in the reset state, determine the output sequence for each of the following input sequences:

(1) $X = 1\ 0\ 0\ 1\ 0\ 0\ 1\ 0\ 0\ 0\ 1\ 1$
(2) $X = 0\ 1\ 1\ 0\ 0\ 0\ 0\ 1\ 1\ 0\ 1\ 1$

**16.10** The 8, 4, −2, −1 BCD code is similar to the 8-4-2-1 BCD code, except that the weights are negative for the two least significant bit positions. For example, 0111 in 8, 4, −2, −1 code represents

$$8 \times 0 + 4 \times 1 + (-2) \times 1 + (-1) \times 1 = 1$$

Design a Mealy sequential circuit to convert 8, 4, −2, −1 code to 8-4-2-1 code. The input and output should be serial with the least significant bit first. The input $X$ represents an 8, 4, −2, −1 coded decimal digit and the output $Z$ represents the corresponding 8-4-2-1 BCD code. After four time steps the circuit should reset to the starting state regardless of the input sequence. Design your circuit using three D flip-flops, NAND gates, and NOR gates. Any solution which is minimal for your state assignment and uses eight or fewer gates is acceptable. (Assign 000 to the reset state.)

*Test Procedure:* First, check out your state table by starting in each state and making sure that the present output and next state are correct for each input. Then, starting in the reset state, determine the output sequence for each of the 10 possible input sequences and make a table.

**16.11** Design a Mealy sequential circuit (Figure 16-27) which adds five to a binary number in the range 0000 through 1010. The input and output should be serial with the least significant bit first. Find a state table with a minimum number of states. Design the circuit using NAND gates, NOR gates, and three D flip-flops. Any solution which is minimal for your state assignment and uses nine or fewer gates and inverters is acceptable. (Assign 000 to the reset state.)

*Test Procedure:* First, check out your state table by starting in each state and making sure that the present output and the next state are correct for each input.

Then, starting in the reset state, determine the output sequence for each of the 11 possible input sequences and make a table.

**16.12** Design a Mealy sequential circuit (Figure 16-27) to convert a 4-bit binary number in the range 0000 through 1010 to its 10's complement. (The 10's complement of a number $N$ is defined as $10 - N$.) The input and output should be serial with the least significant bit first. The input $X$ represents the 4-bit binary number, and the output $Z$ represents the corresponding 10's complement. After four time steps, the circuit should reset to the starting state regardless of the input sequence. Find a state table with a minimum number of states. Design the circuit using NAND gates, NOR gates, and three D flip-flops. Any solution which is minimal for your state assignment and uses nine or fewer gates and inverters is acceptable. (Assign 000 to the reset state.)

*Test Procedure:* First, check out your state table by starting in each state and making sure that the present output and the next state are correct for each input. Then, starting in the reset state, determine the output sequence for each of the 11 possible input sequences and make a table.

**16.13** Design a Mealy sequential circuit which investigates an input sequence $X$ and which will produce an output of $Z = 1$ for any input sequence ending in 1010, provided that the sequence 001 has occurred at least once.

*Example:*

$X = 1\ 0\ 1\ 0\ 0\ 1\ 0\ 1\ 0\ 1\ 0$
$Z = 0\ 0\ 0\ 0\ 0\ 0\ 0\ 0\ 1\ 0\ 1$

Notice that the circuit does not reset to the start state when an output of $Z = 1$ occurs. However, your circuit should have a start state and should be provided with a method of manually resetting the flip-flops to the start state. A minimum solution requires six states. Design your circuit using NAND gates, NOR gates, and three D flip-flops. Any solution which is minimal for your state assignment and uses nine or fewer gates and inverters is acceptable. (Assign 000 to the start state.)

*Test Procedure:* First, check out your state table by starting in each state and making sure that the present output and the next state are correct for each input. Then, starting in the proper initial state, determine the output sequence for the following input sequences:

(1) $X = 1\ 0\ 0\ 1\ 0\ 0\ 1\ 1\ 0\ 1\ 0\ 1$
(2) $X = 1\ 0\ 1\ 0\ 0\ 0\ 1\ 0\ 1\ 0\ 1\ 0$

**16.14** Design a Mealy sequential circuit which investigates an input sequence $X$ and will produce an output of $Z = 1$ whenever the total number of 0's in the sequence is odd, provided that the sequence 01 has occurred at least once.

*Example:*

$X = 1\ 1\ 0\ 0\ 0\ 1\ 1\ 0\ 1\ 0$
$Z = 0\ 0\ 0\ 0\ 0\ 1\ 1\ 0\ 0\ 1$

A minimum solution requires five states. Design your circuit using NAND gates, NOR gates, and three D flip-flops. Your circuit should have a start state and should be provided with a method of manually resetting the flip-flops to the start state. Any solution which is minimal for your state assignment and which uses 11 or fewer gates and inverters is acceptable. (Assign 000 to the start state.)

*Test Procedure:* First, check out your state table by starting in each state and making sure that the present output and the next state are correct for each input. Then, starting in the proper initial state, determine the output sequence for the following input sequences:

(1) $X = 1\ 0\ 0\ 0\ 1\ 1\ 0\ 1\ 0\ 0\ 1$
(2) $X = 0\ 0\ 0\ 0\ 1\ 0\ 1\ 0\ 0\ 0\ 1$

# Additional Problems

**16.15** Draw a block diagram that shows how a ROM and D flip-flops could be connected to realize Table 13-4 (p. 405). Specify the truth table for the ROM using a straight binary state assignment. (Note that a truth table, not a transition table, is to be specified.)

**16.16** The state table of Figure 15-14(a) is to be realized using a PLA and D flip-flops.
(a) Draw a block diagram.
(b) Specify the contents of the PLA in tabular form using the state assignment of Figure 15-14(a). (See Figure 15-15(b) for the $D$ equations.)

**16.17** An iterative circuit has a form similar to Figure 16-6. The output $Z$ is to be 1 if the total number of $X$ inputs that are 1 is an odd number greater than 2.
(a) Draw a state graph for a typical cell.
(b) Derive the equations and a NAND-gate circuit for a typical cell and for the output circuit.
(c) Specify $a_1$ and $b_1$, and simplify the first cell.
(d) Show how a sequential circuit can be constructed using the typical cell and output circuit.

**16.18** Design a sequential circuit having one input and one output that will produce an output of 1 for every second 0 it receives and for every second 1 it receives.

*Example:*

$$X\ (\text{input}) = 0\,1\,1\,0\,1\,1\,1\,0\,0\,0\,0\,1\,0\,1\,1\,0\,0\,1\,0\,1\,1\,0\,1\,0$$
$$Z\ (\text{output}) = 0\,0\,1\,1\,0\,1\,0\,0\,1\,0\,1\,1\,0\,0\,1\,1\,0\,0\,1\,1\,0\,0\,1\,1$$

(a) Design a Mealy sequential circuit using D flip-flops, showing a reduced state graph, and equations for the output and $D$ inputs. It should be a reasonably economical design.
(b) Repeat Part (a) for J-K flip-flops.

(c) Design a Moore sequential circuit using T flip-flops to do the same task, showing a state graph and input equations for a reasonably economical design.

**16.19** Design a sequential circuit to multiply an 8-4-2-1 binary-coded decimal digit by 3 to give a 5-bit binary number. For example, if the input is 0111, the output should be 10101. The input and output to the circuit should be serial with the least significant bit first. Assume that the input will be 0 at the fifth clock time and reset the circuit after the fifth output bit. [*Hint:* As each bit is received, multiply it by 3, giving a product of either 00 or 11. Thus we either output 0 and carry 0 to the next column, or output 1 and carry 1 to the next column. If we carry a 1 to the next column, then the sum of the carry and the next product is either 01 or 100. In this case, we either output 1 and carry 0 or output 0 and carry 10 (2) to the next column. What happens if we carry 10 (2) to the next column?]

(a) Derive a state table with a minimum number of states (3 states).
(b) Design the circuit using J-K flip-flops and NAND and NOR gates.
(c) Design the circuit using a PLA and D flip-flops. Give the PLA table.

**16.20** A Moore sequential circuit has three inputs ($X_2$, $X_1$, and $X_0$) that specify a temperature range in a room. The circuit has two outputs ($I$ and $D$) that control a heater for the room; $I = 1$ causes the heater to increase its heat output, and $D = 1$ causes the heater to decrease its heat output. If the temperature range is 0, 1, or 2 for three successive clock cycles, the circuit generates $I = 1$, and conversely if the temperature range is 5, 6, or 7 for three successive clock cycles, the circuit generates $D = 1$; otherwise, $I = 0$ and $D = 0$.

(a) Construct a state diagram for the circuit.
(b) Encode the states using a one-hot state assignment and derive the D flip-flop input equations and the output equations.
(c) Use a minimum number of D flip-flops and derive the D flip-flop input equations and the output equations.

**16.21** Repeat Problem 16.20 using a Mealy circuit.

**16.22** A Moore sequential circuit has two inputs ($X$ and $Y$) and three outputs ($Z_2$, $Z_1$, and $Z_0$). The outputs are a 1's complement number specifying the number of successive times $X$ and $Y$ have been equal or not equal as follows: In decimal, the outputs are 1, 2, and 3 if $X$ and $Y$ have been equal for one time, two successive times, and three or more successive times, and the outputs are, $-1$, $-2$, and $-3$ if $X$ and $Y$ have been not equal for one time, two successive times, and three or more successive times. Initially, the outputs are all 0.

(a) Construct a state diagram for the circuit.
(b) Encode the states using a one-hot state assignment and derive the D flip-flop input equations and the output equations.
(c) Use a minimum number of D flip-flops and derive the D flip-flop input equations and the output equations.

**16.23** Repeat Problem 16.22 using a Mealy sequential circuit.

**16.24** A Moore sequential circuit has two inputs ($X$ and $Y$) and three outputs ($Z_2$, $Z_1$, and $Z_0$). The outputs are a 2's complement number specifying the number of successive times $X$ and $Y$ have been equal or not equal as follows: In decimal, the outputs are 1, 2, and 3 if $X$ and $Y$ have been equal for one time, two successive times, and three or more successive times, and the outputs are $-1$, $-2$, $-3$, and $-4$ if $X$ and $Y$ have been not equal for one time, two successive times, three successive times, and four or more successive times. Initially, the outputs are all 0.

(a) Construct a state diagram for the circuit.

(b) Encode the states using a one-hot state assignment and derive the D flip-flop input equations and the output equations.

(c) Use a minimum number of D flip-flops and derive the D flip-flop input equations and the output equations.

**16.25** Repeat Problem 16.24 using a Mealy sequential circuit.

**16.26** The block diagram for an elevator controller for a two-floor elevator follows. The inputs $FB_1$ and $FB_2$ are 1 when someone in the elevator presses the first or second floor buttons, respectively. The inputs $CALL_1$ and $CALL_2$ are 1 when someone on the first or second floor presses the elevator call button. The inputs $FS_1$ and $FS_2$ are 1 when the elevator is at the first or second floor landing. The output $UP$ turns on the motor to raise the elevator car; $DOWN$ turns on the motor to lower the elevator. If neither $UP$ nor $DOWN$ is 1, then the elevator will not move. $R_1$ and $R_2$ reset the latches (described below); and when $DO$ goes to 1, the elevator door opens. After the door opens and remains open for a reasonable length of time (as determined by the door controller mechanism), the door controller mechanism closes the door and sets $DC = 1$. Assume that all input signals are properly synchronized with the system clock.

(a) If we were to realize a control circuit that responded to all of the inputs $FB_1$, $FB_2$, $CALL_1$, $CALL_2$, $FS_1$, $FS_2$, and $DC$, we would need to implement logic equations with nine or more variables (seven inputs plus at least two state variables). However, if we combine the signals $FB_i$ and $CALL_i$ into a signal $N_i$ ($i$ = 1 or 2) that indicates that the elevator is needed on the specified floor, we can reduce the number of inputs into the control circuit. In addition, if the signal $N_i$

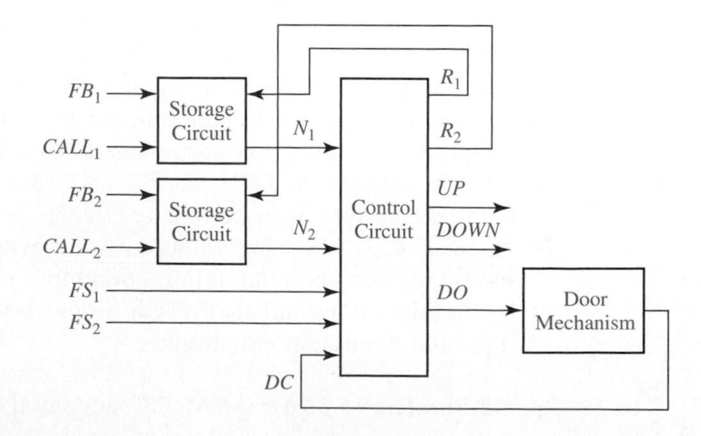

is stored so that a single pulse on $FB_i$ or $CALL_i$ will set $N_i$ to 1 until the control circuit clears it, then the control circuit will be simplified further. Using a D flip-flop and a minimum number of added gates, design a storage circuit that will have an output 1 when either input ($FB_i$ or $CALL_i$) becomes 1 and will stay 1 until reset with a signal $R_i$.

(b) Using the signals $N_1$ and $N_2$ that indicate that the elevator is needed on the first or second floor (to deliver a passenger or pick one up or both), derive a state graph for the elevator controller. (Only four states are needed.)

(c) Realize the storage circuits for $N_1$ and $N_2$ and the state graph.

**16.27** An older model Thunderbird car has three left and three right taillights which flash in unique patterns to indicate left and right turns.

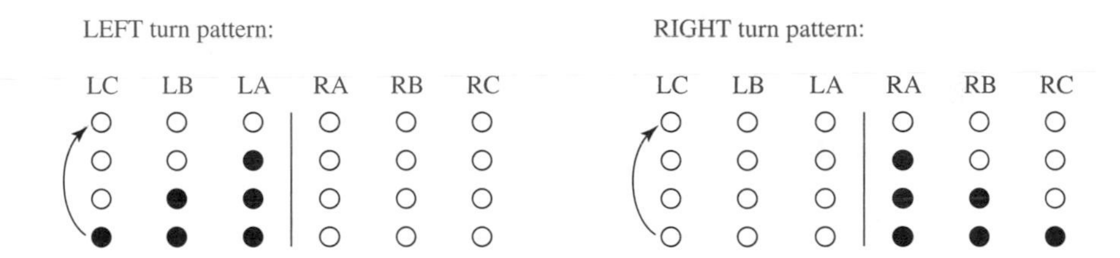


Design a Moore sequential circuit to control these lights. The circuit has three inputs *LEFT*, *RIGHT*, and *HAZ*. *LEFT* and *RIGHT* come from the driver's turn signal switch and cannot be 1 at the same time. As indicated above, when *LEFT* = 1 the lights flash in a pattern *LA* on; *LA* and *LB* on; *LA*, *LB*, and *LC* on; all off; and then the sequence repeats. When *RIGHT* = 1, the light sequence is similar. If a switch from *LEFT* to *RIGHT* (or vice versa) occurs in the middle of a flashing sequence, the circuit should immediately go to the *IDLE* (lights off) state and, then, start the new sequence. *HAZ* comes from the hazard switch, and when *HAZ* = 1, all six lights flash on and off in unison. *HAZ* takes precedence if *LEFT* or *RIGHT* is also on. Assume that a clock signal is available with a frequency equal to the desired flashing rate.

(a) Draw the state graph (eight states).

(b) Realize the circuit using six D flip-flops and make a state assignment such that each flip-flop output drives one of the six lights directly. (Use *LogicAid*.)

(c) Realize the circuit using three D flip-flops, using the guidelines to determine a suitable state assignment. Note the trade-off between more flip-flops and more gates in (b) and (c).

**16.28** Design a sequential circuit to control the motor of a tape player. The logic circuit, shown as follows, has five inputs and three outputs. Four of the inputs are the control buttons on the tape player. The input *PL* is 1 if the play button is pressed, the input *RE* is 1 if the rewind button is pressed, the input *FF* is 1 if the fast forward button is pressed, and the input *ST* is 1 if the stop button is pressed. The fifth input to the control circuit is *M*, which is 1 if the special *music sensor* detects music at the current tape position. The three outputs of the control circuit are *P*, *R*, and *F*, which make the tape play, rewind, and fast forward, respectively, when 1. No more than one output should ever be on at a time; all outputs off cause the motor to stop. The buttons control the tape as follows: If the play button is pressed, the tape player will start playing the tape (output $P = 1$). If the play button is held down and the rewind button is pressed and released, the tape player will rewind to the beginning of the current song (output $R = 1$ until $M = 0$) and then start playing. If the play button is held down and the fast forward button is pressed and released, the tape player will fast forward to the end of the current song (output $F = 1$ until $M = 0$) and then start playing. If rewind or fast forward is pressed while play is released, the tape player will rewind or fast forward the tape. Pressing the stop button at any time should stop the tape player motor.

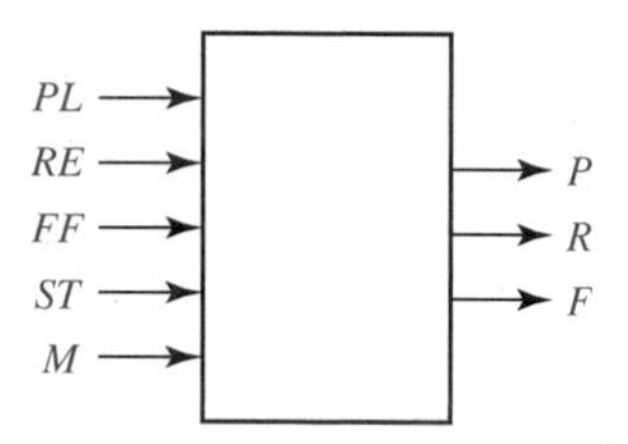


(a) Construct a state graph chart for the tape player control circuit.
(b) Realize the control circuit using a PLA and D flip-flops.

**16.29** An iterative circuit has an output of 1 from the last cell if and only if the input pattern 1011 or 1101 has occurred as inputs to any four adjacent cells in the circuit.
(a) Find a Moore state graph or table with a minimum number of states.
(b) Make a suitable state assignment, and derive one of the equations for a typical cell.
(c) Derive the output equation.

**16.30** An iterative circuit has a form similar to Figure 16-6. The output $Z$ is to be 1 iff at least one of the $X$ inputs is 1, and no group of two or more consecutive 1 inputs occurs.

*Example:*

0 0 1 0 1 0 0 0 1 0 0 gives an output $Z = 1$
0 0 1 0 1 1 0 0 0 0 0 gives an output $Z = 0$

(a) Draw a state graph for a typical cell.
(b) Derive the equations and a NOR-gate circuit for a typical cell and for the output circuit.
(c) Specify $a_1$ and $b_1$, and simplify the first cell.
(d) Show how a sequential circuit can be constructed using the typical cell and output circuit.

# VHDL for Sequential Logic

## Objectives:

1. Represent flip-flops, shift registers, and counters using VHDL processes.
2. Write sequential VHDL statements including if-then-else, case, and wait statements.
3. Explain the sequence of execution for sequential statements and the order in which signals are updated when a process executes.
4. Represent combinational logic using a process.
5. Represent a sequential logic circuit with VHDL code.
   **(a)** Use two processes.
   **(b)** Use logic equations and a process that updates the flip-flops.
   **(c)** Use a ROM and flip-flops.
6. Given VHDL code for sequential logic, draw the corresponding logic circuit.
7. Compile, simulate, and synthesize a sequential logic module.

# Study Guide

1. **1.** Study Section 17.1, *Modeling Flip-Flops Using VHDL Processes.*
   - (a) Under what condition is the expression CLK'event **and** CLK = '0' true?
   - (b) If the first line of a process is Process (St, Q1, V), under what condition will the process execute?
   - (c) In Figure 17-4, if C1 and C3 are false and C2 is true, which statements will execute?
   - (d) What device does the following VHDL code represent? What happens if ClrN = SetN = '0'?

```
process (CLK, ClrN, SetN)
  if ClrN = '0' then Q <= '0';
  elsif SetN = '0' then Q <= '1';
  elsif CLK'event and CLK = '1' then
    Q <= D;
  end if;
end process;
```

   - (e) In Figure 17-6, why are RN and SN tested before CLK? If J = '1', K = '0', and RN changes to '0', and then CLK changes to '0' 10 ns later, what will be the Q output ?
   - (f) In Figure 17-6, if the statements Q <= Qint; and QN <= not Qint; are moved inside the process just before the end-process statement, why will Q and QN have the wrong values?
   - (g) Modify the VHDL code in Figure 17-3 to add a clock enable (CE) to the flip-flop. (*Hint*: **if** CLK'event **and** CLK = '1' **and** ________ .)
   - (h) Work Problem 17.1.

2. **2.** Study Section 17.2, *Modeling Registers and Counters Using VHDL Processes.*
   - (a) Add the necessary VHDL code to Figure 17-9 to make a complete VHDL module.
   - (b) In Figure 17-10, if CLK changes to '1' at time 10 ns, at what time will Q change? (Remember that it takes Δ time to update a signal).
   - (c) What change should be made to Figure 17-10 to cause the register to rotate left one place instead of shifting left? (Do not use shift operators.)
   - (d) In Figure 17-11, what changes would be needed to make the clear asynchronous?

(e) In Figure 17-12, under what conditions does Carry2 = 1?

(f) In Figure 17-11, note that Q is a std_logic vector. Why would the code fail to compile if Q is a bit_vector?

(g) In Figure 17-14, if Qout1 = "1111", Qout2 = "1001", P = T1 = LdN = ClrN = '1', what will Qout1 and Qout2 be after the rising edge of CLK?

(h) If the process in Figure 17-13 is replaced with

```
process (CK)
begin
  if CK'event and CK = '1' then
    if Ld = '1' then Q <= D;
    elsif (P and T) = '1' then Q <= Q + 1;
    elsif Clr = '1' then Q <= "0000";
    end if;
  end if;
end process;
```

Modify Table 17-1 to properly represent the corresponding counter operation.

| Control Signals | | | Next State | | | |
|---|---|---|---|---|---|---|
| Clr | Ld | PT | $Q_3^+$ | $Q_2^+$ | $Q_1^+$ | $Q_0^+$ |
| | | | | | | |

(i) Work Problems 17.2, and 17.3.

**3.** Study Section 17.3, *Modeling Combinational Logic Using VHDL Processes*.

(a) For Figure 17-15, if the circuit is represented by a single sequential statement, make the necessary changes in the VHDL code. Assume that the AND gate delay is negligible and the OR gate delay is 5 ns. (*Hint*: The process sensitivity list should only have three signals on it.)

(b) Work Problem 17.4.

**4.** Study Section 17.4, *Modeling a Sequential Machine*.

(a) If Figure 17-16 implements the state table of Table 17-2, what will NextState and *Z* be if State = S2 and *X* = '1'?

(b) For the VHDL code of Figure 17-17:
(1) Why is the integer range 0 to 6?

(2) Assume that initially Clock = '0', State = 0, and $X$ = '0'. Trace the code to answer the following:
If $X$ changes to '1', what happens?

If CLK then changes to '1', what happens? (*Hint*: Both processes execute.)

Work Problem 17.5.

(c) Explain how the waveform of Figure 17-18 relates to Table 17-2.

Why is there a glitch in the nextstate waveform between next states 0 and 2?

Why does this glitch not cause the state to go to the wrong value?

(d) For the VHDL code of Figure 17-19:
Why do Q1, Q2, and Q3 not appear on the sensitivity list?

If CLK changes from 0 to 1 at time 5 ns, at what time are the new values of Q1, Q2, and Q3 computed? At what time do Q1, Q2, and Q3 change to these new values?

(e) Recall that component instantiation statements are concurrent statements. For the VHDL code of Figure 17-20, if Q1 changes, which of these statements will execute immediately? Relate your answer to the circuit of Figure 16-4.

(f) For Figure 17-22, what value will be read from the ROM array when X = '1' and Q = "010"?

(g) Work Problem 17.6.

**5.** Study Section 17.5, *Synthesis of VHDL Code.*

(a) Implement the following process using only a D-CE flip-flop:

```
process (CLK)
begin
   if CLK'event and CLK = '1' then
      if En = '1' then Q <= A; end if;
   end if;
end process;
```

(b) Implement the same code using a D flip-flop without a clock enable and a MUX.

(c) Implement the following VHDL code using only D-CE flip-flops:

```
signal A: bit_vector(3 downto 0)
-------------------------------------------
process (CLK)
begin
   if CLK'event and CLK = '1' then
      if ASR = '1' then
      A <= A(3)& A(3 downto 1);
      end if;
   end if;
end process;
```

(d) Work Problem 17.7.

6. Study Section 17.6, *More About Processes and Sequential Statements.*

(a) Write an equivalent process that has no sensitivity list on the first line. Use a wait statement instead.

```
process (B, C)
begin
   A <= B or C;
end process;
```

(b) For the following process, if B changes at time 2 ns, at what time does statement (2) execute? (The answer is *not* 7 ns.)

```
process (B, D);
   A <= B after 5 ns;        --(1)
   C <= D;         --(2)
end process;
```

(c) Work Problem 17.8.

7. Complete the assigned lab exercises before you take the test on Unit 17.

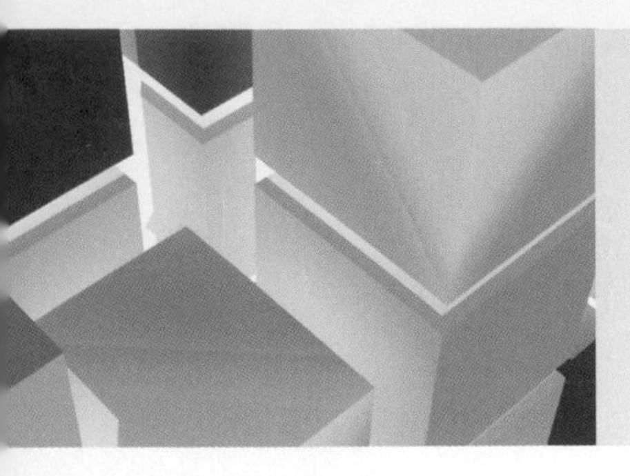

# VHDL for Sequential Logic

In Unit 10 we learned how to represent combinational logic in VHDL by using concurrent signal assignment statements. In this unit, we will learn how to represent sequential logic by using VHDL processes.

## 17.1 Modeling Flip-Flops Using VHDL Processes

A flip-flop can change state either on the rising or on the falling edge of the clock input. This type of behavior is modeled in VHDL by a process. For a simple D flip-flop with a Q output that changes on the rising edge of CLK, the corresponding process is given in Figure 17-1.

The expression in parentheses after the word **process** is called a sensitivity list, and the process executes whenever any signal in the sensitivity list changes. For example, if the process begins with **process**(A, B, C), then the process executes whenever any one of A, B, or C changes. When a process finishes executing, it goes back to the beginning and waits for a signal on the sensitivity list to change again.

In Figure 17-1, whenever CLK changes, the process executes once through and, then, waits at the start of the process until CLK changes again. The **if** statement tests for a rising edge of the clock, and Q is set equal to D when a rising edge occurs. The expression CLK'event (read as clock tick event) is TRUE whenever the signal CLK changes. If CLK = '1' is also TRUE, this means that the change was from '0' to '1', which is a rising edge. If the flip-flop has a delay of 5 ns between the rising edge of the clock and the change in the Q output, we would replace the statement Q <= D; with Q <= D **after** 5 ns; in the process in Figure 17-1.

The statements between **begin** and **end** in a process are called sequential statements. In the process in Figure 17-1, Q <= D; is a sequential statement that only executes

FIGURE 17-1
VHDL Code for a Simple D Flip-Flop

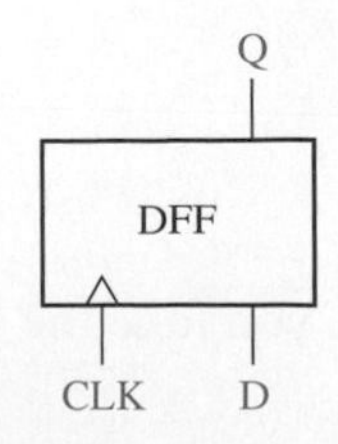


```
process (CLK)
begin
   if CLK'event and CLK = '1' -- rising edge of CLK
      then Q <= D;
   end if;
end process;
```

following the rising edge of CLK. In contrast, the concurrent statement `Q <= D;` executes whenever D changes. If we synthesize the process, the synthesizer infers that Q must be a flip-flop because it only changes on the rising edge of CLK. If we synthesize the concurrent statement `Q <= D;` the synthesizer will simply connect D to Q with a wire or with a buffer.

In Figure 17-1 note that D is not on the sensitivity list because changing D will not cause the flip-flop to change state. Figure 17-2 shows a transparent latch and its VHDL representation. Both G and D are on the sensitivity list because if G = '1', a change in D causes Q to change. If G changes to '0', the process executes, but Q does not change.

**FIGURE 17-2**
VHDL Code for a Transparent Latch

D Q
G

```
process (G,D)
begin
   if G = '1' then Q <= D; end if;
end process;
```

If a flip-flop has an active-low asynchronous clear input (ClrN) that resets the flip-flop independently of the clock, then we must modify the process of Figure 17-1 so that it executes when either CLK or ClrN changes. To do this, we add ClrN to the sensitivity list. The VHDL code for a D flip-flop with asynchronous clear is given in Figure 17-3. Because the asynchronous ClrN signal overrides CLK, ClrN is tested first, and the flip-flop is cleared if ClrN is '0'. Otherwise, CLK is tested, and Q is updated if a rising edge has occurred.

A basic process has the following form:

```
process(sensitivity-list)
begin
  sequential-statements
end process;
```

Whenever one of the signals in the sensitivity list changes, the sequential statements in the process body are executed in sequence one time. The process then goes back to the beginning and waits for a signal in the sensitivity list to change.

In the previous examples, we have used two types of sequential statements—signal assignment statements and **if** statements. The basic **if** statement has the form

```
if condition then
      sequential statements1
   else sequential statements2
end if;
```

**FIGURE 17-3**
VHDL Code for a D Flip-flop with Asynchronous Clear

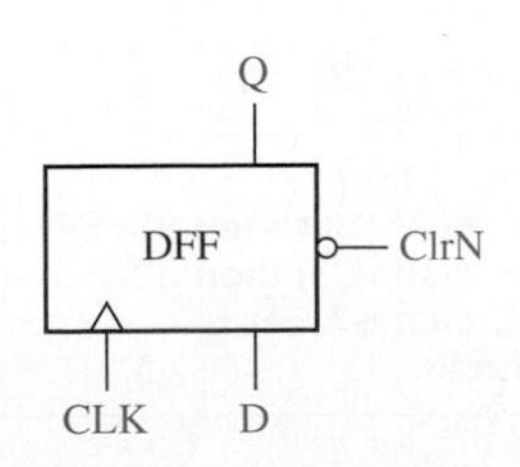


```
process (CLK, ClrN)
begin
   if ClrN = '0' then Q <= '0';
         else if CLK'event and CLK = '1'
              then Q <= D;
              end if;
   end if;
end process;
```

The condition is a Boolean expression which evaluates to TRUE or FALSE. If it is TRUE, sequential statements1 are executed; otherwise, sequential statements2 are executed. VHDL **if** statements are sequential statements that can be used within a process, but they cannot be used as concurrent statements outside of a process. On the other hand, conditional signal assignment statements are concurrent statements that cannot be used within a process.

The most general form of the **if** statement is

```
if condition then
   sequential statements
{elsif condition then
   sequential statements}
   -- 0 or more elsif clauses may be included
[else sequential statements]
end if;
```

The curly brackets indicate that any number of **elsif** clauses may be included, and the square brackets indicate that the **else** clause is optional. The example of Figure 17-4 shows how a flow chart can be represented using nested **if**s or the equivalent using **elsif**s. In this example, C1, C2, and C3 represent conditions that can be TRUE or FALSE, and S1, S2, . . . S8 represent sequential statements. Each **if** requires a corresponding **end if**, but an **elsif** does not.

Next, we will write a VHDL module for a J-K flip-flop (Figure 17-5). This flip-flop has active-low asynchronous preset (SN) and clear (RN) inputs. State changes related to J and K occur on the falling edge of the clock. In this chapter, we use a suffix N to indicate an active-low (negative-logic) signal. For simplicity, we will assume that the condition SN = RN = 0 does not occur.

**FIGURE 17-4** Equivalent Representations of a Flow Chart Using Nested Ifs and Elsifs

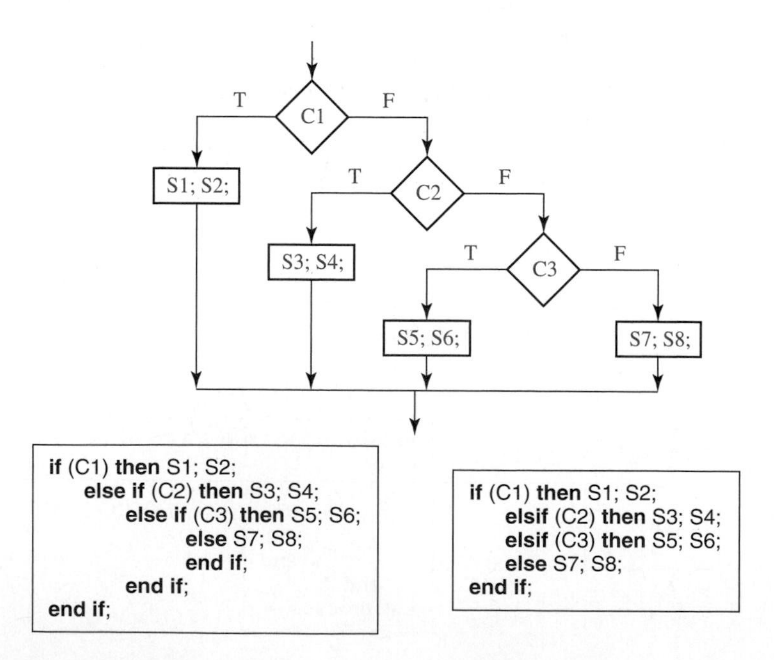

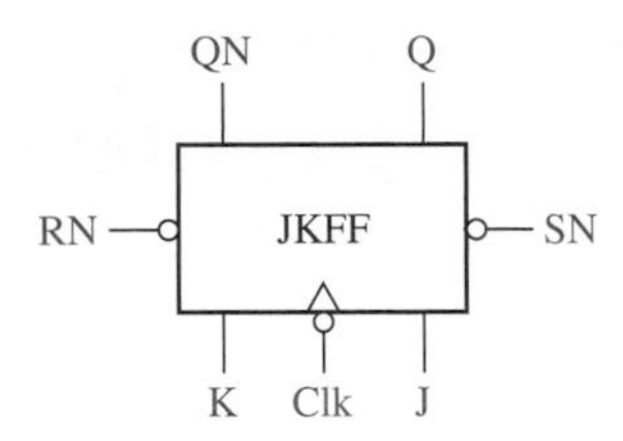


**FIGURE 17-5**
J-K Flip-Flop

The VHDL code for the J-K flip-flop is given in Figure 17-6. The port declaration in the entity defines the input and output signals. Within the architecture we define a signal Qint that represents the state of the flip-flop internal to the module. The two concurrent statements after **begin** transmit this internal signal to the Q and QN outputs of the flip-flop. We do it this way because an output signal in a port cannot appear on the right side of an assignment statement within the architecture. The flip-flop can change state in response to changes in SN, RN, and CLK, so these three signals are in the sensitivity list of the process. Because RN and SN reset and set the flip-flop independently of the clock, they are tested first. If RN and SN are both '1', then we test for the falling edge of the clock. The condition (CLK'event **and** CLK = '0') is TRUE only if CLK has just changed from '1' to '0'. The next state of the flip-flop is determined by its characteristic equation:

$$Q^{+} = JQ' + K'Q$$

The 8-ns delay represents the time it takes to set or clear the flip-flop output after SN or RN changes to 0. The 10-ns delay represents the time it takes for Q to change after the falling edge of the clock.

**FIGURE 17-6**
J-K Flip-Flop Model

```
1   entity JKFF is
2   port (SN, RN, J, K, CLK: in bit;                    --inputs
3      Q, QN: out bit);
4   end JKFF;
5   architecture JKFF1 of JKFF is
6   signal Qint: bit;                                   -- internal value of Q
7   begin
8      Q <= Qint;                                       -- output Q and QN to port
9      QN <= not Qint;
10     process (SN, RN, CLK)
11     begin
12        if RN = '0' then Qint <= '0' after 8 ns;      -- RN = '0' will clear the FF
13        elsif SN = '0' then Qint <= '1' after 8 ns;   -- SN = '0' will set the FF
14        elsif CLK'event and CLK = '0' then            -- falling edge of CLK
15           Qint <= (J and not Qint) or (not K and Qint) after 10 ns;
16        end if;
17     end process;
18  end JKFF1;
```

# 17.2 Modeling Registers and Counters Using VHDL Processes

When several flip-flops change state on the same clock edge, the statements representing these flip-flops can be placed in the same clocked process. Figure 17-7 shows three flip-flops connected as a cyclic shift register. These flip-flops all change state following the rising edge of the clock. We have assumed a 5-ns propagation delay between the clock edge and the output change. Immediately following the clock edge, the three statements in the process execute in sequence with no delay. The new values of the Q's are then scheduled to change after 5 ns. If we omit the delay and replace the sequential statements with

```
Q1 <= Q3;   Q2 <= Q1;   Q3 <= Q2;
```

the operation is basically the same. The three statements execute in sequence in zero time, and, then, the Q's change value after $\Delta$ delay. In both cases the old values of Q1, Q2, and Q3 are used to compute the new values. This may seem strange at first, but that is the way the hardware works. At the rising edge of the clock, all of the D inputs are loaded into the flip-flops, but the state change does not occur until after a propagation delay.

Next we will write structural VHDL code for the cyclic shift register using a D flip-flop as a component. In the writing of structural VHDL code, instantiation statements are used to specify how components are connected together. Components may be declared and defined either in a library or within the architecture part of the VHDL code. Each copy of a component requires a separate instantiation statement to specify how it is connected to other components and to the port inputs and outputs. Instantiation statements are concurrent statements, not sequential statements, and therefore they cannot be used within a process. A component can be as simple as a single gate or as complex as a digital system that contains many internal signals, registers, control circuits, and other components. Each instantiation statement represents a copy of a hardware component. The instantiation statement connects the component inputs and outputs, and the component computes new outputs whenever one of its inputs changes. This is exactly how the real hardware component works. Instantiating a

**FIGURE 17-7**
Cyclic Shift Register

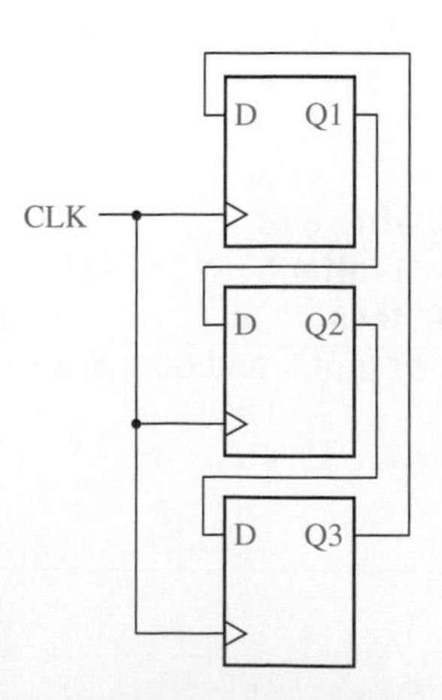


```
process (CLK)
begin
    if CLK'event and CLK = '1' then
        Q1 <= Q3 after 5 ns;
        Q2 <= Q1 after 5 ns;
        Q3 <= Q2 after 5 ns;
    end if;
end process;
```

component is different from calling a function in a computer program. A function returns a new value whenever it is called, but an instantiated component computes a new output value whenever its input changes.

The VHDL code of Figure 17-8 has two modules. The first one models a simple D flip-flop. The second module instantiates three copies of the D flip-flop component to model the cyclic shift register of Figure 17-7. Qout represents a 3-bit output from the register. The internal signals (Q1, Q2, and Q3) that are declared within the architecture are used to connect the flip-flop inputs and outputs. Lines 21, 22, and 23 instantiate three copies of the D flip-flop component. Even though the DFF module has a clock input and internal sequential statements, each instantiation statement is still a concurrent statement and must *not* be placed in a process. If clk changes, this change is passed to the D flip-flop components, and the effect of the clock change is handled within the components.

Figure 17-9 shows a simple register that can be loaded or cleared on the rising edge of the clock. If CLR = 1, the register is cleared, and if Ld = 1, the D inputs are loaded into the register. This register is fully synchronous so that the Q outputs only change in response to the clock edge and not in response to a change in Ld or Clr. In the VHDL code for the register, Q and D are bit vectors dimensioned 3 **downto** 0. Because the register outputs can only change on the rising edge of the clock, CLR is not on the

FIGURE 17-8 Structural VHDL Code for Cyclic Shift Register

```
1     entity DFF is                                       --simple DFF
2        port (D, clk: in bit; q: out bit);
3     end DFF;
4     architecture DFF_simple of DFF is
5     begin
6     process (clk)
7     begin
8        if clk'event and clk = '1' then
9           Q <= D after 5 ns; end if;
10    end process;
11    end DFF_simple;

12    entity cyclicSR is                                  -- 3-bit cyclic shift register
13       port (clk: in bit; Qout: out bit-vector(1 to 3) ) ;
14    end cyclicSR;
15    architecture cyclicSR3 of cyclicSR is
16    component DFF
17       port (D, clk: in bit; Q: out bit);
18    end component;
19    signal Q1, Q2, Q3: bit;
20    begin
21       FF1: DFF port map (Q3, clk, Q1);
22       FF2: DFF port map (Q1, clk, Q2);
23       FF3: DFF port map (Q2, clk, Q3) ;
24       Qout <= Q1&Q2&Q3;
25    end cyclicSR3;
```

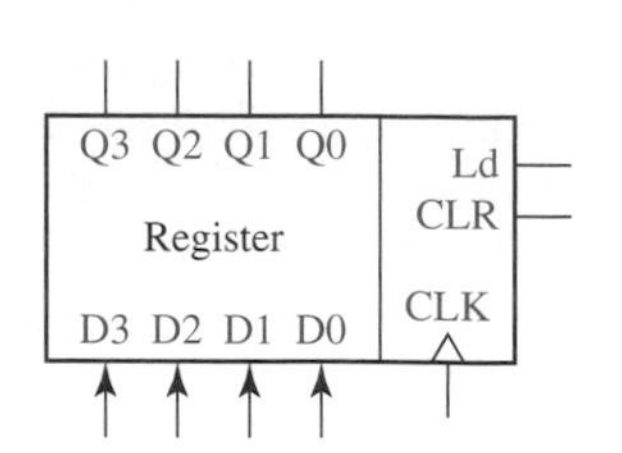


```
process (CLK)
begin
   if CLK'event and CLK = '1' then
       if CLR = '1' then Q <= "0000";
           elsif Ld = '1' then Q <= D;
       end if;
   end if;
end process;
```

**FIGURE 17-9** Register with Synchronous Clear and Load

sensitivity list. It is tested after the rising edge of the clock instead of being tested first as in Figure 17-3. If Clr = Ld = '0', Q does not change. Because Clr is tested before Ld, if Clr = '1', the **elsif** prevents Ld from being tested and Clr overrides Ld.

Next, we will model a left-shift register using a VHDL process. The register in Figure 17-10 is similar to that in Figure 17-9, except we have added a left-shift control input (LS). When LS is '1', the contents of the register are shifted left and the rightmost bit is set equal to Rin. The shifting is accomplished by taking the rightmost three bits of Q, Q(2 **downto** 0) and concatenating them with Rin. For example, if Q = "1101" and Rin = '0', then Q(2 **downto** 0) &Rin = "1010", and this value is loaded back into the Q register on the rising edge of CLK. The code implies that if CLR = Ld = LS = '0', then Q remains unchanged.

Figure 17-11 shows a simple synchronous counter. On the rising edge of the clock, the counter is cleared when ClrN = '0', and it is incremented when ClrN = En = '1'. In this example, the signal Q represents the 4-bit value stored in the counter. Because addition is not defined for bit_vectors, we have declared Q to be of type std_logic_vector. Then, we can increment the counter using the overloaded "+" operator that is defined in the ieee.std_logic_unsigned package. The statement Q <= Q + 1; increments the counter. When the counter is in state "1111", the next increment takes it back to state "0000".

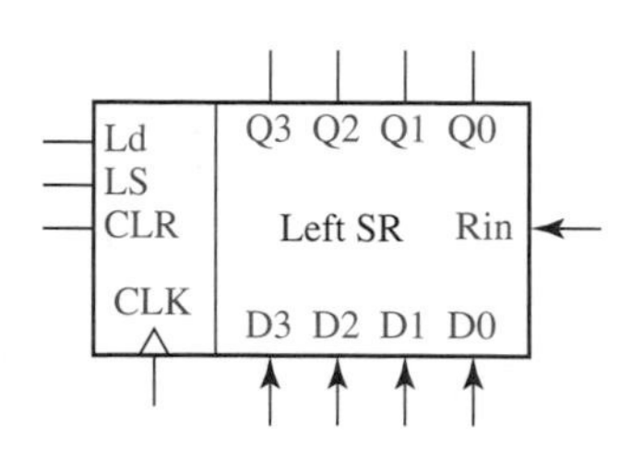


```
process (CLK)
begin
  if CLK'event and CLK = '1' then
    if CLR = '1' then Q <= "0000";
      elsif Ld = '1' then Q <= D;
      elsif LS = '1' then Q <= Q(2 downto 0)& Rin;
    end if;
  end if;
end process;
```

**FIGURE 17-10** Left-Shift Register with Synchronous clear and Load

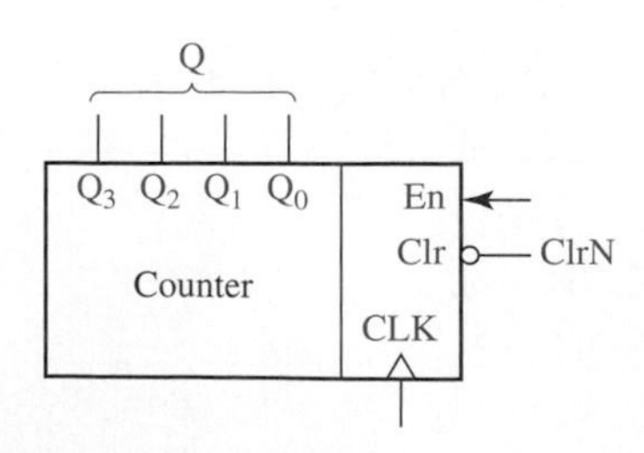


```
signal Q: std_logic_vector(3 downto 0);
------------
process (CLK)
begin
   if CLK'event and CLK = '1' then
       if ClrN = '0' then Q <= "0000";
           elsif En = '1' then Q <= Q + 1;
       end if;
   end if;
end process;
```

**FIGURE 17-11** VHDL Code for a Simple Synchronous Counter

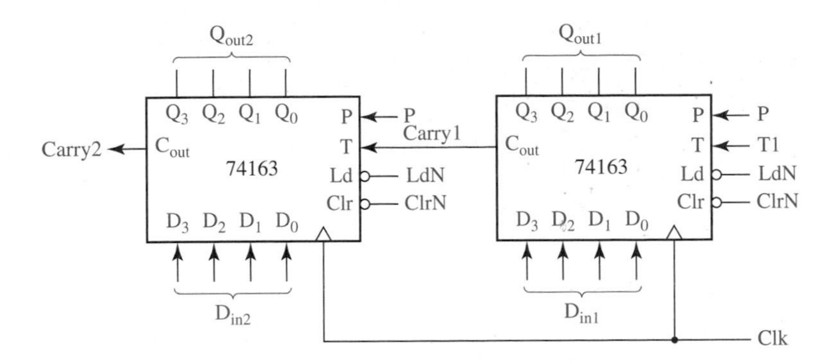


**FIGURE 17-12** Two 74163 Counters Cascaded to Form an 8-Bit Counter

The 74163 (see Figure 17-12) is a 4-bit fully synchronous binary counter which is available in both TTL and CMOS logic families. Although rarely used in new designs at present, it represents a general type of counter that is found in many CAD design libraries. In addition to performing the counting function, it can be cleared or loaded in parallel. All operations are synchronized by the clock, and all state changes take place following the rising edge of the clock input.

This counter has four control inputs: ClrN, LdN, P, and T. Inputs P and T are used to enable the counting function. Operation of the counter is as follows:

1. If ClrN = 0, all flip-flops are set to 0 following the rising clock edge.
2. If ClrN = 1 and LdN = 0, the D inputs are transferred in parallel to the flip-flops following the rising clock edge.
3. If ClrN = LdN = 1 and P = T = 1, the count is enabled and the counter state will be incremented by 1 following the rising clock edge.

If T = 1, the counter generates a carry ($C_{out}$) in state 15, so

$$C_{out} = Q_3\, Q_2\, Q_1\, Q_0\, T$$

Table 17-1 summarizes the operation of the counter. Note that ClrN overrides the load and count functions in the sense that when ClrN = 0, clearing occurs regardless of the values of LdN, P, and T. Similarly, LdN overrides the count function. The ClrN input on the 74163 is referred to as a *synchronous* clear input because it clears the counter in synchronization with the clock, and no clearing can occur if a clock pulse is not present.

The VHDL description of the counter is shown in Figure 17-13. Q represents the four flip-flops that make up the counter. The counter output, $Q_{out}$, changes whenever Q changes. The carry output is computed whenever Q or T changes. The first **if** statement in the process tests for a rising edge of Clk. Because clear overrides load and count, the next **if** statement tests ClrN first. Because load overrides count, LdN is

**TABLE 17-1** 74163 Counter Operation

| Control Signals | | | Next State | | | | |
|---|---|---|---|---|---|---|---|
| ClrN | LdN | *PT* | $Q_3^+$ | $Q_2^+$ | $Q_1^+$ | $Q_0^+$ | |
| 0 | X | X | 0 | 0 | 0 | 0 | (Clear) |
| 1 | 0 | X | $D_3$ | $D_2$ | $D_1$ | $D_0$ | (Parallel load) |
| 1 | 1 | 0 | $Q_3$ | $Q_2$ | $Q_1$ | $Q_0$ | (No change) |
| 1 | 1 | 1 | Present state + 1 | | | | (Increment count) |

FIGURE 17-13
74163 Counter Model

```
-- 74163 FULLY SYNCHRONOUS COUNTER
1   library IEEE;
2   use IEEE.STD_LOGIC_1164.ALL;
3   use IEEE.STD_LOGIC_ARITH.ALL;
4   use IEEE.STD_LOGIC_UNSIGNED.ALL;
5   entity c74163 is
6     port(LdN, ClrN, P, T, ClK: in std_logic;
7          D: in std_logic_vector(3 downto 0);
8          Cout: out std_logic; Qout: out std_logic_vector(3 downto 0) );
9   end c74163;

10  architecture b74163 of c74163 is
11  signal Q: std_logic_vector(3 downto 0);        -- Q is the counter register
12  begin
13    Qout <= Q;
14    Cout <= Q(3) and Q(2) and Q(1) and Q(0) and T;
15    process (Clk)
16    begin
17      if Clk'event and Clk = '1' then            -- change state on rising edge
18        if ClrN = '0' then Q <= "0000";
19          elsif LdN = '0' then Q <= D;
20          elsif (P and T) = '1' then Q <= Q + 1;
21        end if;
22      end if;
23    end process;
24  end b74163;
```

tested next. Finally, the counter is incremented if both P and T are 1. Because Q is type std_logic_vector, we can use the overloaded "+" operator from the ieee.std_logic_unsigned library to add 1 to increment the counter. The expression Q + 1 would not be legal if Q were a bit_vector because addition is not defined for bit_vectors.

To test the counter, we have cascaded two 74163's to form an 8-bit counter (Figure 17-12). When the counter on the right is in state 1111 and T1 = 1, the T input to the left counter is Carry1 = 1. Then, if P = 1, on the next clock the right counter is incremented to 0000 at the same time the left counter is incremented. Figure 17-14 shows the VHDL code for the 8-bit counter. In this code we have used the c74163 model as a component and instantiated two copies of it. For convenience in reading the output, we have defined a signal Count which is the integer equivalent of the 8-bit counter value. The function Conv_integer converts a std_logic_vector to an integer.

The two instantiation statements (lines 21 and 22) connect the inputs and outputs of two copies of the 4-bit counter component. Each of these concurrent statements will execute when one of the counter inputs changes, and then the corresponding counter module computes new values of the counter outputs. Although the 4-bit

FIGURE 17-14
VHDL for 8-Bit Counter

```
-- Test module for 74163 counter
1   library IEEE;
2   use IEEE.STD_LOGIC_1164.ALL;
3   use IEEE.STD_LOGIC_ARITH.ALL;
4   use IEEE.STD_LOGIC_UNSIGNED.ALL;

5   entity c74163test is
6     port(ClrN, LdN, P, T1, Clk: in std_logic;
7       Din1, Din2: in std_logic_vector (3 downto 0);
8       Count: out integer range 0 to 255;
9       Carry2: out std_logic);
10  end c74163test;

11  architecture tester of c74163test is
12    component c74163
13      port(LdN, ClrN, P, T, Clk: in std_logic;
14        D: in std_logic_vector(3 downto 0);
15      Cout: out std_logic; Qout: out std_logic_vector (3 downto 0) );
16    end component;
17    signal Carry1: std_logic;
18    signal Qout1, Qout2: std_logic_vector (3 downto 0);
19  begin
20    ct1: c74163 port map (LdN, ClrN, P, T1, Clk, Din1, Carry1, Qout1);
21    ct2: c74163 port map (LdN, ClrN, P, Carry1, Clk, Din2, Carry2, Qout2);
22    Count <= Conv_integer(Qout2 & Qout1);
23  end tester;
```

counter module (Figure 17-13) contains a process and sequential statements, each statement that instantiates a counter module is nevertheless a concurrent statement and cannot be placed within a process.

## 17.3 Modeling Combinational Logic Using VHDL Processes

Although processes are most useful for modeling sequential logic, they can also be used to model combinational logic. The circuit of Figure 10-1 can be modeled by the process shown in Figure 17-15.

For a combinational process, every signal that appears on the right side of a signal assignment must appear on the sensitivity list. Suppose that initially $A = 1$, and $B = C = D = E = 0$. If B changes to 1 at time = 4 ns, the process executes, and the two sequential assignment statements execute in sequence. The new value

FIGURE 17-15
VHDL Code for
Gate Circuit

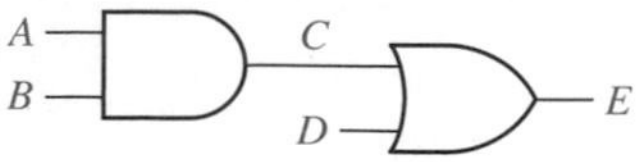

```
process (A, B, C, D)
begin
   C <= A and B after 5 ns;
   E <= C or D after 5 ns;
end process;
```

of C is computed to be '1', and C is scheduled to change 5 ns later. Meanwhile, E is immediately computed using the old value of C, but it does not change because C has not yet changed. After 5 ns, C changes, and because it is on the sensitivity list, the process executes again, and the sequential statements again execute in sequence. This time C does not change, but E is scheduled to change after 5 ns. Because E is not on the sensitivity list, no further execution of the process occurs. The following listing summarizes the operation:

```
time  A  B  C  D  E
 0    1  0  0  0  0
 4    1  1  0  0  0
      process executes (C ← 1 after 5 ns; E ← 0, no change)
 9    1  1  1  0  0
      process executes (C ← 1, no change; E ← 1 after 5 ns)
14    1  1  1  0  1
      no further execution until A, B, C, or D changes
```

In Section 10.2, we modeled a MUX using a conditional signal assignment statement and a selected signal assignment statement. Because these are concurrent statements, they cannot be used inside a process. However, the **case** statement is a sequential statement that can be used to model a MUX within a process. The 4-to-1 MUX of Figure 10-7 can be modeled as follows:

```
signal sel: bit_vector(0 to 1);

-----------------------------------------------------------------------
sel <= A&B;              -- a concurrent statement, outside of the process
process (sel, I0, I1, I2, I3)
begin
  case sel is            -- a sequential statement in the process
    when "00" =>   F <= I0;
    when "01" =>   F <= I1;
    when "10" =>   F <= I2;
    when "11" =>   F <= I3;
    when others => null;          -- required if sel is a std_logic_vector;
                                  -- omit if sel is a bit_vector
  end case;
end process;
```

The **case** statement has the general form:

```
case expression is
   when choice1 => sequential statements1
   when choice2 => sequential statements2
   . . .
   [when others => sequential statements]
end case;
```

The "expression" is evaluated first. If it is equal to "choice1", then "sequential statements1" are executed; if it is equal to "choice2", then "sequential statements2" are executed, etc. All possible values of the expression must be included in the choices. If all values are not explicitly given, a "**when others**" clause is required in the case statement. If no action is specified for the other choices, the clause should be

```
when others => null;
```

## 17.4 Modeling a Sequential Machine

In this section we will discuss several ways of writing VHDL descriptions for sequential machines. First, we will write a behavioral model for a Mealy sequential circuit based on the state table of Table 17-2. This table is the same as Table 16-3 with the states renamed. It represents a BCD to excess-3 code converter with inputs and outputs LSB first.

As shown in Figure 17-16, a Mealy machine consists of a combinational circuit and a state register. The VHDL model of Figure 17-17 uses two processes to represent these two parts of the circuit. Because X and Z are external signals, they are declared in the port. State and Nextstate are internal signals that represent the state and next state of the sequential circuit, so they are declared at the start of the architecture. At the behavioral level, we represent the state and next state of the circuit by integer signals with a range of 0 to 6.

**TABLE 17-2**
State Table for Code Converter

| PS | NS | | Z | |
|---|---|---|---|---|
| | X = 0 | X = 1 | X = 0 | X = 1 |
| S0 | S1 | S2 | 1 | 0 |
| S1 | S3 | S4 | 1 | 0 |
| S2 | S4 | S4 | 0 | 1 |
| S3 | S5 | S5 | 0 | 1 |
| S4 | S5 | S6 | 1 | 0 |
| S5 | S0 | S0 | 0 | 1 |
| S6 | S0 | – | 1 | – |

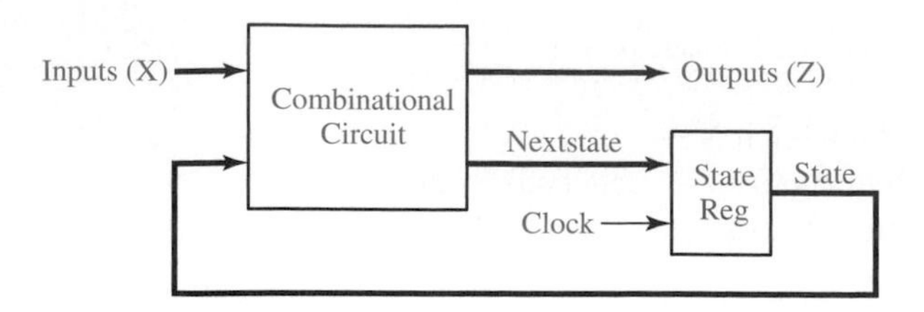


FIGURE 17-16 General Model of Mealy Sequential Machine

The first process represents the combinational circuit of Figure 17-16. Because the circuit outputs, Z and Nextstate, can change when either the State or X changes, the sensitivity list includes both State and X. The case statement tests the value of State, and then for each state, the **if** statement tests X to determine the new values of Z and Nextstate. For state S6, we assigned values to the don't-cares so that Z and Nextstate are independent of X. The second process represents the state register. Whenever the rising edge of the clock occurs, the State is updated to the Nextstate value, so CLK appears in the sensitivity list. A typical sequence of execution for the two processes is as follows:

1. X changes and the first process executes. New values of Z and NextState are computed.
2. The clock falls, and the second process executes. Because CLK = '0', nothing happens.
3. The clock rises, and the second process executes again. Because CLK = '1', State is set equal to the Nextstate value.
4. If State changes, the first process executes again. New values of Z and Nextstate are computed.

A simulator command file which can be used to test Figure 17-17 follows:

```
add wave CLK X State Nextstate Z
force CLK 0 0, 1 100 -repeat 200
force X 0 0, 1 350, 0 550, 1 750, 0 950, 1 1350
run 1600
```

The first command specifies the signals which are to be included in the waveform output. The next command defines a clock with period of 200 ns. CLK is '0' at time 0 ns, '1' at time 100 ns, and repeats every 200 ns. In a command of the form

```
force signal_name v1 t1, v2 t2, . . .
```

signal_name gets the value v1 at time t1, the value v2 at time t2, etc. X is '0' at time 0 ns, changes to '1' at time 350 ns, changes to '0' at time 550 ns, etc. The X input corresponds to the sequence 0010 1001, and only the times at which X changes are

**FIGURE 17-17**
Behavioral Model for Table 17-2

```
-- This is a behavioral model of a Mealy state machine (Table 17-2) based on its state
-- table. The output (Z) and next state are computed before the active edge of the clock.
-- The state change occurs on the rising edge of the clock.

1   entity SM17_2 is
2     port (X, CLK: in bit;
3       Z: out bit);
4   end SM17_2;

5   architecture Table of SM17_2 is
6     signal State, Nextstate: integer range 0 to 6 := 0;
7   begin
8     process(State, X)        -- Combinational Circuit
9     begin
10      case State is
11      when 0 =>
12        if X = '0' then Z <= '1'; Nextstate <= 1;
13        else Z <= '0'; Nextstate <= 2; end if;
14      when 1 =>
15        if X = '0' then Z <= '1'; Nextstate <= 3;
16        else Z <= '0'; Nextstate <= 4; end if;
17      when 2 =>
18        if X = '0' then Z <= '0'; Nextstate <= 4;
19        else Z <= '1'; Nextstate <= 4; end if;
20      when 3 =>
21        if X = '0' then Z <= '0'; Nextstate <= 5;
22        else Z <= '1'; Nextstate <= 5; end if;
23      when 4 =>
24        if X = '0' then Z <= '1'; Nextstate <= 5;
25        else Z <= '0'; Nextstate <= 6; end if;
26      when 5 =>
27        if X = '0' then Z <= '0'; Nextstate <= 0;
28        else Z <= '1'; Nextstate <= 0; end if;
29      when 6 =>
30        Z <= '1'; Nextstate <= 0;
31      end case;
32    end process;

33    process (CLK)       -- State Register
34    begin
35      if CLK'event and CLK = '1' then       -- rising edge of clock
36        State <= Nextstate;
37      end if;
38    end process;
39  end Table;
```

FIGURE 17-18 Waveforms for Figure 17-17

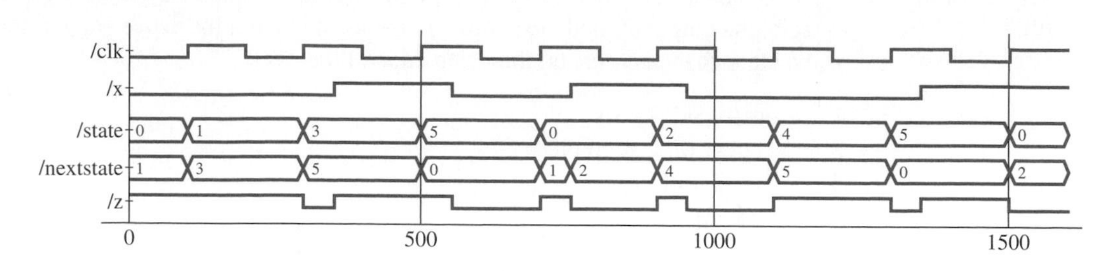


specified. Execution of the preceding command file produces the waveforms shown in Figure 17-18.

The behavioral VHDL model of Figure 17-17 is based on the state table. After we have derived the next-state and output equations from the state table, we can write a data flow VHDL model based on these equations. The VHDL model of Figure 17-19 is based on the next-state and output equations that are derived in Figure 16-3 using the state assignment of Figure 16-2. The flip-flops are updated

FIGURE 17-19 Sequential Machine Model Using Equations

```
-- The following is a description of the sequential machine of Table 17-2 in terms
-- of its next-state equations. The following state assignment was used:
-- S0--> 0; S1--> 4; S2--> 5; S3--> 7; S4--> 6; S5--> 3; S6--> 2

1    entity SM1_2 is
2      port (X, CLK: in bit;
3        Z: out bit);
4    end SM1_2;

5    architecture Equations1_4 of SM1_2 is
6      signal Q1, Q2, Q3: bit;
7    begin
8      process(CLK)
9      begin
10       if CLK'event and CLK = '1' then      -- rising edge of clock
11         Q1 <= not Q2 after 10 ns;
12         Q2 <= Q1 after 10 ns;
13         Q3 <= (Q1 and Q2 and Q3) or (not X and Q1 and not Q3) or
14                    (X and not Q1 and not Q2) after 10 ns;
15       end if;
16     end process;
17     Z <= (not X and not Q3) or (X and Q3) after 20 ns;
18   end Equations1_4;
```

in a process which is sensitive to CLK. When the rising edge of the clock occurs, Q1, Q2, and Q3 are all assigned new values. A 10-ns delay is included to represent the propagation delay between the active edge of the clock and the change of the flip-flop outputs. Even though the assignment statements in the process are executed sequentially, Q1, Q2, and Q3 are all scheduled to be updated at the same time, T + 10 ns, where T is the time at which the rising edge of the clock occurred. Thus, the old value of Q1 is used to compute $Q2^+$, and the old values of Q1, Q2, and Q3 are used to compute $Q3^+$. The concurrent assignment statement for Z causes Z to be updated whenever a change in X or Q3 occurs. The 20-ns delay represents two gate delays.

After we have designed a sequential circuit using components such as gates and flip-flops, we can write a structural VHDL model based on the actual interconnection of these components. Figure 17-20 shows a structural VHDL representation of the circuit of Figure 16-4. Seven NAND gates, three D flip-flops, and one inverter are used. All of these components are defined in a library named

**FIGURE 17-20** Structural Model of Sequential Machine

```
-- The following is a STRUCTURAL VHDL description of the circuit of Figure 16-4.

1   library BITLIB;
2   use BITLIB.bit_pack.all;

3   entity SM17_1 is
4     port (X, CLK: in bit;
5       Z: out bit);
6   end SM17_1;

7   architecture Structure of SM17_1 is
8     signal A1, A2, A3, A5, A6, D3: bit: = '0';
9     signal Q1, Q2, Q3: bit: = '0';
10    signal Q1N, Q2N, Q3N, XN: bit: = '1';
11  begin
12    I1: Inverter port map (X, XN);
13    G1: Nand3 port map (Q1, Q2, Q3, A1);
14    G2: Nand3 port map (Q1, Q3N, XN, A2);
15    G3: Nand3 port map (X, Q1N, Q2N, A3);
16    G4: Nand3 port map (A1, A2, A3, D3);
17    FF1: DFF port map (Q2N, CLK, Q1, Q1N);
18    FF2: DFF port map (Q1, CLK, Q2, Q2N);
19    FF3: DFF port map (D3, CLK, Q3, Q3N);
20    G5: Nand2 port map (X, Q3, A5);
21    G6: Nand2 port map (XN, Q3N, A6);
22    G7: Nand2 port map (A5, A6, Z);
23  end Structure;
```

BITLIB. The component declarations and definitions are contained in a package called bit_pack. The **library** and **use** statements are explained in Section 10.7. Because the NAND gates and D flip-flops are declared as components in bit_pack, they are not explicitly declared in the VHDL code. Because Q1, Q2, and Q3 are initialized to '0', the complementary flip-flop outputs (Q1N, Q2N, and Q3N) are initialized to '1'. G1 is a 3-input NAND gate with inputs Q1, Q2, Q3, and output A1. FF1 is a D flip-flop (see Figure 17-1) with the D input connected to Q2N. All of the gates and flip-flops in the bit_pack have a default delay of 10 ns. Executing the following simulator command file produces the waveforms of Figure 17-21.

```
add wave CLK X Q1 Q2 Q3 Z
force CLK 0 0, 1 100 –repeat 200
force X 0 0, 1 350, 0 550, 1 750, 0 950, 1 1350
run 1600
```

Next, we will implement the state machine of Table 16-6(a) using a ROM, as shown in Figure 16-10. In the VHDL code (Figure 17-22), we have used packages from the IEEE library and IEEE Standard Logic because synthesis tools often use std_logic and std_logic_vector as default types. The constant array ROM1 represents the truth table of Table 16-6(c), which is stored in the ROM. Reading data from the ROM is accomplished by four concurrent statements. First, the ROM address, which is the index into the array, is formed by concatenating X and Q to form a 4-bit vector. The index is converted from a std_logic_vector to an integer by calling the conv_integer function. The ROM1 output is split into the D vector that represents the next state and the Z output. The process updates the state register on the rising edge of the clock.

Next, we will write behavioral VHDL code for the state table given in Table 13-4. We will use a two-process model as we did in Figure 17-17. We will use nested case statements instead of using if-then-else because the state table has more columns. Figure 17-23 shows a portion of the VHDL code for the combinational part of the circuit. The first case statement branches on the state, and the nested case statement for each state defines the Nextstate and outputs by branching on X12 (= X1&X2). The second process (not shown) that updates the state register is identical to the one in Figure 17-17.

**FIGURE 17-21** Waveforms for Figure 16-4

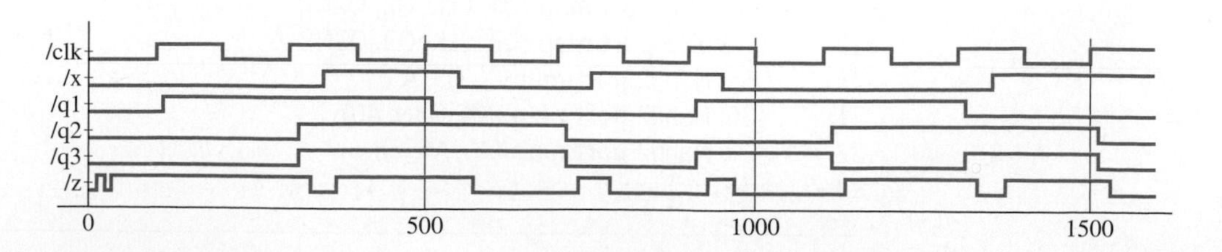

FIGURE 17-22
Sequential Machine Using a ROM

```
1   library IEEE;
2   use IEEE.STD_LOGIC_1164.ALL;
3   use IEEE.STD_LOGIC_ARITH.ALL;
4   use IEEE.STD_LOGIC_UNSIGNED.ALL;

5   entity SM16_6 is
6     Port ( X : in std_logic;
7       CLK : in std_logic;
8       Z : out std_logic);
9   end SM16_5;

10  architecture ROM of SM16_6 is
11  type ROM16X4 is array (0 to 15) of std_logic_vector (0 to 3);
12  constant ROM1: ROM16X4 : = ("1001", "1011", "0100", "0101",
13                              "1101", "0000", "1000", "0000",
14                              "0010", "0100", "1100", "1101",
15                              "0110", "1000", "0000", "0000");
16  signal Q, D: std_logic_vector (1 to 3) := "000";
17  signal Index, Romout: std_logic_vector (0 to 3);
18  begin
19  Index <= X&Q; -- X&Q is a 4-bit vector: X Q1 Q2 Q3
20  Romout <= ROM1 (conv_integer(Index));
            -- this statement reads the output from the ROM
            -- conv_integer converts Index to an Integer
21  Z <= Romout(0);
22  D <= Romout(1 to 3);

23  process (CLK)
24    begin
25      if CLK'event and CLK = '1' then Q <= D; end if;
26    end process;
27  end ROM;
```

A Moore machine can be modeled using two processes just like a Mealy machine. For example, the first row of the Moore table of Table 14-3 could be modeled within the combinational process as follows:

```
case state is
  when 0 =>
    Z <= '0';
    if X = '0' then Nextstate <= 0; else Nextstate <= 1; end if;
  . . .
```

Note that the Z output is specified before X is tested because the Moore output only depends on the state and not on the input.

**FIGURE 17-23**
Partial VHDL Code for the Table of Figure 13-4

```
1    entity Table_13_4 is
2      port(X1, X2, CLK: in bit; Z1, Z2: out bit);
3    end Table_13_4;

4    architecture T1 of Table_13_4 is
5    signal State, Nextstate: integer range 0 to 3: = 0;
6    signal X12: bit_vector(0 to 1);
7    begin
8      X12 <= X1&X2;
9      process(State, X12)
10     begin
11       case State is
12       when 0 =>
13         case X12 is
14           when "00" => Nextstate <= 3; Z1 <= '0'; Z2 <= '0';
15           when "01" => Nextstate <= 2; Z1 <= '1'; Z2 <= '0';
16           when "10" => Nextstate <= 1; Z1 <= '1'; Z2 <= '1';
17           when "11" => Nextstate <= 0; Z1 <= '0'; Z2 <= '1';
18           when others => null;         -- not required since X is a bit_vector
19         end case;
20       when 1 =>                        -- code for state 1 goes here, etc.
```

# 17.5 Synthesis of VHDL Code

The synthesis software for VHDL translates the VHDL code to a circuit description that specifies the needed components and the connections between the components. When writing VHDL code, you should always keep in mind that you are designing hardware, not simply writing a computer program. Each VHDL statement implies certain hardware requirements. So poorly written VHDL code may result in poorly designed hardware. Even if VHDL code gives the correct result when simulated, it may not result in hardware that works correctly when synthesized. Timing problems may prevent the hardware from working properly even though the simulation results are correct.

The synthesis software tries to infer the components needed by "looking" at the VHDL code. In order for code to synthesize correctly, certain conventions must be followed. In order to infer flip-flops or registers that change state on the rising edge of a clock signal, an **if** clause of the form

```
if clock'event and clock = '1' then . . . end if;
```

is required by most synthesizers. For every assignment statement between **then** and **end if** in the preceding statement, a signal on the left side of the assignment will cause creation of a register or flip-flop. The moral to this story is: If you do not want to create unnecessary flip-flops, do not put the signal assignments in a clocked process. If clock' event is omitted, the synthesizer may produce latches instead of flip-flops.

Before synthesis is started, we must specify a target device so that the synthesizer knows what components are available. We will assume that the target is a CPLD or FPGA that has D flip-flops with clock enable (D-CE flip-flops). We will synthesize the VHDL code for a left-shift register (Figure 17-10). Q and D are 4-bit vectors. Because updates to Q follow "CLK'event and CLK = '1' then", this infers that Q must be a register composed of four flip-flops, which we will label Q3, Q2, Q1, and Q0. Because the flip-flops can change state when Clr, Ld, or Ls is '1', we connect the clock enables to an OR gate whose output is Clr + Ld + Ls. Then, we connect gates to the D inputs to select the data to be loaded into the flip-flops. If Clr = 0 and Ld = 1, D is loaded into the register on the rising clock edge. If Clr = Ld = 0 and Ls = 1, then Q2 is loaded into Q3, Q1 is loaded into Q2, etc. Figure 17-24 shows the logic circuit for the first two flip-flops. If Clr = 1, the D flip-flop inputs are 0, and the register is cleared.

A VHDL synthesizer cannot synthesize delays. Clauses of the form "**after** time-expression" will be ignored by most synthesizers, but some synthesizers require that **after** clauses be removed. Although the initial values for signals may be specified in port and signal declarations, these initial values are ignored by the synthesizer. A reset signal should be provided if the hardware must be set to a specific initial state. Otherwise, the initial state of the hardware may be unknown, and the hardware may malfunction. When an integer signal is synthesized, the integer is represented in hardware by its binary equivalent. If the range of an integer is not specified, the synthesizer will assume the maximum number of bits, usually 32. Thus,

```
signal count: integer range 0 to 7;
```

would result in a 3-bit counter, but

```
signal count: integer;
```

could result in a 32-bit counter.

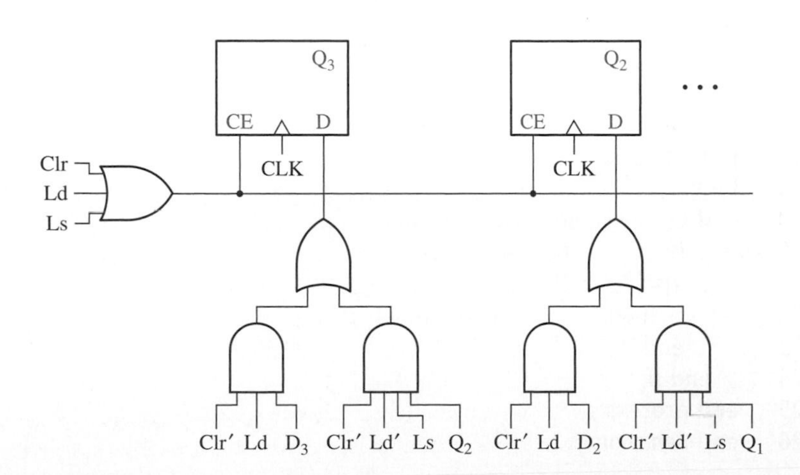


**FIGURE 17-24** Synthesis of VHDL Code From Figure 17-10

VHDL signals retain their current values until they are changed. This can result in the creation of unwanted latches when the code is synthesized. For example, in a combinational process, the statement

**if** X = '1' **then** B <= 1; **end if**;

would create latches to hold the value of B when X changed to '0'. To avoid the creation of unwanted latches in a combinational process, always include an **else** clause in every if statement. For example,

**if** X = '1' **then** B <= 1 **else** B <= 2; **end if**;

would create a MUX to switch the value of B from 1 to 2.

Figure 17-25 shows the VHDL code for a 4-bit adder with accumulator. When the synthesizer analyses this code, it infers the presence of a 4-bit adder with carry in and carry out from line 14. When it analyses the clocked process, it infers from

FIGURE 17-25 VHDL Code and Synthesis Results for 4-Bit Adder with Accumulator

```
1   library IEEE;
2   use IEEE.STD_LOGIC_1164.ALL;
3   use IEEE.STD_LOGIC_UNSIGNED.ALL;

4   entity adder is
5     Port (B: in std_logic_vector(3 downto 0);
6       Ld, Ad, Cin, CLK : in std_logic;
7       Aout : out std_logic_vector(3 downto 0);
8       Cout : out std_logic);
9   end adder;

10  architecture Behavioral of adder is
11  signal A : std_logic_vector(3 downto 0);
12  signal Addout : std_logic_vector(4 downto 0);
13  begin
14    Addout <= ('0' & A) + B + Cin;
15    Cout <= Addout(4);
16    Aout <= A;
17  process(CLK)
18  begin
19    if CLK'event and CLK = '1' then
20      if Ld = '1' then A <= B;
21      elsif Ad = '1'
22        then A <= Addout(3 downto 0);
23      end if;
24    end if;
25  end process;
26  end Behavioral;
```

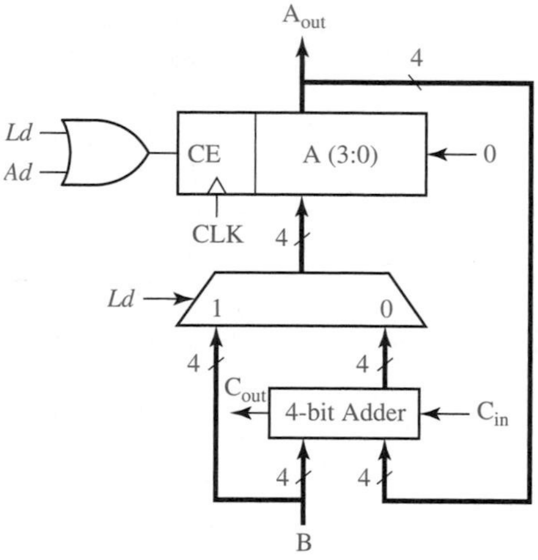

lines 11, 19, and 20 that A is a 4-bit register that changes state on the rising clock edge. It also infers the presence of a 4-wide 2-to-1 multiplexer to select either B or the adder output to load into A. Because A is loaded when $Ld = 1$ or $Ad = 1$, the CE input to the register is $Ld + Ad$. At this point, a block diagram of the synthesized code resembles that shown in Figure 17-25. The synthesizer output is then optimized and fit into a specific target device.

## 17.6 More About Processes and Sequential Statements

An alternative form for a process uses wait statements instead of a sensitivity list. A process cannot have both a wait statement and a sensitivity list. A process with wait statements may have the form

```
process
begin
   sequential-statements
   wait-statement
   sequential-statements
   wait-statement
   . . .
end process;
```

This process will execute the sequential-statements until a wait statement is encountered. Then, it will wait until the specified wait condition is satisfied. It will then execute the next set of sequential-statements until another wait is encountered. It will continue in this manner until the end of the process is reached. Then, it will start over again at the beginning of the process.

Wait statements can be of three different forms:

```
wait on sensitivity-list;
wait for time-expression;
wait until Boolean-expression;
```

The first form waits until one of the signals on the sensitivity list changes. For example, **wait on** A,B,C; waits until A, B, or C changes and, then, execution proceeds. The second form waits until the time specified by time expression has lapsed. If **wait for** 5 ns is used, the process waits for 5 ns before continuing. If **wait for** 0 ns is used, the wait is for one Δ time. Wait statements of the form **wait for** xx ns are useful for writing VHDL code for simulation; however, they should not be used when writing VHDL code for synthesis because they are not synthesizable. For the third form of wait statement, the Boolean expression is evaluated whenever one of the signals in the expression changes, and the process continues execution when the expression evaluates to TRUE. For example,

```
wait until A = B;
```

will wait until either A or B changes. Then, A = B is evaluated, and if the result is TRUE, the process will continue, or else the process will continue to wait until A or B changes again and A = B is TRUE.

After a VHDL simulator is initialized, it executes each process with a sensitivity list one time through, and then waits at the beginning of the process for a change in one of the signals on the sensitivity list. If a process has a wait statement, it will initially execute until a wait statement is encountered. Therefore, the following process is equivalent to the one in Figure 17-15:

```
process
begin
   C <= A and B after 5 ns;
   E <= C or D after 5 ns;
   wait on A, B, C, D;
end process;
```

The wait statement at the end of the process replaces the sensitivity list at the beginning. In this way both processes will initially execute the sequential statements one time and, then, wait until A, B, C, or D changes.

The order in which sequential statements are executed in a process is not necessarily the order in which the signals are updated. Consider the following example:

```
process
begin
   wait until clk'event and clk = '1';
   A <= E after 10 ns;      -- (1)
   B <= F after 5 ns;       -- (2)
   C <= G;                  -- (3)
   D <= H after 5 ns;       -- (4)
end process;
```

This process waits for a rising clock edge. Suppose the clock rises at time = 20 ns. Statements (1), (2), (3), (4) immediately execute in sequence. A is scheduled to change to E at time = 30 ns; B is scheduled to change to F at time = 25 ns; C is scheduled to change to G at time = 20 + $\Delta$ ns; and D is scheduled to change to H at time 25 ns. As simulated time advances, first, C changes. Then, B and D change at time = 25 ns, and finally A changes at time 30 ns. When clk changes to '0', the wait statement is re-evaluated, but it keeps waiting until clk changes to '1', and then the remaining statements execute again.

If several VHDL statements in a process update the same signal at a given time, the last value overrides. For example,

```
process (CLK)
begin
   if CLK'event and CLK = '0' then
      Q <= A; Q <= B; Q <= C;
   end if;
end process;
```

Every time CLK changes from '1' to '0', after $\Delta$ time, Q will change to C.

In this unit, we have introduced processes with sensitivity lists and processes with wait statements. The statements within a process are called sequential statements because they execute in sequence, in contrast with concurrent statements that execute only when a signal on the right-hand side changes. Signal assignment statements can be either concurrent or sequential. However, **if** and **case** statements are always sequential, yet conditional signal assignment statements and selected signal assignment statements can only be concurrent.

# Problems

**17.1** Write VHDL code for a T flip-flop with an active-low asynchronous clear.

**17.2** Write VHDL code for the following right-shift register with synchronous clear.

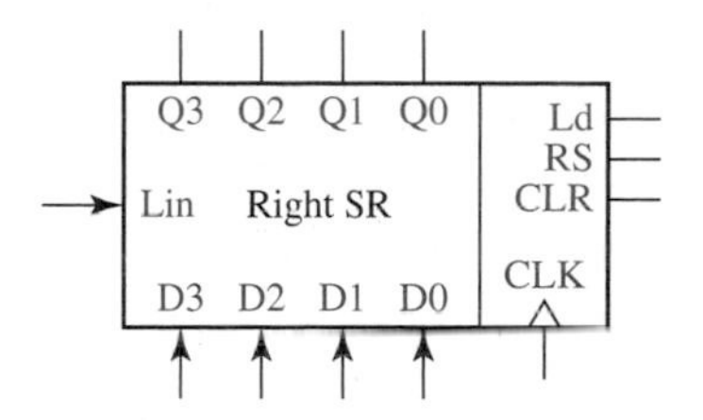


**17.3** A 4-bit up/down binary counter with output Q works as follows: All state changes occur on the rising edge of the CLK input, except the asynchronous clear (ClrN). When ClrN = 0, the counter is reset regardless of the values of the other inputs.

If the LOAD input is 0, the data input D is loaded into the counter.
If LOAD = ENT = ENP = UP = 1, the counter is incremented.
If LOAD = ENT = ENP = 1 and UP = 0, the counter is decremented.
If ENT = UP = 1, the carry output (CO) = 1 when the counter is in state 15.
If ENT = 1 and UP = 0, the carry output (CO) = 1 when the counter is in state 0.

(a) Write a VHDL description of the counter.
(b) Draw a block diagram and write a VHDL description of an 8-bit binary up/down counter that uses two of these 4-bit counters.

**17.4** Represent the given circuit using a process with a case statement.

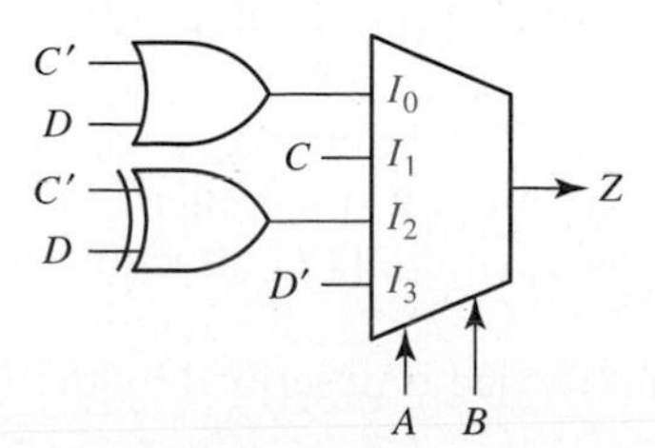

**17.5** Write a VHDL module for the sequential machine of Table 14-1. Use two processes as in Figure 17-17.

**17.6** (a) Draw a block diagram showing how Table 13-4 can be realized using a ROM and D flip-flops (rising-edge trigger).
(b) Write VHDL code for the circuit of part (a). Use a straight binary state assignment and form the ROM address as X1&X2&Q1&Q2.

**17.7** (a) Draw a circuit that implements the following VHDL code using gates and D-CE flip-flops.

```
signal A,B,Q: bit_vector(1 to 2);
---------------------------------
process(CLK)
  if CLK'event and CLK = '0' then
    if LdA = '1' then Q <= A;
      elsif LdB = '1' then Q <= B;
    end if;
  end if;
end process;
```

(b) Show how your circuit can be simplified if LdA = LdB = '1' can never occur. Use MUXes and D-CE flip-flops in your simplified circuit.

**17.8** In the following VHDL process, A, B, C, and D are all integers that have a value of 0 at time = 10 ns. If E changes from '0' to '1' at time 20 ns, specify the time at which each signal will change and the value to which it will change.

```
p1: process
  wait on E;
  A <= 1 after 15 ns;
  B <= A + 1;
  C <= B + 1 after 10 ns;
  D <= B + 2 after 3 ns;
  A <= A + 5 after 15 ns;
  B <= B + 7;
end process p1;
```

**17.9** Write the VHDL code for an S-R flip-flop with a rising-edge clock. Use standard logic, and output 'X' if $S = R =$ '1' at a rising clock edge.

**17.10** Write a VHDL module for a D-G latch, using the code of Figure 17-2. Then, write a VHDL module to implement the D flip-flop shown in Figure 11-15, using two instances of the D-G latch module you wrote.

**17.11** What device is described by the following VHDL code?

```
process(CLK, CLR, PRE)
  if CLR = '1' then Q <= '0';
    elsif PRE = '1' then Q <= '1';
    elsif CLK'event and CLK = '1' and CE = '1' then Q <= D;
  end if;
end process;
```

**17.12** Write the VHDL code for an 8-bit register with data inputs and tri-state outputs. Use control inputs Ld (Load) and En (tri-state output enable).

**17.13** Implement a 4-to-2 priority encoder using **if** and **elsif** statements.

**17.14** Write a VHDL module for a 4-bit comparator. The comparator has two inputs, A and B, which are 4-bit std_logic vectors; and three std_logic outputs, AGB, ALB, and AEB. AGB = '1' if A is greater than B, ALB = '1' if A is less than B, AEB = '1' if A and B are equal.

**17.15** Write the VHDL code for a 6-bit Super-Register with a 3-bit control input A. The register operates according to the following table:

| A | Action |
|---|---|
| 000 | Hold State |
| 001 | Shift Left |
| 010 | Shift Right |
| 011 | Synchronous Clear |
| 100 | Synchronous Preset |
| 101 | Count Up |
| 110 | Count Down |
| 111 | Load |

The register also has a 6-bit output (Q), a 6-bit input (D), a Right-Shift-In input (RSI), and a Left-Shift-In input (LSI). Use a **case** statement.

**17.16** Write VHDL code that will display the value of a BCD input on a seven-segment display. Use a single process with a **case** statement to model this combinational circuit. Refer to Figure 8-14 for a diagram of the seven-segment display.

**17.17** The Mealy and Moore circuits shown both produce an output that is the exclusive-OR of two consecutive inputs. Assume each of the flip-flops has a propagation delay of 10 ns both from the clock edge and from ClrN, and the exclusive-OR gate has a 10 ns propagation delay. ClrN is an asynchronous clear.

(a) Create a VHDL dataflow model for the Mealy circuit. Assign type std_logic to all signals.

(b) Simulate your code for 400 ns, and record the waveforms for the following input patterns:

ClrN 0 at 0 ns, 1 at 20 ns
*x* 1 at 0 ns, 0 at 60 ns, 1 at 140 ns, 0 at 220 ns
CLK symmetrical 80 ns period starting at 0

(c) Repeat Part (a) for the Moore circuit.
(d) Repeat Part (b) for the Moore circuit.
(e) Explain the differences in the outputs from the two circuits. On the waveforms, show the input and output sequences for the two circuits.

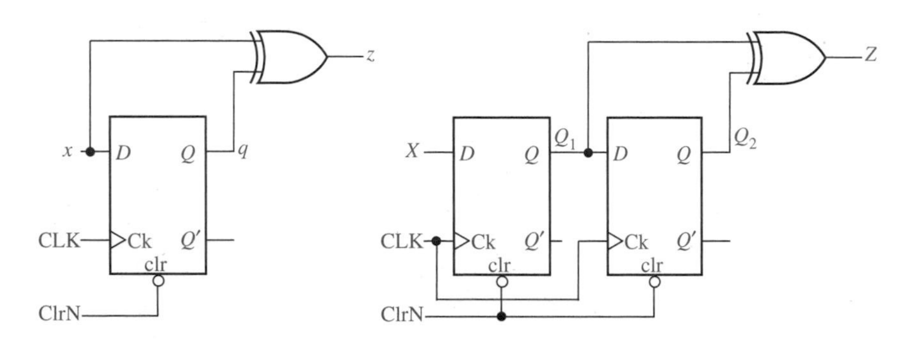


**17.18** A modulo 8 counter cycles through the states $Q_0Q_1Q_2Q_3 = 1000, 1100, 0100, 0110, 0010, 0011, 0001, 1001$. The counter has eight outputs: $Z_0 = 1$ when the counter is in state 1000 and the CLK is 0 and $Z_0 = 0$ otherwise; $Z_1 = 1$ when the counter is in state 1100 and the CLK is 0 and $Z_1 = 0$ otherwise, . . .; $Z_7 = 1$ when the counter is in state 1001 and the CLK is 0 and $Z_7 = 0$ otherwise. The counter has an asynchronous, active-low reset input ClrN.
(a) Derive minimum equations for the counter outputs.
(b) Assume the counter is implemented using D flip-flops. Find minimum input equations for the flip-flops.
(c) Assume the counter is implemented using D-CE flip-flops. Find minimum input equations for the flip-flops.
(d) Write a VHDL behavioral description of the counter. Assume the flip-flops are positive edge triggered.
(e) Write a VHDL dataflow description of the counter using the equations from Part (b). Simulate the counter for a cycle to verify your code.
(f) Write a VHDL dataflow description of the counter using the equations from Part (c). Simulate the counter for a cycle to verify your code.

**17.19** Repeat Problem 17.18 for a modulo 8 counter that cycles through the states $Q_0Q_1Q_2Q_3 = 1000, 1100, 1110, 0110, 0010, 0011, 1011, 1001$.

**17.20** Shown is an iterative circuit for comparing two 4-bit positive numbers. All of the Cmp modules in the circuit are the same. With the proper inputs for ig and ie, the outputs are og = 1 and oe = 0 if the $x$ is larger than $y$, og = 0 and oe = 1 if $x$ and $y$ are equal, and og = 0 and oe = 0 if $y$ is larger than $x$.
(a) Derive the logic equations that describe the Cmp module.
(b) Using your equations from Part (a), write VHDL code that gives a dataflow description of a Cmp module.
(c) Using the VHDL module defined in Part (b), write structural VHDL code that specifies the 4-bit comparator.

(d) Use the Direct VHDL simulator to obtain the signal values for the three input combinations: $x = 0100$ $y = 0011$, $x = 0011$ $y = 0100$, and $x = 0001$ $y = 0001$. Record the waveform report from the simulator.

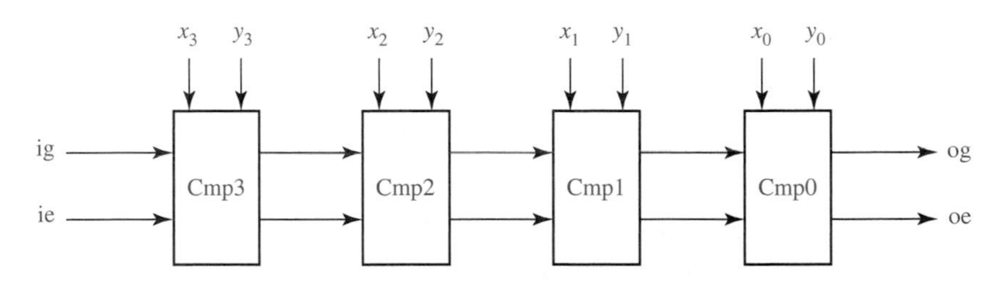


**17.21** The following iterative circuit is a priority selection circuit. When one or more of the inputs is 1, osel = 0 and $y_i = 1$ where $i$ is the largest index such that $x_i = 1$. If none of the inputs is 1, then all outputs are 0 and osel = 1. The four modules in the circuit are identical.

(a) Derive the logic equations that describe the Pr module.

(b) Using your equations from Part (a), write VHDL code that gives a dataflow description of the Pr module.

(c) Using the VHDL module defined in Part (b), write structural VHDL code that specifies the 4-bit priority selector.

(d) Use the Direct VHDL simulator to obtain the signal values for the three input combinations: $x = 1000$, $x = 0111$, and $x = 0000$. Record the waveform report from the simulator.

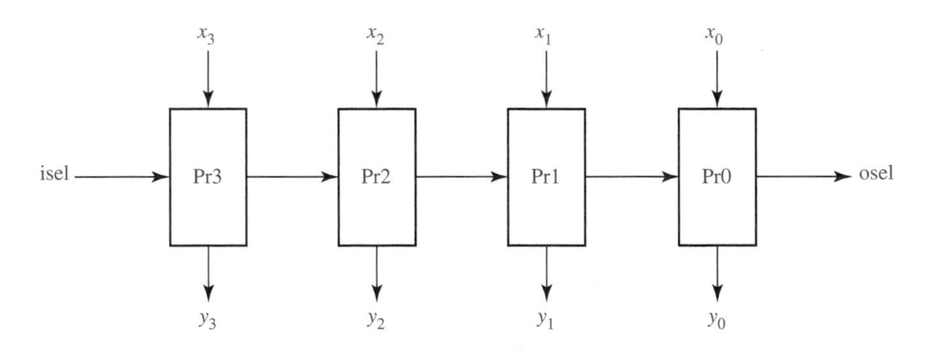


**17.22** A Mealy sequential machine with one input (X) and one output (Z) has the following state table.

| Present State | Next State X = 0 | Next State X = 1 | Z X = 0 | Z X = 1 |
|---|---|---|---|---|
| S0 | S1 | S0 | 0 | 0 |
| S1 | S0 | S2 | 1 | 0 |
| S2 | S3 | S2 | 1 | 1 |
| S3 | S0 | S1 | 0 | 1 |

Write a VHDL module for the sequential machine using a ROM (as in Figure 17-22) and a straight binary assignment.

**17.23** Repeat Problem 17.22 using equations as in Figure 17-19 and using a one-hot state assignment. (*Hint*: It may be easier to do the one-hot state assignment properly if you draw the state graph first.)

**17.24** The following VHDL code is for a 2-to-1 MUX, but it contains mistakes. What are the mistakes?

```
library IEEE;
use IEEE.STD_LOGIC_1164.ALL;
entity mux2 is
   port (d0, d1 : in bit;
   sel : in Boolean;
   z : out bit);
end mux2;

architecture bvhr of mux2 is
   signal muxsel : integer range 0 to 1;
begin
  process(d0, d1, select)
  begin
   muxsel <= 0;
   if sel then muxsel <= muxsel + 1; end if;
   case muxsel is
    when 0 => z <= d0 after 2ns;
    when 1 => z <= d1 after 2ns;
   end case;
  end process;
end bvhr;
```

**17.25** Give the state table implemented by the following VHDL code.

```
entity Problem17_25 is
   port(X, CLK: in bit;
   Z1, Z2: out bit);
end Problem17_25;

architecture Table of Problem17_25 is
   signal State, Nextstate: integer range 0 to 3 := 0;
begin
   process(State, X)        --Combinational Circuit
   begin
      case State is
      when 0 =>
```

```
      if X = '0' then Z1 <= '1'; Z2 <= '0'; Nextstate <= 0;
      else Z1 <= '0'; Z2 <= '0'; Nextstate <= 1; end if;
    when 1 =>
      if X = '0' then Z1 <= '0'; Z2 <= '1'; Nextstate <= 1;
      else Z1 <= '0'; Z2 <= '1'; Nextstate <= 2; end if;
    when 2 =>
      if X = '0' then Z1 <= '0'; Z2 <= '1'; Nextstate <= 2;
      else Z1 <= '0'; Z2 <= '1'; Nextstate <= 3; end if;
    when 3 =>
      if X = '0' then Z1 <= '0'; Z2 <= '0'; Nextstate <= 0;
      else Z1 <= '1'; Z2 <= '0'; Nextstate <= 0; end if;
    end case;
  end process;
  process(CLK)      -- State Register
    begin
      if CLK'event and CLK = '1' then      -- rising edge of clock
      State <= Nextstate;
    end if;
  end process;
end Table;
```

**17.26** Give the state table implemented by the following VHDL code.

```
entity Problem17_26 is
  port(X, CLK: in bit;
    Z: out bit);
end Problem17_26;
architecture Table of Problem17_26 is
  signal State, Nextstate: integer range 0 to 3 := 0;
begin
  process(State, X)      --Combinational Circuit
  begin
    case State is
    when 0 =>  Z <= '1';
      if X = '0' then Nextstate <= 1;
      else Nextstate <= 2; end if;
    when 1 =>  Z <= '0';
      if X = '0' then Nextstate <= 3;
      else Nextstate <= 2; end if;
    when 2 =>  Z <= '0';
      if X = '0' then Nextstate <= 1;
      else Nextstate <= 0; end if;
    when 3 =>  Z <= '0';
      if X = '0' then Nextstate <= 0;
      else Nextstate <= 1; end if;
```

```
      end case;
    end process;
  -- the clocked process goes here, same as in Problem 17.25
  end Table;
```

**17.27** Give the state table implemented by the following VHDL code.

```
entity Problem17_27 is
  port(X1, X2, CLK: in bit;
    Z: out bit);
end Problem17_27;

architecture Table of Problem17_27 is
  signal State, Nextstate: integer range 0 to 2 := 0;
  signal X12: bit_vector(0 to 1);
begin
  X12 <= X1&X2;
  process(State, X12)    --Combinational Circuit
  begin
    case State is
    when 0 =>  Z <= '0';
      case X12 is
        when "00" => Nextstate <= 0;
        when "01" => Nextstate <= 1;
        when "10" => Nextstate <= 2;
        when "11" => Nextstate <= 0;
      end case;
    when 1 =>  Z <= '0';
      case X12 is
        when "00" => Nextstate <= 0;
        when "01" => Nextstate <= 1;
        when "10" => Nextstate <= 2;
        when "11" => Nextstate <= 1;
      end case;
    when 2 =>  Z <= '1';
      case X12 is
        when "00" => Nextstate <= 0;
        when "01" => Nextstate <= 1;
        when "10" => Nextstate <= 2;
        when "11" => Nextstate <= 2;
      end case;
    end case;
  end process;
-- the clocked process goes here, same as in Problem 17.25.
end Table;
```

**17.28** The VHDL specification for a state machine follows. It has one binary input (plus a clock and reset) and one binary output.

(a) Construct a state table for this state machine.

(b) Simulate the circuit for the input sequence xin = 010111011, record the waveform and list the output sequence produced.

(c) Find a minimum row state table that describes this state machine.

(d) What input sequences cause the output to become 1? (*Hint*: The machine recognizes sequences ending in two different patterns.)

```
library IEEE;
use IEEE.STD_LOGIC_1164.ALL;
entity pttrnrcg is
  port (clk, rst, xin : in std_logic;
  zout : out std_logic);
end pttrnrcg;

architecture sttmchn of pttrnrcg is
  type mchnstate is (s1, s2, s3, s4, s5, s6, s7, s8, s9, s10);
  signal state, nextstate: mchnstate;
begin

  cmb_lgc: process(state, xin)
  begin
    case state is
      when s1 =>
        zout <= '0';
        if xin = '0' then nextstate <= s2; else nextstate <= s10; end if;
      when s2 =>
        zout <= '0';
        if xin = '0' then nextstate <= s2; else nextstate <= s3; end if;
      when s3 =>
        zout <= '1';
        if xin = '0' then nextstate <= s4; else nextstate <= s6; end if;
      when s4 =>
        zout <= '0';
        if xin = '0' then nextstate <= s7; else nextstate <= s8; end if;
      when s5 =>
        zout <= '1';
        if xin = '0' then nextstate <= s9; else nextstate <= s10; end if;
      when s6 =>
        zout <= '0';
        if xin = '0' then nextstate <= s9; else nextstate <= s10; end if;
      when s7 =>
        zout <= '0';
        if xin = '0' then nextstate <= s2; else nextstate <= s3; end if;
      when s8 =>
```

```
      zout <= '1';
      if xin = '0' then nextstate <= s4; else nextstate <= s5; end if;
    when s9 =>
      zout <= '0';
      if xin = '0' then nextstate <= s7; else nextstate <= s8; end if;
    when s10 =>
      zout <= '0';
      if xin = '0' then nextstate <= s9; else nextstate <= s10; end if;
  end case;
 end process cmb_lgc;

 stt_trnstn: process(clk,rst)
 begin
  if rst = '1' then
   state <= s1;
   elsif Rising_Edge (clk) then
   state <= nextstate;
  end if;
 end process stt_trnstn;
end sttmchn;
```

**17.29** The VHDL specification for a sequential circuit follows. It has one binary input (plus a clock and reset) and one binary output. Four architectures are given for the sequential circuit.

(a) For each of these architectures, draw the schematic described by the architecture. Use D flip-flops and AND, OR, and NOT gates.

(b) What differences exist in the outputs produced by these architectures?

```
library IEEE;
use IEEE.STD_LOGIC_1164.ALL;
entity diff1 is
 port (clk, rst, xin : in std_logic;
 zout : out std_logic);
end diff1;

architecture df1 of diff1 is
  signal y0,y1,nxty0,nxty1 : std_logic;
begin
 process(y0,y1,xin)
 begin
  zout <= y0 AND (xin XOR y1); nxty0 <= NOT y0; nxty1 <= xin;
 end process;
 process(clk,rst)
 begin
  if rst = '1' then
   y0 <= '0'; y1 <= '0';
   elsif Rising_Edge (clk) then
```

```
      y0 <= nxty0; y1 <= nxty1;
    end if;
  end process;
end df1;

architecture df2 of diff1 is
    signal y0,y1,nxty0,nxty1 : std_logic;
begin
    zout <= y0 AND (xin XOR y1); nxty0 <= NOT y0; nxty1 <= xin;
  process(clk,rst)
  begin
    if rst = '1' then
      y0 <= '0'; y1 <= '0';
      elsif Rising_Edge (clk) then
      y0 <= nxty0; y1 <= nxty1;
    end if;
  end process;
end df2;

architecture df3 of diff1 is
    signal y0,y1 : std_logic;
begin
    zout <= y0 AND (xin XOR y1);
  process(clk,rst)
  begin
    if rst = '1' then
      y0 <= '0'; y1 <= '0';
      elsif Rising_Edge (clk) then
      y0 <= NOT y0; y1 <= xin;
    end if;
  end process;
end df3;

architecture df4 of diff1 is
    signal y0,y1 : std_logic;
begin
  process(clk,rst)
  begin
    if rst = '1' then
      y0 <= '0'; y1 <= '0'; zout <= '0';
      elsif Rising_Edge (clk) then
      y0 <= NOT y0; y1 <= xin; zout <= y0 AND (xin XOR y1);
    end if;
  end process;
end df4;
```

17.30 Write a VHDL module for an 8-bit mask circuit. When the signal Store = 1, the 8-bit input X is stored in an 8-bit mask register M. The 8-bit output Z of the mask circuit is always the AND of the bits of M with the corresponding bits of X. The circuit should also have an asynchronous active-high signal Set, which will set all the bits of M to 1.

17.31 Write a VHDL module for the sequential machine of Table 14-3. Use two processes as in Figure 17-17.

## Simulation Problems

17.A Write a behavioral VHDL module that implements the 8-bit shift register of Figure 12-8. Do not use individual flip-flops in your code. Add an active-low asynchronous reset input, ClrN. Simulate the module to obtain a timing diagram similar to Figure. 12-9. Then, write VHDL code for a 16-bit serial-in, serial-out shift register using two of these modules.

17.B Write a VHDL module for a 4-bit counter with enable that increments by different amounts, depending on the control input C. If En = 0, the counter holds its state. Otherwise, if C = 0, the counter increments by 1 every rising clock edge, and if C = 1, the counter increments by 3 every rising clock edge. The counter also has an active-low asynchronous preset signal, PreN.

17.C Write a VHDL module to implement a counter that counts in the following sequence: 000, 010, 100, 110, 001, 011, 101, 111, (repeat) 000, etc. Use a ROM and D flip-flops.

17.D Write a VHDL module to implement a circuit that can generate a clock signal whose time period is a multiple of the input clock. A control signal F determines the multiplying factor. If F = 0, the output signal has a time period twice that of the input clock. If F = 1, the output signal has a time period three times that of the input clock. The portion of the clock cycle when the clock is 1 may be longer than the portion when it is 0, or vice versa. Use a counter with an active-high synchronous clear input.

17.E Write a VHDL module to implement an 8-bit serial-in, serial-out right-left shift register with inputs RSI, LSI, En, R, and Clk. RSO and LSO are the serial outputs, so they should be the rightmost and leftmost bits of the register. However, the values of the other flip-flops inside the register should not appear on the outputs. When En = 1, at the rising edge of the clock, the register shifts right if R = 1 or left if R = 0. RSI should be the shift-in input if R = 1, and LSI should be the shift-in input if R = 0. When En = 0, the register holds its state. There should also be an asynchronous active-low clear input ClrN.

**17.F** Work Problem 17.E, but change the register to 6 bits, remove the input En, and add an input L. At the rising edge of the clock, if R = 1 and L = 0, the register shifts right. If R = 0 and L = 1, the register shifts left. If R = L = 0 or R = L = 1, the register holds its state.

**17.G** Write a VHDL module for a 6-bit accumulator with carry-in (CI) and carry-out (CO). When Ad = 0, the accumulator should hold its state. When Ad = 1, the accumulator should add the value of the data inputs D (plus CI) to the value already in the accumulator. The accumulator should also have an active-low asynchronous clear signal ClrN.

**17.H** Write a VHDL module for a 4-bit up-down counter. If En = 0, the counter will hold its state. If En = 1, the counter will count up if U = 1 or down if U = 0. The counter should also have an asynchronous active-low clear signal ClrN.

**17.I** Write a VHDL module for a 6-bit up-down counter. If U = 1 and D = 0, the counter will count up, and if U = 0 and D = 1, the counter will count down. If U = D = 0 or U = D = 1, the counter will hold its state. The counter should also have an asynchronous active-low preset signal PreN that sets all flip-flops to 1.

**17.J** Write a VHDL module for a memory circuit. The memory stores four 6-bit words in registers. The output Memout is always the value of the memory register selected by the 2-bit select signal Sel. Use tri-state buffers to connect the register outputs. If Ld = 1, the register specified by Sel will load the value of the 6-bit input signal Memin at the next rising clock edge.

**17.K** Write a VHDL module for the Parallel-in, Parallel-out right-shift register of Figure 12-10, but add an active-low asynchronous clear signal ClrN. Do not use individual flip-flops in your code. Simulate the module to obtain a timing diagram similar to Figure 12-11.

**17.L** Write a VHDL module for an 8-bit accumulator which can also shift the bits in the accumulator register to the left. If Ad = 1, the accumulator should add the value of the data inputs D to the value already in the accumulator. If Ad = 0 and Sh = 1, the bits in the accumulator should shift left (i.e., multiply by 2). If Ad = Sh = 0, the accumulator should hold its state. The accumulator should also have an active-low asynchronous clear signal ClrN. Assume that carry-in and carry-out signals are unnecessary for this application. Use an overloaded "+" operator for addition.

**17.M** Write a VHDL module for an 8-bit accumulator for subtraction, which can also shift the accumulator bits to the right. There are two control inputs, A and B. If A = B = 1, the value of the data inputs D are subtracted from the accumulator. If A = 1 and B = 0, the value of the data inputs D are loaded directly into the register. If A = 0 and B = 1, the accumulator should shift right with zero fill. If A = B = 0, the accumulator should hold its state. Use an overloaded "–" operator for subtraction.

# Circuits for Arithmetic Operations

## Objectives

1. Analyze and explain the operation of various circuits for adding, subtracting, multiplying, and dividing binary numbers and for similar operations.
2. Draw a block diagram and design the control circuit for various circuits for adding, subtracting, multiplying, and dividing binary numbers and for similar operations.

## Study Guide

**1.** Study Section 18.1, *Serial Adder with Accumulator.*

(a) Study Figure 18-2 carefully to make sure you understand the operation of this type of adder. Work out a table similar to Table 18-1 starting with $X = 6$ and $Y = 3$:

| | $X$ | $Y$ | $c_i$ | $s_i$ | $c_i^+$ |
|---|---|---|---|---|---|
| $t_0$ | 0110 | 0011 | | | |
| $t_1$ | | | | | |
| $t_2$ | | | | | |
| $t_3$ | | | | | |
| $t_4$ | | | | | |

(b) What changes would be made in this table if the *SI* input to the addend register (Figure 18-1) was connected to a logic 0 instead of to $y_0$?

(c) Note in Table 18-1 that when the adding has finished, the full adder still generates a sum and a carry output. The full adder consists of combinational logic, so it will still automatically do the work of calculating its outputs even when they are not needed. What bits are added to generate the last values of $s_i$ and $c_i^+$? [See Figure 18-2(e).] Are the last values of $s_i$ and $c_i^+$ useful for anything?

(d) Work Problem 18.3.

**2.** Study Section 18.2, *Design of a Parallel Multiplier.*

(a) For the binary multiplier of Figure 18-7, if the initial contents of the accumulator is 000001101 and the multiplicand is 1111, show the sequence of add and shift signals and the contents of the accumulator at each time step.

(b) For the state diagram of Figure 18-8, what is the maximum number of clock cycles required to carry out the multiplication? The minimum number?

(c) For the state diagram of Figure 18-9(c), assuming the counter sets $K = 1$ when the counter is in state 3 ($11_2$), what is the maximum number of clock cycles required to carry out the multiplication? The minimum number?

(d) For Figure 18-7, how many bits would be required for the product register if the multiplier was 6 bits and the multiplicand was 8 bits?

(e) Work Problems 18.4 and 18.5.

(f) Consider the design of a binary multiplier which multiplies 8 bits by 8 bits to give a 16-bit product. What changes would need to be made in Figure 18-7?

If a multiplier control of the type shown in Figure 18-8 were used, how many states would be required?

If a control of the type shown in Figure 18-9 is used, how many bits should the counter have? $K$ should equal 1 in what state of the counter? How many states will the control state graph have?

(g) Work Programmed Exercise 18.1.

**3.** Study Section 18.3, *Design of a Binary Divider.*

(a) Using the state diagram of Figure 18-11 to determine when to shift or subtract, work through the division example given at the start of this section.

(b) What changes would have to be made in Figure 18-12 if the subtraction was done using full adders rather than full subtracters?

(c) For the block diagram of Figure 18-10, under what conditions will an overflow occur and why?

(d) Work Programmed Exercise 18.2.

(e) Derive the control circuit equations, Equations (18-1).

(f) In Figure 18-13, why is one of the inputs to the bus merger at the 0 input of the MUX set to 1?

(g) For a binary multiplier of the type described in Section 18.2, addition is done before shifting. Division requires a series of shift and subtract operations. Since division is the inverse of multiplication, which operation should be done first, subtract or shift?

(h) Work Problems 18.6, 18.7, and 18.8.

4. Optional simulation exercises:
   (a) Simulate the serial adder of Figure 13-12 and test it.
   (b) Connect two 4-bit shift registers to the inputs of the adder that you simulated in (a) to form a serial adder with accumulator (as in Figure 18-1). Supply the shift signal and clock signal from switches so that a control circuit is unnecessary. Test your adder using the following pairs of binary numbers:

   $$0101 + 0110, 1011 + 1101$$

   (c) Input the control circuit from the equations of Figure 18-4, connect it to the circuit which you built in (b), and test it.
5. When you are satisfied that you can meet all of the objectives, take the readiness test.

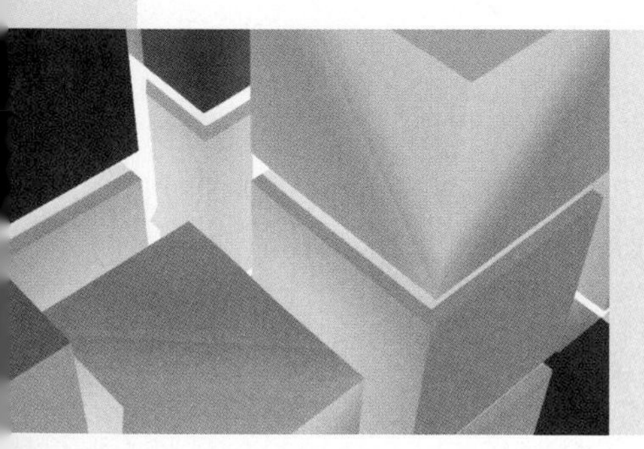

# Circuits for Arithmetic Operations

This unit introduces the concept of using a sequential circuit to control a sequence of operations in a digital system. Such a control circuit outputs a sequence of control signals that cause operations such as addition or shifting to take place at the appropriate times. We will illustrate the use of control circuits by designing a serial adder, a multiplier, and a divider.

## 18.1 Serial Adder with Accumulator

In this section we will design a control circuit for a serial adder with an accumulator. Figure 18-1 shows a block diagram for the adder. Two shift registers are used to hold the 4-bit numbers to be added, *X* and *Y*. The *X* register serves as an

**FIGURE 18-1**
Block Diagram for Serial Adder with Accumulator

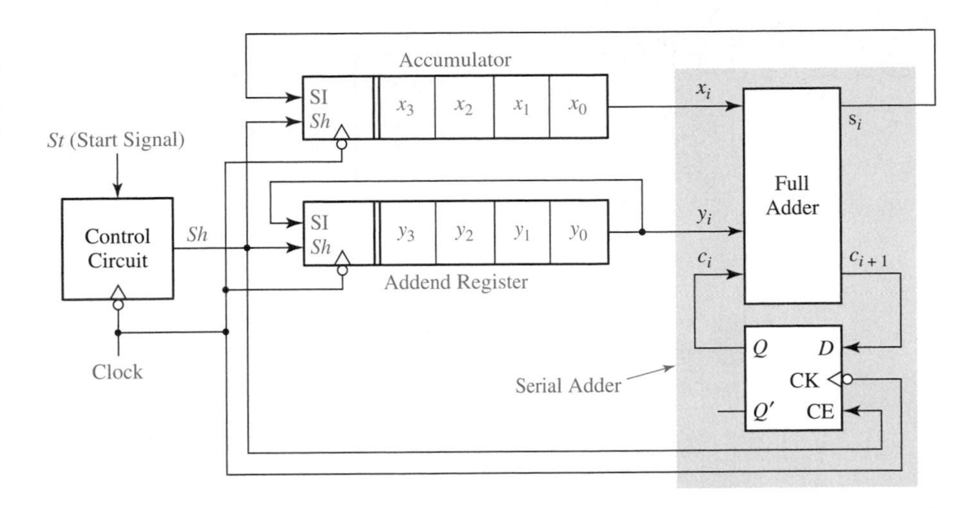


accumulator and the $Y$ register serves as an addend register. When the addition is completed, the contents of the $X$ register are replaced with the sum of $X$ and $Y$. The addend register is connected as a cyclic shift register so that after shifting four times it is back in its original state, and the number $Y$ is not lost. The box at the left end of each shift register shows the inputs: *Sh* (shift signal), SI (serial input), and *Clock*. When $Sh = 1$ and an active clock edge occurs, SI is entered into $x_3$ (or $y_3$) at the same time as the contents of the register are shifted one place to the right. The additional connections required for initially loading the $X$ and $Y$ registers and clearing the carry flip-flop are not shown in the block diagram.

The serial adder, highlighted in blue in the diagram, is the same as the one in Figure 13-12, except the D flip-flop has been replaced with a D flip-flop with clock enable. At each clock time, one pair of bits is added. Because the full adder is a combinational circuit, the sum and carry appear at the full adder output after the propagation delay. When $Sh = 1$, the falling clock edge shifts the sum bit into the accumulator, stores the carry bit in the carry flip-flop, and rotates the addend register one place to the right. Because *Sh* is connected to CE on the flip-flop, the carry is only updated when shifting occurs.

Figure 18-2 illustrates the operation of the adder. Shifting occurs on the falling clock edge when $Sh = 1$. In this figure, $t_0$ is the time before the first shift, $t_1$ is the time after the first shift, $t_2$ is the time after the second shift, etc. Initially, at time $t_0$, the accumulator contains $X$ and the addend register contains $Y$. Because the full adder is a combinational circuit, $x_0$, $y_0$, and $c_0$ are added independently of the clock to form the sum $s_0$ and carry $c_1$. When the first falling clock edge occurs, $s_0$ is shifted into the accumulator and the remaining accumulator digits are shifted one position to the right. The same clock edge stores $c_1$ in the carry flip-flop and rotates the addend register right. The next pair of bits, $x_1$ and $y_1$, are now at the full adder input, and the adder generates the sum and carry, $s_1$

FIGURE 18-2
Operation of Serial Adder

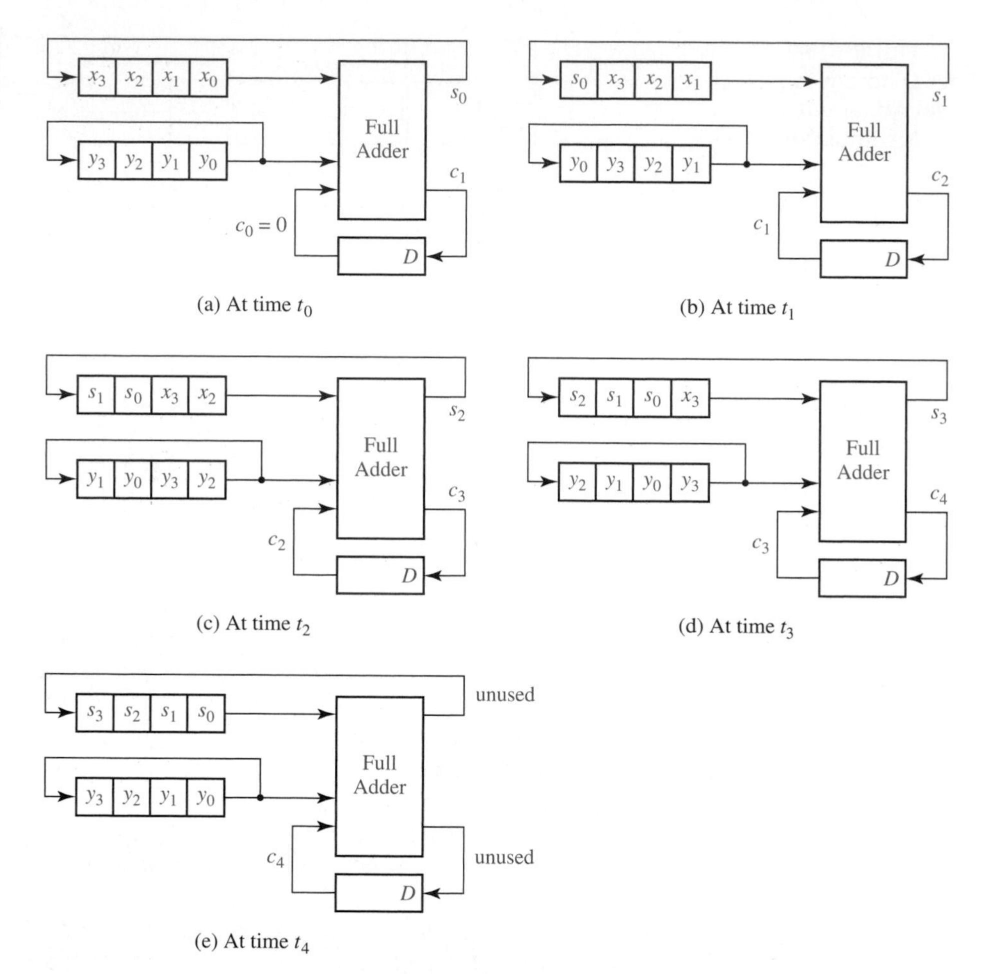


and $c_2$, as seen in Figure 18-2(b). The second falling edge shifts $s_1$ into the accumulator, stores $c_2$ in the carry flip-flop, and cycles the addend register right. Bits $x_2$ and $y_2$ are now at the adder input, as seen in Figure 18-2(c), and the process continues until all bit pairs have been added, as shown in Figure 18-2(e).

Table 18-1 shows a numerical example of the serial adder operation. Initially, the accumulator contains 0101 and the addend register contains 0111. At $t_0$, the full adder computes $1 + 1 + 0 = 10$, so $s_i = 0$ and $c_i^+ = 1$. After the first falling clock

TABLE 18-1
Operation of Serial Adder

| | $X$ | $Y$ | $c_i$ | $s_i$ | $c_i^+$ |
|---|---|---|---|---|---|
| $t_0$ | 0101 | 0111 | 0 | 0 | 1 |
| $t_1$ | 0010 | 1011 | 1 | 0 | 1 |
| $t_2$ | 0001 | 1101 | 1 | 1 | 1 |
| $t_3$ | 1000 | 1110 | 1 | 1 | 0 |
| $t_4$ | 1100 | 0111 | 0 | (1) | (0) |

edge (time $t_1$) the first sum bit has been entered into the accumulator, the carry has been stored in the carry flip-flop, and the addend has been cycled right. After four falling clock edges (time $t_4$), the sum of $X$ and $Y$ is in the accumulator, and the addend register is back to its original state.

The control circuit for the adder must now be designed so that after receiving a start signal, the control circuit will put out four shift signals and then stop. Figure 18-3 shows the state graph and table for the control circuit. The circuit remains in $S_0$ until a start signal is received, at which time the circuit outputs $Sh = 1$ and goes to $S_1$. Then, at successive clock times, three more shift signals are put out. It will be assumed that the start signal is terminated before the circuit returns to state $S_0$ so that no further output occurs until another start signal is received. Dashes appear on the graph because once $S_1$ is reached, the circuit operation continues regardless of the value of $St$. Starting with the state table of Figure 18-3 and using a straight binary state assignment, the control circuit equations are derived in Figure 18-4.

A serial processing unit, such as a serial adder with an accumulator, processes data one bit at a time. A typical serial processing unit (Figure 18-5) has two shift registers. The output bits from the shift register are inputs to a combinational circuit. The combinational circuit generates at least one output bit. This output bit is fed into the input of a shift register. When the active clock edge occurs, this bit is stored in the first bit of the shift register at the same time the register bits are shifted to the right.

The control for the serial processing unit generates a series of shift signals. When the start signal ($St$) is 1, the first shift signal ($Sh$) is generated. If the shift registers

**FIGURE 18-3** State Graph for Serial Adder Control

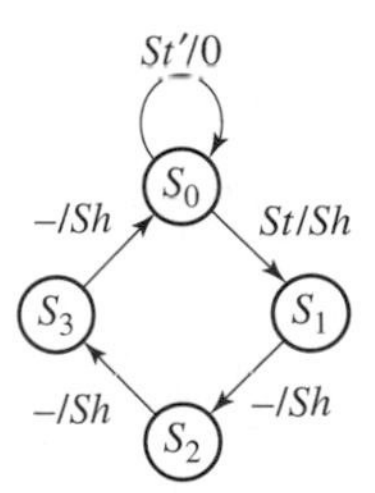


| | Next State $St = 0$ | $1$ | $Sh$ $0$ | $1$ |
|---|---|---|---|---|
| $S_0$ | $S_0$ | $S_1$ | 0 | 1 |
| $S_1$ | $S_2$ | $S_2$ | 1 | 1 |
| $S_2$ | $S_3$ | $S_3$ | 1 | 1 |
| $S_3$ | $S_0$ | $S_0$ | 1 | 1 |

**FIGURE 18-4** Derivation of Control Circuit Equations

| | $AB$ | $A^+B^+$ 0 | 1 |
|---|---|---|---|
| $S_0$ | 00 | 00 | 01 |
| $S_1$ | 01 | 10 | 10 |
| $S_2$ | 10 | 11 | 11 |
| $S_3$ | 11 | 00 | 00 |

| AB \ St | 0 | 1 |
|---|---|---|
| 00 | 0 | 0 |
| 01 | 1 | 1 |
| 11 | 0 | 0 |
| 10 | 1 | 1 |

$A^+$

$D_A = A'B + AB'$
$= A \oplus B$

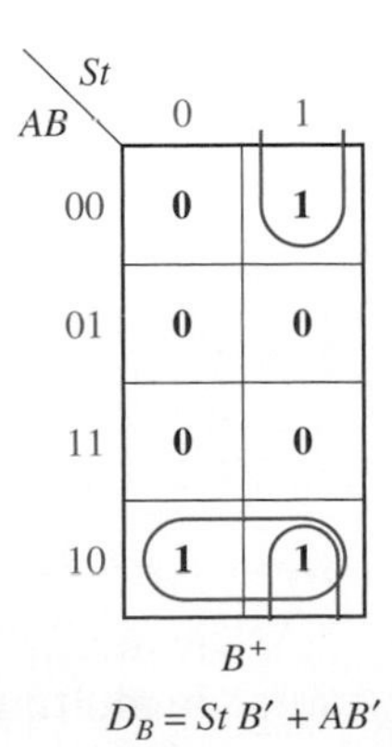


$B^+$

$D_B = St\,B' + AB'$

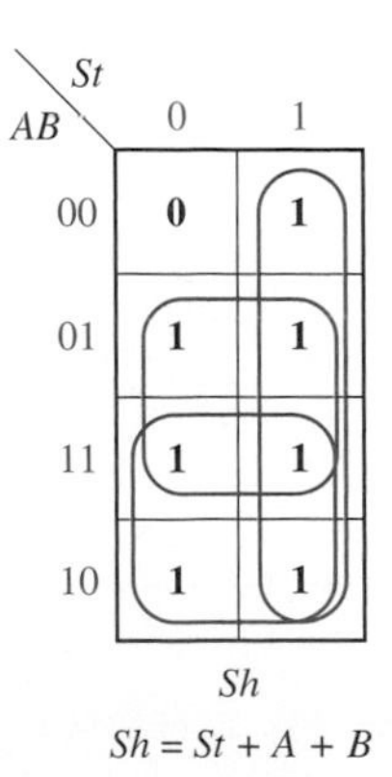


$Sh$

$Sh = St + A + B$

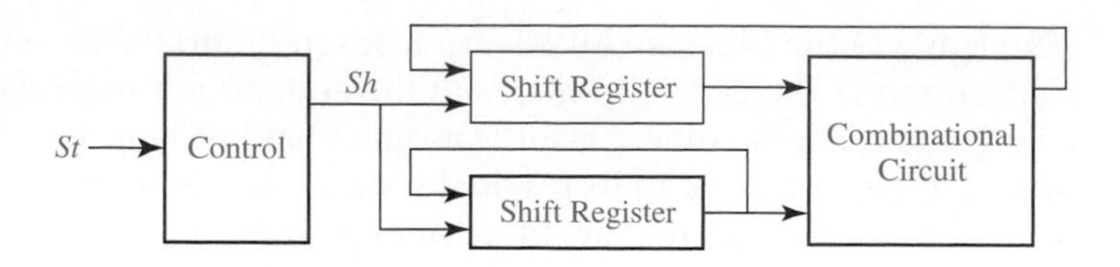


**FIGURE 18-5** Typical Serial Processing Unit

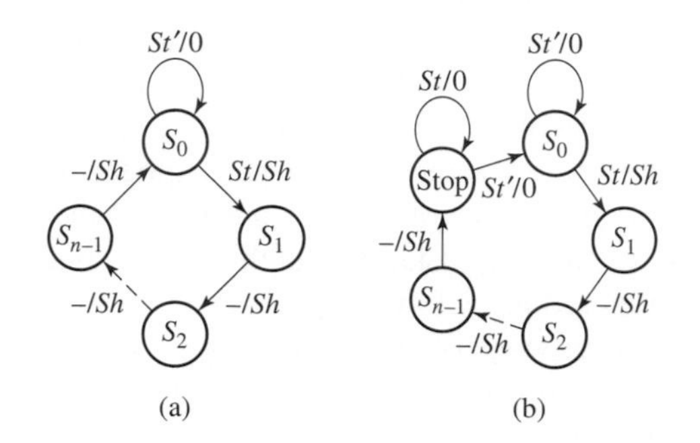


**FIGURE 18-6** State Graphs for Serial Processing Unit

have $n$ bits, then a total of $n$ shift signals must be generated. If *St* is 1 for only one clock time, then the control state graph [Figure 18-6(a)] stops when it returns to state $S_0$. However, if *St* can remain 1 until after the shifting is completed, then a separate stop state is required, as shown in Figure 18-6(b). The control remains in the stop state until *St* returns to 0.

## 18.2 Design of a Parallel Multiplier

Next, we will design a parallel multiplier for positive binary numbers. As illustrated in the example in Section 1.3, binary multiplication requires only shifting and adding. The following example shows how each partial product is added in as soon as it is formed. This eliminates the need for adding more than two binary numbers at a time.

```
Multiplicand ──→      1101    (13)
Multiplier   ──→      1011    (11)
                      ----
                      1101
Partial     }        1101
Products    }       ------
                    100111
                    0000
                    ------
                    100111
                   1101
                  --------
Product      ──→  10001111    (143)
```

The multiplication of two 4-bit numbers requires a 4-bit multiplicand register, a 4-bit multiplier register, and an 8-bit register for the product. The product register serves as an accumulator to accumulate the sum of the partial products. Instead of shifting the multiplicand left each time before it is added, as was done in the previous example, it is more convenient to shift the product register to the right each time. Figure 18-7 shows a block diagram for such a parallel multiplier. As indicated by the arrows on the diagram, 4 bits from the accumulator and 4 bits from the multiplicand register are connected to the adder inputs; the 4 sum bits and the carry output from the adder are connected back to the accumulator. (The actual connections are similar to the parallel adder with accumulator shown in Figure 12-5.) The adder calculates the sum of its inputs, and when an add signal ($Ad$) occurs, the adder outputs are stored in the accumulator by the next rising clock edge, thus causing the multiplicand to be added to the accumulator. An extra bit at the left end of the product register temporarily stores any carry ($C_4$) which is generated when the multiplicand is added to the accumulator.

Because the lower four bits of the product register are initially unused, we will store the multiplier in this location instead of in a separate register. As each multiplier bit is used, it is shifted out the right end of the register to make room for additional product bits.

The Load signal loads the multiplier into the lower four bits of ACC and at the same time clears the upper 5 bits. The shift signal ($Sh$) causes the contents of the product register (including the multiplier) to be shifted one place to the right when the next rising clock edge occurs. The control circuit puts out the proper sequence of add and shift signals after a start signal ($St = 1$) has been received. If the current multiplier bit ($M$) is 1, the multiplicand is added to the accumulator followed by a right shift; if the multiplier bit is 0, the addition is skipped and only the right shift occurs. The multiplication example at the beginning of this section ($13 \times 11$) is reworked below showing the location of the bits in the registers at each clock time.

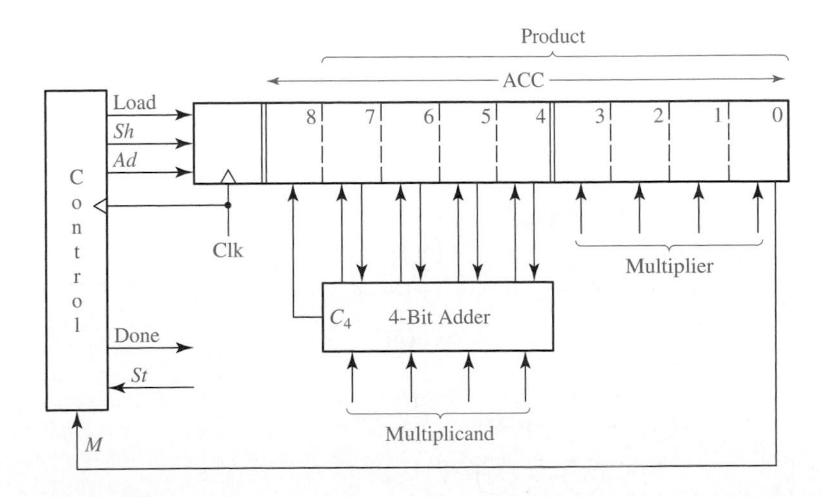


**FIGURE 18-7** Block Diagram for Parallel Binary Multiplier

| | | |
|---|---|---|
| initial contents of product register | 0 0 0 0 0 ¦ 1 0 1 1 ← $M$ | (11) |
| (add multiplicand because $M = 1$) | 1 1 0 1 | (13) |
| after addition | 0 1 1 0 1 ¦ 1 0 1 1 | |
| after shift | 0 0 1 1 0 1 ¦ 1 0 1 ← $M$ | |
| (add multiplicand because $M = 1$) | 1 1 0 1 | |
| after addition | 1 0 0 1 1 1 ¦ 1 0 1 | |
| after shift | 0 1 0 0 1 1 1 ¦ 1 0 ← $M$ | |
| (skip addition because $M = 0$) | | |
| after shift | 0 0 1 0 0 1 1 1 ¦ 1 ← $M$ | |
| (add multiplicand because $M = 1$) | 1 1 0 1 | |
| after addition | 1 0 0 0 1 1 1 1 ¦ 1 | |
| after shift (final answer) | 0 1 0 0 0 1 1 1 1 ¦ | (143) |

dividing line between product and multiplier →

The control circuit must be designed to output the proper sequence of add and shift signals. Figure 18-8 shows a state graph for the control circuit. The notation used on this graph is defined in Section 14.5. $M/Ad$ means if $M = 1$, then the output $Ad$ is 1 (and the other outputs are 0). $M'/Sh$ means if $M' = 1$ ($M = 0$), then the output $Sh$ is 1 (and the other outputs are 0). In Figure 18-8, $S_0$ is the reset state, and the circuit stays in $S_0$ until a start signal ($St = 1$) is received. This generates a Load signal, which causes the multiplier to be loaded into the lower 4 bits of the accumulator (ACC) and the upper 5 bits of ACC to be cleared on the next rising clock edge. In state $S_1$, the low order bit of the multiplier ($M$) is tested. If $M = 1$, an add signal is generated and, then, a shift signal is generated in $S_2$. If $M = 0$ in $S_1$, a shift signal is generated because adding 0 can be omitted. Similarly, in states $S_3$, $S_5$, and $S_7$, $M$ is tested to determine whether to generate an add signal followed by shift or just a shift signal. A shift signal is always generated at the next clock time following an add signal (states $S_2$, $S_4$, $S_6$, and $S_8$). After four shifts have been generated, all four multiplier bits have been processed, and the control circuit goes to a Done state and terminates the multiplication process.

**FIGURE 18-8**
State Graph for Multiplier Control

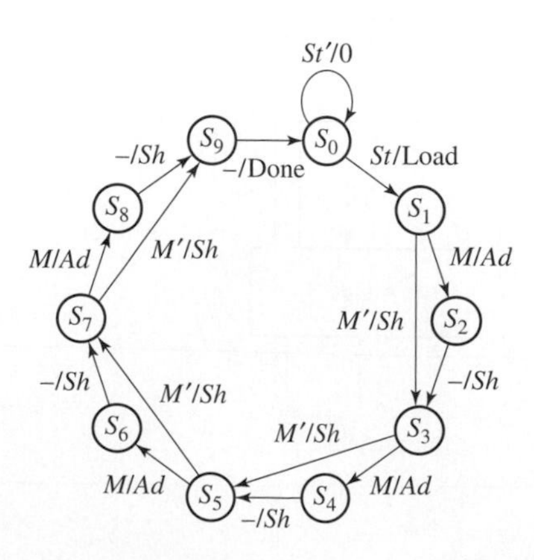

As the state graph indicates, the control performs two functions—generating add or shift signals as needed and counting the number of shifts. If the number of bits is large, it is convenient to divide the control circuit into a counter and an add-shift control, as shown in Figure 18-9(a). First, we will derive a state graph for the add-shift control which tests $M$ and $St$ and outputs the proper sequence of add and shift signals (Figure 18-9(b)). Then, we will add a completion signal ($K$) from the counter which stops the multiplier after the proper number of shifts have been completed. Starting in $S_0$ in Figure 18-9(b), when a start signal ($St = 1$) is received, a Load signal is generated. In state $S_1$, if $M = 0$, a shift signal is generated and the circuit stays in $S_1$. If $M = 1$, an add signal is generated and the circuit goes to state $S_2$. In $S_2$ a shift signal is generated because a shift always follows an add. Back in $S_1$, the next multiplier bit ($M$) is tested to determine whether to shift, or add and then shift. The graph of Figure 18-9(b) will generate the proper sequence of add and shift signals, but it has no provision for stopping the multiplier.

In order to determine when the multiplication is completed, the counter is incremented on the active clock edge each time a shift signal is generated. If the multiplier is $n$ bits, a total of $n$ shifts are required. We will design the counter so that a completion signal ($K$) is generated after $n - 1$ shifts have occurred. When $K = 1$, the circuit should perform one more addition if necessary and then do the final shift. The control operation in Figure 18-9(c) is the same as Figure 18-9(b) as long as $K = 0$. In state $S_1$, if $K = 1$, we test $M$ as usual. If $M = 0$, we output the final shift signal and stop; however, if $M = 1$, we add before shifting and go to state $S_2$. In state $S_2$, if $K = 1$, we output one more shift signal and then go to $S_3$. The last shift signal will reset the counter to 0 at the same time the add-shift control goes to the Done state.

As an example, consider the multiplier of Figure 18-7, but replace the control circuit with Figure 18-9(a). Because $n = 4$, a 2-bit counter is needed, and $K = 1$ when the counter is in state 3 ($11_2$). Table 18-2 shows the operation of the multiplier when 1101 is multiplied by 1011. $S_0$, $S_1$, and $S_2$ represent states of the control circuit [Figure 18-9(c)]. The contents of the product register at each step is the same as given on p. 600.

At time $t_0$ the control is reset and waiting for a start signal. At time $t_1$, the start signal $St = 1$, and a Load signal is generated. At time $t_2$, $M = 1$, so an $Ad$ signal is generated. When the next clock occurs, the output of the adder is loaded into the accumulator and the control goes to $S_2$. At $t_3$, an $Sh$ signal is generated, so, shifting occurs and the counter is incremented at the next clock. At $t_4$, $M = 1$, so $Ad = 1$, and the

**FIGURE 18-9**

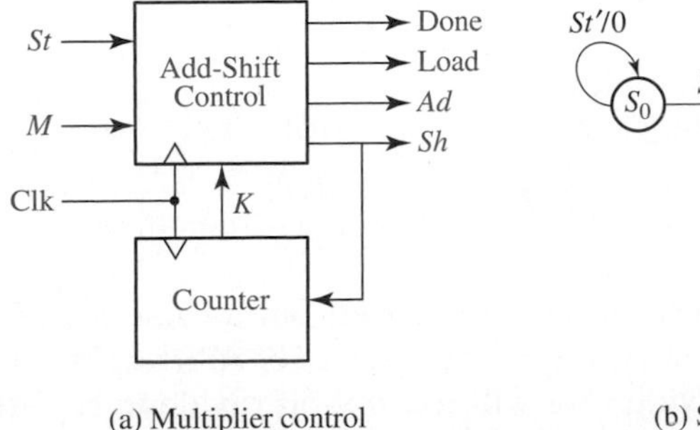


(a) Multiplier control

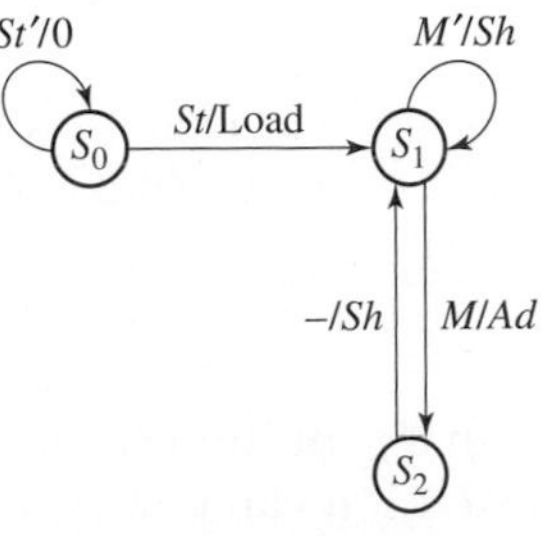


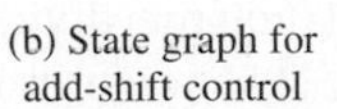
(b) State graph for add-shift control

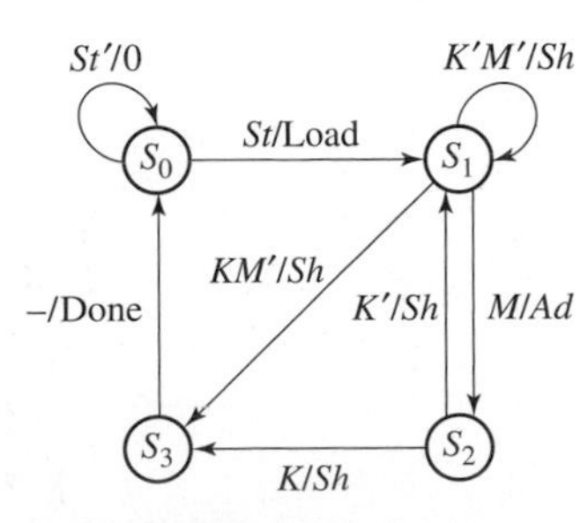


(c) Final state graph for add-shift control

**TABLE 18-2** Operation of a Multiplier Using a Counter

| Time | State | Counter | Product Register | *St* | *M* | *K* | Load | *Ad* | *Sh* | Done |
|---|---|---|---|---|---|---|---|---|---|---|
| $t_0$ | $S_0$ | 00 | 000000000 | 0 | 0 | 0 | 0 | 0 | 0 | 0 |
| $t_1$ | $S_0$ | 00 | 000000000 | 1 | 0 | 0 | 1 | 0 | 0 | 0 |
| $t_2$ | $S_1$ | 00 | 000001011 | 0 | 1 | 0 | 0 | 1 | 0 | 0 |
| $t_3$ | $S_2$ | 00 | 011011011 | 0 | 1 | 0 | 0 | 0 | 1 | 0 |
| $t_4$ | $S_1$ | 01 | 001101101 | 0 | 1 | 0 | 0 | 1 | 0 | 0 |
| $t_5$ | $S_2$ | 01 | 100111101 | 0 | 1 | 0 | 0 | 0 | 1 | 0 |
| $t_6$ | $S_1$ | 10 | 010011110 | 0 | 0 | 0 | 0 | 0 | 1 | 0 |
| $t_7$ | $S_1$ | 11 | 001001111 | 0 | 1 | 1 | 0 | 1 | 0 | 0 |
| $t_8$ | $S_2$ | 11 | 100011111 | 0 | 1 | 1 | 0 | 0 | 1 | 0 |
| $t_9$ | $S_3$ | 00 | 010001111 | 0 | 1 | 0 | 0 | 0 | 0 | 1 |

adder output is loaded into the accumulator at the next clock. At $t_5$ and $t_6$, shifting and counting occurs. At $t_7$, three shifts have occurred and the counter state is 11, so $K = 1$. Because $M = 1$, addition occurs, and the control goes to $S_2$. At $t_8$, $Sh = K = 1$, so at the next clock the final shift occurs, and the counter is incremented back to state 00. At $t_9$, a Done signal is generated.

The multiplier design given here can easily be expanded to 8, 16, or more bits simply by increasing the register size and the number of bits in the counter. The add-shift control would remain unchanged.

## 18.3 Design of a Binary Divider

We will consider the design of a parallel divider for positive binary numbers. As an example, we will design a circuit to divide an 8-bit dividend by a 4-bit divisor to obtain a 4-bit quotient. The following example illustrates the division process:

```
                          1010  quotient
divisor   1101  /     10000111  dividend
                      1101
                       0111
                       0000
                        1111
(135 ÷ 13 = 10 with     1101
  a remainder of 5)      0101
                         0000
                         0101  remainder
```

Just as binary multiplication can be carried out as a series of add and shift operations, division can be carried out by a series of subtraction and shift operations. To construct the divider, we will use a 9-bit dividend register and a 4-bit divisor register, as shown in Figure 18-10. During the division process, instead of

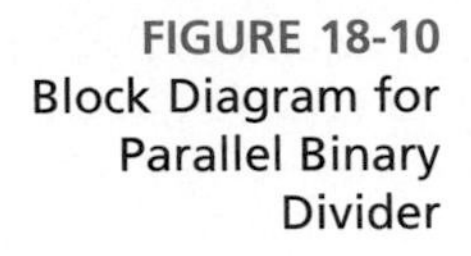

**FIGURE 18-10** Block Diagram for Parallel Binary Divider

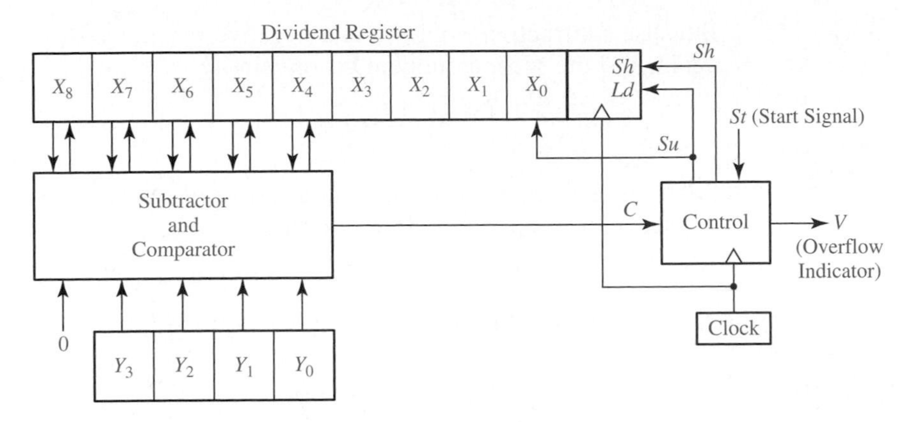


shifting the divisor to the right before each subtraction as shown in the preceding example, we will shift the dividend to the left. Note that an extra bit is required on the left end of the dividend register so that a bit is not lost when the dividend is shifted left. Instead of using a separate register to store the quotient, we will enter the quotient bit-by-bit into the right end of the dividend register as the dividend is shifted left. Circuits for initially loading the dividend into the register will be added later.

The preceding division example (135 divided by 13) is now reworked, showing the location of the bits in the registers at each clock time. Initially, the dividend and divisor are entered as follows:

```
0 1 0 0 0 0 1 1 1
  1 1 0 1
```

Subtraction cannot be carried out without a negative result, so we will shift before we subtract. Instead of shifting the divisor one place to the right, we will shift the dividend one place to the left:

```
                  ← Dividing line between dividend and quotient
1 0 0 0 0 1 1 1 | 0
  1 1 0 1       |  ↖ Note that after the shift, the rightmost position
                     in the dividend register is "empty".
```

Subtraction is now carried out, and the first quotient digit of 1 is stored in the unused position of the dividend register:

```
0 0 0 1 1 1 1 1 | 1 ←——— first quotient digit
```

Next, we shift the dividend one place to the left:

```
0 0 1 1 1 1 1 | 1 0
  1 1 0 1     |
```

Because subtraction would yield a negative result, we shift the dividend to the left again, and the second quotient bit remains 0:

```
0 1 1 1 1 1 ¦ 1 0 0
  1 1 0 1   ¦
```

Subtraction is now carried out, and the third quotient digit of 1 is stored in the unused position of the dividend register:

```
0 0 0 1 0 1 ¦ 1 0 1  <——————  third quotient digit
```

A final shift is carried out and the fourth quotient bit is set to 0:

```
0 0 1 0 1 ¦ 1 0 1 0
remainder ¦ quotient
```

The final result agrees with that obtained in the first example. Note that in the first step the leftmost 1 in the dividend is shifted left into the leftmost position ($X_8$) in the $X$ register. If we did not have a place for this bit, the division operation would have failed at this step because $0000 < 1101$. However, by keeping the leftmost bit in $X_8$, $10000 \geq 1101$, and subtraction can occur.

If as a result of a division operation, the quotient would contain more bits than are available for storing the quotient, we say that an overflow has occurred. For the divider of Figure 18-10 an overflow would occur if the quotient is greater than 15, because only 4 bits are provided to store the quotient. It is not actually necessary to carry out the division to determine if an overflow condition exists, because an initial comparison of the dividend and divisor will tell if the quotient will be too large. For example, if we attempt to divide 135 by 7, the initial contents of the registers would be:

```
0 1 0 0 0 0 1 1 1
  0 1 1 1
```

Because subtraction can be carried out with a nonnegative result, we should subtract the divisor from the dividend and enter a quotient bit of 1 in the rightmost place in the dividend register. However, we cannot do this because the rightmost place contains the least significant bit of the dividend, and entering a quotient bit here would destroy that dividend bit. Therefore, the quotient would be too large to store in the 4 bits we have allocated for it, and we have detected an overflow condition. In general, for Figure 18-10, if initially $X_8X_7X_6X_5X_4 \geq Y_3Y_2Y_1Y_0$ (i.e., if the left five bits of the dividend register exceed or equal the divisor), the quotient will be greater than 15 and an overflow occurs. Note that if $X_8X_7X_6X_5X_4 \geq Y_3Y_2Y_1Y_0$, the quotient is

$$\frac{X_8\,X_7\,X_6\,X_5\,X_4\,X_3\,X_2\,X_1\,X_0}{Y_3\,Y_2\,Y_1\,Y_0} \geq \frac{X_8\,X_7\,X_6\,X_5\,X_4\,0000}{Y_3\,Y_2\,Y_1\,Y_0} = \frac{X_8\,X_7\,X_6\,X_5\,X_4 \times 16}{Y_3\,Y_2\,Y_1\,Y_0} \geq 16$$

The operation of the divider can be explained in terms of the block diagram of Figure 18-10. A shift signal ($Sh$) will shift the dividend one place to the left on the next rising clock edge. Because the subtracter is a combinational circuit, it computes

$X_8X_7X_6X_5X_4 - Y_3Y_2Y_1Y_0$, and this difference appears at the subtracter output after a propagation delay. A subtract signal ($Su$) will load the subtracter output into $X_8X_7X_6X_5X_4$ and set the quotient bit (the rightmost bit in the dividend register) to 1 on the next rising clock edge. To accomplish this, $Su$ is connected to both the $Ld$ input on the shift register and the data input on flip-flop $X_0$. If the divisor is greater than the five leftmost dividend bits, the comparator output is $C = 0$; otherwise, $C = 1$. The control circuit generates the required sequence of shift and subtract signals. Whenever $C = 0$, subtraction cannot occur without a negative result, so a shift signal is generated. Whenever $C = 1$, a subtract signal is generated, and the quotient bit is set to one.

Figure 18-11 shows the state diagram for the control circuit. When a start signal ($St$) occurs, the 8-bit dividend and 4-bit divisor are loaded into the appropriate registers. If $C$ is 1, the quotient would require five or more bits. Because space is only provided for a 4-bit quotient, this condition constitutes an overflow, so the divider is stopped, and the overflow indicator is set by the $V$ output. Normally, the initial value of $C$ is 0, so a shift will occur first, and the control circuit will go to state $S_2$. Then, if $C = 1$, subtraction occurs. After the subtraction is completed, $C$ will always be 0, so the next active clock edge will produce a shift. This process continues until four shifts have occurred, and the control is in state $S_5$. Then, a final subtraction occurs if $C = 1$, and no subtraction occurs if $C = 0$. No further shifting is required, and the control goes to the stop state. For this example, we will assume that when the start signal ($St$) occurs, it will be 1 for one clock time, and, then, it will remain 0 until the control circuit is back in state $S_0$. Therefore, $St$ will always be 0 in states $S_1$ through $S_5$.

We will now design the control circuit using a one-hot assignment (see Section 15.9) to implement the state graph. One flip-flop is used for each state with $Q_0 = 1$ in $S_0$, $Q_1 = 1$ in $S_1$, $Q_2 = 1$ in $S_2$, etc. By inspection, the next-state and output equations are

$$Q_0^+ = St'Q_0 + CQ_1 + Q_5 \qquad Q_1^+ = StQ_0 \qquad (18\text{-}1)$$
$$Q_2^+ = C'Q_1 + CQ_2 \qquad Q_3^+ = C'Q_2 + CQ_3$$
$$Q_4^+ = C'Q_3 + CQ_4 \qquad Q_5^+ = C'Q_4$$
$$Load = St\,Q_0 \qquad V = CQ_1$$
$$Sh = C'(Q_1 + Q_2 + Q_3 + Q_4) = C'(Q_0 + Q_5)'$$
$$Su = C(Q_2 + Q_3 + Q_4 + Q_5) = C(Q_0 + Q_1)'$$

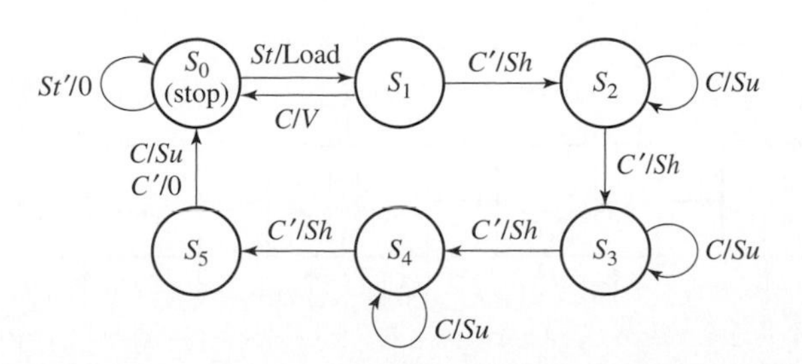


**FIGURE 18-11** State Graph for Divider Control Circuit

Because there are three arrows leading into $S_0$, $Q_0^+$ has three terms. The equation for $Sh$ has been simplified by noting that if the circuit is in state $S_1$ or $S_2$ or $S_3$ or $S_4$, it is not in state $S_0$ or $S_5$.

The subtracter in Figure 18-10 can be constructed using five full subtracters, as shown in Figure 18-12. Because the subtracter is a combinational circuit, whenever the numbers in the divisor and dividend registers change, these changes will propagate to the subtracter outputs. The borrow signal will propagate through the full subtracters before the subtracter output is transferred to the dividend register. If the last borrow signal ($b_9$) is 1, this means that the result is negative. Hence, if $b_9$ is 1, the divisor ($Y_3Y_2Y_1Y_0$) is greater than $X_8X_7X_6X_5X_4$, and $C = 0$. Therefore, $C = b_9'$, and a separate comparator circuit is unnecessary. Under normal operating conditions (no overflow) for this divider, we can also show that $C = d_8'$. At any subtraction step, because the divisor is only four bits, $d_8 = 1$ would allow a second subtraction without shifting. However, this can never occur because the quotient digit cannot be greater than 1. Therefore, if subtraction is possible, $d_8$ will always be 0 after the subtraction, so $d_8 = 0$ implies $X_8X_7X_6X_5X_4$ is greater than $Y_3Y_2Y_1Y_0$ and $C = d_8'$.

The block diagram of Figure 18-10 does not show how the dividend is initially loaded into the $X$ register. This can be accomplished by adding a MUX at the $X$ register inputs, as shown in Figure 18-13. This diagram uses bus notation to avoid drawing multiple wires. When several busses are merged together to form a single bus, a *bus merger* is used. For example, the symbol

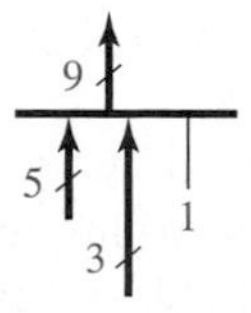


means that the 5-bit subtracter output is merged with bits $X_3X_2X_1$ and a logic 1 to form a 9-bit bus. Thus, the MUX output will be $d_8d_7d_6d_5d_4X_3X_2X_11$ when Load = 0.

Similarly, the symbol

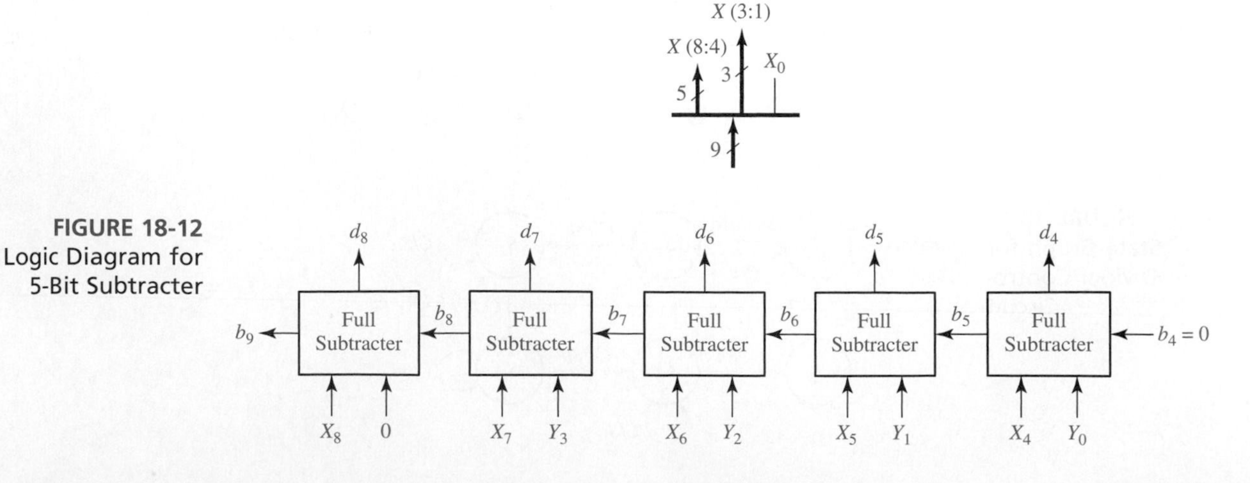


**FIGURE 18-12** Logic Diagram for 5-Bit Subtracter

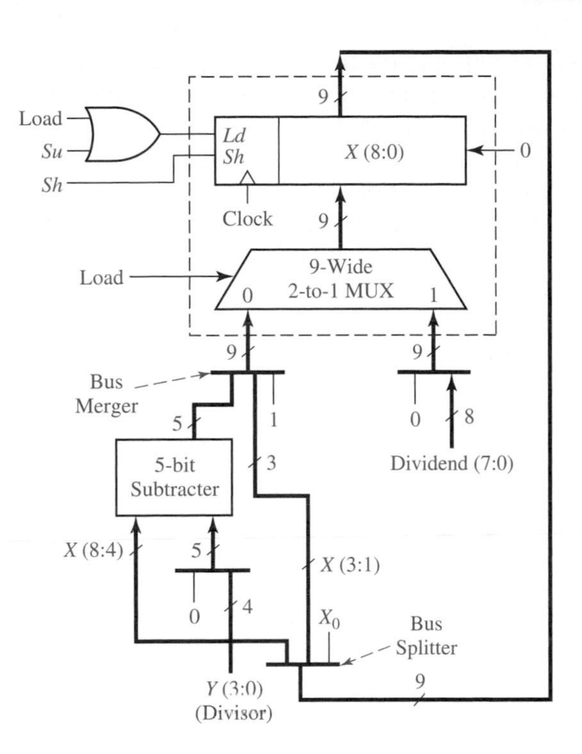


**FIGURE 18-13**
Block Diagram for Divider Using Bus Notation

represents a *bus splitter* that splits the 9 bits from the $X$ register into $X_8X_7X_6X_5X_4$ and $X_3X_2X_1$; $X_0$ is not used. Bus mergers and splitters do not require any actual hardware; they are just a symbolic way of showing bus connections.

The $X$ register is a left-shift register with parallel load capability, similar to the register in Figure 12-10. On the rising clock edge, it is loaded when Ld = 1 and shifted left when $Sh$ = 1. Because the register must be loaded with the dividend when Load = 1 and with the subtracter output when $Su$ = 1, Load and $Su$ are ORed together and connected to the Ld input. The MUX selects the dividend (preceded by a 0) when Load = 1. When Load = 0, it selects the bus merger output which consists of the subtracter output, $X_3X_2X_1$, and a logic 1. When $Su$ = 1 and the clock rises, this MUX output is loaded into $X$. The net result is that $X_8X_7X_6X_5X_4$ gets the subtracter output, $X_3X_2X_1$ is unchanged, and $X_0$ is set to 1.

# Programmed Exercise 18.1

Cover the lower part of each page with a sheet of paper and slide it down as you check your answers. Write your answer in the space provided before looking at the correct answers.

This exercise concerns the design of a circuit which forms the 2's complement of a 16-bit binary number. The circuit consists of three main components—a 16-bit shift register which initially holds the number to be complemented, a control circuit, and a counter which counts the number of shifts. The control circuit processes the number in the shift register one bit at a time and stores the 2's complement back in the shift register. Draw a block diagram of the circuit. Show the necessary inputs and outputs for the control circuit including a start signal ($N$) which is used to initiate the 2's complement operation.

**Answer**

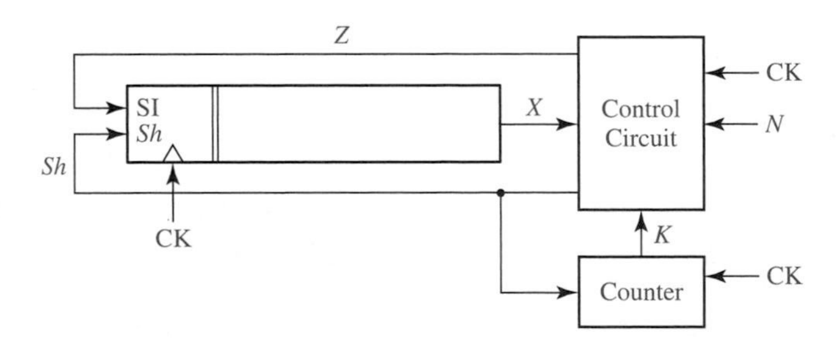


State a rule for forming the 2's complement which is appropriate for use with the preceding block diagram.

**Answer** Starting with the least significant bit, complement all of the bits to the left of the first 1.

Draw a state graph for the control circuit (three states) which implements the preceding rule. The 2's complement operation should be initiated when $N = 1$. (Assume that $N$ will be 1 for only one clock time.) When drawing your graph, do not include any provision for stopping the circuit. (In the next step you will be asked to add the signal $K$ to your state graph so that the circuit will stop after 16 shifts.) Explain the meaning of each state in your graph.

**Answer**

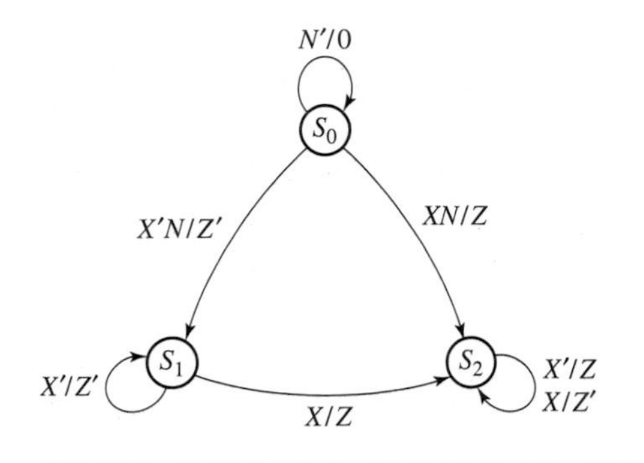


The counter will generate a completion signal ($K$) when it reaches state 15. Modify your state graph so that when $K = 1$, the circuit will complete the 2's complement operation and return to the initial state. Also, add the $Sh$ output in the appropriate places.

**Answer** Check the input labels on all arrows leaving each state of your graph. Make sure that two of the labels on arrows leaving a given state cannot have the value 1 at the same time. Make any necessary corrections to your graph, and then check your final answer.

**Final Answer**

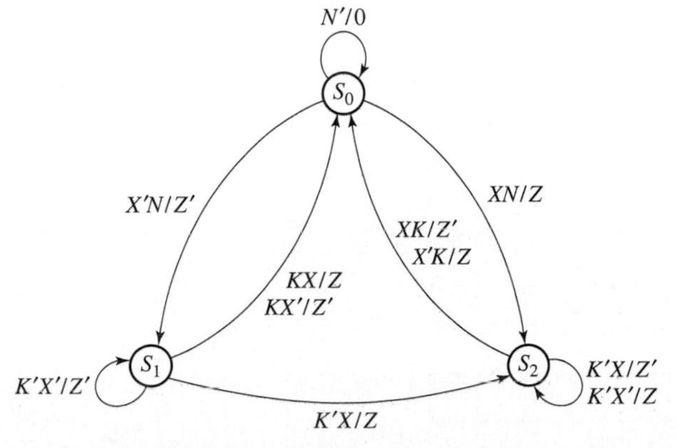


(*Note:* $Sh$ should be added to the graph everywhere $Z$ or $Z'$ appears.)

# Programmed Exercise 18.2

This exercise concerns the design of a binary divider to divide a 6-bit number by a 3-bit number to find a 3-bit quotient. The right 3 bits of the dividend register should be used to store the quotient. Draw a block diagram for the divider. Omit the signals required to initially load the dividend register and assume the dividend is already loaded.

**Answer**

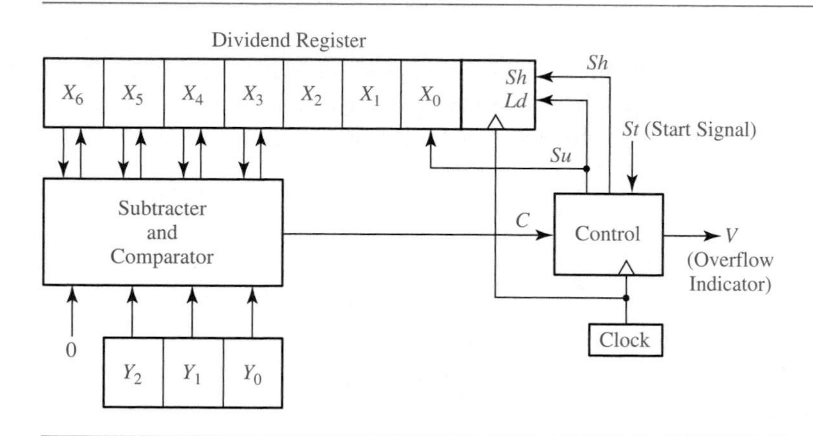


If the contents of the dividend register is initially 0100010 and the divisor is 110, show the contents of the dividend register after each of the first three rising clock edges. Also, indicate whether a shift or a subtraction should occur next.

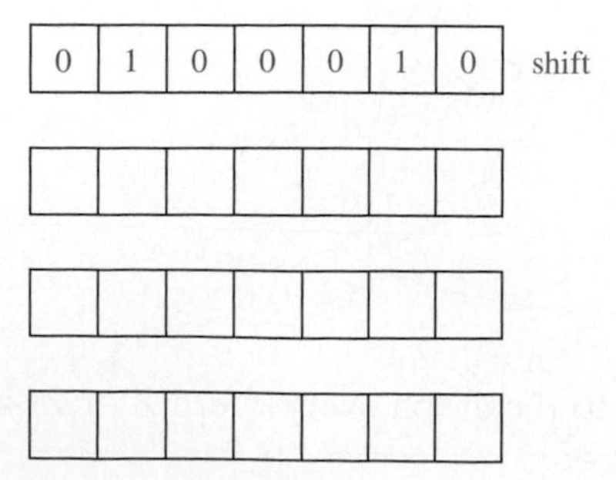

**Answer**

| | | | | | | | |
|---|---|---|---|---|---|---|---|
| 0 | 1 | 0 | 0 | 0 | 1 | 0 | shift |
| 1 | 0 | 0 | 0 | 1 | 0 | 0 | subtract |
| 0 | 0 | 1 | 0 | 1 | 0 | 1 | shift |
| 0 | 1 | 0 | 1 | 0 | 1 | 0 | shift |

Now, show the remaining steps in the computation and check your answer by converting to decimal.

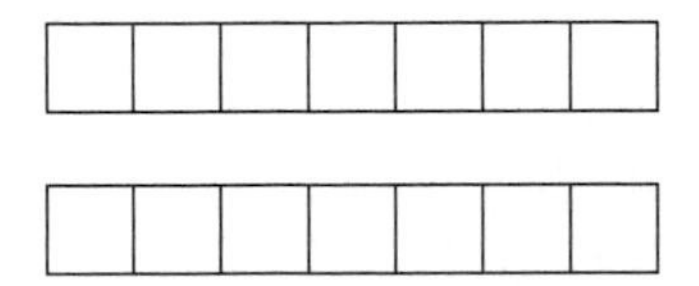

**Answer**

| | | | | | | | |
|---|---|---|---|---|---|---|---|
| 1 | 0 | 1 | 0 | 1 | 0 | 0 | subtract |
| 0 | 1 | 0 | 0 | 1 | 0 | 1 | (finished) |

If the dividend register initially contained 0011001 and the divisor is 010, can division take place? Explain.

**Answer** No. Because 011 > 010, subtraction should occur first, but there is no place to store the quotient bit. In other words, the quotient would be greater than three bits, so an overflow would occur.

Draw a state graph for the divider which will produce the necessary sequence of *Su* and *Sh* signals. Assume that the comparator output is $C = 1$ if the upper four bits of the dividend register is greater than the divisor. Include a stop state in your graph which is different than the reset state. Assume that the start signal (*St*) will remain 1 until the division is completed. The circuit should go to the stop state when division is complete or when an overflow is detected. The circuit should then reset when $St = 0$.

**Answer**

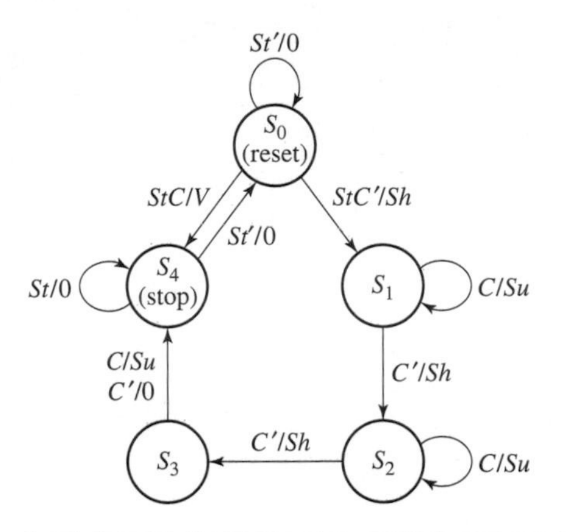


# Problems

**18.3** Design a serial subtracter with accumulator for 5-bit binary numbers. Assume that negative numbers are represented by 2's complement. Use a circuit of the form of Figure 18-1, except implement a serial subtracter using a D-CE flip-flop and any kind of gates. Give the state graph for the control circuit. Assume that *St* will remain 1 until the subtraction is complete, and the circuit will not reset until *St* returns to 0.

**18.4** Design a parallel binary multiplier which multiplies two 3-bit binary numbers to form a 6-bit product. This multiplier is to be a combinational circuit consisting of an array of full adders and AND gates (no flip-flops). Demonstrate that your circuit works by showing all of the signals which are present when 111 is multiplied by 111. (*Hint*: The AND gates can be used to multiply by 0 or 1, and the full adders can be used to add 2 bits plus a carry. Six full adders are required.)

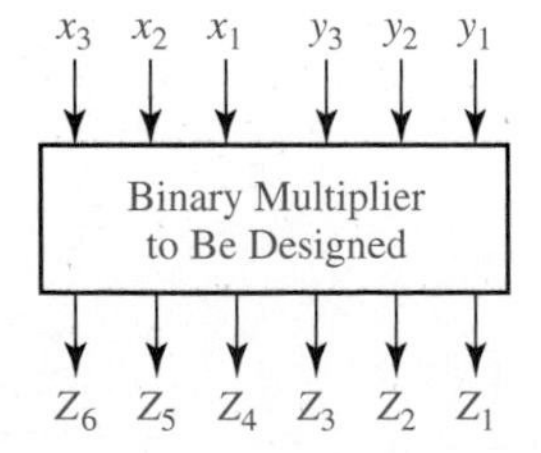


**18.5** The binary multiplier of Figure 18-7 has been redesigned so that whenever addition occurs the multiplier bit ($M$) will be set to 0. Specifically, the *Ad* signal is now connected to a synchronous clear input on only the rightmost flip-flop of the product register of Figure 18-7. Thus, if $M$ is 1 at a given clock time and addition takes place, $M$ will be 0 at the next clock time. Now, we can always add when $M = 1$ and always shift when $M = 0$. This means that the control circuit does not have to change state when $M = 1$, and the number of states can be reduced from ten to six. Draw the resulting state graph for the multiplier control with six states.

**18.6** In order to allow for a larger number of bits, the control circuit of the binary divider (Figure 18-10) is to be redesigned so that it uses a separate counter and a subtract-shift control which is analogous to Figure 18-9(a). Draw the state graph for the subtract-shift control.

**18.7** Below is the block diagram of a divider which will divide a 5-bit binary number $X_4X_3X_2X_1X_0$ by a 5-bit binary number $Y_4Y_3Y_2Y_1Y_0$. Initially, the 5-bit dividend is loaded into bits $X_4$ through $X_0$, and 0's are loaded into bits $X_9$ through $X_5$. Because of its design, overflow will only occur if the divisor is 0. This divider operates similarly to the one given in Figures 18-10 and 18-11, except for the starting placement of the dividend.

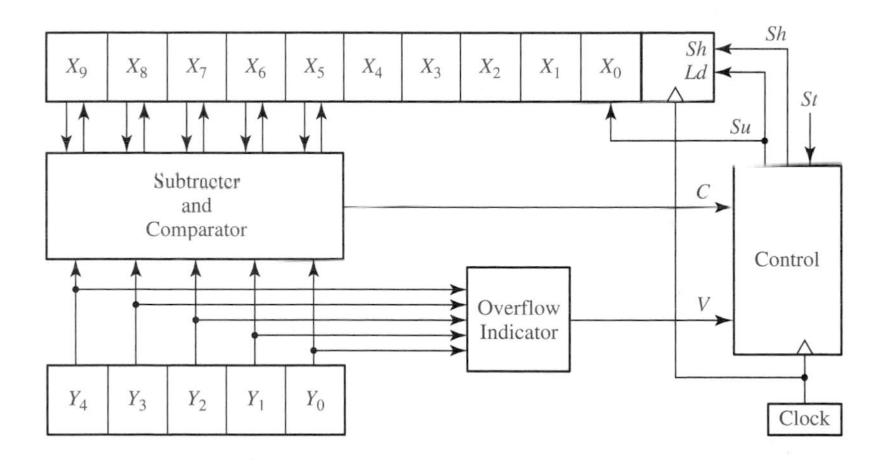


(a) Give the equation for the overflow signal $V$, generated by the overflow indicator.

(b) Illustrate the operation of the divider when 26 is divided by 5. Specify the sequence of $Su$ and $Sh$ outputs and the contents of the dividend register, and specify the quotient and the remainder.

(c) Draw the state graph for the control circuit. If there is an overflow, the circuit should remain in the starting state. Otherwise, when $St = 1$, the circuit should begin operation. Assume that $St$ will be 1 for only one clock cycle.

(d) In Figure 18-10, the subtracter-comparator and the dividend register have one more bit on the left than the divisor register. Why is that not necessary here?

**18.8** A serial logic unit has two 8-bit shift registers, $X$ and $Y$, shown as follows. Inputs $K_1$ and $K_2$ determine the operation to be performed on $X$ and $Y$. When $St = 1$, $X$ and $Y$ are shifted into the logic circuit one bit at a time and replaced with new values. If $K_1K_2 = 00$, $X$ is complemented and $Y$ is unchanged. If $K_1K_2 = 01$, $X$ and $Y$

are interchanged. If $K_1K_2 = 10$, $Y$ is set to 0 and each bit of $X$ is replaced with the exclusive-OR of the corresponding bits of $X$ and $Y$, that is, the new $x_i$ is $x_i \oplus y_i$. If $K_1K_2 = 11$, $X$ is unchanged and $Y$ is set to all 1's.

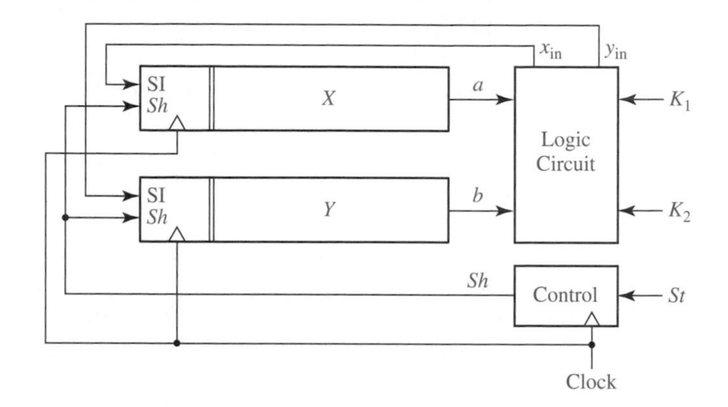


(a) Derive logic equations for $x_{in}$ and $y_{in}$.
(b) Derive a state graph for the control circuit. Assume that once $St$ is set to 1 it will remain 1 until all 8 bits have been processed. Then, $St$ will be changed back to 0 some time before the start of the next computation cycle.
(c) Realize the logic circuit using two 4-to-1 multiplexers and a minimum number of added gates.

**18.9** A circuit for adding one to the contents of a shift register has the following form:

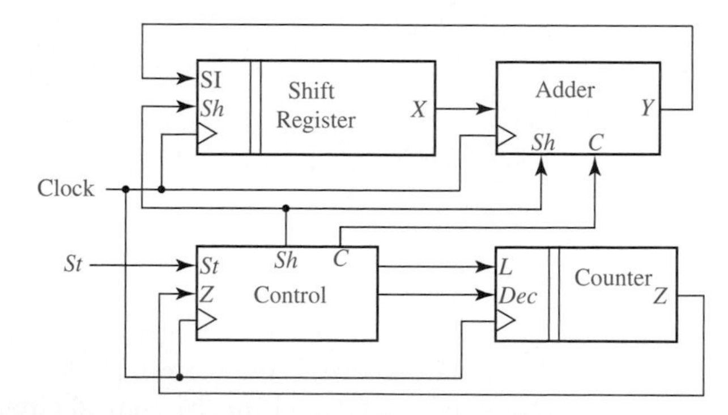


The adder circuit has an internal flip-flop that can be used to store a carry from the adder operation. The control unit has a counter available to determine when the add operation is complete. The counter input $L$ enables a parallel load of the counter with the length of the shift register. The counter input $Dec$ causes the counter to decrement. The counter output $Z$ becomes 1 when the counter value is zero. When $St$ becomes 1, the control unit generates $Sh$ and $C$ the required number of times to cause 1 to be added to the shift register contents. The control unit also generates the signals $L$ and $Dec$ to control the counter.

Design the adder and the control unit, using D flip-flops and NOR gates.

**18.10** Repeat Problem 18.9 so that 2 is added to the shift register contents rather than 1.

**18.11** Repeat Problem 18.9 so that 3 is added to the shift register contents rather than 1.

**18.12** A sequential circuit receives decimal numbers encoded in BCD one digit (4 bits) at a time, starting with the least significant digit. The circuit outputs are the 10's complement of the input number, also encoded in BCD least significant digit first. Input decimal numbers are separated by one or more inputs of all 1's, during which the circuit outputs all 1's. Once valid BCD digits of a new number start, the circuit resumes computing and outputting the 10's complement of the new number.

(a) Construct a state table and output table for the circuit. (Two states are sufficient.)
(b) Realize the circuit using a minimum number of flip-flops.

**18.13** Repeat Problem 18.12 assuming the decimal digits are encoded in excess-3 and the separator between decimal numbers is all 0's, which produces all 0's on the outputs.

**18.14** A circuit that adds one to the contents of a shift register has the following form:

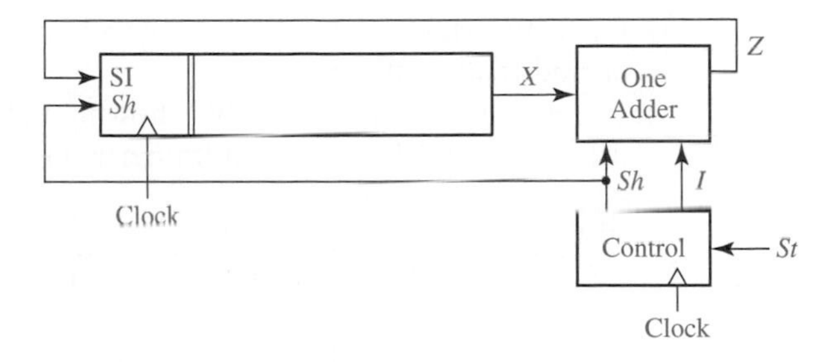


The control circuit outputs $I$, which should set the ONE ADDER to the proper initial state, and then outputs $Sh$ to the shift register the required number of times. Design the box labeled "ONE ADDER" using NOR gates and a D flip-flop with preset and clear inputs.

EXAMPLE:
Contents of shift register before: 000001011
Contents of shift register after $I$ and 9 $Sh$ outputs: 000001100

**18.15** (a) Draw a block diagram for a parallel multiplier that can multiply two binary numbers, where the multiplier is 3 bits and the multiplicand is 4 bits. Use an 8-bit shift register along with other necessary blocks.
(b) Draw a state graph for the multiplier control.
(c) Illustrate the operation of the multiplier when 11 is multiplied by 5. Specify the sequence of add and shift outputs generated by the control circuit and specify the contents of the 8-bit register at each clock time.
(d) Draw the logic diagram for the multiplier using an 8-bit shift register of the form of Figure 12-10, a 4-bit adder, three J-K flip-flops, and any necessary gates.

**18.16** Work Problem 18.15 if the multiplier is 3 bits and the multiplicand is 5 bits, and show 20 multiplied by 6. Use a 9-bit shift register similar to Figure 12-10, five full adders, three D flip-flops, and a PLA for Part (d). Show the PLA table.

**18.17** The block diagram for a parallel multiplier for positive binary numbers follows. The counter counts the number of shifts and outputs a signal $K = 1$ after two shifts.

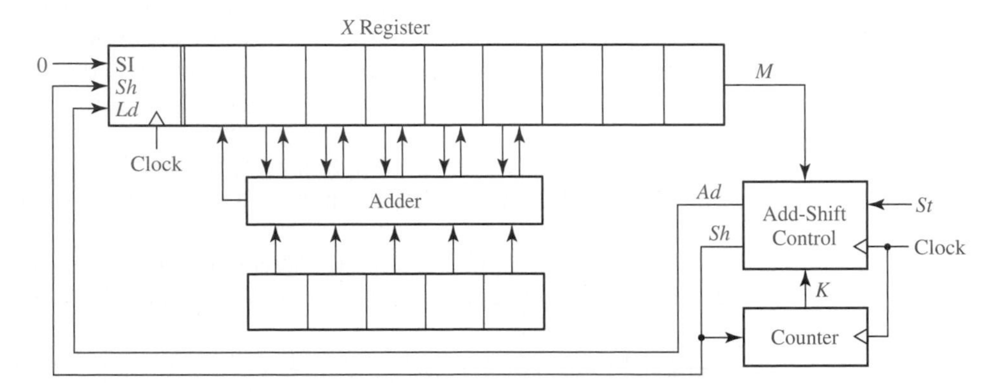


(a) Draw the state graph for the control circuit. Assume that *St* is 1 for one clock period to start the multiplier.
(b) Complete the following table showing the operation of the circuit if the multiplicand is 11001 and the multiplier is 111:

| State | Counter | X | St | M | K | Ad | Sh |
|---|---|---|---|---|---|---|---|
| $S_0$ | 00 | 000000111 | 1 | 1 | 0 | | |

**18.18** Design a binary divider which divides a 7-bit dividend by a 2-bit divisor to give a 5-bit quotient. The system has an input *St* that starts the division process.
(a) Draw a block diagram for the subtracter-comparator. You may use full adders or full subtracters.
(b) Draw a block diagram for the rest of the system (do not show the adders or subtracters in the subtracter-comparator block).
(c) Draw the state graph for the control circuit. Assume that the start signal (*St*) is present for one clock period.
(d) Give the contents of the dividend register and the value of *C* at each time step if initially the dividend is 01010011 and the divisor is 11.

**18.19** (a) Draw a block diagram for a parallel divider that is capable of dividing a positive 6-bit binary number by a positive 4-bit binary number to give a 2-bit quotient. Use a dividend register, a divisor register, a subtracter-comparator block, and a control block.

(b) Draw a state graph for the control circuit. Assume that the start signal *St* remains 1 for one or more clock times after the division is complete, and *St* must be set to 0 to reset the circuit.

(c) Show how the subtracter-comparator could be realized using full adders and inverters.

(d) Show the contents of the registers and the value of *C* after each time step if initially the dividend is 101101 and the divisor is 1101.

**18.20** Design a controller for an odd-parity generator. The circuit should transmit 7 bits from a shift register onto the output *X*. Then, on the next clock cycle, the eighth value of *X* should be chosen to make the number of 1's be odd. In other words, the last value of *X* should be 1 if there was an even number of 1's in the shift register, so that the 8-bit output word will have odd parity. (Parity was discussed in Section 13.1.) The circuit is shown. *K* will be 1 when the counter reaches 111.

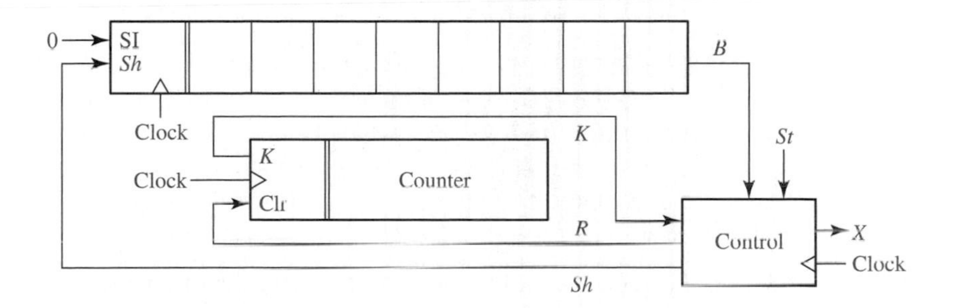


(a) Give the state graph for the control circuit. Assume $St = 1$ for one clock cycle (three states).

(b) Implement the controller using D flip-flops and any necessary gates. Use a one-hot state assignment.

**18.21** Design a serial logic unit to multiply a 6-bit number *X* by $-1$. Assume negative numbers are represented by their 2's complements. Recall that one way to find the 2's complement is to invert all of the bits to the left of the rightmost 1. If the number is $-32 = 100000$, there is no 6-bit 2's complement representation of $+32$, so an error signal *Er* should be generated.

(a) Give a block diagram for the circuit, using a control block, a 6-bit right-shift register, and a 3-bit counter. The controller has inputs *St*, *K*, and $S_O$, and outputs *Er*, *Clr*, *Sh*, and SI. The shift register is like the register of Figure 12-7, but it has 6 bits. The counter has a *Clr* input and an output *K* which is 1 when the counter reaches 6. Assume the shift register contains *X* at the beginning of the operation. The shift register should contain $-X$ when the operation is complete.

(b) Give the state graph for the control circuit. Be sure the circuit will work properly when taking the 2's complement of 0. $(0 \times -1 = 0.)$

(c) Implement the controller using a one-hot state assignment and D flip-flops.

**18.22** A serial Boolean logic unit has two 16-bit shift registers, $A$ and $B$. A control signal ($C$) is used to select the Boolean operation to be performed. If $C = 0$, the contents of $A$ are serially replaced by the bit-by-bit Boolean AND of $A$ and $B$. If $C = 1$, the contents of $A$ are serially replaced by the bit-by-bit exclusive-OR of $A$ and $B$. After the numbers have been placed in $A$ and $B$, and $C$ is set to 0 or 1, a start signal ($St$) sets the circuit in operation. A counter is used to count the number of shifts. When the counter reaches state 15, it outputs a signal $K = 1$, which causes the control circuit to stop after one more shift. Assume that $St$ remains 1 and $C$ does not change until the operation is completed. The control then remains in the stop state until $St$ is changed back to 0.

(a) Draw a block diagram of the system, which includes the shift registers, the counter, the control circuit, and a logic circuit that generates the serial input (SI) to the $A$ register.
(b) Draw a state graph for the control circuit (three states).
(c) Design the control circuit using a PLA and D flip-flops.
(d) Design a logic circuit that generates SI.

**18.23** Repeat 18.22, but assume that $St = 1$ for only one clock cycle, and that $C$ may change during the operation of the circuit. Therefore, the circuit should operate according to what the value of $C$ was when $St = 1$. Use a one-hot state assignment for (c). [*Hint:* $C$ should be an input to the control circuit, and you will need another output of the control circuit to take the place of $C$ in the logic circuit of Part (d) of 18.22.]

**18.24** A serial logic unit consists of a 4-bit shift register $X$ and a control unit. The control unit has a start input ($St$), a shift output ($Sh$), and an output $M$ which is the serial input to the shift register. In addition, signals $C_1$ and $C_2$ are used to select the logic operation performed on the shift register. When $St = 1$, then
If $C_1C_2 = 00$, the contents of register $X$ is serially replaced by all 0's.
If $C_1C_2 = 01$, the contents of register $X$ is serially replaced by all 1's.
If $C_1C_2 = 11$, the contents of register $X$ is serially replaced by its bit-by-bit complement. Assume that $C_1C_2$ does not change until the selected operation is complete.

(a) Draw a block diagram for the system.
(b) Specify the state graph for the control unit. Assume that $St$ stays 1 for one clock period.
(c) Design the control unit (not the shift register) using J-K flip-flops and any kind of gates. Also, design the logic inside the control unit which generates the serial input $M$ to the shift register. [*Hint:* $M$ depends only on $C_1$, $C_2$, and $X$.]

**18.25** Design a circuit which sets a specified number of bits on the right side of a shift register to 0. The number of bits to be set to 0 is in register $N$ before the start of the operation. When $St = 1$, the controller should shift right $N$ times, and then shift left $N$ times. The counter only counts down, and $K = 1$ when the counter reaches 000.

(a) Give the circuit. Use a control block, a 3-bit $N$ register, a 3-bit down counter with load input ($Ld$) and $K$ output (which is 1 when the counter reaches 000), and an 8-bit right/left shift register which functions according to the table in Problem 12.3 (except that it has 8 bits). Note that the counter does not count up, so you will have to load $N$ into the counter twice. The controller has inputs $St$ and $K$, and outputs $A$, $B$, and $Ld$.

(b) Give the state graph for the control circuit. Assume $St = 1$ for one clock period.

(c) Implement the controller using two D flip-flops. Use a straight binary assignment.

**18.26** Design a controller for the circuit of Problem 12.35 that will add three numbers. Assume each number (including the first one) appears on the 8-bit input data line for two consecutive clock cycles. You may *not* assume that the registers begin with a value of 0. When $St = 1$, the first input appears on the input data line for that clock cycle and the next one. The circuit should halt when the answer goes into the accumulator, and output a signal Done $= 1$. Done should remain 1 until $St$ returns to 0. You may assume $St = 1$ for enough time for the operation to complete. Give the block diagram and the state graph (seven states), but you do not need to implement the state graph.

**18.27** The given multiplier uses only counters to multiply a 4-bit multiplicand by a 4-bit multiplier to obtain an 8-bit product. This Ultra-Slow Multiplier is based on the principle that multiplication is repeated addition and that addition is repeated incrementing. The multiplier works as follows: When the $St$ signal is received, the 8-bit up counter is cleared, $N_1$ is loaded into 4-bit counter $A$, and $N_2$ is loaded into 4-bit counter $B$. Then, the controller decrements $A$ and increments the up counter until $A$ reaches zero. When $A$ reaches zero, $B$ is decremented and $A$ is reloaded with $N_1$. Then, the process is repeated until $B$ reaches zero. When $B$ reaches zero, the 8-bit up counter contains the product.

(a) Draw the state graph for the controller. Assume $St = 1$ for only one clock period.

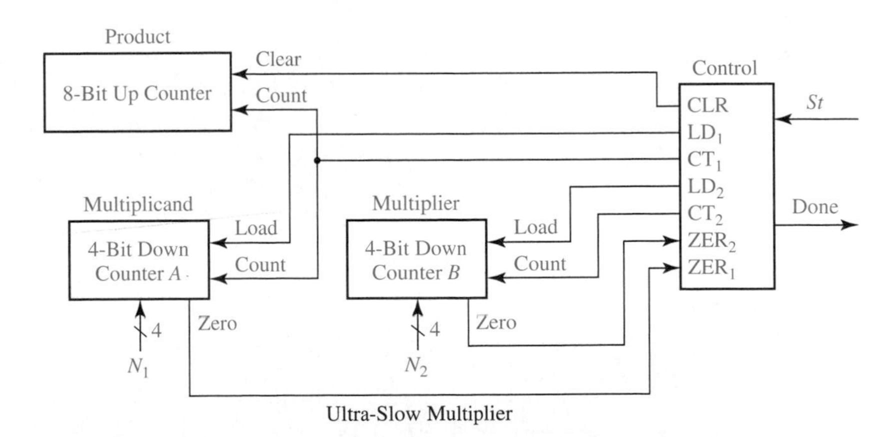

Ultra-Slow Multiplier

(b) Realize the state graph using one or two J-K flip-flops and a minimum number of gates.

(c) If the multiplier is $N_1$ and the multiplicand is $N_2$, how many clock periods does it take for the Ultra-Slow Multiplier to calculate the product?

**18.28** The following circuit is a multiplier circuit for 4-bit positive numbers. Multiplication is performed by adding the multiplicand to a partial product while decrementing the multiplier. This is continued until the multiplier is decremented to zero. (If the multiplier is initially zero, no additions are done.) When the start input ($S$) changes to 1, the multiplicand and multiplier are available; the multiplier circuit loads them into A Reg and B Counter, respectively. The partial product register, implemented in two parts (PU and PL), is cleared, as is the carry-out FF for the adder (C FF). To avoid having an adder twice as long as the operands, the addition of the multiplicand to the partial product is done in two steps: First, the multiplicand is added to the lower half of the partial product; second, the carry from the first addition is added to the upper half of the partial product. The multiplier in the B counter is decremented for each addition, and the additions continue until the multiplier has been decremented to zero. Then the done signal (D) is generated, with the product available in the partial product register; D remains asserted and the product available until S returns to 0.

The control signals that the controller must generate are

| | |
|---|---|
| LB | Load B Counter |
| DB | Decrement B Counter |
| CP | Clear PU and PL |
| LPU | Load PU |
| LPL | Load PL |
| LA | Load A Reg |
| MS | MUX Select Signal |
| EA | Signal ANDed with A Reg output |
| CC | Clear C FF |
| D | Done |

The input signals to the controller are start, S, and BZ; BZ = 0 when the B counter is zero.

(a) Determine the contents of the partial product register for each addition step when the multiplicand is 1011 and the multiplier is 0101.
(b) Draw a state graph for the controller. (Four states are sufficient.)
(c) Realize the controller using D FFs and a one-hot state assignment. Give the next-state equations and the controller output equations.
(d) Realize the controller using a minimum number of D FFs.

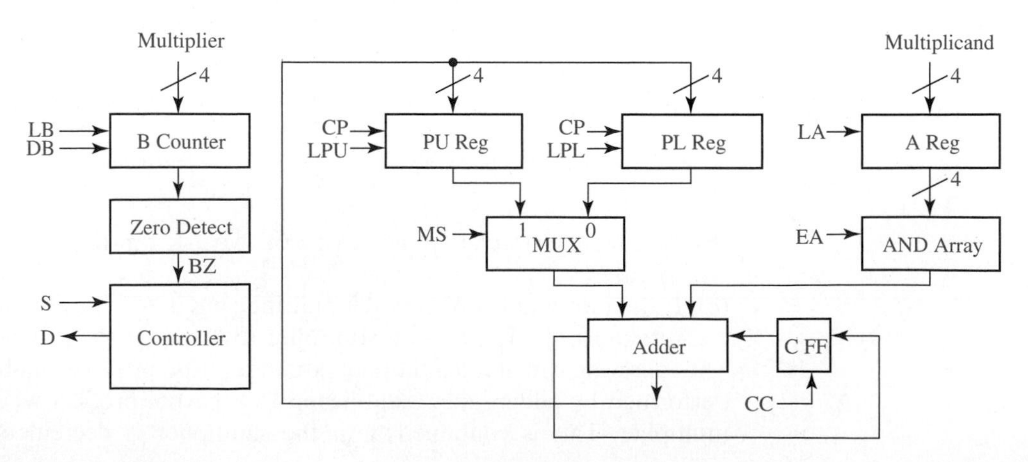

**18.29** A few modifications of the circuit of Problem 18.28 are necessary so that it will multiply 2's complement numbers. For example, the Controller must have inputs that are the sign bits of the multiplier and multiplicand; the B Counter must be able to increment a negative multiplier to 0; and the AND Array must be changed so that its outputs can be the Multiplicand, all 0's or all 1's.

(a) Redesign the multiplier so that it can multiply 2's complement numbers using these suggested modifications.
(b) Determine the contents of the partial product register for each addition step when the multiplicand is 1011 and the multiplier is 0101. Repeat when the multiplicand is 0101 and the multiplier is 1011.
(c) Draw a state graph for the controller. (At most, five states are required.)
(d) Realize the controller using D FFs and a one-hot state assignment. Give the next-state equations and the controller output equations.
(e) Realize the controller using a minimum number of D FFs.

**18.30** The Ultra-Slow Divider, shown in the following block diagram, works on a principle similar to the Ultra-Slow Multiplier in Problem 18.27. When the *St* signal is received, the 8-bit down counter is loaded with the dividend ($N_1$), the 4-bit down counter is loaded with the divisor ($N_2$), and the 4-bit quotient up counter is cleared. The dividend counter and the divisor counter are decremented together, and every time the 4-bit divisor counter reaches zero, it is reloaded with the divisor and the quotient up counter is incremented. When the dividend counter reaches zero, the process terminates and the quotient counter contains the result.

(a) Draw the state graph for the controller.
(b) Realize the state graph using one or two D flip-flops and a minimum number of gates.
(c) If the dividend is $N_1$ and the divisor is $N_2$, how many clock cycles does it take to calculate the quotient?
(d) How can you tell if an overflow occurs during division?
(e) What will happen in your circuit if the divisor is zero?

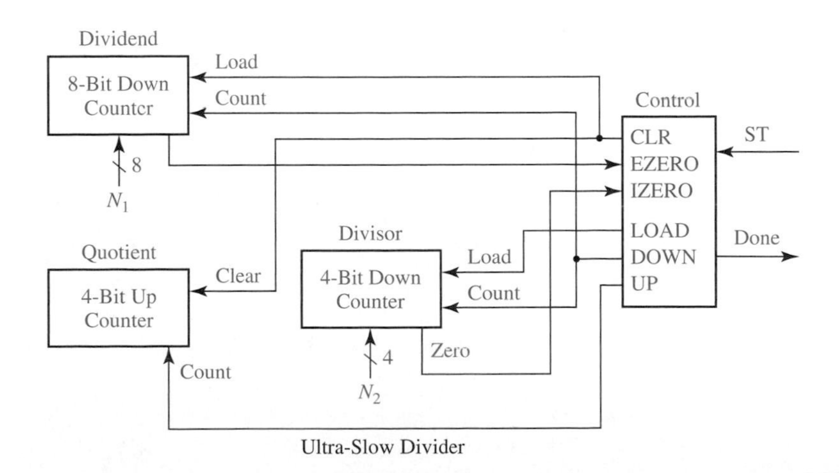


Ultra-Slow Divider

**18.31** This problem involves the design of a circuit that finds the integer part of the square root of an 8-bit unsigned binary number $N$ using the method of subtracting out odd integers. To find the square root of $N$, we subtract 1, then 3, then 5, etc., until we can no longer subtract without the result going negative. The number of times we subtract is equal to the integer part of the square root of $N$. For example, to find $\sqrt{27}$: $27 - 1 = 26$; $26 - 3 = 23$; $23 - 5 = 18$; $18 - 7 = 11$; $11 - 9 = 2$; $2 - 11$ (cannot subtract). Because we subtracted five times, $\sqrt{27} = 5$. Note that the final odd integer is $11_{10} = 1011_2$, and this consists of the square root ($101_2 = 5_{10}$) followed by a 1.

# State Machine Design with SM Charts

## Objectives

1. Explain the different parts of an SM chart.
2. Given the input sequence to a state machine, determine the output sequence from its SM chart and construct a timing diagram.
3. Convert a state graph to an SM chart.
4. Construct an SM chart for the control circuit for a multiplier, divider, or other simple digital system.
5. Determine the next-state and output equations for a state machine by tracing link paths on its SM chart.
6. Realize an SM chart using a PLA or ROM and flip-flops.

# Study Guide

1. Study Section 19.1, *State Machine Charts.*

   (a) For the example of Figure 19-2, if $X_1 = 0$ and $X_2 = 1$ when the machine is in state $S_1$, specify the values of all of the outputs and the exit path number.

   (b) For Figures 19-6(a) and (b), trace the link paths and determine the outputs when $X_1 = X_3 = 1$.

   (c) Verify that the SM chart and state graph of Figure 19-7 are equivalent.

   (d) Construct a timing chart for Figure 19-7(b) when the input sequence is $X = 0, 1, 1, 0$.

   (e) Work Problems 19.1, 19.2, and 19.3.

2. Study Section 19.2, *Derivation of SM Charts.*

   (a) Using the SM chart of Figure 19-9 to determine when to subtract and when to shift for the binary divider of Figure 18-10, show the contents of the dividend register at each time step when 28 is divided by 5.

   (b) Compare the SM chart of Figure 19-10 with the state graph of Figure 18-9(c) and verify that in each state they will generate the same outputs when the inputs are the same.

   (c) Compare the flowchart for the dice game (Figure 19-12) with the SM chart (Figure 19-13). Note that the Roll Dice box on the flowchart requires two states to implement on the SM chart. In the first state, the machine waits for the roll button to be pressed; in the second state, it generates a roll signal which lasts until the roll button is released. In state $S_1$ 3 variables are tested; if they are all 0, *Sp* is generated so that the sum will be stored in the point register at the same time the transition from $S_1$ to $S_4$ occurs.

   (d) Work Problems 19.4, 19.5, and 19.6.

3. Study Section 19.3, *Realization of SM Charts.*

   (a) For Figure 19-7(b) find simplified equations for $A^+$ and $B^+$.

   (b) Verify Tables 19-1 and 19-2. For Table 19-2, why is $Sp = 1$ only in row 4, and $Win = 1$ in both rows 7 and 8?

   (c) Expand row 16 of Table 19-2 to give the corresponding rows of the ROM table.

   (d) Work Problems 19.7, 19.8, 19.9, and 19.10.

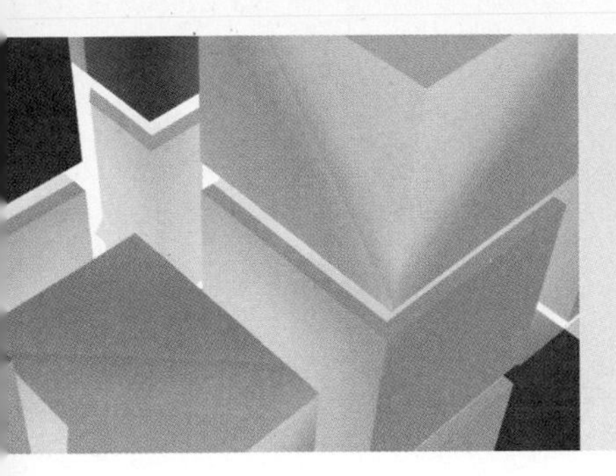

# State Machine Design with SM Charts

Another name for a sequential circuit is an algorithmic state machine or simply a state machine. These names are often used when the sequential circuit is used to control a digital system that carries out a step-by-step procedure or algorithm. The state graphs in Figures 18-3, 18-8, 18-9, and 18-11 define state machines for controlling adders, multipliers, and dividers. As an alternative to using state graphs, a special type of flowchart, called a state machine flowchart or SM chart, may be used to describe the behavior of a state machine. This unit describes the properties of SM charts and how they are used in the design of state machines.

## 19.1 State Machine Charts

Just as flowcharts are useful in software design, flowcharts are useful in the hardware design of digital systems. In this section we introduce a special type of flowchart called a state machine flowchart, or SM chart for short. SM charts are also called ASM (algorithmic state machine) charts. We will see that the SM chart offers several advantages. It is often easier to understand the operation of a digital system by inspection of the SM chart instead of the equivalent state graph. A given SM chart can be converted into several equivalent forms, and each form leads directly to a hardware realization.

An SM chart differs from an ordinary flowchart in that certain specific rules must be followed in constructing the SM chart. When these rules are followed, the SM chart is equivalent to a state graph, and it leads directly to a hardware realization. Figure 19-1 shows the three principal components of an SM chart. The state of the system is represented by a *state box*. The state box contains a state name, and it may contain an *output list*. A *state code* may be placed outside the box at the top. A *decision box* is represented by a diamond-shaped symbol with true and false branches. The *condition* placed in the box is a Boolean expression that is evaluated to determine which branch to take. The *conditional output box*, which has curved ends, contains a *conditional output list*. The conditional outputs depend on both the state of the system and the inputs.

FIGURE 19-1 Components of an SM Chart

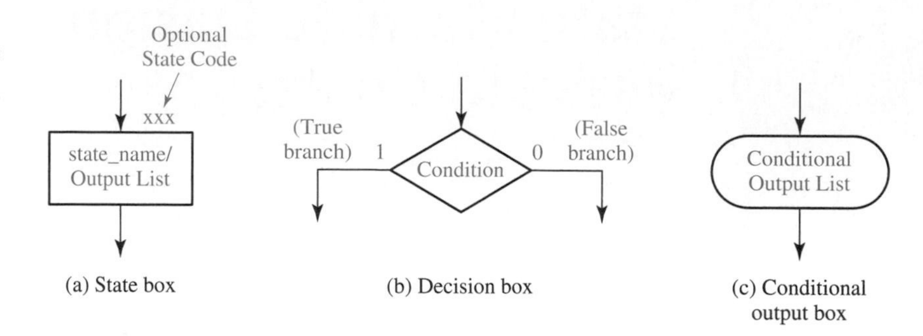


An SM chart is constructed from *SM blocks*. Each SM block (Figure 19-2) contains exactly one state box together with the decision boxes and conditional output boxes associated with that state. An SM block has exactly one *entrance path* and one or more *exit paths*. Each SM block describes the machine operation during the time that the machine is in one state. When a digital system enters the state associated with a given SM block, the outputs on the output list in the state box become true. The conditions in the decision boxes are evaluated to determine which path (or paths) is (are) followed through the SM block. When a conditional output box is encountered along such a path, the corresponding conditional outputs become true. A path through an SM block from entrance to exit is referred to as a *link path*.

For the example of Figure 19-2, when state $S_1$ is entered, outputs $Z_1$ and $Z_2$ become 1. If inputs $X_1$ and $X_2$ are both equal to 0, $Z_3$ and $Z_4$ are also 1, and at the end of the state time, the machine goes to the next state via exit path 1. On the other hand, if $X_1 = 1$ and $X_3 = 0$, the output $Z_5$ is 1, and an exit to the next state will occur via exit path 3.

A given SM block can generally be drawn in several different forms. Figure 19-3 shows two equivalent SM blocks. In both Figure 19-3(a) and (b), the output $Z_2 = 1$ if $X_1 = 0$; the next state is $S_2$ if $X_2 = 0$ and $S_3$ if $X_2 = 1$.

FIGURE 19-2 Example of an SM Block

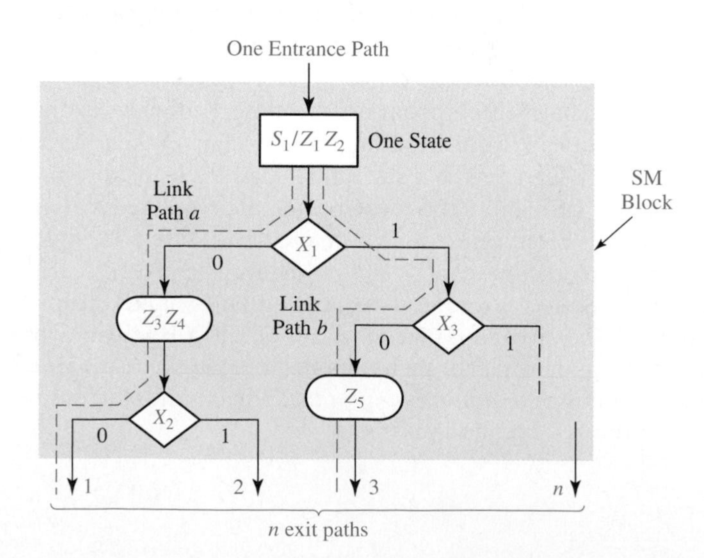

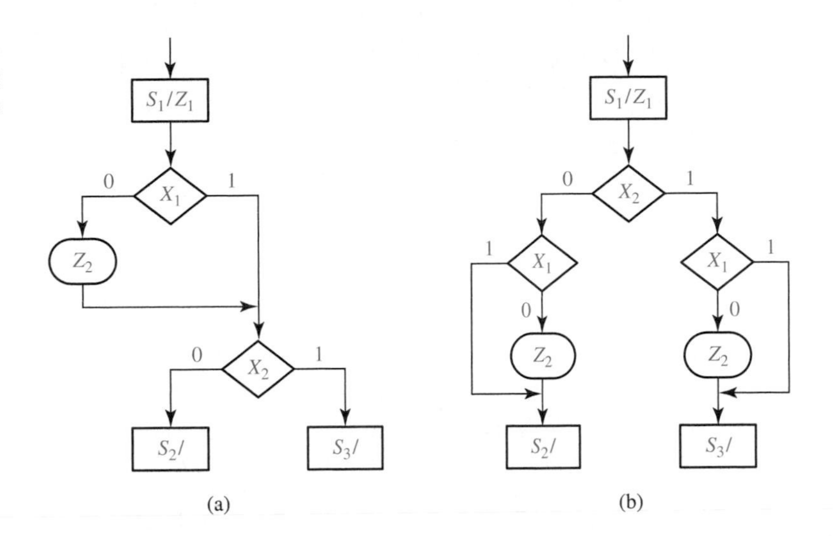


FIGURE 19-3 Equivalent SM Blocks

The SM chart of Figure 19-4(a) represents a combinational circuit because there is only one state and no state change occurs. The output is $Z_1 = 1$ if $A + BC = 1$; or else $Z_1 = 0$. Figure 19-4(b) shows an equivalent SM chart in which the input variables are tested individually. The output is $Z_1 = 1$ if $A = 1$ or if $A = 0, B = 1$, and $C = 1$. Hence,

$$Z_1 = A + A'BC = A + BC$$

which is the same output function realized by the SM chart of Figure 19-4(a).

Certain rules must be followed when constructing an SM block. First, for every valid combination of input variables, there must be exactly one exit path defined. This is necessary because each allowable input combination must lead to a single next state. Second, no internal feedback within an SM block is allowed. Figure 19-5 shows an incorrect and correct way of drawing an SM block with feedback.

As shown in Figure 19-6(a), an SM block can have several parallel paths which lead to the same exit path, and more than one of these paths can be active at the same time. For example, if $X_1 = X_2 = 1$ and $X_3 = 0$, the link paths marked with

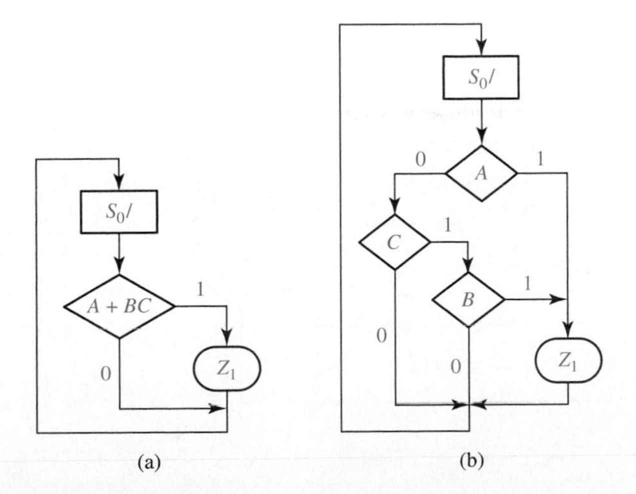


FIGURE 19-4 Equivalent SM Charts for a Combinational Circuit

FIGURE 19-5
SM Block with Feedback

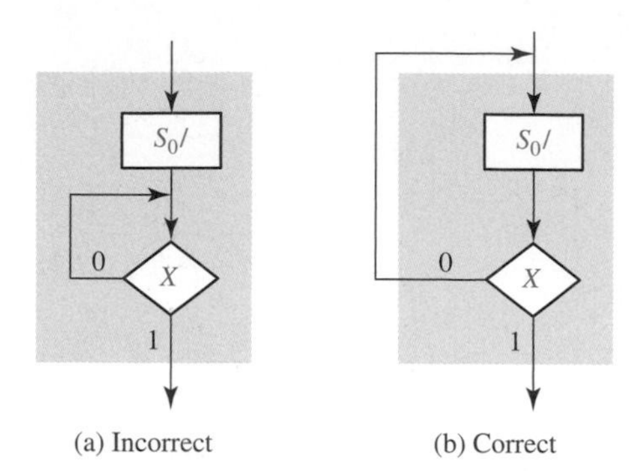


(a) Incorrect (b) Correct

dashed lines are active, and the outputs $Z_1$, $Z_2$, and $Z_3$ will be 1. Although Figure 19-6(a) would not be a valid flowchart for a program for a serial computer, it presents no problems for a state machine implementation. The state machine can have a multiple-output circuit that generates $Z_1$, $Z_2$, and $Z_3$ at the same time. Figure 19-6(b) shows a serial SM block, which is equivalent to Figure 19-6(a). In the serial block only one active link path between entrance and exit is possible.

For any combination of input values the outputs will be the same as in the equivalent parallel form. The link path for $X_1 = X_2 = 1$ and $X_3 = 0$ is shown with a dashed line, and the outputs encountered on this path are $Z_1$, $Z_2$, and $Z_3$. Regardless of whether the SM block is drawn in serial or parallel form, all of the tests take place within one clock time.

A state graph for a sequential machine is easy to convert to an equivalent SM chart. The state graph of Figure 19-7(a) has both Moore and Mealy outputs. The equivalent SM chart has three blocks—one for each state. The Moore outputs ($Z_a$, $Z_b$ and $Z_c$) are placed in the state boxes because they do not depend on the

FIGURE 19-6
Equivalent SM Blocks

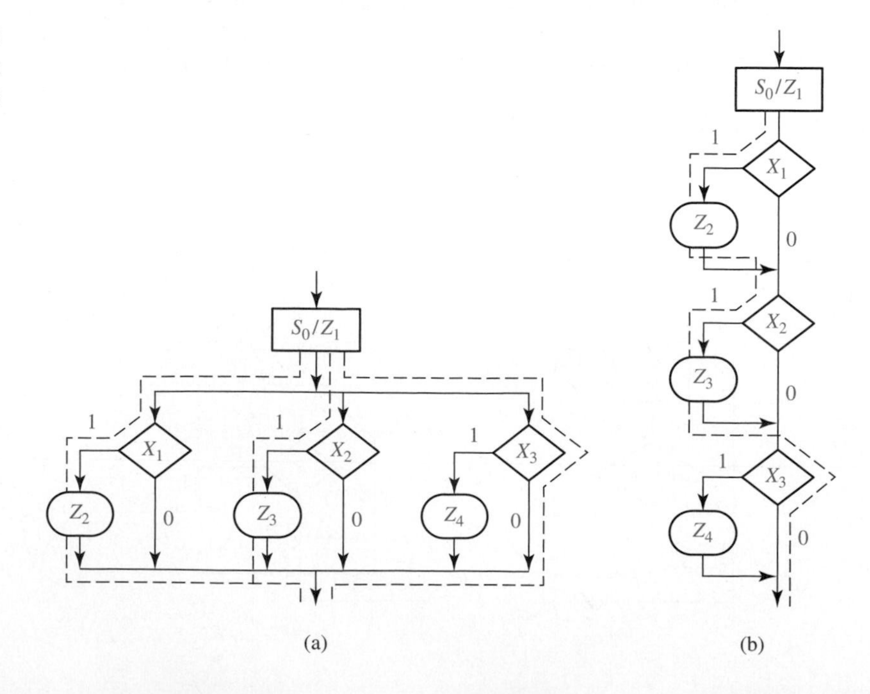


(a) (b)

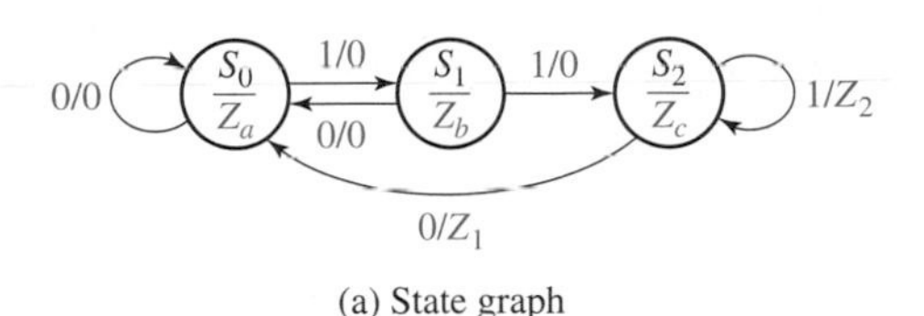

(a) State graph

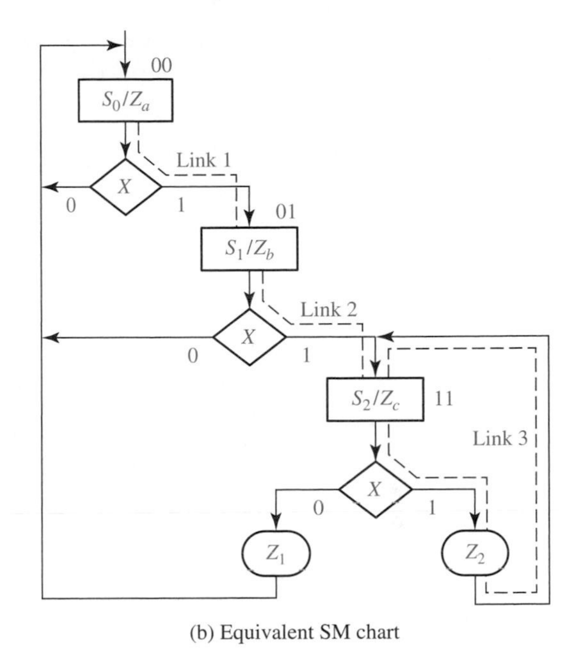

(b) Equivalent SM chart

**FIGURE 19-7** Conversion of a State Graph to an SM Chart

input. The Mealy outputs ($Z_1$ and $Z_2$) appear in conditional output boxes because they depend on both the state and input. In this example, each SM block has only one decision box because only one input variable must be tested. For both the state graph and SM chart, $Z_c$ is always 1 in state $S_2$. If $X = 0$ in state $S_2$, $Z_1 = 1$ and the next state is $S_0$. If $X = 1$, $Z_2 = 1$ and the next state is $S_2$.

Figure 19-8 shows a timing chart for the SM chart of Figure 19-7 with an input sequence $X = 1, 1, 1, 0, 0, 0$. In this example, all state changes occur immediately after the rising edge of the clock. Because the Moore outputs ($Z_a$, $Z_b$ and $Z_c$) depend on the state, they can only change immediately following a state change. The Mealy outputs

**FIGURE 19-8** Timing Chart for Figure 19-7

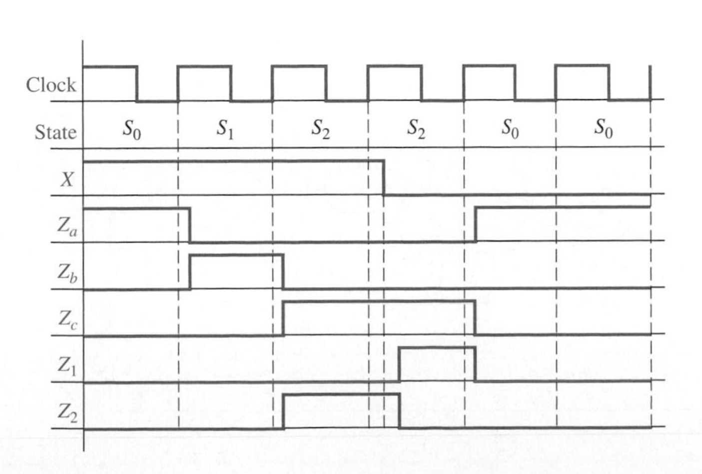

($Z_1$ and $Z_2$) can change immediately after a state change or an input change. In any case, all outputs will have their correct value during the active edge of the clock.

## 19.2 Derivation of SM Charts

The method used to derive an SM chart for a sequential control circuit is similar to that used to derive the state graph. First, we should draw a block diagram of the system we are controlling. Next, we should define the required input and output signals to and from the control circuit. Then, we can construct an SM chart that tests the input signals and generates the proper sequence of output signals.

In this section we will give several examples of SM charts. The first example is an SM chart for control of the parallel binary divider, as shown in Figure 18-10. As described in Section 18.3, binary division requires a series of subtract and shift operations. Derivation of an SM chart to generate the proper sequence of subtract and shift signals is very similar to derivation of the state graph of Figure 18-11. For the SM chart of Figure 19-9, $S_0$ is the starting state. In $S_0$, the start signal ($St$) is tested,

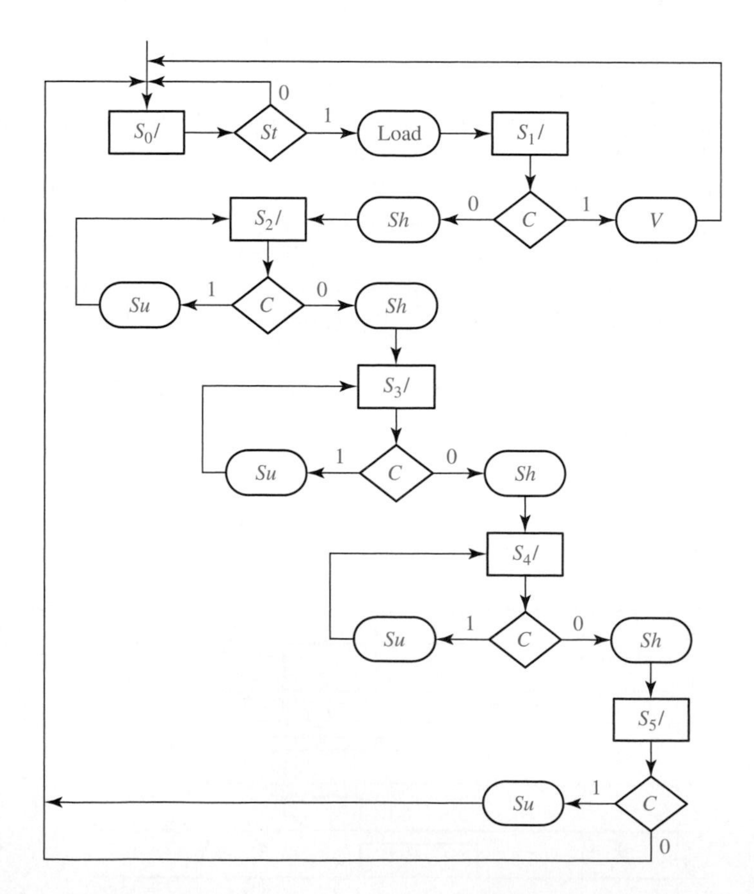


**FIGURE 19-9**
SM Chart for Binary Divider

and if $St = 1$, the *Load* signal is turned on and the next state is $S_1$. In $S_1$, the compare signal ($C$) is tested. If $C = 1$, the quotient would be larger than 4 bits, so an overflow signal ($V = 1$) is generated and the state changes back to $S_0$. If $C = 0$, $Sh$ becomes 1, so at the next clock the dividend is shifted to the left and the state changes to $S_2$. $C$ is tested again in state $S_2$. If $C = 1$, subtraction is possible, so $Su$ becomes 1 and no state change occurs. If $C = 0$, $Sh = 1$, and the dividend is shifted as the state changes to $S_3$. The action in states $S_3$ and $S_4$ is identical to that in state $S_2$. In state $S_5$ the next state is always $S_0$, and $C = 1$ causes subtraction to occur.

Next, we will derive the SM chart for the multiplier control of Figure 18-9(a). This control generates the required sequence of add and shift signals for a binary multiplier of the type shown in Figure 18-7. The counter counts the number of shifts and outputs $K = 1$ just before the last shift occurs. The SM chart for the multiplier control (Figure 19-10) corresponds closely to the state graph of Figure 18-9(c). In state $S_0$, when the start signal $St$ is 1, Load is turned on and the next state is $S_1$. In $S_1$, the multiplier bit $M$ is tested to determine whether to add or shift. If $M = 1$, an add signal is generated and the next state is $S_2$. If $M = 0$, no addition is required, so a shift signal is generated and $K$ is tested. If $K = 1$, the circuit goes to the Done state, $S_3$, at the time of the last shift; otherwise, the next state is $S_1$. In $S_2$ a shift signal is generated because a shift must always follow an add, and $K$ is tested to determine the next state.

As a third example of SM chart construction, we will design an electronic dice game. Figure 19-11 shows the block diagram for the dice game. Two counters are used to simulate the roll of the dice. Each counter counts in the sequence 1, 2, 3, 4, 5, 6, 1, 2, . . . . Thus, after the "roll" of the dice, the sum of the values in the two counters will be in the range 2 through 12.

**FIGURE 19-10**
SM Chart for Binary Multiplier

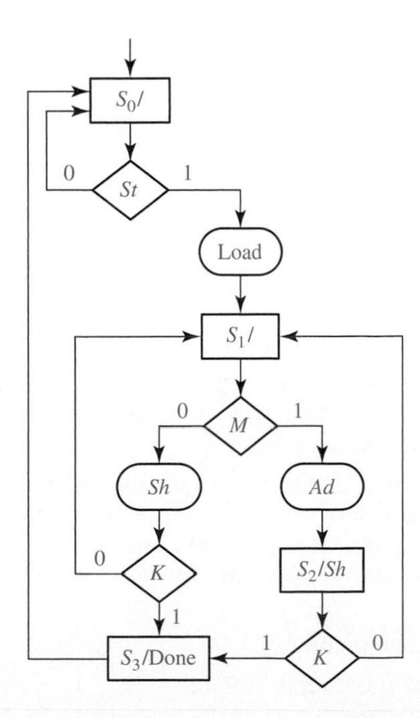

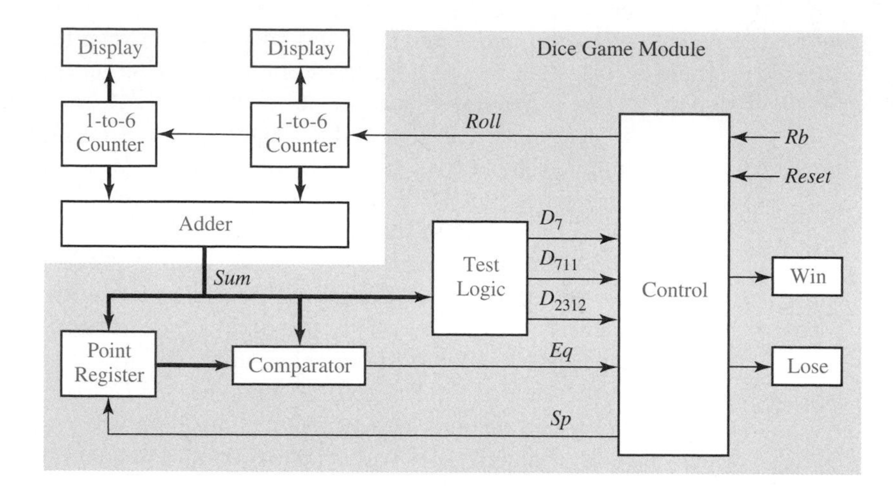


**FIGURE 19-11** Block Diagram for Dice Game

The rules of the game are as follows:

1. After the first roll of the dice, the player wins if the sum is 7 or 11. He loses if the sum is 2, 3, or 12. Otherwise, the sum which he obtained on the first roll is referred to as his point, and he must roll the dice again.
2. On the second or subsequent roll of the dice, he wins if the sum equals his point, and he loses if the sum is 7. Otherwise, he must roll again until he finally wins or loses.

The inputs to the dice game come from two push buttons, *Rb* (roll button) and Reset. Reset is used to initiate a new game. When the roll button is pushed, the dice counters count at a high speed, so the values cannot be read on the display. When the roll button is released, the values in the two counters are displayed and the game can proceed. Because the button is released at a random time, this simulates a random roll of the dice. If the Win light or Lose light is not on, the player must push the roll button again. We will assume that the push buttons are properly debounced and that the changes in *Rb* are properly synchronized with the clock. Methods for debouncing and synchronization were discussed previously.

Figure 19-12 shows a flowchart for the dice game. After rolling the dice, the sum is tested. If it is 7 or 11, the player wins; if it is 2, 3, or 12, he loses. Otherwise, the sum is saved in the point register, and the player rolls again. If the new sum equals the point, he wins; if it is 7, he loses. Otherwise, he rolls again. After winning or losing, he must push Reset to begin a new game.

The components for the dice game shown in the block diagram (Figure 19-11) include an adder which adds the two counter outputs, a register to store the point, test logic to determine conditions for win or lose, and a control circuit. The input signals to the control circuit are defined as follows:

$D_7 = 1$ if the sum of the dice is 7

$D_{711} = 1$ if the sum of the dice is 7 or 11

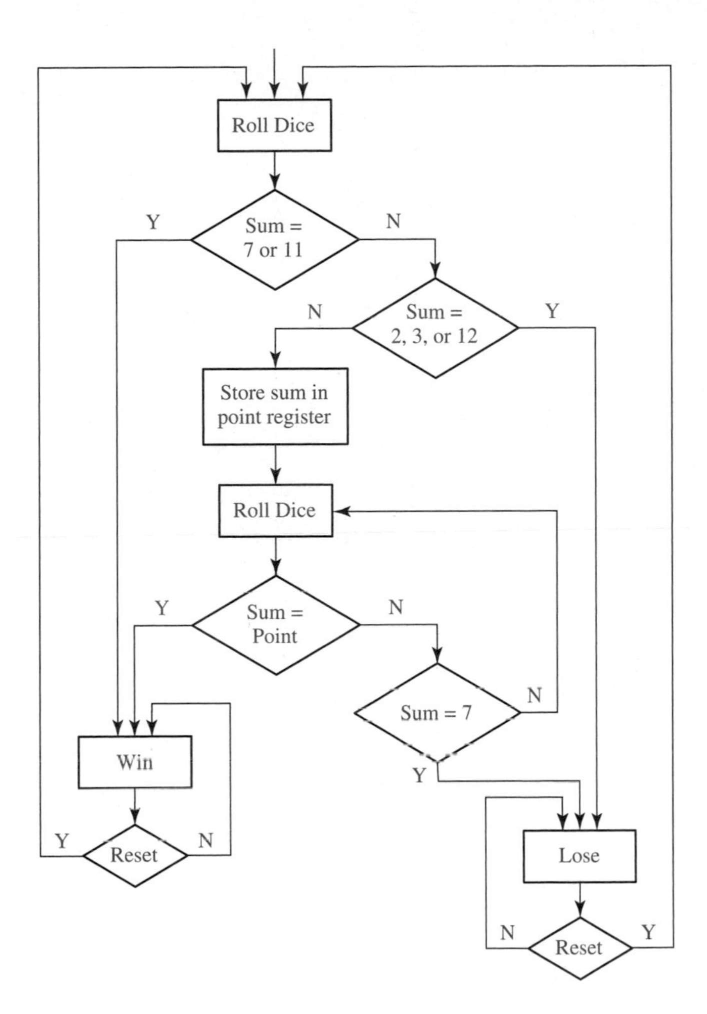


**FIGURE 19-12** Flowchart for Dice Game

$D_{2312}$ = 1 if the sum of the dice is 2, 3, or 12

$Eq$ = 1 if the sum of the dice equals the number stored in the point register

$Rb$ = 1 when the roll button is pressed

$Reset$ = 1 when the reset button is pressed

The outputs from the control circuit are defined as follows:

$Roll$ = 1 enables the dice counters

$Sp$ = 1 causes the sum to be stored in the point register

$Win$ = 1 turns on the win light

$Lose$ = 1 turns on the lose light

FIGURE 19-13
SM Chart for Dice Game

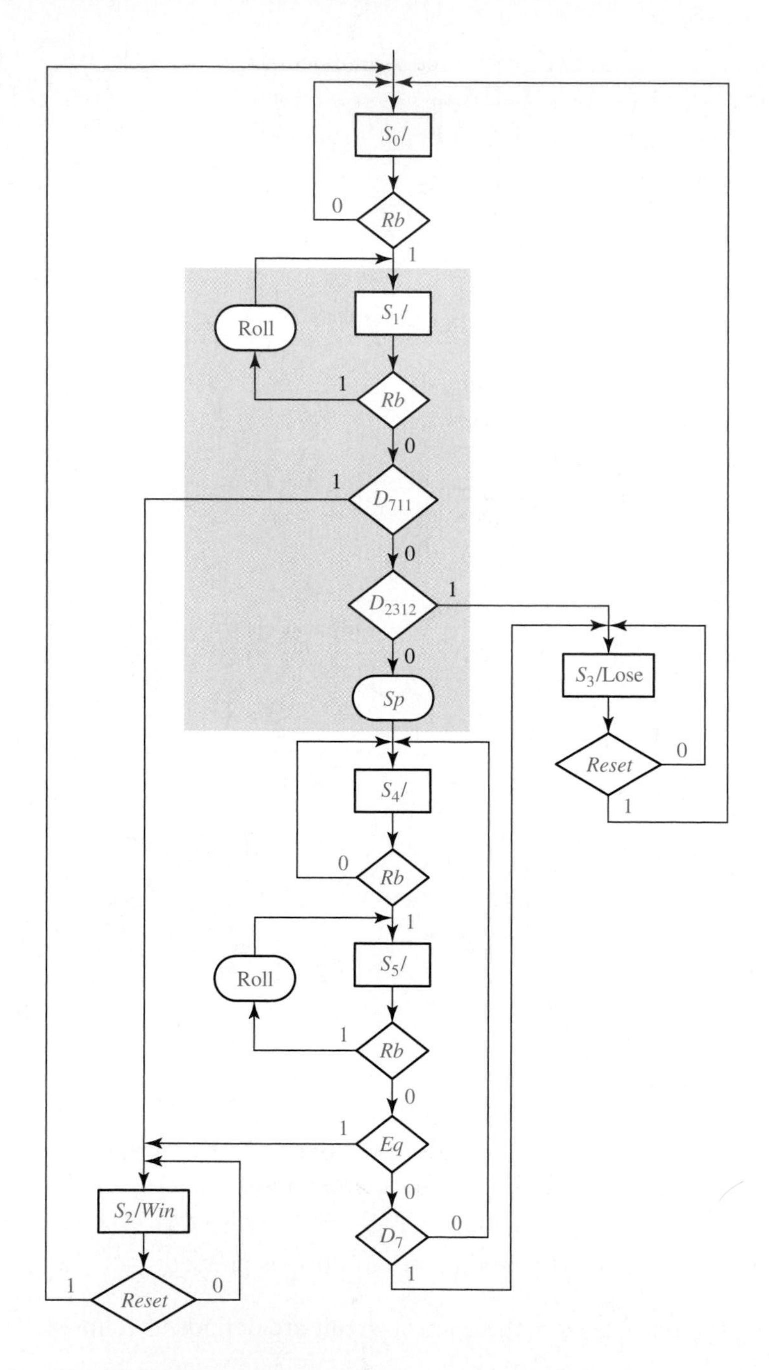


We can now convert the flowchart for the dice game to an SM chart for the control circuit using the defined control signals. Figure 19-13 shows the resulting SM chart. The control circuit waits in state $S_0$ until the roll button is pressed ($Rb = 1$). Then, it goes to state $S_1$, and the roll counters are enabled as long as $Rb = 1$. As soon as the roll button is released ($Rb = 0$), $D_{711}$ is tested. If the sum is 7 or 11, the circuit goes to state $S_2$ and turns on the Win light; otherwise, $D_{2312}$ is tested. If the sum is 2, 3, or 12, it goes to state $S_3$ and turns on the Lose light; otherwise, the signal $Sp$ becomes 1, and the sum is stored in the point register. It then enters $S_4$ and waits for the player to "roll the dice" again. In $S_5$, after the roll button is released, if $Eq = 1$, the sum equals the point and state $S_2$ is entered to indicate a win. If $D_7 = 1$, the sum is 7 and $S_3$ is entered to indicate a loss. Otherwise, the control returns to $S_4$ so that the player can roll again. When in $S_2$ or $S_3$, the game is reset to $S_0$ when the Reset button is pressed.

Instead of using an SM chart, we could construct an equivalent state graph from the flowchart. Figure 19-14 shows a state graph for the dice game controller. The state graph has the same states, inputs, and outputs as the SM chart. The arcs have been labeled consistently with the rules for proper alphanumeric state graphs given in Section 14.5. Thus, the arcs leaving state $S_1$ are labeled $Rb$, $Rb'D_{711}$, $Rb'D'_{711}D_{2312}$, and $Rb'D'_{711}D'_{2312}$. With these labels, only one next state is defined for each combination of input values. Note that the structure of the SM chart automatically defines only one next state for each combination of input values.

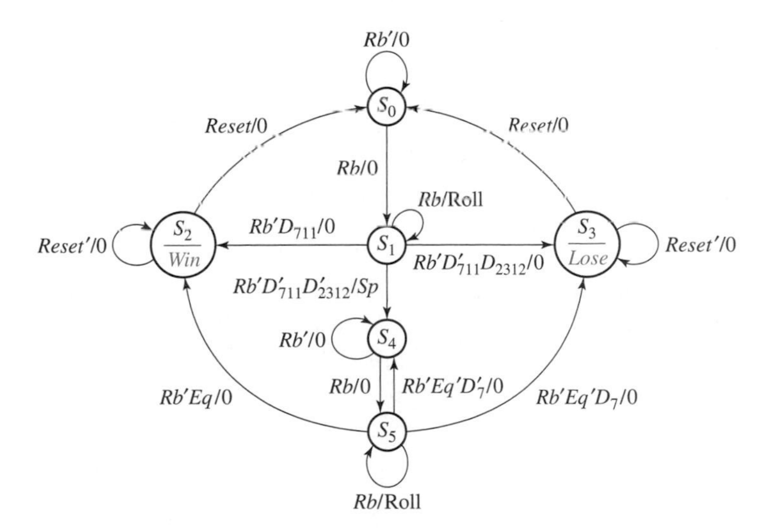


FIGURE 19-14 State Graph for Dice Game Controller

## 19.3 Realization of SM Charts

The methods used to realize SM charts are similar to the methods used to realize state graphs. As with any sequential circuit, the realization will consist of a combinational subcircuit together with flip-flops for storing the state of the circuit (see Figure 13-17). In some cases, it may be possible to identify equivalent states in an SM chart and eliminate redundant states using the same method as was used for reducing state tables. However, an SM chart is usually incompletely specified in the sense that all inputs are not tested in every state, which makes the reduction procedure more difficult. Even if the number of states in an SM chart can be reduced, it is not always desirable to do so because combining states may make the SM chart more difficult to interpret.

Before deriving next-state and output equations from an SM chart, a state assignment must be made. The best way of making the assignment depends on how the SM chart is realized. If gates and flip-flops (or the equivalent PLD realization) are used, the guidelines for state assignment given in Section 15.8 may be useful.

As an example of realizing an SM chart, consider Figure 19-7(b). We have made the state assignment $AB = 00$ for $S_0$, $AB = 01$ for $S_1$, and $AB = 11$ for $S_2$. After a state assignment has been made, output and next-state equations can be read directly from the SM chart. Because the Moore output $Z_a$ is 1 only in state 00, $Z_a = A'B'$. Similarly, $Z_b = A'B$ and $Z_c = AB$. The conditional output $Z_1 = ABX'$ because the only link path through $Z_1$ starts with $AB = 11$ and takes the $X = 0$ branch. Similarly, $Z_2 = ABX$. There are three link paths (labeled link 1, link 2, and link 3), which terminate in a state that has $B = 1$. Link 1 starts with a present state $AB = 00$, takes the $X = 1$ branch, and terminates on a state in which $B = 1$. Therefore, the next state of $B$ ($B^+$) equals 1 when $A'B'X = 1$. Link 2 starts in state 01, takes the $X = 1$ branch, and ends in state 11, so $B^+$ has a term $A'BX$. Similarly, $B^+$ has a term $ABX$ from link 3. The next-state equation for $B$ thus has three terms corresponding to the three link paths:

$$B^+ = \underbrace{A'B'X}_{\text{link 1}} + \underbrace{A'BX}_{\text{link 2}} + \underbrace{ABX}_{\text{link 3}}$$

Similarly, two link paths terminate in a state with $A = 1$, so

$$A^+ = A'BX + ABX$$

These output and next-state equations can be simplified with a Karnaugh map, using the unused state assignment ($AB = 10$) as a don't-care condition.

As illustrated, the next-state equation for a flip-flop $Q$ can be derived from the SM chart as follows:

1. Identify all of the states in which $Q = 1$.
2. For each of these states, find all of the link paths that lead into the state.
3. For each of these link paths, find a term that is 1 when the link path is followed. That is, for a link path from $S_i$ to $S_j$, the term will be 1 if the machine is in state $S_i$ and the conditions for exiting to $S_j$ are satisfied.
4. The expression for $Q^+$ (the next state of $Q$) is formed by ORing together the terms found in step 3.

Next, we will implement the multiplier control SM chart of Figure 19-10 using a PLA and two D flip-flops connected, as shown in Figure 19-15. The PLA has five inputs and six outputs. We will use a straight binary state assignment ($S_0 = 00$, $S_1 = 01$, etc.). Each row in the PLA table (Table 19-1) corresponds to one of the link paths in the SM chart. Because $S_0$ has two exit paths, the table has two rows for present state $S_0$. Because only $St$ is tested in $S_0$, $M$ and $K$ are don't-cares as indicated by dashes. The first row corresponds to the $St = 0$ exit path, so the next state is 00 and all outputs

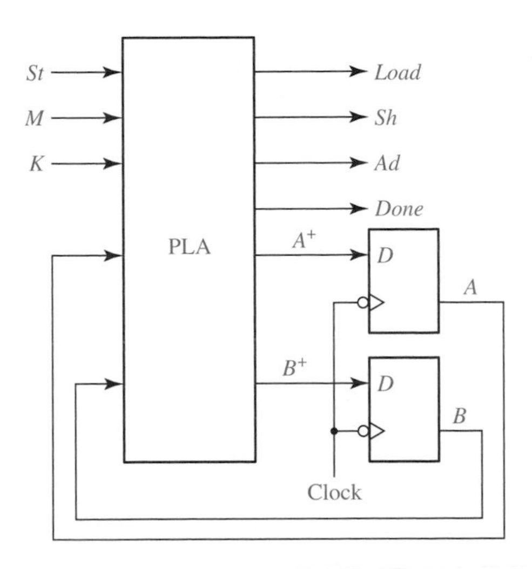


**FIGURE 19-15** Realization of Figure 19-10 Using a PLA and Flip-Flops

are 0. In the second row, $St = 1$, so the next state is 01 and the other PLA outputs are 1000. Because *St* is not tested in states $S_1$, $S_2$, and $S_3$, *St* is a don't-care in the corresponding rows. The outputs for each row can be filled in by tracing the corresponding link paths on the SM chart. For example, the link path from $S_1$ to $S_2$ passes through conditional output *Ad* when $M = 1$, so $Ad = 1$ in this row. Because $S_2$ has a Moore output *Sh*, $Sh = 1$ in both of the rows for which $AB = 10$.

The SM chart for the dice game controller can be implemented using a PLA and three D flip-flops, as shown in Figure 19-16. The PLA has nine inputs and seven outputs, which are listed at the top of Table 19-2. In state $ABC = 000$, the next state is $A^+B^+C^+ = 000$ or 001, depending on the value of *Rb*. Because state 001 has four exit paths, the PLA table has four corresponding rows. When *Rb* is 1, *Roll* is 1 and there is no state change. When $Rb = 0$ and $D_{711}$ is 1, the next state is 010. When $Rb = 0$ and $D_{2312} = 1$, the next state is 011. For the link path from state 001 to 100, *Rb*, $D_{711}$, and $D_{2312}$ are all 0, and *Sp* is a conditional output. This path corresponds to row 4 of the PLA table, which has $Sp = 1$ and $A^+B^+C^+ = 100$. In state 010, the *Win* signal is always on, and the next state is 010 or 000, depending on the value of *Reset*.

**TABLE 19-1** PLA Table for Multiplier Control

| Present State | PLA Inputs | | | | | PLA Outputs | | | | | |
|---|---|---|---|---|---|---|---|---|---|---|---|
| | $A$ | $B$ | $St$ | $M$ | $K$ | $A^+$ | $B^+$ | $Load$ | $Sh$ | $Ad$ | $Done$ |
| $S_0$ | 0 | 0 | 0 | - | - | 0 | 0 | 0 | 0 | 0 | 0 |
| | 0 | 0 | 1 | - | - | 0 | 1 | 1 | 0 | 0 | 0 |
| $S_1$ | 0 | 1 | - | 0 | 0 | 0 | 1 | 0 | 1 | 0 | 0 |
| | 0 | 1 | - | 0 | 1 | 1 | 1 | 0 | 1 | 0 | 0 |
| | 0 | 1 | - | 1 | - | 1 | 0 | 0 | 0 | 1 | 0 |
| $S_2$ | 1 | 0 | - | - | 0 | 0 | 1 | 0 | 1 | 0 | 0 |
| | 1 | 0 | - | - | 1 | 1 | 1 | 0 | 1 | 0 | 0 |
| $S_3$ | 1 | 1 | - | - | - | 0 | 0 | 0 | 0 | 0 | 1 |

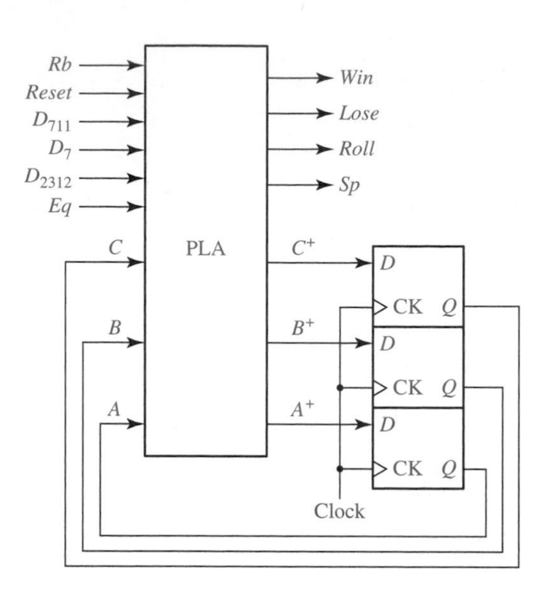


**FIGURE 19-16** PLA Realization of Dice Game Controller

**TABLE 19-2** PLA Table for Dice Game

| | *ABC* | *Rb* | *Reset* | $D_7$ | $D_{711}$ | $D_{2312}$ | *Eq* | $A^+B^+C^+$ | *Win* | *Lose* | *Roll* | *Sp* |
|---|---|---|---|---|---|---|---|---|---|---|---|---|
| 1 | 000 | 0 | – | – | – | – | – | 000 | 0 | 0 | 0 | 0 |
| 2 | 000 | 1 | – | – | – | – | – | 001 | 0 | 0 | 0 | 0 |
| 3 | 001 | 1 | – | – | – | – | – | 001 | 0 | 0 | 1 | 0 |
| 4 | 001 | 0 | – | – | 0 | 0 | – | 100 | 0 | 0 | 0 | 1 |
| 5 | 001 | 0 | – | – | 0 | 1 | – | 011 | 0 | 0 | 0 | 0 |
| 6 | 001 | 0 | – | – | 1 | – | – | 010 | 0 | 0 | 0 | 0 |
| 7 | 010 | – | 0 | – | – | – | – | 010 | 1 | 0 | 0 | 0 |
| 8 | 010 | – | 1 | – | – | – | – | 000 | 1 | 0 | 0 | 0 |
| 9 | 011 | – | 1 | – | – | – | – | 000 | 0 | 1 | 0 | 0 |
| 10 | 011 | – | 0 | – | – | – | – | 011 | 0 | 1 | 0 | 0 |
| 11 | 100 | 0 | – | – | – | – | – | 100 | 0 | 0 | 0 | 0 |
| 12 | 100 | 1 | – | – | – | – | – | 101 | 0 | 0 | 0 | 0 |
| 13 | 101 | 0 | – | 0 | – | – | 0 | 100 | 0 | 0 | 0 | 0 |
| 14 | 101 | 0 | – | 1 | – | – | 0 | 011 | 0 | 0 | 0 | 0 |
| 15 | 101 | 0 | – | – | – | – | 1 | 010 | 0 | 0 | 0 | 0 |
| 16 | 101 | 1 | – | – | – | – | – | 101 | 0 | 0 | 1 | 0 |
| 17 | 110 | – | – | – | – | – | – | --- | – | – | – | – |
| 18 | 111 | – | – | – | – | – | – | --- | – | – | – | – |

Similarly, *Lose* is always on in state 011. In state 101, $A^+B^+C^+ = 010$ if $Eq = 1$; otherwise, $A^+B^+C^+ = 011$ or 100, depending on the value of $D_7$. States 110 amd 111 are unused, so all inputs and outputs are don't-cares in these states.

If a ROM is used instead of a PLA, the PLA table must be expanded to $2^9 = 512$ rows. To expand the table, the dashes in each row must be replaced with all possible combinations of 0's and 1's. For example, row 5 would be replaced with the following 8 rows:

| | | | | | | | | | | | | | |
|---|---|---|---|---|---|---|---|---|---|---|---|---|---|
| 001 | 0 | **0** | **0** | 0 | 1 | **0** | 0 | 1 | 1 | 0 | 0 | 0 | 0 |
| 001 | 0 | **0** | **0** | 0 | 1 | **1** | 0 | 1 | 1 | 0 | 0 | 0 | 0 |
| 001 | 0 | **0** | **1** | 0 | 1 | **0** | 0 | 1 | 1 | 0 | 0 | 0 | 0 |
| 001 | 0 | **0** | **1** | 0 | 1 | **1** | 0 | 1 | 1 | 0 | 0 | 0 | 0 |
| 001 | 0 | **1** | **0** | 0 | 1 | **0** | 0 | 1 | 1 | 0 | 0 | 0 | 0 |
| 001 | 0 | **1** | **0** | 0 | 1 | **1** | 0 | 1 | 1 | 0 | 0 | 0 | 0 |
| 001 | 0 | **1** | **1** | 0 | 1 | **0** | 0 | 1 | 1 | 0 | 0 | 0 | 0 |
| 001 | 0 | **1** | **1** | 0 | 1 | **1** | 0 | 1 | 1 | 0 | 0 | 0 | 0 |

The added entries have been printed in boldface.

The dice game controller can also be realized using a PAL. The required PAL equations can be derived from Table 19-2 using the method of map-entered variables (Section 6.5) or using a CAD program such as *LogicAid*. Figure 19-17 shows maps for $A^+$, $B^+$, and *Win*. Because $A^+$, $B^+$, $C^+$, and *Rb* have assigned values in most of the rows of the table, these four variables are used on the map edges, and the remaining variables are entered within the map. $E_1$, $E_2$, $E_3$, and $E_4$ on the maps represent the expressions given below the maps.

The resulting equations are

$$A^+ = A'B'C{\cdot}Rb'D'_{711}D'_{2312} + AC' + A{\cdot}Rb + A{\cdot}D'_7\,Eq'$$
$$B^+ = A'B'C{\cdot}Rb'(D_{711} + D_{2312}) + B{\cdot}Reset' + AC{\cdot}Rb'\,(Eq + D_7)$$
$$C^+ = B'{\cdot}Rb + A'B'C{\cdot}D'_{711}D_{2312} + BC{\cdot}Reset' + AC{\cdot}D_7Eq'$$
$$Win = BC'$$
$$Lose = BC$$
$$Roll = B'C{\cdot}Rb$$
$$Sp = A'B'C{\cdot}Rb'D'_{711}D'_{2312} \qquad (19\text{-}1)$$

These equations can also be derived using *LogicAid* or another CAD program. The entire dice game, including the control circuit, can be implemented using a small CPLD or FPGA. Implementation using VHDL is described in Section 20.4.

**FIGURE 19-17** Maps Derived from Table 19-2

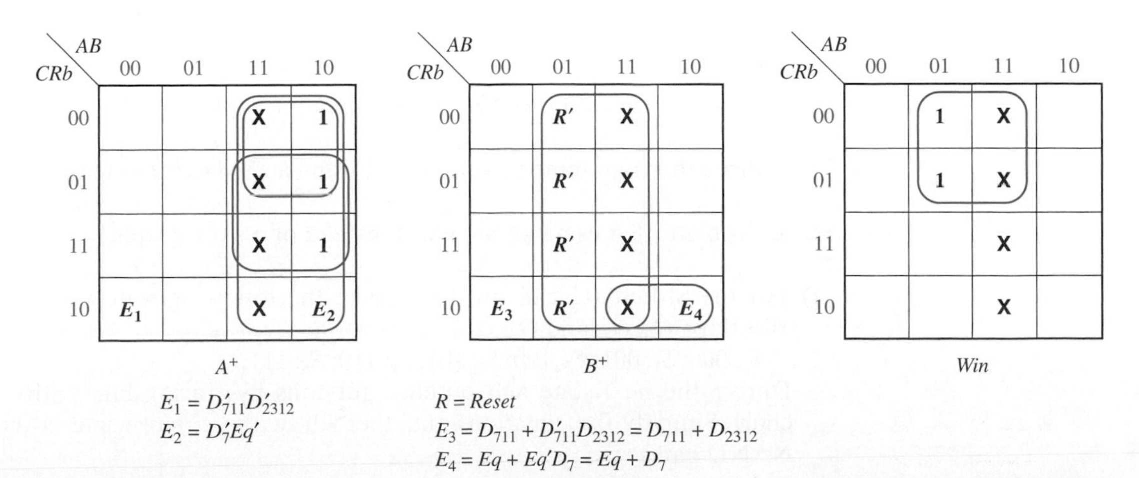

This unit has illustrated one way of realizing an SM chart using a PLA or ROM. Alternative procedures are available which make it possible to reduce the size of the PLA or ROM by adding some components to the circuit. These methods are generally based on transformation of the SM chart to different forms and encoding the inputs or outputs of the circuit.

# Problems

**19.1** Construct an SM block that has three input variables ($D$, $E$, $F$), four output variables ($P$, $Q$, $R$, $S$), and two exit paths. For this block, output $P$ is always 1, and $Q$ is 1 iff $D = 1$. If $D$ and $F$ are 1 or if $D$ and $E$ are 0, $R = 1$ and exit path 2 is taken. If ($D = 0$ and $E = 1$) or ($D = 1$ and $F = 0$), $S = 1$ and exit path 1 is taken.

**19.2** Convert the state graph of Figure 13-11 to an SM chart.

**19.3** Complete the following timing diagram for the SM chart of Figure 19-10. Assume $St = 1$.

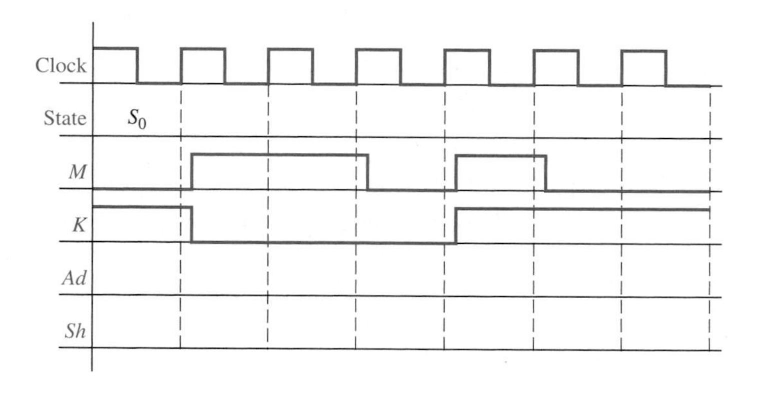


**19.4** Solve Problem 18.5 using an SM chart instead of a state graph.

**19.5** Work through Programmed Exercise 18.1 using an SM chart instead of a state graph.

**19.6** Solve Problem 18.6 using an SM chart instead of a state graph.

**19.7** (a) For the SM chart of Figure 19-9, make the following state assignment for the flip-flops $Q_0$, $Q_1$, and $Q_2$:

$S_0$: 000; $S_1$: 001; $S_2$: 100; $S_3$: 101; $S_4$: 110; $S_5$: 111.

Derive the next-state and output equations by tracing link paths on the SM chart. Simplify the equations and, then, draw the circuit using D flip-flops and NAND gates.

(b) Repeat for the SM chart of Figure 19-10, using the following state assignment for flip-flops $Q_0$ and $Q_1$: $S_0$: 00; $S_1$: 01; $S_2$:11; $S_3$: 10.

**19.8** (a) Write the next-state and output equations for the dice game by tracing link paths on the SM chart (Figure 19-13). Use a straight binary assignment.
(b) Design the block labeled "Test Logic" on Figure 19-11.

**19.9** Realize the SM chart of Figure 19-7(b) using a PLA and two D flip-flops. Draw the block diagram and give the PLA table.

**19.10** For the following SM chart:
(a) Draw a timing chart that shows the clock, the state ($S_0$, $S_1$, or $S_2$), the inputs $X_1$ and $X_2$, and the outputs. Assume that $X_3 = 0$ and the input sequence for $X_1X_2$ is 01, 00, 10, 11, 01, 10. Assume that all state changes occur on the rising edge of the clock, and the inputs change between clock pulses.
(b) Using a straight binary assignment, derive the next-state and output equations by tracing link paths. Simplify these equations using the don't-care state ($AB = 11$) and draw the corresponding circuit.
(c) Realize the chart using a PLA and D flip-flops. Give the PLA table.
(d) If a ROM is used instead of a PLA, what size ROM is required? Give the first five rows of the ROM table.

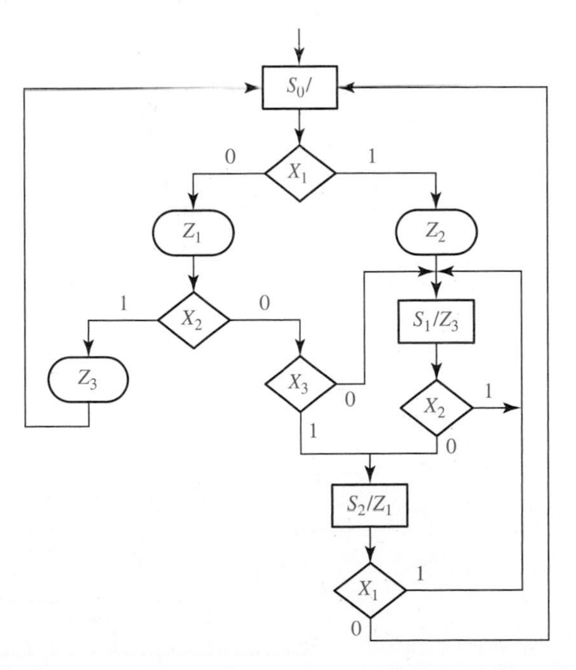


**19.11** Construct an SM block that has three input variables ($A$, $B$, and $C$), four outputs ($W$, $X$, $Y$, and $Z$), and two exit paths. For this block, output $Z$ is always 1, and $W$ is 1 iff $A$ and $B$ are both 1. If $C = 1$ and $A = 0$, $Y = 1$ and exit path 1 is taken. If $C = 0$ or $A = 1$, $X = 1$ and exit path 2 is taken.

**19.12** Convert the state graphs of Figures 14-4 and 14-6 to SM charts. Use conditional outputs for Figure 14-4.

**19.13** Convert the state graph of Figure 13-15 to an SM chart. Test only one variable in each decision box. Try to minimize the number of decision boxes.

**19.14** (a) Construct an SM chart for a Moore sequential circuit with a single input and a single output such that the output is 1 if and only if the input has been 1 for at least three consecutive clock times.
(b) Use a one-hot state assignment for the sequential circuit and derive the next-state and output equations.
(c) Make a state assignment for the sequential circuit using a minimum number of state variables and derive the next-state equation and output equations directly from the equations for the one-hot assignment.
(d) Simplify the next-state equations found in Part (c).

**19.15** (a) Construct an SM chart for the controller in Problem 18.21.
(b) Implement the controller using two D flip-flops and derive minimum two-level NAND gate logic for the flip-flop input equations and the output equations. (Assign 00 to the initial state, 01 to the state reachable from the initial state, and 11 to the third state.)
(c) Implement the controller using a one-hot state assignment. Again use D flip-flops and two-level NAND gate logic for the flip-flop input equations and the output equations.
(d) Implement the controller using two D flip-flops with a 2-to-4 decoder connected to the D flip-flops outputs and two-level NAND gate logic connected to the decoder outputs for the flip-flop input equations and the output equations. (Use the same 00, 01, 11 state assignment.)

**19.16** Convert the state graph shown in Figure 18-8 to an SM chart.

**19.17** Complete the following timing diagram for the SM chart of Figure 19-9.

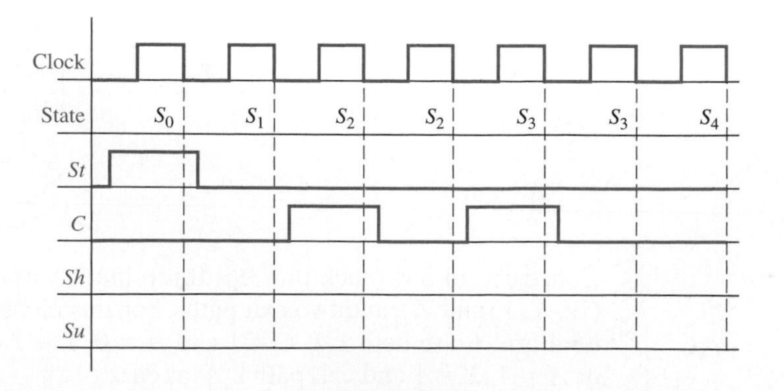

**19.18** Realize the SM chart of Figure 19-10 using a PLA and two D flip-flops. Draw the block diagram and give the PLA table. Use the same state assignment as in Problem 19.7(b).

**19.19** Work Problem 19.10 for the following SM chart and the input sequence $X_1X_2X_3 =$ 011, 101, 111, 010, 110, 101, 001.

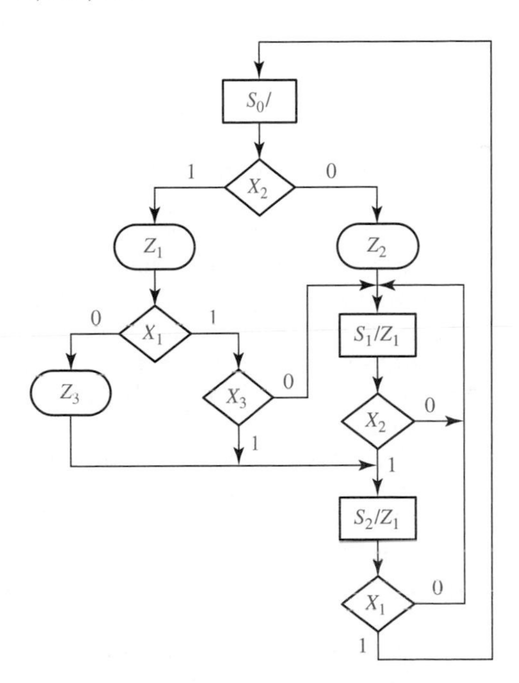


**19.20** A sequential circuit has an input ($s$) and two outputs ($z_1$ and $z_2$). When $s$ changes from 0 to 1, the circuit repeats the following pattern 12 times: $z_1\ z_2 = 10, 01$, i.e., $z_1$ is 1 for one clock period followed by $z_2$ is 1 for one clock period repeated 12 times; otherwise, both $z_1$ and $z_2$ are 0. After the 24 output patterns, the circuit waits until $s$ returns to 0, if it hasn't already, and then the operation can repeat. The sequential circuit is to be designed in two parts: (1) a four-state controller and (2) a 4-bit parallel loading counter. The counter diagram is shown. When LDN is 0, the parallel inputs are loaded into the counter. When LDN is 1 and CE is 1, the counter increments. When LDN is 1 and CE is 0, the counter does not change state. The output TC is 1 when the counter value is decimal 15.

(a) Construct an SM chart for the controller. You need to specify the signals between the controller and the counter.

(b) Using a one-hot state assignment, write the next-state and output equations for the controller.

(c) Make a state assignment for the controller using two state variables. Assign 00 to the initial state and make the other assignments so that only one variable changes during each state change. Derive the next-state and output equations for the controller.

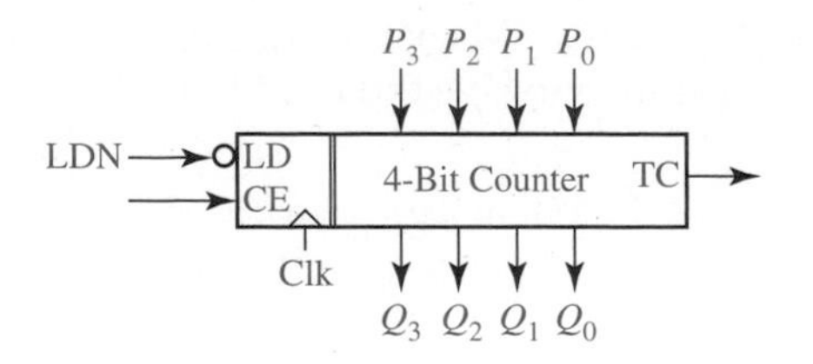


**19.21** Work Problem 18.28, Parts (a), (b) and (c), but use an SM chart instead of a state graph. For Part (d), design the controller using a minimum number of D flip-flops, a decoder, and NAND gates.

**19.22** The following circuit is a multiplier for 8-bit, unsigned (positive) numbers. When the start input ($S$) changes to 1, the multiplicand and multiplier are available on the input lines, and this signals the controller to begin the multiplication process. Upon completion, the product is available in the lower 8 bits of the 9-bit PU register combined with the 8-bit PL register. Assume $S$ remains 1 until the $D$ signal is generated. Then the circuit holds $D$ and the product until $S$ is returned to 0.

The data path portion of the circuit has the following components:

(a) 8-bit A register for holding the multiplicand
(b) 8-bit B register for holding the multiplier
(c) 9-bit PU register for accumulating the upper part of the product
(d) 8-bit PL register for the lower part of the product
(e) 8-bit adder with inputs from A reg and 8 bits of PU; sum and carry-out are inputs of PU reg; carry-in is 0

The control section contains a 3-bit counter ($C$) with an all 1's detection circuit connected to its outputs. The inputs to the controller are S, the least significant bit of $B$ (B0) and the output of the all 1's detection circuit (C1). The outputs of the controller are $D$, and all of the control signals for the registers. The control signals for the registers are

(a) LA, load A
(b) LB, load B
(c) SB, shift B right with B0 connected to the shift-in bit
(d) CC, clear C
(e) IC, increment C
(f) CP, clear PU and PL
(g) LPU, load PU
(h) SP, shift PU and PL right; the shift-in bit of PU is 0 and the shift-in bit of PL is the least significant bit of PU

Construct an SM chart for the controller using a minimum number of states. Do *not* modify the data path portion of the circuit.

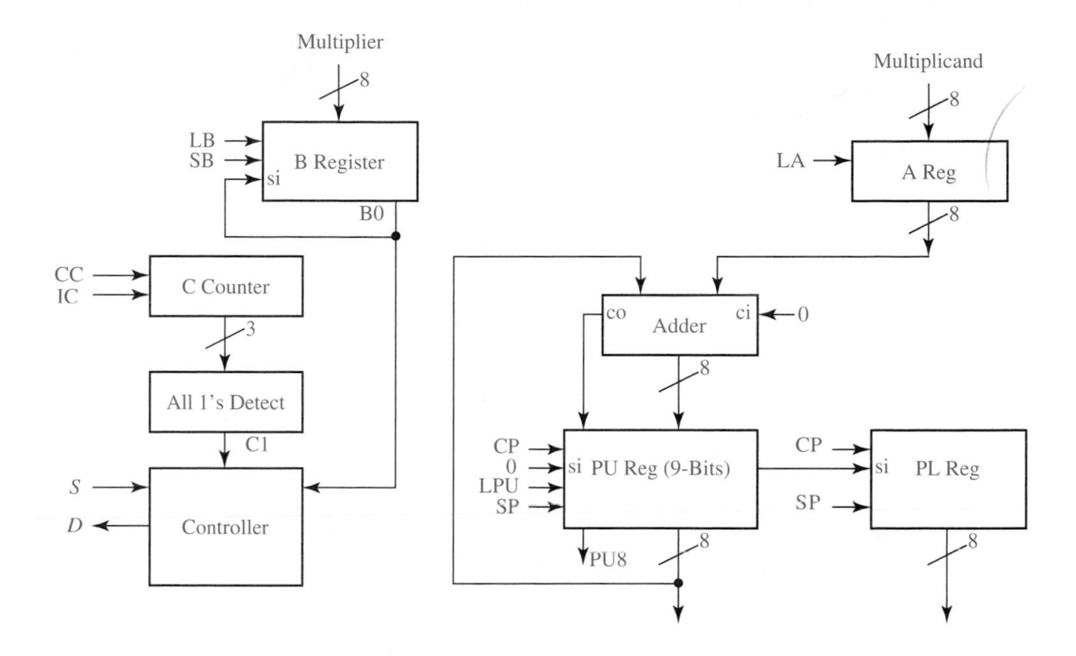


**19.23** (a) Derive an SM chart for the Ultra-Slow Divider in Problem 18.30.
(b) Realize the SM chart in (a) using a PLA and D flip-flops.

**19.24** (a) Derive an SM chart for the elevator controller in Problem 16.26.
(b) Realize the SM chart in (a) using a PLA and D flip-flops.

**19.25** Derive an SM chart for the Thunderbird taillight controller in Problem 16.27.

**19.26** (a) Derive the SM chart for the tape player controller of Problem 16.28.
(b) Realize the control circuit using a PLA and D flip-flops.

**19.27** Convert the state graph of Figure 13-9 to an SM chart.

# VHDL for Digital System Design

## Objectives

1. Given a block diagram and a state graph for a digital system's control unit of the type discussed in Unit 18, write behavioral VHDL code for the system. Use one clocked process.
2. Compile and simulate the VHDL code you wrote for Objective 1.
3. Write synthesizable VHDL code for the system using control signals. Use two processes, one for the combinational logic and one for updating registers.
4. Compile, simulate, and synthesize the VHDL code you wrote for Objective 3.
5. Write a VHDL test bench to test a VHDL module.

# Study Guide

1. Study Section 20.1, *VHDL Code for a Serial Adder*

   (a) In Figure 20-1:
   Which statements represent the full adder?

   Why are concurrent statements used for the full adder instead of a clocked process?

   Which VHDL statements are used to shift the X and Y register? Why are these statements in the clocked process?

   (b) What change is required in the VHDL code if all register updates occur on the rising edge of the clock instead of the falling edge?

2. Study Section 20.2, *VHDL Code for a Binary Multiplier*

   (a) In Figure 20-2: Why are Mplier, Mcand, and ACC declared as type std_logic_vector instead of bit_vector?

   After what state change does Done change from '1' to '0' ?

   When adding Mcand to ACC(7 **downto** 4), why is '0' concatenated to ACC (7 **downto** 4)? (line 29)

   What docs the notation **when** 2 | 4 | 6 | 8 mean? (line 34)

   (b) In Figure 20-3:
   Why is the initial value of ACC "UUUUUUUUU" ?

   When should the product be read?

(c) If the signal X, of type std_logic_vector(8 **downto** 0), is "111001101" initially, what is X after the execution of the following for loop? How long does it take?

```
for i in 5 downto 0 loop
   X <= X(7 downto 0) & X(8);
   wait for 10 ns;
end loop;
```

(d) In Figure 20-5, on lines 29 and 30, when i = 2 and Done = '1', what are the values of Mcand and Mplier? What is the value of Product if the multiplier is working properly?

(e) In Figure 20-7:
Which statement represents the adder?

Why are Load, Ad, Sh, and Done set to '0' (line 22) before the case statement?

Write a single VHDL statement that will clear ACC(8 **downto** 4) and load ACC(3 **downto** 0) with Mplier, so that lines 39 and 40 can be replaced with a single line.

Why is addout loaded into ACC in the second process instead of the first process?

In Figure 20-2 we set `Done <= '1'` in a concurrent statement (line 42), and not after `when 9 =>` on line 37 in the process. Why?

In Figure 20-7 line 33, we set `Done <= '1'` after `when 9 =>`, which is unlike what we did in Figure 20-2. Why is this correct in this case?

(f) In Figure 20-9:
When does the statement in line 22 execute?

If Sh = '1', which statements execute following a rising clock edge?

If the clock rises at $t = 10$ ns, at what time are A, B, Count, and State updated? Explain why A and B are shifted as a unit even though the statements for updating A and B execute in sequence (line 51).

(g) In Figure 20-10 at time = 60 ns, explain the contents of the registers after the rising clock edge. [*Hint*: Refer to Figure 18-9(c) to determine what happens in state 2 when K = '0'. Convert hexadecimal to binary and shift the binary before converting back to hexadecimal.] Repeat for time = 140 ns, noting that M = '1' before the rising clock edge.

(h) Read Appendix C, *Tips for Writing Synthesizable VHDL Code.*

(i) Work Problems 20.1, 20.2, 20.3, and 20.4.

**3.** Read Section 20.3, *VHDL Code for a Binary Divider*

(a) In Figure 20-11:
If Dividend(8 **downto** 4) >= Divisor, what is the value of Subout(4)?

If Dividend(8 **downto** 4) < Divisor, what is the value of Subout(4)?

Why is C equal to not Subout(4)?

(b) Work Problems 20.5, 20.6, and 20.7.

**4.** Read Section 20.4, *VHDL Code for a Dice Game Simulator.*
Work Problem 20.8.

**5.** Read Section 20.5, *Concluding Remarks.*
By looking at the VHDL code for the dice game, determine the minimum number of flip-flops required. Verify this against the value given in Table 20-1.

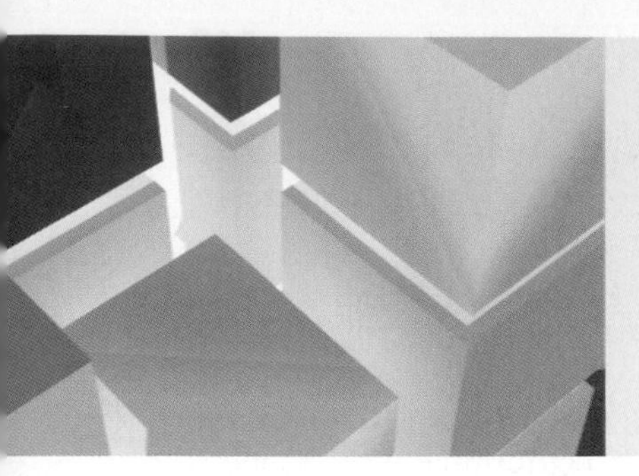

# VHDL for Digital System Design

In this chapter, we illustrate the use of VHDL in the design of digital systems. Several of the examples are based on the multiplier and divider designs developed in Unit 18. We will use VHDL to describe a digital system at the behavioral level, so we can simulate the system to check out the algorithms used and to make sure that the sequence of operations is correct. We can then define the required control signals and the actions performed by these signals. Next, we write a VHDL description of the system in terms of the control signals and verify its correct operation by simulation. We can then synthesize our design and download it to a CPLD or FPGA for final testing.

## 20.1 VHDL Code for a Serial Adder

First, we will write VHDL code that represents the serial adder with accumulator shown in Figure 18-1. The operation of the adder is explained in Section 18.1. In Figure 18-1, if $Sh = 1$, the carry from the full adder is stored in the flip-flop at the same time the registers are shifted on the falling edge of the clock.

Figure 20-1 shows VHDL code for the serial adder. Provision for loading the X and Y registers and clearing the carry flip-flop (Ci) is not included in this code; however, the VHDL simulator can be used to initialize X, Y, and Ci for testing the code. The code is based on the state graph for the controller shown in Figure 18-3. We have used two processes to represent the state machine in a manner similar to the state machine model of Figure 17–17. The first process (lines 18–28) executes whenever state or St changes, and it generates the NextState and Sh signals. The second process (lines 29–38) updates the state after the falling edge of the clock. At the same time, if Sh = '1' the registers are shifted, and the carry is stored in the flip-flop (lines 33–36). The full adder is implemented using concurrent statements for the sum and carry (lines 15–16). This is appropriate because the full adder uses combinational logic that does not require a clock. Because std_logic and std_logic vectors are used in the code, the library and use statements (lines 1 and 2) are required. These statements could be omitted if bits and bit_vectors were used instead.

FIGURE 20-1
VHDL Code for Figure 18-1

```
1   library IEEE;
2   use IEEE.STD_LOGIC_1164.all;
3   entity serial is
4     Port (St: in std_logic;
5       Clk: in std_logic;
6       Xout: out std_logic_vector(3 downto 0));
7   end serial;
8   architecture Behavioral of serial is
9     signal X, Y: std_logic_vector(3 downto 0);
10    signal Sh: std_logic;
11    signal Ci, Ciplus: std_logic;
12    signal Sumi: std_logic;
13    signal State, NextState: integer range 0 to 3;        -- 4 states
14    begin
15      Sumi <= X(0) xor Y(0) xor Ci;                        -- full adder
16      Ciplus <= (Ci and X(0)) or (Ci and Y(0)) or (X(0) and Y(0));
17      Xout <= X;
18    process (State, St)
19    begin
20      case State is
21        when 0 =>
22          if St = '1' then Sh <= '1'; NextState <= 1;
23          else Sh <= '0'; NextState <= 0; end if;
24        when 1 =>  Sh <= '1'; NextState <= 2;
25        when 2 =>  Sh <= '1'; NextState <= 3;
26        when 3 =>  Sh <= '1'; NextState <= 0;
27        end case;
28    end process;
29    process (clk)
30      begin
31      if clk'event and clk = '0' then
32      State <= NextState;                                  -- update state register
33        if Sh = '1' then
34          X <= Sumi & X(3 downto 1);                       -- shift Sumi into X register
35          Y <= Y (0) & Y(3 downto 1);                      -- rotate right Y register
36          Ci <= Ciplus; end if;                            -- store next carry
37        end if;
38    end process;
39  end Behavioral;
```

# 20.2 VHDL Code for a Binary Multiplier

In Section 18.2, we designed a multiplier for unsigned binary numbers. In this section we will show several ways of writing VHDL code to describe the multiplier operation. As indicated in Figure 18-7, 4 bits from the accumulator (ACC) and 4 bits from the multiplicand register are connected to the adder inputs; the 4 sum bits and the carry output from the adder are connected back to the accumulator. When an add signal ($Ad$) occurs, the adder outputs are loaded into the accumulator by the next clock pulse, thus, causing the multiplicand to be added to the accumulator. An extra bit at the left end of the product register temporarily stores any carry which is generated when the multiplicand is added to the accumulator. When a shift signal ($Sh$) occurs, all 9 bits of ACC are shifted right by the next clock pulse. See Section 18.2 for a more detailed explanation of the multiplier operation.

We will write a behavioral VHDL model for the multiplier (Figure 20-2) based on the block diagram of Figure 18-7 and the state graph of Figure 18-8. This model will allow us to check out the basic design of the multiplier and the multiplication algorithm before proceeding with a more detailed design. Because the control circuit has ten states, we have declared an integer in the range 0 to 9 for the state

**FIGURE 20-2** Behavioral VHDL Code for Multiplier of Figure 18-7

```
-- This is a behavioral model of a multiplier for unsigned binary numbers. It multiplies a 4-bit
-- multiplicand by a 4-bit multiplier to give an 8-bit product. The maximum number of clock
-- cycles needed for a multiply is 10.
1    library IEEE;
2    use IEEE.STD_LOGIC_1164. ALL;
3    use IEEE.STD_LOGIC_ARITH. ALL;
4    use IEEE.STD_LOGIC_UNSIGNED. ALL;

5    entity mult4X4 is
6      port (Clk, St: in std_logic;
7      Mplier, Mcand : in std_logic_vector(3 downto 0);
8      Done: out std_logic;
9      Product: out std_logic_vector (7 downto 0));
10   end mult4X4;

11   architecture behave1 of mult4X4 is
12     signal State: integer range 0 to 9;
13     signal ACC: std_logic_vector(8 downto 0);        -- accumulator
14     alias M: std_logic is ACC(0);                     -- M is bit 0 of ACC
15     begin
16       Product <= ACC (7 downto 0);
17       process (Clk)
18       begin
19         if Clk'event and Clk = '1' then               -- executes on rising edge of clock
```

FIGURE 20-2 (Continued)

```
20         case State is
21         when 0 =>                                  --initial State
22           if St = '1' then
23             ACC(8 downto 4) <= "00000";            -- clear upper ACC
24             ACC(3 downto 0) <= Mplier;             -- load the multiplier
25             State <= 1;
26           end if;
27         when 1 | 3 | 5 | 7 =>                      -- "add/shift" State
28           if M = '1' then                          -- Add multiplicand to ACC
29             ACC(8 downto 4) <= ('0'& ACC(7 downto 4)) + Mcand;
30             State <= State + 1;
31           else ACC <= '0' & ACC(8 downto 1);       -- Shift accumulator right
32             State <= State + 2;
33           end if;
34         when 2 | 4 | 6 | 8 =>                      -- "shift" State
35             ACC <= '0' & ACC(8 downto 1);          -- Right shift
36             State <= State + 1;
37         when 9 =>                                  -- end of cycle
38             State <= 0;
39         end case;
40         end if;
41       end process;
42       Done <= '1' when State = 9 else '0';
43     end behave1;
```

signal (line 12). The signal ACC represents the 9-bit accumulator output (line 13). The signals ACC, Mcand, and Mplier are declared as type std_logic_vector so that the overloaded "+" operator can be used for addition. The statement "**alias** M: std_logic **is** ACC(0);" allows us to use the name M in place of ACC(0). The product is set equal to the lower 8 bits of ACC in a concurrent statement (line 16).

Because all register operations and state changes occur on the rising edge of the clock, we will use a process that executes when Clk changes. The case statement specifies the actions to be taken in each state. In state 0, if St = '1' the multiplier is loaded into the accumulator at the same time the state changes to 1 (lines 21–26). From the state graph, we see that the same operations occur in states 1, 3, 5, and 7. The notation "**when** 1 | 3 | 5 | 7 =>" means when state is 1 or 3 or 5 or 7, the statements that follow will execute. When M = '1', the expression

'0'& ACC(7 **downto** 4) + Mcand

computes the adder output, which is loaded into ACC (lines 28–29). At the same time, the circuit goes to the next state in sequence (2, 4, 6, or 8). If M = '0', ACC is shifted to the right by loading ACC with '0' concatenated with the upper 8 bits of ACC (line 31). At the same time the state changes to 3, 5, 7, or 9 (the present state + 2). In states 2, 4, 6, or 8 ACC is shifted to the right, and state changes to the next state in sequence (lines 34–36).

The Done signal needs to be turned on only in state 9. If we had used the statement "**when** 9 => State <= 0; Done <= '1' ", Done would be turned on at the same time the State changed to 0. This is too late because we want Done to turn on when the State becomes 9. Furthermore, if Done <= '1' were included in the clocked process, a synthesizer would infer that we wanted to store Done in a flip-flop. Because we do not want to do this, we use a separate concurrent assignment statement. This statement is placed outside the process so that Done will be updated whenever the State changes.

Before continuing with the design, we will test the behavioral level VHDL code to make sure that the algorithm is correct and consistent with the hardware block diagram. At early stages of testing, we will want a step-by-step printout to verify the internal operations of the multiplier and to aid in debugging if required. When we think that the multiplier is functioning properly, we will only want to look at the final product output so that we can quickly test a large number of cases.

Figure 20-3 shows the command file and test results for multiplying 13 × 11. A clock is defined with a 20-ns period. The St signal is turned on at 2 ns and turned off one clock

**FIGURE 20-3** Command File and Simulation Results for (13 by 11)

```
-- command file to test multiplier
view list
add list CLK St State ACC done product
force st 1 2, 0 22
force clk 1 0, 0 10 –repeat 20
force Mcand 1101
force Mplier 1011
run 200
```

| ns | delta | clk | St | State | ACC | done | product |
|---|---|---|---|---|---|---|---|
| 0 | +1 | 1 | U | 0 | UUUUUUUUU | 0 | UUUUUUUU |
| 2 | +0 | 1 | 1 | 0 | UUUUUUUUU | 0 | UUUUUUUU |
| 10 | +0 | 0 | 1 | 0 | UUUUUUUUU | 0 | UUUUUUUU |
| 20 | +2 | 1 | 1 | 1 | 000001011 | 0 | 00001011 |
| 22 | +0 | 1 | 0 | 1 | 000001011 | 0 | 00001011 |
| 30 | +0 | 0 | 0 | 1 | 000001011 | 0 | 00001011 |
| 40 | +2 | 1 | 0 | 2 | 011011011 | 0 | 11011011 |
| 50 | +0 | 0 | 0 | 2 | 011011011 | 0 | 11011011 |
| 60 | +2 | 1 | 0 | 3 | 001101101 | 0 | 01101101 |
| 70 | +0 | 0 | 0 | 3 | 001101101 | 0 | 01101101 |
| 80 | +2 | 1 | 0 | 4 | 100111101 | 0 | 00111101 |
| 90 | +0 | 0 | 0 | 4 | 100111101 | 0 | 00111101 |
| 100 | +2 | 1 | 0 | 5 | 010011110 | 0 | 10011110 |
| 110 | +0 | 0 | 0 | 5 | 010011110 | 0 | 10011110 |
| 120 | +2 | 1 | 0 | 7 | 001001111 | 0 | 01001111 |
| 130 | +0 | 0 | 0 | 7 | 001001111 | 0 | 01001111 |
| 140 | +2 | 1 | 0 | 8 | 100011111 | 0 | 00011111 |
| 150 | +0 | 0 | 0 | 8 | 100011111 | 0 | 00011111 |
| 160 | +2 | 1 | 0 | 9 | 010001111 | 1 | 10001111 |
| 170 | +0 | 0 | 0 | 9 | 010001111 | 1 | 10001111 |
| 180 | +0 | 1 | 0 | 0 | 010001111 | 0 | 10001111 |
| 190 | +0 | 0 | 0 | 0 | 010001111 | 0 | 10001111 |
| 200 | +0 | 1 | 0 | 0 | 010001111 | 0 | 10001111 |

period later. By inspection of the state graph, the multiplication requires at most ten clocks, so the run time is set at 200 ns. The simulator output corresponds to the example given on page 600. Note that when Done = '1', the final product is $10001111_2 = 143$.

To thoroughly test the multiplier, we need to run additional tests, including special cases and limiting cases. Test values for the multiplicand and multiplier should include 0, maximum values, and smallest nonzero values. We will write VHDL code to test the multiplier by supplying a sequence of values for the multiplicand and multiplier. VHDL code that is written to test another VHDL module is often referred to as a *test bench*. Figure 20-4 shows how the test bench is connected to the multiplier module. The test bench generates the Clk and St signals as well as supplying values of Mplier and Mcand to the Multiplier module. The Multiplier module, in turn, sends the Done signal and the Product values back to the test bench. Using the VHDL test bench is analogous to having a hardware tester sitting on a work bench and plugging in the multiplier module into a test socket to test it.

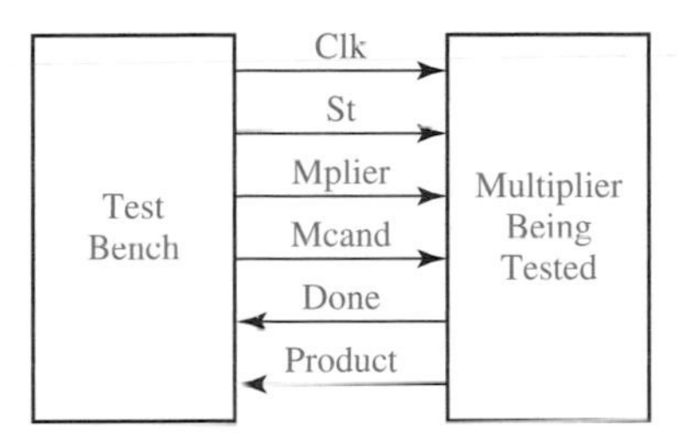


**FIGURE 20-4** Test Bench for Multiplier

We will use a **for loop** within the test bench code. The syntax for a VHDL for loop statement is

```
[loop-label:] for index in range loop
   sequential statements
end loop [loop-label];
```

The index is an integer variable that is defined only within the loop. This variable must *not* be explicitly declared because it is automatically declared by the compiler. When the loop is entered, the index is initialized to the first value in the range, and the sequential statements in the loop are executed. Then, the index is incremented (or decremented) to the next value, and the statements are executed again. This continues until the index equals the last value in the range, at which point the statements are executed for the last time and the loop exits. The for loop statement is a sequential statement that can be used within a process.

The VHDL code listing for the test bench is given in Figure 20-5. The test bench code is intended for simulation purposes only and does not have to be synthesizable. The port declaration has been omitted from the entity (lines 5–6) because we plan to use internal signals to connect the Multiplier to the test bench. The Multiplier module (mult4X4) is declared as a component within the architecture (lines 8–14). The multiplicand and multiplier test values are placed in constant arrays dimensioned 1 to N (lines 16–18). Because we are using six pairs of values, the constant N is set to 6 (line 15). The internal signals in the test bench are declared in lines 19–22. For convenience, we have used the same signal names as used in the component declaration, although we do not have to do this. At the start of the architecture body, we use a component instantiation statement to connect the Multiplier module to the test bench signals

FIGURE 20-5
Test Bench for Multiplier

```
1   library IEEE;
2   use IEEE.STD_LOGIC_1164. ALL;
3   use IEEE.STD_LOGIC_ARITH. ALL;
4   use IEEE.STD_LOGIC_UNSIGNED. ALL;
5   entity testmult is
6   end testmult;
7   architecture test1 of testmult is
8   component mult4X4
9   port (CLK: in std_logic;
10     St: in std_logic;
11     Mplier, Mcand: in std_logic_vector(3 downto 0);
12     Product: out std_logic_vector(7 downto 0);
13     Done: out std_logic);
14  end component;
15  constant N: integer:= 6;
16  type arr is array(1 to N) of std_logic_vector(3 downto 0);
17  constant Mcandarr: arr:= ("1011", "1101", "0001", "1000", "1111", "1101");
18  constant Mplierarr: arr:= ("1101", "1011", "0001", "1000", "1111","0000");
19  signal CLK: std_logic: = '0';
20  signal St, Done: std_logic;
21  signal Mplier, Mcand: std_logic_vector(3 downto 0);
22  signal Product: std_logic_vector(7 downto 0);
23  begin
24     mult1: mult4X4 port map(CLK, St, Mplier, Mcand, Product, Done);
25     CLK <= not CLK after 10 ns;   -- clock has 20 ns period
26     process
27     begin
28        for i in 1 to N loop
29           Mcand <= Mcandarr(i);
30           Mplier <= Mplierarr(i);
31           St <= '1';
32           wait until CLK = '1' and CLK'event;
33           St <= '0';
34           wait until done = '1' ;
35           wait until CLK = '1' and CLK'event;
36        end loop;
37     end process;
38  end test1;
```

(line 24). The port map lists the test signals in the same order as in the component port. The next statement generates a CLK signal with a half period of 10 ns.

The process contains a for loop that reads values from the multiplicand and multiplier arrays and then sets the start signal to '1' (lines 29–31). After the next rising clock edge, the start signal is turned off. Meanwhile, the multiplication is taking place within the Multiplier module. When the multiplication is complete, the multiplier turns on the Done signal. Done is turned off at the same time the multiplier control goes back to $S_0$. The test bench process waits for Done = '1' and then waits for the next rising clock edge before looping back to read new values of Mcand and Mplier and restart the multiplication. After N times through the loop, the test is complete.

Figure 20-6 shows the command file for executing the test bench code and the simulator output. In the add list command line, "–Notrigger Mplier Mcand product" together with "–Trigger done" causes the output to be displayed only when the Done signal changes. Without the –NOtrigger and –Trigger, the output would be displayed every time any signal on the list changed. We have annotated the simulator output to interpret the test results.

Next, we will model the same multiplier using two processes (see Figure 20-7). This model is based on Figures 17-16 and 17-17. The first process represents the

**FIGURE 20-6**
Command File and Simulation of Multiplier

```
-- Command file to test multiplier
view list
add list –NOtrigger Mplier Mcand product – Trigger done
run 1320 ns
```

| ns | +delta | mcand | mplier | product | done | |
|---|---|---|---|---|---|---|
| 0 | +0 | UUUU | UUUU | UUUUUUUU | U | |
| 0 | +1 | 1011 | 1101 | UUUUUUUU | 0 | |
| 150 | +2 | 1011 | 1101 | 10001111 | 1 | 11 × 13 = 143 |
| 170 | +2 | 1011 | 1101 | 10001111 | 0 | |
| 330 | +2 | 1101 | 1011 | 10001111 | 1 | 13 × 11 = 143 |
| 350 | +2 | 1101 | 1011 | 10001111 | 0 | |
| 470 | +2 | 0001 | 0001 | 00000001 | 1 | 1 × 1 = 1 |
| 490 | +2 | 0001 | 0001 | 00000001 | 0 | |
| 610 | +2 | 1000 | 1000 | 01000000 | 1 | 8 × 8 = 64 |
| 630 | +2 | 1000 | 1000 | 01000000 | 0 | |
| 810 | +2 | 1111 | 1111 | 11100001 | 1 | 15 × 15 = 225 |
| 830 | +2 | 1111 | 1111 | 11100001 | 0 | |
| 930 | +2 | 1101 | 0000 | 00000000 | 1 | 13 × 0 = 0 |

**FIGURE 20-7**
Two-Process VHDL Model for Multiplier

```
-- This is a behavioral model of a multiplier for unsigned binary numbers. It multiplies a 4-bit
-- multiplicand by a 4-bit multiplier to give an 8-bit product. The maximum number of clock cycles
-- needed for a multiply is 10.

1     library IEEE;
2     use IEEE.STD_LOGIC_1164. all;
3     use IEEE.STD_LOGIC_ARITH. all;
4     use IEEE.STD_LOGIC_UNSIGNED. all;
```

FIGURE 20-7 (Continued)

```
5   entity mult4X4 is
6     port (Clk, St: in std_logic;
7       Mplier, Mcand: in std_logic_vector(3 downto 0);
8       Product: out std_logic_vector(7 downto 0);
9       Done: out std_logic);
10  end mult4X4;

11  architecture control_signals of mult4X4 is
12    signal State, Nextstate: integer range 0 to 9;
13    signal ACC: std_logic_vector(8 downto 0);         -- accumulator
14    alias M: std_logic is ACC(0);                      -- M is bit 0 of ACC
15    signal addout: std_logic_vector(4 downto 0);      -- adder output including carry
16    signal Load, Ad, Sh: std_logic;
17    begin
18      Product <= ACC(7 downto 0);
19      addout <= ('0' & ACC(7 downto 4)) + Mcand;
          -- uses "+" operator from the ieee._std_logic_unsigned package
20      process(State, St, M)
21      begin
22        Load <= '0'; Ad <= '0'; Sh <= '0'; Done <= '0';
23        case State is
24        when 0 =>
25          if St = '1' then Load <= '1'; Nextstate <= 1;
26          else Nextstate <= 0; end if;
27        when 1 | 3 | 5 | 7 =>                         -- "add/shift" State
28          if M = '1' then Ad <= '1';                  -- Add multiplicand
29          Nextstate <= State + 1;
30          else Sh <= '1'; Nextstate <= State + 2; end if;
31        when 2 | 4 | 6 | 8 = >                        -- "shift" State
32          Sh <= '1'; Nextstate <= State + 1;
33        when 9 => Done <= '1'; Nextstate <= 0;
34        end case;
35      end process;

36      process (Clk)                                  -- Register update process
37      begin
38        if Clk'event and Clk = '1' then              -- executes on rising edge of clock
39          if Load = '1' then ACC(8 downto 4) <= "00000";
40            ACC(3 downto 0) <= Mplier; end if; -- load the multiplier
41          if Ad = '1' then ACC(8 downto 4) <= addout; end if;
42          if Sh = '1' then ACC <= '0' & ACC(8 downto 1); end if;
                            --Shift accumulator right
43          State <= Nextstate;
44        end if;
45      end process;
46  end control_signals;
```

combinational circuit that generates control signals and next-state information. The second process updates all of the registers on the rising edge of the clock. This model corresponds more closely to the actual hardware than the one-process model of Figure 20-2, and the control signals Ld, Sh, and Ad, as well as the adder output, appear explicitly in the code. The port declaration is the same for the two models, but the architectures are different.

Because the adder is a combinational circuit, we can define the adder output in a concurrent statement (line 19). This 5-bit output includes the 4 sum bits and the carry. It is efficient to represent the combinational part of the sequential control circuit by a process with a case statement (lines 20–35). This process executes whenever State (S) or M changes, and it computes the values of Nextstate, Load, St, Ad, and Done. The four control signals are set to '0' in line 22, and then they are set to '1' as required in the case statement. This technique avoids the necessity of setting these signals to '0' in each state and in each else clause where they are not set to '1'. At first glance, setting a signal to '0' and '1' at the same instant of time appears to be a conflict. However, when two sequential statements in a process both change the same signal at the same time, the value assigned by the second statement to execute overrides the value assigned by the first statement. The case statement determines the values of Nextstate and the control signals. For example, when state is 1, 3, 5, or 7, if M = '1', the Ad signal is turned on and the Nextstate is the present state plus 1. However, no registers can change until the next active clock edge.

All register updates occur in the second process after the rising edge of Clk. If Load = '1', Mplier is loaded into the lower ACC and the upper ACC is cleared (lines 38–40). If Ad = '1', the adder output is loaded into the upper ACC (line 41). If Sh = '1' ACC is shifted to the right (line 42). The state register is always updated (line 43).

Because the entity is the same for both multipliers, we can use the same test bench to test the second multiplier as we did for the first one, and we should obtain the same test results.

Next, we will write VHDL code for a binary multiplier that multiplies two 8-bit numbers to give a 16-bit product. For the control circuit, we will use an add-shift control with a counter, as shown in Figure 18-9, instead of using a state graph with more states. Figure 20-8 shows the block diagram for the 8 × 8 multiplier. This is of the

**FIGURE 20-8** Block Diagram for 8 × 8 Binary Multiplier

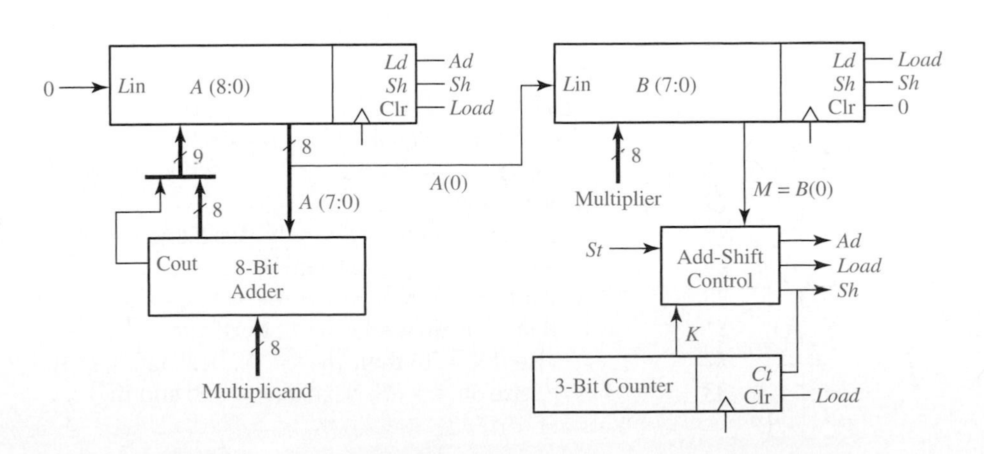

same form as Figure 18-7 except the ACC register has been split into two registers—*A* and *B*. *A* is the 9-bit accumulator register, and *B* initially holds the 8-bit multiplier. When the multiplication is complete, the 16-bit product is in A(7 **downto** 0)& B. The controller generates signals *Load*, *Sh*, and *Ad*. The *Load* signal clears *A*, loads the multiplier into *B*, and clears the shift counter. The *Sh* signal shifts both *A* and *B* together and increments the counter. The *Ad* signal loads the adder outputs into *A*.

The VHDL code for the 8 × 8 multiplier (Figure 20-9) is based on the block diagram of Figure 20-8 and the state graph of Figure 18-9(c). The entity and signal declarations are similar to those used in the previous examples except for the

**FIGURE 20-9** VHDL Code for Multiplier with Shift Counter

```
1   library IEEE;
2   use IEEE.STD_LOGIC_1164. all;
3   use IEEE.STD_LOGIC_ARITH. all;
4   use IEEE.STD_LOGIC_UNSIGNED. all;

5   entity mult8X8 is
6     Port (Clk, St: in std_logic;
7     Mplier, Mcand: in std_logic_vector(7 downto 0);
8     Done: out std_logic;
9     Product: out std_logic_vector(15 downto 0));
10  end mult8X8;

11  architecture Behavioral of mult8X8 is
12    signal State, NextState: integer range 0 to 3;
13    signal count: std_logic_vector(2 downto 0) := "000";   -- 3-bit counter
14    signal A: std_logic_vector(8 downto 0);                -- accumulator
15    signal B: std_logic_vector(7 downto 0);
16    alias M: std_logic is B(0);                            -- M is bit 0 of B
17    signal addout: std_logic_vector(8 downto 0);
18    signal K, Load, Ad, Sh: std_logic;
19    begin
20      Product <= A(7 downto 0) & B;                  -- 16-bit product is in A and B
21      addout <= '0' & A(7 downto 0) + Mcand;  -- adder output is 9 bits including carry
22      K <= '1' when count = 7 else '0';
23      process (St, State, K, M)
24      begin
25        Load <= '0'; Sh <= '0'; Ad <= '0'; Done <= '0';
            -- control signals are '0' by default
26        case State is
27        when 0 =>
28          if St = '1' then Load <= '1'; NextState <= 1;
29          else NextState <= 0; end if;
30        when 1 =>
31          if M = '1' then Ad <= '1'; NextState <= 2;
32          else if K = '0' then Sh <= '1'; NextState <= 1;
33            else Sh <= '1'; NextState <= 3; end if;
```

FIGURE 20-9 (Continued)

```
34              end if;
35          when 2 =>
36              if K = '0' then Sh <= '1'; NextState <= 1;
37              else Sh <= '1'; NextState <= 3; end if;
38          when 3 =>
39              Done <= '1'; NextState <= 0;
40          end case;
41      end process;
42        process (Clk)
43        begin
44          if Clk'event and Clk = '1' then          -- update registers on rising edge of Clk
45            if load = '1' then
46              A <= "000000000"; Count <= "000";    -- clear A and counter
47              B <= Mplier;
48            end if;                                -- load multiplier
49            if Ad = '1' then A <= addout; end if;
50            if Sh = '1' then
51              A <= '0' & A(8 downto 1); B <= A(0)& B(7 downto 1);
                      -- right shift A and B
52              count <= count + 1;                  -- increment counter
53                      -- uses "+" operator from ieee_std_logic_unsigned package
54            end if;
55            State <= NextState;
56          end if;
57      end process;
58    end Behavioral;
```

number of bits. The signal count in line 13 represents the 3-bit counter. Line 21 implements the 8-bit adder using a concurrent statement, and line 22 sets K to 1 when the count is 7. The first process (lines 23–40) represents the combinational part of the state machine. It generates control signals Ad, Load, and Sh whenever the inputs state, St, M, and K change.

To make sure that the code will synthesize properly, we have included an else clause in each **if** statement so that the NextState is properly defined, regardless whether the condition is TRUE or FALSE. For example, in lines 28–29 NextState is 1 or 0 depending on the value of St. For simulation purposes, we could omit the else clause because a VHDL signal holds its value until it is explicitly changed. However, if we omitted the else clause, most synthesizers would generate an unnecessary latch.

The second process updates the registers on the rising edge of Clk. In lines 45–48, if Load = '1', the counter is cleared when the multiplier is loaded. In lines 50–53, if Sh = '1', the counter is incremented when the A-B registers are shifted. In a clocked process, the **if** statements do not need else clauses because all registers hold their current values until changed.

FIGURE 20-10 Command File and Simulation of 8 × 8 Multiplier

```
add wave clk st state count a b done product
force st 1 2, 0 22
force clk 1 0, 0 10 –repeat 20
force mcand 00001011
force mplier 00001101
run 280
```

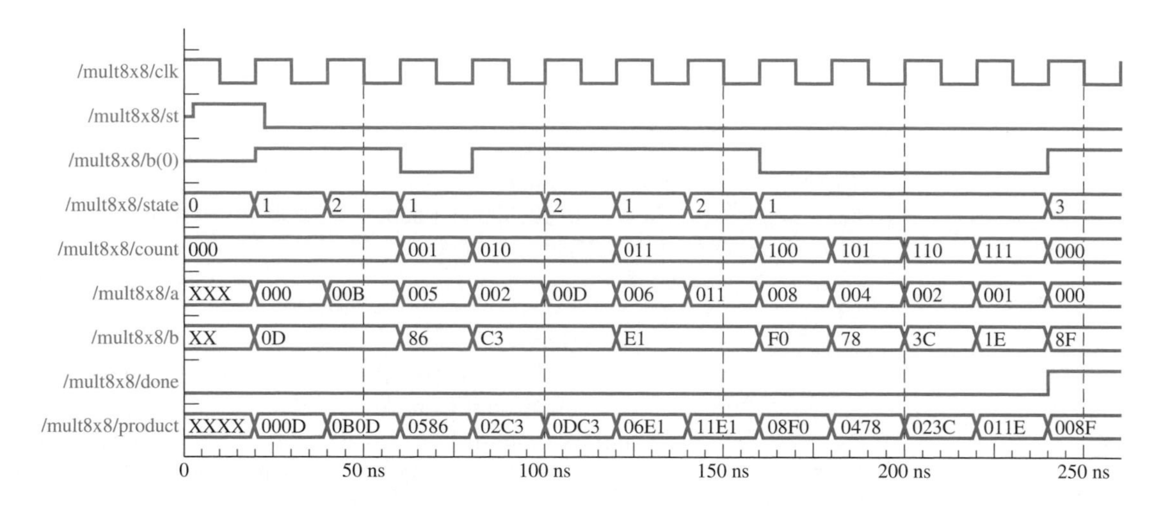


Figure 20-10 shows a simulator command file used to test the multiplier with inputs 11 × 13. The A and B register values and the product are shown in hexadecimal on the resulting waveforms. The current multiplier bit (M) is the same as b(0). Note that at time = 240 ns, the state changes to 3, the Done signal is turned on, and the final product is the correct answer, $8F_{16} = 143_{10}$.

## 20.3 VHDL Code for a Binary Divider

In Section 18.3 we designed a parallel divider for positive binary numbers that divides an 8-bit dividend by a 4-bit divisor to obtain a 4-bit quotient. Figure 20-11 shows VHDL code for the divider based on the block diagram of Figure 18-10 and

FIGURE 20-11 VHDL Code for Divider

```
1   library IEEE;
2   use IEEE.STD_LOGIC_1164. all;
3   use IEEE.STD_LOGIC_ARITH. all;
4   use IEEE.STD_LOGIC_UNSIGNED. all;

5   entity Divider is
6   Port (Dividend_in: in std_logic_vector(7 downto 0);
7      Divisor: in std_logic_vector(3 downto 0);
```

FIGURE 20-11 (Continued)

```
8      St, Clk: in std_logic;
9      Quotient: out std_logic_vector(3 downto 0);
10     Remainder: out std_logic_vector(3 downto 0);
11     Overflow: out std_logic);
12   end Divider;

13   architecture Behavioral of Divider is
14   signal State, NextState: integer range 0 to 5;
15   signal C, Load, Su, Sh: std_logic;
16   signal Subout: std_logic_vector(4 downto 0);
17   signal Dividend: std_logic_vector(8 downto 0);
18   begin
19     Subout <= Dividend(8 downto 4) – ('0' & divisor);
20     C <= not Subout (4);
21     Remainder <= Dividend(7 downto 4);
22     Quotient <= Dividend(3 downto 0);

23   State_Graph: process (State, St, C)
24     begin
25       Load <= '0'; Overflow <= '0'; Sh <= '0'; Su <= '0';
26     case State is
27       when 0= >
28         if (St = '1') then Load <= '1'; NextState <= 1;
29         else NextState <= 0; end If;
30       when 1 = >
31         if (C = '1') then Overflow <= '1'; NextState <= 0;
32         else Sh <= '1'; NextState <= 2; end if;
33       when 2 | 3 | 4 = >
34         if (C = '1') then Su <= '1'; NextState <= State;
35         else Sh <= '1'; NextState <= State + 1; end if;
36       when 5 = >
37         if (C = '1') then Su <= '1'; end if;
38         NextState <= 0;
39     end case;
40   end process State_Graph;

41   Update: process (Clk)
42     begin
43     if Clk'event and Clk = '1' then   -- rising edge of Clk
44       State <= NextState;
45       if Load = '1' then Dividend <= '0' & Dividend_in; end if;
46       if Su = '1' then Dividend(8 downto 4) <= Subout; Dividend(0) <= '1'; end if;
47       if Sh = '1' then Dividend <= Dividend(7 downto 0) & '0'; end if;
48     end if;
49   end process update;
50   end Behavioral;
```

the state graph of Figure 18-11. A concurrent statement (line 19) computes the subtracter output, subout, using an overloaded "−"operator. Then, line 20 computes C as the complement of the high order bit of the subtracter output (see Section 18.3 for justification).

The first process (lines 23–40) represents the combinational part of the sequential circuit. It computes the values of NextState and the control signals whenever state, St, or C changes. As in the other examples, line 24 sets the control signals to '0', and these signals are set to '1' as required within the case statement. The second process (lines 41–49) updates the state and dividend registers on the rising edge of the clock. If Ld = '1', the 9-bit dividend register is loaded with '0' followed by the 8-bit dividend (line 45). If Su = '1', the subtracter output is loaded into the upper part of the dividend register and the quotient bit is set to '1' (line 46). If Sh = '1', the dividend register is shifted left (line 47).

## 20.4 VHDL Code for a Dice Game Simulator

In this section we will write behavioral VHDL code for the dice game described in Section 19.2. The code in Figure 20-12 is based on the block diagram for the DiceGame Module in Figure 19-11 and the SM chart of Figure 19-13. The two counters and the adder will be placed in a separate module, so the input to this module is the sum of the two counters, which represents the roll of the dice. This sum must be in the range 2 to 12 as declared in line 3. The Point register is a signal with the same range (line 8). We will use a two-process model for the dice game. The first process represents the combinational logic for the controller. Whenever the inputs Rb, Reset, Sum, or State change, this process computes new values for NextState, for the control signals (Sp and Roll), and for the outputs (Win and Lose). The case statement tests the state, and in each state nested if-then-else (or **elsif** ) statements implement the conditional tests. In State 1 the Roll signal is turned on when Rb is 1. If all conditions test FALSE, Sp is set to 1, and the next state is 4. In the second process, the state is updated after the rising edge of the clock (line 38), and if Sp is 1, the sum is stored in the point register (line 39).

FIGURE 20-12
VHDL Code for Dice Game Controller

```
1   entity DiceGame is
2     port (Rb, Reset, CLK: in bit;
3       Sum: in integer range 2 to 12;
4       Roll, Win, Lose: out bit);
5   end DiceGame;

6   architecture DiceBehave of DiceGame is
7     signal State, NextState: integer range 0 to 5;
8     signal Point: integer range 2 to 12;
9     signal Sp: bit;
10  begin
```

FIGURE 20-12 (Continued)

```
11       process(Rb, Reset, Sum, State)
12         begin
13           Sp <= '0'; Roll <= '0'; Win <= '0'; Lose <= '0';
14           case State is
15             when 0 => if Rb = '1' then NextState <= 1; else NextState <= 0; end if;
16             when 1 =>
17               if Rb = '1' then Roll <= '1'; NextState <= 1;
18               elsif Sum = 7 or Sum = 11 then NextState <= 2;
19               elsif Sum = 2 or Sum = 3 or Sum = 12 then NextState <= 3;
20               else Sp <= '1'; NextState <= 4;
21               end if;
22             when 2 => Win <= '1';
23               if Reset = '1' then NextState <= 0; else NextState <= 2; end if;
24             when 3 => Lose <= '1';
25               if Reset = '1' then NextState <= 0; else NextState <= 3; end if;
26             when 4 => if Rb = '1' then NextState <= 5; else NextState <= 4; end if;
27             when 5 =>
28               if Rb = '1' then Roll <= '1'; NextState <= 5;
29                 elsif Sum = Point then NextState <= 2;
30                 elsif Sum = 7 then NextState <= 3;
31                 else NextState <= 4;
32               end if;
33           end case;
34         end process;
35         process(CLK)
36         begin
37           if CLK'event and CLK = '1' then
38             State <= NextState;
39             if Sp = '1' then Point <= Sum; end if;
40           end if;
41         end process;
42     end DiceBehave;
```

To complete the VHDL implementation of the dice game we will add a module with two counters, which count from 1 to 6, and an adder as shown in Figure 20-13. The counters are initialized to 1 so that the sum of the two dice will always be in the range 2 through 12. When Cnt1 is in state 6, the next clock sets it to state 1, and Cnt2 is incremented (or Cnt2 is set to 1 if it is in state 6). The concurrent statement in line 19 implements the adder.

The main module shown in Figure 20-14 connects the DiceGame and Counter modules together. The architecture starts with two component declarations (lines 6–14). The internal signals that connect the two modules, roll1 and sum1, are declared in lines 15 and 16. The two components are instantiated in lines 18 and 19. These statements connect the two components to each other and to the port signals.

FIGURE 20-13
Counter Module for Dice Game

```
1   entity Counter is
2   port(Clk, Roll: in bit;
3     Sum: out integer range 2 to 12);
4   end Counter;

5   architecture Count of Counter is
6   signal Cnt1,Cnt2: integer range 1 to 6 := 1;
7   begin
8     process (Clk)
9     begin
10      if Clk'event and Clk = '1' then
11        if Roll = '1' then
12          if Cnt1 = 6 then Cnt1 <= 1; else Cnt1 <= Cnt1 + 1; end if;
13          if Cnt1 = 6 then
14            if Cnt2 = 6 then Cnt2 <= 1; else Cnt2 <= Cnt2 + 1; end if;
15          end if;
16        end if;
17      end if;
18  end process;
19  Sum <= Cnt1 + Cnt2;
20  end Count;
```

FIGURE 20-14
Main Module for Dice Game

```
1   entity Game is
2     port (Rb, Reset, Clk: in bit;
3       Win, Lose: out bit);
4   end Game;

5   architecture Play1 of Game is
6     component Counter
7       port(Clk, Roll: in bit;
8         Sum: out integer range 2 to 12);
9     end component;
10    component DiceGame
11      port (Rb, Reset, Clk: in bit;
12        Sum: in integer range 2 to 12;
13        Roll, Win, Lose: out bit);
14    end component;
15  signal roll1: bit;
16  signal sum1: integer range 2 to 12;
17  begin
18    Dice: Dicegame port map(Rb, Reset, Clk, sum1, roll1, Win, Lose);
19    Count: Counter port map(Clk, roll1, sum1);
20  end Play1;
```

# 20.5 Concluding Remarks

Except for the test bench, all of the VHDL code in this chapter is synthesizable. The synthesis results depend on the target device and synthesizer that is used. Most synthesizers offer the choices of optimizing for area, for speed, or for something in between. Optimizing for area implies fewer macrocells or function generators are used, resulting in a smaller area used on the IC chip. Optimizing for speed means reducing the delay times along the various paths so that a higher clock speed may be used. This often results in using more components and a larger chip area.

Table 20-1 shows some typical synthesis results for five VHDL code examples from this chapter when the optimize for area option was chosen. Results shown here are for Xilinx CoolRunner CPLDs and for the Xilinx Spartan and Spartan II FPGAs. The Xilinx XST synthesizer was used for CoolRunner, and the FPGA Express synthesizer was used for Spartan. In all cases, the number of flip-flops is minimum and the same for the different devices. For CPLDs, the most important factors in determining the required chip area are the number of macrocells and the number of product terms, and the optimizer attempts to minimize these. For FPGAs, the optimizer attempts to reduce chip area by minimizing the required number of logic cells (CLBs, or slices). Each CLB or slice contains two four-input function generators (also called lookup tables or LUTs) and two flip-flops. Most designs require more function generators than flip-flops, so a key factor in optimizing for area is to reduce the number of four-input function generators (LUTs).

In this text we have introduced the basic VHDL features needed to write synthesizable code. In most examples, we have related the VHDL code to the actual hardware that it represents. In Unit 10, we used concurrent statements to represent combinational logic. In Unit 17, we used sequential statements in a process to represent sequential logic and also to represent combinational logic. In this chapter we wrote VHDL code to describe small synchronous digital systems based on their block diagrams and state graphs.

In the example of Figure 20-2, we wrote a behavioral model for a multiplier using a single process to update the state and the registers on the rising clock edge. When a single process is used, it is often necessary to add concurrent statements for the combinational outputs (the Done signal, for example) to assure proper timing. The two-process model, used in the example of Figure 20-7, is closer to the actual

**TABLE 20-1** Synthesis Results (Optimized for Area)

| Device | | Multiplier Fig. 20-2 | Multiplier Fig. 20-7 | Multiplier Fig. 20-9 | Divider Fig. 20-11 | Dice Game Fig. 20-12 + |
|---|---|---|---|---|---|---|
| | Flip-Flops | 13 | 13 | 22 | 12 | 13 |
| CoolRunner CPLD | Macrocells | 18 | 19 | 32 | 18 | 24 |
| | Product terms | 63 | 61 | 108 | 70 | 72 |
| Spartan FPGA | 4-Input LUTs | 38 | 32 | 36 | 23 | 31 |
| | CLBs | 20 | 18 | 19 | 14 | 16 |
| Spartan II FPGA | 4-Input LUTs | 30 | 30 | 35 | 30 | 30 |
| | Slices | 16 | 15 | 19 | 16 | 19 |

hardware in that it explicitly generates control signals in a combinational process and then uses these signals to control register updates in a clocked process. We generally prefer the two-process model because it introduces fewer timing problems. This is particularly important in large systems where the operation of a number of modules must be properly coordinated.

When writing VHDL code for synthesis, you must constantly keep in mind that you are designing hardware, not simply writing a computer program. Every VHDL statement that you write implies certain hardware. Poorly written VHDL code may result in excessive amounts of hardware when synthesized, and the hardware may malfunction because of timing problems. Simulation plays an important role in digital design using VHDL. Functional simulation before synthesis is important to make sure that the hardware performs the intended functions and that the basic design is sound. However, just because the code simulates correctly does not mean that the code will synthesize and implement correctly. Review of the reports generated by the synthesizer may reveal problems such as generation of unintended latches. After the code is implemented, a timing simulation of the actual hardware is desirable. This type of simulation may reveal timing problems in the design, and it will help to determine the maximum clock speed. Debugging using a simulator is generally much easier than using the actual hardware because the internal signals within the hardware are generally not available for observation.

Appendix B summarizes the syntax for all VHDL statements used in the text. VHDL has many other features that are not discussed in this text. VHDL variables, as distinguished from signals, have not been introduced because VHDL code using variables may have timing problems when synthesized. Other useful features of VHDL include procedures, functions, attributes, generics, and generate. These features are described in references [1], [2], [3], [13], and [14].

# Problems

**20.1** In Figure 20-7, if St changes from '0' to '1' at time 2 ns, and a rising edge of Clk occurs at 10 ns, in what sequence do the VHDL statements execute? (*Hint*: The first process executes more than one time.)

**20.2** Write VHDL code for the 16-bit 2's complementer described in Programmed Exercise 18.1. Use two processes.

**20.3** Modify the VHDL code of Figure 20-7 to implement the multiplier of Problem 18.5. You may refer to the answer to Problem 18.5 for the state graph of the control unit.

**20.4** Write a test bench to test the BCD-to-excess-3 code converter of Table 17-2. Test all 10 BCD digits in order, using an input stream consisting of a single constant vector (which should begin "000010000100 . . ."). Note that the order of bits is least significant bit first, as in Section 16.2. (Table 16-3 is the same as Table 17-2, but with the states named differently.) Define an expected output vector ("110000101010..."). Set an error flag to '1' if the actual output does not match the expected output.

**20.5** For the following VHDL code, draw a block diagram of the corresponding hardware and a state graph for the controller. If MplierData is 0101 and McandData is 1001 at the first clock edge when Start is 1, how many clock cycles will it take for Done to become 1, and what will the value of Product be when Done becomes 1?

```
library IEEE;
use IEEE.STD_LOGIC_1164.ALL;
use IEEE.STD_LOGIC_ARITH.ALL;
use IEEE.STD_LOGIC_UNSIGNED.ALL;

entity olorin is
  Port ( Clk, Start: in std_logic;
    McandData, MplierData: in std_logic_vector(3 downto 0);
    Done: out std_logic;
    Product: out std_logic_vector(7 downto 0));
end olorin;

architecture Behavioral of olorin is
  signal Init, K, Add: std_logic;
  signal Sum, Accumulator: std_logic_vector(7 downto 0);
  signal Mcand, Mplier: std_logic_vector(3 downto 0);
  signal State, NextState: integer range 0 to 2;
begin
  Sum <= Accumulator + Mcand;
  K <= not Mplier(3) and not Mplier(2) and not Mplier(1) and not Mplier(0);
  Product <= Accumulator;

Process(State, Start, K)
begin
  Init <= '0'; Add <= '0'; Done <= '0';
  case state is
  when 0 =>
    if Start = '1' then Init <= '1'; NextState <= 1;
    else NextState <= 0; end if;
  when 1 =>
    if K = '1' then Done <= '1'; NextState <= 2;
    else Add <= '1'; NextState <= 1; end if;
  when 2 =>
    if Start = '1' then Done <= '1'; NextState <= 2;
    else NextState <= 0; end if;
  end case;
end process;
```

```
Process(Clk)
begin
   if Clk'event and Clk = '1' then
      State <= NextState;
      If Init = '1' then Mcand <= McandData; Mplier <= MplierData;
                     Accumulator <= "00000000"; end if;
      If Add = '1' then Accumulator <= Sum; Mplier <= Mplier – 1; end if;
   end if;
end process;
end Behavioral;
```

---

**20.6** A digital system consists of three registers and two adders, as shown in the following figure. An input bus is used to load the registers in sequence $A$, $B$, and $C$. The sum of $A$, $B$, and $C$ is then loaded into $A$. Write VHDL code that describes the system.

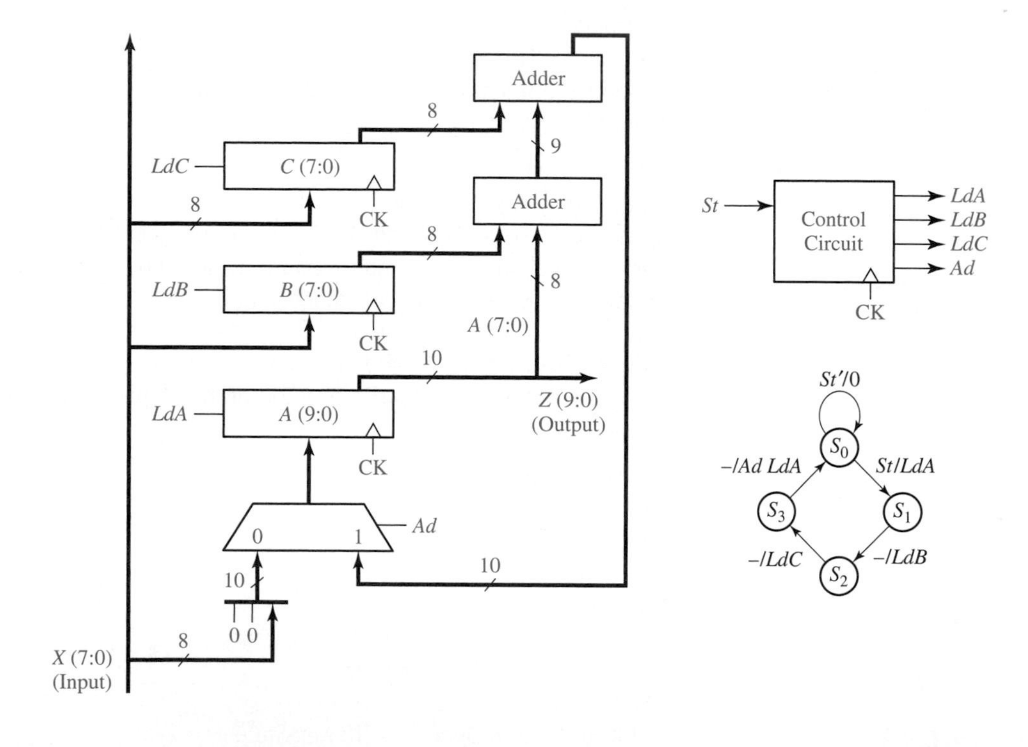


**20.7** Modify the VHDL code of Figure 20-11 to use a counter as in Figure 20-9. You may refer to the answer to Problem 18.6 for the state graph for the control unit.

**20.8** Write a test bench for the DiceGame controller of Figure 20-12. Use the following test sequence for sum: 7, 11, 2, 4, 7, 5, 6, 7, 6, 8, 9, 6.

**20.9** Modify the VHDL code of Figure 20-11 to implement the divider of Problem 18.7.

**20.10** Consider the multiplier of Problem 18.28.
(a) Write VHDL code that describes the multiplier.
(b) Write a test bench that tests the code of Part (a). The test cases should include at least the following: zero multiplier, maximum multiplier and maximum multiplicand, and maximum multiplier and zero multiplicand.

**20.11** Repeat Problem 20.10 for the multiplier of Problem 18.29. In addition to the test cases listed in Problem 20.10, the test cases should include combinations of maximum and minimum (signed) values of the multiplicand and multiplier.

## Lab Design Problems

Each of these problems is designed to fit on a small CPLD or FPGA circuit board that has at least eight input switches, two pushbuttons, and eight LEDs. Carry out the following steps for your assigned digital system design problem:

1. Draw a block diagram of the system showing registers, adders, MUXes, and other components. Define the necessary control signals. Specify the sizes of registers, adders, etc. Provide an active-high asynchronous reset for your design.
2. Draw a state graph for the control circuit.
3. Based on the results of Steps 1 and 2, write a behavioral VHDL description of the system. Use one clocked process to update the state and the registers as in Figure 20-2. Compile and simulate your code.
4. Based on the results of Steps 1 and 2, write a VHDL description of the system using control signals and two processes as in Figure 20-7. Use one combinational process to generate the next-state and control signals. Use a clocked process to update the state and other registers. Compile and simulate your code.
5. Synthesize your VHDL code from Step 4, download it to a CPLD or FPGA board, test it, and then demonstrate its operation.

**20.A** Design a divider for unsigned binary numbers that will divide a 7-bit dividend by a 4-bit divisor to give a 3-bit quotient. Assume that the start signal (St) is 1 for exactly one clock time. When St = 1, the dividend register should be loaded from the input bus. On the next clock cycle, the divisor register should be loaded from the same input bus. Then, if the quotient would require more than 3 bits, an overflow would occur, so V should be set to 1, and the controller should go back to the reset state. Otherwise, the controller should generate the appropriate sequence of shift and subtract signals and turn on a done signal when division is complete. Use an 8-bit dividend register and store the quotient in the lower 3 bits of the register.

**20.B** Same as 20.A, except use a 3-bit divisor and a 4-bit quotient. (An overflow would occur if the quotient would require more than 4 bits.)

**20.C** Same as 20.A, except use an 8-bit dividend, a 3-bit divisor, a 5-bit quotient, and a 9-bit dividend register. (An overflow would occur if the quotient would require more than 5 bits.)

**20.D** Same as 20.A, except use an 8-bit dividend, a 5-bit divisor, a 3-bit quotient, and a 9-bit dividend register. (An overflow would occur if the quotient would require more than 3 bits.)

**20.E** Design a multiplier for unsigned binary numbers that will multiply a 3-bit multiplicand by a 4-bit multiplier to give a 7-bit product. Assume that the start signal (St) is 1 for exactly one clock time. When St = 1, the multiplier register should be loaded. After loading the multiplier, load the multiplicand into a separate register on the next clock, and then proceed with the multiplication. Both the multiplier and multiplicand should come from the same input bus. Inputs to this bus should come from switches on the FPGA board. Use an 8-bit accumulator register. The controller should generate the appropriate sequence of add and shift signals and turn on a done signal when multiplication is complete.

**20.F** Same as 20.E except use a 4-bit multiplicand, a 3-bit multiplier, and a 7-bit product.

**20.G** Same as 20.E except use a 5-bit multiplicand, a 3-bit multiplier, an 8-bit product, and a 9-bit accumulator.

**20.H** Same as 20.E except use a 3-bit multiplicand, a 5-bit multiplier, an 8-bit product, and a 9-bit accumulator.

**20.I** Design an 8-bit serial adder with accumulator for signed binary numbers similar to Figure 18-1, except provide for loading the registers and clearing the carry flip-flop. Represent signed negative numbers in 2's complement. Assume that the start signal (St) is 1 for exactly one clock time. When St = 1, the accumulator register should be parallel loaded from a bus. Then, at the next clock the addend register should be loaded from the same bus. When addition is completed, output a Done signal for one clock time. Output an overflow signal if a 2's complement overflow occurs. Design the control circuit using a 3-bit counter and a state graph with four states.

**20.J** Same as 20.I except change 8-bit to 7-bit.

**20.K** Same as 20.I except design a serial subtracter instead of an adder.

**20.L** Same as 20.I except design a serial subtracter instead of an adder and change 8-bit to 7-bit.

**20.M** Design a divider for unsigned binary numbers that divides a 16-bit dividend by an 8-bit divisor to give an 8-bit quotient. Use a 17-bit dividend register and store the quotient in the lower 8 bits of the register. Also, use a 4-bit counter to count the number of shifts, together with a subtract-shift controller.

The following instructions only apply to 20.N, 20.O, 20.P and 20.Q:

1. Use an active-high asynchronous reset to reset the circuit at any time.
2. When you press start and then clock the circuit, the multiplicand should be loaded in some internal register.

3. On the next clock cycle, the multiplier should be loaded into another internal register.
4. Note that both the multiplicand and the multiplier should be loaded from the *same* 8 switches on the board. Use the least significant bits of the 8 switches to enter the multiplicand and the multiplier.
5. Once they are loaded, the circuit should cycle through the states until the final answer is calculated.
6. Once the product is calculated, the state should not change, and a done signal should be set to high (and remain high).

**20.N** Design a multiplier for unsigned binary numbers that will multiply a 6-bit multiplicand by a 7-bit multiplier to give a 13-bit result. Assume that the start signal (St) is 1 for exactly one clock time. When St = 1, the multiplier and multiplicand should be loaded in sequence. Use a 14-bit accumulator. The controller should generate the appropriate sequence of add and shift signals.

**20.O** Work Problem 20.N except use a 7-bit multiplicand and a 6-bit multiplier.

**20.P** Work Problem 20.N except use a 8-bit multiplicand and a 5-bit multiplier.

**20.Q** Work Problem 20.N except use a 5-bit multiplicand and a 8-bit multiplier.

**20.R** Design a divider for unsigned binary numbers that will divide a 6-bit dividend by a 4-bit divisor to give a 6-bit quotient. An asynchronous reset must be used to reset the circuit. Assume that the start signal (St) is 1 for exactly one clock cycle time. When St = 1, the dividend should be loaded from the input bus. On the next clock cycle the divisor should be loaded from the same input bus. Then if the divisor is 0, an overflow will occur, the V signal should be set and the controller should go back to the reset state. Otherwise, the controller should generate the appropriate sequence of shift and subtract signals and then turn on a done signal. Use an *11-bit* dividend register and store the quotient in the lower bits. You may consult Problem 18.7 as a reference. You need to show only the quotient on the FPGA LEDs.

**20.S** Work Problem 20.R except use a 7-bit dividend and a 3-bit divisor to give a 7-bit quotient.

**20.T** Design an arithmetic unit that computes $W = X*Y + Z$, where $X$, $Y$ and $Z$ are all 4-bit unsigned numbers. $X$, $Y$ and $Z$ should be read sequentially from the same input bus. Assume that the start signal (St) is 1 exactly for one clock cycle. When St is '1', in the first clock cycle, the Multiplier ($X$) should be loaded from the bus. In the second clock cycle, the Multiplicand ($Y$) should be loaded from the same bus. Finally, in the third clock cycle, Z (the term to be added) should be loaded. Then the state machine should multiply $X$ by $Y$. Use a 9-bit accumulator, and design the multiplier without using a counter. Use the overloaded addition operator to add. Use a second adder to add $Z$ to $X*Y$ and store the result in the accumulator using a fourth load signal.

**20.U** Same as Problem 20.T, except that $X$ is a 5-bit number, $Y$ is a 3-bit number, and $Z$ is a 5-bit number. Use a 9-bit accumulator.

**20.V** Same as Problem 20.T, except that $X$ is a 3-bit number, $Y$ is a 5-bit number, and $Z$ is a 5-bit number. Use a 9-bit accumulator.

**20.W** Same as Problem 20.T, except that $X$ is a 6-bit number, $Y$ is a 2-bit number, and $Z$ is a 6-bit number. Use a 9-bit accumulator.

# MOS and CMOS Logic

Most integrated circuits designed today use MOS or CMOS logic. MOS logic is based on the use of MOSFETs (metal-oxide-semiconductor field-effect transistors) as switching elements. Figure A-1 shows the symbols used to represent MOSFETs. The substrate (or body) is a thin slice of silicon. The gate is a thin metallic layer deposited on the substrate and insulated from it by a thin layer of silicon dioxide. A voltage applied to the gate is used to control the flow of current between the drain and source.

In normal operation of an *n*-channel MOSFET, shown in Figure A-1(a), a positive voltage ($V_{DS}$) is applied between the drain and source. If the gate voltage ($V_{GS}$) is 0, there is no channel between the drain and source and no current flows. When $V_{GS}$ is positive and exceeds a certain threshold, an *n*-type channel is formed between the drain and source, which allows current to flow from $D$ to $S$. Operation of a *p*-channel MOSFET is similar, except $V_{DS}$ and $V_{GS}$ are negative. When $V_{GS}$ assumes a negative value less than the threshold, a *p*-type channel is formed between drain and source, which allows current to flow from $S$ to $D$.

The symbol in Figure A-1(c) may be used to represent either a *p*- or *n*-channel MOSFET. When this symbol is used, it is generally understood that the substrate is connected to the most positive circuit voltage for *p*-channel MOSFETs (or the most negative for *n*-channel). If the power supply voltage is $V_{DD}$, we will use positive logic

**FIGURE A-1**
MOSFET Symbols

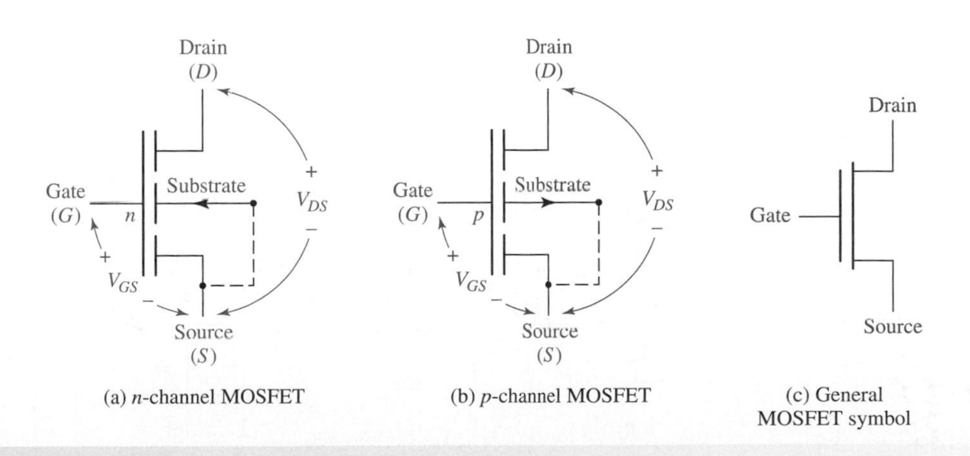

(0 volts = logic 0 and $V_{DD}$ volts = logic 1) for *n*-channel MOS circuits and negative logic ($V_{DD}$ volts = logic 0 and 0 volts = logic 1) for *p*-channel MOS circuits. Using this convention, a logic 1 applied to the gate will switch the MOSFET to the ON state (low resistance between drain and source), and a logic 0 will switch it to the OFF state (high resistance between drain and source).

Figure A-2(a) shows a MOS inverter. When a logic 0 is applied to the gate, the MOSFET is in a high-resistance or OFF state, and the output voltage is approximately $V_{DD}$. When a logic 1 is applied to the gate, the MOSFET switches to a low-resistance or ON state, the output is connected to ground, and the output voltage is approximately 0.

Thus, the operation of the MOSFET is analogous to the operation of a switch in Figure A-2(b) which is open when $V_{in}$ is a logic 0 and closed when $V_{in}$ is a logic 1. In Figure A-2(d), a second MOSFET serves as a load resistor. The geometry of this MOSFET and the gate voltage $V_{GG}$ are chosen so that its resistance is high compared with the ON resistance of the lower MOSFET so that the switching operation of Figure A-2(d) is essentially the same as Figure A-2(a).

As shown in Figure A-3, MOSFETs can be connected in parallel or series to form NOR or NAND gates. In Figure A-3(a), a logic 1 applied to $A$ or $B$ turns on the corresponding transistor and $F$ becomes 0. Thus $F' = A + B$ and $F = (A + B)'$, which is the NOR function. In Figure A-3(c), a logic 1 applied to the $A$ and $B$ inputs turns on both transistors and $F$ becomes 0. In this case $F' = AB$ and $F = (AB)'$, which is the NAND function. More complex functions can be realized by using series-parallel combinations of MOSFETs. For example, the circuit of Figure A-3(e) performs the exclusive-OR function. The output of this circuit has a conducting path to ground, and $F = 0$ if $A$ and $B$ are both 1 or if $A'$ and $B'$ are both 1. Thus, $F' = AB + A'B'$ and $F = A'B + AB' = A \oplus B$. $A'$ and $B'$ are generated by inverters as in Figure A-2(d).

CMOS (complementary MOS) logic performs logic functions using a combination of *p*-channel and *n*-channel MOSFETs. Compared with TTL or other bipolar transistor technologies, CMOS has the advantage of much lower power consumption. Figure A-4(a) shows a CMOS inverter built from a *p*-channel and an *n*-channel MOSFET. When 0 volts (logic 0) is applied to the gate inputs, the *p*-channel transistor ($Q_1$) is on and the *n*-channel transistor ($Q_2$) is off, so the output is $+V$ (logic 1). When $+V$ (logic 1) is applied to the gate inputs, $Q_1$ is off and $Q_2$ is on, so the output is 0 volts (logic 0).

**FIGURE A-2**
**MOS Inverter**

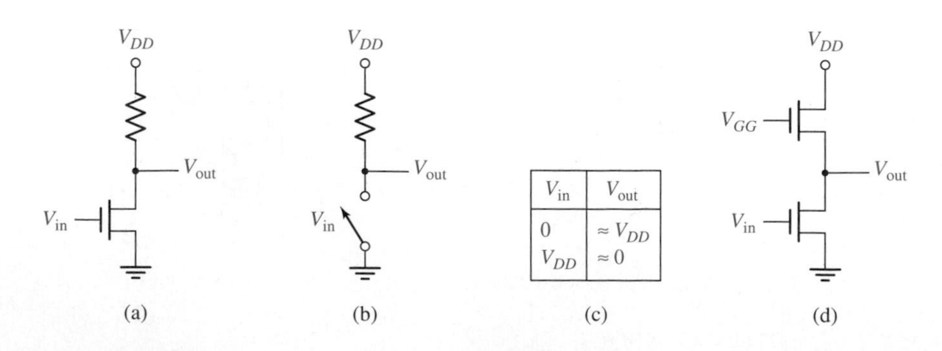

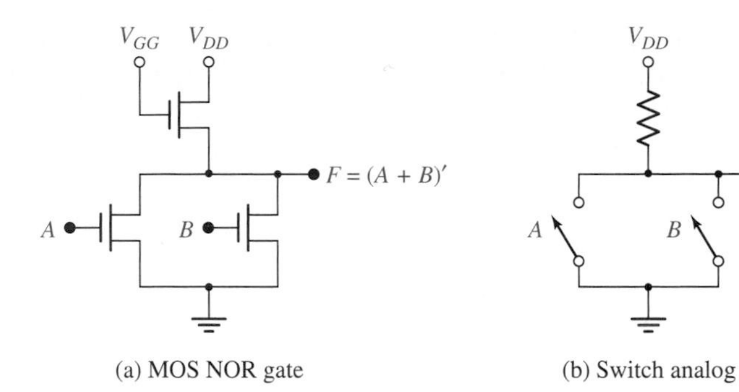


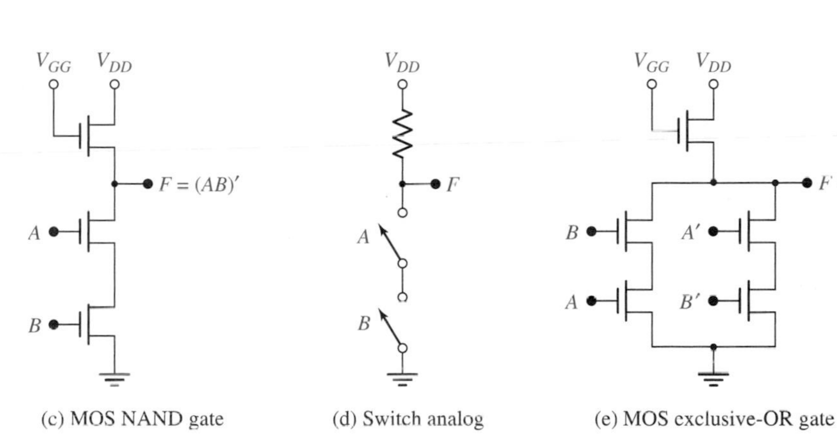


**FIGURE A-3** MOS Gates

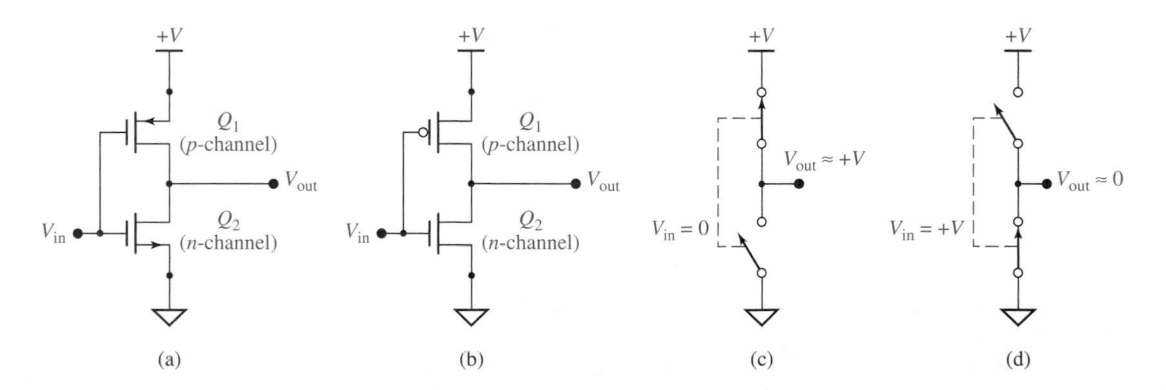


**FIGURE A-4** CMOS Inverter

In the remainder of this discussion we will use a bubble at the MOSFET gate input to indicate a *p*-channel transistor, which is turned on by a logic 0. No bubble at the gate input indicates an *n*-channel transistor, which is turned on by a logic 1. Figure A-4(b) shows the CMOS inverter using this bubble notation. The switch analog in Figure A-4(c) illustrates the operation of the inverter when the inverter input is 0. $Q_1$ is on and $Q_2$ is off as indicated by the closed and open switches. When the input is $+V$(logic 1), $Q_1$ is off and $Q_2$ is on, as indicated by the open and closed switches in Figure A-4(d). The following table summarizes the operation:

| $V_{in}$ | $V_{out}$ | $Q_1$ | $Q_2$ |
|---|---|---|---|
| 0 | +V | ON | OFF |
| +V | 0 | OFF | ON |

Figure A-5 shows a CMOS NAND gate. If $A$ or $B$ is 0 volts, then $Q_1$ or $Q_2$ is ON while $Q_3$ or $Q_4$ is off, and the output is $+V$. If $A$ and $B$ are both $+V$, then $Q_3$ and $Q_4$ are both ON while $Q_1$ and $Q_2$ are off, and the output is 0 volts. If 0 volts represents a logic 0 and $+V$ represents a logic 1, this gate performs the NAND function, as indicated by the truth table of Figure A-5(b).

**FIGURE A-5**
CMOS NAND Gate

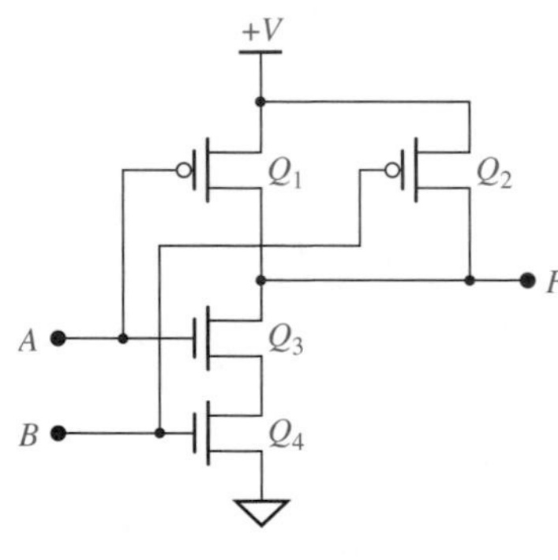


(a) Circuit diagram

| $A$ | $B$ | $F$ | $Q_1$ | $Q_2$ | $Q_3$ | $Q_4$ |
|---|---|---|---|---|---|---|
| 0 | 0 | $+V$ | ON | ON | OFF | OFF |
| 0 | $+V$ | $+V$ | ON | OFF | OFF | ON |
| $+V$ | 0 | $+V$ | OFF | ON | ON | OFF |
| $+V$ | $+V$ | 0 | OFF | OFF | ON | ON |

(b) Truth table

Figure A-6 shows a CMOS NOR gate. If $A = 1\,(+V)$, $Q_1$ is off and $Q_4$ is on, $F = 0$. Likewise, if $B = 1$, $Q_2$ is off and $Q_3$ is on, so $F = 0$. Because $F = 0$ when $A$ or $B$ is 1, $F' = A + B$, and $F = (A + B)'$, which is the NOR function.

**FIGURE A-6**
CMOS NOR Gate

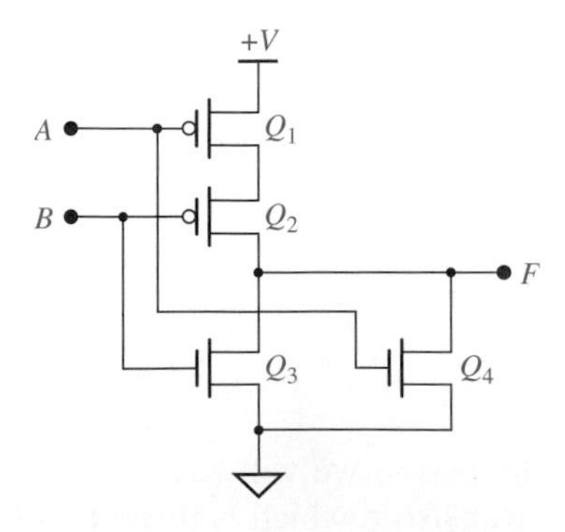


A $p$-channel and $n$-channel transistor pair can be connected to form a CMOS transmission gate (TG) as shown in Figure A-7. The two enable inputs are normally complements so that when $En = 1$, both transistors are enabled and a low impedance path connects $A$ and $B$. When $En = 0$, points $A$ and $B$ are disconnected. In other words, the

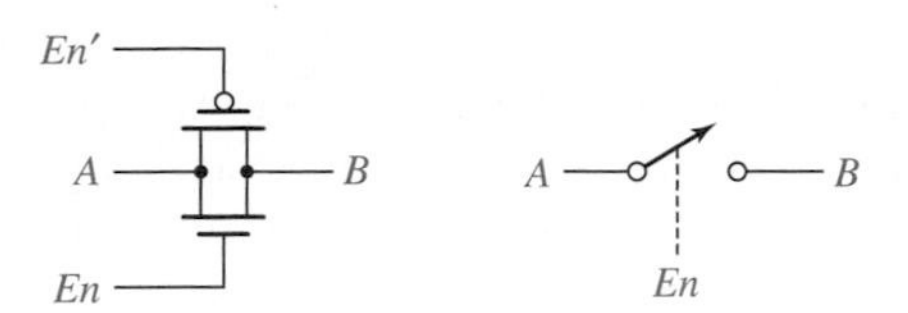


**FIGURE A-7** CMOS Transmission Gate and Switch Analog

transmission gate acts like a switch that is closed when $En = 1$ and open when $En = 0$. Two transistors are used because the $p$-channel transistor does a good job of transmitting a logic 1 and the $n$-channel transistor does a good job of transmitting a logic 0. The 2-to-1 multiplexer of Figure 9-1 can be constructed from two TGs and an inverter, as shown in Figure A-8. When $A = 0$, the upper TG is enabled so that $I_0$ is connected to $F$; when $A = 1$, the lower TG is enabled so that $I_1$ is connected to $F$.

A CMOS gated D latch, as shown in Figure A-9(a), is easily constructed using two TGs and two inverters. The switch analogs of Figures A-9(b) and (c) represent the

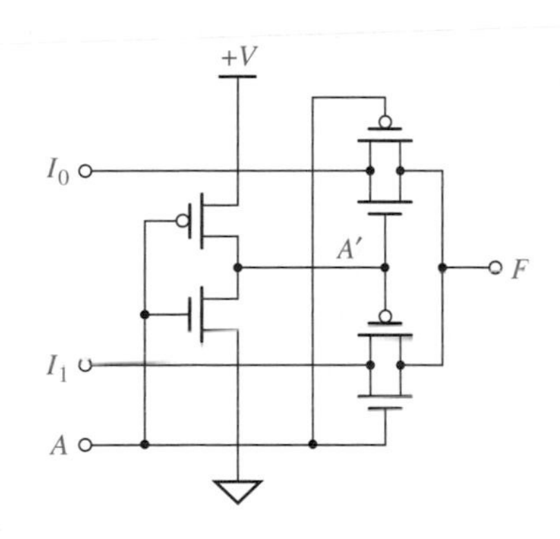


**FIGURE A-8** CMOS Multiplexer

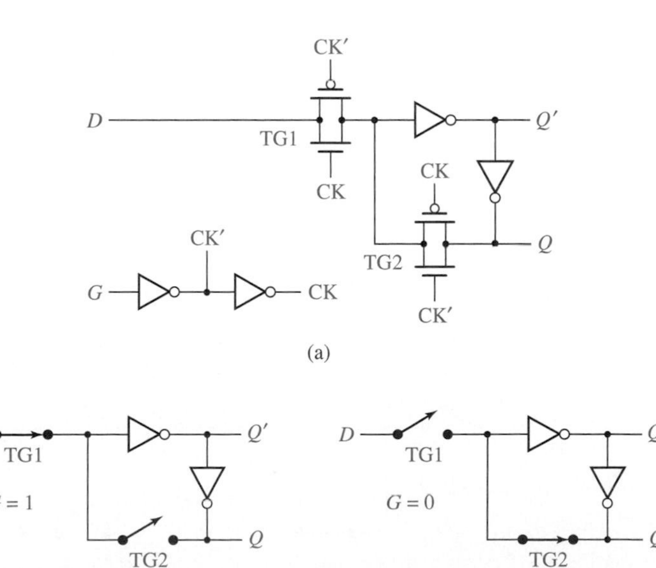


**FIGURE A-9** CMOS Latch and Switch Analogs

TGs by switches. When $G = 1$, CK = 1 and $TG_1$ is closed. Therefore, the latch is transparent, and D is transmitted through the inverters to the $Q$ output. When $G = 0$, $TG_2$ is closed, and the data in the latch is stored in the closed loop of the two inverters. That is, if $Q = 0$, it is still 0 after going through the two inverters, and if $Q = 1$, it is still 1 after going through the two inverters. Because $TG_1$ is open, the data does not change when $D$ changes, and the latch holds the stored value of $Q$.

A CMOS falling-edge-triggered D flip-flop, similar to the type shown in Figure 11-15, can be constructed from two CMOS latches [Figure A-10(a)]. The switch analogs of Figure A-10(b) and (c) illustrate the flip-flop operation. When Clock is 1, the input latch is transparent and the output latch holds the current value of $Q$. When Clock goes to 0, the input latch holds its value, which is transmitted through the output latch to $Q$. Thus, $Q$ can only change states following the falling edge of Clock.

The technology for implementing a CMOS integrated circuit continues to improve, resulting in smaller transistors, lower voltage levels, faster operation, and very high density logic. When no inputs are changing, the static power dissipation is very low. When the CMOS gates are switching, the power dissipation is proportional to the switching frequency. Thus, the power dissipation at a switching frequency of 100 MHz is ten times that at 10 MHz.

FIGURE A-10 Falling-Edge-Triggered D Flip-Flop

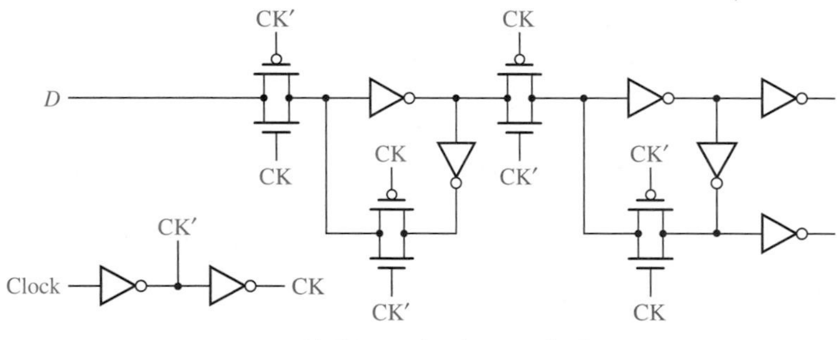


(a) Construction from two latches

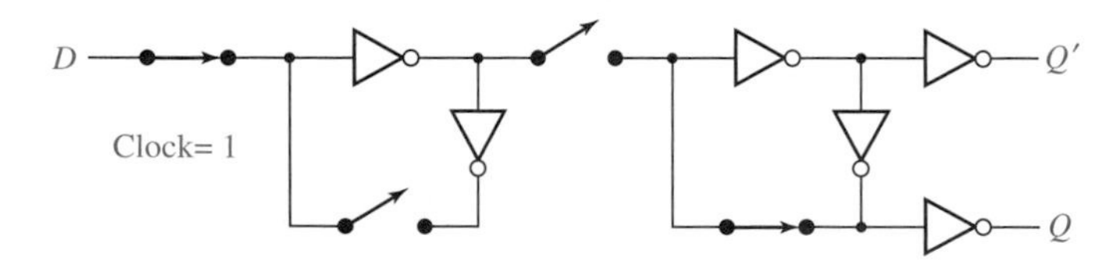


(b) Switch analog for Clock = 1

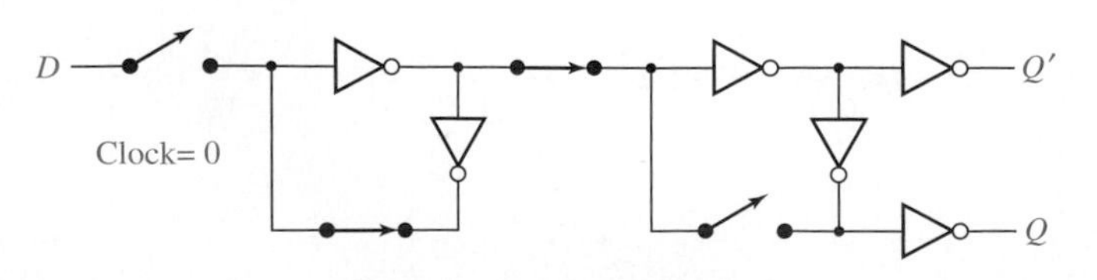


(c) Switch analog for Clock = 0

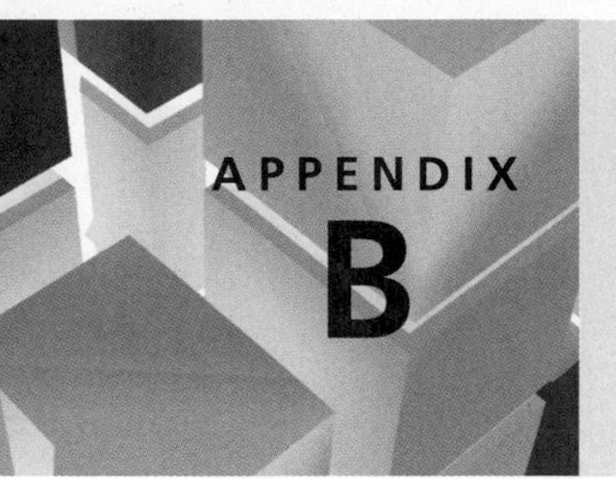

# APPENDIX B VHDL Language Summary

Reserved words are in boldface type. Square brackets enclose optional items. Curly brackets enclose items which are repeated zero or more times. A vertical bar (|) indicates or.

*Disclaimer:* This VHDL summary is not complete and contains some special cases. Only VHDL statements used in this text are listed. For a complete description of VHDL syntax, refer to references [1] and [2].

*entity declaration*

```
entity entity-name is
   port(interface-signal-declaration);
end [entity] [entity-name];
```

*interface-signal declaration*

```
list-of-interface-signals: mode type [:= initial-value]
{; list-of-interface-signals: mode type [:= initial-value]}
```

*Note:* A signal can be of mode **in**, **out**, **inout**, or **buffer**.

*architecture declaration*

```
architecture  architecture-name of entity-name is
[declarations]              -- declare internal signals here
begin
     architecture-body
end [architecture] [architecture-name];
```

*Note:* The architecture body may contain component-instantiation statements, processes, assignment statements, procedure calls, etc.

*integer type declaration*

```
type type_name is range integer_range;
```

*signal declaration*

```
signal  list-of-signal-names : type_name [ := initial_value ];
```

*constant declaration*

```
constant  constant_name : type_name := constant_value;
```

*alias declaration*
**alias** identifier [:identifier-type] **is** item-name;

*Note:* Item-name can be a constant, signal, variable, file, type name, etc.

*array type and object declaration*
**type** array_type_name **is array** index_range **of** element_type;
**signal** | **constant** array_name: array_type_name [ := initial_values ];

*component declaration*
**component** component-name
[**generic** (list-of-generics-and-their types);]
**port** (list-of-interface-signals-and-their-types);
**end component;**

*component instantiation* (concurrent statement)
label: component-name
[**generic map** (generic-association-list;)]
**port map** (list-of-actual-signals);

*Note:* Use **open** if a component output has no connection.

*signal assignment statement* (sequential or concurrent statement)
**signal** <= [**transport**] expression [**after** delay_time];

*Note:* If concurrent, the signal value is recomputed every time a change occurs on the right-hand side. If [**after** delay-time] is omitted, the signal is updated after Δ time. If [**transport**] is omitted, an inertial delay is assumed.

*conditional assignment statement* (concurrent statement only)
**signal** <= expression1 **when** condition1
**else** expression2 **when** condition2
...
[**else** expression];

*selected signal assignment statement* (concurrent statement only)
**with** expression **select**
**signal** <= expression1 [**after** delay_time1] **when** choice1,
expression2 [**after** delay_time2] **when** choice2,
...
[expression [**after** delay_time] **when others**];

*process statement (with sensitivity list)*
[process-label:] **process** (sensitivity-list)
[declarations] --signal declarations not allowed
**begin**
sequential statements
**end process** [process-label];

*Note:* This form of process is executed initially and thereafter only when an item on the sensitivity list changes value. The sensitivity list is a list of signals. No wait statements are allowed.

*process statement (without sensitivity list)*

```
[process-label:] process
      [declarations]              --signal declarations not allowed
begin
      sequential statements
end process [process-label];
```

*Note:* This form of process must contain one or more wait statements. It starts execution immediately and continues until a wait statement is encountered.

*wait statements*

```
wait on sensitivity-list;
wait until Boolean-expression;
wait for time-expression;
```

*if statement* (sequential statement only)

```
if condition then
      sequential statements
{elsif condition then
      sequential statements }  -- 0 or more elsif clauses may be included
[else  sequential statements]
end if;
```

*case statement* (sequential statement only)

```
case expression is
     when choice1 => sequential statements
     when choice2 => sequential statements
     ...
     [when others => sequential statements]
end case;
```

*for loop statement* (sequential statement only)

```
[loop-label:]  for index in range loop
      sequential statements
end loop [loop-label];
```

*Note:* You may use **exit** to exit the current loop.

*report declaration*

```
report string-expression
    [severity severity-level];
```

## VHDL Libraries and Packages

VHDL libraries and packages are used to extend the functionality of VHDL by defining types, functions, components, and overloaded operators. The syntax for libraries and packages is as follows:

*library declaration*

```
library list-of-library names;
```

*use statement*

use **library**_name.package_name.item; (.item may be **.all**)

*package declaration*

**package** package-name **is**
package declarations
**end** [**package**][package-name];

*package body*

**package body** package-name **is**
package body declarations
**end** [**package body**][package name];

When working with bits and bit_vectors, you may use the following declarations:

**library** BITLIB;
**use** BITLIB.bit_pack.**all;**

The bit_pack package includes functions and components that work with signals of type bit and bit_vector. For example, the function call vec2int(A) converts a bit_vector A to an integer. The CD contains a complete listing of bit_pack.

When working with std_logic and std_logic_vectors, the following declarations are required:

**library** IEEE;
**use** IEEE.std_logic_1164.**all;**

The std_logic_1164 package defines the types std_logic and std_logic_vector, a resolution function for these types, conversion functions, and overloaded operators for logic operations. It does not define overloaded operators for arithmetic operations.

In order to perform arithmetic operations on std_logic_vectors, you may add the declaration

**use** IEEE.std_logic_unsigned.**all;**

Although this package is found in the IEEE library, it was written by Synopsis and it is not an IEEE standard. This package treats std_logic_vectors as if they were unsigned numbers and provides overloaded arithmetic operators for +, −, *, =, /=, >, >=, <, and <=. For "+" and "−" if the left operand is a std_logic_vector, the right operand can be the same type, integer type, or std_logic type. For the comparison operators, the right operand can be a std_logic_vector or an integer. The function call CONV_INTEGER(A) converts a std_logic_vector A to an integer.

As an alternative to using std_logic_vectors and the overloaded operators defined in the std_logic_unsigned package, type unsigned may be used. Unsigned type is defined in the Synopsis package std_logic_arith and in the IEEE package numeric_std. To use the former, add the declaration

**use** IEEE.std_logic_arith.**all;**

A vector of type unsigned is similar to a std_logic_vector in that it is an array of std_logic bits, but it has its own overloaded arithmetic operators. Operators for +, −, *, =, /=, >, >=, <, and <= are defined in the std_logic_arith package for various combinations of left and right operands. Unfortunately, logic operators AND, OR, and NOT are not defined for unsigned vectors in this package, so C <= A + B;

works for unsigned vectors, but C <= A **and** B; is not allowed without calling type conversion functions. Some type conversion functions available in this package are as follows:

conv_integer(A)   converts an unsigned vector A to an integer
conv_std_logic_vector(A)   converts an unsigned vector A to a std_logic_vector
conv_unsigned(B, N)   converts an integer B to an unsigned vector of length N

Conversion of a std_logic_vector to unsigned is not defined.

The IEEE numeric_std package, which actually is an IEEE standard, overcomes a number of the deficiencies in the std_logic_arith package. The statement

**use** IEEE.numeric_std.**all;**

invokes this package. It defines unsigned type and overloaded operators for arithmetic and comparison operations in a way similar to the std_logic_arith package, but in addition it defines overloaded operators for logic operations on unsigned vectors. Useful conversion functions in the package include

TO_INTEGER(A)   converts an unsigned vector A to an integer
TO_UNSIGNED(B, N)   converts an integer to an unsigned vector of length N

The only significant deficiency is that this package does not define an overloaded operator for adding a std_logic bit to an unsigned type. Thus, a statement of the form

sum <= A + B + carry;

is not allowed when carry is of type std_logic. The carry must be converted to an integer before it can be added to the unsigned vector A + B.

We have used the std_logic_unsigned package in many examples in this book because it is easy to use. For complex VHDL projects, we recommend using the numeric_std package. Most VHDL simulators and synthesizers work well with either package.

APPENDIX C

# Tips for writing Synthesizable VHDL Code

One of our goals throughout this text is to write VHDL code that not only simulates correctly but also synthesizes correctly to implement hardware that works correctly. First and foremost, always remember that when you write VHDL code you are not writing a computer program; you are describing hardware. If you are designing a multiplier for binary numbers, do not simply write a program to multiply binary numbers. Instead think in terms of what registers are required and what sequence of operations on these registers will produce the desired result.

VHDL code that simulates correctly will not always implement correctly in hardware. A frequent cause of problems is the creation of unintended latches. Even though code simulates correctly, the presence of latches may cause timing problems when the code is actually implemented in hardware. After synthesizing your code, check the synthesis report to make sure no latches are present. If latches are present, check your code for the following:

1. Counters, shift registers, flip-flops, and other devices that change state in response to a clock edge must be updated only in a clocked process. The state of these devices should never be changed in a combinational process or in a concurrent statement. All state changes for a given device must be specified within the same process.
   *Example:* count <= count + 1; should not appear in a combinational process. When this statement, which increments a counter, is placed in a clocked process, any statement that clears the counter must be placed in the same process.
2. If a combinational process sets control signals to '1' at various places in a case statement, all of these signals should be set to '0' *before* the start of the case statement.
3. For every **if** statement in a combinational process, check each signal that is assigned a value in the **then** clause. If such a signal is not assigned a value in step 2, then make sure that it is assigned a value in the **else** clause.
   *Example:* **if** St = '1' **then** nextstate <=1; load <='1'; **end if;** will create a latch because nextstate is not defined when St ='0'. To eliminate the latch write **if** St = '1' **then** nextstate <= 1; load <= '1'; **else** nextstate <= 0; **end if;** This assumes that load is set to '0' in step (2).

Do not attempt to set the same signal to two different values in two different processes or in a process and in a concurrent statement.

*Example* A <= '0'; is a concurrent statement, and A <= B; is another concurrent statement or a sequential statement in a process. These statements can attempt to set A to two different values at the same time. If A and B are bit signals, when you try to simulate, you will get an error message that a signal has multiple drivers. That means a conflict exists because A could be driven to '0' and to '1' at the same time. If A and B are std_logic, the conflict still exists, but you will not get the error message. Instead, during simulation A will assume the value 'X' (unknown) if the simulator tries to set A to '0' and '1' at the same time. In both cases, the code will not synthesize properly because it does not correspond to any real hardware.

Also consider the following example:

```
-- Example of what NOT TO DO: output A is assigned values
-- in a concurrent statement and in a processes.

entity two_drivers is
   port (B,clk,reset : in bit; A : out bit);
end two_drivers;
architecture arch of two_drivers is
begin
   A <= '0' when reset = '0';
   process (clk)
   begin
      if clk'event and clk = '0' then
         A <= B; end if;
   end process;
end arch;
```

In this example, A is supposed to represent a flip-flop that is reset to '0' when the signal reset is '0' and set equal to B on the falling clock edge. Although this code has correct syntax, it will not simulate properly because the two statements that change A occur as a concurrent statement and also as a sequential statement in a process so that A has two drivers. If the signals are std_logic instead of bits, A will assume a value of 'X' at times during the simulation. The code will not synthesize because all statements that change the output of flip-flop A must be placed in the same process. This also would apply if A were a register or a counter.

## Excercise

Change the preceding code so that the reset signal will work properly.

---

An easy way to write synthesizable VHDL code to perform arithmetic operations is to represent binary numbers as std_logic_vectors so that overloaded operators can be used. This is explained on pages 305–306 of the text.

Overloaded + and – operators cannot be used with bit vectors. If you use overloaded operators with std_logic_vectors in your VHDL code, place the following declarations at the start of your code:

```
library IEEE;                     -- this library contains several useful
                                  --   packages
use IEEE.std_logic_1164.all;      -- this package defines std_logic,
                                  --   std_logic_vectors and logic
                                  --   operations on these types
use IEEE.std_logic_unsigned.all;  -- this package defines overloaded
                                  --   operators for std_logic_vectors
```

Remember that the VHDL operators +, −, and & have the same precedence and will be applied from left to right as they appear in a VHDL statement.

Thus A <= B + C&D; is treated as A <= (B+C)&D;

If you want to do concatenation first, you must use parentheses.

A <= B + (C&D);

# Proofs of Theorems

## Finding Essential Prime Implicants

Section 5.4 presents a method for finding all of the essential prime implicants which is based on finding adjacent 1's on a Karnaugh map. The validity of the method is based on the following theorem:

If a given minterm $m_j$ of $F$ and all of its adjacent minterms are covered by a single term $p_j$, then $p_j$ is an essential prime implicant of $F$.

*Proof:*

1. Assume $p_j$ is *not a prime implicant.* Then, it can be combined with another term $p_k$ to eliminate some variable $x_i$ and form another term which does not contain $x_i$. Therefore, $x_i = 0$ in $p_j$ and $x_i = 1$ in $p_k$, or vice versa. Then, $p_k$ covers a minterm $m_k$ which differs from $m_j$ only in the variable $x_i$. This means that $m_k$ is adjacent to $m_j$, but $m_k$ is not covered by $p_j$. This contradicts the original assumption that all minterms adjacent to $m_j$ are covered by $p_j$; therefore, $p_j$ is a prime implicant.
2. Assume $p_j$ is *not essential.* Then, there is another prime implicant $p_h$ which covers $m_j$. Because $p_h$ is not contained in $p_j$, $p_h$ must contain at least one minterm which is adjacent to $m_j$ and not covered by $p_j$. This is a contradiction, so $p_j$ must be essential.

## State Equivalence Theorem

The methods for determining state equivalence presented in Unit 15 are based on Theorem 15.1:

Two states $p$ and $q$ of a sequential network are equivalent if and only if for every single input $x$, the outputs are the same and the next states are equivalent.

*Proof:* We must prove both part 1, the "if" part of the theorem, and part 2, the "only if" part.

1. Assume that $\lambda(p, x) = \lambda(q, x)$ and $\delta(p, x) \equiv \delta(q, x)$ for every input $x$. Then, from Definition 15.1, for every input sequence $\underline{X}$,

$$\lambda[\delta(p, x), \underline{X}] = \lambda[\delta(q, x), \underline{X}].$$

For the input sequence $\underline{Y} = x$ followed by $\underline{X}$, we have

$$\lambda(p, \underline{Y}) = \lambda(p, x) \text{ followed by } \lambda[\delta(p, x), \underline{X})]$$
$$\lambda(q, \underline{Y}) = \lambda(q, x) \text{ followed by } \lambda[\delta(q, x), \underline{X})]$$

Hence, $\lambda(p, \underline{Y}) = \lambda(q, \underline{Y})$ for every input sequence $\underline{Y}$, and $p \equiv q$ by Definition 15.1.

2. Assume that $p \equiv q$. Then, by Definition 15.1, $\lambda(p, \underline{Y}) = \lambda(q, \underline{Y})$ for every input sequence $\underline{Y}$. Let $\underline{Y} = x$ followed by $\underline{X}$. Then,

$$\lambda(p, x) = \lambda(q, x) \text{ and } \lambda[\delta(p, x), \underline{X}] = \lambda[\delta(q, x), \underline{X}]$$

for every sequence $\underline{X}$. Hence, from Definition 15.1, $\delta(p, x) \equiv \delta(q, x)$.

APPENDIX E

# Answers to Selected Study Guide Questions and Problems

## UNIT 1 Study Guide Answers

2. (e) Two of the rows are:

| | | | |
|---|---|---|---|
| 1110 | 16 | 14 | *E* |
| 1111 | 17 | 15 | *F* |

3. (b)

$$
\begin{aligned}
1100_2 - 101_2 &= [1 \times 2^3 + 1 \times 2^2 + 0 \times 2^1 + 0 \times 2^0] \\
&\quad - [\qquad\qquad 1 \times 2^2 + 0 \times 2^1 + 1 \times 2^0] \quad \text{note borrow from column 1} \\
&= [1 \times 2^3 + 1 \times 2^2 + (0 - 1) \times 2^1 + (10 + 0) \times 2^0] \\
&\quad - [\qquad\qquad 1 \times 2^2 + \quad 0 \times 2^1 + \quad 1 \times 2^0] \quad \text{note borrow from column 2} \\
&= [1 \times 2^3 + (1 - 1) \times 2^2 + (10 - 1) \times 2^1 + 10 \times 2^0] \\
&\quad - [\qquad\qquad 1 \times 2^2 + \quad 0 \times 2^1 + \quad 1 \times 2^0] \quad \text{note borrow from column 3} \\
&= [(1 - 1) \times 2^3 + (10 - 0) \times 2^2 + 1 \times 2^1 + 10 \times 2^0] \\
&\quad - [\qquad\qquad\qquad 1 \times 2^2 + 0 \times 2^1 + 1 \times 2^0] \\
&= [\quad 0 \times 2^3 + \qquad 1 \times 2^2 + 1 \times 2^1 + 1 \times 2^0] = 111_2
\end{aligned}
$$

5. (f) sign & mag: −0, 2's comp: −32, 1's comp: −31

(g) Overflow occurs when adding $n$-bit numbers and the result requires $n + 1$ bits for proper representation. You can tell that an overflow has occurred when the sum of two positive numbers is negative or the sum of two negative numbers is positive.
A carry out of the last bit position does *not* indicate that an overflow has occurred.

6. (a)

| | | | |
|---|---|---|---|
| BCD: | 0001 | 1000 | 0111 |
| excess-3: | 0100 | 1011 | 1010 |
| 6-3-1-1: | 0001 | 1011 | 1001 |
| 2-out-of-5: | 00101 | 10100 | 10010 |

## UNIT 1 Answers to Problems

1.1 (a) $2F5.40_{16} = 001011110101.01000000_2$

(b) $7B.2B_{16} = 01111011.00101011_2$

(c) $164.E3_{16} = 000101100100.11100011_2$

(d) $427.8_{16} = 010000100111.1000_2$

1.2 (a) $7261.3_8 = 3761.4_{10}$, $EB1.6_{16} = 3761.4_{10}$
(b) $2635.6_8 = 1437.8_{10}$, $59D.C_{16} = 1437.8_{10}$

1.3 $3252.1002_6$

1.4 (a) $5B1.1C_{16}$ (b) $010110110001.00011100_2 = 2661.070_8$
(c) $112301.0130_4$ (d) $3564.6_{10}$

1.5 (a) Add: 11001. Subtract: 0101. Multiply: 10010110.
(b) Add: 1010011. Subtract: 011001. Multiply: 11000011110.
(c) Add: 111010. Subtract: 001110. Multiply: 1100011000.

1.6

| (a) | (b) | (c) |
|---|---|---|
| 1111 | 111 1 | 11111 1 |
| 11110100 | 1110110 | 10110010 |
| − 1000111 | − 111101 | − 111101 |
| 10101101 | 0111001 | 01110101 |

1.7 2's complement:

| (a) | (b) | (c) |
|---|---|---|
| 010101 | 110010 | 100111 |
| + 001011 | + 100000 | + 010010 |
| 100000 | (1) 010010 | 111001 |
| **OVERFLOW!** | **OVERFLOW!** | |

| (d) | (e) |
|---|---|
| 110100 | 110101 |
| + 001101 | + 101011 |
| (1) 000001 | (1) 100000 |

1's complement:

| (a) | (b) | (c) |
|---|---|---|
| 010101 | not assigned | 100110 |
| + 001011 | because –32 cannot | + 010010 |
| 100000 | be represented | 111000 |
| **OVERFLOW!** | in 6 bits | |

| (d) | (e) |
|---|---|
| 110011 | 110100 |
| + 001101 | + 101010 |
| (1) 000000 | (1) 011110 |
| + 1 | + 1 |
| 000001 | 011111 |
| | **OVERFLOW!** |

1.8 For a word length of $N$, the range of 2's complement numbers that can be represented is $-2^{N-1}$ to $2^{N-1} - 1$.
So, for a word length of 8, the range is $-2^7$ to $2^7 - 1$, or $-128$ *to* $127$. Because 1's complement has a "negative zero" (11111111) in addition to zero (00000000), the values that can be represented range from $-(2^7 - 1)$ to $2^7 - 1$, or $-127$ *to* $127$.

1.9

| Dec. | 7-3-2-1 |
|---|---|
| 0 | 0000 |
| 1 | 0001 |
| 2 | 0010 |
| 3 | 0011 or 0100 |
| 4 | 0101 |
| 5 | 0110 |
| 6 | 0111 |

| 3 | 6 | 5 | 9 |
|---|---|---|---|
| 0011 or 0100 | 0111 | 0110 | 1010 |

7 1000
8 1001
9 1010

## UNIT 2 Study Guide Answers

2. (d) 1; 0; 1; 1 (e) 1, 1; 0, 0; 0; 1
3. (a) four variables, 10 literals (d) $F = (A'B)'$ (e) $F = (A + B')C$
   (f) *Circuit* should have two OR gates, three AND gates, and three inverters.
4. (b) $A, 0, 0, A; A, 1, A, 1$
6. (c) $Z = ABC$
7. (a) Sum of products
   Neither
   Product of sums (Here, $A$ and $B'$ are each considered to be separate terms in the product.)
   Neither
   (b) Fewer terms are generated.
   (c) $D[A + B'(C + E)] = D(A + B')(A + C + E)$
8. (a) $AE + B'C' + C'D$ (b) $C'DE + AB'CD'E$
10. (a) $a' + b + c$ (b) $ab'c'd$ (c) $a(b' + c')$
    (d) $(a + b)(c' + d')$ (e) $a' + b(c + d')$

## UNIT 2 Answers to Problems

2.1 (a) $X(X' + Y) = XX' + XY = 0 + XY = XY$
   (b) $X + XY = X(1 + Y) = X(1) = X$
   (c) $XY + XY' = X(Y + Y') = X(1) = X$
   (d) $(A + B)(A + B') = AA + AB' + AB + BB' = A + AB' + AB + BB'$
   $= A(1 + B + B') + 0 = A(1) = A$

2.2

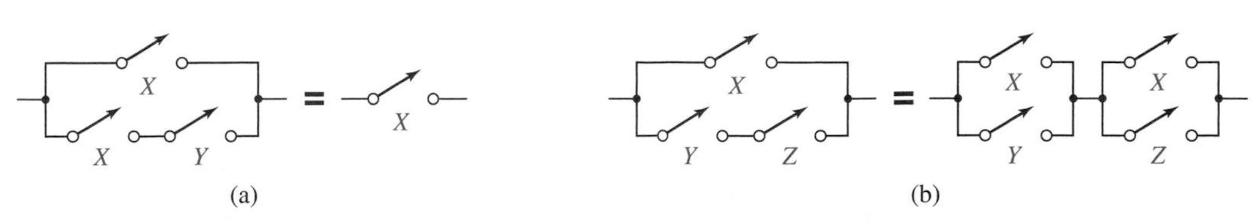


2.3 (a) 1 (Theorem 5)
   (b) $CD + AB'E$ (Theorem 8D) (technically, we also used Theorem 3D)
   (c) $AF$ (Theorem 9) (d) $C + D'B + A'$ (Theorem 11D)
   (e) $A'B + D$ (Theorem 10D) (f) $A + BC + DE + F$ (Theorem 11D)

2.4 (a) $F = A + E + BCD$ (one AND gate and one OR gate with three inputs)
   (b) $Y = A + B$

2.5 (a) $ACD' + BE$ (b) $A'B' + A'D' + C'B' + C'D'$

2.6 (a) $(A + C')(A + D')(B + C')(B + D')$ (b) $X(W + Z)(W + Y)$
   (c) $(A' + E)(B + E)(C + E)(A' + D + F)(B + D + F)(C + D + F)$

(d) $Z(W' + X)(Q' + W' + Y)$ (e) $(A' + D')(C + D')$
(f) $(A + B + D)(A + C + D)(A + B + E)(A + C + E)$

2.7

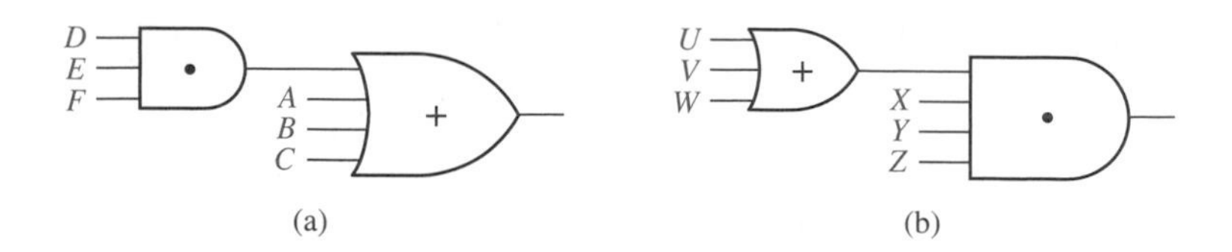


(a) (b)

2.8 (a) $ABC + ABD'$ (b) $A'B' + A'CD'$ (c) $A'BC'$

2.9 (a) $F = A'B$ (b) $G = T'$

## UNIT 3 Study Guide Answers

1. (b) $(b' + d)(b + a)(b + c)$ $(a + d)(b + d)(a' + b' + c)$
   (c) $w'y' + x'y'z' + xy + wyz$
5. (b) $A'B'C + BC'D' + AB'D' + BCD$
   (c) Add $BCD$; eliminate $A'BD, ABC$

## UNIT 3 Answers to Problems

3.6 (a) $WY'X + WY'Z' + W'X'Y + W'X'Z$ (b) $A'D + AC$

3.7 (a) $(C' + D)(C + D' + B')$
(b) $(D' + A' + B')(D' + C + B')(D + A + C')(D + A' + B)$

3.8 $F = (AB) \oplus [(A \equiv D) + D] = A' + BD' + B'D$

3.9 No. Consider $A = 1, B = 1, C = 0$ or $A = 1, B = 0, C = 1$.

3.10 (a) $W'X + WY'Z + WYZ'$ (b) $BD + A'BC + AB' + AC'$
(c) $(A + C + D)(A' + C' + D')(B + C' + D)$

3.11 $AE' + AC' + B' + CD' + D'E'$

3.12 $A'CD'E + A'B'D' + ABCE + ABD = A'CD'E + BCD'E + A'B'D' + ABCE + ABD = A'B'D' + ABD + BCD'E$

## UNIT 4 Study Guide Answers

2. (d) $ab'c'd$ (e) $a + b + c' + d'$
   (g) $(a + b' + c)(a' + b + c')(a' + b' + c)(a' + b' + c')$
3. (c) $m_0 + m_1 + m_3 + m_4 = \Sigma\, m(0, 1, 3, 4)$ $M_2M_5M_6M_7 = \Pi\, M(2, 5, 6, 7)$
4. (b) $m_{19}$ (c) $A'BCD'E$
   (e) $M_{19}$ (f) $(A + B' + C' + D + E')$
5. (a) 65536
   (d) $(a_0m_0 + a_1m_1 + a_2m_2 + a_3m_3)(b_0m_0 + b_1m_1 + b_2m_2 + b_3m_3) = \ldots = a_0b_0m_0 + a_1b_1m_1 + a_2b_2m_2 + a_3b_3m_3$
   (f) $f = \Pi\, M(2, 5, 6)$ $f' = \Sigma\, m(2, 5, 6) = \Pi\, M(0, 1, 3, 4, 7)$
6. (b) $\Sigma\, m(0, 5) + \Sigma\, d(1, 3, 4)$

## UNIT 4 Answers to Problems

4.1 (a) $U$: Safe unlocked, $J$: Mr. Jones present, $E$: Mr. Evans present, $B$: Normal business hours, $S$: Security guard present

$$U = (J + E)BS$$

(b) $O$: Wear overshoes, $A$: You are outside, $R$: Raining heavily, $S$: Wearing suede shoes, $M$: Mother tells you to

$$O = ARS + M$$

(c) $L$: Laugh at joke, $F$: It is funny, $G$: Good taste, $O$: Offensive, $P$: Told by professor

$$L = FGO' + PO'$$

(d) $D$: Elevator door opens, $S$: Elevator is stopped, $F$: Level with floor, $T$: Timer expired, $B$: Button pressed

$$D = SFT' + SFB$$

4.2 (a) $Y = A'B'C'D'E' + AB'C'D'E' + ABC'D'E'$ or $Y = C'$
(b) $Z = ABC'D'E' + ABCD'E' + ABCDE'$ or $Z = BE'$

4.3 $F_1 + F_2 = \Sigma\, m\,(0, 3, 4, 5, 6, 7)$; General rule: $F_1 + F_2$ is the sum of all minterms which are present in either $F_1$ or $F_2$, because $F_1 + F_2 = \Sigma\, a_i m_i + \Sigma\, b_i m_i = \Sigma\,(a_i + b_i)m_i$

4.4 (a) 16
(b) $F(x, y) = 0, x'y', x'y, x', xy', y', x'y + xy', x' + y', xy, x'y' + xy, y, x' + y, x, x + y', x + y, 1$

4.5

| A | B | C | D | E | F | |
|---|---|---|---|---|---|---|
| 0 | 0 | 0 | 1 | 1 | X | |
| 0 | 0 | 1 | X | X | 1 | |
| 0 | 1 | 0 | X | X | X | |
| 0 | 1 | 1 | X | X | 1 | or 1 1 X |
| 1 | 0 | 0 | X | 0 | 0 | |
| 1 | 0 | 1 | X | X | 1 | |
| 1 | 1 | 0 | X | X | X | |
| 1 | 1 | 1 | X | 0 | 0 | or 0 X 0 |

4.6 (a) $F = A'B' + AB\ (d_1 = 1, d_5 = 0)$ (b) $G = C\ (d_2 = 0, d_6 = 0)$

4.7 (a) $\Sigma\, m\,(1, 2, 4)$ (b) $\Pi\, M\,(0, 3, 5, 6, 7)$

4.8 (a) $F = A'B'C'D' + A'B'C'D + A'B'CD' + A'B'CD + A'BC'D' + A'BC'D + A'BCD' + AB'C'D' + AB'C'D + ABC'D'$
$F = \Sigma\, m\,(0, 1, 2, 3, 4, 5, 6, 8, 9, 12)$
(b) $F = (A + B' + C' + D')(A' + B + C' + D)(A' + B + C' + D')(A' + B' + C + D')(A' + B' + C' + D)(A' + B' + C' + D')$
$F = \Pi\, M\,(7, 10, 11, 13, 14, 15)$

4.9 (a) $F = \Sigma\, m\,(0, 1, 4, 5, 6)$ (b) $F = \Pi\, M\,(2, 3, 7)$
(c) $F' = \Sigma\, m\,(2, 3, 7)$ (d) $F' = \Pi\, M\,(0, 1, 4, 5, 6)$

4.10 (a) $F = \Sigma\, m\,(1, 4, 5, 6, 7, 10, 11)$
(b) $F = \Pi\, M\,(0, 2, 3, 8, 9, 12, 13, 14, 15)$
(c) $F' = \Sigma\, m\,(0, 2, 3, 8, 9, 12, 13, 14, 15)$
(d) $F' = \Pi\, M\,(1, 4, 5, 6, 7, 10, 11)$

4.11 (a) $d_i = x_i \oplus y_i \oplus b_i$
$b_{i+1} = b_i x_i' + x_i' y_i + b_i y_i$

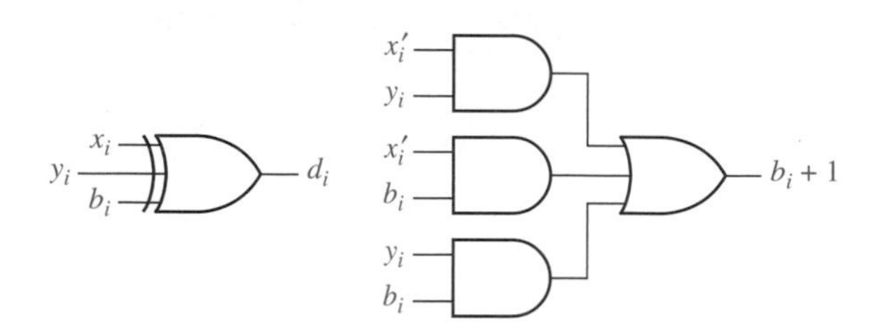


(b) $d_i = s_i$, $b_{i+1}$ is the same as $c_{i+1}$ with $x_i$ replaced by $x_i'$

4.12

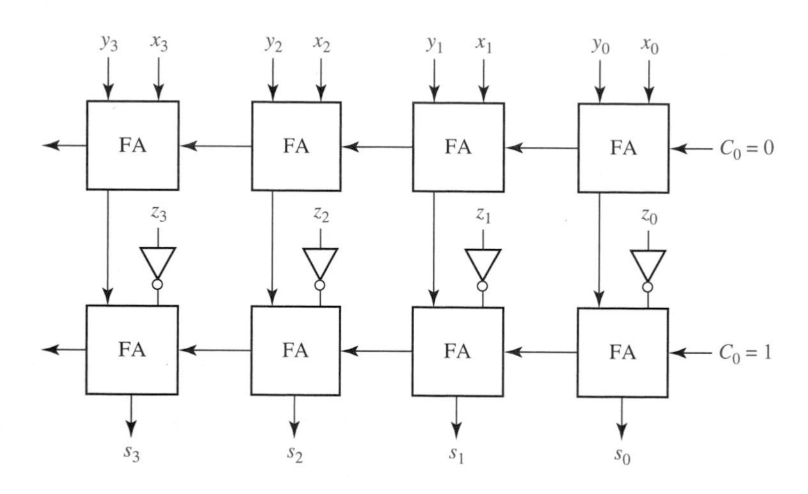


## UNIT 5 Study Guide Answers

3. (d) 6, 10, 12, 15; 0, 12, 9, 10
   (g) $f_1 = a'b + bc' + a'cd + ac'd$ $\quad f_2 = b'c + cd + a'bd + ab'd'$
4. (a) $a'b'd', b'c'd', ac'd', ac'd$, also $a'b'cd$, and all the other minterms.
   (b) $AB'C'$ and $AC'D$ are prime implicants.
5. (a) 4 (c) We cannot determine if $B'C'$ is essential.
   (f) Yes (i) $A'D'$ because of $m_4$, $B'D'$ because of $m_{10}$
6. (b) $A'D'$ is not essential because all of its minterms are covered by other prime implicants. $BC'$ is essential because of $m_{13}$. $B'CD$ is essential because of $m_{11}$. Minimum sum $= B'CD + BC' + BD' + A'B'$.
   (d) $A'C' + ACD + \{AB \text{ or } BC'\}$
8. (a) $F = AB'D' + B'D'E' + A'BDE$ (b) 8, 16, 25, 26, 28
   (d) $P_1 + P_2 + P_3 + P_4 + BCDE + AC'E$
   (f) $AC'E' + A'DE + ACE + B'CE + (AB'C \text{ or } ADE' \text{ or } ACD \text{ or } AB'E')$

## UNIT 5 Answers to Problems

5.3 (a) $f = bc' + a'c' + ab'c$ (b) $f = e'f' + d'e' + d'f'$
(c) $f = r' + t'$ (d) $f = y + x'z + xz'$

5.4 (a)

| CD \ AB | 00 | 01 | 11 | 10 |
|---|---|---|---|---|
| 00 | **1** (0) | **1** (4) | **1** (12) | **1** (8) |
| 01 | **0** (1) | **0** (5) | **1** (13) | **0** (9) |
| 11 | **1** (3) | **0** (7) | **1** (15) | **1** (11) |
| 10 | **1** (2) | **1** (6) | **1** (14) | **1** (10) |

(b) $F = D' + B'C + AB$
(c) $F = (A + B' + D')(B + C + D')$

5.5 (a)

| $C_1$ | $C_2$ | $X_1$ | $X_2$ | $Z$ |
|---|---|---|---|---|
| 0 | 0 | 0 | 0 | 0 |
| 0 | 0 | 0 | 1 | 1 |
| 0 | 0 | 1 | 0 | 1 |
| 0 | 0 | 1 | 1 | 1 |
| 0 | 1 | 0 | 0 | 0 |
| 0 | 1 | 0 | 1 | 1 |
| 0 | 1 | 1 | 0 | 1 |
| 0 | 1 | 1 | 1 | 0 |
| 1 | 0 | 0 | 0 | 0 |
| 1 | 0 | 0 | 1 | 0 |
| 1 | 0 | 1 | 0 | 0 |
| 1 | 0 | 1 | 1 | 1 |
| 1 | 1 | 0 | 0 | 1 |
| 1 | 1 | 0 | 1 | 0 |
| 1 | 1 | 1 | 0 | 0 |
| 1 | 1 | 1 | 1 | 1 |

(b) $Z = C_1' X_1' X_2 + C_1' X_1 X_2' + C_1 X_1 X_2 + C_1 C_2 X_1' X_2' + \{C_1' C_2' X_2$ **or** $C_1' C_2' X_1$ **or** $C_2' X_1 X_2\}$

5.6 (a) $f = \underline{a'd} + \underline{a'b'c'} + \underline{b'cd} + \underline{abd'} + \{a'bc$ **or** $bcd'\}$
$a'd \rightarrow m_5; a'b'c' \rightarrow m_0; b'cd \rightarrow m_{11}; abd' \rightarrow m_{12}$
(b) $f = \underline{bd} + \underline{a'c} + \underline{b'd'} + \{a'b$ **or** $a'd'\}$
$bd \rightarrow m_{13}, m_{15}; a'c \rightarrow m_3; b'd' \rightarrow m_8, m_{10}$
(c) $f = \underline{c'd'} + \underline{a'd'} + \underline{b'}$
$c'd' \rightarrow m_{12}; a'd' \rightarrow m_6; b' \rightarrow m_{10}, m_{11}$

5.8 (a) $f = a'bc' + ac'd + b'cd'; f = (b' + c')(c' + d')(a + b + c)(a' + c + d)$
(b) $f = a'b'd + bc'd' + cd; f = (b + d)(b' + d')(a' + c)\{(b' + c')$ **or** $(c' + d)\}$

5.10 (a) $C'D'E' \rightarrow m_{16}, m_{24}; A'CE' \rightarrow m_{14}; ACE \rightarrow m_{31}; A'B'DE \rightarrow m_3$
(b) $A'B'DE, A'D'E', CD'E, A'CE', ACE, A'B'C, B'CE, C'D'E', A'CD'$

5.11 $f = \underline{(a + b + c + d)}\ \underline{(a + b' + e')}\ \underline{(a' + d' + e)}\ \underline{(a' + b + c')}\ \underline{(a + c + e')}$
$(c + d + e')\{(a' + b' + c + d)$ **or** $(a' + b' + c + e)\}$

5.12 (a) $F = \Pi\, M(0, 1, 9, 12, 13, 14)$
$$F = (A + B + C + D)(A + B + C + D')(A' + B' + C + D)$$
$$(A' + B' + C + D')(A' + B' + C' + D)(A' + B + C + D')$$
(b) $F' = A'B'C' + ABD' + AC'D$
(c) $F = (A + B + C)(A' + B' + D)(A' + C + D')$

5.13 $F = A'C' + B'C + ACD' + BC'D$
Minterms $m_0, m_1, m_2, m_3, m_4, m_5, m_7, m_8, m_{10}$, and $m_{11}$ can be made don't-cares individually and will not change the given expression.

## UNIT 6 Study Guide Answers

2. (f) (2,6)
3. (a) $m_0 - a'b'c'$ $(m_0, m_1) - a'b'$
$m_1 - a'b'c$ $(m_1, m_5) - b'c$ prime
$m_5 - ab'c$ $(m_5, m_7) - ac$
$m_7 - abc$
(d) $A'B'C'$ and $ABC$ are not prime implicants.
4. (b) $a'c'd', bc', ab'c$
5. (b) $F = bd + a'b, F = bd + bc', F = bc' + a'b, F = a'b + c'd$

## UNIT 6 Answers to Problems

6.2 (a)

| | |
|---|---|
| $a'c'd$ | (1,5) |
| $b'c'd$ | (1,9) |
| $a'bd$ | (5,7) |
| $ab'd$ | (9,11) |
| $abd'$ | (12,14) |
| $bcd$ | (7,15) |
| $acd$ | (11,15) |
| $abc$ | (14,15) |

(b)

| | |
|---|---|
| $a'b'c'$ | (0,1) |
| $b'c'd'$ | (0,8) |
| $ab'd'$ | (8,10) |
| $acd'$ | (10,14) |
| $a'd$ | (1,3,5,7) |
| $bc$ | (6,7,14,15) |

6.3 (a) $f = a'c'd + ab'd + abd' + bcd$ **or** $f = b'c'd + a'bd + abd' + acd$

(b) $$f = a'd + bc + \left\{\begin{array}{c} a'b'c' + ab'd' \\ \text{OR} \\ b'c'd' + acd' \\ \text{OR} \\ b'c'd' + ab'd' \end{array}\right\}$$

6.4 $f = b'cd' + bc' + a'd + (a'b \text{ OR } a'c)$ [1 other solution]

6.5 Prime implicants: $ab, c'd, ac', bc', ad, bd$
$F = ab + c'd$ **or** $F = ab + ac'$ **or** $F = ab + ad$ **or** $F = ac' + ad$ **or** $F = ac' + bd$ **or**
$F = ad + bc'$

6.6 (a) $F = A'B + A'C'D' + AB'D + A'C'E + BCDE$
(b) $Z = A'B' + ABD + EB'C' + EA'C + FAB + GBD$ [several other solutions]

## UNIT 7 Study Guide Answers

1. (b) $Z_1$: six gates, 13 inputs, four levels $Z_2$: five gates, 11 inputs, five levels
(d)

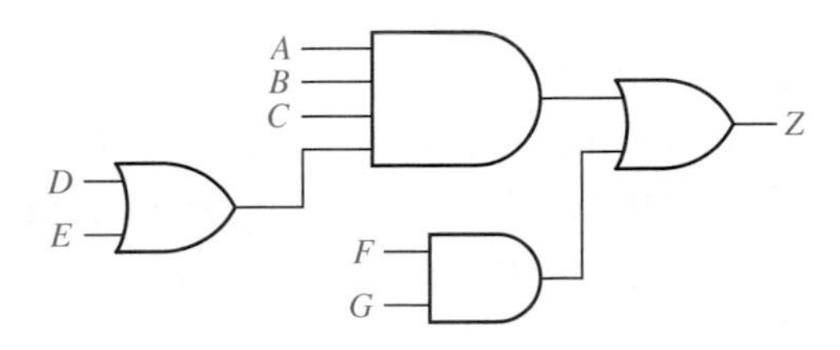


2. (a) 0; 1; 1,1,1; 0,0,0
6. (a) (1) No (2) Yes (3) No (b) because $C$ requires no gate
(c) five gates, 10 inputs; using common gate: four gates, nine inputs
(d) $F_1 = \underline{a'cd} + \underline{acd} + ab'c'$; $F_2 = \underline{a'cd} + \underline{bcd} + a'bc' + acd'$;
$F_3 = \underline{bcd} + \underline{acd} + a'c'd$

## UNIT 7 Answers to Problems

7.1 (a) $f = (a + b)(a' + b')(a + c + d')(a' + c' + d')$
OR $f = (a + b)(a' + b')(a + c + d')(b + c' + d')$
OR $f = (a + b)(a' + b')(b' + c + d')(a' + c' + d')$
OR $f = (a + b)(a' + b')(b' + c + d')(b + c' + d')$
(b) $f = a'b(c + d') + ab'(c' + d')$

7.2 (a) $Z = (C' + E')(AD + B) + A'D'E'$ (four levels, 13 inputs)
(b) $Z = (B(C + D) + A)(E + FG)$ (four levels, 12 inputs)

7.3 AND-OR: $F = a'bd + ac'd$; OR-AND: $F = d(a' + c')(a + b)$

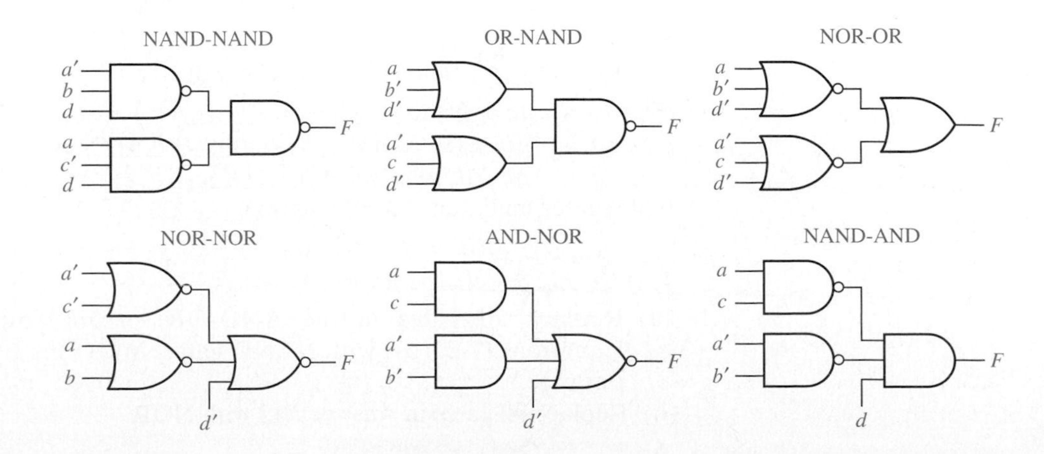

7.4 $F = BC'(A + D) + AB'C$ (three levels, four gates, 10 inputs)

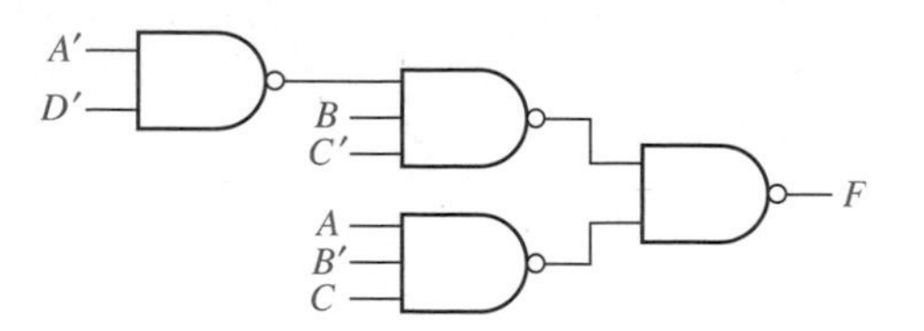


7.5 $Z = (A + C + D)(A' + B'C'D')$ (convert circuit to four NOR gates)

7.6 $Z = A(BC + D) + C'D$ (convert circuit to five NAND gates)

7.7 $Z = E(A + B(D + CF))$ (convert circuit to five NOR gates)

7.8 (a)

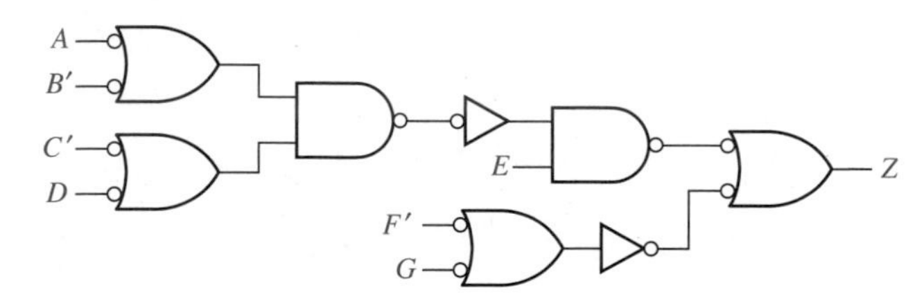


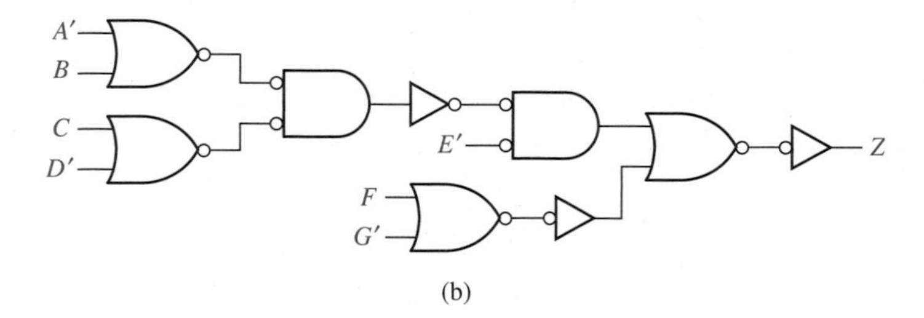


(b)

7.9 $f_1 = \underline{acd'} + ad + \underline{a'b'd}$; $f_2 = a'd' + \underline{a'b'd} + \underline{acd'}$ (six gates, 16 inputs)

7.10 $f_1 = ab'd + \underline{b'cd} + a'bd'$ $f_2 = \underline{ab'c} + b'cd' + bc'd' + \{ac'd'$ **or** $ab'd'\}$
$f_3 = \underline{ab'c} + \underline{b'cd} + a'bc$ (11 gates, 34 inputs)

7.11 $F_1 = (a + c)(a + b')\,\underline{(a' + b' + c)}\underline{(a' + b + c')}$
$F_2 = (a + c')(b' + c + d)\underline{(a' + b' + c)}\underline{(a' + b + c')}$ **or**
$(a + c')(a + b' + d)\underline{(a' + b' + c)}\underline{(a' + b + c')}$
(eight gates minimum, 23 gate inputs)

7.12 $f_1 = \underline{(a + b + c)}(b' + d)$ $f_2 = \underline{(a + b + c)}\underline{(b' + c + d)}(a' + c)$
$f_3 = \underline{(b' + c + d)}(a + c)(b + c')$

7.13 (a) Replace all gates in the AND-OR *circuit* which corresponds to Equations (7-23(b)) with NAND gates. Invert the $c$ input to the $f_2$ output gate.
(b) Replace all gates in Answer 7.12 with NOR.

## UNIT 8 Study Guide Answers

3.

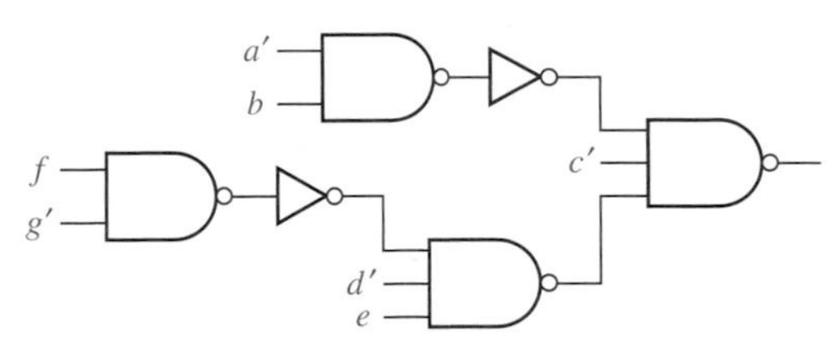


4. (a) Factor out the expression such that the number of inputs on each gate is less than or equal to the maximum allowed. This will result in the addition of more levels of logic.
   (b) Yes.
   (c) Even if the two-level expressions had common terms, most of these common terms would be lost when the expressions are factored.
5. (a) $B'$ goes to 0 at 80 ns. $Z$ goes to 1 at 50 ns and goes to 0 at 110 ns.
6. (a) $y_1$ goes to 1 at 15 ns. $y_2$ goes to 0 at 30 ns. $Z$ goes to 1 at 25 ns and, then, goes to 0 at 40 ns.
   (c) A pair of adjacent 1's corresponding to $a'bc$ and $abc$ are not in the same loop in the Karnaugh map, but $a'bc$ and $a'bc'$ are both in $a'b$. Without the map, when $b = c = 1$ and $a$ changes from 0 to 1, $a'b$ may go to 0 before $ac$ becomes 1. But when $a = 0, b = 1$, and $c$ changes from 1 to 0, $a'b$ remains 1.
   (g) The application of DeMorgan's laws to convert a circuit from one form to another will not introduce any hazards.
7. (b) If $G = 0$, gate 4 is faulty. If $G = 1$, gate 1 is faulty.

## UNIT 8 Answers to Problems

8.1

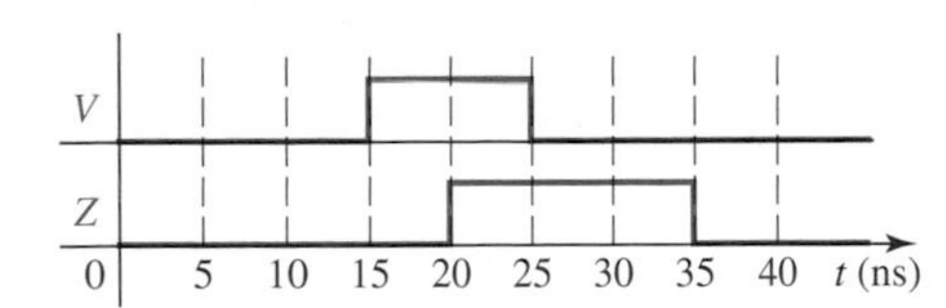


8.2 (a) $F = A'C'D' + BC'D + AC$ (hazards are 1101 ↔ 1111 and 0100 ↔ 0101 [static 1])
**OR** $F = (A' + C + D)(B + C + D')(A + C')$ (hazards are 0001 ↔ 0011 and 1000 ↔ 1001 [static 0])
(b) $F^t = A'C'D' + BC'D + AC + \underline{A'BC'} + \underline{ABD}$
(c) $F^t = (A' + C + D)(B + C + D')(A + C')\underline{(A' + B + C)}\,\underline{(A + B + D')}$

8.3 (a) Glitch in output of $G$ occurs between 6 ns and 7 ns (static 1-hazard).
(b) Modified equation to avoid hazards: $G = A'C'D + BC + A'BD$

8.4 $A = 1$ $E = \text{X}$
$B = \text{Z}$ $F = 0$
$C = \text{X}$ $G = 0$
$D = 1$ $H = \text{X}$

8.5 Gate 3 is connected incorrectly or is malfunctioning.

## UNIT 9 Study Guide Answers

2. (a)

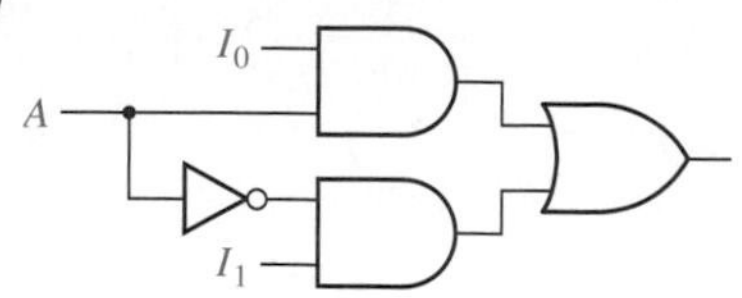


(b) $Z = A'C'I_0 + A'CI_1 + AC'I_2 + ACI_3$

(c) Before $C$ changes, $Z = I_4$, and after $C$ changes, $Z = I_5$.

(d)

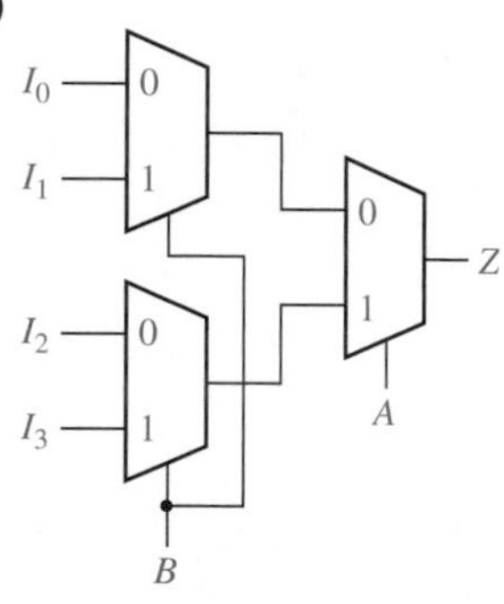


(e) MUX inputs: $I_0 = B, I_1 = B'$, control $= A$

3. (f) AND gate inputs are $A'B'$, $A'B$, $AB'$, and $AB$

4. (a) Inputs $BCD$; $A = 0$

5. (b) 32 words $\times$ 4 bits; 1024 $\times$ 8 (c) 16 words $\times$ 5 bits; 16 $\times$ 10

6. (a) Four inputs, seven terms, three outputs

(b) Four inputs, four terms, three outputs

(c)

| A | B | C | D | $F_1$ | $F_2$ | $F_3$ |
|---|---|---|---|---|---|---|
| 1 | 1 | - | - | 1 | 0 | 1 |
| 1 | - | 1 | 1 | 1 | 1 | 0 |
| 1 | 1 | 0 | - | 0 | 1 | 0 |
| 0 | - | 1 | 1 | 0 | 1 | 1 |

(f) When $ABC = 010$, $F_0F_1F_2F_3 = 0111$.

8. (c) $f = c'(d' + a) + c(a'b' + bd)$

(d)

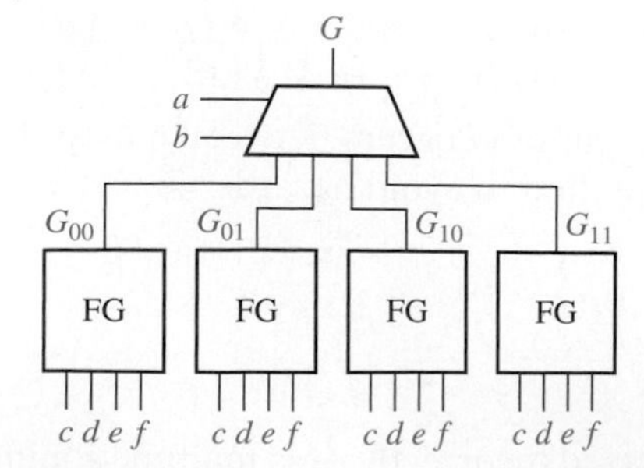

## UNIT 9 Answers to Problems

9.1 (a)

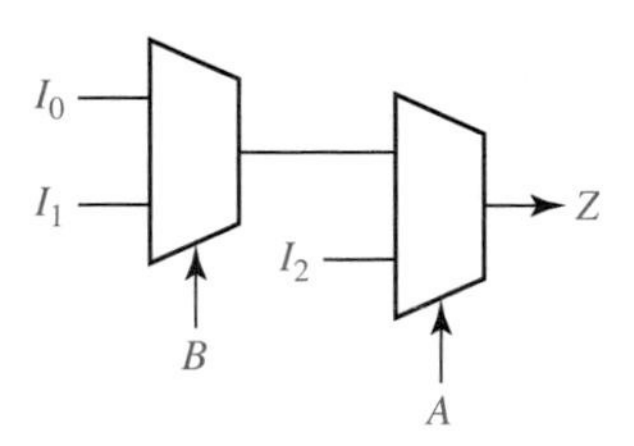


(b)

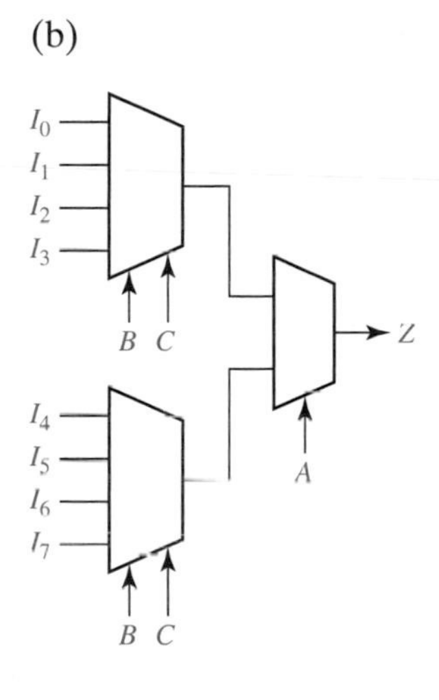


(c)

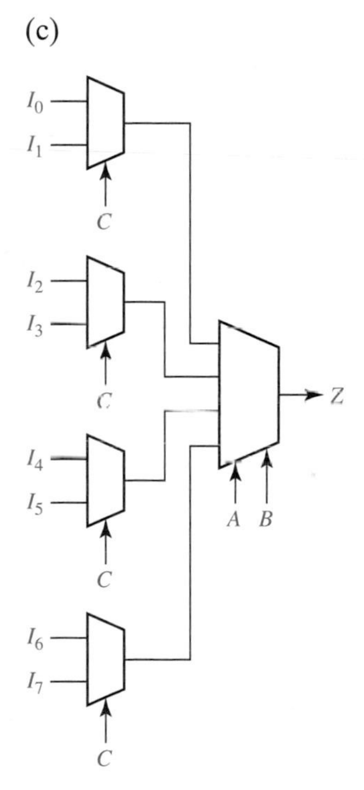


9.2

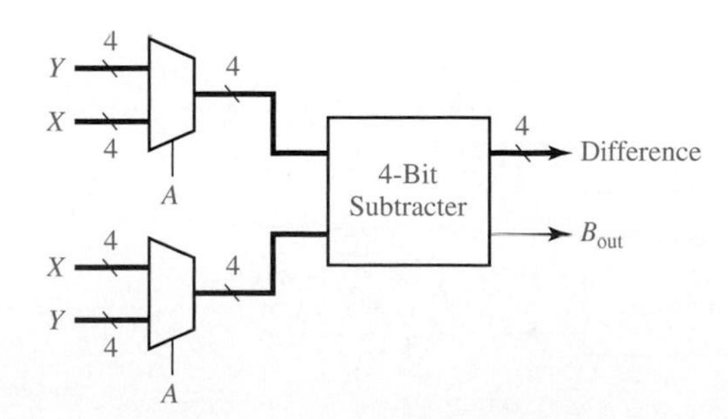

9.3

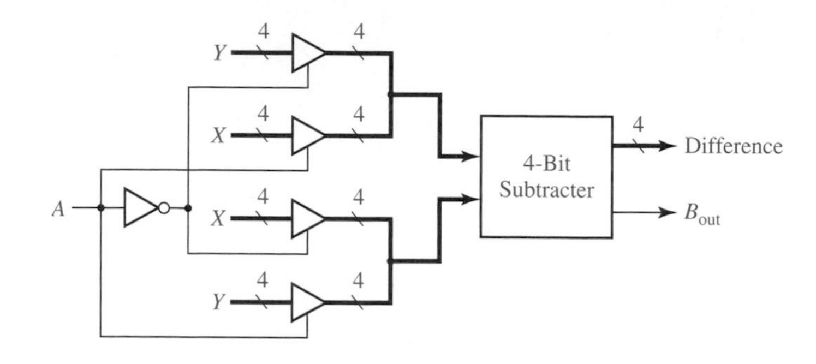


9.4 (a)

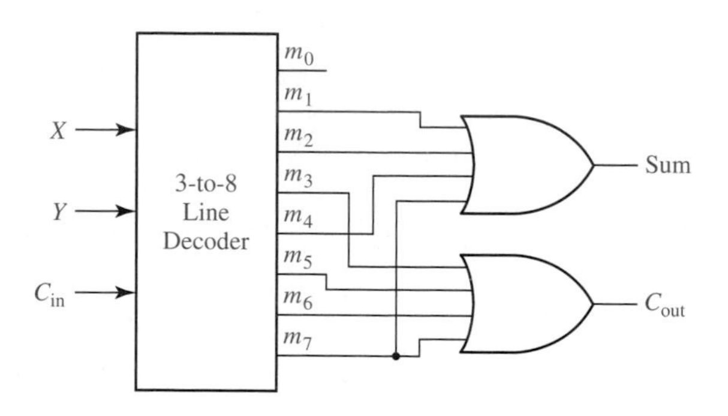


(b)

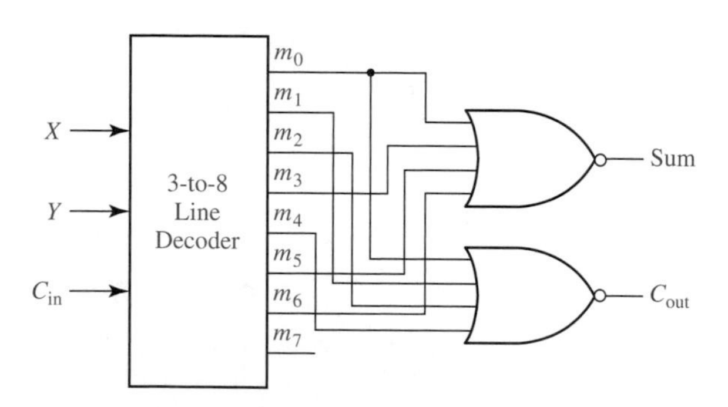


9.5

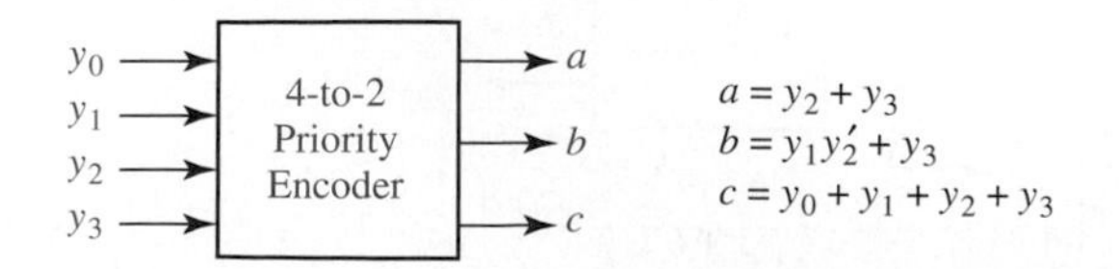


$a = y_2 + y_3$
$b = y_1 y_2' + y_3$
$c = y_0 + y_1 + y_2 + y_3$

9.6

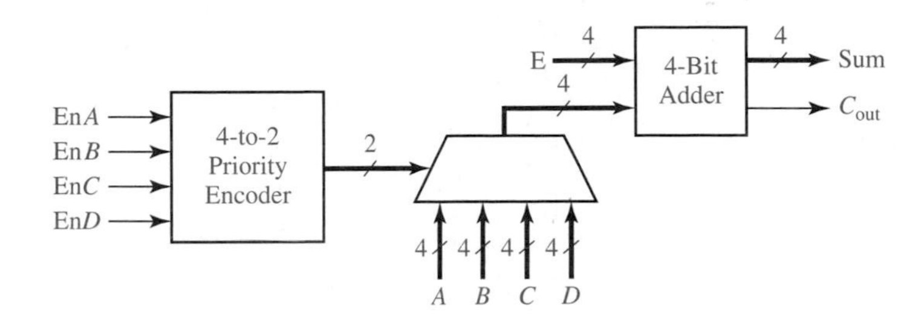


9.7 Block diagram for a Gray code adder:

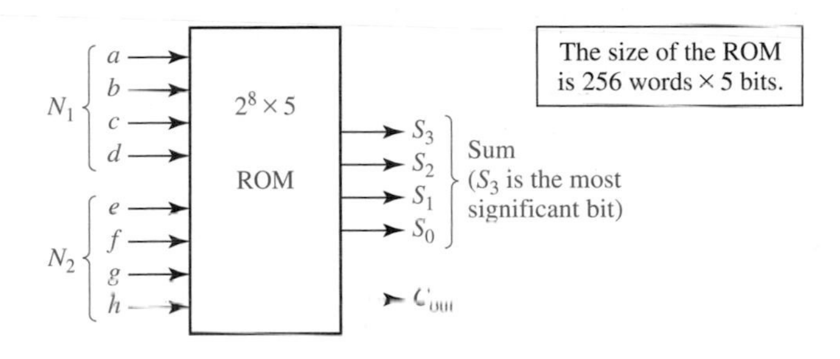


Partial Truth Table

| | a | b | c | d | e | f | g | h | $S_3$ | $S_2$ | $S_1$ | $S_0$ | $C_{out}$ |
|---|---|---|---|---|---|---|---|---|---|---|---|---|---|
| (0 + 0 = 0) | 0 | 0 | 0 | 0 | 0 | 0 | 0 | 0 | 0 | 0 | 0 | 0 | 0 |
| (1 + 2 = 3) | 0 | 0 | 0 | 1 | 0 | 0 | 1 | 1 | 0 | 0 | 1 | 0 | 0 |
| (5 + 7 = 12) | 1 | 1 | 1 | 0 | 1 | 0 | 1 | 1 | 0 | 0 | 1 | 1 | 1 |
| (8 + 9 = 17) | 1 | 0 | 0 | 1 | 1 | 0 | 0 | 0 | 1 | 0 | 1 | 1 | 1 |

9.8 (a)

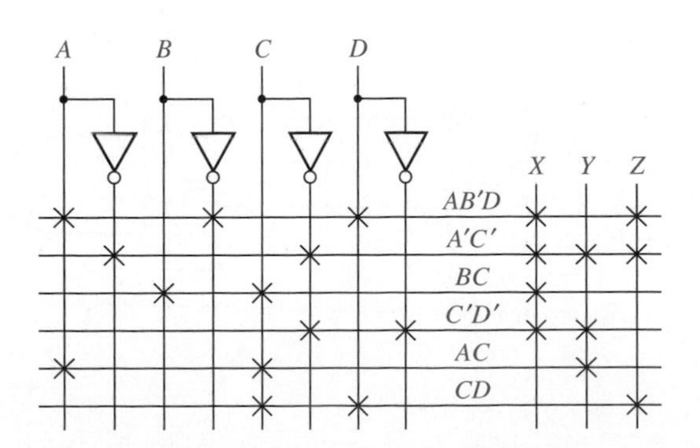

(b) Truth Table for the ROM

| *A* | *B* | *C* | *D* | *X* | *Y* | *Z* |
|---|---|---|---|---|---|---|
| 0 | 0 | 0 | 0 | 1 | 1 | 1 |
| 0 | 0 | 0 | 1 | 1 | 1 | 1 |
| 0 | 0 | 1 | 0 | 0 | 0 | 0 |
| 0 | 0 | 1 | 1 | 0 | 0 | 1 |
| 0 | 1 | 0 | 0 | 1 | 1 | 1 |
| 0 | 1 | 0 | 1 | 1 | 1 | 1 |
| 0 | 1 | 1 | 0 | 1 | 0 | 0 |
| 0 | 1 | 1 | 1 | 1 | 0 | 1 |
| 1 | 0 | 0 | 0 | 1 | 1 | 0 |
| 1 | 0 | 0 | 1 | 1 | 0 | 1 |
| 1 | 0 | 1 | 0 | 0 | 1 | 0 |
| 1 | 0 | 1 | 1 | 1 | 1 | 1 |
| 1 | 1 | 0 | 0 | 1 | 1 | 0 |
| 1 | 1 | 0 | 1 | 0 | 0 | 0 |
| 1 | 1 | 1 | 0 | 1 | 1 | 0 |
| 1 | 1 | 1 | 1 | 1 | 1 | 1 |

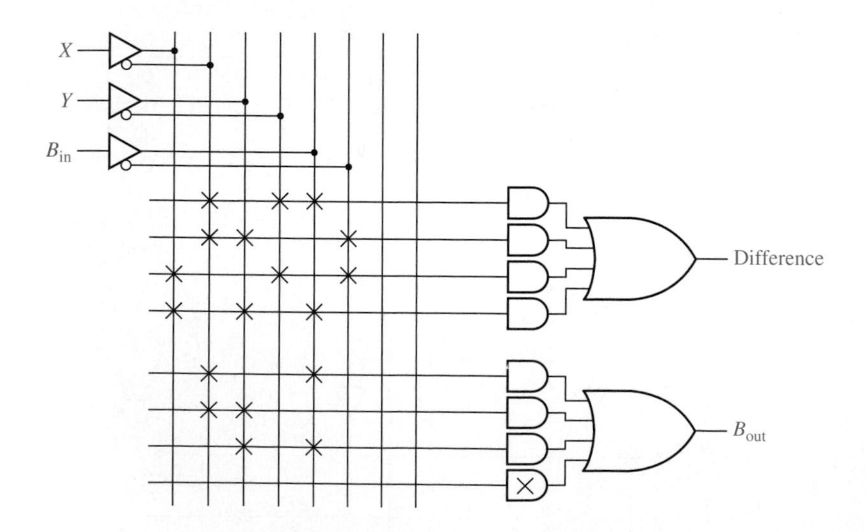

9.10 (a) $A_4 = W' + X'Y' \quad A_3 = WX'Y' \quad A_2 = W'X + XZ + XY$
$A_1 = W'Y + WXY'Z' + YZ \quad A_0 = W'Z + WXZ' + X'Y'Z + WYZ'$

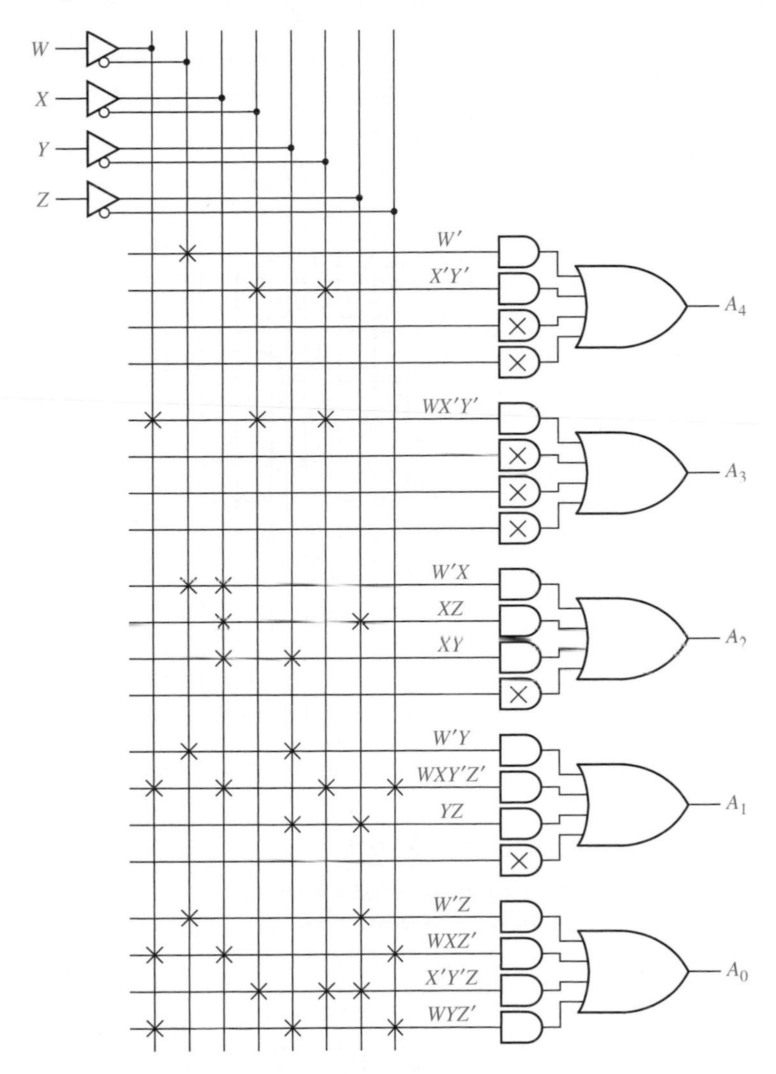


(b)

| $W$ | $X$ | $Y$ | $Z$ | $A_4$ | $A_3$ | $A_2$ | $A_1$ | $A_0$ |
|---|---|---|---|---|---|---|---|---|
| 0 | - | - | - | 1 | 0 | 0 | 0 | 0 |
| - | 0 | 0 | - | 1 | 0 | 0 | 0 | 0 |
| 1 | 0 | 0 | - | 0 | 1 | 0 | 0 | 0 |
| 0 | 1 | - | - | 0 | 0 | 1 | 0 | 0 |
| - | 1 | - | 1 | 0 | 0 | 1 | 0 | 0 |
| - | 1 | 1 | - | 0 | 0 | 1 | 0 | 0 |
| 0 | - | 1 | - | 0 | 0 | 0 | 1 | 0 |
| 1 | 1 | 0 | 0 | 0 | 0 | 0 | 1 | 0 |
| - | - | 1 | 1 | 0 | 0 | 0 | 1 | 0 |
| 0 | - | - | 1 | 0 | 0 | 0 | 0 | 1 |

*(continued)*

| | | | | | | | | | |
|---|---|---|---|---|---|---|---|---|---|
| 1 | 1 | - | 0 | | 0 | 0 | 0 | 0 | 1 |
| - | 0 | 0 | 1 | | 0 | 0 | 0 | 0 | 1 |
| 1 | - | 1 | 0 | | 0 | 0 | 0 | 0 | 1 |

9.11 (a) Not inverting, three AND gates. Inverting, $F = ac + b'c'd$, two AND gates.
(b) Not inverting, two AND gates. Inverting, $F = ad + ac + bd + bc$, four AND gates.

9.12 (b)

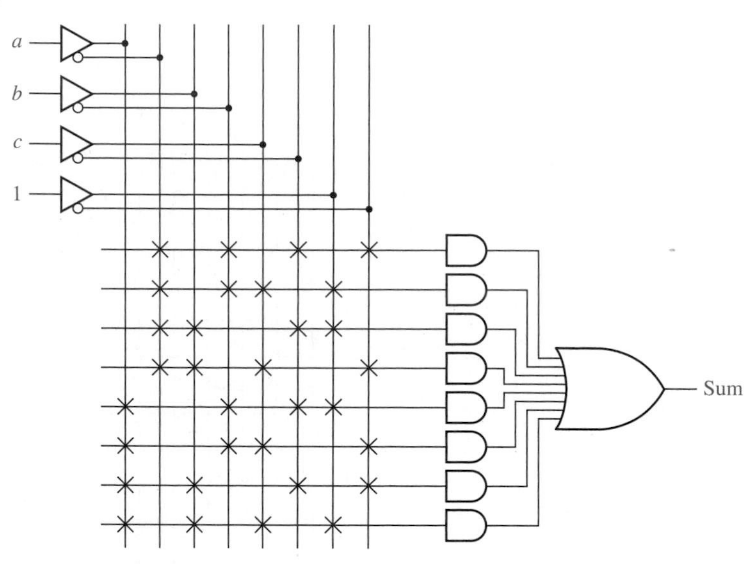


9.13 $F = b'(ade' + a'cd'e) + b\,((c'd'e + a'cd'e) + ac'de')$

## UNIT 10 Study Guide Answers

1. (b) Both statements execute at 5 ns. C and D are updated at $5 + \Delta$ ns.
(c) M <= **not** M **after** 5 ns;

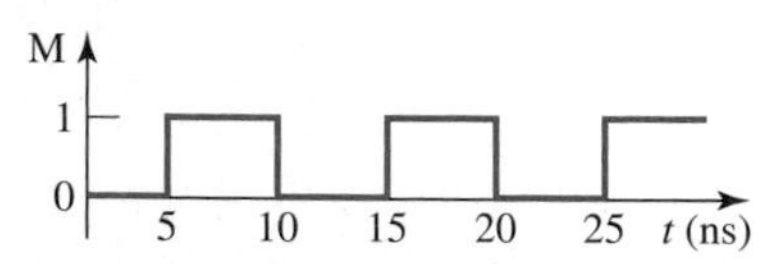


(d) A <= (**not** B **and** C) or (B **and not** C);

2. (a)

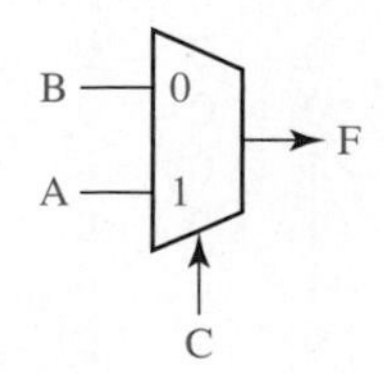


(b) F <= '1' **when** A&B = "00" **else** '0' **when** A&B = "01" **else** C;

(c)
```
AB <= A&B;
with AB select
  F <= '1' when "00", '0' when "01", C when "10", C when "11";
```

3. (c) Change all (3 **downto** 0) to (4 **downto** 0). Change (3 **downto** 1) to (4 **downto** 1). Add another instance of a full adder –
FA4: FullAdder **port map** (A(4), B(4), C(4), Co, S(4));
Change Co in FA3 to C(4).

(f)
```
architecture ckt of fig8_5 is
signal G1: bit;
begin
  G1 <= A and B after 20 ns;
  G2 <= G1 nor C after 20 ns;
end ckt;
```

5. (a) **not** (A&B **xor** "10") **not** (A&B) **xor** "10"
(b) The given statement will keep executing over and over again.

7. (a) A = '1', B = 'X', C = '0', D = '1', E = 'X', F = 'Z'
(b) If *F* is of type bit, compiler will log an error.
If *F* is std_logic, it will be 0 for 2 ns and, then, become X.
(c) Addout = 10011, Sum = 0011, Cout = 1
(d)
```
Addout <= ('0' & A) + ("000" & B);
Sum <= Addout(5 downto 0);
Cout <= Addout(6);
```
(e)

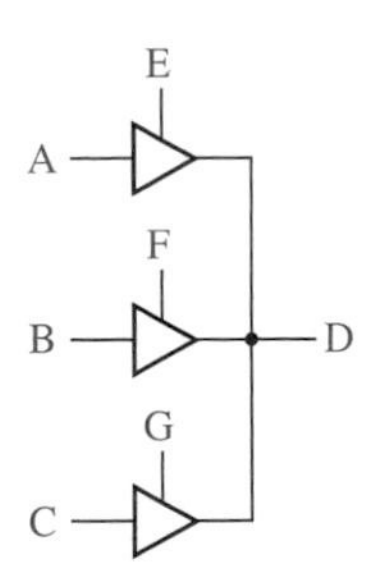


## UNIT 10 Answers to Problems

10.1 (a) F <= **not** A **and** B **and** C; G <= D **and not** E; N <= F **xor** G; I <= **not** N;
(b) I <= **not** (( **not** A **and** B **and** C) **xor** (D **and not** E));

10.2

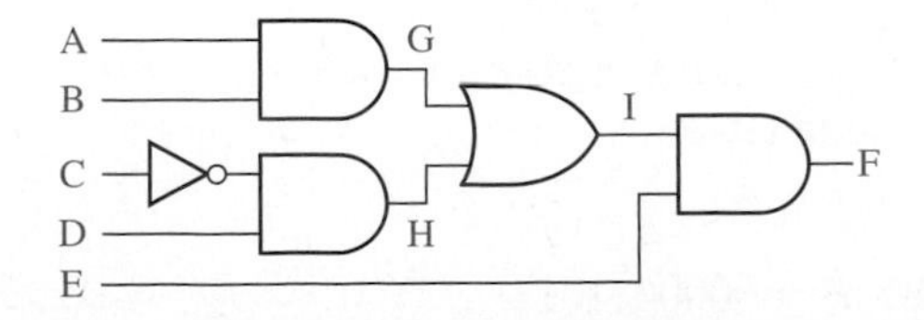

10.3 (a) (b)

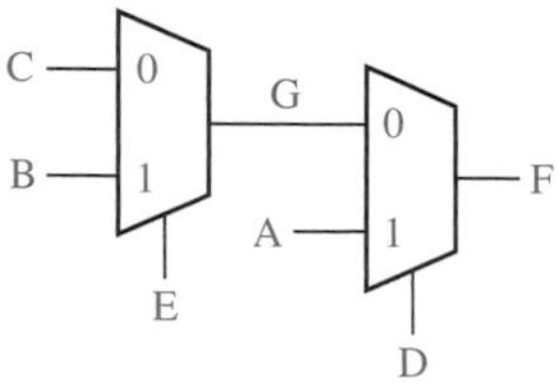

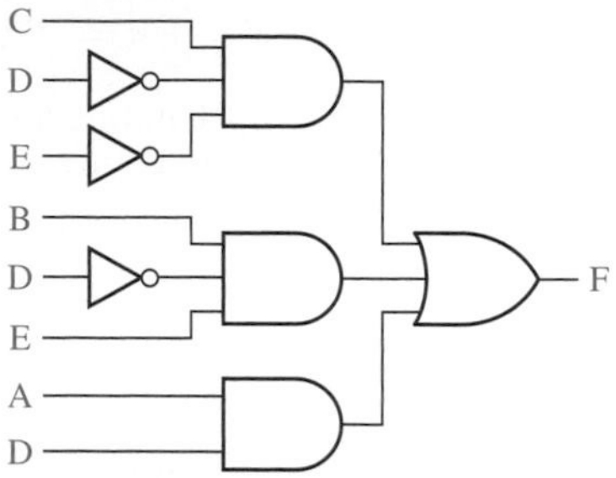

10.4

```
entity quad_mux is
   port (X, Y: in bit_vector(3 downto 0);
      A: in bit;
      Z: out bit_vector(3 downto 0));
end quad_mux;
architecture equations of quad_mux is
begin
   Z <= X when A = '0' else Y;
end equations;
```

10.5

```
entity ROM is
   port (A, B, Cin: in bit;
      Sum, Cout: out bit);
end ROM;
architecture table of ROM is
   type ROM8_2 is array(0 to 7) of bit_vector(1 downto 0);
   constant ROM1: ROM8_2 := ("00", "01", "01", "10", "01", "10", "10","11");
   signal index: integer range 0 to 7;
   signal S: bit_vector(1 downto 0);
   begin
      index <= vec2int(A&B&Cin);
      S <= ROM1(index);
      Sum <= S(0);
      Cout <= S(1);
end table;
```

10.6 (a) $F = 000001101$

(b) The expression evaluates to TRUE.

10.7

```
entity average is
   port (a, b, c, d: in std_logic_vector(15 downto 0);
   f: out std_logic_vector(15 downto 0));
end average;
architecture behavioral of average is
   signal sum: std_logic_vector(17 downto 0);
   begin
   sum <= ("00" & a) + b + c + d;
   f <= sum (17 downto 2) + sum (1);
end behavioral;
```

10.8

```
Bus <= A when EnA = '1' else "ZZZZ";
Bus <= B when EnB = '1' else "ZZZZ";
Bus <= C when EnC = '1' else "ZZZZ";
Bus <= D when EnD = '1' else "ZZZZ";
```

10.9 (a)

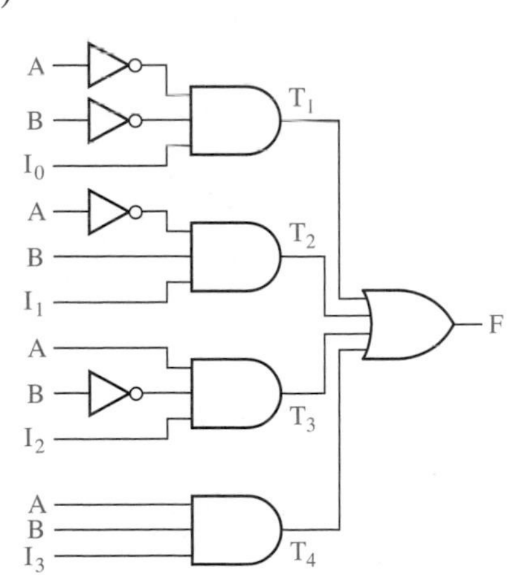

(b)

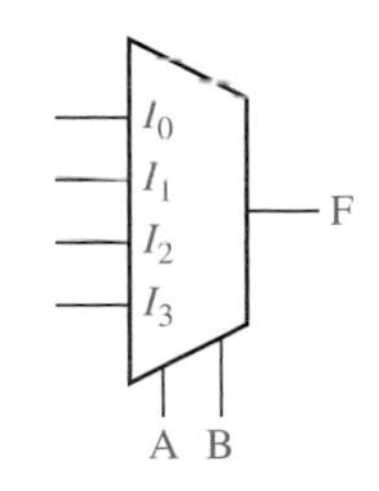


```
sel <= A&B;
with sel select
F <= I0 when "00", I1 when "01",
       I2 when "10", I3 when "11";
```

## UNIT 11 Study Guide Answers

1. Left inverter has a 1 output; right inverter has a 0 output.
2. (b) $P = Q = 0$ (c) $S$ and $R$ cannot both be 1 simultaneously.
3. (c)

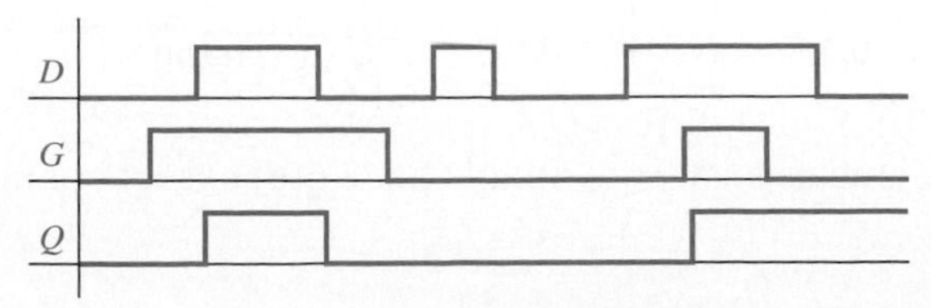

4. (b) $Q$ changes to 1 after first rising clock edge and back to 0 after third rising clock edge.
   (d) Hold time violation ($D$ is not stable for 2 ns after second falling clock edge.)
   (e) = 7 ns.
5. (c) For a rising-edge-triggered flip-flop, the value of the inputs is sensed at the rising edge of the clock, and inputs can change when the clock is low. For a master-slave flip-flop, if the inputs change when the clock is low, the flip-flop outputs may be incorrect.
6. (c)

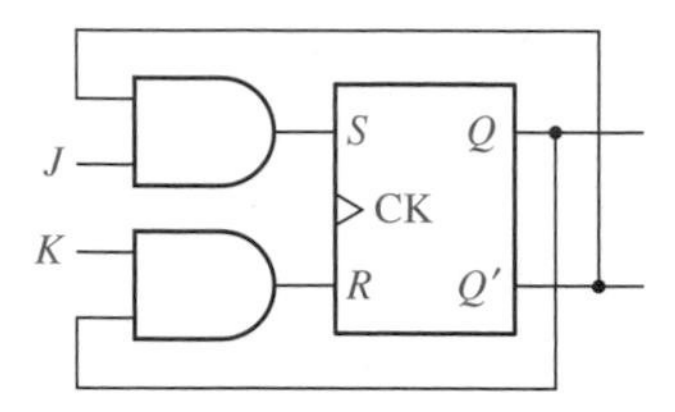


7. (b) $Q$ changes its value at times 1 and 2.
8. (b)

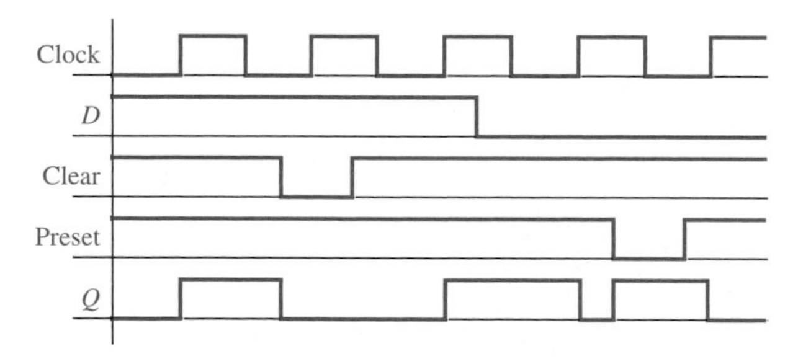


   (c) If CLK = 1, it will produce a falling edge at the clock input of the D flip-flop, causing the output to change. If CLK = 0, only the rising edge is affected, so the output does not change at the wrong time.
   En cannot be changed when the clock is 1.
   The flip-flops in Figures 11-27(b) and (c) can only change on the falling edge of the clock.
   (d)

| CK | D | CE | ClrN | $Q^+$ |
|---|---|---|---|---|
| x | x | x | 0 | 0 |
| x | x | 0 | 1 | Q (no change) |
| ↓ | 0 | 1 | 1 | 0 |
| ↓ | 1 | 1 | 1 | 1 |
| 0,1,↑ | x | 1 | 1 | Q (no change) |

9. (b) $S = Q'T$, $R = QT$
   Same as answer to Study Guide 6(c) except connect $J$ and $K$ and label it $T$.

## UNIT 11 Answers to Problems

11.1

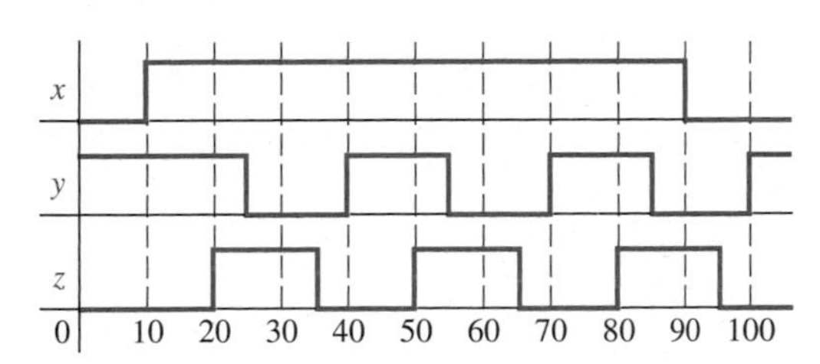


11.2 (a) $R = 1$ and $H = 0$ cannot occur at the same time.

(b)

| R | H | Q | $Q^+$ |
|---|---|---|---|
| 0 | 0 | 0 | 0 |
| 0 | 0 | 1 | 0 |
| 0 | 1 | 0 | 0 |
| 0 | 1 | 1 | 1 |
| 1 | 0 | 0 | X |
| 1 | 0 | 1 | X |
| 1 | 1 | 0 | 1 |
| 1 | 1 | 1 | 1 |

$Q^+ = R + HQ$

(c)

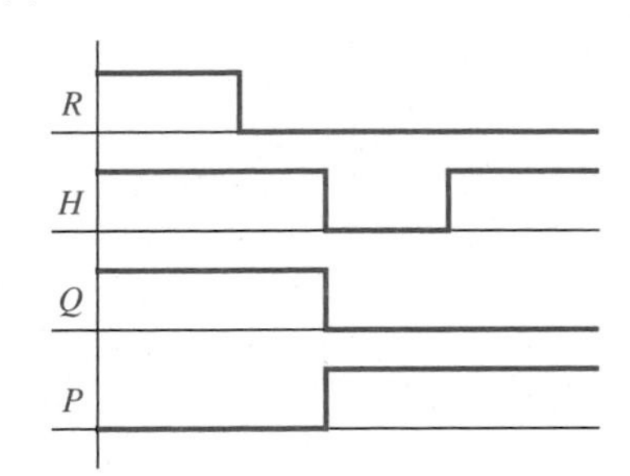


11.3

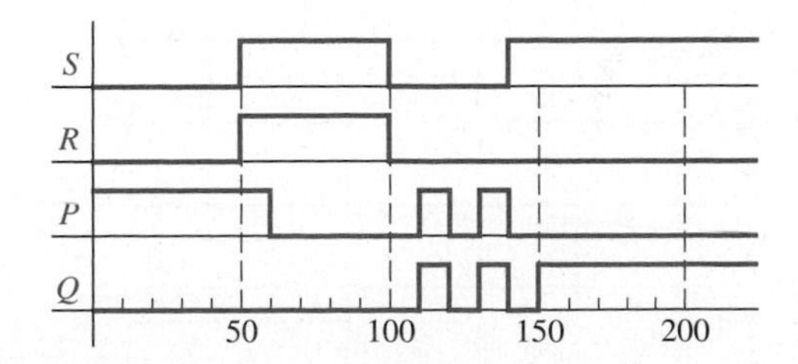

11.4

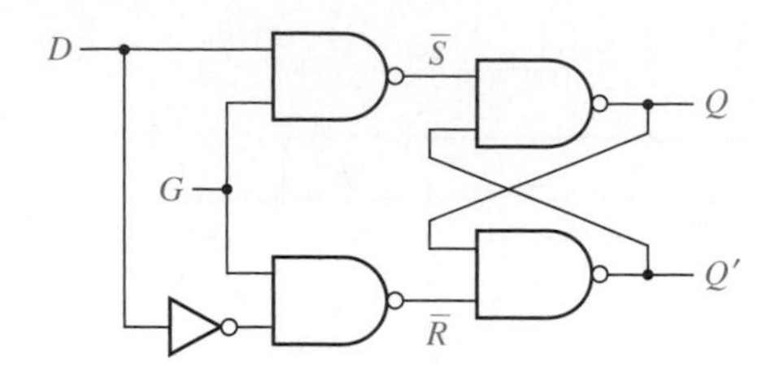


11.5 Connect the clock directly to the input $G_1$ and connect the clock to $G_2$ through an inverter.

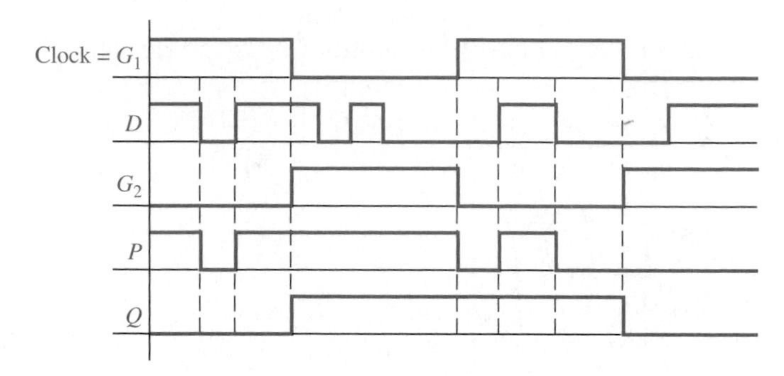


11.6 (a) $Q^+ = SR' + R'Q$ (b)

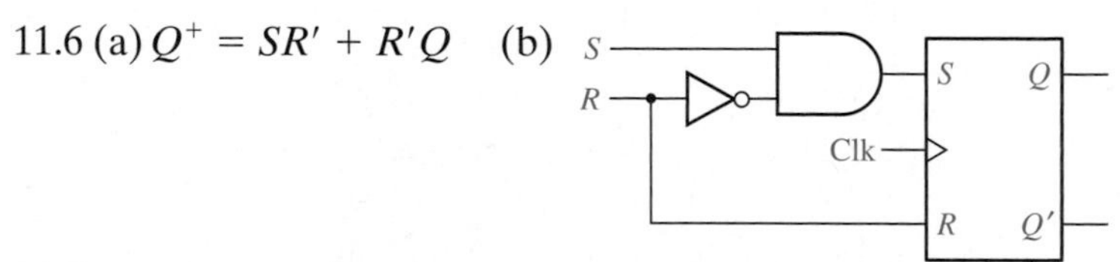


11.7

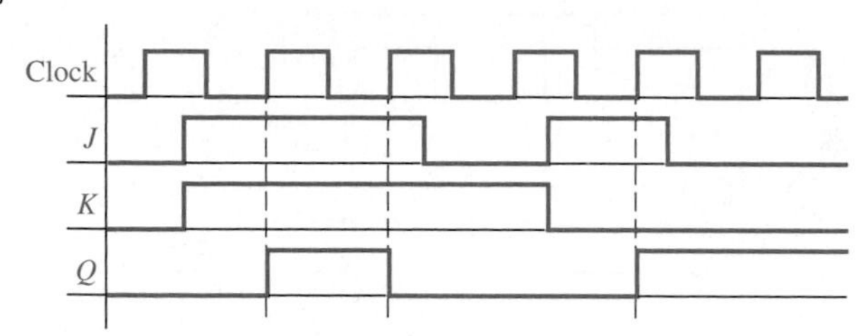


11.8

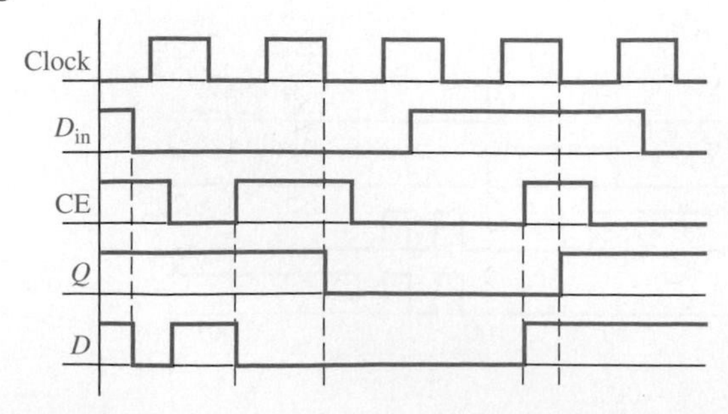

11.9 (a)

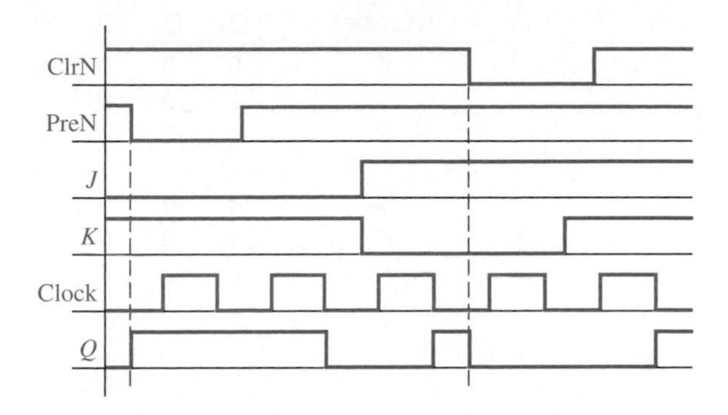


(b)

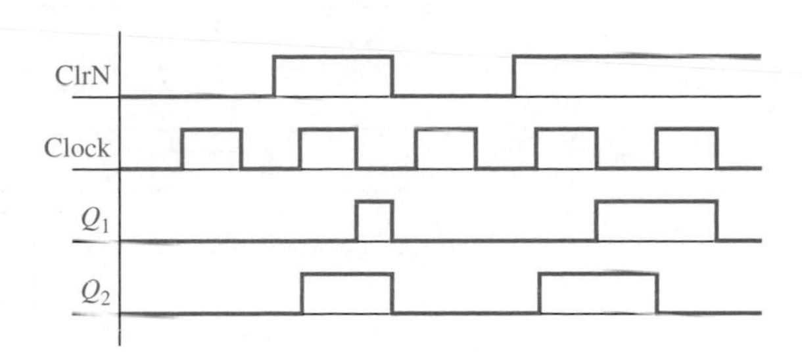


11.10 (a)

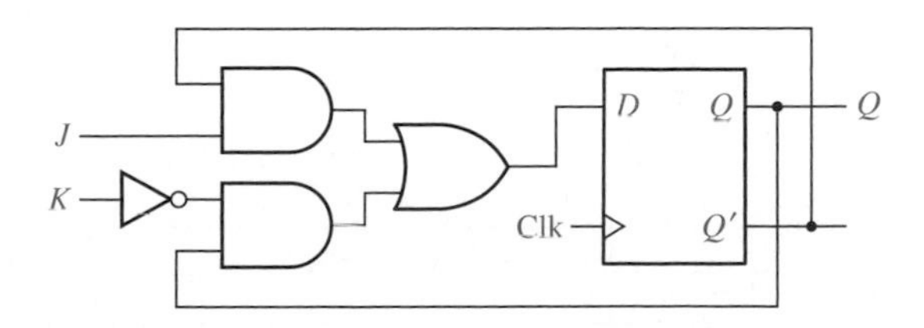


(b)

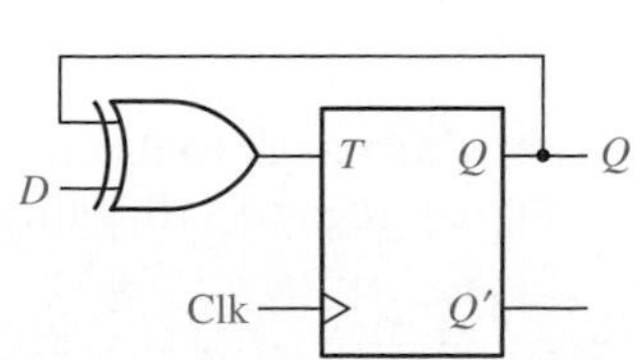


(c)

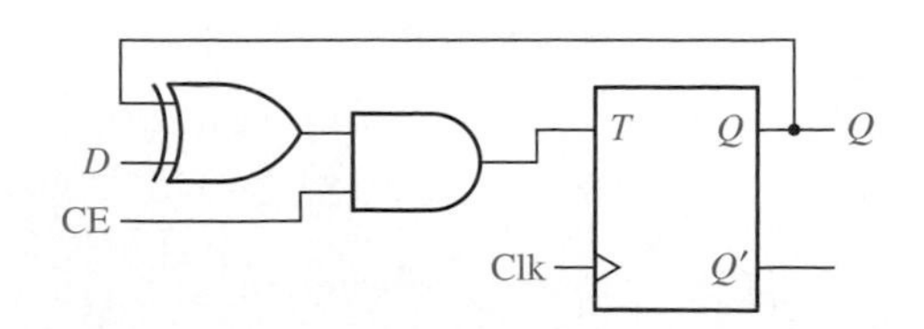


## UNIT 12 Study Guide Answers

1. (a) $G = 0, H = 249; G = 0, H = 70; G = 118, H = 118; G = 91, H = 118; G = 91, H = 118$
2. (b) $S_0$ is 1 between the rising edges of clocks 10 and 11, and also 1 between the rising edges of clocks 14 and 16.

| Clock Cycle Number | State of Shift Register When Clock = 1 $Q_7$ | $Q_6$ | $Q_5$ | $Q_4$ | $Q_3$ | $Q_2$ | $Q_1$ | $Q_0$ |
|---|---|---|---|---|---|---|---|---|
| 1 | 0 | 0 | 0 | 0 | 0 | 0 | 0 | 0 |
| 2 | 0 | 0 | 0 | 0 | 0 | 0 | 0 | 0 |
| 3 | 1 | 0 | 0 | 0 | 0 | 0 | 0 | 0 |
| - | - | - | - | - | - | - | - | - |
| 14 | 0 | 0 | 0 | 0 | 0 | 0 | 1 | 1 |
| 15 | 0 | 0 | 0 | 0 | 0 | 0 | 0 | 1 |
| 16 | 0 | 0 | 0 | 0 | 0 | 0 | 0 | 0 |

3. (b)

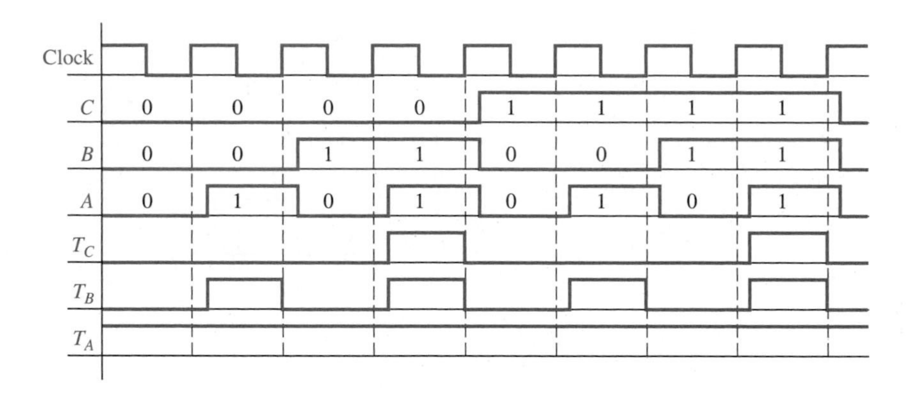


(d)

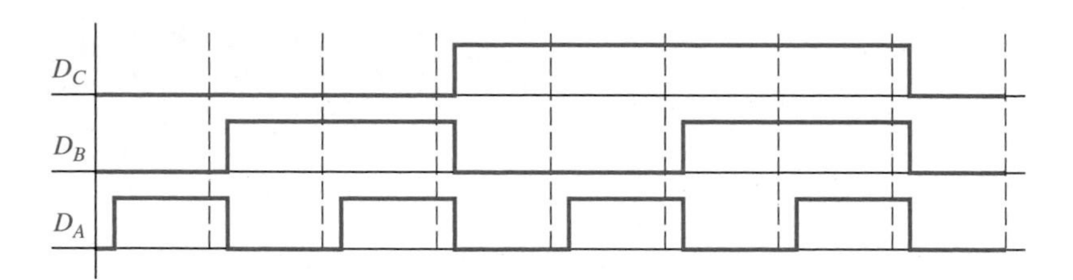


(f) State 101 goes to 110, which goes to 011.
(g) State 001 goes to 100; 101 goes to 110, which goes to 011.

4. (e)

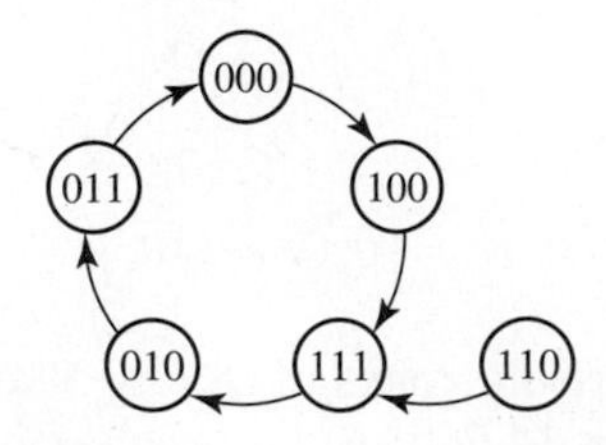


(k) $F_1 = 1 \qquad F_2 = 0$

## UNIT 12 Answers to Problems

12.1

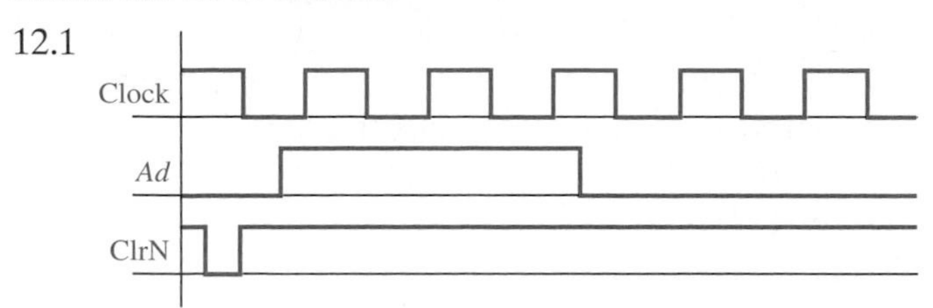


12.2

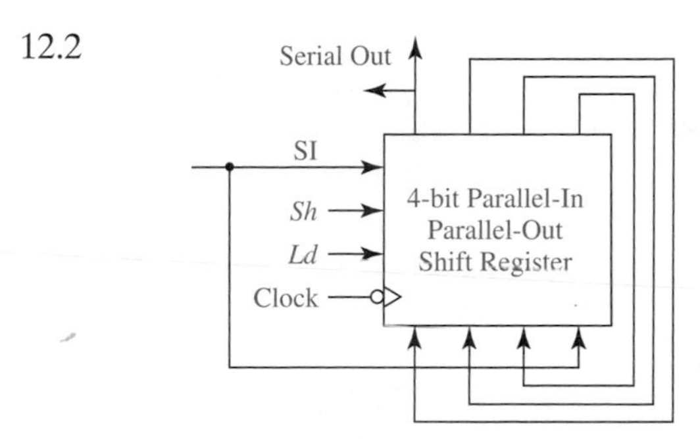


For a right shift, $Sh = 1, Ld = 0$ or 1. For a left shift, $Sh = 0, Ld = 1$.

12.3

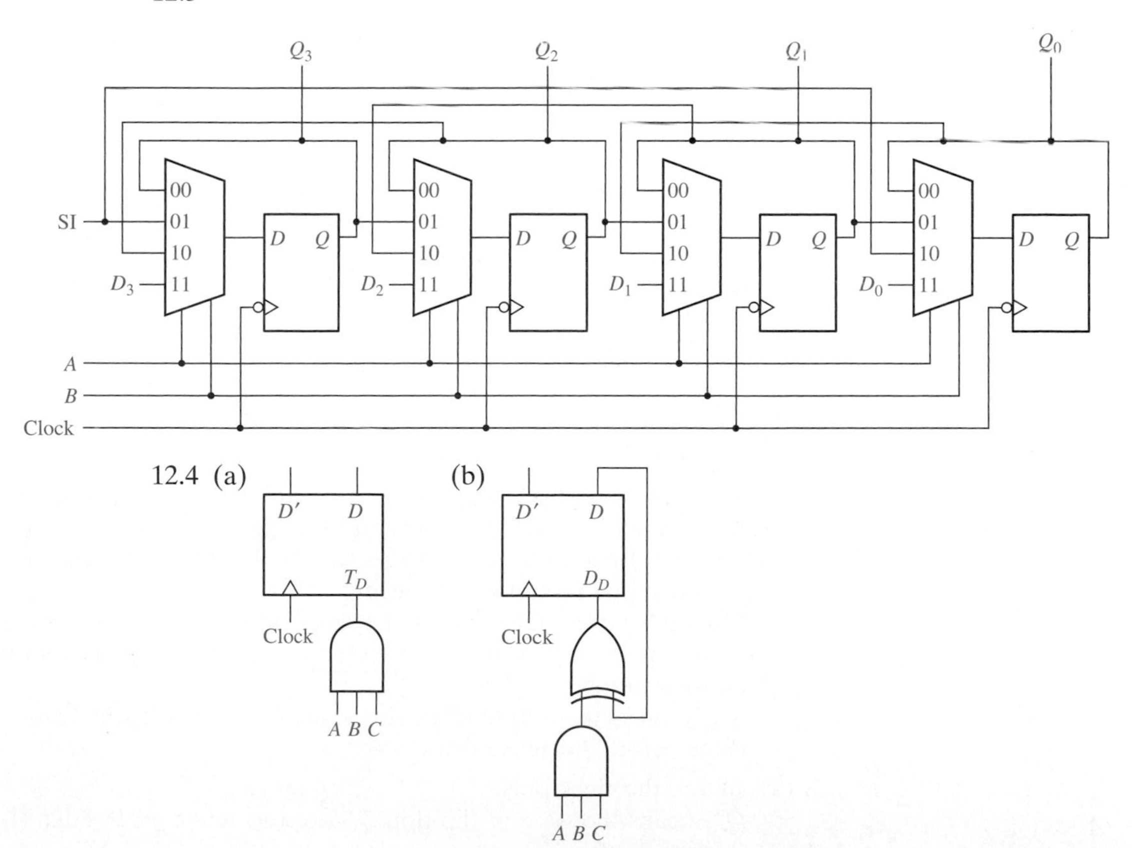

12.5 $D_D = D^+ = D'CBA + DC' + DB' + DA'$ $\quad D_B = B^+ = B'A + BA'$
$D_C = C^+ = C'BA + CB' + CA'$ $\quad D_A = A^+ = A'$

12.6 Uses three flip-flops: $Q_3\, Q_2\, Q_1$
Many correct solutions are possible. One is:
$D_3 = Q_1 + Q_2Q_3'$ $\quad D_2 = Q_2'Q_3$
$D_1 = Q_1'Q_3'$

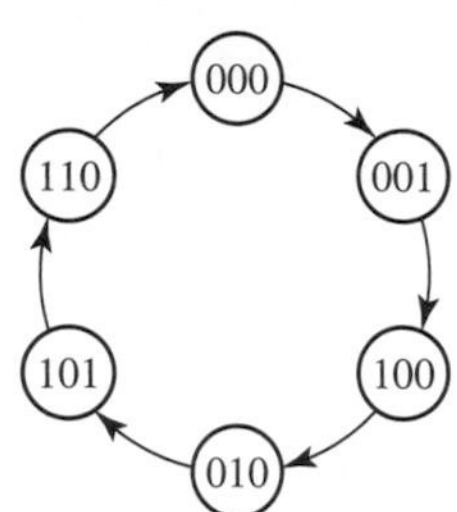


12.7 (a) $D_C = CA + BA'$ $\quad D_B = C' + BA'$
$D_A = B'A' + CB + C'B'$
If $CBA = 000$, next state is 011.
(b) $T_C = B'A' + C'A'$ $\quad T_B = C'B' + CBA$
$T_A = CB' + CA' + C'BA$
If $CBA = 000$, next state is 110.

12.8 (a) $J_C = A'$ $\quad J_B = C'$ $\quad J_A = C$
$K_C = B'A'$ $\quad K_B = CA$ $\quad K_A = CB' + C'B$
If $CBA = 000$, next state is 110.
(b) $S_C = BA'$ $\quad S_B = C'$ $\quad S_A = CA'$
$R_C = B'A'$ $\quad R_B = CA$ $\quad R_A = CB'A + C'B$
If $CBA = 000$, next state is 010.

12.9 (a)

| Q | Q⁺ | M | N |
|---|---|---|---|
| 0 | 0 | 0 | X |
| 0 | 1 | 1 | X |
| 1 | 0 | X | 0 |
| 1 | 1 | X | 1 |

(b) $M_C = B$ $\quad M_B = C'A$ $\quad M_A = C'$
$N_C = A$ $\quad N_B = C'$ $\quad N_A = C' + B$

## UNIT 13 Study Guide Answers

2. (a) Mealy: output a function of both input and state
Moore: output a function of state only
(b) Before the active clock edge
After the active clock edge
When the flip-flops change state
When the flip-flops change state or when the inputs change
(c) Immediately preceding the active clock edge
(d) Mealy: False outputs can appear when the state has changed to its next value, but the input has not yet changed.
Moore: No false outputs occur because output is not a function of input.
Changing the inputs at the same time the state change occurs will eliminate false outputs.
No, because the output of the first Mealy circuit will still change to its final value before the active clock edge.

3. (a) Before the clock pulse
$Q^+$ means the state of flip-flop $Q$ after the active clock edge (i.e., the next state of flip-flop $Q$).

(c) Mealy: output associated with transitions between states
Moore: output associated with state
(d) Present: Before the active clock edge
Next: after the clock pulse
(e) Output depends only on the state and not on the input.

4. (a) 1101 (c) 1001
(e)

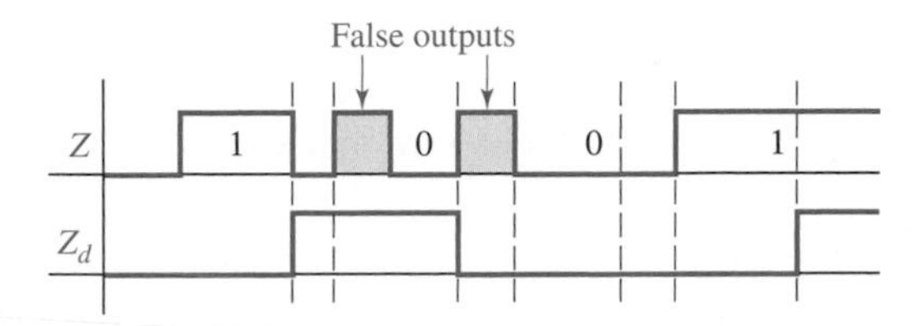


5. (a)

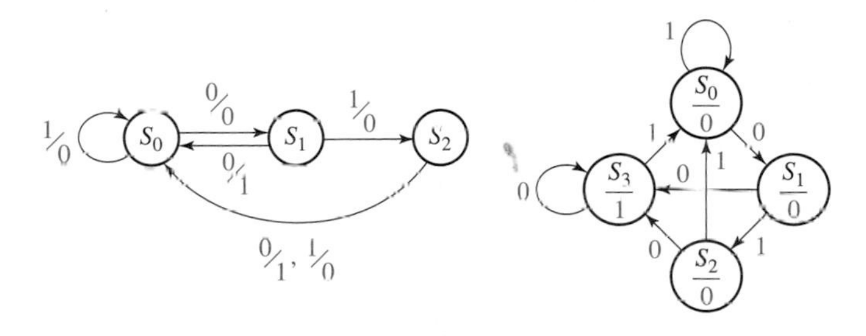


(c)

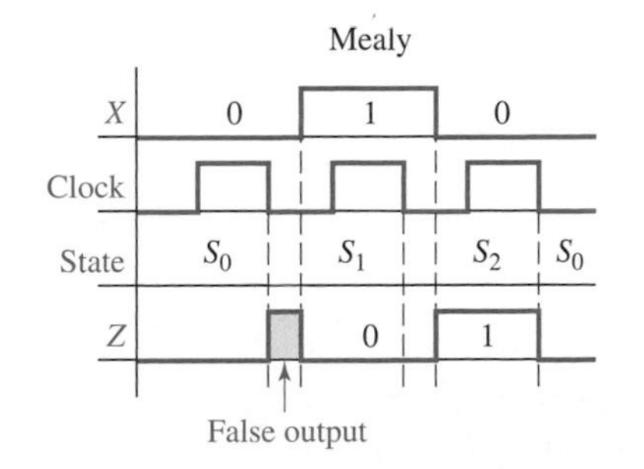


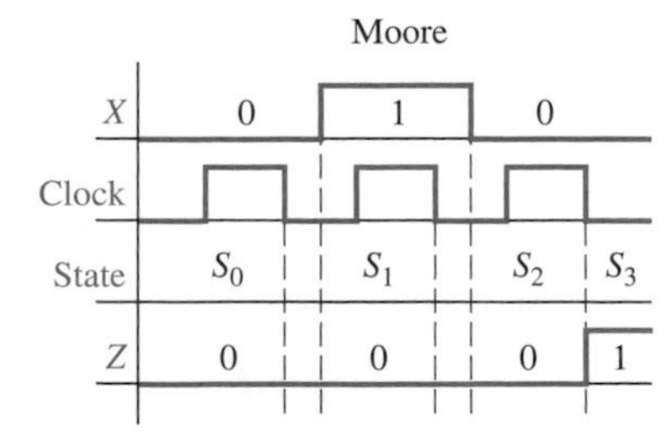


6. (a)

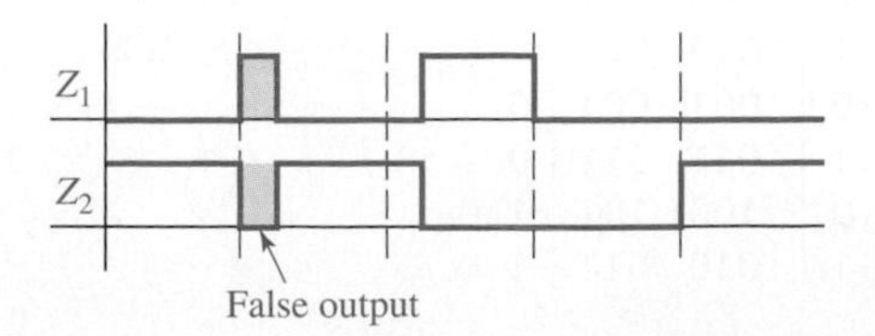


(g) $\delta(S_3, 1) = S_2, \lambda(S_3, 1) = 0, \delta(S_1, 2) = S_2, \lambda(S_1, 2) = 3$

## UNIT 13 Answers to Problems

13.2

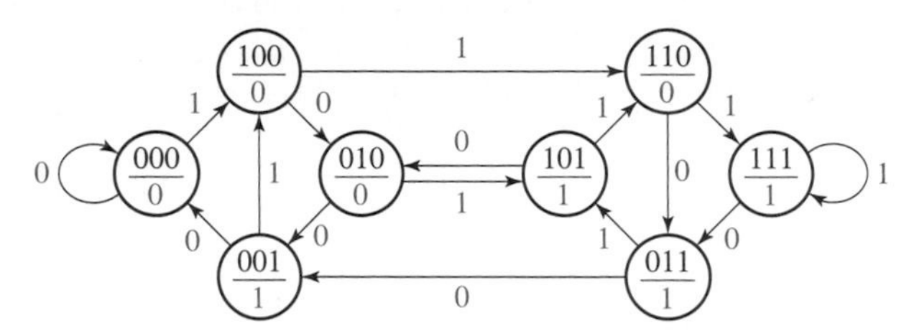


This is a Moore machine.

13.3 (a) $A^+ = A(B' + X) + A'(BX' + B'X)$ $\quad B^+ = AB'X + B(A' + X')$

| Present State $AB$ | Next State ($A^+B^+$) $x = 0$ | $x = 1$ | $Z$ |
|---|---|---|---|
| 00 | 00 | 10 | 0 |
| 01 | 11 | 01 | 0 |
| 11 | 01 | 10 | 1 |
| 10 | 10 | 11 | 0 |

This is a Moore machine.

(b) $Z = (0)00101$

(c)

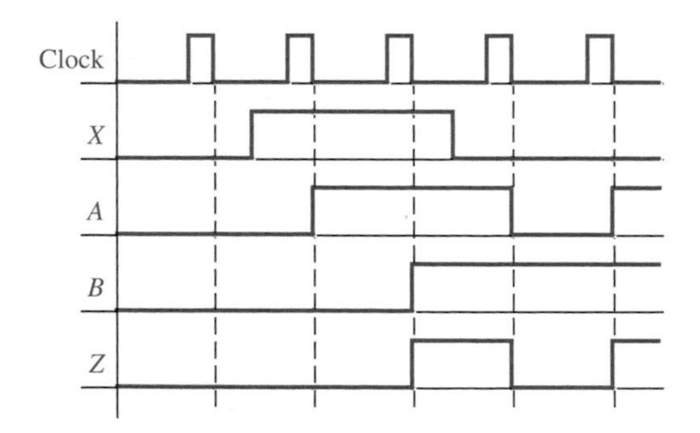


13.4 (a)

| | $Q_1Q_2Q_3$ | $Q_1^+Q_2^+Q_3^+$ $X = 0$ | 1 | $Z$ 0 | 1 |
|---|---|---|---|---|---|
| $S_0$ | 0 0 0 | 001 | 001 | 0 | 1 |
| $S_1$ | 0 0 1 | 011 | 011 | 0 | 1 |
| $S_2$ | 0 1 0 | 100 | 101 | 1 | 0 |
| $S_3$ | 0 1 1 | 010 | 011 | 1 | 0 |
| $S_4$ | 1 0 0 | 001 | 001 | 0 | 1 |
| $S_5$ | 1 0 1 | 011 | 011 | 0 | 1 |
| $S_6$ | 1 1 0 | 100 | 101 | 1 | 0 |
| $S_7$ | 1 1 1 | 010 | 011 | 1 | 0 |

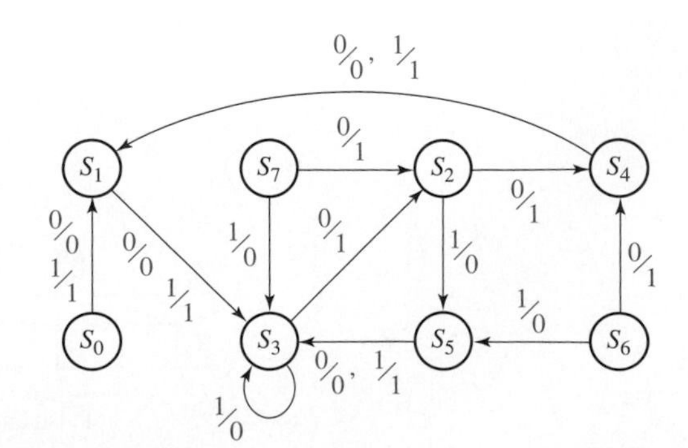

(b)

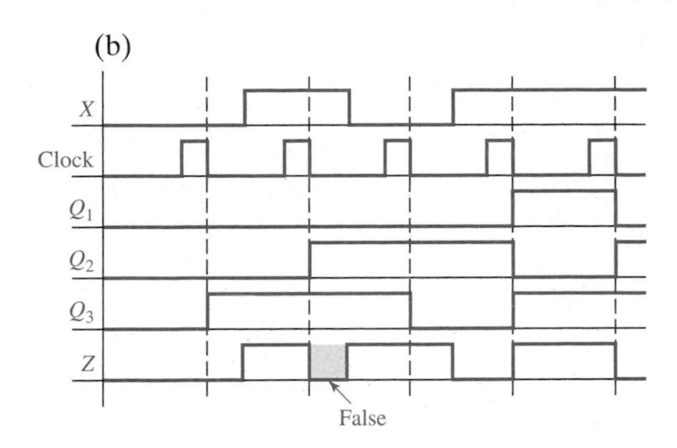


(c) From diagram: 0 1(0) 1 0 1
From graph: 0 1 1 0 1
(same except for false output)

(d) Change input on falling edge of clock

13.5 (a) Mealy Machine

(b)

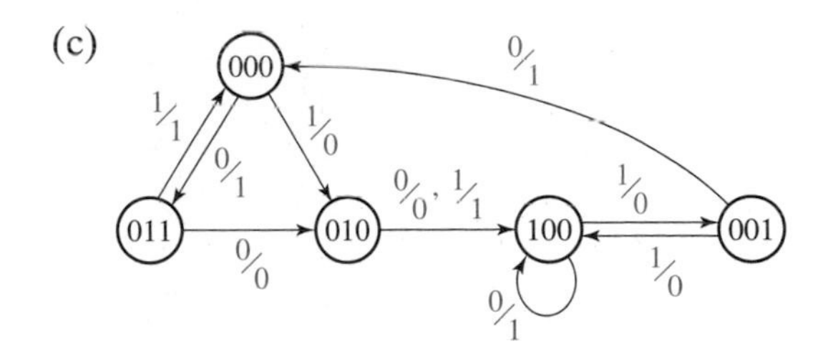

| ABC | $A^+B^+C^+$ $X = 0$ | $X = 1$ | $Z$ $X = 0$ | $X = 1$ |
|---|---|---|---|---|
| 000 | 011 | 010 | 1 | 0 |
| 001 | 000 | 100 | 1 | 0 |
| 010 | 100 | 100 | 0 | 1 |
| 011 | 010 | 000 | 0 | 1 |
| 100 | 100 | 001 | 1 | 0 |

(c)

(d)

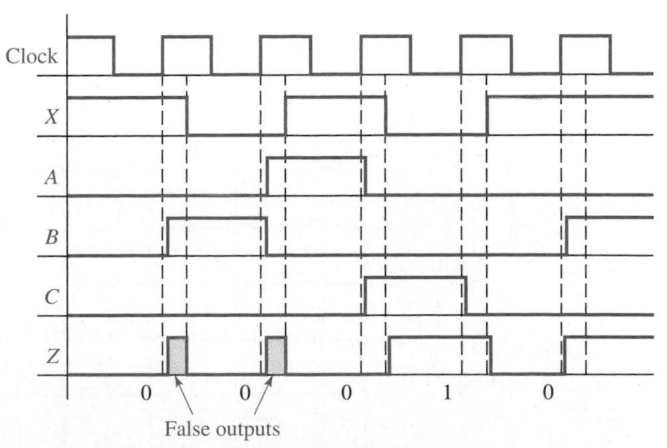


Correct output sequence: 00010

13.6 (a) 14 ns

(b)

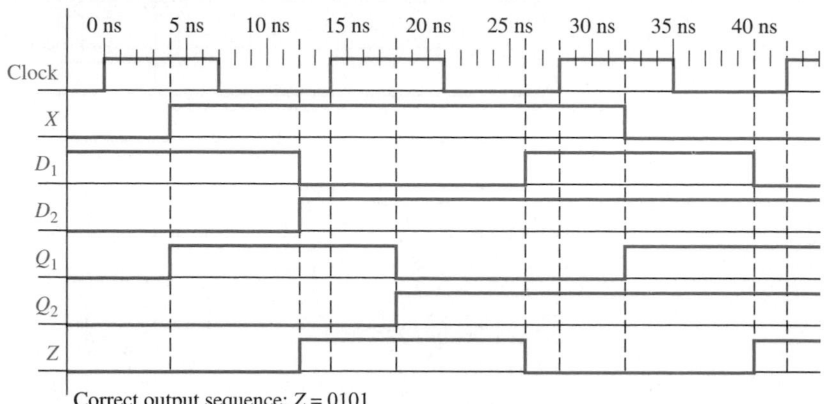


(c)

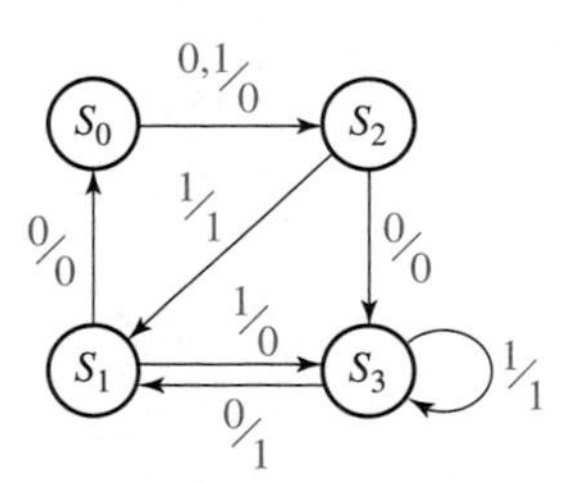


| | Next State | | Z | |
|---|---|---|---|---|
| | $X = 0$ | $X = 1$ | $X = 0$ | $X = 1$ |
| $S_0$ | $S_2$ | $S_2$ | 0 | 0 |
| $S_1$ | $S_0$ | $S_3$ | 0 | 0 |
| $S_2$ | $S_3$ | $S_1$ | 0 | 1 |
| $S_3$ | $S_1$ | $S_3$ | 1 | 1 |

## UNIT 14 Study Guide Answers

1. (b) last row: 11 10 01 0 1

   (c) $J_A = BX'$ $K_A = X \oplus B$ $J_B = A + X$ $K_B = A$ $Z = AB'$

8. (a)

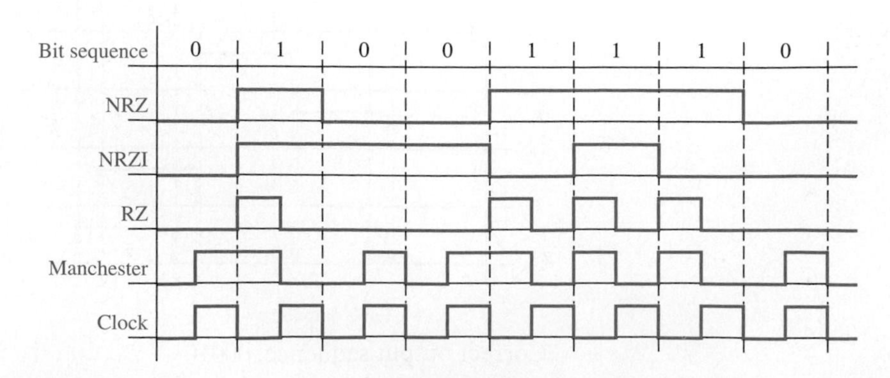

9. (b) Change $N$ to $NS'$; add loop to $S_3$: $S'N'/0$

| | NS = 00 | 01 | 10 | 11 | Z 00 | 01 | 10 | 11 |
|---|---|---|---|---|---|---|---|---|
| $S_3$ | $S_3$ | $S_5$ | $S_1$ | $S_5$ | 0 | 0 | 1 | 0 |

## UNIT 14 Answers to Problems

14.4

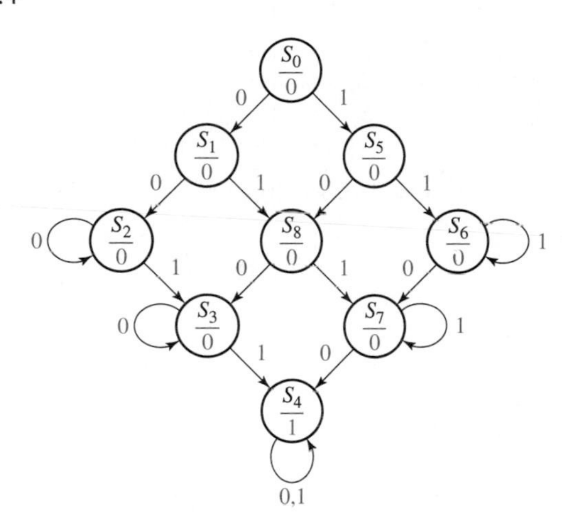


14.5

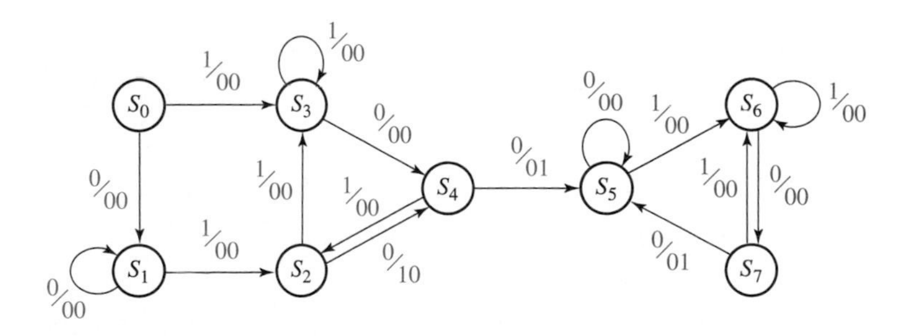


| | X = 0 | X = 1 | $Z_1Z_2$ X = 0 | X = 1 |
|---|---|---|---|---|
| $S_0$ | $S_1$ | $S_3$ | 00 | 00 |
| $S_1$ | $S_1$ | $S_2$ | 00 | 00 |
| $S_2$ | $S_4$ | $S_3$ | 10 | 00 |
| $S_3$ | $S_4$ | $S_3$ | 00 | 00 |
| $S_4$ | $S_5$ | $S_2$ | 01 | 00 |
| $S_5$ | $S_5$ | $S_6$ | 00 | 00 |
| $S_6$ | $S_7$ | $S_6$ | 00 | 00 |
| $S_7$ | $S_5$ | $S_6$ | 01 | 00 |

14.6

| | $X_1X_2 =$ 00 | 01 | 10 | 11 | $Z$ |
|---|---|---|---|---|---|
| $S_0$ | $S_0$ | $S_1$ | $S_3$ | $S_2$ | 0 |
| $S_1$ | $S_0$ | $S_1$ | $S_3$ | $S_2$ | 0 |
| $S_2$ | $S_4$ | $S_1$ | $S_3$ | $S_2$ | 0 |
| $S_3$ | $S_4$ | $S_1$ | $S_3$ | $S_2$ | 0 |
| $S_4$ | $S_4$ | $S_5$ | $S_7$ | $S_6$ | 1 |
| $S_5$ | $S_0$ | $S_5$ | $S_7$ | $S_6$ | 1 |
| $S_6$ | $S_4$ | $S_5$ | $S_7$ | $S_6$ | 1 |
| $S_7$ | $S_0$ | $S_5$ | $S_7$ | $S_6$ | 1 |

(a 4-state solution is also possible)

14.7 (a)

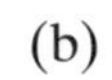


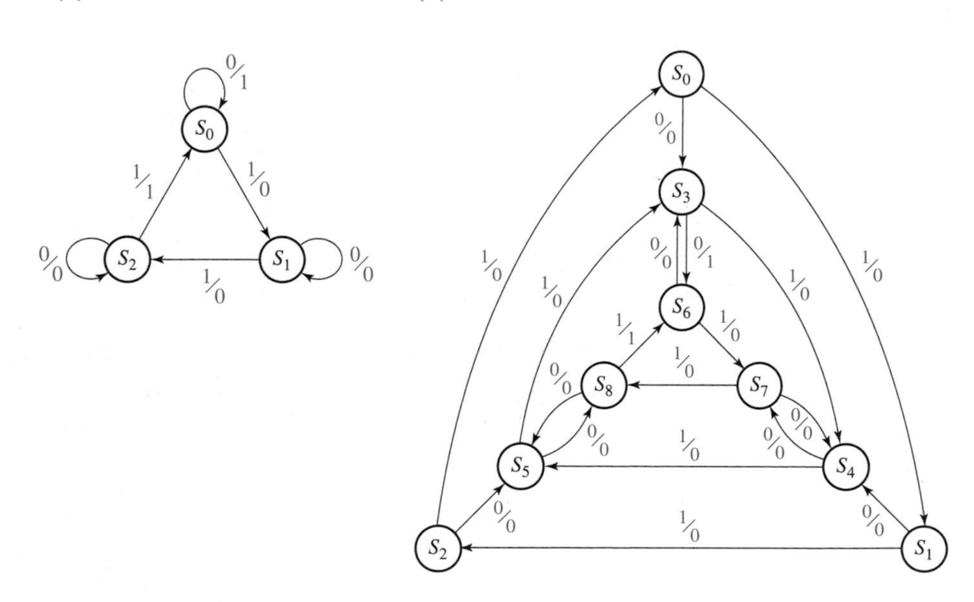


14.8

(a)

| | $X_1X_2 =$ 00 | 01 | 10 | 11 | $Z_1Z_2$ 00 | 01 | 10 | 11 |
|---|---|---|---|---|---|---|---|---|
| $S_0$ | $S_1$ | $S_2$ | $S_3$ | $S_4$ | 00 | 00 | 00 | 00 |
| $S_1$ | $S_1$ | $S_2$ | $S_3$ | $S_4$ | 00 | 10 | 10 | 10 |
| $S_2$ | $S_1$ | $S_2$ | $S_3$ | $S_4$ | 01 | 00 | 10 | 10 |
| $S_3$ | $S_1$ | $S_2$ | $S_3$ | $S_4$ | 01 | 01 | 00 | 10 |
| $S_4$ | $S_1$ | $S_2$ | $S_3$ | $S_4$ | 01 | 01 | 01 | 00 |

(b)

| | $X_1X_2 =$ 00 | 01 | 10 | 11 | $Z_1Z_2$ |
|---|---|---|---|---|---|
| $S_0$ | $S_1$ | $S_4$ | $S_7$ | $S_{10}$ | 00 |
| $S_1$ | $S_1$ | $S_3$ | $S_6$ | $S_9$ | 00 |
| $S_2$ | $S_1$ | $S_3$ | $S_6$ | $S_9$ | 01 |
| $S_3$ | $S_2$ | $S_4$ | $S_6$ | $S_9$ | 10 |
| $S_4$ | $S_2$ | $S_4$ | $S_6$ | $S_9$ | 00 |
| $S_5$ | $S_2$ | $S_4$ | $S_6$ | $S_9$ | 01 |
| $S_6$ | $S_2$ | $S_5$ | $S_7$ | $S_9$ | 10 |
| $S_7$ | $S_2$ | $S_5$ | $S_7$ | $S_9$ | 00 |
| $S_8$ | $S_2$ | $S_5$ | $S_7$ | $S_9$ | 01 |
| $S_9$ | $S_2$ | $S_5$ | $S_8$ | $S_{10}$ | 10 |
| $S_{10}$ | $S_2$ | $S_5$ | $S_8$ | $S_{10}$ | 00 |

14.9

(a)

| | $X = 0$ | 1 | $X = 0$ | 1 |
|---|---|---|---|---|
| $S_0$ | $S_0$ | $S_1$ | 0 | 1 |
| $S_1$ | $S_1$ | $S_0$ | 1 | 0 |

(b)

| | $X = 0$ | 1 | |
|---|---|---|---|
| $S_0$ | $S_0$ | $S_1$ | 0 |
| $S_1$ | $S_1$ | $S_0$ | 1 |

(c, d)

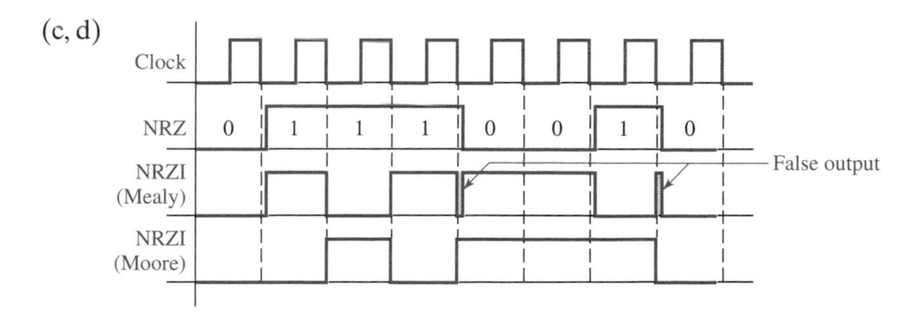


14.10

| | Next State 000 | 001 | 010 | 011 | 100 | 101 | 110 | 111 | Output ($DEF$) 000 | 001 | 010 | 011 | 100 | 101 | 110 | 111 |
|---|---|---|---|---|---|---|---|---|---|---|---|---|---|---|---|---|
| $S_0$ | $S_0$ | $S_0$ | $S_0$ | $S_0$ | $S_1$ | $S_1$ | $S_0$ | $S_0$ | 100 | 100 | 100 | 100 | 010 | 010 | 001 | 001 |
| $S_1$ | $S_1$ | $S_0$ | $S_1$ | $S_0$ | $S_1$ | $S_0$ | $S_1$ | $S_0$ | 110 | 000 | 110 | 000 | 101 | 000 | 101 | 000 |

For $S_0$:

$A' + AB + AB' = A' + A = 1$

$A' \cdot AB = 0;\ A' \cdot AB' = 0;$

$AB \cdot AB' = 0$

For $S_1$:

$A'C' + AC' + C = C' + C = 1$

$A'C' \cdot AC' = 0;$

$A'C' \cdot C = 0;\ AC' \cdot C = 0$

14.11

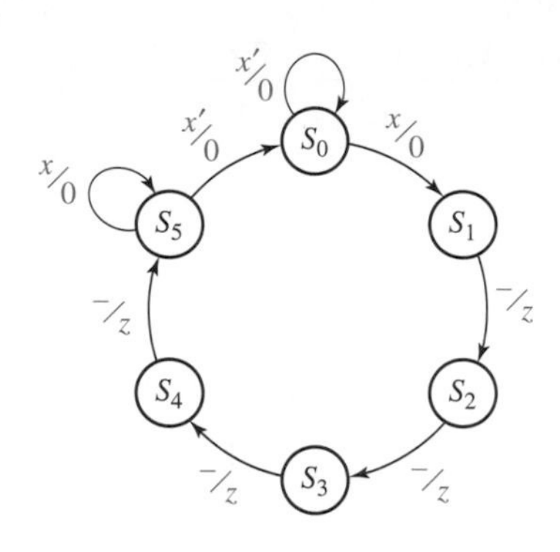


## UNIT 15 Study Guide Answers

2. (b) $\lambda(\mathrm{p}, 01) = 00$ and $\lambda(\mathrm{q}, 01) = 01$; therefore, $\mathrm{p} \not\equiv \mathrm{q}$
   (c) No. You would have to try an infinite number of test sequences to be sure the *circuits* were equivalent.
   (d) $S_2 \equiv S_3$ iff $S_1 \equiv S_5$ and $S_4 \equiv S_2$. But $S_1 \not\equiv S_5$ because the outputs are different. Therefore, $S_2 \not\equiv S_3$.

3. (a)

| | | | | | |
|---|---|---|---|---|---|
| *b* | X | | | | |
| *c* | X | *a-d* *a-b* | | | |
| *d* | *a-g* *b-f* | X | X | | |
| *f* | X | *a-g* | *a-d* *b-g* | X | |
| *g* | X | *a-f* | *a-d* *b-f* | X | ✓ |
| | *a* | *b* | *c* | *d* | *f* |

(b) $f \equiv g$
(c) $a \equiv c, b \equiv d, b \equiv e, d \equiv e$

| | 0 | 1 | 0 | 1 |
|---|---|---|---|---|
| a | b | a | 0 | 1 |
| b | b | b | 0 | 0 |

7. (b) $Z = X'AB' + XA'BC'$

8. (b) Interchanging columns or complementing columns does not affect circuit cost for symmetric flip-flops.
   (c) Complementing columns (to make the first row all 0's) does not change the cost of the circuit.
   (f) Numbering columns from left to right, column 3 is same as column 4, column 2 is column 5 complemented, column 1 is column 6 complemented.
9. (e) $D_1 = XQ_1' + XQ_3 + Q_2Q_3' + X'Q_1Q_2'$ or $D_1 = XQ_1' + XQ_2 + Q_2'Q_3 + X'Q_1Q_3'$, $D_2 = Q_3, D_3 = X'Q_3 + XQ_2Q_3' + (Q_1'Q_3 \text{ or } Q_1'Q_2)$
   $Z = XQ_2Q_3 + X'Q_2'Q_3 + X'Q_2Q_3'$
   (f) $J_1 = X, K_1 = X'Q_2Q_3 + XQ_2'Q_3'$
11. (b) $Q_5^+ = XQ_2 + X'YQ_2 + X'Q_5$

(c) $Q_2^+ = Q_1M$
$Q_3^+ = Q_2K + Q_1KM'$
(d) $Ad = Q_1M$
$Done = Q_3$

## UNIT 15 Answers to Problems

15.1 (a)

| Present State | Next State X = 0 | X = 1 | Output X = 0 | X = 1 |
|---|---|---|---|---|
| A | A | C | 1 | 0 |
| B | C | F | 0 | 0 |
| C | B | A | 0 | 0 |
| F | B | F | 1 | 0 |

(b)

| Input | 1 | 0 | 0 |
|---|---|---|---|
| Output(from B) | 0 | 1 | 0 |
| Output (from G) | 0 | 1 | 1 |

15.2

| Present State | Next State X = 0 | X = 1 | Output |
|---|---|---|---|
| a | c | c | 1 |
| c | d | f | 0 |
| d | f | a | 1 |
| f | c | d | 0 |

15.3 (a) No, states $S_2$ and $S_4$ have no corresponding states in Mr. Ipflop's design.
(b) Because there is no way of reaching $S_2$ and $S_4$ by starting from $S_0$, the two *circuits* would perform the same.

15.4 (a) $D = X_1'X_2Q' + X_1X_2'Q' + X_2'X_3Q + X_2X_3'Q$
(b) $S = X_1'X_2Q' + X_1X_2'Q'$
$R = X_2'X_3'Q + X_2X_3Q$

15.5 (a) Only *one* assignment—000 011 101 OR 001 010 100 etc.

(b)

| | | | | | | | | | | |
|---|---|---|---|---|---|---|---|---|---|---|
| 000 | 000 | 000 | 000 | 000 | 000 | 000 | 000 | 000 | 000 | |
| 001 | 001 | 001 | 001 | 001 | 001 | 001 | 001 | 001 | 001 | etc. |
| 010 | 010 | 010 | 010 | 011 | .011 | 011 | 011 | 110 | 110 | |
| 100 | 101 | 110 | 111 | 100 | 101 | 110 | 111 | 010 | 011 | |

15.6 (a)

| $BC \backslash A$ | 0 | 1 |
|---|---|---|
| 00 | $S_1$ | $S_2$ |
| 01 | $S_4$ | $S_3$ |
| 11 | $S_6$ | $S_5$ |
| 10 | $S_7$ | $S_8$ |

$Z = A$

(b)

| $BC \backslash A$ | 0 | 1 |
|---|---|---|
| 00 | $S_1$ | $S_8$ |
| 01 | $S_7$ | $S_5$ |
| 11 | $S_3$ | $S_4$ |
| 10 | $S_2$ | $S_6$ |

$D_A = A^+ = A'B' + XA' + X'AC'$
$D_B = B^+ = \text{etc.}$
$D_C = C^+ = \text{etc.}$

15.7 (a)

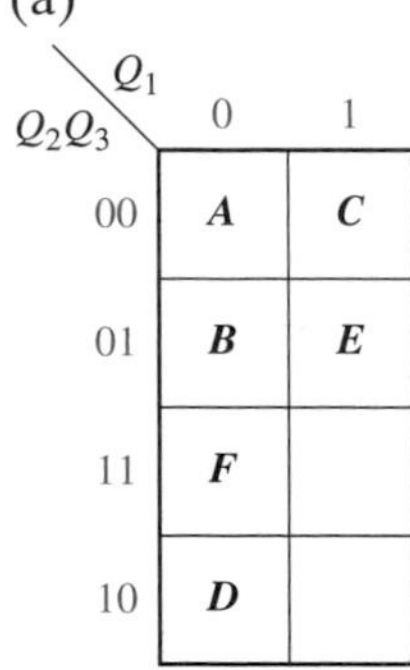


(b) $D_1 = XQ_1$
$D_2 = Q_1'Q_3' + X'Q_1'$
$D_3 = X'Q_1'Q_3' + XQ_1'Q_3 + \{XQ_2' \text{ or } Q_2Q_3\}$
$Z = XQ_1$

15.8 (a) $A = 00, B = 01, C = 10, D = 11$
(b) $T_1 = X_1'X_2Q_2' + X_1'Q_1Q_2 + X_1Q_1'Q_2 + X_1X_2'$
$T_2 = X_1Q_1'Q_2' + X_1Q_1Q_2$
$Z_1 = X_1Q_2, Z_2 = X_1'Q_1 + Q_1Q_2'$

15.9

| | | $Q_1$ | $Q_2$ | $Q_3$ |
|---|---|---|---|---|
| Assign | $S_0$ | 1 | 0 | 0 |
| | $S_1$ | 0 | 1 | 0 |
| | $S_2$ | 0 | 0 | 1 |

$D_1 = X'Q_1 + XY'Q_3$
$D_2 = XQ_1 + YQ_3 + X'Q_2$
$D_3 = XQ_2 + X'Y'Q_3$
$P = XQ_1 + Y'Q_3 + XQ_2$
$S = X'Q_1 + XY'Q_3$

## UNIT 16 Study Guide Answers

1. (a) Because the input sequences are listed in reverse order.
2. (b) $m$ leads, where $2^{m-1} < n \leq 2^m$
3. (b) 64 words $\times$ 7 bits
   (d) $Z = 0, D_1 = 1, D_2 = 1, D_3 = 0;$ $Q_1Q_2Q_3 = 110$
8. (a) Yes
   (c) Yes
9. (a) After the clock (when the state has just changed) and before the input is set to its new value, the output may be wrong (false output).
   (b) No, because the output is always correct before the active clock edge.

## UNIT 16 Answers to Problems

16.15

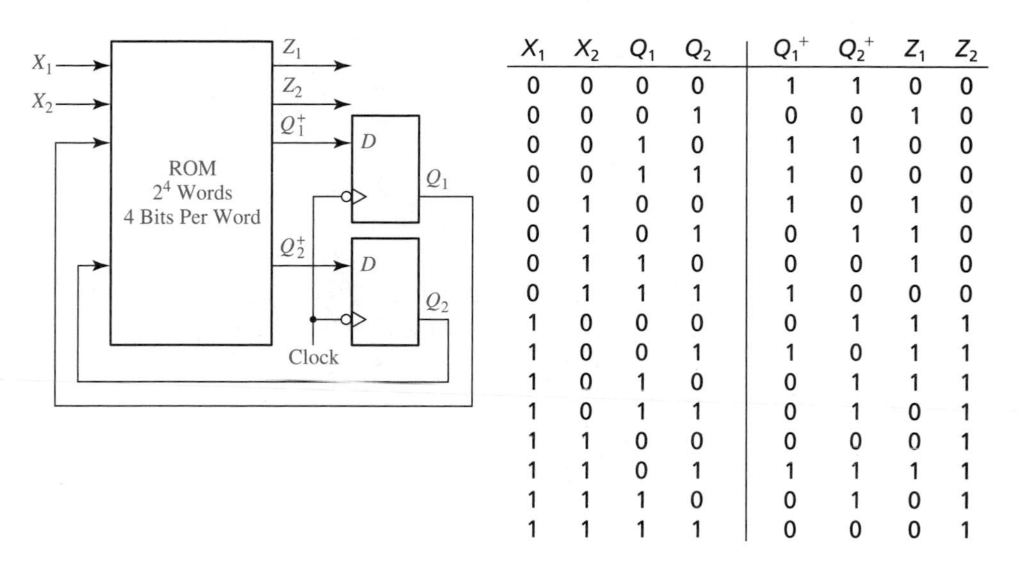


| $X_1$ | $X_2$ | $Q_1$ | $Q_2$ | $Q_1^+$ | $Q_2^+$ | $Z_1$ | $Z_2$ |
|---|---|---|---|---|---|---|---|
| 0 | 0 | 0 | 0 | 1 | 1 | 0 | 0 |
| 0 | 0 | 0 | 1 | 0 | 0 | 1 | 0 |
| 0 | 0 | 1 | 0 | 1 | 1 | 0 | 0 |
| 0 | 0 | 1 | 1 | 1 | 0 | 0 | 0 |
| 0 | 1 | 0 | 0 | 1 | 0 | 1 | 0 |
| 0 | 1 | 0 | 1 | 0 | 1 | 1 | 0 |
| 0 | 1 | 1 | 0 | 0 | 0 | 1 | 0 |
| 0 | 1 | 1 | 1 | 1 | 0 | 0 | 0 |
| 1 | 0 | 0 | 0 | 0 | 1 | 1 | 1 |
| 1 | 0 | 0 | 1 | 1 | 0 | 1 | 1 |
| 1 | 0 | 1 | 0 | 0 | 1 | 1 | 1 |
| 1 | 0 | 1 | 1 | 0 | 1 | 0 | 1 |
| 1 | 1 | 0 | 0 | 0 | 0 | 0 | 1 |
| 1 | 1 | 0 | 1 | 1 | 1 | 1 | 1 |
| 1 | 1 | 1 | 0 | 0 | 1 | 0 | 1 |
| 1 | 1 | 1 | 1 | 0 | 0 | 0 | 1 |

16.16 (a) Same as Figure 16-10 with ROM replaced by PLA.

(b)

| X | A | B | C | Z | $D_A$ | $D_B$ | $D_C$ |
|---|---|---|---|---|---|---|---|
| 0 | - | - | - | 0 | 1 | 0 | 0 |
| 0 | - | - | 0 | 0 | 0 | 1 | 0 |
| - | 0 | - | 1 | 0 | 0 | 1 | 0 |
| - | 0 | 1 | - | 0 | 0 | 1 | 0 |
| - | 1 | - | - | 0 | 0 | 0 | 1 |
| 1 | - | 0 | - | 0 | 0 | 0 | 1 |
| 0 | 1 | 0 | 1 | 1 | 0 | 0 | 0 |
| 1 | 0 | 1 | 0 | 1 | 0 | 0 | 0 |

16.17 (a)

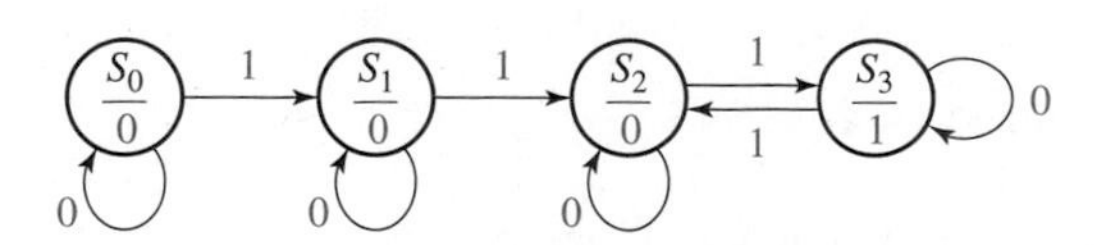


(b) $a_{i+1} = a_i + x_i b_i = [a_i' (x_i b_i)']'$ $\qquad$ $b_{i+1} = x_i b_i' + x_i' b_i = [(x_i b_i')' (x_i' b_i)']'$

$z = a_{n+1} b_{n+1}$

(c) $a_1 = b_1 = 0, a_2 = 0, b_2 = x_1$

(d) Similar to Figure 16-9 with one output

## UNIT 17 Study Guide Answers

1. (a) When a falling edge of CLK occurs
   (b) Whenever there is a change in St or $Q_1$ or $V$
   (c) Statements $S_3$ and $S_4$ will execute
   (d) The code represents a rising-edge-triggered D flip-flop with asynchronous, active-low clear and preset. Because ClrN has higher priority, $Q$ will be set to '0' when both ClrN and SetN are '0'.
   (e) $SN$ and $RN$ override $J$ and $K$, including at a clock edge. $Q$ will be '0'.
   (f) They will get the old value of $Q_{int}$ because of the 8-ns delay.
2. (a)
```
entity register is
port (CLK, CLR, Ld: in bit; D: in bit_vector (3 downto 0);
   Q: out bit_vector (3 downto 0));
end register;

architecture eqn of register is
begin
   -- Given code here
end eqn;
```
   (b) Q will change at time 10 + $\Delta$ ns
   (c) On line 6 of the VHDL code, make this change—
```
elsif LS = '1' then Q <= Q (2 downto 0) & Q (3);
```
   (d) Change the indicated lines of code as shown—
```
line 3: process (ClrN, CLK)
lines 5 to 7: if ClrN = '0' then Q <= "0000";
              elsif CLK'event and CLK = '1' then
                 if En = '1' then Q <= Q + 1;
```
   (e) When Carry1 = '1' and Qout2 = "1111"
   (f) No overloaded "+" operator is defined for bit_vectors
   (g) After the rising edge, Qout1 = "0000" and Qout2 = "1010"
   (h)

| Control Signals | | | Next State | | | |
|---|---|---|---|---|---|---|
| ClrN | LdN | PT | $Q_3^+$ | $Q_2^+$ | $Q_1^+$ | $Q_0^+$ |
| X | 0 | X | $D_3$ | $D_2$ | $D_1$ | $D_0$ |
| X | 1 | 1 | Present state + 1 | | | |
| 0 | 1 | 0 | 0 | 0 | 0 | 0 |
| 1 | 1 | 0 | $Q_3$ | $Q_2$ | $Q_1$ | $Q_0$ |

3. (a)
```
process (A, B, D)
begin
   E <= (A and B) or D after 5 ns;
end process;
```
4. (a) Nextstate = S4 and Z = 1
   (b) Because there are only seven states. Also, specifying the range restricts the number of bits used to represent the integer

When X changes to 1, Nextstate = 2, and Z = 0. Then, when CLK changes to 1, State = 2, Nextstate = 4, and Z = 1

(c) The glitch occurs because the change in state and change in the value of X a little while later causes process (State, X) to execute two times, thereby updating the value of Nextstate two times. This glitch does not affect the 'state' because the state will not be updated until the next positive clock edge.

(d) Because Q1, Q2, and Q3 must be updated only on the CLK edge, the other signals must not appear on the sensitivity list. The new values will be computed at 5 ns, and the values are updated at 15 ns.

(e) The statements of lines 13, 14, and 18 will execute.

(f) ROM output = 1100

5. (a) Connect En to CE and A to D

(b) See Figure 11-27(c) (change to rising-edge trigger)

(c) Use four D-CE flip-flops. Connect ASR to every CE input, D3 to Q3, D2 to Q3, D1 to Q2, and D0 to Q1. Label Q3 . . . Q0 as A(3) . . . A(0).

6. (a)

```
process
begin
   A <= B or C;
   wait on B, C;
end process;
```

(b) 2 ns (Both sequential statements execute immediately with no delay.)

## UNIT 17 Answers to Problems

### 17.1 Code to implement a T flip-flop

```
entity tff is
   port (t, clk, clrn: in bit;
            q, qn: out bit);
end tff;
architecture eqn of tff is
signal qint: bit; -- Internal value of q
begin
q <= qint;
qn <= not qint;
process (clk, clrn)
begin
   if clrn = '0' then qint <= '0';
   elsif clk'event and clk = '1' then
      qint <= (t and not qint) or (not t and qint);
   end if;
end process;
end eqn;
```

## 17.2 Right-shift register with synchronous reset

```
entity rsr is
   Port (clk, clr, ld, rs, lin: in bit;
         d: in bit_vector(3 downto 0);
         q: out bit_vector(3 downto 0));
end rsr;

architecture eqn of rsr is
signal qint: bit_vector(3 downto 0);
begin
q <= qint;

process (clk)
begin
   if clk'event and clk = '1' then
      if clr = '1' then qint <= "0000";
      elsif ld = '1' then qint <= d;
      elsif rs = '1' then qint <= lin & qint(3 downto 1);
      end if;
   end if;
end process;
end eqn;
```

## 17.3 (a) 4-bit binary up/down counter

```
entity updown is
   Port (clrn, clk, load, ent, enp, up: in std_logic;
         d: in std_logic_vector(3 downto 0);
         q: out std_logic_vector(3 downto 0);
         co: out std_logic);
end updown;

architecture eqn of updown is
signal qint: std_logic_vector(3 downto 0) := "0000";
begin
q <= qint;
co <= (qint(3) and qint(2) and qint(1) and qint(0) and ent and up)
        or (not qint(3) and not qint(2) and not qint(1) and not qint(0)
        and ent and not up);

process (clrn, clk)
begin
   if clrn = '0' then qint <= "0000";
   elsif clk'event and clk = '1' then
   if load = '0' then qint <= d;
      elsif (ent and enp and up) = '1' then qint <= qint + 1;
```

```
      elsif (ent and enp and not up) = '1' then qint <= qint - 1;
      end if;
    end if;
  end process;
end eqn;
```

17.3 (b) 8-bit binary up/down counter. (For block diagram, connect the Carry-out of the first counter to ENT of the second.)

```
entity updown8bit is
Port (clrn, clk, load, ent, enp, up: in std_logic;
          d: in std_logic_vector(7 downto 0);
          q: out std_logic_vector(7 downto 0);
          co: out std_logic);
end updown8bit;
architecture structure of updown8bit is
component updown is
Port (clrn, clk, load, ent, enp, up: in std_logic;
          d: in std_logic_vector(3 downto 0);
          q: out std_logic_vector(3 downto 0);
          co: out std_logic);
end component;
signal co1: std_logic;
signal q1,q2: std_logic_vector(3 downto 0);
begin
  c1: updown port map (clrn, clk, load, ent, enp, up, d(3 downto 0),q1,co1);
  c2: updown port map (clrn, clk, load, co1, enp, up, d(7 downto 4),q2, co);
  q <= q2 & q1;
end structure;
```

17.4 MUX with a and b as control inputs

```
entity mymux is
Port (a, b, c, d: in bit;
      z: out bit);
end mymux;
architecture eqn of mymux is
signal sel: bit_vector(1 downto 0);
begin
sel <= a & b;
process (a, b, c, d)
begin
  case sel is
    when "00" => z <= not c or d;
    when "01" => z <= c;
```

```
        when "10" => z <= not c xor d;
        when "11" => z <= not d;
    end case;
end process;
end eqn;
```

17.5 Implements the state machine of Table 14-1

```
entity sm1 is
Port (x, clk: in bit;
        z: out bit);
end sm1;
architecture table of sm1 is
signal State, Nextstate: integer range 0 to 2 := 0;
begin
process (State, x)
begin
    case State is
    when 0 =>
        if x = '0' then Nextstate <= 0; else Nextstate <= 1; end if;
        z <= '0';
    when 1 =>
        if x = '0' then Nextstate <= 2; else Nextstate <= 1; end if;
        z <= '0';
    when 2 =>
        if x = '0' then Nextstate <= 0; z <= '0';
        else Nextstate <= 1; z <= '1'; end if;
    end case;
end process;
process (clk)
begin
if clk'event and clk = '0' then
    State <= Nextstate;
end if;
end process;
end table;
```

17.6 (a) See Figure 13-17, with $m = 2, n = 2$, and $k = 2$.
(b) Implements the state machine of Table 13-4

```
library BITLIB;
use BITLIB.bit_pack.all;
entity sm is
Port (x1, x2, clk: in bit;
    z1,z2: out bit);
```

```
end sm;
architecture Behavioral of sm is
type rom16_4 is array (0 to 15) of bit_vector(3 downto 0);
-- Input is in the order X1 X2 Q1 Q2
-- Output in order Q1 Q2 Z1 Z2
constant myrom: rom16_4 := ("1100", "0010", "1100", "1000", "1010", "0110","0010",
          "1000", "0111", "1011", "0111", "0101", "0001", "1111", "0101", "0001");
signal index: integer range 0 to 15;
signal q1,q2: bit;
signal rom_out: bit_vector(3 downto 0);
begin
index <= vec2int(x1&x2&q1&q2);
rom_out <= myrom(index);
z1 <= rom_out(1);
z2 <= rom_out(0);
process(clk)
begin
  if clk'event and clk = '1' then
    q1 <= rom_out(3);
    q2 <= rom_out(2);
  end if;
end process;
end Behavioral;
```

17.7 (a) There are two D-CE flip-flops. For each, CE = LdA + LdB.
D1 = LdA A1 + LdA′ LdB B1, D2 = LdA A2 + LdA′ LdB B2.

(b) CE does not change. For each D input, replace the gates with a 2-to-1 MUX, with LdA as the control input, and B and A as the data inputs for 0 and 1, respectively. (Alternately, use LdB as the control input, and swap A and B on the data inputs.)

17.8 All statements execute at time = 20 ns
A becomes 1 at 35 ns (not the final value)
B becomes 1 at 20 ns + Δ (not the final value)
C becomes 1 at 30 ns
D becomes 2 at 23 ns
A becomes 5 at 35 ns (overrides the previous value)
B becomes 7 at 20 ns + Δ (overrides the previous value)

## UNIT 18 Study Guide Answers

1. (a)

| | $X$ | $Y$ | $c_i$ | $s_i$ | $c_i^+$ |
|---|---|---|---|---|---|
| $t_0$ | 0110 | 0011 | 0 | 1 | 0 |
| $t_1$ | 1011 | 1001 | 0 | 0 | 1 |
| $t_2$ | 0101 | 1100 | 1 | 0 | 1 |
| $t_3$ | 0010 | 0110 | 1 | 1 | 0 |
| $t_4$ | 1001 | 0011 | 0 | (0) | (1) |

(b) $Y$ would fill up with 0's from the left: 0011, 0001, 0000, 0000, 0000.
(c) $S_0$ and $Y_0$, no.

2. (a)

| | | | | | | | | | |
|---|---|---|---|---|---|---|---|---|---|
| add | 0 | 0 | 0 | 0 | 0 | 1 | 1 | 0 | 1 |
| | | 1 | 1 | 1 | 1 | | | | |
| shift | 0 | 1 | 1 | 1 | 1 | 1 | 1 | 0 | 1 |
| shift | 0 | 0 | 1 | 1 | 1 | 1 | 1 | 1 | 0 |
| add | 0 | 0 | 0 | 1 | 1 | 1 | 1 | 1 | 1 |
| | | 1 | 1 | 1 | 1 | | | | |
| shift | 1 | 0 | 0 | 1 | 0 | 1 | 1 | 1 | 1 |
| add | 0 | 1 | 0 | 0 | 1 | 0 | 1 | 1 | 1 |
| | | 1 | 1 | 1 | 1 | | | | |
| shift | 1 | 1 | 0 | 0 | 0 | 0 | 1 | 1 | 1 |
| | 0 | 1 | 1 | 0 | 0 | 0 | 0 | 1 | 1 |

(b) 10, 6. (c) 10, 6. (d) 15 bits
(f) Product register has 17 bits. Adder is 8 bits wide, multiplicand has 8 bits. 18 states. 3-bit counter, $K = 1$ when counter is in state 7 ($111_2$), control graph unchanged.

3. (b) Change $Y$ to 2's complement by inverting each bit and adding 1 (by setting the carry input of the first full adder to 1). Also change $C$ so that it is equal to the carry out of the last full adder.
(c) An overflow will occur if $X_8X_7X_6X_5X_4 \geq Y_3Y_2Y_1Y_0$, because subtraction is possible but there is no place to store the quotient bit, since there are only 4 bits available to store the quotient.
(f) To set the quotient bit to 1.

## UNIT 18 Answers to Problems

18.3

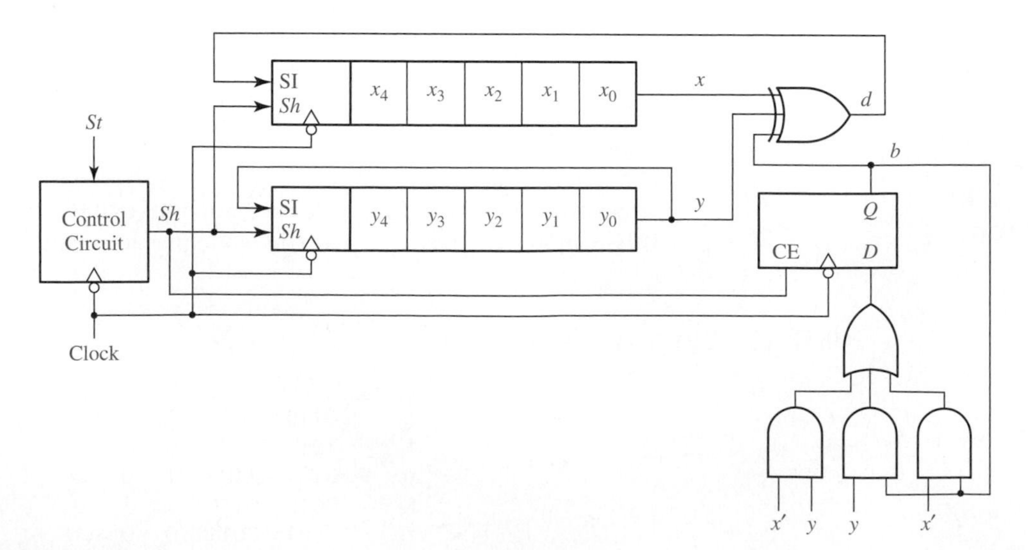

18.4

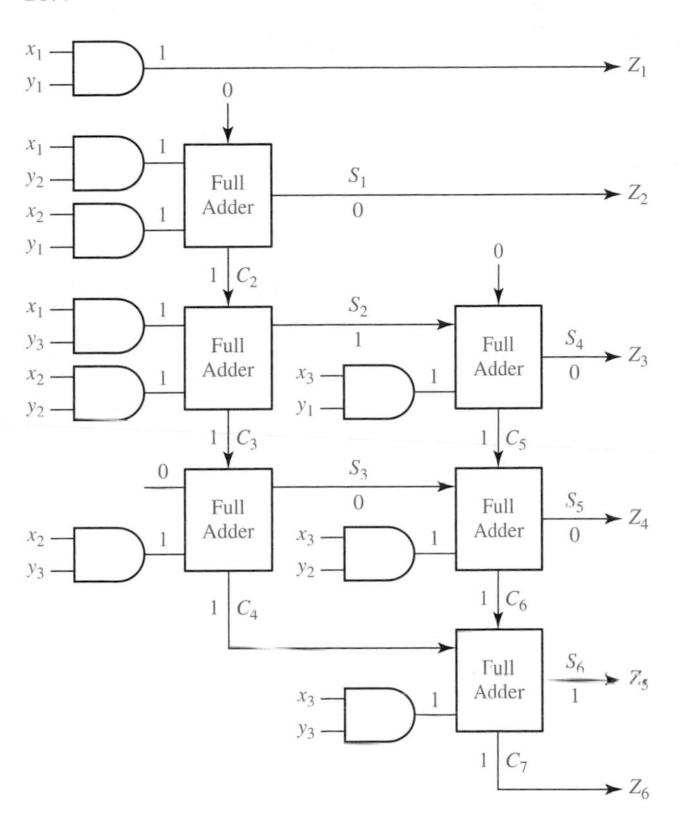


18.5 18.6

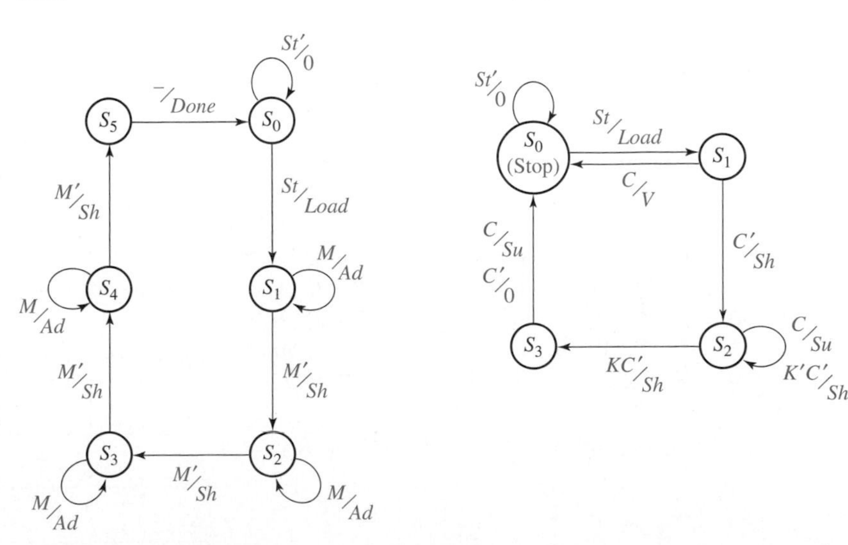


18.7 (a) $V = y_0'y_1'y_2'y_3'y_4' = (y_0 + y_1 + y_2 + y_3 + y_4)'$

(b)

| (0 | 0 | 1 | 0 | 1) | | | | | | *Sh* | *Su* |
|---|---|---|---|---|---|---|---|---|---|---|---|
| 0 | 0 | 0 | 0 | 0 | 1 | 1 | 0 | 1 | 0 | 1 | 0 |
| 0 | 0 | 0 | 0 | 1 | 1 | 0 | 1 | 0 | 0 | 1 | 0 |
| 0 | 0 | 0 | 1 | 1 | 0 | 1 | 0 | 0 | 0 | 1 | 0 |
| 0 | 0 | 1 | 1 | 0 | 1 | 0 | 0 | 0 | 0 | 0 | 1 |
| 0 | 0 | 0 | 0 | 1 | 1 | 0 | 0 | 0 | 1 | 1 | 0 |
| 0 | 0 | 0 | 1 | 1 | 0 | 0 | 0 | 1 | 0 | 1 | 0 |
| 0 | 0 | 1 | 1 | 0 | 0 | 0 | 1 | 0 | 0 | 0 | 1 |
| 0 | 0 | 0 | 0 | 1 | 0 | 0 | 1 | 0 | 1 | | |

remainder = 1 quotient = 5

(c)

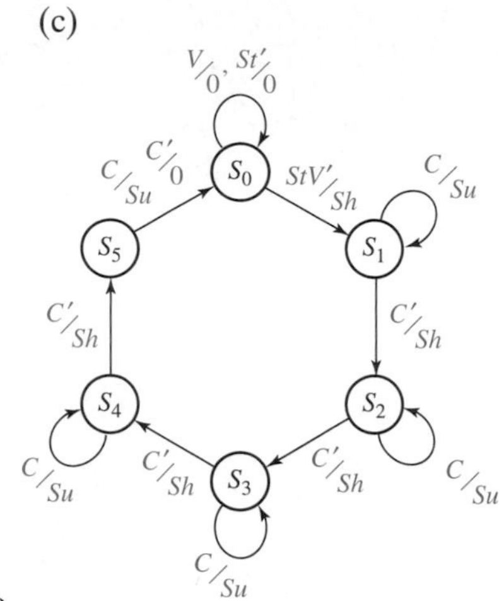


(d) After five shifts, the division is complete.

18.8 (a) $x_{in} = K_1'K_2'a' + K_1'K_2b + K_1K_2'(a \oplus b) + K_1K_2a$

$y_{in} = K_1'K_2'b + K_1'K_2a + K_1K_2' \cdot 0 + K_1K_2 \cdot 1$

(b) Use the state graph of Figure 18-6(b), with nine states total.

(c)

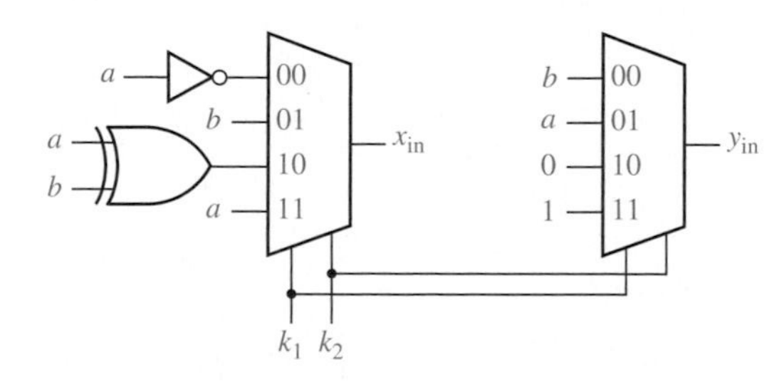


## UNIT 19 Study Guide Answers

1. (b) $Z_1, Z_2, Z_4$ (for both charts) (d)
2. (a)

| | | | | | | | | | |
|---|---|---|---|---|---|---|---|---|---|
| 0 | 0 | 0 | 0 | 1 | 1 | 1 | 0 | 0 | C = 0, *Sh* |
| 0 | 0 | 0 | 1 | 1 | 1 | 0 | 0 | 0 | C = 0, *Sh* |
| 0 | 0 | 1 | 1 | 1 | 0 | 0 | 0 | 0 | C = 1, *Su* |
| 0 | 0 | 0 | 1 | 0 | 0 | 0 | 0 | 1 | C = 0, *Sh* |
| 0 | 0 | 1 | 0 | 0 | 0 | 0 | 1 | 0 | C = 0, *Sh* |
| 0 | 1 | 0 | 0 | 0 | 0 | 1 | 0 | 0 | C = 1, *Su* |
| 0 | 0 | 0 | 1 | 1 | 0 | 1 | 0 | 1 | (result) |

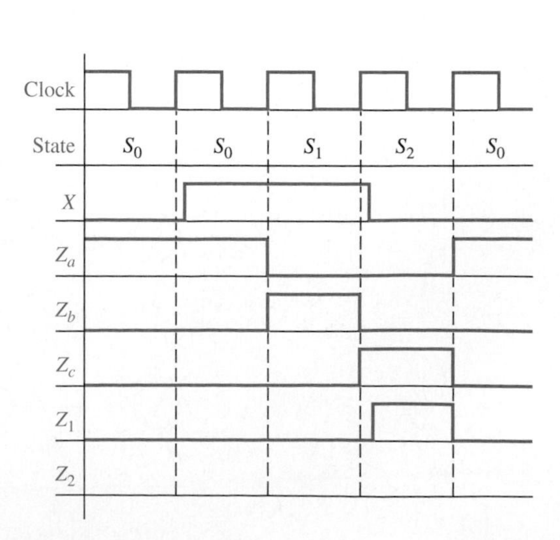


3. (a) $A^+ = BX \quad B^+ = A'X + BX$

## UNIT 19 Answers to Problems

19.1

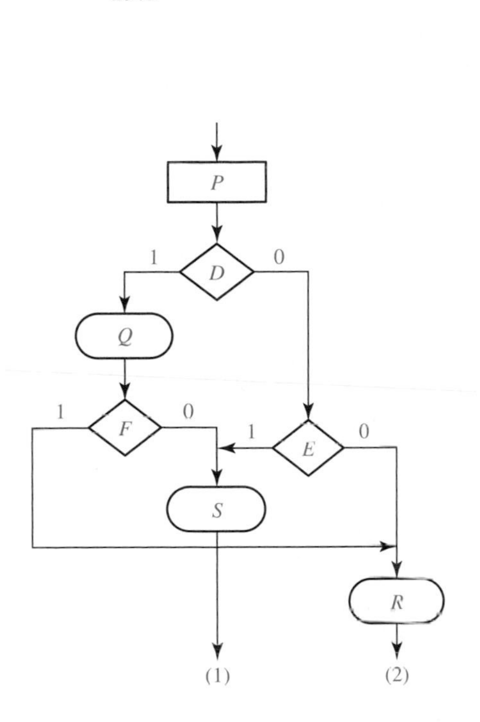


19.2

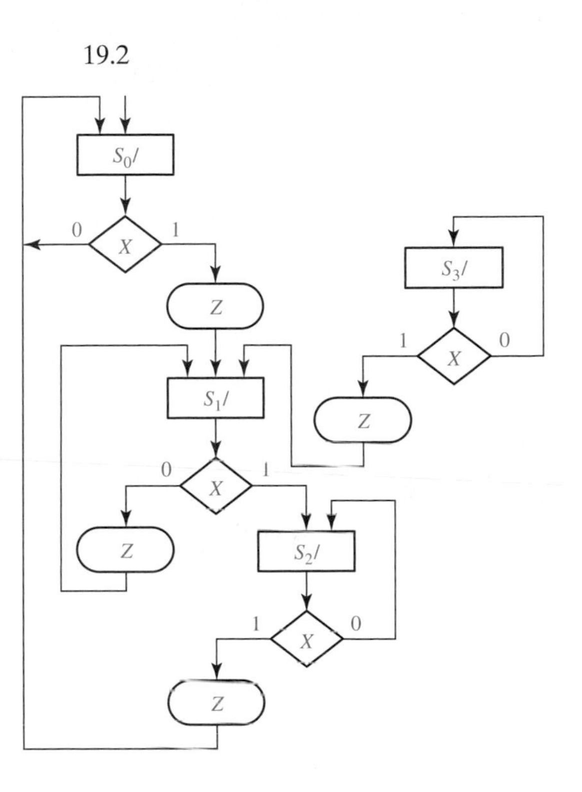


19.3

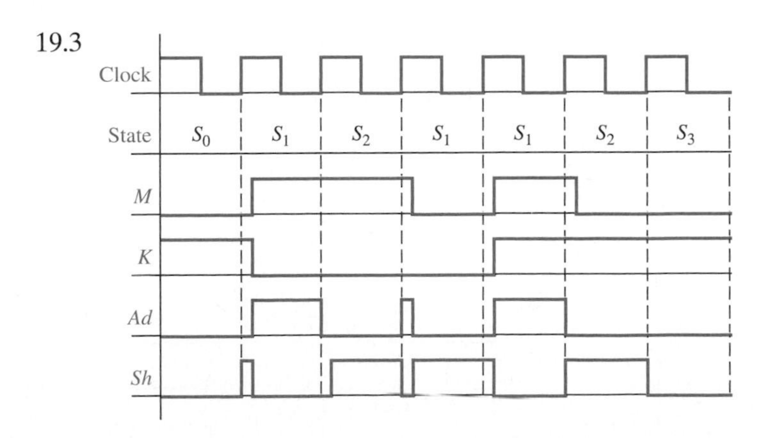

19.4

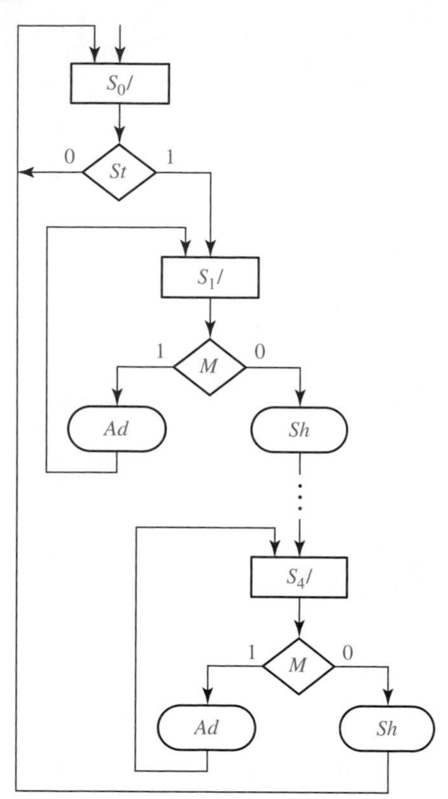

$S_0/$
0
St
1
$S_1/$
1
M
0
Ad
Sh
$S_4/$
1
M
0
Ad
Sh


19.5

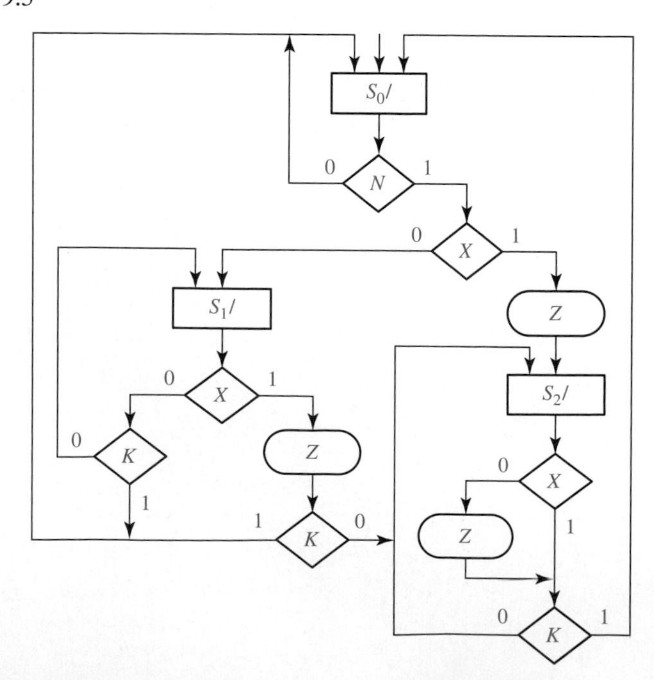

$S_0/$
0
N
1
0
X
1
Z
$S_1/$
0
X
1
0
K
Z
1
1
K
0
$S_2/$
0
X
Z
1
0
K
1

19.6

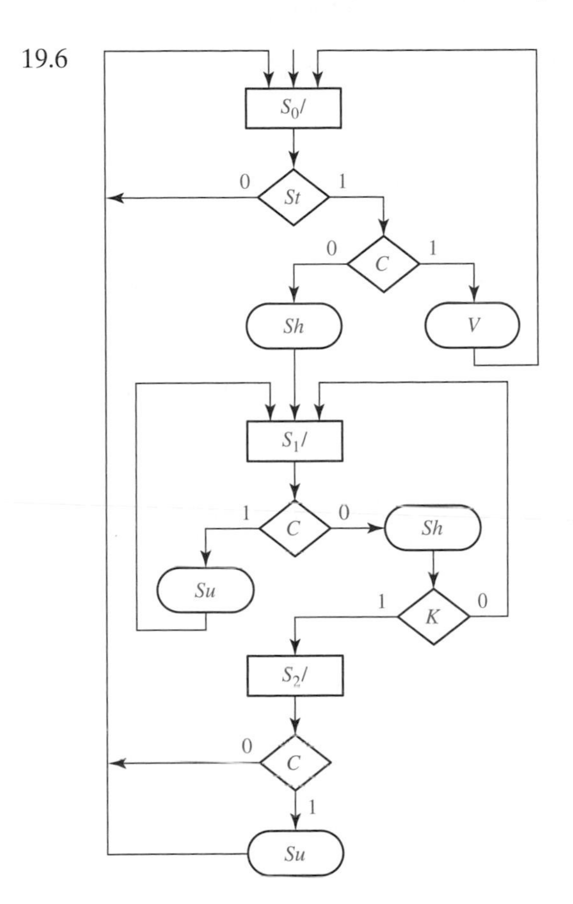


19.7 (a) $Q_0^+ = Q_0'Q_1'Q_2C' + Q_0Q_1' + Q_0Q_2'$

$Q_1^+ = Q_0Q_1'Q_2C' + Q_0Q_1Q_2'$

$Q_2^+ = Q_0'Q_1'Q_2'St + Q_0Q_2'C' + Q_0Q_1'Q_2C$

$Load = Q_0'Q_1'Q_2'St$, $Su = Q_0C$, $Sh = Q_1'Q_2C' + Q_0Q_2'C'$, $V = Q_0'Q_1'Q_2C$

(These equations could be further simplified using don't-cares.)

(b) $Q_0^+ = Q_0'Q_1M + Q_0'Q_1M'K + Q_0Q_1K$

$Q_1^+ = Q_0'Q_1'St + Q_0'Q_1M'K' + Q_0'Q_1M + Q_0Q_1K'$

$Load = Q_0'Q_1'St$, $Sh = Q_0Q_1 + Q_0'Q_1M'$, $Ad = Q_0'Q_1M$, $\text{Done} = Q_0Q_1'$

19.8 (a) $A^+ = A'B'C \cdot Rb'D_{711}'D_{2312}' + AB'C' + AB' \cdot Rb + AB'Eq'D_7'$

$B^+ = A'B'C \cdot D_{711} + A'B \cdot Reset' + AB'C \cdot Rb'Eq'D_7'$

$C^+ = A'B'Rb + A'BC \cdot Reset' + B'C'Rb + AB'C \cdot Rb'Eq'D_7$

$Roll = B'C \cdot Rb$  $Sp = A'B'C \cdot Rb'D_{711}'D_{2312}'$

$Win = A'BC'$  $Lose = A'BC$

(These equations could be further simplified using don't-cares.)

(b) If the input from the adder is $S_3S_2S_1S_0$, then the equations realized by the test logic block are

$D_7 = S_2S_1S_0$  $D_{711} = S_1S_0(S_2 + S_3)$  $D_{2312} = S_3'S_2' + S_3S_2$

19.9 (a) $A^+ = BX$  $Z_a = A'B'$  $Z_1 = ABX'$

$B^+ = A'X + BX$  $Z_b = A'B$  $Z_2 = ABX$

$Z_c = AB$

| $X$ | $A$ | $B$ | $A^+$ | $B^+$ | $Z_a$ | $Z_b$ | $Z_c$ | $Z_1$ | $Z_2$ |
|---|---|---|---|---|---|---|---|---|---|
| 1 | - | 1 | 1 | 1 | 0 | 0 | 0 | 0 | 0 |
| 1 | 0 | - | 0 | 1 | 0 | 0 | 0 | 0 | 0 |
| - | 0 | 0 | 0 | 0 | 1 | 0 | 0 | 0 | 0 |
| - | 0 | 1 | 0 | 0 | 0 | 1 | 0 | 0 | 0 |
| - | 1 | 1 | 0 | 0 | 0 | 0 | 1 | 0 | 0 |
| 0 | 1 | 1 | 0 | 0 | 0 | 0 | 0 | 1 | 0 |
| 1 | 1 | 1 | 0 | 0 | 0 | 0 | 0 | 0 | 1 |

19.10 (a)

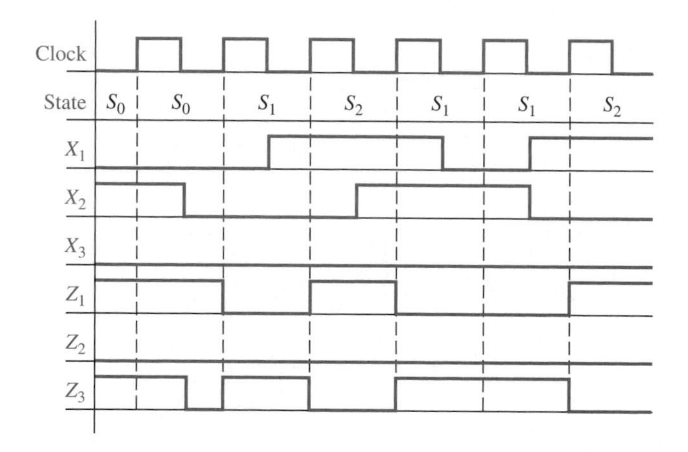


(b) $D_1 = Q_1'X_1'X_2'X_3 + Q_2X_2'$ $\quad D_2 = Q_2'X_1 + Q_1'Q_2'X_2'X_3' + Q_2X_2$

$Z_1 = Q_2'X_1' + Q_1$ $\quad Z_2 = Q_1'Q_2'X_1$ $\quad Z_3 = Q_1'X_1'X_2 + Q_2$

(c)

| $Q_1$ | $Q_2$ | $X_1$ | $X_2$ | $X_3$ | $D_1$ | $D_2$ | $Z_1$ | $Z_2$ | $Z_3$ |
|---|---|---|---|---|---|---|---|---|---|
| 0 | 0 | 1 | - | - | 0 | 1 | 0 | 1 | 0 |
| 0 | 0 | 0 | 1 | - | 0 | 0 | 1 | 0 | 1 |
| 0 | 0 | 0 | 0 | 1 | 1 | 0 | 1 | 0 | 0 |
| 0 | 0 | 0 | 0 | 0 | 0 | 1 | 1 | 0 | 0 |
| 0 | 1 | - | 0 | - | 1 | 0 | 0 | 0 | 1 |
| 0 | 1 | - | 1 | - | 0 | 1 | 0 | 0 | 1 |
| 1 | 0 | 0 | - | - | 0 | 0 | 1 | 0 | 0 |
| 1 | 0 | 1 | - | - | 0 | 1 | 1 | 0 | 0 |

(d) $2^5 \times 5$ ROM

| $Q_1$ | $Q_2$ | $X_1$ | $X_2$ | $X_3$ | $D_1$ | $D_2$ | $Z_1$ | $Z_2$ | $Z_3$ |
|---|---|---|---|---|---|---|---|---|---|
| 0 | 0 | 0 | 0 | 0 | 0 | 1 | 1 | 0 | 0 |
| 0 | 0 | 0 | 0 | 1 | 1 | 0 | 1 | 0 | 0 |
| 0 | 0 | 0 | 1 | 0 | 0 | 0 | 1 | 0 | 1 |
| 0 | 0 | 0 | 1 | 1 | 0 | 0 | 1 | 0 | 1 |
| 0 | 0 | 1 | 0 | 0 | 0 | 1 | 0 | 1 | 0 |

## UNIT 20 Study Guide Answers

1. (a) lines 15 and 16
   The full adder is combinational logic.
   Lines 34 and 35, which are in clocked process because it is a clocked register
   (b) In line 31, change clk = '0' to clk = '1'.

2. (a) So we can use the overloaded "+" operator
   The change from state 9 to state 0
   To make the result be 5 bits
   Lines 35 and 36 will execute when State is 2, 4, 6, or 8.
   (b) ACC is uninitialized and is not loaded until St = '1' at a rising clock edge. When Done = 1, i.e., in state 9, 160-180 ns.
   (c) $X = 101111001$, 60 ns
   (d) Mcand = 1101, Mplier = 1011, and product is 10001111 = 143
   (e) Line 19
   To avoid having to set them to 0 in each case where they are not 1. When they are set to 1, it overrides line 22 because these are sequential statements.
   ACC <= "00000" & Mplier;
   Because it is a clocked register that is updated on the rising clock edge.
   The process executes on the rising clock edge, and when state is 9 at the rising clock edge, it is too late; the state is about to change to 0.
   The process of lines 20–34 is not clocked; it executes when State changes to 9.
   (f) Whenever the value of count changes.
   Lines 51 and 52.
   10 ns + $\Delta$.
   Sequential statements execute in 0 time, so A and B update simultaneously.
   (g) At time 60 ns, we are in state 2 when K = 0, so Sh = 1. So A = $00B_{16}$ = $000001011_2$ shifts to the right to become $005_{16} = 000000101_2$. At time 140 ns, we are in state 1 and M = 1, so Ad = 1. So we add the multipicand, $000001011_2$, to A = $006_{16} = 000000110_2$ to get $011_{16} = 000010001_2$.

3. (a) 0; 1
   C should be 1 iff we can subtract, i.e., Dividend(8 **downto** 4) > Divisor.

## UNIT 20 Answers to Problems

20.1 First process executes at t = 2 ns. Lines 22–25 execute.
Second process executes at t = 10 ns. Lines 38–40 and 43 execute.
Because the state changes, first process executes again at 10 + $\Delta$ ns. Lines 22–23 and lines 27–30 execute.

20.2

```
entity complementer is
   Port (clk, n: in std_logic;
   Regout: out std_logic_vector(15 downto 0));
end complementer;
architecture Behavioral of complementer is
signal State, NextState: integer range 0 to 2 := 0;
signal count: std_logic_vector(3 downto 0) := "0000";--4-bit counter
```

```
signal X, Z, Sh: std_logic;
signal K: std_logic := '0';
signal Reg: std_logic_vector(15 downto 0);
begin
    Regout <= Reg;
    X <= Reg(0);
    K <= '1' when count = "1111" else '0';
    process (State, X, N, K)
    begin
        case State is
        when 0 =>
            if N = '0' then NextState <= 0; Sh <= '0'; Z <= '0';
            elsif X = '1' then NextState <= 2; Sh <= '1'; Z <= '1';
         else NextState <= 1; Sh <= '1'; Z <= '0'; end if;
        when 1 => Sh <= '1';
          if K = '1' then NextState <= 0;
                if X = '1' then Z <= '1';
                else Z <= '0'; end if;
          elsif X = '0' then NextState <= 1; Z <= '0';
          else NextState <= 2; Z <= '1'; end if;
        when 2 => Sh <= '1';
            if K = '1' then NextState <= 0;
               if X = '1' then Z <= '0';
               else Z <= '1'; end if;
            elsif X = '0' then NextState <= 2; Z <= '1';
            else NextState <= 2; Z <= '0'; end if;
        end case;
    end process;
    process (clk)
    begin
        if clk'event and clk = '1' then
            if Sh = '1' then Reg <= Z & Reg(15 downto 1);
            count <= count + 1; end if;
            State <= NextState;
        end if;
    end process;
end Behavioral;
```

20.4

```
entity test is
end test;
architecture Behavioral of test is
component sm17_2 is
  Port (x,clk: in std_logic;
        z: out std_logic);
end component;
constant N: integer:= 40;
```

```
signal flag: std_logic:= '0';
signal clk: std_logic:= '1';
signal x,z: std_logic;
constant x_seq: std_logic_vector(1 to 40) :=
                    ("0000100001001100001010100110111000011001");
constant z_seq: std_logic_vector(1 to 40) :=
                    ("1100001010100110111000011001010111010011");
begin
  sm1: sm17_2 port map(x,clk,z);
  clk <= not clk after 10 ns;    -- clock has 20 ns period
  process
  begin
    for i in 1 to N loop
      x <= x_seq(i);
      wait for 5 ns;    -- wait for z to become stable
      if z = z_seq(i) then flag <= '0'; else flag <= '1'; end if;
      wait until clk'event and clk = '1';
      wait for 5 ns;
    end loop;
  end process;
end Behavioral;
```

20.5

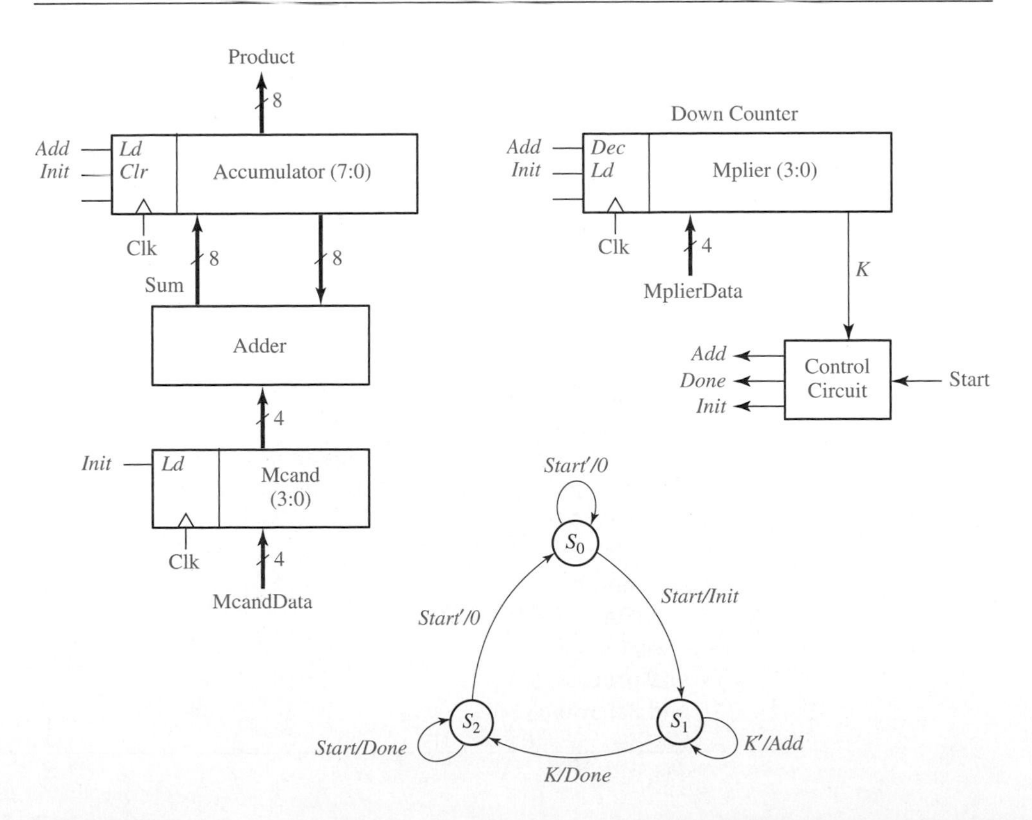

20.6

```
entity prob20_6 is
   Port (st, clk: in std_logic;
            X: in std_logic_vector(7 downto 0);
            Z: out std_logic_vector(9 downto 0));
end prob20_6;
architecture Behavioral of prob20_6 is
   signal State, NextState: integer range 0 to 3 := 0;
   signal lda, ldb, ldc, ad: std_logic;
   signal B, C: std_logic_vector(7 downto 0);
   signal A: std_logic_vector(9 downto 0);
   signal sumAB: std_logic_vector(8 downto 0);
   signal sumABC: std_logic_vector(9 downto 0);
begin
   sumAB <= ("0"&A(7 downto 0)) + B;
   sumABC <= ("0"&sumAB) + C;
   Z <= A;
process (st, State)
begin
   lda <= '0'; ldb <= '0'; ldc <= '0'; ad <= '0';
   case State is
      when 0 =>
         if st = '1' then lda <= '1'; NextState <= 1;
         else NextState <= 0; end if;
      when 1 =>
         ldb <= '1'; NextState <= 2;
      when 2 =>
         ldc <= '1'; NextState <= 3;
      when 3 =>
         ad <= '1'; lda <= '1'; NextState <= 0;
   end case;
end process;
process(clk)
begin
   if clk'event and clk = '1' then
      if lda = '1' then
         if ad = '1' then A <= sumABC;
         else A <= ("00" & X); end if;
      elsif ldb = '1' then B <= X;
      elsif ldc = '1' then C <= X;
      end if;
      State <= NextState;
   end if;
end process;
end Behavioral;
```

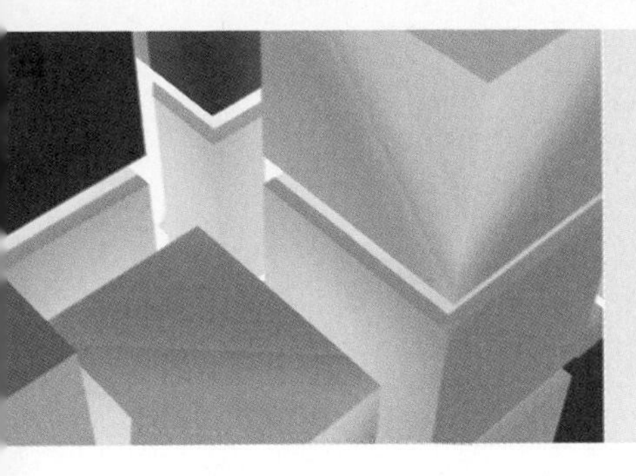

# References


1. Ashenden, Peter J. *The Designer's Guide to VHDL*, 2nd ed. San Francisco: Morgan Kaufmann Publishers, 2002.
2. Bhasker, J. *A Guide to VHDL Syntax*. Upper Saddle River, NJ: Prentice-Hall, 1995.
3. Bhasker, J. *VHDL Primer*, 3rd ed. Upper Saddle River, NJ: Prentice-Hall, 1999.
4. Brayton, Robert, et al. *Logic Minimization Algorithms for VLSI Synthesis*. Secaucus, NJ: Springer, 1984.
5. Givone, Donald D. *Digital Principles and Design*. New York: McGraw-Hill, 2003.
6. Katz, Randy H. and Gaetano Borriello. *Contemporary Logic Design*, 2nd ed. Upper Saddle River, NJ: Prentice Hall, 2004.
7. Mano, M. Morris. *Digital Design*, 3rd ed. Upper Saddle River, NJ: Prentice Hall, 2001.
8. Mano, M. Morris and Charles R. Kime. *Logic and Computer Design Fundamentals*, 4th ed. Old Tappan, NJ: Pearson Prentice Hall, 2008.
9. Marcovitz, Alan B. *Introduction to Logic Design,* 2nd ed. New York: McGraw-Hill, 2002.
10. McCluskey, Edward J. *Logic Design Principles*. Upper Saddle River, NJ: Prentice Hall, 1986.
11. Miczo, Alexander. *Digital Logic Testing and Simulation*, 2nd ed. New York: John Wiley & Sons, Ltd West Sussex, England, 2003.
12. Patt, Yale N. and Sanjay J. Patel. *Introduction to Computing Systems: From Bits and Gates to C and Beyond*, 2nd ed. New York: McGraw-Hill, 2004.
13. Roth, Charles H. Jr. and Lizy Kurian John. *Digital Systems Design Using VHDL*, 2nd ed. Toronto, Ontario: Thomson, 2008.
14. Rushton, Andrew. *VHDL for Logic Synthesis*, 2nd ed. West Sussex, England: John Wiley & Sons, Ltd, 1998.
15. Wakerly, John F. *Digital Design Principles & Practices*, 4th ed. Upper Saddle River, NJ: Prentice Hall, 2006.
16. Weste, Neil and Kaamran Eshraghian. *Principles of CMOS VLSI Design*, 2nd ed. Reading, MA: Addison-Wesley, 1993.

# Index

# Description of the CD

The CD that accompanies this text contains three programs that are useful in the computer-aided design and simulation of digital logic—LogicAid, SimUaid, and DirectVHDL-PE. Principal features of these programs are listed below. User manuals for LogicAid and SimUaid are provided on the CD in PDF format. The user manuals for DirectVHDL, which are provided in the form of HTML help files, will be installed when you run setup from the DirectVHDL directory.

**LogicAid Features:**

- Logic functions may be input in the following forms: sum-of-products, product-of-sums, truth table, PLA table, Karnaugh map, minterm or maxterm expansion
- Choice of logic simplification algorithms provides for finding a fast solution or all minimum solutions
- Sequential logic may be input as Mealy or Moore state tables, state graphs (with either binary or alphanumeric input/output), or SM charts
- Reduces state tables to a minimum number of rows and derives flip-flop input equations for D, T, J-K, and S-R flip-flops
- Creates JEDEC files for programming 22V10 PALs
- Tutorial aids include Karnaugh map tutor, state table checker, and partial graph checker

**SimUaid Features:**

- Friendly user interface allows easy placement and wiring of components
- Available devices include basic gates, flip-flops, switches, probes, registers, counters, adders, multiplexers, decoders, 7-segment indicators, clocks, tri-state buffers, and state machines
- Four-valued logic simulation (0, 1, X, Z)
- Displays all device inputs and outputs for ease of signal tracing and debugging
- Probe placement automatically sets up waveform display
- Live simulation mode allows immediate observation of response to input switch changes
- Synchronous simulation allows stepping one clock period at a time
- Asynchronous simulation allows stepping until a signal changes
- Converts a circuit diagram to synthesizable VHDL code

**DirectVHDL Features:**

- Edits, compiles, and simulates VHDL code
- Easy to learn user interface
- VHDL editor highlights syntax errors as you type
- Simulator displays waveforms and listing output
- Command interface allows forcing input values interactively or from a command file
- Compatible with IEEE Standard 1076-1993 VHDL